AF894327

Surveys of Modern Mathematics
Volume 9

# Compressible Flow and Euler's Equations

Demetrios Christodoulou
*Eidgenössische Technische Hochschule (ETH) Zürich*

Shuang Miao
*Department of Mathematics, University of Michigan*

International Press
www.intlpress.com

HIGHER EDUCATION PRESS

Surveys of Modern Mathematics, Volume 9
Compressible Flow and Euler's Equations

Demetrios Christodoulou
Eidgenössische Technische Hochschule (ETH) Zürich

Shuang Miao
Department of Mathematics, University of Michigan



ISBN: 978-1-57146-297-8

Printed in the United States of America.

18 17 16 15 14 1 2 3 4 5 6 7 8 9

# Surveys of Modern Mathematics

www.intlpress.com/smm

# Introduction

The equations describing the motion of a perfect fluid were first formulated by Euler in 1752 (see [9], [10]), based, in part, on the earlier work of D. Bernoulli [1]. These equations were among the first partial differential equations to be written down, preceded, it seems, only by D'Alembert's 1749 formulation [8] of the 1-dimensional wave equation describing the motion of a vibrating string in the linear approximation. In contrast to D'Alembert's equation, however, we are still, after the lapse of two and a half centuries, far from having achieved an adequate understanding of the observed phenomena which are supposed to lie within the domain of validity of Euler's equations.

The phenomena displayed in the interior of a fluid fall into two broad classes, the phenomena of sound, the linear theory of which is acoustics, and the phenomena of vortex motion. The sound phenomena depend on the compressibility of a fluid, while the vortex phenomena occur even in a regime where the fluid may be considered to be incompressible. The formation of shocks, the subject of the present monograph, belongs to the class of sound phenomena, but lies in nonlinear regime, beyond the range covered by the linear acoustics.

Let us make a short review of the history of the study of the sound phenomena in fluids, in particular the phenomena of the formation of shocks in the nonlinear regime. At the time when the equations of the fluid motion were formulated, thermodynamics was in its infancy, however, it was already clear that the local state of a fluid as seen by a comoving observer is determined by two thermodynamic variables, say pressure and temperature. Of these, only pressure entered the equations of motion, while the equations involve also the density of the fluid. Density was already known to be a function of pressure and temperature for a given type of fluid. However, in the absence of an additional equation, the system of equations at the time of Euler, which consisted of the momentum equations and the equation of continuity, was underdetermined, except in the incompressible limit. The additional equation was supplied by Laplace in 1816 [13] in the form of what was later to be called adiabatic condition, and allowed him to make the first correct calculation of the sound speed.

The first work on the formation of shocks was done by Riemann in 1858 [17]. Riemann considered the case of isentropic flow with plane symmetry, where the equations of fluid mechanics reduces to a system of conservation laws for two unknowns and with two independent variables, a single space coordinate and time. He introduced for such systems the so-called Riemann invariants, and with the help of these showed that solutions which arise from smooth initial conditions develop infinite gradients in finite time.

In 1865 the concept of entropy was introduced into theoretical physics by Clausius [7], and the adiabatic condition was understood to be the requirement that the entropy of each fluid element remains constant during its evolution.

The first general result on the formation of singularity in 3-dimensional fluids was obtained by Sideris in 1985 [18]. Sideris considered the compressible Euler equations in the case of a classical ideal gas with adiabatic index $\gamma > 1$ and with initial data which coincide with those of a constant state outside a ball. The assumptions of his theorem on the initial data were that there is an annular region bounded by the sphere outside which the constant state holds, and a concentric sphere in its interior, such that a certain integral in this annular region of $\rho - \rho_0$, the departure of the density $\rho$ from its value $\rho_0$ in the constant state, is positive, while another integral in the same region of $\rho v^r$, the radial momentum density, is non-negative. These integrals involve kernels which are functions of the distance from the center. It is also assumed that everywhere in the annular region the specific entropy $s$ is not less than its value $s_0$ in the constant state. The conclusion of the theorem is that the maximal time interval of existence of a smooth solution is finite. The chief drawback of this theorem is that it tells us nothing about the nature of the breakdown. Also the method relies the strict convexity of the pressure as a function of density displayed by the equation of state of an ideal gas, and does not extend to more general equation of state.

The most recent and complete results on the formation of shocks in three dimensional fluids were obtained by Christodoulou in 2007 [5]. Christodoulou considered the relativistic Euler equations in three space dimensions for a perfect fluid with an arbitrary equation of state. He considered the regular initial data on a spacelike hyperplane $\Sigma_0$ in Minkowski spacetime which outside a sphere coincide with the data corresponding to a constant state. He considered the restriction of the initial data to the exterior of a concentric sphere in $\Sigma_0$ and the maximal classical development of this data. Under suitable restriction on the size of the departure of the initial data from those of the constant state, he proved certain theorems which give a complete description of the maximal classical development. In particular, theorems give a detailed description of the geometry of the boundary of the domain of the maximal classical solution and a detailed analysis of the behavior of the solution at this boundary.

The aim of the present monograph is to derive analogous results for the classical, non-relativistic, compressible Euler's equations taking the data to be irrotational and isentropic, and to give a proof of these results which is considerably simpler and completely self-contained. The present monograph in fact not only gives simpler proofs but also sharpens some of the results. In addition the present monograph explains in depth the ideas on which the approach is based. Finally certain geometric aspects which pertain only to the non-relativistic theory are discussed.

We shall presently explain the basis of the approach of the present monograph (also of the previous one dealing with the relativistic case). This basic idea can be thought as an extension of the method of Riemann invariants combined with the method of the partial hodograph transformation, to the case of more than one space dimension. We first recall some basic facts about Riemann invariants. In the case of one space dimension, the isentropic Euler system reads

$$\partial_t \rho + \partial_x(\rho v) = 0$$

$$\partial_t v + v\partial_x v = -\frac{1}{\rho}\partial_x p$$

and it can be written as a single equation of the velocity potential $\phi$:

$$(g^{-1})^{\mu\nu}\partial_\mu\partial_\nu\phi = 0$$

or, setting $\psi_\mu = \partial_\mu\phi$,

$$(g^{-1})^{\mu\nu}\partial_\mu\psi_\nu = 0$$

where $g$ is the following Lorentzian metric on the spacetime manifold $\mathcal{M}$:

$$g = -\eta^2 dt^2 + (dx - vdt)^2.$$

Here

$$v = -\psi_x$$

is the fluid velocity and

$$h = \psi_t - \frac{1}{2}\psi_x^2$$

is the enthalpy. The pressure $p$ is a given function of $h$ and

$$\rho = dp/dh$$

while the sound speed $\eta$ is given by

$$\eta^2 = dp/d\rho$$

Riemann invariants are the functions $R_1, R_2$ defined on the cotangent space $T_p^*\mathcal{M}$, which are the two functionally independent solutions of the following eikonal equation:

$$g_{\mu\nu}\frac{\partial R}{\partial\psi_\mu}\frac{\partial R}{\partial\psi_\nu} = 0$$

i.e.

$$-\eta^2(\frac{\partial R}{\partial\psi_t})^2 + (\frac{\partial R}{\partial\psi_x} + \psi_x\frac{\partial R}{\partial\psi_t})^2 = 0$$

We define the vectorfields $N_1, N_2$ on $\mathcal{M}$ by

$$N_1 := \frac{\partial R_1}{\partial \psi_\mu} \frac{\partial}{\partial x^\mu}, \quad N_2 := \frac{\partial R_2}{\partial \psi_\mu} \frac{\partial}{\partial x^\mu}$$

These are null vectorfields with respect to $g$. We choose $R_1$ and $R_2$ so that the integral curves of $N_1$ and $N_2$ are the incoming and outgoing null curves respectively. Then $R_1$, $R_2$ as functions on $\mathcal{M}$ satisfy

$$N_2 R_1 = 0, \quad N_1 R_2 = 0 \tag{1}$$

Let us introduce the acoustical coordinates $(t, u)$ so that $u$ is constant along the outgoing null curves. Then the vectorfields

$$L = \frac{\partial}{\partial t}, \quad \underline{L} = \eta^{-1}\kappa L + 2T$$

where

$$T = \frac{\partial}{\partial u} \quad \text{and} \quad \kappa = -\frac{\partial x}{\partial u} \tag{2}$$

are null vectorfields with respect to $g$, and the integral curves of $L$ and $\underline{L}$ are outgoing and incoming null curves respectively. Therefore $L$ and $\underline{L}$ are colinear to $N_2$ and $N_1$ respectively. Therefore Equations (1) are equivalent to

$$L R_1 = 0, \quad \underline{L} R_2 = 0 \tag{3}$$

To write down explicit expressions for $R_1, R_2$ we use, instead of $(\psi_t, \psi_x)$, the following variables in $T_p^*\mathcal{M}$:

$$h = \psi_t - \frac{1}{2}\psi_x^2, \quad v = -\psi_x$$

Then for any function $f = f(h, v)$ defined on $T_p^*\mathcal{M}$, we have

$$\frac{\partial f}{\partial \psi_x} + \psi_x \frac{\partial f}{\partial \psi_t} = -\frac{\partial f}{\partial v}$$

Let us introduce a function $r = r(h)$ by

$$\frac{dr}{dh} = \frac{1}{\eta}, \quad r(0) = 0$$

Then

$$R_1 = r + v, \quad R_2 = r - v$$

are the two functionally independent solutions of the eikonal equation, therefore the two Riemann invariants. From the first equation of (3), we know that

$$R_1 = R_1(u)$$

is determined by initial data, while to obtain $R_2$, we consider the second equation of (3), namely the equation

$$\eta^{-1}\kappa\frac{\partial R_2}{\partial t}+2\frac{\partial R_2}{\partial u}=0$$

Here $\kappa$ enters, which is defined by (2). To obtain an equation for $\kappa$ we consider the following equations:

$$\frac{\partial x}{\partial t}=c_+,\quad c_+=v+\eta$$

We shall derive an equation for $\kappa$. From (3) we have

$$\frac{\partial\kappa}{\partial t}=-\frac{\partial c_+}{\partial u}=-\frac{1}{2}\frac{\partial R_1}{\partial u}+\frac{1}{2}\frac{\partial R_2}{\partial u}-\frac{d\eta}{dh}\frac{\partial h}{\partial u}$$

while

$$\frac{\partial h}{\partial u}=\frac{dh}{dr}\frac{\partial r}{\partial u}=\frac{1}{2}\eta(\frac{\partial R_1}{\partial u}+\frac{\partial R_2}{\partial u})$$

Substituting the above we obtain

$$\frac{\partial\kappa}{\partial t}=\frac{1}{2}(-1-\eta\frac{d\eta}{dh})\frac{\partial R_1}{\partial u}+\frac{1}{2}(1-\eta\frac{d\eta}{dh})\frac{\partial R_2}{\partial u}$$

Since

$$\frac{\partial R_1}{\partial u}=\frac{2}{\eta}\frac{\partial h}{\partial u}-\frac{\partial R_2}{\partial u}$$

we have

$$\frac{\partial\kappa}{\partial t}=\frac{1}{2\eta}(-2-2\eta\frac{d\eta}{dh})\frac{\partial h}{\partial u}+\frac{\partial R_2}{\partial u}$$

Let us define

$$H=-2h-\eta^2$$

Noting that by the second equation of (3)

$$\frac{\partial R_2}{\partial u}=-\frac{\kappa}{2\eta}\frac{\partial R_2}{\partial t}$$

We conclude that $\kappa$ satisfies the following equation:

$$\frac{\partial\kappa}{\partial t}=m'+\kappa e'$$

with

$$m'=\frac{1}{2\eta}\frac{dH}{dh}\frac{\partial h}{\partial u},\quad e'=-\frac{1}{2\eta}\frac{\partial R_2}{\partial t}$$

The main idea in the 1-dimensional case is that $R_1, R_2$ as well as the rectangular coordinate $x$ are smooth functions of $(t,u)$. The partial hodograph transformation is the transformation

$$(t,u) \longmapsto (t,x)$$

from acoustical to rectangular coordinates. The Jacobian is

$$\frac{\partial(t,x)}{\partial(t,u)} = \begin{vmatrix} 1 & 0 \\ v+\eta & -\kappa \end{vmatrix} = -\kappa$$

and vanishes when $\kappa$ vanishes. This means $R_1, R_2$ are not smooth in $(t,x)$ when shocks form.

In the case of more than one space dimension, in particular, the case of three space dimensions, we do not have Riemann invariants. We work instead with the first order variations, which are defined through the *variation fields*:

$$\partial_\mu, \quad \overset{\circ}{R}_i = \epsilon_{ijk} x^j \frac{\partial}{\partial x^k}, \quad x^\mu \partial_\mu - I; \quad \mu = 0,1,2,3, \quad 1 \leqslant i,j,k \leqslant 3$$

Here $I$ is the multiplication operator by 1. These fields are the generators of the subgroup of the scale-extended Galilean group, the invariance group of the compressible Euler system, which leaves the constant state invariant. These first order variations satisfy the linear wave equation

$$\square_{\tilde{g}} \psi = 0 \tag{4}$$

where $\tilde{g}$ is the conformal acoustical metric:

$$\tilde{g} = \Omega g, \quad \Omega = \frac{\rho}{\eta}, \quad g = -\eta^2 dt^2 + \sum_i (dx^i - v^i dt)^2$$

Like the equations satisfied by $R_1, R_2$ in the case of one space dimension, Equation (4) does not depend on the Galilean structure, but depends only on the properties of $(\mathcal{M}, g)$ as a Lorentzian manifold. Actually, it depends more sensitively on the conformal class of $(\mathcal{M}, g)$. This is similar to the fact that in the case of one space dimension, null curves depend only on the conformal class of the acoustical metric. Also like in the case of one space dimension, we shall work in the *acoustical coordinates* $(t, u, \vartheta)$. Here $u$ is the *acoustical function* in $\mathcal{M}$, whose level sets $C_u$ are outgoing null hypersurfaces. The level curves of $\vartheta \in \mathbb{S}^2$ on each $C_u$ are the generators of $C_u$. Like in the case of one space dimension, we denote by $L$ the tangent vectorfield of the generator of $C_u$ parametrized by $t$, and $\underline{L}$ the incoming null normals of $S_{t,u} := C_u \bigcap \Sigma_t$, whose definitions are formally the same as one space dimensional case. We also denote by $T$ the tangent vectorfield of the inward normal curves to the $S_{t,u}$ in $\Sigma_t$ parametrized by $u$. Here $\Sigma_t$ is the level set of the function $t$, which is isometric to the Euclidean space.

To obtain a fundamental energy estimate for this linear equation in $(t, u, \vartheta)$, we need two *multiplier vectorfields* $K_0, K_1$. These are non-spacelike and future-directed with respect to $g$ (a requirement which actually depends only on the conformal class of $g$). They are linear combinations of $L$ and $\underline{L}$, with coefficients which are smooth in $(t, u, \vartheta)$. The concept of multiplier vectorfields originates from Noether's theorem [14] on conserved currents. A modern more general treatment of *compatible currents* is found in [4]. In order to obtain higher order energy estimates, we consider the $n$th order variations by applying a string of *commutation vectorfields* of length $n-1$ to the first order variations:

$$T, \quad Q := (1+t)L, \quad R_i = \Pi \mathring{R}_i$$

Here $\Pi$ is the orthogonal projection from $T_p\Sigma_t$ to $T_pS_{t,u}$. Then we obtain an inhomogeneous wave equation for the $n$th order variation $\psi_n$:

$$\Box_{\tilde{g}}\psi_n = \rho_n \tag{5}$$

where $\rho_n$ is determined by the deformation tensors of the commutation vectorfields. Actually $\rho_n$ depends on up to the $(n-1)$th order derivatives of the deformation tensors, and $\rho_1 = 0$. The use of commutation fields originates in [12]. Multiplier fields and commutation fields on general curved spacetimes have first been used in [6].

After we solve (5) in the acoustical coordinates, we need to go back to the original rectangular coordinates. Again we must consider the inverse of the transformation:

$$(t, u, \vartheta) \longmapsto (t, x)$$

where $\vartheta \in \mathbb{S}^2$ and $x \in \mathbb{R}^3$. This is what replaces the partial hodograph transformation in higher dimensions. The Jacobian of this transformation is

$$\kappa\sqrt{\det \not{g}} \tag{6}$$

Here $\not{g}$ is the induced acoustical metric on $S_{t,u}$. So we consider the system satisfied by the rectangular coordinates on each $C_u$:

$$\frac{\partial x^i}{\partial t} = L^i = -\eta\hat{T}^i - \psi_i$$

which is a fully nonlinear system for $x^i$. Here $\hat{T}$ is the inward unit normal of $S_{t,u}$ in $\Sigma_t$, whose expression is the ratio of a homogeneous quadratic polynomial in $\frac{\partial x^i}{\partial \vartheta^A}$ to the square root of a homogeneous quartic polynomial in $\frac{\partial x^i}{\partial \vartheta^A}$. The estimates of the derivatives of $x^i$ reduce to the estimates of the derivatives of $\chi$ and $\mu$. These are defined as follows:

$$2\chi = \not{\mathcal{L}}_L \not{g}, \quad \mu = \eta\kappa$$

where $\kappa$ is the magnitude of $T$. Thus $\chi$ is the second fundamental form of $S_{t,u}$ in $C_u$. Finally the way we estimate $\chi, \mu$ is to study the geometric structure equations of the foliation of $\mathcal{M}$ by surfaces $S_{t,u}$.

To summarize, in the case of $n$ space dimensions, the role of Riemann invariants is played by the first order variations $\psi$, which shall be proved to be smooth functions of $(t, u, \vartheta)$. Moreover, we shall show that the $x^i$ are also smooth functions of $(t, u, \vartheta)$. This shall be done through estimates on $\chi$ and $\mu$ based on the geometric structure equations. One of these equations is the same as in the one space dimensional case:

$$L\mu = m + \mu e$$

where

$$m = \frac{1}{2}\frac{dH}{dh}Th, \quad H = -2h - \eta^2$$

and

$$e = \frac{1}{2\eta^2}(\frac{\rho}{\rho'})'Lh + \frac{1}{\eta}\hat{T}^i(L\psi_i)$$

As a consequence of these facts, the boundary of the maximal classical development contains a singular part $\mathcal{H}$, where the Jacobian (6) vanishes. Since $\sqrt{\det \not{g}}$ is, by virtue of the estimates for $\chi$, bounded from below by a positive constant, $\kappa$, equivalently $\mu$, vanishes on $\mathcal{H}$. Thus, the inverse of the transformation

$$(t, u, \vartheta) \longmapsto (t, x)$$

is not differentiable at $\mathcal{H}$. Therefore the $\psi$ are not differentiable with respect to the rectangular coordinates at $\mathcal{H}$. Nevertheless, $\mathcal{H}$, the zero-level set of $\mu$, a smooth function of $(t, u, \vartheta)$, is a smooth hypersurface in $\mathcal{M}$ relative to the differential structure induced by the acoustical coordinates. This is because we can show that $L\mu$ is bounded from above by a strictly negative function at $\mathcal{H}$, therefore $\mathcal{H}$ is a non-critical level set of $\mu$.

The first main result in the present monograph can be thought as an existence theorem (**Theorem 17.1**) for the nonlinear wave equation of the velocity potential

$$(g^{-1})^{\mu\nu}\partial_\mu\partial_\nu\phi = 0$$

with small initial data:

*The solution of the above nonlinear wave equation can be extended smoothly to the boundary of the maximal solution in acoustical coordinates $(t, u, \vartheta)$, and the solution is also smooth in rectangular coordinates before $\mu$ becomes 0, since the differential structures induced by acoustical coordinates and rectangular coordinates are equivalent to each other when $\mu > 0$.*

Theorem 17.1 also gives a lower bound for the time when the shock forms (i.e. $\mu = 0$), and the energy estimates for solution as well as various geometric quantities associated to the acoustical spacetime $(\mathcal{M}, g)$.

Based on this existence theorem, we find some conditions on initial data which guarantee the formation of shocks in finite time (see **Theorem 18.1**). The conditions are imposed on a $\Sigma_t$-integral of the following function:

$$(1 - u + t)\underline{L}\psi_0 - \psi_0 \tag{7}$$

where

$$\psi_0 := \partial_t \phi$$

is one of the first order variations. The principal part of function (7), namely, $(1-u+t)\underline{L}\psi_0$ determines the properties of function $m$, which, in turn, determines the formation of shocks. Moreover, the spherical mean of function (7) on $S_{t,u}$ satisfies an ordinary differential inequality in the parameter $t$. Then we can connect the properties of $m$ near the point where the shocks form and the properties of $m$ on the initial hypersurface $\Sigma_0$ by using this ordinary differential inequality. Then the necessary properties of $m$ then follow from its properties on the initial hypersurface $\Sigma_0$.

Also based on the existence theorem, we can give a geometric description of the boundary of the maximal classical solution in acoustical differential structure (**Proposition 19.1**):

*The boundary contains a regular part* $\underline{C}$*, which is an incoming null hypersurface in* $(\mathcal{M}, g)$*, and a singular part* $\mathcal{H}$*, on which the function* $\mu$ *vanishes.* $\mathcal{H}$ *is a spacelike hypersurface in* $(\mathcal{M}, g)$*, and it has the common past boundary with* $\underline{C}$*, denoted by* $\partial_{-}\mathcal{H}$*, which is a 2-dimensional spacelike surface in* $(\mathcal{M}, g)$*. However, the singular boundary, from the intrinsic point of view, is a null hypersurface in* $(\mathcal{M}, g)$*, on which the acoustical metric* $g$ *degenerates in acoustical coordinates.*

The corresponding description of the singular boundary in the standard differential structure (that is, in rectangular coordinates) is given in **Proposition 19.3**.

Moreover, we establish a trichotomy theorem (**Theorem 19.1**) describing the behavior of the past-directed null geodesics initiated at the singular boundary:

*For each point* $q$ *of the singular boundary, the intersection of the past null geodesic conoid of* $q$ *with any* $\Sigma_t$ *in the past of* $q$ *splits into three parts, the parts corresponding to the outgoing and to the incoming sets of null geodesics ending at* $q$ *being embedded discs with a common boundary, an embedded circle, which corresponds to the set of the remaining null geodesics ending at* $q$*. All outgoing null geodesics ending at* $q$ *have the same tangent vector at* $q$*.*

Finally, considering the transformation from one acoustical function to another, we show that the foliations corresponding to different families of outgoing null hypersurfaces have equivalent geometric properties and degenerate in precisely the same way on the same singular boundary (See **Proposition 19.2**).

Let us now give an outline of the present monograph. The first four chapters concern the geometric set up. Then in Chapter 5, we obtain energy estimates for the linear wave equation associated to the conformal acoustical metric. Chapter 6 deals with the preliminary estimates for the deformation tensor of the commutation vectorfields, the precise estimates of which are given in Chapters 10 and 11. We also introduce the basic bootstrap assumptions on variations as well as on $\chi$ and $\mu$ in Chapters 5 and 6. Chapters 8 and 9 are crucial in the whole work, because it is here that estimates for $\chi$ and $\mu$ are derived which do not lose derivatives, thus allowing us to close the bootstrap argument. Chapter 8 concerns the estimates for the top order spatial derivatives of $\chi$. In fact only the top order angular derivatives are involved. While Chapter 9 concerns the estimates for the top order spatial derivatives of $\mu$. In Chapter 12, based on the bootstrap assumptions on variations, we recover the bootstrap assumptions on $\chi$ and $\mu$, except **C1**, **C2** and **C3**, which are recovered in Chapter 13. In Chapter 14, based on a crucial lemma (**Lemma 8.11**) established in Chapter 8, we estimate the *borderline* contribution from the top order spatial derivatives of $\chi$ and $\mu$. Then in Chapter 15, we obtain the energy estimates for the top order variations. These are allowed to blow up as shocks begin to form. We then revisit the lower order energy estimates and show that the estimates of each preceding order blow up successively more slowly until we finally reach energy estimates of a certain order which do not blow up at all. These bounded energy estimates allow us to close the bootstrap argument (See Chapters 16 and 17).

In regard to the notational conventions, Latin indices take the values $1, 2, 3$, while Greek indices take the values $0, 1, 2, 3$. Repeated indices are meant to be summed, unless otherwise specified.

---

Work supported by ERC Advanced Grant 246574 *Partial Differential Equations of Classical Physics.*

# Contents

# 1

# Compressible Flow and Non-linear Wave Equations

## 1.1 Euler's Equations

The mathematical description of the state of a moving fluid is determined by the distribution of the fluid velocity $\mathbf{v} = \mathbf{v}(\mathbf{x}, t)$ and of any two thermodynamic quantities pertaining to the fluid, for instance the pressure $p = p(\mathbf{x}, t)$ and the mass density $\rho = \rho(\mathbf{x}, t)$. All the thermodynamics are determined by the value of any two of them, together with the equation of state. The equation of motion of a perfect fluid can be derived as follows.

We begin with the equation which expresses the conservation of mass. We consider the domain $\Omega(t)$ in $\mathbb{R}^3$ occupied by some fluid at time $t$. The mass of fluid in $\Omega(t)$ is $\int_{\Omega(t)} \rho dV$. This integral should not depend on $t$ due to the mass conservation. So we get

$$\frac{d}{dt} \int_{\Omega(t)} \rho dV = 0 \tag{1.1}$$

This means

$$\int_{\Omega(t)} \frac{\partial \rho}{\partial t} dV + \int_{\partial\Omega(t)} \rho \mathbf{v} \cdot \mathbf{n} dS = 0 \tag{1.2}$$

where $\mathbf{n}$ is the out normal of $\partial\Omega(t)$. By Green's formula, we have

$$\int_{\Omega(t)} (\frac{\partial \rho}{\partial t} + \mathrm{div}(\rho \mathbf{v})) dV = 0 \tag{1.3}$$

Since $\Omega(t)$ is arbitrary, we have

$$\frac{\partial \rho}{\partial t} + \mathrm{div}(\rho \mathbf{v}) = 0 \tag{1.4}$$

This is the equation of continuity.

Then we consider the conservation of momentum. By Newton's law,

$$\frac{d\mathbf{P}}{dt} = \mathbf{F} \tag{1.5}$$

where $\mathbf{P}$ is the momentum, $\mathbf{F}$ is the force. In our case, Equation (1.5) becomes

$$\frac{d}{dt}\int_{\Omega(t)} \rho \mathbf{v} dV = -\int_{\partial\Omega(t)} p\mathbf{n} dS \tag{1.6}$$

i.e.

$$\frac{d}{dt}\int_{\Omega(t)} \rho \mathbf{v} dV = -\int_{\Omega(t)} \nabla p dV \tag{1.7}$$

A component of this equation is

$$\frac{d}{dt}\int_{\Omega(t)} \rho v^i dV = -\int_{\Omega(t)} \frac{\partial p}{\partial x^i} dV \tag{1.8}$$

Then we get

$$\int_{\Omega(t)} (\frac{\partial(\rho v^i)}{\partial t} + \mathrm{div}(\rho v^i \mathbf{v}))dV = -\int_{\Omega(t)} \frac{\partial p}{\partial x^i} dV \tag{1.9}$$

Since $\Omega(t)$ is arbitrary, as well as the equation of continuity, we get

$$\frac{\partial v^i}{\partial t} + \mathbf{v}\cdot\nabla v^i = -\frac{1}{\rho}\frac{\partial p}{\partial x^i} \tag{1.10}$$

In a perfect fluid, heat exchange between different parts of the fluid is absent. So the motion is adiabatic throughout the fluid.

In the adiabatic motion the entropy of any particle of fluid remains constant as that particle moves about in space. Denoting by $s$ the entropy per unit mass. We can express the condition for adiabatic motion as

$$\frac{ds}{dt} = 0 \tag{1.11}$$

which can be written as

$$\frac{\partial s}{\partial t} + \mathbf{v}\cdot\nabla s = 0 \tag{1.12}$$

## 1.2 Irrotational Flow and the Nonlinear Wave Equation

The adiabatic condition may take a much simpler form. If the entropy is constant throughout the fluid at some initial instant, it remains everywhere the same constant value at all times. In this case, we can write the adiabatic condition simply as

$$s = \mathrm{const} \tag{1.13}$$

Such a motion is said to be isentropic. We may in this case put (1.10) into a different form. To do this, we employ the thermodynamic relation

$$dh = Vdp + \theta ds \tag{1.14}$$

where $h$ is the enthalpy, defined in terms of the energy per unit mass:

$$h = e + pV$$

where $V = \frac{1}{\rho}$ is the specific volume, and $\theta$ is the temperature. Since $s$ is constant, we have

$$dh = V dp = \frac{1}{\rho} dp \tag{1.15}$$

So (1.10) becomes

$$\frac{\partial v^i}{\partial t} + \mathbf{v} \cdot \nabla v^i = -\frac{\partial h}{\partial x^i} \tag{1.16}$$

This implies

$$\frac{\partial \omega}{\partial t} + \mathbf{v} \cdot \nabla \omega = \omega \cdot \nabla \mathbf{v} - (\operatorname{div} \mathbf{v})\omega \tag{1.17}$$

where $\omega = \nabla \times \mathbf{v}$.

If $\omega \mid_{t=0} = 0$, then by (1.17) $\omega \equiv 0$. A flow for which $\omega = \nabla \times \mathbf{v} \equiv 0$ in all space and time is called an irrotational flow. In this case, since $\mathbb{R}^3$ is simply-connected, there exists a function $\phi$, such that $\mathbf{v} = -\nabla\phi$. Equation (1.16) becomes

$$\frac{\partial^2 \phi}{\partial x^i \partial t} - \sum_j \frac{\partial \phi}{\partial x^j} \frac{\partial^2 \phi}{\partial x^j \partial x^i} = \frac{\partial h}{\partial x^i} \tag{1.18}$$

i.e.

$$\frac{\partial}{\partial x^i} \left( \frac{\partial \phi}{\partial t} - \frac{1}{2} |\nabla \phi|^2 - h \right) = 0 \tag{1.19}$$

Since $\phi$ is defined up to a constant which may depend on $t$, without loss of generality, we may set

$$h = \frac{\partial \phi}{\partial t} - \frac{1}{2} |\nabla \phi|^2 \tag{1.20}$$

Then only the continuity equation remains. While $\frac{dh}{dp} = V > 0$, then by inverse function theorem, $p$ can be viewed as a function of $h$. This is because $h$ can be viewed as a function of $p$ and $s$, and in the isentropic case, $h$ is a function of $p$. Then $\rho = \frac{dp}{dh}$, $\rho$ is also a function of $h$.

The continuity equation is

$$\frac{\partial \rho}{\partial t} - \operatorname{div}(\rho \nabla \phi) = 0 \tag{1.21}$$

i.e.

$$\frac{d\rho}{dh} \frac{\partial}{\partial t} \left( \frac{\partial \phi}{\partial t} - \frac{1}{2} \sum_i \left( \frac{\partial \phi}{\partial x^i} \right)^2 \right) - \sum_j \frac{d\rho}{dh} \frac{\partial}{\partial x^j} \left( \frac{\partial \phi}{\partial t} - \frac{1}{2} \sum_i \left( \frac{\partial \phi}{\partial x^i} \right)^2 \right) \frac{\partial \phi}{\partial x^j} - \rho \Delta \phi = 0 \tag{1.22}$$

$$\frac{\partial^2 \phi}{\partial t^2} - 2 \sum_i \frac{\partial \phi}{\partial x^i} \frac{\partial^2 \phi}{\partial t \partial x^i} + \sum_{ij} \frac{\partial \phi}{\partial x^i} \frac{\partial \phi}{\partial x^j} \frac{\partial^2 \phi}{\partial x^i \partial x^j} - \eta^2 \Delta \phi = 0 \tag{1.23}$$

where we have used the fact that $\frac{dp}{dh} = \rho$ and the definition of $\eta$, the sound speed, which is the following:

$$\eta^2 = \eta^2(h) := (\frac{\partial p}{\partial \rho})_s \tag{1.24}$$

We define the function

$$H = H(h) := -\eta^2 - 2h \tag{1.25}$$

From (1.24) we know that

$$\eta^{-2} = (\frac{d\rho}{dp})_s = -\frac{1}{V^2}(\frac{dV}{dp})_s \tag{1.26}$$

Since $V = (\frac{dh}{dp})_s$, by direct calculation we obtain

$$\frac{dH}{dh} = -\frac{\eta^4}{V^3}(\frac{d^2V}{dp^2})_s \tag{1.27}$$

So $\frac{dH}{dh}$ vanishes if and only if $V$ is linear in $p$. This fact will be used later. The meaning of the function $H$ will become apparent in the sequel.

Equation (1.21) is the Euler-Lagrange equation of the Lagrangian

$$L = p(h) \tag{1.28}$$

In general, $L = L(t, \mathbf{x}; \phi, \frac{\partial \phi}{\partial t}, \nabla\phi)$, and the Euler-Lagrange equation is

$$\sum_\alpha \frac{\partial}{\partial x^\alpha}(\frac{\partial L(t, \mathbf{x}; \phi, \frac{\partial \phi}{\partial t}, \nabla\phi)}{\partial u_\alpha}) = \frac{\partial L}{\partial q}(t, \mathbf{x}; \phi, \frac{\partial \phi}{\partial t}, \nabla\phi) \tag{1.29}$$

where $q = \phi$, $x^0 = t$, $u_0 = \frac{\partial \phi}{\partial t}$, $u_i = \frac{\partial \phi}{\partial x^i}$, $i = 1, 2, 3$.

In the case of $L = p(h)$, then

$$\frac{\partial L}{\partial u_\alpha} = \frac{dp}{dh}\frac{\partial h}{\partial u_\alpha} = \rho\frac{\partial h}{\partial u_\alpha} \tag{1.30}$$

where $h = u_0 - \frac{1}{2}\sum_i (u_i)^2$.

So we have

$$\frac{\partial h}{\partial u_0} = 1, \frac{\partial h}{\partial u_i} = -u_i, \frac{\partial h}{\partial q} = 0$$

Then the Euler-Lagrange equation is

$$\frac{\partial \rho}{\partial t} - \sum_i \frac{\partial}{\partial x^i}(\rho u_i) = 0 \tag{1.31}$$

This is Equation (1.21).

## 1.3 The Equation of Variations and the Acoustical Metric

Next, we shall discuss the linearized equation corresponding to (1.21).

Let $\phi$ be a given solution of (1.21) and let $\phi_\tau : \tau \in I$, $I$ an open interval of the real line containing 0, be a differentiable 1-parameter family of solution such that $\phi_0 = \phi$. Then

$$\psi = (\frac{d\phi_\tau}{d\tau})_{\tau=0} \tag{1.32}$$

is a variation of $\phi$ through solutions.

Consider the Lagrangian of the unknown $\psi$:

$$L_\tau[\psi] := L[\phi + \tau\psi] \tag{1.33}$$

Then the linearized Lagrangian of $L[\phi]$ is

$$\dot{L}[\psi] := \frac{1}{2}\frac{d^2 L_\tau[\psi]}{d\tau^2}|_{\tau=0} \tag{1.34}$$

In this case, $L_\tau[\psi] = p(h_\tau)$, where

$$h_\tau = \frac{\partial}{\partial t}(\phi + \tau\psi) - \frac{1}{2}\sum_i(\frac{\partial}{\partial x^i}(\phi + \tau\psi))^2$$

By direct calculation, we have

$$\frac{d^2 L_\tau}{d\tau^2}[\psi] \,|_{\tau=0}= \frac{d^2 p}{dh^2}(h)(\frac{\partial\psi}{\partial t} - \sum_i \frac{\partial\phi}{\partial x^i}\frac{\partial\psi}{\partial x^i})^2 + \frac{dp}{dh}(h)(-\sum_i(\frac{\partial\psi}{\partial x^i})^2) \tag{1.35}$$

i.e.

$$\frac{d^2 L_\tau}{d\tau^2}[\psi] \,|_{\tau=0}= \frac{d\rho}{dh}(h)(\frac{\partial\psi}{\partial t} - \sum_i \frac{\partial\phi}{\partial x^i}\frac{\partial\psi}{\partial x^i})^2 - \rho(\sum_i(\frac{\partial\psi}{\partial x^i})^2) \tag{1.36}$$

i.e.

$$\frac{d^2 L_\tau}{d\tau^2}[\psi] \,|_{\tau=0}= \rho(\eta^{-2}(\partial_t\psi - \sum_i \partial_i\phi\partial_i\psi)^2 - \sum_i(\partial_i\psi)^2) \tag{1.37}$$

i.e.

$$\frac{d^2 L_\tau}{d\tau^2}[\psi] \,|_{\tau=0}= -\rho(g^{-1})^{\mu\nu}\partial_\mu\psi\partial_\nu\psi \tag{1.38}$$

where the metric $g$ is

$$g = -\eta^2 dt^2 + \sum_i(dx^i - v^i dt)^2 \tag{1.39}$$

and

$$g^{-1} = -\eta^{-2}(\frac{\partial}{\partial t} + v^i\frac{\partial}{\partial x^i}) \otimes (\frac{\partial}{\partial t} + v^j\frac{\partial}{\partial x^j}) + \sum_i \frac{\partial}{\partial x^i} \otimes \frac{\partial}{\partial x^i} \tag{1.40}$$

Here, $v^i = -\partial_i\phi$, and we have used the fact

$$\frac{\rho'}{\rho} = \frac{d\rho}{dh}\frac{1}{\rho} = \frac{d\rho}{dp}\frac{dp}{dh}\frac{1}{\rho} = \eta^{-2}$$

Consider the conformal metric

$$\tilde{g}_{\mu\nu} = \Omega g_{\mu\nu} \tag{1.41}$$

which satisfy the following conditions:

$$\rho(g^{-1})^{\mu\nu}\partial_\mu\psi\partial_\nu\psi = (\tilde{g}^{-1})^{\mu\nu}\partial_\mu\psi\partial_\nu\psi\sqrt{-\det\tilde{g}} \tag{1.42}$$

Since $\sqrt{-\det g} = \eta$, we get $\sqrt{-\det\tilde{g}} = \Omega^2\eta$, we have $\Omega = \frac{\rho}{\eta}$.

Since the linearized equation of (1.21) is the Euler-Lagrange equation of the linearized Lagrangian $\dot{L}[\psi]$, the linearized equation of (1.21) is

$$\Box_{\tilde{g}}\psi = 0 \tag{1.43}$$

Since $\rho_0, \eta_0$, which are the mass density and the sound speed corresponding to the constant state, are constants, and the Euler-Lagrangian equations are not affected if the Lagrangian is multiplied by a constant, we may choose

$$\Omega = \frac{\rho/\rho_0}{\eta/\eta_0} \tag{1.44}$$

so that $\Omega = 1$ in the constant state.

We may in fact choose the unit of time in relation to the unit of length so that

$$\eta_0 = 1$$

and we may further choose the unit of mass in relation to the unit of length so that

$$\rho_0 = 1$$

These choices shall be understood from the next chapter to the end of the book.

## 1.4 The Fundamental Variations

Now if $\phi$ is a solution of (1.21) and $f_\tau$ is a 1-parameter subgroup of the isometry group of $\mathbb{E}^3$, that is $\mathbb{R}^3$ with the standard Euclidean metric

$$\sum_i (dx^i)^2$$

or the translation on $t$-direction, then for each $\tau$,

$$\phi_\tau = \phi \circ f_\tau \tag{1.45}$$

is also a solution. It follows that

$$\psi = (\frac{d(\phi \circ f_\tau)}{d\tau})_{\tau=0} = X\phi \tag{1.46}$$

satisfies the linear equation (1.43).

$X$ can be the following:

$$\Omega_{ij} := x^i \partial_j - x^j \partial_i, \quad 1 \leqslant i < j \leqslant 3 \tag{1.47}$$

and

$$T_\mu := \frac{\partial}{\partial x^\mu}, \quad \mu = 0, 1, 2, 3 \tag{1.48}$$

which are the generators of $f_\tau$.

Consider now the scaling group for Equation (1.21). It is easy to see that the physical dimension of $\phi$ is $\mathrm{L}^2\mathrm{T}^{-1}$, so we consider the transformation, with constants $a, b > 0$:

$$\tilde{\phi}(t, \mathbf{x}) := \frac{b^2}{a}\phi(\frac{t}{a}, \frac{\mathbf{x}}{b}) \tag{1.49}$$

A direct calculation implies

$$\tilde{h}(t, \mathbf{x}) := \frac{\partial \tilde{\phi}}{\partial t} - \frac{1}{2}|\nabla \tilde{\phi}|^2 = \frac{b^2}{a^2}h(\frac{t}{a}, \frac{\mathbf{x}}{b}) \tag{1.50}$$

In general, $p$ is an arbitrary function of $h$:

$$p = p(h) \tag{1.51}$$

so

$$\frac{dp}{dh} = \rho > 0 \quad \text{i.e} \quad p' > 0 \quad \text{and also} \quad p > 0 \tag{1.52}$$

and

$$\frac{dp}{d\rho} = \eta^2 = \frac{dp}{dh}\frac{dh}{d\rho} = \rho / \frac{d^2p}{dh^2} > 0 \tag{1.53}$$

Under the transformation (1.49), we have

$$\tilde{p}(t, \mathbf{x}) = p(\tilde{h}(t, \mathbf{x})) = p(\frac{b^2}{a^2}h(\frac{t}{a}, \frac{\mathbf{x}}{b})) \tag{1.54}$$

and then

$$\tilde{\rho}(t, \mathbf{x}) = \frac{d\tilde{p}}{d\tilde{h}} = p'(\frac{b^2}{a^2}h(\frac{t}{a}, \frac{\mathbf{x}}{b})) = \rho(\frac{b^2}{a^2}h(\frac{t}{a}, \frac{\mathbf{x}}{b})) \tag{1.55}$$

Now we can transform the Euler-Lagrange equation (1.21) under the transformation (1.49).

We can see that in general, only in the case $a = b$ the Euler-Lagrange equation is invariant under the transformation (1.49). But still, there are some special cases that we have 2-dimensional scaling group.

First we consider the case

$$p(h) = \frac{1}{2}h^2 \tag{1.56}$$

Note that this function $p(h)$ satisfies (1.51)—(1.53).

Now we may take $a \neq b$:

$$\tilde{h}(t, \mathbf{x}) = \frac{b^2}{a^2} h(\frac{t}{a}, \frac{\mathbf{x}}{b})$$

Then

$$p(\tilde{h}(t, \mathbf{x})) = \frac{1}{2}\tilde{h}^2(t, \mathbf{x}) = \frac{b^4}{a^4}\frac{1}{2}h^2(\frac{t}{a}, \frac{\mathbf{x}}{b}) = \frac{b^4}{a^4}p(h(\frac{t}{a}, \frac{\mathbf{x}}{b})) \tag{1.57}$$

That is

$$\tilde{p}(t, \mathbf{x}) = \frac{b^4}{a^4} p(\frac{t}{a}, \frac{\mathbf{x}}{b}) \tag{1.58}$$

and also

$$\tilde{\rho}(t, \mathbf{x}) = \tilde{h}(t, \mathbf{x}) = \frac{b^2}{a^2} h(\frac{t}{a}, \frac{\mathbf{x}}{b}) = \frac{b^2}{a^2}\rho(\frac{t}{a}, \frac{\mathbf{x}}{b}) \tag{1.59}$$

So now we have

$$\begin{aligned} &\frac{\partial \tilde{\rho}}{\partial t} - \operatorname{div}(\tilde{\rho}\nabla\tilde{\phi}) \\ &= \frac{b^2}{a^3}\frac{\partial \rho}{\partial t} - \frac{b^2}{a^2}\frac{1}{a}\rho\Delta\phi - \frac{b}{a^2}\nabla\rho \cdot \frac{b}{a}\nabla\phi = \frac{b^2}{a^3}(\frac{\partial\rho}{\partial t} - \operatorname{div}(\rho\nabla\phi)) \end{aligned} \tag{1.60}$$

This means Equation (1.21) is invariant under the transformation (1.49). So in this case, the scaling group is 2-dimensional.

More generally, we can consider the case

$$p(h) = Ch^\alpha, \alpha > 1, C > 0, h > 0 \tag{1.61}$$

Obviously in this case the function $p(h)$ satisfies (1.51)—(1.53) and we have

$$p = C(\frac{\rho}{\alpha C})^{\frac{\alpha}{\alpha - 1}} \tag{1.62}$$

If we set

$$\gamma = \frac{\alpha}{\alpha - 1}, \quad k = \frac{C}{(\alpha C)^{\frac{\alpha}{\alpha - 1}}}$$

then

$$p = k\rho^\gamma \tag{1.63}$$

with $\gamma > 1$ and $k$ depending on $\gamma$.

This is the polytropic case. We can calculate as above to see that in this case, we also have the 2-dimensional scaling group. Also, one can easily see that we have the 2-dimensional scaling group only when the function $p$ is a homogeneous function of $h$.

In present book, we shall consider the general case $p = p(h)$, so the scaling group is 1-dimensional. In this case, we consider the 1-parameter group of dilations of $\mathbb{R} \times \mathbb{E}^3$, given by

$$(t, \mathbf{x}) \mapsto (e^\tau t, e^\tau \mathbf{x}) \tag{1.64}$$

where $\tau \in \mathbb{R}$.

Let $\phi$ be a solution of (1.21) and define for each $\tau \in \mathbb{R}$ the function

$$\phi_\tau := e^{-\tau} \phi(e^\tau t, e^\tau \mathbf{x}) \tag{1.65}$$

As we have seen before, $\phi_\tau$ is also a solution of (1.21). Then

$$\psi = (\frac{d\phi_\tau}{d\tau})|_{\tau=0} := S\phi \tag{1.66}$$

satisfies the linear equation (1.43). Here $S$ is the differential operator

$$S = D - I, \quad D = x^\mu \frac{\partial}{\partial x^\mu} \tag{1.67}$$

# 2

# The Basic Geometric Construction

## 2.1 Null Foliation Associated with the Acoustical Metric

### 2.1.1 Galilean Spacetime

Since the Galilean spacetime is the frame of classical fluid mechanics, we shall mention some basic facts about it.

A *Galilean spacetime* is a quadruple $(\mathcal{G}, \mathcal{A}, \pi, e)$ where

(1) $\mathcal{G}$ is a 4-dimensional affine space;

(2) $\mathcal{A}$ is a 1-dimensional affine space with unit displacement, which represents time;

(3) The map $\pi\colon \mathcal{G} \to \mathcal{A}$ is a surjective affine map;

So $\Sigma_t := \pi^{-1}(t)$ is an affine subspace of $\mathcal{G}$ for each $t \in \mathcal{A}$, and we can introduce a Euclidean metric $e_t$ on each $\Sigma_t$. The $\Sigma_t$ is the hypersurfaces of absolute simultaneity.

(4) A *Galilean frame* is an affine foliation of $\mathcal{G}$ by a family of parallel lines transversal to $\Sigma_t$. A Galilean frame represents a family of inertial observers at rest relative to each other. The metric $e$ satisfies the condition that parallel transport along the lines of a Galilean frame induces an isometry between $(\Sigma_{t_1}, e_{t_1})$ and $(\Sigma_{t_2}, e_{t_2})$ for any $t_1, t_2 \in \mathcal{A}$.

Two Galilean spacetimes, $(\mathcal{G}, \mathcal{A}, \pi, e)$ and $(\mathcal{G}', \mathcal{A}', \pi', e')$ are equivalent if and only if there is a pair $(I, J)$, where $I$ is an affine spaces isomorphism of $\mathcal{G}$ onto $\mathcal{G}'$, and $J$ is an affine spaces isomorphism of $\mathcal{A}$ onto $\mathcal{A}'$ such that

(1) $\pi' \circ I = J \circ \pi$ ;

(2) $J$ takes the unit displacement of $\mathcal{A}$ into the unit displacement of $\mathcal{A}'$ ;

(3) $I|_{\Sigma_t}$ is an isometry of $(\Sigma_t, e_t)$ onto $(\Sigma'_{t'}, e'_{t'})$, for each $t \in \mathcal{A}$. Here $\Sigma_t = \pi^{-1}(t)$, $\Sigma'_{t'} = \pi'^{-1}(t')$, and $t' = J(t)$.

Classical mechanics including classical continuum mechanics are invariant under such Galilean transformations. We have used in fact this Galilean invariance to define the first order variations in Chapter 1.

However, in this book, we shall adopt a different point of view in analyzing the equation of variations. We shall retain from the Galilean spacetime $\mathcal{G}$ only its differential manifold structure $\mathbb{R}^4$, a manifold endowed with the acoustical metric $g$. We call the pair $(\mathbb{R}^4, g)$ the acoustical spacetime, which is a Lorentzian manifold. Two such Lorentzian manifolds $(\mathcal{M}, g)$ and $(\mathcal{M}', g')$ are equivalent if and only if there is a diffeomorphism $f$ of $\mathcal{M}'$ onto $\mathcal{M}$ such that $g' = f^*g$, i.e. $(\mathcal{M}', g')$ is isometric to $(\mathcal{M}, g)$, through an arbitrary diffeomorphism $f$.

### 2.1.2 Null Foliation and Acoustical Coordinates

The initial data for the equations of motion

$$\frac{\partial \rho}{\partial t} + \operatorname{div}(\rho \mathbf{v}) = 0 \tag{2.1}$$

$$\frac{\partial \mathbf{v}}{\partial t} + \mathbf{v} \cdot \nabla \mathbf{v} = -\frac{1}{\rho} \nabla p \tag{2.2}$$

$$\frac{\partial s}{\partial t} + \mathbf{v} \cdot \nabla s = 0 \tag{2.3}$$

is to be given on the hyperplane $\Sigma_0$ and consists in specification of the triplet $(p, s, \mathbf{v})$. We assume that there is a sphere $S_{0,0}$ outside which the initial data coincide with those of a constant state, that is we have

$$p = p_0, \quad s = s_0, \quad \mathbf{v} = 0 \tag{2.4}$$

By choosing the unit of length equal to the radius of $S_{0,0}$, we then take $S_{0,0}$ to be the unit sphere centered at the origin in $\Sigma_0$. We consider an annular interior neighborhood of $S_{0,0}$ in $\Sigma_0$:

$$\Sigma_0^{\epsilon_0} = \{\mathbf{x} \in \Sigma_0 : 1 - \epsilon_0 \leqslant r(\mathbf{x}) \leqslant 1\} \tag{2.5}$$

where $r$ is the Euclidean distance function from the origin in $\Sigma_0$ and $\epsilon_0$ is a positive constant, satisfying the condition

$$\epsilon_0 \leqslant \frac{1}{2} \tag{2.6}$$

We define in $\Sigma_0$ the function

$$u = 1 - r \tag{2.7}$$

which on $\Sigma_0$ minus the origin is a smooth function without critical points, vanishing on $S_{0,0}$ and increasing inward. For each value of $u$ in the closed interval $[0, \epsilon_0]$, the corresponding level set $S_{0,u}$ of $u$ is a sphere of radius $1 - u$ in the interval $[1 - \epsilon_0, 1]$, and we have

$$\Sigma_0^{\epsilon_0} = \bigcup_{u \in [0, \epsilon_0]} S_{0,u} \tag{2.8}$$

In the case that the initial data is irrotational and isentropic outside $S_{0,\epsilon_0}$, the data in the exterior of $S_{0,\epsilon_0}$ in $\Sigma_0$ specify the pair $(\phi, \partial_t \phi)$ in this region. This is initial data for the nonlinear wave equation (1.23) in the region in question and we have

$$\phi = 0, \quad \partial_0 \phi = h_0 \tag{2.9}$$

in the exterior of $S_{0,0}$ in $\Sigma_0$, where $h_0$ is the enthalpy of constant state. As we have seen in Chapter 1, the enthalpy $h$ is defined up to an additive constant, we may use this freedom to set

$$h_0 = 0$$

To any given initial data set as above there corresponds a unique *maximal solution* of Equations (2.1)—(2.3), of the nonlinear wave equation (1.23) in the irrotational isentropic case. The notion of maximal solution or maximal development of an initial data set is the following. Given an initial data set, the local existence theorem asserts the existence of a *development* of this set, namely of a domain $\mathcal{D}$ in Galilean spacetime, whose past boundary is the domain of the initial data, and of a solution defined in $\mathcal{D}$ and taking the given data at the past boundary, such that if we consider any point $p \in \mathcal{D}$ and any curve issuing at $p$ with the property that its tangent vector at any point $q$ belongs to $\bar{I}_q^-$, the closure of the past component of the double cone defined by $g_q$, the acoustical metric at $q$, then the curve terminates in the past at a point of the domain of the initial data. The local uniqueness theorem asserts that if $(\mathcal{D}_1, (p_1, s_1, \mathbf{v}_1))$ and $(\mathcal{D}_2, (p_2, s_2, \mathbf{v}_2))$ are two developments of the same initial data set ($(\mathcal{D}_1, \phi_1)$ and $(\mathcal{D}_2, \phi_2)$ in the irrotational isentropic case), then $(p_1, s_1, \mathbf{v}_1)$ coincides with $(p_2, s_2, \mathbf{v}_2)$ in $\mathcal{D}_1 \bigcap \mathcal{D}_2$ ($\phi_1$ coincides with $\phi_2$ in $\mathcal{D}_1 \bigcap \mathcal{D}_2$ in the irrotational isentropic case). It follows that the union of all developments of a given initial data set is itself a development, the unique *maximal development* of the initial data set. We consider, in the domain of maximal solution, the family $\{C_u : u \in [0, \epsilon_0]\}$ of outgoing characteristic hypersurfaces corresponding to the family $\{S_{0,u} : u \in [0, \epsilon_0]\}$:

$$C_u \bigcap \Sigma_0 = S_{0,u}, \quad \forall u \in [0, \epsilon_0] \tag{2.10}$$

Each bicharacteristic generator of each $C_u$ is to extend in the domain of maximal solution as long as it remains on the boundary of the domain of dependence of the exterior of the surface $S_{0,u}$ in $\Sigma_0$. If we denote by $W_{\epsilon_0}$ the spacetime domain:

$$W_{\epsilon_0} = \bigcup_{u \in [0, \epsilon_0]} C_u \tag{2.11}$$

then the domain $M_{\epsilon_0}$ of the maximal solution corresponding to the given data set is the union of $W_{\epsilon_0}$ with the domain in Galilean spacetime bounded by the exterior of the unit sphere $S_{0,0}$ in $\Sigma_0$ and by the outgoing characteristic hypersurface $C_0$ corresponding to $S_{0,0}$. By the domain of dependence theorem the solution coincides in $M_{\epsilon_0} \setminus W_{\epsilon_0}$ with the constant state. This implies that $C_0$ is a complete cone, each of its bicharacteristic generators extending to infinity in the parameter $t$. We extend the function $u$ to $W_{\epsilon_0}$ by requiring that its level sets are precisely the outgoing characteristic hypersurfaces $C_u$. The function $u$ is then the solution of the equation

$$(g^{-1})^{\mu\nu}\partial_\mu u \partial_\nu u = 0 \tag{2.12}$$

We shall call such a function an *acoustical function*. The vector field $\hat{L}$ given by

$$\hat{L} = -(g^{-1})^{\mu\nu}\partial_\nu u \tag{2.13}$$

is then a future-directed null geodesic vectorfield with respect to the Lorentzian metric $g$. Its integral curves are the bicharacteristic generators of each $C_u$. Now the parametrization of these given by $\hat{L}$ is affine; however, for reasons which shall become apparent in the following we wish the generators to be parametrized by the function $t$ instead. For this reason we choose the colinear vectorfield

$$L = \mu\hat{L} \tag{2.14}$$

where the proportionality factor $\mu$, a positive function, is chosen so that

$$Lt = 1 \tag{2.15}$$

Thus

$$\frac{1}{\mu} = -(g^{-1})^{\mu\nu}\partial_\mu t \partial_\nu u \tag{2.16}$$

For each $u \in [0, \epsilon_0]$ there is a greatest lower bound $t_*(u)$ of the extent of the generators of $C_u$, in the parameter $t$, in the domain of the maximal solution. This $t_*(u)$ is either a positive real number or $\infty$. According to the above we have $t_*(0) = \infty$. Let us denote

$$t_{*\epsilon_0} = \inf_{u\in[0,\epsilon_0]} t_*(u) \tag{2.17}$$

In the following we shall confine attention to the spacetime domain

$$W^*_{\epsilon_0} = \bigcup_{(t,u)\in[0,t_{*\epsilon_0})\times[0,\epsilon_0]} S_{t,u} \tag{2.18}$$

where

$$S_{t,u} = C_u \bigcap \Sigma_t \tag{2.19}$$

Define in $W^*_{\epsilon_0}$ the vectorfield $T$ such that it is tangential to $\Sigma_t$, orthogonal to the surfaces $\{S_{t,u} : u \in [0, \epsilon_0]\}$ in metric $g$, and it verifies

$$Tu = 1 \tag{2.20}$$

Consider the commutator

$$\Lambda = [L, T] \tag{2.21}$$

From (2.12)—(2.14), we know

$$Lu = 0 \tag{2.22}$$

and by the fact that $T$ is tangential to $\Sigma_t$, we have

$$Tt = 0 \tag{2.23}$$

then from (2.15), (2.20), we get

$$\Lambda u = \Lambda t = 0 \tag{2.24}$$

Therefore $\Lambda$ is tangential to $S_{t,u}$. From (2.14) and (2.22), we know that

$$g(L, L) = 0 \tag{2.25}$$

i.e. $L$ is a null vector with respect to the acoustical metric $g$. Also, from (2.13), (2.14) and (2.20),

$$g(L, T) = \mu g(\hat{L}, T) = -\mu T u = -\mu \tag{2.26}$$

Since the restriction of $g$ to $\Sigma_t$ is the Euclidean metric, $T$ is a spacelike vector with respect to $g$ and we can write

$$g(T, T) = \kappa^2 \tag{2.27}$$

where $\kappa$ is a positive function, the Euclidean magnitude of $T$. Let $N$ be any vector tangent to one of $C_u$, then we have

$$N u = 0 \tag{2.28}$$

i.e.

$$g(L, N) = 0 \tag{2.29}$$

So for any $X$ tangent to one of $S_{t,u}$,

$$g(L, X) = 0 \tag{2.30}$$

Since by definition, $X$ is also $g$ orthogonal to $T$, it follows that

$$T_p W^*_{\epsilon_0} = \Pi_p \oplus T_p S_{t,u} \tag{2.31}$$

where $\Pi_p$ is spanned by $L$ and $T$. The metric $g$ in $\Pi_p$ is given by (2.25)—(2.27) in terms of $\mu$ and $\kappa$. By (2.31) $g$ is then completely specified once we give the metric $\not{g}$ induced on $S_{t,u}$:

$$\not{g}(X, Y) = g(X, Y), \quad \forall X, Y \in T_p S_{t,u} \tag{2.32}$$

For each $u \in [0, \epsilon_0]$ the generator of $C_u$ defines a smooth mapping of $S_{0,u}$ to $S_{t,u}$ for each $t \in [0, t_{*\epsilon_0})$. While each $S_{0,u}$ is diffeomorphic to $S^2$. Therefore, we get a diffeomorphism from $S_{t,u}$ to $S^2$:

$$p \mapsto \vartheta = \vartheta(p) \tag{2.33}$$

where $\vartheta \in S^2$, $p \in S_{t,u}$. If local coordinates $(\vartheta^1, \vartheta^2)$ are chosen on $S^2$, this diffeomorphism defines local coordinates on $S_{t,u}$ for every $(t, u) \in [0, t_{*\epsilon_0}) \times [0, \epsilon_0]$. Since the diffeomorphism from $S_{0,u}$ to $S^2$ is arbitrary, the diffeomorphism (2.33) is arbitrary as it may be composed on the left by a transformation of the form

$$\vartheta \mapsto \tilde{\vartheta} = \tilde{\vartheta}(u, \vartheta) \tag{2.34}$$

The local coordinates $(\vartheta^1, \vartheta^2)$, together with the functions $(t, u)$ define a complete system of local coordinates $(t, u, \vartheta^1, \vartheta^2)$ for $W^*_{\epsilon_0}$. We shall call these *acoustical coordinates* and we shall derive an expression for $g$ in $W^*_{\epsilon_0}$ in these coordinates.

First, the integral curves of $L$ are the lines of constant $\vartheta$ and $u$, parametrized by $t$. Therefore

$$L = \frac{\partial}{\partial t} \tag{2.35}$$

Next, by (2.20), (2.23), we have

$$T = \frac{\partial}{\partial u} - \Xi \tag{2.36}$$

where $\Xi$ is a vectorfield tangential to $S_{t,u}$. Thus $\Xi$ can be expanded in terms of the frame field $(\frac{\partial}{\partial \vartheta^A}, A = 1, 2)$ and we have

$$\Xi = \Xi^A \frac{\partial}{\partial \vartheta^A} \tag{2.37}$$

By (2.21), (2.35), (2.36) we have

$$[L, \Xi] = -\Lambda \tag{2.38}$$

or, in terms of components,

$$\frac{\partial \Xi^A}{\partial t} = -\Lambda^A \tag{2.39}$$

We can set $\Xi = 0$ along any one of the hypersurfaces $\Sigma_t$. However the non-vanishing of $\Lambda$ forbids setting $\Xi = 0$ everywhere. With the components

$$\not{g}_{AB} = \not{g}(\frac{\partial}{\partial \vartheta^A}, \frac{\partial}{\partial \vartheta^B}) \tag{2.40}$$

where $A, B = 1, 2$, we have

$$g = -2\mu du dt + \kappa^2 du^2 + \not{g}_{AB}(d\vartheta^A + \Xi^A du)(d\vartheta^B + \Xi^B du) \tag{2.41}$$

We define in $W^*_{\epsilon_0}$ the vectorfield $B$ by the conditions that it be orthogonal to $\Sigma_t$ with respect to $g$ and that it verifies

$$Bt = 1 \tag{2.42}$$

Since $\Sigma_t$ is spacelike relative to $g$, $B$ is future-directed timelike relative to $g$. Thus, there is a positive function $\alpha$ such that

$$g(B, B) = -\alpha^2 \tag{2.43}$$

In fact, we have

$$B^\mu = -\alpha^2 (g^{-1})^{\mu\nu} \partial_\nu t \tag{2.44}$$

and

$$\alpha^{-2} = -(g^{-1})^{\mu\nu} \partial_\mu t \partial_\nu t = \eta^{-2} \tag{2.45}$$

Since $\alpha$ and $\eta$ are both positive, $\alpha = \eta$. Then we have

$$B^0 = -\alpha^2(g^{-1})^{00} = 1, \quad B^i = -\alpha^2(g^{-1})^{i0} = -\alpha^2(-\eta^{-2})v^i = v^i \tag{2.46}$$

So

$$B = \frac{\partial}{\partial t} + v^i \frac{\partial}{\partial x^i} \tag{2.47}$$

and in irrotational isentropic case,

$$B = (1, -\nabla\phi)$$

Now at any point $p$, $B \in \Pi_p$, a timelike plane relative to $g$. Therefore, $B$ is a linear combination of $L$ and $T$. From (2.35), (2.36), (2.42), and the fact that $B$ is future-directed timelike relative to $g$, we have

$$B = L + fT \tag{2.48}$$

where $f$ is a positive function. Taking $g$-inner product of (2.48) with $T$, using (2.26) and the fact that $g(B,T) = 0$, we have

$$\mu = f\kappa^2 \tag{2.49}$$

On the other hand, $L$ is null relative to $g$, we have

$$0 = g(B,B) + f^2 g(T,T) = -\alpha^2 + f^2\kappa^2 \tag{2.50}$$

Since $f$ is positive, we have

$$f = \frac{\alpha}{\kappa} \tag{2.51}$$

Then we get

$$\mu = \alpha\kappa \tag{2.52}$$

Let us introduce the vectorfield $\hat{T} := \kappa^{-1}T$. Then from (2.48), (2.51), we have

$$L = B - \alpha\hat{T} \tag{2.53}$$

By (2.46), (2.47), we have

$$L = \frac{\partial}{\partial t} - (\alpha\hat{T}^i - v^i)\frac{\partial}{\partial x^i} \tag{2.54}$$

in the irrotational isentropic case:

$$L = \frac{\partial}{\partial t} - (\alpha\hat{T}^i + \partial_i\phi)\frac{\partial}{\partial x^i} \tag{2.55}$$

We finally calculate the Jacobian of the mapping:

$$(t, u, \vartheta^1, \vartheta^2) \mapsto (x^0, x^1, x^2, x^3) \tag{2.56}$$

Since $x^0 = t$, we have

$$\frac{\partial x^0}{\partial t} = 1, \quad \frac{\partial x^0}{\partial u} = \frac{\partial x^0}{\partial \vartheta^A} = 0 \tag{2.57}$$

where $A = 1, 2$. Also, since

$$\frac{\partial x^\mu}{\partial t} = L^\mu \tag{2.58}$$

we have

$$\frac{\partial x^i}{\partial t} = L^i \tag{2.59}$$

where $i = 1, 2, 3$. Similarly, we have

$$\frac{\partial x^i}{\partial u} = T^i + \xi^i \tag{2.60}$$

where

$$\xi^i = \Xi^A X^i_A, \quad X^i_A = \frac{\partial x^i}{\partial \vartheta^A} \tag{2.61}$$

where $i = 1, 2, 3; A = 1, 2$.

Then we get the Jacobian determinant of (2.56):

$$\Delta = \begin{vmatrix} 1 & 0 & 0 & 0 \\ L^1 & T^1 + \xi^1 & X^1_1 & X^1_2 \\ L^2 & T^2 + \xi^2 & X^2_1 & X^2_2 \\ L^3 & T^3 + \xi^3 & X^3_1 & X^3_2 \end{vmatrix} \tag{2.62}$$

We have

$$\Delta = \begin{vmatrix} T^1 + \xi^1 & X^1_1 & X^1_2 \\ T^2 + \xi^2 & X^2_1 & X^2_2 \\ T^3 + \xi^3 & X^3_1 & X^3_2 \end{vmatrix} = \dot{\Delta} + \ddot{\Delta} \tag{2.63}$$

where

$$\dot{\Delta} = \begin{vmatrix} T^1 & X^1_1 & X^1_2 \\ T^2 & X^2_1 & X^2_2 \\ T^3 & X^3_1 & X^3_2 \end{vmatrix} \tag{2.64}$$

and

$$\ddot{\Delta} = \begin{vmatrix} \xi^1 & X^1_1 & X^1_2 \\ \xi^2 & X^2_1 & X^2_2 \\ \xi^3 & X^3_1 & X^3_2 \end{vmatrix} \tag{2.65}$$

Now from (2.61), we have

$$\ddot{\Delta} = \sum_{A=1}^{2} \Xi^A \begin{vmatrix} X^1_A & X^1_1 & X^1_2 \\ X^2_A & X^2_1 & X^2_2 \\ X^3_A & X^3_1 & X^3_2 \end{vmatrix} = 0 \tag{2.66}$$

then we can simply write

$$\Delta = (T, X_1, X_2) \tag{2.67}$$

i.e.

$$\Delta = |T||X_1 \wedge X_2| \tag{2.68}$$

where the magnitude is with respect to $\bar{g}$, the restriction of $g$ on $\Sigma_t$, which is the Euclidean metric. Then we know that

$$|X_1 \wedge X_2| = \sqrt{|X_1|^2|X_2|^2 - (X_1, X_2)^2} = \sqrt{\det \not{g}} \tag{2.69}$$

then we get

$$\Delta = \kappa\sqrt{\det \not{g}} \tag{2.70}$$

## 2.2 A Geometric Interpretation for Function *H*

Recall the definition of $H$ in Chapter 1:

$$H = -2h - \eta^2 \tag{2.71}$$

Here we just consider the case $H$ = const. Actually, we just need to consider the case $H = 0$ due to the following transformation:

$$\tilde{\phi} := \phi + \frac{k}{2}t, \quad \tilde{h} := h + \frac{k}{2}, \quad \tilde{H} := H - k, \quad k \text{ is the “const”} \tag{2.72}$$

which leaves the non-linear wave equation (1.23) invariant.

Since

$$H = 0$$

implies

$$h = -\frac{1}{2}\eta^2$$

in this case the acoustical metric is

$$g = 2\frac{\partial\phi}{\partial t}(dt)^2 + 2\sum_i \frac{\partial\phi}{\partial x^i}dx^i dt + \sum_i (dx^i)^2$$

i.e.

$$g = 2d\phi dt + \sum_i (dx^i)^2 \tag{2.73}$$

Introducing a new coordinate $s$, we consider the Lorentzian manifold $(\mathbb{R}^{1+4}, \tilde{g})$ where

$$\tilde{g} = 2dsdt + \sum_i (dx^i)^2 \tag{2.74}$$

So the manifold $(\mathbb{R} \times \mathbb{E}^3, g)$ can be realized as a submanifold of $(\mathbb{R}^{1+4}, \tilde{g})$ with the induced metric, in fact as the graph over the null hyperplane:

$$s = \phi(\mathbf{x}, t) \tag{2.75}$$

Also, by direct calculation, we find

$$\det g = 2h = -\eta^2 < 0 \tag{2.76}$$

which implies that this submanifold is a time-like hypersurface in $(\mathbb{R}^{1+4}, \tilde{g})$.

Moreover, the manifold $(\mathbb{R} \times \mathbb{E}^3, g)$ is a minimal submanifold of $(\mathbb{R}^{1+4}, \tilde{g})$. Since in general

$$\frac{d\rho}{dp} = \eta^{-2} \quad \text{and} \quad \frac{d\rho}{dp} = \frac{1}{\rho}\frac{d\rho}{dh}$$

In the present case by (2.71) we have

$$\frac{1}{\rho}\frac{d\rho}{dh} = -\frac{1}{2h}$$

Hence we can solve this ODE to obtain

$$\rho = \frac{C}{\sqrt{-2h}} = \frac{C}{\eta} \tag{2.77}$$

where $C$ is a positive constant. The equation

$$\frac{dp}{dh} = \rho$$

gives

$$p = p_0 - C\sqrt{-2h}$$

By (2.77) we can write

$$p = p_0 - C^2 V \quad \text{or} \quad p = p_0 - C\eta \tag{2.78}$$

Thus, the action in a domain $\Omega$ in $\mathbb{R} \times \mathbb{E}^3$ is

$$\int_\Omega p dt d^3x = p_0 \int_\Omega dt d^3x - C \int_\Omega \eta dt d^3x \tag{2.79}$$

while

$$\int_\Omega \eta dt d^3x = \int_\Omega \sqrt{-\det g}\, dt d^3x \tag{2.80}$$

is the volume of $\Omega$ with respect to $g$.

It follows that the Euler-Lagrange equation coincides with that corresponding to the Lagrangian $\sqrt{-\det g}$. But the Euler-Lagrange equation corresponding to the volume of a domain as the action in the domain is the minimal surface equation.

# 3

# The Acoustical Structure Equations

## 3.1 The Acoustical Structure Equations

Define $k$ by

$$2\alpha k = \bar{\mathcal{L}}_B g \tag{3.1}$$

where $\bar{\mathcal{L}}$ is the restriction of $\mathcal{L}$ to $\Sigma_t$.

If $X$, $Y$ are two vectors tangent to $\Sigma_t$ at a point, we then have

$$\alpha k(X, Y) = g(D_X B, Y) = g(D_Y B, X) \tag{3.2}$$

where $k$ is the second fundamental form of $\Sigma_t$ relative to $g$ and $D$ is the covariant derivative associated to $g$. Recall from Chapter 2, we have

$$B = \partial_0 + v^i \partial_i \tag{3.3}$$

Then

$$2\alpha k_{ij} = (\bar{\mathcal{L}}_B g)_{ij} = (B\bar{g}_{ij}) + \bar{g}_{im}\partial_j v^m + \bar{g}_{jm}\partial_i v^m \tag{3.4}$$

Since $\bar{g}_{ij} = \delta_{ij}$, we have

$$2\alpha k_{ij} = \partial_j v^i + \partial_i v^j = -2\partial_i \psi_j \tag{3.5}$$

The last equality above is in the irrotational isentropic case.

Any vector $X$ tangent to $C_u$ at a point $p$ can be uniquely decomposed into a vector colinear to $L$ and a vector tangent to $S_{t,u}$:

$$X = c_X L + \Pi X \tag{3.6}$$

If $X$, $Y$ are tangent to $C_u$, then $g(X, Y) = \not{g}(\Pi X, \Pi Y)$.

Let $X_A = \frac{\partial}{\partial \vartheta^A}$, given any $Z \in T_p C_u$, we can expand $\Pi Z = Z^A X_A$. Taking inner product with $X_B$:

$$\not{g}(X_B, \Pi Z) = \not{g}_{AB} Z^A \tag{3.7}$$

Define $\chi$ by

$$2\chi = \not{\mathcal{L}}_L g \tag{3.8}$$

Here, $\not{\mathcal{L}}$ is the restriction of $\mathcal{L}$ to $S_{t,u}$.

If $X$, $Y$ are tangent to $S_{t,u}$, then we have

$$\chi(X,Y) = g(D_X L, Y) = g(D_Y L, X) \tag{3.9}$$

Since $\hat{L} = \mu^{-1}L$ is an affinely parametrized geodesic field, we have

$$D_L L = \mu D_{\hat{L}}(\mu \hat{L}) = L(\mu)\hat{L} = \mu^{-1}(L\mu)L \tag{3.10}$$

Denoting $\chi_{AB} = \chi(X_A, X_B)$. We shall derive a propagation equation for $\chi_{AB}$ along $C_u$. First we have

$$L\chi_{AB} = g(D_L D_{X_A} L, X_B) + g(D_{X_A} L, D_L X_B) \tag{3.11}$$

Now $D_L X_A - D_{X_A} L = [L, X_A] = 0$, if we denote by $R$ the curvature of $g$, we have

$$D_L D_{X_A} L - D_{X_A} D_L L = R(L, X_A)L$$

where the curvature transformation is defined by

$$R(X,Y)Z = D_X D_Y Z - D_Y D_X Z - D_{[X,Y]} Z$$

Then

$$g(D_L D_{X_A} L, X_B) = \mu^{-1}(L\mu)\chi_{AB} - \alpha_{AB} \tag{3.12}$$

where $\alpha_{AB} = R(X_A, L, X_B, L)$. Recall that the curvature tensor is defined through the curvature transformation as follows:

$$R(W, Z, X, Y) := g(W, R(X,Y)Z) \tag{3.13}$$

The second term on the right hand side of (3.11) is equal to $g(D_{X_A} L, D_{X_B} L)$.

Since $L$ is null with respect to $g$, then for any vectorfield $X$, $D_X L$ is $g$-orthogonal to $L$, therefore tangential to $C_u$. Thus, setting $W_A = D_{X_A} L$, $A = 1,2$, we have

$$g(W_A, W_B) = \not{g}(\Pi W_A, \Pi W_B) \tag{3.14}$$

We now expand $\Pi W_A$ in the basis $X_B, B = 1,2$. The coefficient of $X_B$ in this expansion is

$$(\not{g}^{-1})^{BC} \not{g}(\Pi W_A, X_C) = (\not{g}^{-1})^{BC} g(D_{X_A} L, X_C) = (\not{g}^{-1})^{BC} \chi_{AC} \tag{3.15}$$

hence

$$\not{g}(\Pi W_A, \Pi W_B) = \not{g}(\chi_A^C X_C, \chi_B^D X_D) = \chi_A^C \chi_{BC}$$

Here the capital indices are raised and lowered with respect to $\not{g}_{AB}$. Thus

$$\chi_A^B = (\not{g}^{-1})^{BC} \chi_{AC}$$

So we get the propagation equation

$$L\chi_{AB} = \mu^{-1}(L\mu)\chi_{AB} + \chi_A^C\chi_{BC} - \alpha_{AB} \tag{3.16}$$

From this, we can get a propagation equation for $\text{tr}\chi$. We denote by $S_{\mu\nu}$ the Ricci tensor:

$$S_{\mu\nu} := (g^{-1})^{\kappa\lambda}R_{\mu\kappa\nu\lambda} \tag{3.17}$$

We can express $(g^{-1})^{\kappa\lambda}$ in terms of the frame $L, T, X_1, X_2$. We have

$$(g^{-1})^{\mu\nu} = -\alpha^{-2}L^\mu L^\nu - \mu^{-1}(L^\mu T^\nu + L^\nu T^\mu) + (\not{g}^{-1})^{AB}X_A^\mu X_B^\nu \tag{3.18}$$

From (3.18), we have

$$\text{tr}\alpha = (\not{g}^{-1})^{AB}R_{\mu\kappa\nu\lambda}X_A^\mu L^\kappa X_B^\nu L^\lambda = (g^{-1})^{\mu\nu}R_{\mu\kappa\nu\lambda}L^\kappa L^\lambda = S_{\kappa\lambda}L^\kappa L^\lambda \tag{3.19}$$

On the other hand,

$$L(\text{tr}\chi) = L[(\not{g}^{-1})^{AB}\chi_{AB}] = (\not{g}^{-1})^{AB}L(\chi_{AB}) + \chi_{AB}L[(\not{g}^{-1})^{AB}] \tag{3.20}$$

This equals

$$(\not{g}^{-1})^{AB}L(\chi_{AB}) - \chi_{AB}(\not{g}^{-1})^{AC}(\not{g}^{-1})^{BD}L(\not{g}_{CD}) \tag{3.21}$$

i.e.

$$(\not{g}^{-1})^{AB}L(\chi_{AB}) - 2\chi_{AB}(\not{g}^{-1})^{AC}(\not{g}^{-1})^{BD}\chi_{CD} \tag{3.22}$$

where we have used the following fact:

$$L[(\not{g}^{-1})^{AB}] = -(\not{g}^{-1})^{AC}(\not{g}^{-1})^{BD}L(\not{g}_{CD})$$

This is simply the derivative of the reciprocal of a non-degenerate matrix:

$$\frac{dM^{-1}}{dt} = -M^{-1}\frac{dM}{dt}M^{-1}$$

Since we have

$$2\chi_{AB} = L(\not{g}_{AB}) \tag{3.23}$$

then taking trace with (3.16), we get

$$L(\text{tr}\chi) = \mu^{-1}(L\mu)\text{tr}\chi - |\chi|^2_{\not{g}} - S(L, L) \tag{3.24}$$

We shall write down the Gauss and Codazzi equations of the embedding of $S_{t,u}$ in the acoustical spacetime. First, we consider $S_{t,u}$ as a submanifold of $\Sigma_t$. In this case, the normal vectorfield is $T$.

We define $\theta$, the second fundamental form of $S_{t,u}$ relative to $\Sigma_t$, by

$$2\kappa\theta := \not{\mathcal{L}}_T\bar{g} \tag{3.25}$$

For any two vectors tangent to $S_{t,u}$ at a point, we have

$$\kappa\theta(X, Y) = \bar{g}(\bar{D}_X T, Y) = \bar{g}(\bar{D}_Y T, X) \tag{3.26}$$

Denote by $\not{k}$, the restriction of $k$, the second fundamental form of $\Sigma_t$ in spacetime manifold, to $TS_{t,u}$, then by (2.53), we have

$$\chi = \alpha(\not{k} - \theta) \tag{3.27}$$

The Gauss equation of $S_{t,u}$ in $\Sigma_t$ is expressed in terms of $(X_1, X_2)$ by

$$\not{R}_{ABCD} - \theta_{AC}\theta_{BD} + \theta_{AD}\theta_{BC} = 0 \tag{3.28}$$

$\not{R}$ is the curvature tensor of $\not{g}$. Since $\dim S_{t,u} = 2$, we have

$$\not{R}_{ABCD} = K(\not{g}_{AC}\not{g}_{BD} - \not{g}_{AD}\not{g}_{BC}) \tag{3.29}$$

where $K$ is the Gauss curvature of $S_{t,u}$. Then we have

$$\theta_{AC}\theta_{BD} - \theta_{AD}\theta_{BC} = K(\not{g}_{AC}\not{g}_{BD} - \not{g}_{AD}\not{g}_{BC}) \tag{3.30}$$

contracting this with $\frac{1}{2}(\not{g}^{-1})^{AC}(\not{g}^{-1})^{BD}$:

$$\frac{1}{2}(\mathrm{tr}_{\not{g}}\theta)^2 - \frac{1}{2}|\theta|^2_{\not{g}} = K \tag{3.31}$$

Also from the Gauss equations of the embedding of $\Sigma_t$ in the whole manifold, we have

$$\not{k}_{AC}\not{k}_{BD} - \not{k}_{BC}\not{k}_{AD} = R_{ABCD} \tag{3.32}$$

In view of the symmetries of the curvature tensor we can express

$$R_{ABCD} = \rho\epsilon_{AB}\epsilon_{CD} \tag{3.33}$$

where $\epsilon_{AB}$ are the components of the area form of $(S_{t,u}, \not{g})$:

$$\epsilon_{AB} = \sqrt{\det \not{g}}[AB] \tag{3.34}$$

Contracting (3.33) with $\frac{1}{2}(\not{g}^{-1})^{AC}(\not{g}^{-1})^{BD}$ and taking into account the fact that

$$(\not{g}^{-1})^{BD}\epsilon_{AB}\epsilon_{CD} = \not{g}_{AC} \tag{3.35}$$

we obtain

$$\frac{1}{2}(\mathrm{tr}_{\not{g}}\not{k})^2 - \frac{1}{2}|\not{k}|^2_{\not{g}} = \rho \tag{3.36}$$

This is the Gauss equation of $S_{t,u}$ in acoustical spacetime.

We then derive the Codazzi equation: Let $X, Y, Z$ be vectorfields tangent to $S_{t,u}$, we can extend them to $C_u$ by pushing forward by the flow of $L$. They are then tangential to each of the sections $S_{t,u}$, $t \in [0, t_{*\varepsilon_0})$. We have

$$X(\chi(Y,Z)) = X(g(D_Y L, Z)) = g(D_X D_Y L, Z) + g(D_Y L, D_X Z) \tag{3.37}$$

On the other hand,

$$(\not{D}_X\chi)(Y,Z) = X(\chi(Y,Z)) - \chi(\not{D}_X Y, Z) - \chi(Y, \not{D}_X Z) \tag{3.38}$$

Here $\not{D}$ is the covariant derivative operator on the $S_{t,u}$ associated to the induced metric $\not{g}$. Hence

$$(\not{D}_X\chi)(Y,Z) = g(D_X D_Y L, Z) + g(D_Y L, D_X Z) - \chi(\not{D}_X Y, Z) - \chi(Y, \not{D}_X Z) \quad (3.39)$$

Exchanging $X, Y$, and noting that

$$D_X D_Y L - D_Y D_X L = R(X,Y)L + D_{[X,Y]}L \quad (3.40)$$

while $g(R(X,Y)L, Z) = R(Z,L,X,Y)$ and $\not{D}_X Y - \not{D}_Y X = [X,Y]$, we obtain

$$\begin{aligned}
&(\not{D}_X\chi)(Y,Z) - (\not{D}_Y\chi)(X,Z) \\
&= R(Z,L,X,Y) + g(D_{[X,Y]}L, Z) + g(D_Y L, D_X Z) - g(D_X L, D_Y Z) \\
&\quad - \chi(\not{D}_X Y, Z) + \chi(\not{D}_Y X, Z) - \chi(Y, \not{D}_X Z) + \chi(X, \not{D}_Y Z) \\
&= R(Z,L,X,Y) + g(D_Y L, D_X Z) - g(D_X L, D_Y Z) - \chi(Y, \not{D}_X Z) + \chi(X, \not{D}_Y Z)
\end{aligned}$$

Define the 1-form $\zeta$ on $S_{t,u}$, by

$$\zeta(X) = g(D_X L, T) \quad (3.41)$$

for any vector $X$ tangent to $S_{t,u}$ at some point. We shall derive an expression for $\zeta$. For this, we set $D_X L = xL + yT + z^A X_A$. Then

$$g(D_X L, T) = -\mu x + \kappa^2 y, \quad g(D_X L, L) = -\mu y, \quad g(D_X L, X_A) = z^B \not{g}_{AB} \quad (3.42)$$

Solving for the coefficients $x, y, z^A$, we obtain

$$D_X L = -\mu^{-1}\zeta(X)L + \chi \cdot X \quad (3.43)$$

where $\not{g}(\chi \cdot X, Y) = \chi(X,Y)$. Then

$$\begin{aligned}
&g(D_X L, D_Y Z) \\
&= \mu^{-1}\zeta(X)\chi(Y,Z) + g(\chi \cdot X, D_Y Z) \\
&= \mu^{-1}\zeta(X)\chi(Y,Z) + \chi(X, \not{D}_Y Z)
\end{aligned} \quad (3.44)$$

Substituting this into the expression above, we get

$$\begin{aligned}
&(\not{D}_X\chi)(Y,Z) - (\not{D}_Y\chi)(X,Z) \\
&= R(Z,L,X,Y) - \mu^{-1}(\zeta(X)\chi(Y,Z) - \zeta(Y)\chi(X,Z))
\end{aligned} \quad (3.45)$$

In this equation, we set $X = X_A, Y = X_B, Z = X_C$. Defining the 1-form $\beta$ on the $S_{t,u}$ by

$$R(X_C, L, X_A, X_B) = \beta_C \epsilon_{AB} \quad (3.46)$$

we obtain the Codazzi equation:

$$\not{D}_A \chi_{BC} - \not{D}_B \chi_{AC} = \beta_C \epsilon_{AB} - \mu^{-1}(\zeta_A \chi_{BC} - \zeta_B \chi_{AC}) \quad (3.47)$$

Denoting by $\not{\text{curl}}\chi$ the 1-form on $S_{t,u}$ with components:

$$\not{\text{curl}}\chi_C := \frac{1}{2}\epsilon^{AB}(\not{D}_A \chi_{BC} - \not{D}_B \chi_{AC}) \quad (3.48)$$

where $\epsilon^{AB} = (\not{g}^{-1})^{AC}(\not{g}^{-1})^{BD}\epsilon_{CD} = (\sqrt{\det \not{g}})^{-1}[AB]$, we can write the Codazzi equation as follows:

$$\not{\text{curl}}\chi = \beta - \mu^{-1}\zeta \wedge \chi \tag{3.49}$$

Here $(\zeta \wedge \chi)_C = \frac{1}{2}\epsilon^{AB}(\zeta_A\chi_{BC} - \zeta_B\chi_{AC})$, and this equation has the whole content of Equation (3.47). Contracting (3.47) with $(\not{g}^{-1})^{AC}$, we obtain the contracted form of the Codazzi equation

$$\not{\text{div}}\chi - \not{d}\text{tr}\chi = \beta^* - \mu^{-1}(\zeta \cdot \chi - \zeta\text{tr}\chi) \tag{3.50}$$

which is equivalent to Equation (3.49). Here $(\zeta \cdot \chi)_B = (\not{g}^{-1})^{AC}\zeta_A\chi_{BC}, \beta^*_B = (\not{g}^{-1})^{AC}\beta_C\epsilon_{AB}$.

We shall now derive an expression for $\zeta$, from the expression for $L$,

$$\zeta(X) = g(D_X B, T) - \alpha\kappa^{-1}g(D_X T, T) - g(T,T)X(\alpha\kappa^{-1}) \tag{3.51}$$

While $g(T,T) = \kappa^2$, we have $g(D_X T, T) = \kappa X\kappa$, so this reduces to

$$\zeta(X) = g(D_X B, T) - \kappa X\alpha \tag{3.52}$$

Define 1-form $\varepsilon$ on $S_{t,u}$ by $\kappa\varepsilon(X) = k(X,T)$. Then

$$\zeta = \kappa(\alpha\varepsilon - \not{d}\alpha) \tag{3.53}$$

We now define $S_{t,u}$ 1-form

$$\eta := \zeta + \not{d}\mu \tag{3.54}$$

Obviously, we have

$$\eta(X) = -g(D_X T, L) \tag{3.55}$$

We can express the commutator $\Lambda$ of $L$ and $T$, a vectorfield tangential to $S_{t,u}$, in terms of $\zeta$ and $\eta$. We have

$$\Lambda = \Lambda^A X_A, \quad \Lambda^A \not{g}_{AB} = g(X_B, D_L T) - g(X_B, D_T L) \tag{3.56}$$

and

$$g(X_B, D_L T) = -g(D_L X_B, T) = -g(D_{X_B} L, T) = -\zeta_B \tag{3.57}$$

To compute $g(X_B, D_T L)$, we express $L$ in terms of the geodesic vectorfield $\hat{L}$. We have

$$\begin{aligned} g(X_B, D_T L) &= g(X_B, D_T(\mu\hat{L})) = \mu g(X_B, D_T\hat{L}) \\ &= \mu g(T, D_{X_B}\hat{L}) = \mu g(T, D_{X_B}(\mu^{-1}L)) \end{aligned} \tag{3.58}$$

Here we have used the fact that $\hat{L}^\mu = -(g^{-1})^{\mu\nu}\partial_\nu u$, that is, $\hat{L}$ is gradient of a function, so we can exchange $T$ and $X_B$. This equals

$$g(T, D_{X_B}L) - \mu^{-1}(X_B\mu)g(T,L) = \zeta_B + X_B\mu = \eta_B \tag{3.59}$$

So we get

$$\Lambda^A = -(\not{g}^{-1})^{AB}(\zeta_B + \eta_B) \tag{3.60}$$

We shall express $D_T L$ and $D_L T$ in the frames $(L, T, X_1, X_2)$. First, by (3.59) and (3.60),

$$g(D_T L, X_B) = \eta_B \tag{3.61}$$

and

$$g(D_T L, T) = g(D_L T, T) = L(\frac{1}{2}\kappa^2), \quad g(D_T L, L) = 0 \tag{3.62}$$

It follows that

$$D_T L = (\not{g}^{-1})^{AB} \eta_B X_A - \alpha^{-1}(L\kappa)L \tag{3.63}$$

Also,

$$g(D_L T, X_B) = -g(D_L X_B, T) = -g(D_{X_B} L, T) = -\zeta_B \tag{3.64}$$

and

$$g(D_L T, L) = g(D_T L, L) = 0, \quad g(D_L T, T) = \frac{1}{2}L(\kappa^2) \tag{3.65}$$

where we have used the fact that $[L, T]$ is tangential to $S_{t,u}$. We thus obtain

$$D_L T = -\not{g}^{AB} \zeta_B X_A - \alpha^{-1}(L\kappa)L \tag{3.66}$$

Next, we compute $D_{X_A} T$. Set

$$D_{X_A} T = a_A L + b_A T + c_A^B X_B \tag{3.67}$$

Taking inner product with $L$:

$$-\mu b_A = g(D_{X_A} T, L) = -\eta_A \Rightarrow b_A = \mu^{-1}\eta_A \tag{3.68}$$

Taking inner product with $T$:

$$-\mu a_A + \kappa^2 b_A = g(D_{X_A} T, T) = \frac{1}{2} X_A(\kappa^2) \tag{3.69}$$

Thus we obtain

$$\begin{aligned} a_A &= \alpha^{-2}\eta_A - \alpha^{-1} X_A(\kappa) \\ &= \alpha^{-2}(\eta_A - X_A(\mu) + \kappa X_A(\alpha)) \\ &= \alpha^{-2}(\zeta_A + \kappa X_A(\alpha)) = \alpha^{-1}\kappa\varepsilon_A \end{aligned} \tag{3.70}$$

Finally, taking inner product with $X_C$:

$$c_A^B \not{g}_{BC} = g(D_{X_A} T, X_C) = \kappa\theta_{AC} \tag{3.71}$$

We conclude that

$$D_{X_A} T = \alpha^{-1}\kappa\varepsilon_A L + \mu^{-1}\eta_A T + \kappa\theta_{AB}(\not{g}^{-1})^{BC} X_C \tag{3.72}$$

In the following, we will compute $D_T T$. We may decompose

$$D_T T = \bar{D}_T T + aB \tag{3.73}$$

where $\bar{D}$ is the covariant derivative operator on the $\Sigma_t$, associated to the induced metric $\bar{g}$, which, as we have seen, coincides with the Euclidean metric. Taking inner product with $B$:

$$\begin{aligned} -\alpha^2 a &= g(D_T T, B) = g(D_T T, L) + \alpha\kappa^{-1} g(D_T T, T) \\ &= -T\mu - g(T, D_T L) + \alpha\kappa^{-1} T(\frac{1}{2}\kappa^2) \end{aligned} \tag{3.74}$$

where we have used the fact that $g(D_T T, L) = T(g(T,L)) - g(T, D_T L)$. The right-hand side of (3.74) equals

$$-T\mu - \kappa L\kappa + \alpha T\kappa = -\kappa(T\alpha + L\kappa) \tag{3.75}$$

Using the fact that $[L,T]$ is tangential to $S_{t,u}$, so $g(T, D_T L) = g(T, D_L T) = \frac{1}{2}L(\kappa^2)$.

We may use the rectangular coordinates on $\Sigma_t$, the induced metric $\bar{g}$ being Euclidean. For any pair of vectorfields $X, Y$ tangential to $\Sigma_t$ we then have, in terms of components in rectangular coordinates,

$$(\bar{D}_X Y)^i = X^j \partial_j Y^i \tag{3.76}$$

So we have

$$(\bar{D}_T T)^i = T^j \bar{D}_j T^i = T^j \partial_j T^i \tag{3.77}$$

and

$$\bar{g}_{ij} T^j = \kappa^2 \partial_i u, \quad \bar{g}_{ij} = \delta_{ij} \Rightarrow T^i = \kappa^2 \partial_i u \tag{3.78}$$

So,

$$(\bar{D}_T T)^i = \kappa^2 \partial_j u \partial_j (\kappa^2 \partial_i u) = \kappa^2 \partial_j u \partial_j \kappa^2 \partial_i u + \kappa^4 \partial_j u \partial_j \partial_i u \tag{3.79}$$

By (3.78), $\sum_i (\partial_i u)^2 = \kappa^{-2}$, then we have

$$(\bar{D}_T T)^i = \kappa^{-2} T^i T^j \partial_j(\kappa^2) + \frac{1}{2}\kappa^4 \partial_i(\kappa^{-2}) \tag{3.80}$$

This equals

$$\kappa^{-2} T^i T^j \partial_j(\kappa^2) - \frac{1}{2}\partial_i(\kappa^2) \tag{3.81}$$

$$\Rightarrow \bar{D}_T T = \frac{1}{2}\kappa^{-2} T(\kappa^2) T - \frac{1}{2}\not{g}^{AB} X_B(\kappa^2) X_A \tag{3.82}$$

In view of (3.74), (3.75) and (3.82), we conclude that

$$D_T T = \kappa\alpha^{-2}(T\alpha + L\kappa)L + (\alpha^{-1}(T\alpha + L\kappa) + \kappa^{-1}T\kappa)T - \frac{1}{2}(\not{g}^{-1})^{AB} X_B(\kappa^2) X_A$$

In regard to (3.74),

$$g(D_T T, B) = -g(T, D_T B) = -\alpha k(T,T) \tag{3.83}$$

Comparing with (3.75), we see that

$$L(\frac{1}{2}\kappa^2) = -\kappa T\alpha + \alpha k(T,T) \tag{3.84}$$

Using this equation, we shall derive a propagation equation along the generator of $C_u$ for $\mu$:

$$L\mu = \kappa L\alpha + \alpha L\kappa = \kappa L\alpha - \alpha T\alpha + \alpha\mu k(\hat{T},\hat{T}) \tag{3.85}$$

First,

$$\alpha\mu k(\hat{T},\hat{T}) = -\mu\hat{T}^i\hat{T}^j\partial_i\psi_j = -\alpha\hat{T}^i(T\psi_i) \tag{3.86}$$

We have used the fact that $\alpha k_{ij} = -\partial_i\psi_j$ (see (3.5)).

To compute the first two terms on the right hand side of (3.85), we shall use $\alpha d\alpha = \frac{1}{2}d\alpha^2 = \frac{1}{2}d\eta^2$. While

$$\eta^2 = \frac{\rho}{\rho'} \Rightarrow d\eta^2 = (\frac{\rho}{\rho'})' dh \tag{3.87}$$

It follows that

$$\kappa L\alpha - \alpha T\alpha = \frac{1}{2}\frac{\mu}{\alpha^2}(\frac{\rho}{\rho'})' Lh - \frac{1}{2}(\frac{\rho}{\rho'})' Th \tag{3.88}$$

Substituting (3.88) and (3.86) into (3.85), we get

$$L\mu = \frac{1}{2}\frac{\mu}{\alpha^2}(\frac{\rho}{\rho'})' Lh - \frac{1}{2}(\frac{\rho}{\rho'})' Th - \alpha\hat{T}^i(T\psi_i) \tag{3.89}$$

Recalling that

$$L = \partial_t - (\alpha\hat{T}^i + \psi_i)\partial_i \Rightarrow -\alpha\hat{T}^i = L^i + \psi_i \tag{3.90}$$

we express

$$-\alpha\hat{T}^i(T\psi_i) = L^i(T\psi_i) + \sum_i \psi_i T\psi_i = T^i(L\psi_i) - Th = \alpha^{-1}\mu\hat{T}^i(L\psi_i) - Th \tag{3.91}$$

We thus obtain the following propagation equation for $\mu$:

$$L\mu = m + \mu e \tag{3.92}$$

where

$$m = \frac{1}{2}\frac{dH}{dh}Th, \quad H = -2h - \eta^2 \tag{3.93}$$

and

$$e = \frac{1}{2\eta^2}(\frac{\rho}{\rho'})' Lh + \eta^{-1}\hat{T}^i(L\psi_i) \tag{3.94}$$

To complete the set of connection coefficients of the frame field $(L, T, X_1, X_2)$, we should compute $D_{X_A}X_B$. We decompose

$$D_{X_A}X_B = \not{D}_{X_A}X_B + a_{AB}L + b_{AB}T \tag{3.95}$$

then

$$-\mu b_{AB} = g(D_{X_A}X_B, L) = -\chi_{AB} \tag{3.96}$$

and

$$-\mu a_{AB} + \kappa^2 b_{AB} = g(T, D_{X_A} X_B) = -\kappa\theta_{AB} \tag{3.97}$$

By $\chi = \alpha(\not{k} - \theta)$, we get

$$b_{AB} = \mu^{-1}\chi_{AB}, \quad a_{AB} = \alpha^{-1}\not{k}_{AB} \tag{3.98}$$

So we get the following table:

$$D_L L = \mu^{-1}(L\mu)L \tag{3.99}$$
$$D_T L = (\not{g}^{-1})^{AB}\eta_B X_A - \alpha^{-1}(L\kappa)L \tag{3.100}$$
$$D_L T = -(\not{g}^{-1})^{AB}\zeta_B X_A - \alpha^{-1}(L\kappa)L \tag{3.101}$$
$$D_{X_A} L = -\mu^{-1}\zeta_A L + \chi_A^B X_B \tag{3.102}$$
$$D_T T = \alpha^{-2}\kappa(T\alpha + L\kappa)L + (\alpha^{-1}L\kappa + \mu^{-1}T\mu)T - \frac{1}{2}(\not{g}^{-1})^{AB} X_B(\kappa^2) X_A \tag{3.103}$$
$$D_{X_A} T = \alpha^{-1}\kappa\varepsilon_A L + \mu^{-1}\eta_A T + \kappa\theta_{AB}(\not{g}^{-1})^{BC} X_C \tag{3.104}$$
$$D_L X_A = D_{X_A} L \tag{3.105}$$
$$D_{X_A} X_B = \not{D}_{X_A} X_B + \alpha^{-1}\not{k}_{AB} L + \mu^{-1}\chi_{AB} T \tag{3.106}$$

We now investigate the connection between the Lie derivative of $\chi$ with respect to $T$ and the derivative of $\eta$ tangential to $S_{t,u}$.

First, we extend $\chi$ from $TS_{t,u}$ for each $t, u$ to $TW^*_{\epsilon_0}$ by the conditions:

$$\chi(X, L) = \chi(X, T) = 0 \tag{3.107}$$

We define $\not{\mathcal{L}}_T \chi$ to be the restriction of $\mathcal{L}_T \chi$ to $S_{t,u}$:

$$(\not{\mathcal{L}}_T \chi)(X_A, X_B) = (D_T \chi)(X_A, X_B) + \chi(X_A, D_{X_B} T) + \chi(X_B, D_{X_A} T) \tag{3.108}$$

Due to the extension of $\chi$ and using (3.99)—(3.106), this equals

$$(D_T \chi)(X_A, X_B) + \kappa\theta_{BC}\chi_A^C + \kappa\theta_{AC}\chi_B^C \tag{3.109}$$

To compute $(D_T \chi)(X_A, X_B)$:

$$(D_T \chi)(X_A, X_B) = T(\chi_{AB}) - \chi(\Pi D_T X_A, X_B) - \chi(\Pi D_T X_B, X_A) \tag{3.110}$$

where $\Pi$ is the $g$-projection to $S_{t,u}$. Now

$$T(\chi_{AB}) = T(g(D_{X_A} L, X_B)) = g(D_T D_{X_A} L, X_B) + g(D_{X_A} L, D_T X_B) \tag{3.111}$$

While

$$\begin{aligned} & g(D_T D_{X_A} L, X_B) \\ &= g(D_{X_A} D_T L, X_B) + g(D_{[T, X_A]} L, X_B) + R(X_B, L, T, X_A) \end{aligned} \tag{3.112}$$

From (3.100),

$$g(D_{X_A}D_T L, X_B) = g(D_{X_A}(\eta^C X_C - \alpha^{-1}(L\kappa)L), X_B) \tag{3.113}$$

This equals

$$-\alpha^{-1}(L\kappa)\chi_{AB} + X_A(\eta_B) - \eta(\not{D}_{X_A}X_B) = -\alpha^{-1}(L\kappa)\chi_{AB} + \not{D}_A\eta_B \tag{3.114}$$

Since $[T, X_A]$ is tangential to $S_{t,u}$, we have

$$\begin{aligned} g(D_{[T,X_A]}L, X_B) &= \chi([T, X_A], X_B) \\ &= \chi(\Pi D_T X_A, X_B) - \chi(\Pi D_{X_A}T, X_B) \qquad (3.115) \\ &= \chi(\Pi D_T X_A, X_B) - \kappa\theta_{AC}\chi_B^C \qquad (3.116) \end{aligned}$$

Also,

$$g(D_{X_A}L, D_T X_B) = -\mu^{-1}\zeta_A g(L, D_T X_B) + g(\chi_{AC}X^C, \Pi D_T X_B) \tag{3.117}$$

which equals

$$\mu^{-1}\zeta_A\eta_B + \chi(X_A, \Pi D_T X_B) \tag{3.118}$$

So we obtain

$$\begin{aligned} &(D_T\chi)(X_A, X_B) \\ &= \not{D}_A\eta_B + \mu^{-1}\zeta_A\eta_B - \alpha^{-1}(L\kappa)\chi_{AB} - \kappa\theta_{AC}\chi_B^C - R(X_A, T, X_B, L) \qquad (3.119) \end{aligned}$$

The symmetric part of this equation is

$$\begin{aligned} (D_T\chi)(X_A, X_B) &= \frac{1}{2}(\not{D}_A\eta_B + \not{D}_B\eta_A) + \frac{1}{2}\mu^{-1}(\zeta_A\eta_B + \zeta_B\eta_A) \qquad (3.120) \\ &- \alpha^{-1}(L\kappa)\chi_{AB} - \frac{1}{2}\kappa(\theta_{AC}\chi_B^C + \theta_{BC}\chi_A^C) \qquad (3.121) \\ &- \frac{1}{2}(R(X_A, T, X_B, L) + R(X_B, T, X_A, L)) \qquad (3.122) \end{aligned}$$

The antisymmetric part of this equation contains no new information.

Denoting

$$(\not{\mathcal{L}}_T\chi)(X_A, X_B) = \not{\mathcal{L}}_T\chi_{AB} \tag{3.123}$$

our conclusion is

$$\begin{aligned} \not{\mathcal{L}}_T\chi_{AB} &= \frac{1}{2}(\not{D}_A\eta_B + \not{D}_B\eta_A) + \frac{1}{2}\mu^{-1}(\zeta_A\eta_B + \zeta_B\eta_A) \qquad (3.124) \\ &- \alpha^{-1}(L\kappa)\chi_{AB} + \frac{1}{2}\kappa(\theta_{AC}\chi_B^C + \theta_{BC}\chi_A^C) - \gamma_{AB} \qquad (3.125) \end{aligned}$$

where

$$\gamma_{AB} = \frac{1}{2}(R(X_A, T, X_B, L) + R(X_B, T, X_A, L)) \tag{3.126}$$

We shall also use the null frame $(L, \underline{L}, X_1, X_2)$, where

$$\underline{L} = \alpha^{-1}\kappa L + 2T \tag{3.127}$$

One can check that $\underline{L}$ is an incoming future directed null vector satisfying the following condition:

$$g(L,\underline{L}) = -2\mu \tag{3.128}$$

The inverse of the acoustical metric is expressed in this frame by

$$(g^{-1})^{\mu\nu} = -\frac{1}{2\mu}(L^\mu \underline{L}^\nu + L^\nu \underline{L}^\mu) + (\not{g}^{-1})^{AB} X_A^\mu X_B^\nu \tag{3.129}$$

From (3.99)—(3.106), we have the following table:

$$D_L L = \mu^{-1}(L\mu)L \tag{3.130}$$

$$D_{\underline{L}} L = -L(\alpha^{-1}\kappa)L + 2\eta^A X_A \tag{3.131}$$

$$D_{X_A} L = -\mu^{-1}\zeta_A L + \chi_A^B X_B \tag{3.132}$$

$$D_L \underline{L} = -2\zeta^A X_A \tag{3.133}$$

$$D_{\underline{L}} \underline{L} = (\mu^{-1}\underline{L}\mu + L(\alpha^{-1}\kappa))\underline{L} - 2\mu(\not{g}^{-1})^{AB} X_B(\alpha^{-1}\kappa)X_A \tag{3.134}$$

$$D_{X_A} \underline{L} = \mu^{-1}\eta_A \underline{L} + \underline{\chi}_A^B X_B \tag{3.135}$$

$$D_L X_A = D_{X_A} L \tag{3.136}$$

$$D_{X_A} X_B = \not{D}_{X_A} X_B + \frac{1}{2}\mu^{-1}\underline{\chi}_{AB} L + \frac{1}{2}\mu^{-1}\chi_{AB}\underline{L} \tag{3.137}$$

Similarly, we have

$$\underline{\chi} = \kappa(\not{k} + \theta) \tag{3.138}$$

This completes the exposition of the acoustical structure equations.

We shall presently derive expressions for the operators $\Box_g$ and $\Box_{\tilde{g}}$. We have

$$\Box_g f = \mathrm{tr}(D^2 f) \tag{3.139}$$

where tr denotes the trace with respect to $g$ and $D^2 f$ denotes the Hessian of $f$ with respect to $g$. In view of the expression (3.129) for $g^{-1}$, we have

$$\Box_g f = (g^{-1})^{\mu\nu}(D^2 f)_{\mu\nu} = -\mu^{-1}(D^2 f)_{\underline{L}L} + (\not{g}^{-1})^{AB}(D^2 f)_{AB} \tag{3.140}$$

Also we can consider the Hessian of the restriction of $f$ to $S_{t,u}$ with respect to the induced metric $\not{g}$:

$$\not{D}^2 f \tag{3.141}$$

then the operator $\not{\Delta} = \Delta_{\not{g}}$ is given by

$$\not{\Delta} = \mathrm{tr}(\not{D}^2 f) \tag{3.142}$$

Here tr is the trace with respect to $\not{g}$. In terms of the frame, we have

$$(D^2 f)_{AB} = X_A(X_B f) - (D_{X_A} X_B)f \tag{3.143}$$

and

$$(\not{D}^2 f)_{AB} = X_A(X_B f) - (\not{D}_{X_A} X_B)f \tag{3.144}$$

By (3.137), we have

$$(D^2 f)_{AB} = (\not{D}^2 f)_{AB} - \frac{1}{2}\mu^{-1}\underline{\chi}_{AB}(Lf) - \frac{1}{2}\mu^{-1}\chi_{AB}(\underline{L}f) \tag{3.145}$$

Hence

$$(\not{g}^{-1})^{AB}(D^2 f)_{AB} = \not{\Delta} f - \frac{1}{2}\mu^{-1}\mathrm{tr}\,\underline{\chi}(Lf) - \frac{1}{2}\mu^{-1}\mathrm{tr}\,\chi(\underline{L}f) \tag{3.146}$$

Also, by (3.133),

$$(D^2 f)_{\underline{L}L} = L(\underline{L}f) - (D_L\underline{L})f = L(\underline{L}f) + 2\zeta\cdot \not{d}f \tag{3.147}$$

So we get the following expressions for $\Box_g$:

$$\Box_g f = \not{\Delta} f - \frac{1}{2}\mu^{-1}\mathrm{tr}\underline{\chi}(Lf) - \frac{1}{2}\mu^{-1}\mathrm{tr}\chi(\underline{L}f) - \mu^{-1}L(\underline{L}f) - 2\mu^{-1}\zeta\cdot\not{d}f \tag{3.148}$$

Since the conformal acoustical metric $\tilde{g}$ is given by

$$\tilde{g}_{\mu\nu} = \Omega g_{\mu\nu} \tag{3.149}$$

we then get

$$\begin{aligned}\Box_{\tilde{g}} &= \Omega^{-1}\Box_g f + \Omega^{-2}(g^{-1})^{\mu\nu}\partial_\mu\Omega\partial_\nu f\\ &= \Omega^{-1}\Box_g f + \Omega^{-2}\frac{d\Omega}{dh}(g^{-1})^{\mu\nu}\partial_\mu h\partial_\nu f\end{aligned} \tag{3.150}$$

In view of (3.129), we have

$$(g^{-1})^{\mu\nu}\partial_\mu h\partial_\nu f = -\frac{1}{2}\mu^{-1}(\underline{L}h)(Lf) - \frac{1}{2}\mu^{-1}(\underline{L}f)(Lh) + \not{d}h\cdot\not{d}f \tag{3.151}$$

Thus, setting

$$\nu = \frac{1}{2}(\mathrm{tr}\,\chi + \Omega^{-1}\frac{d\Omega}{dh}(Lh)) \tag{3.152}$$

$$\underline{\nu} = \frac{1}{2}(\mathrm{tr}\,\underline{\chi} + \Omega^{-1}\frac{d\Omega}{dh}(\underline{L}h)) \tag{3.153}$$

we obtain the formula

$$\Omega\Box_{\tilde{g}} f = \not{\Delta} f - \mu^{-1}L(\underline{L}f) - \mu^{-1}(\nu\underline{L}f + \underline{\nu}Lf) - 2\mu^{-1}\zeta\cdot\not{d}f + \Omega^{-1}\frac{d\Omega}{dh}\not{d}h\cdot\not{d}f \tag{3.154}$$

## 3.2 The Derivatives of the Rectangular Components of $L$ and $\hat{T}$

We shall now draw the conclusion that the derivatives of $L^i$, $\hat{T}^i$ , the rectangular components of $L$ and $\hat{T}$, with respect to $L$, $T$ and $X_A$ are regular as $\mu \to 0$. In the following we shall denote by $\nabla$ the covariant differentiation with respect to the natural connection associated to the affine space $\mathcal{G}$, the Galilean spacetime. The $t$ and $x^i, i = 1,2,3$ being linear coordinates on $\mathcal{G}$, $\nabla$ amounts simply to partial differentiation in these coordinates. We have

$$D_\mu W^\nu = \nabla_\mu W^\nu + \Gamma^\nu_{\mu\lambda} W^\lambda \tag{3.155}$$

$$\Gamma^\nu_{\mu\lambda} = (g^{-1})^{\nu\kappa} \Gamma_{\mu\lambda\kappa} \tag{3.156}$$

$$\Gamma_{\mu\lambda\kappa} = \frac{1}{2}(\partial_\mu g_{\lambda\kappa} + \partial_\lambda g_{\mu\kappa} - \partial_\kappa g_{\mu\lambda}) \tag{3.157}$$

In the irrotational and isentropic case, recall that $v^i = -\partial_i \phi$, then the acoustical metric can be written as follows:

$$g = -\eta^2 dt^2 + \sum_i (dx^i + \partial_i \phi dt)^2 \tag{3.158}$$

By direct calculation, we have the Christoffel symbols:

$$\Gamma_{000} = \frac{1}{2}\partial_0(-\eta^2 + |\mathbf{v}|^2) \tag{3.159}$$

$$\Gamma_{0i0} = \frac{1}{2}\partial_i(-\eta^2 + |\mathbf{v}|^2) \tag{3.160}$$

$$\Gamma_{ij0} = \partial_i \partial_j \phi \tag{3.161}$$

$$\Gamma_{00k} = \partial_0 \partial_k \phi - \frac{1}{2}\partial_k(-\eta^2 + |\mathbf{v}|^2) \tag{3.162}$$

$$\Gamma_{i0k} = \Gamma_{ijk} = 0 \tag{3.163}$$

Consider now the vectorfield

$$L(L^\mu)\partial_\mu \tag{3.164}$$

Since $L(L^0) = 0$, it can be expanded as

$$a_L \hat{T} + b_L \tag{3.165}$$

where $b_L \in TS_{t,u}$. On the other hand, (3.165) is also $\nabla_L L$. That is $\nabla_L L = a_L \hat{T} + b_L$.

Taking inner product with $\hat{T}$:

$$a_L = g(\nabla_L L, \hat{T}) \tag{3.166}$$

Writing

$$b_L = b_L^A X_A \tag{3.167}$$

and taking inner product with $X_B$:

$$g(\nabla_L L, X_B) = b_L^A \not{g}_{AB} \tag{3.168}$$

while

$$g(\nabla_L L, \hat{T}) = g(D_L L, \hat{T}) - \Gamma_{\alpha\beta\nu} L^\alpha L^\beta \hat{T}^\nu \tag{3.169}$$

From (3.92) and (3.99),

$$g(D_L L, \hat{T}) = -\kappa^{-1}(L\mu) = -\kappa^{-1}m - \alpha e \tag{3.170}$$

Also,

$$\Gamma_{\alpha\beta\nu}L^\alpha L^\beta \hat{T}^\nu = \Gamma_{00k}\hat{T}^k = \hat{T}(\psi_0) - \tfrac{1}{2}\hat{T}(-\eta^2 + |\mathbf{v}|^2) = -\tfrac{1}{2}\tfrac{dH}{dh}\hat{T}(h) \tag{3.171}$$

$$\Rightarrow g(\nabla_L L, \hat{T}) = -\alpha e \tag{3.172}$$

The $\kappa^{-1}$ term cancels! From (3.94), we have

$$a_L = -\alpha e = -\frac{1}{2\alpha}L(\eta^2) - \hat{T}^i(L\psi_i) \tag{3.173}$$

Since $g(D_L L, X_B) = 0$, we have

$$\begin{aligned} g(\nabla_L L, X_B) &= -\Gamma_{\alpha\beta\nu}L^\alpha L^\beta X_B^\nu = -\Gamma_{00k}X_B^k \\ &= -X_B(\psi_0) + \frac{1}{2}X_B(-\eta^2 + |\mathbf{v}|^2) \\ &= \frac{1}{2}\frac{dH}{dh}X_B(h) \end{aligned} \tag{3.174}$$

So

$$b_L^A = (\not{g}^{-1})^{AB} g(\nabla_L L, X_B) = \frac{1}{2}\frac{dH}{dh}X^A(h) \tag{3.175}$$

Consider next the vectorfield $L(\hat{T}^\mu)\partial_\mu$. Also it can be expanded as

$$\nabla_L \hat{T} = p_L \hat{T} + q_L \tag{3.176}$$

Obviously,

$$p_L = g(\nabla_L \hat{T}, \hat{T}) \tag{3.177}$$

Writing

$$q_L = q_L^A X_A \tag{3.178}$$

and taking inner product with $X_B$:

$$q_L^A \not{g}_{AB} = g(\nabla_L \hat{T}, X_B) \tag{3.179}$$

Since $g(D_L \hat{T}, \hat{T}) = \frac{1}{2}L(g(\hat{T}, \hat{T})) = 0$, we have

$$g(\nabla_L \hat{T}, \hat{T}) = -\Gamma_{\alpha\beta\nu}L^\alpha \hat{T}^\beta \hat{T}^\nu = 0 \Rightarrow p_L = 0 \tag{3.180}$$

Also we have

$$\begin{aligned} g(\nabla_L \hat{T}, X_B) &= g(D_L \hat{T}, X_B) - \Gamma_{\alpha\beta\nu}L^\alpha \hat{T}^\beta X_B^\nu \\ &= g(D_L \hat{T}, X_B) = \kappa^{-1} g(D_L T, X_B) \\ &= -\kappa^{-1}\zeta^A \not{g}_{AB} \end{aligned} \tag{3.181}$$

$$\begin{aligned}\Rightarrow q_L^A &= -\kappa^{-1}\zeta^A = -\alpha\varepsilon^A + X^A(\alpha)\\ &= -\alpha k_{ij}X^{Ai}\hat{T}^j + X^A(\alpha)\\ &= -\hat{T}^j(X^A\psi_j) + X^A(\alpha)\end{aligned} \tag{3.182}$$

This is regular as $\mu \to 0$. Consider the vectorfield

$$T(L^\mu)\partial_\mu \tag{3.183}$$

Also, it can be expanded as

$$\nabla_T L = a_T\hat{T} + b_T \tag{3.184}$$

$$\Rightarrow a_T = g(\nabla_T L, \hat{T}), \quad b_T = b_T^A X_A, \quad b_T^A \not{g}_{AB} = g(\nabla_T L, X_B) \tag{3.185}$$

We have $g(\nabla_T L, \hat{T}) = g(D_T L, \hat{T}) - \Gamma_{\alpha\beta\nu}T^\alpha L^\beta \hat{T}^\nu$. From (3.100), we have

$$g(D_T L, \hat{T}) = g(-\alpha^{-1}(L\kappa)L, \hat{T}) = L\kappa \tag{3.186}$$

While $-\Gamma_{\alpha\beta\nu}T^\alpha L^\beta \hat{T}^\nu = 0$, we get

$$a_T = L\kappa \tag{3.187}$$

also

$$g(\nabla_T L, X_B) = g(D_T L, X_B) - \Gamma_{\alpha\beta\nu}T^\alpha L^\beta X_B^\nu = g(D_T L, X_B) = \eta_B \tag{3.188}$$

$$\Rightarrow b_T^A = \eta^A \tag{3.189}$$

Consider finally the vectorfield

$$T(\hat{T}^\mu)\partial_\mu \tag{3.190}$$

It can be expanded as

$$\nabla_T \hat{T} = p_T\hat{T} + q_T \tag{3.191}$$

$$\Rightarrow p_T = g(\nabla_T\hat{T}, \hat{T}) = g(D_T\hat{T}, \hat{T}) - \Gamma_{\alpha\beta\nu}T^\alpha \hat{T}^\beta \hat{T}^\nu = 0 \tag{3.192}$$

Writing $q_T = q_T^A X_A$, we have

$$q_T^A \not{g}_{AB} = g(\nabla_T\hat{T}, X_B) = g(D_T\hat{T}, X_B) - \Gamma_{\alpha\beta\nu}T^\alpha \hat{T}^\beta X_B^\nu \tag{3.193}$$

This equals

$$g(D_T\hat{T}, X_B) = \kappa^{-1}g(D_T T, X_B) = -\frac{1}{2\kappa}(\not{g}^{-1})^{AC}X_C(\kappa^2)\not{g}_{AB} = -X_B(\kappa) \tag{3.194}$$

We proceed to derive expressions for

$$\nabla_{X_A} L = X_A(L^\mu)\partial_\mu \tag{3.195}$$

and

$$\nabla_{X_A}\hat{T} = X_A(\hat{T}^\mu)\partial_\mu \tag{3.196}$$

First, we expand

$$\nabla_{X_A} L = \not{a}_A \hat{T} + \not{b}_A \tag{3.197}$$

then we have

$$\begin{aligned} \not{a}_A &= g(\nabla_{X_A} L, \hat{T}) = g(D_{X_A} L, \hat{T}) - \Gamma_{\alpha\beta\nu} X_A^\alpha L^\beta \hat{T}^\nu \\ &= g(D_{X_A} L, \hat{T}) = \kappa^{-1} \zeta_A \end{aligned} \tag{3.198}$$

This is regular as $\mu \to 0$ and we have used (3.102).

On the other hand

$$\begin{aligned} \not{g}_{BC} \not{b}_A^B &= g(\nabla_{X_A} L, X_C) = g(D_{X_A} L, X_C) - \Gamma_{\alpha\beta\nu} X_A^\alpha L^\beta X_C^\nu \\ &= g(D_{X_A} L, X_C) = \chi_{AC} \end{aligned} \tag{3.199}$$

Finally

$$\nabla_{X_A} \hat{T} = \not{p}_A \hat{T} + \not{q}_A \tag{3.200}$$

$$\Rightarrow \not{p}_A = g(\nabla_{X_A} \hat{T}, \hat{T}) = g(D_{X_A} \hat{T}, \hat{T}) - \Gamma_{\alpha\beta\nu} X_A^\alpha \hat{T}^\beta \hat{T}^\nu = 0 \tag{3.201}$$

$$\begin{aligned} \not{q}_A^B \not{g}_{BC} &= g(\nabla_{X_A} \hat{T}, X_C) = g(D_{X_A} \hat{T}, X_C) - \Gamma_{\alpha\beta\nu} X_A^\alpha \hat{T}^\beta X_C^\nu \\ &= \kappa^{-1} g(D_{X_A} T, X_C) = \theta_{AC} \end{aligned} \tag{3.202}$$

# 4

# The Acoustical Curvature

We presently derive expressions for the curvature tensor of the acoustical metric $g$.

In this chapter, we need the following expressions of Christoffel symbols derived in Chapter 3:

$$
\begin{aligned}
\Gamma_{000} &= \frac{1}{2}\partial_0(-\eta^2 + |\mathbf{v}|^2) \\
\Gamma_{0i0} &= \frac{1}{2}\partial_i(-\eta^2 + |\mathbf{v}|^2) \\
\Gamma_{ij0} &= \partial_i \partial_j \phi \\
\Gamma_{00k} &= \frac{1}{2}(2\partial_0 \partial_k \phi - \partial_k(-\eta^2 + |\mathbf{v}|^2)) \\
\Gamma_{i0k} &= \Gamma_{ijk} = 0
\end{aligned}
\tag{4.1}
$$

## 4.1 Expressions for Curvature Tensor

We have the expressions for the curvature tensor:

$$R_{\mu\nu\alpha\beta} = R^{(2)}_{\mu\nu\alpha\beta} + R^{(1)}_{\mu\nu\alpha\beta} \tag{4.2}$$

where

$$R^{(2)}_{\mu\nu\alpha\beta} = \frac{1}{2}(\partial_\alpha \partial_\nu g_{\beta\mu} + \partial_\beta \partial_\mu g_{\alpha\nu} - \partial_\alpha \partial_\mu g_{\beta\nu} - \partial_\beta \partial_\nu g_{\alpha\mu}) \tag{4.3}$$

$$R^{(1)}_{\mu\nu\alpha\beta} = -(g^{-1})^{\kappa\lambda}(\Gamma_{\alpha\mu\kappa}\Gamma_{\beta\nu\lambda} - \Gamma_{\beta\mu\kappa}\Gamma_{\alpha\nu\lambda}) \tag{4.4}$$

From these expressions and those for the Christoffel symbols, we have

$$R_{ijkl} = R^{(1)}_{ijkl} = \eta^{-2}(\partial_i \partial_k \phi \partial_j \partial_l \phi - \partial_i \partial_l \phi \partial_j \partial_k \phi) \tag{4.5}$$

Also, we have

$$R_{0i0j} = R^{(2)}_{0i0j} + R^{(1)}_{0i0j} \tag{4.6}$$

where

$$R^{(2)}_{0i0j} = \frac{1}{2}(\partial_0\partial_i g_{0j} + \partial_0\partial_j g_{0i} - \partial^2_0 g_{ij} - \partial_i\partial_j g_{00}) = \partial^2_{ij}(\partial_0\phi - \frac{1}{2}\sum_k(\partial_k\phi)^2 + \frac{1}{2}\eta^2) \quad (4.7)$$

i.e.

$$R^{(2)}_{0i0j} = \partial^2_{ij}h + \frac{1}{2}\partial^2_{ij}\eta^2 = -\frac{1}{2}\partial^2_{ij}H = -\frac{1}{2}\frac{dH}{dh}\partial^2_{ij}h - \frac{1}{2}\frac{d^2H}{dh^2}\partial_i h\partial_j h \quad (4.8)$$

So the real principal part of $R_{0i0j}$ is $-\frac{1}{2}\frac{dH}{dh}\partial^2_{ij}h$. We can write this in terms of $D^2_{ij}h$, we have

$$R_{0i0j} = R^{[P]}_{0i0j} + R^{[N]}_{0i0j} \quad (4.9)$$

where

$$R^{[P]}_{0i0j} = -\frac{1}{2}\frac{dH}{dh}D^2_{ij}h \quad (4.10)$$

Since

$$R^{(2)}_{0i0j} = -\frac{1}{2}\frac{dH}{dh}D^2_{ij}h - \frac{1}{2}\frac{dH}{dh}\Gamma^{\alpha}_{ij}\partial_\alpha h - \frac{1}{2}\frac{d^2H}{dh^2}\partial_i h\partial_j h$$

we have

$$R^{[N]}_{0i0j} = -\frac{1}{2}\frac{d^2H}{dh^2}\partial_i h\partial_j h - \frac{1}{2}\frac{dH}{dh}\Gamma^{\alpha}_{ij}\partial_\alpha h + R^{(1)}_{0i0j} \quad (4.11)$$

where

$$R^{(1)}_{0i0j} = \eta^{-2}(\frac{1}{2}\partial_0(-\eta^2 + |\mathbf{v}|^2)\partial_i\partial_j\phi - \frac{1}{4}\partial_i(-\eta^2 + |\mathbf{v}|^2)\partial_j(-\eta^2 + |\mathbf{v}|^2)) \quad (4.12)$$
$$-\eta^{-2}\partial_k\phi(\partial_i\partial_j\phi)\frac{1}{2}(2\partial_0\partial_k\phi - \partial_k(-\eta^2 + |\mathbf{v}|^2))$$

Finally, we have the component

$$R_{0ijk} = R^{(2)}_{0ijk} + R^{(1)}_{0ijk} \quad (4.13)$$

and

$$R^{(2)}_{0ijk} = \frac{1}{2}(\partial^2_{ij}g_{0k} + \partial^2_{0k}g_{ij} - \partial^2_{0j}g_{ki} - \partial^2_{ki}g_{0j}) = 0 \quad (4.14)$$

while

$$R^{(1)}_{0ijk} = -(g^{-1})^{\sigma\lambda}(\Gamma_{j0\sigma}\Gamma_{ik\lambda} - \Gamma_{k0\sigma}\Gamma_{ij\lambda}) \quad (4.15)$$

Hence

$$R_{0ijk} = R^{(1)}_{0ijk} = \eta^{-2}(\frac{1}{2}\partial_j(-\eta^2 + |\mathbf{v}|^2)\partial^2_{ik}\phi - \frac{1}{2}\partial_k(-\eta^2 + |\mathbf{v}|^2)\partial^2_{ij}\phi) \quad (4.16)$$

Now we have calculated all the components of the curvature tensor, and we shall express all the components in a more systematic way. To do this, we need to introduce the following notations:

$$\begin{aligned}\tau_\mu &= \partial_\mu h \\ v_{\mu\nu} &= D_\mu \tau_\nu = (D^2 h)_{\mu\nu} \\ w_{\mu\nu} &= \partial_\mu \psi_\nu = w_{\nu\mu}\end{aligned}$$

First, since only the component $R_{0i0j}$ has the principal part, we have

$$R^{[P]}_{\mu\nu\alpha\beta} = \frac{1}{2}\frac{dH}{dh}(\delta^0_\mu\delta^0_\beta v_{\alpha\nu} + \delta^0_\alpha\delta^0_\nu v_{\mu\beta} - \delta^0_\nu\delta^0_\beta v_{\alpha\mu} - \delta^0_\alpha\delta^0_\mu v_{\nu\beta}) \tag{4.17}$$

The lower order terms are much more complicated. First from (4.5),

$$R^{[N]}_{ijkl} = \eta^{-2}(w_{ik}w_{jl} - w_{il}w_{jk}) \tag{4.18}$$

From (4.11) and (4.12),

$$\begin{aligned}R^{[N]}_{0i0j} = &-\frac{1}{2}\frac{d^2H}{dh^2}\tau_i\tau_j - \frac{1}{2}\frac{dH}{dh}(\Gamma^0_{ij}\tau_0 + \Gamma^k_{ij}\tau_k) \\ &+ \eta^{-2}(\frac{1}{2}\partial_0(-\eta^2 + |\mathbf{v}|^2)w_{ij} - \frac{1}{4}\partial_i(-\eta^2 + |\mathbf{v}|^2)\partial_j(-\eta^2 + |\mathbf{v}|^2)) \\ &- \eta^{-2}\tau_k(w_{ij})\frac{1}{2}(2w_{0k} - \partial_k(-\eta^2 + |\mathbf{v}|^2))\end{aligned}$$

From (4.1),

$$\begin{aligned}\Gamma^0_{ij} &= (g^{-1})^{00}\Gamma_{ij0} = -\eta^{-2}\omega_{ij} \\ \Gamma^k_{ij} &= (g^{-1})^{k0}\Gamma_{ij0} = -\eta^{-2}v^k\omega_{ij}\end{aligned}$$

Also, since

$$h = \psi_0 - \frac{1}{2}|\mathbf{v}|^2 \quad \text{and} \quad H = -2h - \eta^2$$

we have

$$-\eta^2 + |\mathbf{v}|^2 = H + 2\psi_0 \tag{4.19}$$

Substituting in the expression for $R^{[N]}_{0i0j}$, we obtain, in view of the fact that $\psi_k = -v^k$,

$$\begin{aligned}R^{[N]}_{0i0j} = &-\frac{1}{2}(\frac{d^2H}{dh^2} + \frac{1}{2}\eta^{-2}(\frac{dH}{dh})^2)\tau_i\tau_j \\ &+ \frac{1}{2}\eta^{-2}\frac{dH}{dh}(2\tau_0 w_{ij} - \tau_i w_{j0} - \tau_j w_{i0}) \\ &+ \eta^{-2}(w_{00}w_{ij} - w_{i0}w_{j0})\end{aligned} \tag{4.20}$$

(4.16) reads

$$R^{[N]}_{0ijk} = \frac{1}{2}\eta^{-2}(\partial_j(-\eta^2 + |\mathbf{v}|^2)w_{ik} - \partial_k(-\eta^2 + |\mathbf{v}|^2)w_{ij})$$

Substituting gives

$$R^{[N]}_{0ijk} = \frac{1}{2}\eta^{-2}\frac{dH}{dh}(\tau_j w_{ik} - \tau_k w_{ij}) + \eta^{-2}(w_{j0}w_{ik} - w_{k0}w_{ij}) \tag{4.21}$$

Finally, the results for $R^{[N]}_{ijkl}$, $R^{[N]}_{0ijk}$ and $R^{[N]}_{0i0j}$, combine to produce the formula for $R^{[N]}_{\mu\nu\alpha\beta}$:

$$R^{[N]}_{\mu\nu\alpha\beta} = \eta^{-2}A_{\mu\nu\alpha\beta} + \frac{1}{2}\eta^{-2}\frac{dH}{dh}C_{\mu\nu\alpha\beta} - \frac{1}{2}H_2 B_{\mu\nu\alpha\beta} \tag{4.22}$$

where

$$A_{\mu\nu\alpha\beta} = w_{\mu\alpha}w_{\nu\beta} = w_{\mu\beta}w_{\nu\alpha}$$
$$B_{\mu\nu\alpha\beta} = (\delta^0_\mu\tau_\nu - \delta^0_\nu\tau_\mu)(\delta^0_\alpha\tau_\beta - \delta^0_\beta\tau_\alpha)$$

and, with

$$\xi_{\mu\alpha\beta} = \delta^0_\alpha w_{\beta\mu} - \delta^0_\beta w_{\alpha\mu}$$

we have

$$C_{\mu\nu\alpha\beta} = \tau_\mu\xi_{\nu\alpha\beta} - \tau_\nu\xi_{\mu\alpha\beta} + \tau_\alpha\xi_{\beta\mu\nu} - \tau_\beta\xi_{\alpha\mu\nu}$$

Also, the function $H_2$ is defined by

$$H_2 = \frac{d^2H}{dh^2} + \frac{1}{2}\eta^{-2}(\frac{dH}{dh})^2$$

The expressions $A_{\mu\nu\alpha\beta}, B_{\mu\nu\alpha\beta}, C_{\mu\nu\alpha\beta}$ all possess the algebraic properties of the curvature tensor, namely the antisymmetry in the first as well as the second pair of indices, and the cyclic property, which imply the symmetry under exchange of the first with the second pair of indices. In particular the cyclic property of $C_{\mu\nu\alpha\beta}$ follows from that of $\xi_{\mu\alpha\beta}$:

$$\xi_{\mu\alpha\beta} + \xi_{\alpha\beta\mu} + \xi_{\beta\mu\alpha} = 0$$

## 4.2 Regularity for the Acoustical Structure Equations as $\mu \to 0$

First we compute $\alpha_{AB}$, which appears in the propagation equation of $\chi_{AB}$. By definition,

$$\begin{aligned}\alpha_{AB} &= R_{\mu\nu\alpha\beta}X^\mu_A L^\nu X^\alpha_B L^\beta \\ &= R_{i0j0}X^i_A X^j_B + R_{ijkl}X^i_A L^j X^k_B L^l + R_{i0jk}X^i_A X^j_B L^k + R_{i0jk}X^j_A X^i_B L^k\end{aligned}$$

From (3.106), the principal part of $\alpha_{AB}$ is

$$\begin{aligned}
\alpha_{AB}^{[P]} &= R_{0i0j}^{[P]} X_A^i X_B^j = -\frac{1}{2}\frac{dH}{dh} D^2 h(X_A, X_B) \\
&= -\frac{1}{2}\frac{dH}{dh}(X_A(X_B h) - (D_{X_A} X_B) h) \\
&= -\frac{1}{2}\frac{dH}{dh}(\not{D}^2 h(X_A, X_B) - \alpha^{-1} \not{k}_{AB} L h - \mu^{-1} \chi_{AB} T h)
\end{aligned} \tag{4.23}$$

So the singular term as $\mu \to 0$ in $\alpha_{AB}^{[P]}$ is

$$\frac{1}{2\mu}\frac{dH}{dh}\chi_{AB} T(h) \tag{4.24}$$

For the lower order term, we have

$$\alpha_{AB}^{[N]} = \eta^{-2}\alpha_{AB}^{[A]} + \frac{1}{2}\eta^{-2}\frac{dH}{dh}\alpha_{AB}^{[C]} - \frac{1}{2}H_2 \alpha_{AB}^{[B]} \tag{4.25}$$

and

$$\begin{aligned}
\alpha_{AB}^{[A]} &= (w_{\mu\alpha} w_{\nu\beta} - w_{\mu\beta} w_{\nu\alpha}) X_A^\mu L^\nu X_B^\alpha L^\beta = \not{\psi}_{AB} w_{LL} - (\not{\psi}_L)_A (\not{\psi}_L)_B \\
\alpha_{AB}^{[B]} &= (\delta_\mu^0 \tau_\nu - \delta_\nu^0 \tau_\mu)(\delta_\alpha^0 \tau_\beta - \delta_\beta^0 \tau_\alpha) X_A^\mu L^\nu X_B^\alpha L^\beta = \not{\tau}_A \not{\tau}_B \\
\alpha_{AB}^{[C]} &= (\tau_\mu \xi_{\nu\alpha\beta} - \tau_\nu \xi_{\mu\alpha\beta} + \tau_\alpha \xi_{\beta\mu\nu} - \tau_\beta \xi_{\alpha\mu\nu}) X_A^\mu L^\nu X_B^\alpha L^\beta \\
&= 2\tau_L \not{\psi}_{AB} - \not{\tau}_A (\not{\psi}_L)_B - \not{\tau}_B (\not{\psi}_L)_A
\end{aligned} \tag{4.26}$$

So $\alpha_{AB}^{[C]}$ is regular as $\mu \to 0$. The only singular term in $\alpha_{AB}$ is (4.24).

Recall the propagation equation for $\chi$ in Chapter 3:

$$L\chi_{AB} = \mu^{-1}(L\mu)\chi_{AB} + \chi_A^C \chi_{BC} - \alpha_{AB} = e\chi_{AB} + \mu^{-1} m \chi_{AB} + \chi_A^C \chi_{BC} - \alpha_{AB} \tag{4.27}$$

We have used the propagation equation for $\mu$

$$L\mu = m + \mu e \tag{4.28}$$

where

$$m = \frac{1}{2}\frac{dH}{dh} T(h), \quad e = \frac{1}{2\eta}\frac{d(\eta^2)}{dh} L(h) + \frac{1}{\eta}\hat{T}^i (L\psi_i) \tag{4.29}$$

Then we know that the right hand side of (4.27) is regular as $\mu \to 0$.

Next, we will analyze the regularity of the right hand side of Codazzi equation and Gauss equation.

Recall the Codazzi equation (3.50):

$$(\not{d}\mathrm{iv}\chi)_A - \not{d}_A \mathrm{tr}\chi = \beta_A^* - \mu^{-1}(\zeta_B \chi_A^B - \zeta_A \mathrm{tr}\chi) \tag{4.30}$$

Here $\beta_B^* = (\not{g}^{-1})^{AC} \beta_C \epsilon_{AB}$.

Recalling (3.53) and the definition of $k_{ij}$, we know that $\mu^{-1}\zeta_A$ is regular as $\mu \to 0$. Therefore the regularity of the right hand side of (4.30) reduces to the regularity of

$$R(X_C, L, X_A, X_B) = \beta_C \epsilon_{AB} \tag{4.31}$$

We have

$$\epsilon_{AB}\beta_C^{[P]} = R^{[P]}_{\mu\nu\alpha\beta} X_C^\mu L^\nu X_A^\alpha X_B^\beta = 0 \tag{4.32}$$

and

$$\beta_A^{[N]} = \eta^{-2}\beta_A^{[A]} + \frac{1}{2}\eta^{-2}\frac{dH}{dh}\beta_A^{[C]} - \frac{1}{2}H_2\beta_A^{[B]} \tag{4.33}$$

where

$$\begin{aligned} \epsilon_{AB}\beta_C^{[A]} &= \psi_{CA}(\psi_L)_B - \psi_{CB}(\psi_L)_A \\ \epsilon_{AB}\beta_C^{[B]} &= 0 \\ \epsilon_{AB}\beta_C^{[C]} &= -\not{t}_A\psi_{BC} + \not{t}_B\psi_{AC} \end{aligned}$$

So $\beta_A$ is regular as $\mu \to 0$.

We turn to Gauss equation (3.36). Since

$$R_{ABCD} = \rho\epsilon_{AB}\epsilon_{CD} \tag{4.34}$$

it suffice to prove that $R_{ABCD}$ is regular as $\mu \to 0$. We have

$$R^{[P]}_{ABCD} = R^{[P]}_{\mu\nu\alpha\beta} X_A^\mu X_B^\nu X_C^\alpha X_D^\beta = 0 \tag{4.35}$$

and

$$R^{[N]}_{ABCD} = -\eta^{-2}R^{[A]}_{ABCD} + \frac{1}{2}\eta^{-2}\frac{dH}{dh}R^{[C]}_{ABCD} - \frac{1}{2}H_2R^{[B]}_{ABCD} \tag{4.36}$$

where

$$\begin{aligned} R^{[A]}_{ABCD} &= \psi_{AC}\psi_{BD} - \psi_{AD}\psi_{BC} \\ R^{[B]}_{ABCD} &= R^{[C]}_{ABCD} = 0 \end{aligned}$$

$\rho$ is regular as $\mu \to 0$.

Finally, we shall discuss about (3.124)—(3.125). As we have seen, $\mu^{-1}\zeta$ is regular as $\mu \to 0$. We shall thus discuss about $\gamma_{AB}$. We have

$$\gamma^{[P]}_{AB} = \frac{1}{2}(R^{[P]}(X_A, T, X_B, L) + R^{[P]}(X_B, T, X_A, L)) \tag{4.37}$$

and

$$\gamma^{[N]}_{AB} = \eta^{-2}\gamma^{[A]}_{AB} + \frac{1}{2}\eta^{-2}\frac{dH}{dh}\gamma^{[C]}_{AB} - \frac{1}{2}H_2\gamma^{[B]}_{AB} \tag{4.38}$$

where

$$\gamma^{[A]}_{AB} = \frac{\kappa}{2}(2\psi_{AB}w_{L\hat{T}} - (\psi_L)_A(\psi_{\hat{T}})_B - (\psi_L)_B(\psi_{\hat{T}})_A)$$

$$\gamma_{AB}^{[B]} = 0$$

$$\gamma_{AB}^{[C]} = \frac{1}{2}(2\tau_T \not{\psi}_{AB} - \kappa \not{\tau}_A (\not{\psi}_{\hat{T}})_B - \kappa \not{\tau}_B (\not{\psi}_{\hat{T}})_A)$$

$\gamma_{AB}$ is regular as $\mu \to 0$.

Note that all the components of $\tau$ in the $(L, T, X_1, X_2)$ frame i.e. $\tau_L, \tau_T, \not{\tau}_A$ are regular as $\mu \to 0$. Also all the components of $w$ in the $(L, \hat{T}, X_1, X_2)$ frame i.e. $w_{LL}, w_{L\hat{T}}, (\not{\psi}_{\hat{T}})_A, \not{\psi}_{AB}$, except $w_{\hat{T}\hat{T}}$ are regular as $\mu \to 0$. The last is $\kappa^{-1} w_{T\hat{T}}$ and $w_{T\hat{T}}$ is regular as $\mu \to 0$.

## 4.3 A Remark

We end this chapter with a remark. By (4.15), we can see that if $H = \text{const}$, then the principal part of the acoustical curvature tensor vanishes. This fact can be also obtained by the geometric interpretation of $H$ in Chapter 2. By the Gauss equation,

$$R_{\alpha\beta\mu\nu} - \tilde{k}_{\alpha\mu}\tilde{k}_{\beta\nu} + \tilde{k}_{\beta\mu}\tilde{k}_{\alpha\nu} = \tilde{R}_{\alpha\beta\mu\nu} \tag{4.39}$$

where $R$ and $\tilde{R}$ are the curvature tensor of $(\mathbb{R} \times \mathbb{E}^3, g)$ and $(\mathbb{R}^{1+4}, \tilde{g})$ respectively. And $\tilde{k}$ is the second fundamental form of $(\mathbb{R} \times \mathbb{E}^3, g)$ in $(\mathbb{R}^{1+4}, \tilde{g})$, which contains only the first derivatives of the acoustical metric. Now since $(\mathbb{R}^{1+4}, \tilde{g})$ is a flat spacetime, $R$ contains only the first derivative of the acoustical metric, i.e. the principal part of $R$ vanishes.

# 5

# The Fundamental Energy Estimate

## 5.1 Bootstrap Assumptions. Statement of the Theorem

In this chapter we consider the wave equation

$$\Box_{\tilde{g}}\psi = 0 \tag{5.1}$$

in a spacetime whose metric is the conformal acoustical metric $\tilde{g}$. As we have seen in Chapter 1, this equation is satisfied by a variation of the wave function $\phi$ through solutions. $\tilde{g}$ is related to the acoustical metric $g$ by

$$\tilde{g}_{\mu\nu} = \Omega g_{\mu\nu} \tag{5.2}$$

where

$$\Omega = \frac{\rho/\rho_0}{\eta/\eta_0} = \frac{\rho}{\eta} \tag{5.3}$$

where $\rho_0$ and $\eta_0$ are the mass density and sound speed corresponding to the constant state which we can both set equal to unity by appropriate choice of units. More generally, we consider the inhomogeneous wave equation

$$\Box_{\tilde{g}}\psi = \rho \tag{5.4}$$

where the metric $\tilde{g}$, $\psi$ and $\rho$ are all defined in the domain $M_{\epsilon_0}$ of the maximal solution.

In application to first order estimates we shall take each $\psi$ to be one of

$$\psi_1 = \partial_0\phi - h_0 = \psi_0; \quad \partial_i\phi, i = 1,2,3; \quad \Omega_{ij}\phi, i < j = 1,2,3; \quad S\phi \tag{5.5}$$

Remember in Chapter 2 we have set

$$h_0 = 0$$

We know that each $\psi_1$ vanishes in the exterior of the outgoing characteristic hypersurface $C_0$. In application to higher order estimates we shall take each $\psi$ to be one of

$$\psi_n = Y_{i_1} \cdots Y_{i_{n-1}} \psi_1 \tag{5.6}$$

where $\psi_1$ is any one of the above first order variations and the indices $i_1, \cdots, i_{n-1}$ take values in the set $1, 2, 3, 4, 5$, with $\{Y_i : i = 1, 2, 3, 4, 5\}$ the set of vectorfields to be discussed in the next chapter. Each of the $\psi_n$ also vanishes in the exterior of $C_0$.

We shall therefore assume in general that $\psi$ vanishes in the exterior of $C_0$. We may then confine attention to the spacetime domain $W_{\epsilon_0}$. For any $s \in (0, t_{*\epsilon_0})$, we set

$$W^s_{\epsilon_0} = \bigcup_{(t,u)\in[0,s]\times[0,\epsilon_0]} S_{t,u} \tag{5.7}$$

In particular we have

$$\bigcup_{s\in(0,t_{*\epsilon_0})} W^s_{\epsilon_0} = W^*_{\epsilon_0} \tag{5.8}$$

Everything which follows depends on the geometric construction in Chapter 2. We note that the characteristic hypersurfaces $C_u$ depend only on the conformal class of the metric $g$. Thus, the 2-parameter foliation of $W^*_{\epsilon_0}$ given by the surfaces $S_{t,u}$ likewise depends only on the conformal class of $g$. The geometric properties of this foliation may be referred either to $g$ or to the conformal metric $\tilde{g}$. Since we are studying the wave equation in the metric $\tilde{g}$, it would appear more natural if we refer these properties to $\tilde{g}$. However, our bootstrap argument is constructed relative to $g$, and we introduce additional assumptions referring to the conformal factor $\Omega$.

The bootstrap assumptions are the followings:

There is a positive constant $C$ independent of $s$ such that in $W^*_{\epsilon_0}$,

$$\mathbf{A1}: C^{-1} \leqslant \Omega \leqslant C$$
$$\mathbf{A2}: C^{-1} \leqslant \eta \leqslant C$$
$$\mathbf{A3}: \mu \leqslant C(1 + \log(1 + t))$$

and

$$\mathbf{B1}: C^{-1}(1+t)^{-1} \leqslant \nu \leqslant C(1+t)^{-1}$$
$$\mathbf{B2}: |\underline{\nu}| \leqslant C(1+t)^{-1}(1 + \log(1+t))^4$$
$$\mathbf{B3}: |\hat{\chi}| \leqslant C(1+t)^{-1}(1 + \log(1+t))^{-2}$$
$$\mathbf{B4}: |\hat{\underline{\chi}}| \leqslant C(1+t)(1 + \log(1+t))^{-6}$$

Here,

$$\nu = \frac{1}{2}\tilde{\mathrm{tr}}\,\chi = \frac{1}{2}(\mathrm{tr}\,\chi + L\log\Omega) \tag{5.9}$$

$$\underline{\nu} = \frac{1}{2}\tilde{\mathrm{tr}}\,\underline{\chi} = \frac{1}{2}(\mathrm{tr}\,\underline{\chi} + \underline{L}\log\Omega) \tag{5.10}$$

$\hat{\chi}$ and $\hat{\underline{\chi}}$ are the trace-free part of $\chi$ and $\underline{\chi}$ respectively, and the pointwise norms of the tensors on $S_{t,u}$ are with respect to the induced metric $\not{g}$.

Next, we have the assumptions:

$$\mathbf{B5}: |L \log \Omega| \leqslant C(1+t)^{-1}(1+\log(1+t))^{-2}$$
$$\mathbf{B6}: |\underline{L} \log \Omega| \leqslant C(1+t)(1+\log(1+t))^{-6}$$

and

$$\mathbf{B7}: |\zeta+\eta| \leqslant C(1+t)^{-1}(1+\log(1+t))$$
$$\mathbf{B8}: |\not{d}(\eta^{-1}\kappa)| \leqslant C(1+t)^{-1}(1+\log(1+t))$$
$$\mathbf{B9}: |L\eta| \leqslant C(1+t)^{-1}(1+\log(1+t))^2$$
$$\mathbf{B10}: |L(\eta^{-1}\kappa)| \leqslant C(1+t)^{-1}(1+\log(1+t))^3$$
$$\mathbf{B11}: |\underline{L}(\eta^{-1}\kappa)| \leqslant C(1+t)(1+\log(1+t))^{-2}$$

as well as

$$\mathbf{B12}: |L\nu+\nu^2| \leqslant C(1+t)^{-2}(1+\log(1+t))^{-2}$$
$$\mathbf{B13}: |\underline{L}\nu| \leqslant C(1+t)^{-2}(1+\log(1+t))^3$$
$$\mathbf{B14}: |\not{d}\nu| \leqslant C(1+t)^{-2}(1+\log(1+t))^{\frac{1}{2}}$$

The next set of bootstrap assumptions concerns the behavior of the function $\mu$. In the following we denote by $f_+$ and $f_-$ respectively the positive and negative parts of the function $f$:

$$f_+ = \max\{f(x), 0\}, f_- = \min\{f(x), 0\} \tag{5.11}$$

$$\mathbf{C1}: \mu^{-1}(L\mu)_+ \leqslant (1+t)^{-1}(1+\log(1+t))^{-1} + A(t)$$

where $A(t)$ is a nonnegative function such that

$$\int_0^s A(t)dt \leqslant C \tag{5.12}$$

($C$ is independent of $s$.)

$$\mathbf{C2}: \mu^{-1}(L\mu + \underline{L}\mu)_+ \leqslant B(t)$$

where $B(t)$ is a nonnegative function such that

$$\int_0^s (1+t)^{-2}(1+\log(1+t))^4 B(t)dt \leqslant C \tag{5.13}$$

($C$ is also independent of $s$.)

Moreover, denoting by $\mathcal{U}$ the region

$$\mathcal{U} = \{x \in W^*_{\epsilon_0} : \mu < 1/4\} \tag{5.14}$$

we have

$$\mathbf{C3}: L\mu \leqslant -C^{-1}(1+t)^{-1}(1+\log(1+t))^{-1}$$

in $\mathcal{U}\bigcap W^s_{\epsilon_0}$.

The final set of bootstrap assumptions concerns the existence of a function $\omega$ verifying the following conditions:

$$\mathbf{D1}: C^{-1}(1+t) \leqslant \omega \leqslant C(1+t)$$

$$\mathbf{D2}: |L\omega - \nu\omega| \leqslant C(1+\log(1+t))^{-2}$$

$$\mathbf{D3}: |\underline{L}\omega| \leqslant C(1+\log(1+t))^{3}$$

$$\mathbf{D4}: |\not{d}\omega| \leqslant C(1+\log(1+t))^{\frac{1}{2}}$$

$$\mathbf{D5}: \int_0^s \Big(\int_0^{\epsilon_0} \sup_{S_{t,u}}(\mu|\Box_{\tilde{g}}\omega|)du\Big)dt \leqslant C(1+\log(1+s))^4$$

Here we note that by (5.3) **A1** is equivalent to the following modulo **A2**:

$$\mathbf{A1}': C^{-1} \leqslant \rho \leqslant C$$

The main theorem in this chapter is

**Theorem 5.1** Let $\psi$ be a solution of the wave equation associated to the metric $\tilde{g}$, defined in $M_{\epsilon_0}$ and vanishing in the exterior of $C_0$. Suppose that assumptions **A1** – **A3**, **B1** – **B14**, **C1** – **C3**, **D1** – **D5** hold in $W^s_{\epsilon_0}$, for some $s \in (0, t_{*\epsilon_0}]$. Let us denote

$$D_0 = \int_0^{\epsilon_0} \Big(\int_{S_{0,u}} ((L\psi)^2 + (\underline{L}\psi)^2 + |\not{d}\psi|^2)d\mu_{\not{g}}\Big)du \tag{5.15}$$

Then there exist constants $C$ independent of $s$ such that

**(i)** $\displaystyle\sup_{t\in[0,s]} \int_0^{\epsilon_0} \Big(\int_{S_{t,u}} (\mu(1+\mu)((L\psi)^2 + |\not{d}\psi|^2) + (\underline{L}\psi)^2)d\mu_{\not{g}}\Big)du \leqslant CD_0$

**(ii)** $\displaystyle\int_{S_{t,u}} \psi^2 d\mu_{\not{g}} \leqslant C\epsilon_0 D_0$

**(iii)** $\displaystyle\sup_{u\in[0,\epsilon_0]} \int_0^s \Big(\int_{S_{t,u}} ((1+\mu)(L\psi)^2 + \mu|\not{d}\psi|^2)d\mu_{\not{g}}\Big)dt \leqslant CD_0$

**(iv)** $\displaystyle\sup_{t\in[0,s]} (1+\log(1+t))^{-4}(1+t)^2 \int_0^{\epsilon_0} \Big(\int_{S_{t,u}} \mu((L\psi+\nu\psi)^2 + |\not{d}\psi|^2)d\mu_{\not{g}}\Big)du \leqslant CD_0$

**(v)** $\displaystyle\sup_{u\in[0,\epsilon_0]} (1+\log(1+t))^{-4} \int_0^s (1+t)^2\Big(\int_{S_{t,u}} (L\psi+\nu\psi)^2 d\mu_{\not{g}}\Big)dt \leqslant CD_0$

**(vi)** $\displaystyle\int_{\mathcal{U}\bigcup W^s_{\epsilon_0}} (1+t)(1+\log(1+t))^{-1}|\not{d}\psi|^2 d\mu_{\not{g}}dudt \leqslant CD_0(1+\log(1+s))^4$

## 5.2 The Multiplier Fields $K_0$ and $K_1$. The Associated Energy-Momentum Density Vectorfields

We begin with the energy-momentum-stress tensor $\tilde{T}_{\mu\nu}$ associated to the function $\psi$ through the metric $\tilde{g}$:

$$\tilde{T}_{\mu\nu} := \partial_\mu\psi\partial_\nu\psi - \frac{1}{2}\tilde{g}_{\mu\nu}(\tilde{g}^{-1})^{\kappa\lambda}\partial_\kappa\psi\partial_\lambda\psi$$
$$= \partial_\mu\psi\partial_\nu\psi - \frac{1}{2}g_{\mu\nu}(g^{-1})^{\kappa\lambda}\partial_\kappa\psi\partial_\lambda\psi = T_{\mu\nu} \tag{5.16}$$

We have

$$\tilde{D}^\mu\tilde{T}_{\mu\nu} := (\tilde{g}^{-1})^{\mu\lambda}\tilde{D}_\mu\tilde{T}_{\lambda\nu} = \partial_\nu\psi\Box_{\tilde{g}}\psi$$

where $\tilde{D}$ is the covariant derivative operator associated to the metric $\tilde{g}_{\mu\nu}$. Thus, for a solution of (5.1), $\tilde{T}_{\mu\nu}$ is divergence-free with respect to the metric $\tilde{g}$, while for a solution of the inhomogeneous wave equation (5.4),

$$\tilde{D}^\mu\tilde{T}_{\mu\nu} = \rho\partial_\nu\psi \tag{5.17}$$

We consider the future-directed, time-like vectorfield $K_0$:

$$K_0 = (1 + \alpha^{-1}\kappa)L + \underline{L} \tag{5.18}$$

Also, given a function $\omega$ satisfying **D1**—**D5** we consider the future-directed, null vectorfield $K_1$:

$$K_1 = (\omega/\nu)L \tag{5.19}$$

At this point, we discuss why we choose these two multiplier fields. A multiplier should be non-spacelike and future-directed so that it gives positive energies. Also the multiplier should be smooth with respect to the basis $(\frac{\partial}{\partial t}, \frac{\partial}{\partial u}, \frac{\partial}{\partial \vartheta^A})$ in acoustical coordinates $(t, u, \vartheta^A)$, because only the smooth vectorfields give error terms which can be estimated, due to their expressions in terms of deformation tensors. So in the flat spacetime, $\partial_t$ is a natural choice. While in the present case, $L + \underline{L}$ fits the above requirements. Indeed $L + \underline{L}$ is the analogue of time translation, i.e. $\partial_t$ in flat spacetime. However, in contrast to this, $K_0$ becomes null when $\mu \to 0$. On the other hand, before we set $\eta_0 = 1$, $L$ and $\underline{L}$ have different physical dimensions, $L$ being $\mathrm{T}^{-1}$ and $\underline{L}$ being $\mathrm{L}^{-1}$. However, after we set $\eta_0 = 1$, the unit of time is the time required for sound to travel the unit length in the constant state. Therefore time is the length and $L$, $\underline{L}$ are both $\mathrm{L}^{-1}$. To deal with the growth of $\mu$, we need to plus the term $\alpha^{-1}\kappa L$. This is how we choose $K_0$.

Concerning the choice of $K_1$, we remind you the generator of the inverted time translation in Minkowski spacetime:

$$K = u^2\underline{L} + \underline{u}^2 L$$

where

$$L = \partial_t + \partial_r, \quad \underline{L} = \partial_t - \partial_r$$

and

$$\underline{u} = t + r, \quad u = t - r$$

Now at the present case, we want to choose a similar vectorfield. The analogue of $t - r$ is our acoustical function $u$, which is bounded in $W^*_{\epsilon_0}$. While when $t$ is large,

$\underline{u} \sim t$. Since the first term in the expression for $K$ is contained in $K_0$, so by **B1** and **D1**, it is reasonable to choose

$$K_1 = (\omega/\nu)L$$

We will see later more clearly why we exactly choose this form.

We denote by $\tilde{\pi}_0$ and $\tilde{\pi}_1$ the Lie derivatives of the conformal metric $\tilde{g}$ with respect to the vectorfields $K_0$ and $K_1$:

$$\tilde{\pi}_0 = \mathcal{L}_{K_0}\tilde{g}, \quad \tilde{\pi}_1 = \mathcal{L}_{K_1}\tilde{g} \tag{5.20}$$

To $K_0$ we associate the vectorfield

$$\tilde{P}_0^\mu = -\tilde{T}_\nu^\mu K_0^\nu \tag{5.21}$$

where

$$\tilde{T}_\nu^\mu = (\tilde{g}^{-1})^{\mu\kappa} T_{\kappa\nu}$$

To $K_1$ we associate the vectorfield

$$\tilde{P}_1^\mu = -\tilde{T}_\nu^\mu K_1^\nu - (\tilde{g}^{-1})^{\mu\nu}(\omega\psi\partial_\nu\psi - \frac{1}{2}\psi^2\partial_\nu\omega) \tag{5.22}$$

Let

$$\tilde{T}^{\mu\nu} = (\tilde{g}^{-1})^{\nu\lambda}\tilde{T}_\lambda^\mu = (\tilde{g}^{-1})^{\mu\kappa}(\tilde{g}^{-1})^{\nu\lambda}\tilde{T}_{\kappa\lambda} \tag{5.23}$$

For any vectorfield $X$, we have, by the virtue of the symmetry of $\tilde{T}^{\mu\nu}$,

$$\begin{aligned}\tilde{D}_\mu(\tilde{T}_\nu^\mu X^\nu) &= (\tilde{D}_\mu\tilde{T}_\nu^\mu)X^\nu + \tilde{T}^{\mu\lambda}(\tilde{g}_{\lambda\nu}\tilde{D}_\mu X^\nu) \\ &= \rho X\psi + \frac{1}{2}\tilde{T}^{\mu\lambda}(\tilde{g}_{\lambda\nu}\tilde{D}_\mu X^\nu + \tilde{g}_{\mu\nu}\tilde{D}_\lambda X^\nu) \\ &= \rho X\psi + \frac{1}{2}\tilde{T}^{\mu\lambda}\mathcal{L}_X\tilde{g}_{\mu\lambda}\end{aligned} \tag{5.24}$$

Thus, we have

$$\tilde{D}_\mu\tilde{P}_0^\mu = -\rho K_0\psi - \frac{1}{2}\tilde{T}^{\mu\nu}\tilde{\pi}_{0,\mu\nu} := \tilde{Q}_0 \tag{5.25}$$

Also, we have

$$\tilde{D}_\mu\tilde{P}_1^\mu = -\rho(K_1\psi + \omega\psi) - \frac{1}{2}\tilde{T}^{\mu\nu}\tilde{\pi}_{1,\mu\nu} - \omega(\tilde{g}^{-1})^{\mu\nu}\partial_\mu\psi\partial_\nu\psi + \frac{1}{2}\psi^2\Box_{\tilde{g}}\omega$$

Taking into account the fact that

$$\text{tr}\,\tilde{T} := \tilde{g}_{\mu\nu}\tilde{T}^{\mu\nu} = -(\tilde{g}^{-1})^{\mu\nu}\partial_\mu\psi\partial_\nu\psi$$

and introducing

$$\tilde{\pi}_1' = \tilde{\pi}_1 - 2\omega\tilde{g} \tag{5.26}$$

we can write the above in the form

$$\tilde{D}_\mu \tilde{P}_1^\mu = -\rho(K_1\psi + \omega\psi) - \frac{1}{2}\tilde{T}^{\mu\nu}\tilde{\pi}'_{1,\mu\nu} + \frac{1}{2}\psi^2\Box_{\tilde{g}}\omega := \tilde{Q}_1 \tag{5.27}$$

Consider now the equation

$$\tilde{D}_\mu \tilde{P}^\mu = \tilde{Q} \tag{5.28}$$

for an arbitrary vectorfield $\tilde{P}$ and function $\tilde{Q}$. In arbitrary local coordinates we have

$$\begin{aligned}\tilde{D}_\mu \tilde{P}^\mu &= \frac{1}{\sqrt{-\det \tilde{g}}}\partial_\mu(\sqrt{-\det \tilde{g}}\tilde{P}^\mu)\\ &= \frac{1}{\Omega^2\sqrt{-\det g}}\partial_\mu(\Omega^2\sqrt{-\det g}\tilde{P}^\mu) = \Omega^{-2}D_\mu P^\mu\end{aligned}$$

where

$$P^\mu = \Omega^2 \tilde{P}^\mu \tag{5.29}$$

Hence, with

$$Q = \Omega^2\tilde{Q} \tag{5.30}$$

Equation (5.28) is equivalent to the equation

$$D_\mu P^\mu = Q \tag{5.31}$$

We may express (5.31) in acoustical coordinates $(t, u, \vartheta^1, \vartheta^2)$. Expanding $P$ in the associated coordinate frame field,

$$\left(\frac{\partial}{\partial t}, \frac{\partial}{\partial u}, \frac{\partial}{\partial \vartheta^1}, \frac{\partial}{\partial \vartheta^2}\right)$$

$$P = P^t\frac{\partial}{\partial t} + P^u\frac{\partial}{\partial u} + \sum_A (P^\vartheta)^A \frac{\partial}{\partial \vartheta^A}$$

and noting that, by (2.41), we have

$$\sqrt{-\det g} = \mu\sqrt{\det \not{g}} \tag{5.32}$$

Equation (5.31) takes the form

$$\frac{1}{\sqrt{\det \not{g}}}\left(\frac{\partial}{\partial t}(\mu\sqrt{\det \not{g}}P^t) + \frac{\partial}{\partial u}(\mu\sqrt{\det \not{g}}P^u)\right) + \not{\mathrm{div}} M = \mu Q \tag{5.33}$$

where $M$ is the vectorfield on $S^2$:

$$M = \mu\sum_A (P^\vartheta)^A\frac{\partial}{\partial \vartheta^A}$$

This follows from the fact that

$$\not{\mathrm{div}} M = \frac{1}{\sqrt{\det \not{g}}}\frac{\partial}{\partial \vartheta^A}(\sqrt{\det \not{g}}M^A)$$

We integrate (5.33) on $S^2$ with respect to the measure

$$d\mu_{\not g} = \sqrt{\det \not g} d\vartheta^1 d\vartheta^2 \tag{5.34}$$

to obtain the equation

$$\frac{\partial}{\partial t}\Big(\int_{S_{t,u}} \mu P^t d\mu_{\not g}\Big) + \frac{\partial}{\partial u}\Big(\int_{S_{t,u}} \mu P^u d\mu_{\not g}\Big) = \int_{S_{t,u}} \mu Q d\mu_{\not g} \tag{5.35}$$

Replacing $(t,u)$ by $(t',u')$ and integrating with respect to $(t',u')$ on $[0,t] \times [0,u]$, we obtain, under the hypothesis that $P$ vanishes in the closure of the exterior of $C_0$,

$$\mathcal{E}^u(t) - \mathcal{E}^u(0) + \mathcal{F}^t(u) = \int_{W_u^t} Q d\mu_g \tag{5.36}$$

Here $\mathcal{E}^u$ is the "energy":

$$\mathcal{E}^u(t) = \int_{\Sigma_t^u} \mu P^t d\mu_{\not g} du' = \int_0^u \Big(\int_{S_{t,u'}} \mu P^t d\mu_{\not g}\Big) du' \tag{5.37}$$

and $\mathcal{F}^t$ is the "flux":

$$\mathcal{F}^t(u) = \int_{C_u^t} \mu P^u d\mu_{\not g} dt' = \int_0^t \Big(\int_{S_{t',u}} \mu P^u d\mu_{\not g}\Big) dt' \tag{5.38}$$

In obtaining (5.36) we have used the fact that

$$\int_{W_u^t} Q d\mu_g = \iint_{[0,t]\times[0,u]} \Big(\int_{S_{t',u'}} \mu Q d\mu_{\not g}\Big) du' dt' \tag{5.39}$$

In the above, $\Sigma_t^u$ is the annular region:

$$\Sigma_t^u = \bigcup_{u' \in [0,u]} S_{t,u'} \tag{5.40}$$

in the spacelike hyperplane $\Sigma_t$, $C_u^t$ is the characteristic hypersurface $C_u$ truncated from above by $\Sigma_t$:

$$C_u^t = \bigcup_{t' \in [0,t]} S_{t',u} \tag{5.41}$$

and $W_u^t$ is the spacetime domain:

$$W_u^t = \bigcup_{(t',u') \in [0,t]\times[0,u]} S_{t',u'} \tag{5.42}$$

bounded by the characteristic hypersurfaces $C_u$ and $C_0$ and the spacelike hypersurfaces $\Sigma_t$ and $\Sigma_0$.

Now we expand $P$ in the frame field $(L, \underline{L}, X_1, X_2)$:

$$P = P^L L + P^{\underline{L}} \underline{L} + \sum_A P^A X_A$$

From the conclusion of Chapters 2 and 3,

$$L = \frac{\partial}{\partial t} \tag{5.43}$$

$$\underline{L} = \eta^{-1}\kappa \frac{\partial}{\partial t} + 2\left(\frac{\partial}{\partial u} - \Xi^A \frac{\partial}{\partial \vartheta^A}\right) \tag{5.44}$$

$$X_A = \frac{\partial}{\partial \vartheta^A} \tag{5.45}$$

By direct calculations, we get

$$P^t = P^L + \eta^{-1}\kappa P^{\underline{L}} \tag{5.46}$$

$$P^u = 2P^{\underline{L}} \tag{5.47}$$

$$(P^\vartheta)^A = P^A - 2P^{\underline{L}} \Xi^A \tag{5.48}$$

We shall write down the expressions for the energy and flux integrals corresponding to $P_0$ and $P_1$.

We begin with the components of the energy-momentum-stress tensor in the null frame $(L, \underline{L}, X_1, X_2)$:

$$\begin{aligned}
&T_{LL} = (L\psi)^2, \quad T_{\underline{L}\underline{L}} = (\underline{L}\psi)^2, \quad T_{L\underline{L}} = \mu|\not{d}\psi|^2 \\
&T_{LA} = (L\psi)(\not{d}_A\psi), \quad T_{\underline{L}A} = (\underline{L}\psi)(\not{d}_A\psi) \\
&T_{AB} = (\not{d}_A\psi)(\not{d}_B\psi) - \frac{1}{2}\not{g}_{AB}(-\mu^{-1}(L\psi)(\underline{L}\psi) + |\not{d}\psi|^2)
\end{aligned} \tag{5.49}$$

Here, $\not{d}_A\psi = X_A\psi$, and the pointwise norms of tensors on $S_{t,u}$ are with respect to the induced metric $\not{g}$.

We also give the components of the tensor $T^{\mu\nu}$. From (3.130), we have

$$\begin{aligned}
T^{LL} &= \frac{(\underline{L}\psi)^2}{4\mu^2}, \quad T^{\underline{L}\underline{L}} = \frac{(L\psi)^2}{4\mu^2}, \quad T^{L\underline{L}} = \frac{|\not{d}\psi|^2}{4\mu} \\
T^{LA} &= -\frac{1}{2\mu}(\underline{L}\psi)(\not{d}^A\psi), \quad T^{\underline{L}A} = -\frac{1}{2\mu}(L\psi)(\not{d}^A\psi) \\
T^{AB} &= \frac{1}{2\mu}(\not{g}^{-1})^{AB}(L\psi)(\underline{L}\psi) + ((\not{d}^A\psi)(\not{d}^B\psi) - \frac{1}{2}(\not{g}^{-1})^{AB}|\not{d}\psi|^2)
\end{aligned} \tag{5.50}$$

where $\not{d}^A\psi = (\not{g}^{-1})^{AB}\not{d}_B\psi$.

We consider $P_0$. From (5.21) and (5.29) we have

$$P_0^\mu = \Omega^2 \tilde{P}_0^\mu = -\Omega^2 \tilde{T}^\mu_\nu K_0^\nu = -\Omega T^\mu_\nu K_0^\nu \tag{5.51}$$

We get

$$P_0^{\underline{L}} = -\Omega T^{\underline{L}}_L (1 + \eta^{-1}\kappa) - \Omega T^{\underline{L}}_{\underline{L}} = \frac{\Omega}{2\mu}((1 + \eta^{-1}\kappa)T_{LL} + T_{L\underline{L}})$$

$$P_0^L = -\Omega T^L_L (1 + \eta^{-1}\kappa) - \Omega T^L_{\underline{L}} = \frac{\Omega}{2\mu}((1 + \eta^{-1}\kappa)T_{L\underline{L}} + T_{\underline{L}\underline{L}})$$

By (3.91), we have

$$\begin{aligned} P_0^{\underline{L}} &= \frac{\Omega}{2\mu}((1+\eta^{-1}\kappa)(L\psi)^2 + \mu|\not{d}\psi|^2) \\ P_0^{L} &= \frac{\Omega}{2\mu}((1+\eta^{-1}\kappa)\mu|\not{d}\psi|^2 + (\underline{L}\psi)^2) \end{aligned} \tag{5.52}$$

So we get

$$\mathcal{E}_0^u(t) = \int_{\Sigma_t^u} \frac{\Omega}{2}(\eta^{-1}\kappa(1+\eta^{-1}\kappa)(L\psi)^2 + (\underline{L}\psi)^2 + (1+2\eta^{-1}\kappa)\mu|\not{d}\psi|^2)d\mu_{\not{g}}du' \tag{5.53}$$

$$\mathcal{F}_0^t(u) = \int_{C_u^t} \Omega((1+\eta^{-1}\kappa)(L\psi)^2 + \mu|\not{d}\psi|^2)d\mu_{\not{g}}dt' \tag{5.54}$$

We now consider $P_1$. By some similar calculation we have

$$P_1^\mu = -\Omega(T_\nu^\mu K_1^\nu + (g^{-1})^{\mu\nu}(\omega\psi\partial_\nu\psi - \frac{1}{2}\psi^2\partial_\nu\omega)) \tag{5.55}$$

Then we get

$$\begin{aligned} P_1^{\underline{L}} &= \frac{\Omega}{2\mu}(\omega\nu^{-1}T_{LL} + \omega\psi(L\psi) - \frac{1}{2}\psi^2(L\omega)) \\ P_1^{L} &= \frac{\Omega}{2\mu}(\omega\nu^{-1}T_{L\underline{L}} + \omega\psi(\underline{L}\psi) - \frac{1}{2}\psi^2(\underline{L}\omega)) \end{aligned}$$

Then from (5.49) we get

$$\begin{aligned} P_1^{\underline{L}} &= \frac{\Omega}{2\mu}(\omega\nu^{-1}(L\psi)^2 + \omega\psi(L\psi) - \frac{1}{2}\psi^2(L\omega)) \\ P_1^{L} &= \frac{\Omega}{2\mu}(\omega\nu^{-1}\mu|\not{d}\psi|^2 + \omega\psi(\underline{L}\psi) - \tfrac{1}{2}\psi^2(\underline{L}\omega)) \end{aligned} \tag{5.56}$$

and

$$\mathcal{E}_1^u(t) = \int_{\Sigma_t^u} \frac{\Omega}{2}(\omega\nu^{-1}(\eta^{-1}\kappa(L\psi)^2 + \mu|\not{d}\psi|^2) + \omega\psi(\eta^{-1}\kappa(L\psi) + (\underline{L}\psi)) - \frac{1}{2}\psi^2(\eta^{-1}\kappa(L\omega) + (\underline{L}\omega)))d\mu_{\not{g}}du' \tag{5.57}$$

$$\mathcal{F}_1^t(u) = \int_{C_u^t} \Omega(\omega\nu^{-1}(L\psi)^2 + \omega\psi(L\psi) - \frac{1}{2}\psi^2(L\omega))d\mu_{\not{g}}dt' \tag{5.58}$$

We actually define the flux integral associated to the vectorfield $K_1$ to be

$$\mathcal{F}_1'^t(u) = \int_{C_u^t} \Omega\omega\nu^{-1}(L\psi + \nu\psi)^2 d\mu_{\not{g}}dt' \tag{5.59}$$

By a direct calculation, we deduce

$$\mathcal{F}_1^t(u) - \mathcal{F}_1'^t(u) = -\int_{C_u^t} \frac{1}{2}\Omega(L(\omega\psi^2) + 2\nu\omega\psi^2)d\mu_{\not g}dt' \tag{5.60}$$

Now we have

$$\not{\mathcal{L}}_L d\mu_{\not g} = \mathrm{tr}\chi d\mu_{\not g} \tag{5.61}$$

In view of (5.9), we have for an arbitrary function $f$ defined in $W_{\epsilon_0}^*$

$$\frac{\partial}{\partial t}(\int_{S_{t,u}} \Omega f d\mu_{\not g}) = \int_{S_{t,u}} \Omega(Lf + 2\nu f)d\mu_{\not g} \tag{5.62}$$

Setting $f = \frac{1}{2}\omega\psi^2$, we have from (5.60),

$$\begin{aligned}\mathcal{F}_1^t(u) - \mathcal{F}_1'^t(u) &= -\int_0^t \frac{\partial}{\partial t'}(\int_{S_{t',u}} \frac{1}{2}\Omega\omega\psi^2 d\mu_{\not g})dt' \\ &= -\int_{S_{t,u}} \frac{1}{2}\Omega\omega\psi^2 d\mu_{\not g} + \int_{S_{0,u}} \frac{1}{2}\Omega\omega\psi^2 d\mu_{\not g}\end{aligned} \tag{5.63}$$

Consider next the energy integral (5.57) and (5.58). We actually define the energy integral associated to the vectorfield $K_1$ to be

$$\mathcal{E}_1'^u(t) = \int_{\Sigma_t^u} \frac{\Omega}{2}\omega\nu^{-1}(\eta^{-1}\kappa(L\psi + \nu\psi)^2 + \mu|\not d\psi|^2)d\mu_{\not g}du' \tag{5.64}$$

Taking into account the fact that

$$\underline{L} - \eta^{-1}\kappa L = 2T \tag{5.65}$$

we find

$$\begin{aligned}&\mathcal{E}_1^u(t) - \mathcal{E}_1'^u(t) \\ &= \int_{\Sigma_t^u} \frac{\Omega}{2}(2\omega\psi(T\psi) - \frac{1}{2}(\eta^{-1}\kappa L\omega + \underline{L}\omega + 2\eta^{-1}\kappa\nu\omega)\psi^2)d\mu_{\not g}du'\end{aligned} \tag{5.66}$$

From the definition of $\theta$, we have

$$\not{\mathcal{L}}_T d\mu_{\not g} = \kappa \mathrm{tr}\,\theta d\mu_{\not g} \tag{5.67}$$

From (5.67), we know that for an arbitrary function $f$ defined in $W_{\epsilon_0}^*$:

$$\begin{aligned}\frac{\partial}{\partial u}(\int_{S_{t,u}} \Omega f d\mu_{\not g}) &= \int_{S_{t,u}} \not{\mathcal{L}}_T(\Omega f d\mu_{\not g}) \\ &= \int_{S_{t,u}} \Omega(Tf + (\kappa \mathrm{tr}\theta + T(\log\Omega))f)d\mu_{\not g}\end{aligned} \tag{5.68}$$

Setting $f = \frac{1}{2}\omega\psi^2$ yields

$$\int_{\Sigma_t^u} \frac{\Omega}{2}(2\omega\psi(T\psi) + (T\omega)\psi^2 + (\kappa \text{tr}\theta + T(\log\Omega))\omega\psi^2) d\mu_{\not g} du'$$
$$= \int_0^u \frac{\partial}{\partial u'}(\int_{S_{t,u'}} \frac{\Omega}{2}\omega\psi^2 d\mu_{\not g}) du' = \int_{S_{t,u}} \frac{1}{2}\Omega\omega\psi^2 d\mu_{\not g}$$

Here we have used the fact that $\psi$ vanishes on $S_{t,0} \subset C_0$. Comparing this with (5.66) and using (5.65), we get

$$\mathcal{E}_1^u(t) - \mathcal{E}_1'^u(t) = \int_{S_{t,u}} \frac{1}{2}\Omega\omega\psi^2 d\mu_{\not g} - I^u(t) \tag{5.69}$$

where

$$I^u(t) = \int_{\Sigma_t^u} \frac{1}{2}\Omega(2T\omega + \eta^{-1}\kappa(L\omega + (\nu + \eta \text{tr}\theta)\omega) + \omega T(\log\Omega))\psi^2 d\mu_{\not g} du'$$

Taking into account the fact that

$$\text{tr}\chi = \eta(\text{tr}\not k - \text{tr}\theta), \quad \text{tr}\underline{\chi} = \kappa(\text{tr}\not k + \text{tr}\theta) \tag{5.70}$$

we have

$$\text{tr}\underline{\chi} = \eta^{-1}\kappa(\text{tr}\chi + 2\eta \text{tr}\theta)$$

Then we get

$$I^u(t) = \int_{\Sigma_t^u} \frac{1}{2}\Omega(\underline{L}\omega + \underline{\nu}\omega)\psi^2 d\mu_{\not g} du' \tag{5.71}$$

So (5.69) takes the form

$$\mathcal{E}_1^u(t) - \mathcal{E}_1'^u(t) = \int_{S_{t,u}} \frac{1}{2}\Omega\omega\psi^2 d\mu_{\not g} - \int_{\Sigma_t^u} \frac{1}{2}\Omega(\underline{L}\omega + \underline{\nu}\omega)\psi^2 d\mu_{\not g} du' \tag{5.72}$$

Substituting this and (5.63) into (5.36), the surface integrals cancel and we get

$$\mathcal{E}_1'^u(t) + \mathcal{F}_1'^t(u) = I^u(t) - I^u(0) + \mathcal{E}_1'^u(0) + \int_{W_u^t} Q_1 d\mu_g \tag{5.73}$$

Also for $K_0$ we have

$$\mathcal{E}_0^u(t) + \mathcal{F}_0^t(u) = \mathcal{E}_0^u(0) + \int_{W_u^t} Q_0 d\mu_g \tag{5.74}$$

Obviously, by **A1** and **A2** there is a positive constant $C$ such that

$$C^{-1}\mathcal{E}_0^u(t) \leqslant \int_0^u (\int_{S_{t,u'}} (\mu(1+\mu)((L\psi)^2 + |\not d\psi|^2) + (\underline{L}\psi)^2) d\mu_{\not g}) du' \leqslant C\mathcal{E}_0^u(t) \tag{5.75}$$

and

$$C^{-1}\mathcal{F}_0^t(u) \leqslant \int_0^t (\int_{S_{t',u}} ((1+\mu)(L\psi)^2 + \mu|\not d\psi|^2) d\mu_{\not g}) dt' \leqslant C\mathcal{F}_0^t(u) \tag{5.76}$$

By **A1**, **A2** and **B1**, **D1** there is a positive constant $C$ such that

$$C^{-1}\mathcal{E}_1^{\prime u}(t) \leqslant (1+t)^2 \int_0^u \left( \int_{S_{t,u'}} \mu((L\psi + \nu\psi)^2 + |\not{d}\psi|^2) d\mu_{\not{g}} \right) du' \leqslant C\mathcal{E}_1^{\prime u}(t) \quad (5.77)$$

and

$$C^{-1}\mathcal{F}_1^{\prime t}(u) \leqslant \int_0^t (1+t')^2 \left( \int_{S_{t',u}} (L\psi + \nu\psi)^2 d\mu_{\not{g}} \right) dt' \leqslant C\mathcal{F}_1^{\prime t}(u) \quad (5.78)$$

**Lemma 5.1** There is a numerical constant $C$ such that for all $u \in [0, \epsilon_0]$,

$$\int_{S_{t,u}} \psi^2 d\mu_{\not{g}} \leqslant \epsilon_0 C \mathcal{E}_0^u(t)$$

*Proof*. We shall use the acoustical coordinates $(t, u, \vartheta^1, \vartheta^2)$. Now on a given hyperplane $\Sigma_t$ we may set $\Xi = 0$, then on this $\Sigma_t$ we have

$$T = \frac{\partial}{\partial u}$$

hence, in view of the fact that $\psi$ vanishes on $C_0$,

$$\psi(t, u, \vartheta) = \int_0^u (T\psi)(t, u', \vartheta) du' \quad (5.79)$$

It follows that

$$\begin{aligned} \int_{S_{t,u}} \psi^2 d\mu_{\tilde{\not{g}}} &= \int_{S^2} \psi^2(t,u,\vartheta) d\mu_{\tilde{\not{g}}}(t,u,\vartheta) \\ &= \int_{S^2} \left( \int_0^u (T\psi)(t,u',\vartheta) du' \right)^2 d\mu_{\tilde{\not{g}}}(t,u,\vartheta) \\ &\leqslant \epsilon_0 \int_{S^2} \left( \int_0^u (T\psi)^2(t,u',\vartheta) du' \right) d\mu_{\tilde{\not{g}}}(t,u,\vartheta) \end{aligned} \quad (5.80)$$

In view of the fact $\tilde{\not{g}} = \Omega \not{g}$,

$$\not{\mathcal{L}}_T d\mu_{\tilde{\not{g}}} = \Omega \not{\mathcal{L}}_T d\mu_{\not{g}} + T(\Omega) d\mu_{\not{g}} = (\kappa \mathrm{tr}\, \theta + T \log \Omega) d\mu_{\tilde{\not{g}}} \quad (5.81)$$

From (5.9) and (5.10) we have

$$\kappa \mathrm{tr}\, \theta + T \log \Omega = -\eta^{-1}\kappa\nu + \underline{\nu} \quad (5.82)$$

Thus

$$\not{\mathcal{L}}_T d\mu_{\tilde{\not{g}}} = (-\eta^{-1}\kappa\nu + \underline{\nu}) d\mu_{\tilde{\not{g}}} \quad (5.83)$$

From **B1**, **B2** and **A2**, **A3** we have

$$|-\eta^{-1}\kappa\nu + \underline{\nu}| \leqslant C(1+t)^{-1}(1+\log(1+t))^4 \leqslant C' \quad (5.84)$$

Integrate (5.83) along $T$ on $\Sigma_t$ and using the fact $\epsilon_0 \leqslant \frac{1}{2}$ we get

$$C^{-1} \leqslant \frac{d\mu_{\tilde{\not{g}}(t,u,\vartheta)}}{d\mu_{\tilde{\not{g}}(t,0,\vartheta)}} \leqslant C \tag{5.85}$$

for all $(u, \vartheta) \in [0, \epsilon_0] \times S^2$. Then (5.80) is bounded by

$$C\epsilon_0 \int_0^u \left( \int_{S^2} (T\psi)^2(t, u', \vartheta) d\mu_{\tilde{\not{g}}}(t, u', \vartheta) \right) du' \leqslant C\epsilon_0 \int_0^u \left( \int_{S_{t,u'}} (T\psi)^2 d\mu_{\not{g}} \right) du' \tag{5.86}$$

Obviously

$$(T\psi)^2 \leqslant C((\underline{L}\psi)^2 + \eta^{-2}\kappa^2(L\psi)^2) \tag{5.87}$$

The result follows. □

We now introduce the following quantities which are non-decreasing functions of $t$ at each $u$:

$$\bar{\mathcal{E}}_0^u(t) = \sup_{t' \in [0,t]} \mathcal{E}_0^u(t') \tag{5.88}$$

$$\mathcal{F}_0^t(u) \tag{5.89}$$

$$\bar{\mathcal{E}}_1'^u(t) = \sup_{t' \in [0,t]} (1 + \log(1 + t'))^{-4} \mathcal{E}_1'^u(t') \tag{5.90}$$

$$\bar{\mathcal{F}}_1'^t(u) = \sup_{t' \in [0,t]} (1 + \log(1 + t'))^{-4} \mathcal{F}_1'^{t'}(u) \tag{5.91}$$

Obviously, $\bar{\mathcal{E}}_0^u(t)$, $\bar{\mathcal{E}}_1'^u(t)$ are also non-decreasing functions of $u$ at each $t$. The statements (**i**)—(**v**) of Theorem 5.1 is equivalent to

$$\bar{\mathcal{E}}_0^{\epsilon_0}(s), \ \sup_{u \in [0,\epsilon_0]} \mathcal{F}_0^s(u), \bar{\mathcal{E}}_1'^{\epsilon_0}(s), \ \sup_{u \in [0,\epsilon_0]} \bar{\mathcal{F}}_1'^s(u) \leqslant C D_0 \tag{5.92}$$

In view of (5.73) and (5.74), to prove the theorem we must estimate properly the spacetime integrals

$$\int_{W_u^t} Q_0 d\mu_g, \quad \int_{W_u^t} Q_1 d\mu_g \tag{5.93}$$

in terms of $\bar{\mathcal{E}}_0^u(t), \mathcal{F}_0^t(u), \bar{\mathcal{E}}_1'^u(t), \bar{\mathcal{F}}_1'^t(u)$. When we estimate the error spacetime integral we shall see the reason why we allow the growth like $(1 + \log(1 + t))^4$ for $\mathcal{E}_1'^u(t)$ and $\mathcal{F}_1'^t(u)$.

## 5.3 The Error Integrals

From (5.25) and (5.30), we have

$$\begin{aligned} Q_0 &= -\Omega^2 \rho K_0 \psi - \frac{1}{2} T^{\mu\nu} \tilde{\pi}_{0,\mu\nu} \\ &= Q_{0,0} + Q_{0,1} + Q_{0,2} + Q_{0,3} + Q_{0,4} + Q_{0,5} + Q_{0,6} + Q_{0,7} \end{aligned} \tag{5.94}$$

where

$$Q_{0,0} = -\Omega^2 \rho K_0 \psi \tag{5.95}$$

$$Q_{0,1} = -\frac{1}{2} T^{LL} \tilde{\pi}_{0,LL} = -\frac{1}{8} \mu^{-2} (\underline{L}\psi)^2 \tilde{\pi}_{0,LL} \tag{5.96}$$

$$Q_{0,2} = -\frac{1}{2} T^{\underline{LL}} \tilde{\pi}_{0,\underline{LL}} = -\frac{1}{8} \mu^{-2} (L\psi)^2 \tilde{\pi}_{0,\underline{LL}} \tag{5.97}$$

$$Q_{0,3} = -T^{L\underline{L}} \tilde{\pi}_{0,L\underline{L}} = -\frac{1}{4} \mu^{-1} |\not{d}\psi|^2 \tilde{\pi}_{0,L\underline{L}} \tag{5.98}$$

$$Q_{0,4} = -T^{LA} \tilde{\pi}_{0,LA} = \frac{1}{2} \mu^{-1} (\underline{L}\psi)(\not{d}^A \psi) \tilde{\pi}_{0,LA} \tag{5.99}$$

$$Q_{0,5} = -T^{\underline{L}A} \tilde{\pi}_{0,\underline{L}A} = \frac{1}{2} \mu^{-1} (L\psi)(\not{d}^A \psi) \tilde{\pi}_{0,\underline{L}A} \tag{5.100}$$

and

$$-\frac{1}{2} T^{AB} \tilde{\pi}_{0,AB} = Q_{0,6} + Q_{0,7}$$

where

$$Q_{0,6} = -\frac{1}{2} (\not{d}^A \psi \not{d}^B \psi - \frac{1}{2} (\not{g}^{-1})^{AB} |\not{d}\psi|^2) \hat{\tilde{\not{\pi}}}_{0,AB} \tag{5.101}$$

$$Q_{0,7} = -\frac{1}{4} \mu^{-1} (L\psi)(\underline{L}\psi) \mathrm{tr} \tilde{\not{\pi}}_0 \tag{5.102}$$

Here $\tilde{\not{\pi}}_0$ denotes the restriction to $S_{t,u}$ of $\tilde{\pi}_0$, and $\hat{\tilde{\not{\pi}}}_0$ denotes the trace-free part of $\tilde{\not{\pi}}_0$ and we have made use of the trace-free nature of

$$\not{d}^A \psi \not{d}^B \psi - \frac{1}{2} (\not{g}^{-1})^{AB} |\not{d}\psi|^2$$

We have

$$\tilde{\pi}_0 - \Omega \pi_0 + (K_0 \Omega) g$$

Using (3.130)—(3.137) we have

$$\tilde{\pi}_{0,LL} = 0 \tag{5.103}$$

$$\tilde{\pi}_{0,\underline{LL}} = -4\Omega\mu(\underline{L}(\eta^{-1}\kappa) - (\eta_0^{-1} + \alpha^{-1}\kappa) L(\eta^{-1}\kappa)) \tag{5.104}$$

$$\tilde{\pi}_{0,L\underline{L}} = -2\Omega\mu(\mu^{-1}(K_0\mu) + (K_0 \log \Omega) + 2L(\eta^{-1}\kappa)) \tag{5.105}$$

$$\tilde{\pi}_{0,LA} = -2\Omega(\zeta_A + \eta_A) \tag{5.106}$$

$$\tilde{\pi}_{0,\underline{L}A} = 2\Omega((\eta_0^{-1} + \alpha^{-1}\kappa)(\zeta_A + \eta_A) - \mu \not{d}_A(\eta^{-1}\kappa)) \tag{5.107}$$

$$\hat{\tilde{\not{\pi}}}_{0,AB} = 2\Omega((\eta_0^{-1} + \eta^{-1}\kappa)\hat{\chi}_{AB} + \hat{\underline{\chi}}_{AB}) \tag{5.108}$$

$$\mathrm{tr}\tilde{\not{\pi}}_0 = 4\Omega((\eta_0^{-1} + \eta^{-1}\kappa)\nu + \underline{\nu}) \tag{5.109}$$

From (5.27) and (5.30) we have

$$\begin{aligned} Q_1 &= -\Omega^2 \rho K_1 \psi - \Omega^2 \rho \omega \psi - \frac{1}{2} T^{\mu\nu} \tilde{\pi}'_{1,\mu\nu} + \frac{1}{2} \Omega^2 \psi^2 \Box_{\tilde{g}} \omega \\ &= Q_{1,0} + Q_{1,1} + Q_{1,2} + Q_{1,3} + Q_{1,4} + Q_{1,5} + Q_{1,6} + Q_{1,7} + Q_{1,8} \end{aligned} \tag{5.110}$$

where

$$Q_{1,0} = -\Omega^2 \rho K_1 \psi - \Omega^2 \rho \omega \psi \tag{5.111}$$

$$Q_{1,1} = -\frac{1}{2} T^{LL} \tilde{\pi}'_{1,LL} = -\frac{1}{8}\mu^{-2}(\underline{L}\psi)^2 \tilde{\pi}'_{1,LL} \tag{5.112}$$

$$Q_{1,2} = -\frac{1}{2} T^{\underline{L}\underline{L}} \tilde{\pi}'_{1,\underline{L}\underline{L}} = -\frac{1}{8}\mu^{-2}(L\psi)^2 \tilde{\pi}'_{1,\underline{L}\underline{L}} \tag{5.113}$$

$$Q_{1,3} = -T^{L\underline{L}} \tilde{\pi}'_{1,L\underline{L}} = -\frac{1}{4}\mu^{-1}|\not{d}\psi|^2 \tilde{\pi}'_{1,L\underline{L}} \tag{5.114}$$

$$Q_{1,4} = -T^{LA} \tilde{\pi}'_{1,LA} = \frac{1}{2}\mu^{-1}(\underline{L}\psi)(\not{d}^A\psi)\tilde{\pi}'_{1,LA} \tag{5.115}$$

$$Q_{1,5} = -T^{\underline{L}A} \tilde{\pi}'_{1,\underline{L}A} = \frac{1}{2}\mu^{-1}(L\psi)(\not{d}^A\psi)\tilde{\pi}'_{1,\underline{L}A} \tag{5.116}$$

and

$$-\frac{1}{2} T^{AB} \tilde{\pi}'_{1,AB} = Q_{1,6} + Q_{1,7}$$

where

$$Q_{1,6} = -\frac{1}{2}(\not{d}^A\psi \not{d}^B\psi - \frac{1}{2}(\not{g}^{-1})^{AB}|\not{d}\psi|^2)\hat{\tilde{\not{\pi}}}'_{1,AB} \tag{5.117}$$

$$Q_{1,7} = -\frac{1}{4}\mu^{-1}(L\psi)(\underline{L}\psi)\mathrm{tr}\tilde{\not{\pi}}'_1 \tag{5.118}$$

Also,

$$Q_{1,8} = \frac{1}{2}\Omega^2\psi^2\Box_{\tilde{g}}\omega \tag{5.119}$$

The symbols are similar as (5.101) and (5.102).

Since $\tilde{\pi}_1 = \Omega\pi_1 + (K_1\Omega)g$, we have $\tilde{\pi}'_1 = \Omega(\pi_1 + (K_1 \log\Omega - 2\omega)g)$. So we get the following:

$$\tilde{\pi}'_{1,LL} = 0 \tag{5.120}$$

$$\tilde{\pi}'_{1,\underline{L}\underline{L}} = -4\Omega\mu(\underline{L}(\nu^{-1}\omega) - \nu^{-1}\omega L(\eta^{-1}\kappa)) \tag{5.121}$$

$$\tilde{\pi}'_{1,L\underline{L}} = -2\Omega\mu(\mu^{-1}K_1\mu + L(\nu^{-1}\omega) - 2\omega + K_1\log\Omega) \tag{5.122}$$

$$\tilde{\pi}'_{1,LA} = 0 \tag{5.123}$$

$$\tilde{\pi}'_{1,\underline{L}A} = 2\Omega(\nu^{-1}\omega(\zeta_A + \eta_A) - \mu\not{d}_A(\nu^{-1}\omega)) \tag{5.124}$$

$$\hat{\tilde{\not{\pi}}}'_{1,AB} = 2\Omega\nu^{-1}\omega\hat{\chi}_{AB} \tag{5.125}$$

$$\mathrm{tr}\tilde{\not{\pi}}'_1 = 0 \tag{5.126}$$

At this point, we can see more clearly why we choose

$$K_1 = (\omega/\nu)L$$

To see this, let us first set

$$K_1 = fL$$

where $f$ is a function to be determined.

Recall (5.26),

$$\tilde{\pi}'_{1,AB} = \tilde{\pi}_{1,AB} - 2\omega\tilde{\not{g}}_{AB}$$

while

$$\begin{aligned}\tilde{\pi}_{1,AB} &= f(\tilde{g}(D_A L, e_B) + \tilde{g}(D_B L, e_A)) = 2f\tilde{\chi}_{AB}\\ &= 2f\hat{\tilde{\chi}}_{AB} + f\mathrm{tr}\tilde{\chi}\tilde{\not{g}}_{AB}\end{aligned}$$

Since

$$\mathrm{tr}\ \tilde{\chi} = 2\nu$$

we have

$$\mathrm{tr}\ \tilde{\not{\pi}}'_1 = 4(\nu f - \omega)$$

So if we choose

$$f = \omega/\nu$$

then

$$\mathrm{tr}\ \tilde{\not{\pi}}'_1 = 0$$

Hence $Q_{1,7}$ vanishes. This fact is very important, otherwise, the growth of $\omega$ would prevent us from appropriately estimating this term.

## 5.4 The Estimates for the Error Integrals

We now consider the spacetime integral of $Q_0$.

In view of (5.32), we have for an arbitrary function $f$,

$$\int_{W_u^t} \mu^{-1} f d\mu_g = \int_0^t (\int_{\Sigma_{t'}^u} f)dt' = \int_0^u (\int_{C_{u'}^t} f)du' \tag{5.127}$$

where

$$\int_{\Sigma_t'^u} f = \int_0^u (\int_{S_{t',u'}} f d\mu_{\not{g}})du', \quad \int_{C_{u'}^t} f = \int_0^t (\int_{S_{t',u'}} f d\mu_{\not{g}})dt' \tag{5.128}$$

In view of the fact that $\rho$ vanishes and (5.103),

$$Q_{0,0} = Q_{0,1} = 0 \tag{5.129}$$

By **B10**, **B11** and **A1**,

$$\mu^{-1}|\tilde{\pi}_{0,\underline{LL}}| \leqslant C(1+t)(1+\log(1+t))^{-2} \tag{5.130}$$

hence, from (5.97), we have

$$\mu|Q_{0,2}| \leqslant C(1+t)(1+\log(1+t))^{-2}(L\psi)^2 \tag{5.131}$$

Writing

$$(L\psi)^2 \leqslant 2(L\psi+\nu\psi)^2 + 2(\nu\psi)^2 \tag{5.132}$$

then

$$\int_{W_u^t} |Q_{0,2}| d\mu_g \leqslant C(J_0 + J_1) \tag{5.133}$$

where

$$J_0 = \int_0^t (1+t')(1+\log(1+t'))^{-2}\left(\int_{\Sigma_{t'}^u} (\nu\psi)^2\right)dt' \tag{5.134}$$

$$J_1 = \int_0^t (1+t')(1+\log(1+t'))^{-2}\left(\int_{\Sigma_{t'}^u} (L\psi+\nu\psi)^2\right)dt' \tag{5.135}$$

By **B1**,

$$J_0 \leqslant C\int_0^t (1+t')^{-1}(1+\log(1+t'))^{-2}\left(\int_{\Sigma_{t'}^u} \psi^2\right)dt' \tag{5.136}$$

Using Lemma 5.1 we obtain

$$J_0 \leqslant C\int_0^t (1+t')^{-1}(1+\log(1+t'))^{-2}\mathcal{E}_0^u(t')dt' \tag{5.137}$$

On the other hand,

$$J_1 = \int_0^t f_0(t')\frac{dg(t')}{dt'}dt' \tag{5.138}$$

where

$$g(t) = \int_0^t (1+t')^2\left(\int_{\Sigma_{t'}^u} (L\psi+\nu\psi)^2\right)dt' \tag{5.139}$$

and

$$f_0(t) = (1+t)^{-1}(1+\log(1+t))^{-2} \tag{5.140}$$

Integrating by parts:

$$J_1 = f_0(t)g(t) - \int_0^t g(t')\frac{df_0(t')}{dt'}dt' \tag{5.141}$$

By (5.76),

$$\int_0^t \int_{S_{t',u}} (L\psi)^2 d\mu_{\not g} dt' \leqslant C\mathcal{F}_0^t(u) \tag{5.142}$$

Hence

$$\int_0^t \left(\int_{\Sigma_{t'}^u} (L\psi)^2\right)dt' \leqslant C\int_0^u \mathcal{F}_0^t(u')du' \tag{5.143}$$

Also by Lemma 5.1 and **B1**,

$$\int_0^t \int_{\Sigma_{t'}^u} (\nu\psi)^2 d\mu_{\not g} dt' \leqslant C\epsilon_0 \int_0^t (1+t')^{-2} \sup_{u'\in[0,u]} (\int_{S_{t',u'}} \psi^2 d\mu_{\not g}) dt'$$
$$\leqslant C\epsilon_0^2 \int_0^t (1+t')^{-2} \mathcal{E}_0^u(t') dt' \tag{5.144}$$

Set

$$F(t,u) = (\int_{\Sigma_t^u} (L\psi + \nu\psi)^2)^{\frac{1}{2}} \tag{5.145}$$

We then have

$$g(t) = \int_0^t (1+t')^2 F^2(t',u) dt' \tag{5.146}$$

So

$$f_0(t)g(t) \leqslant (1+\log(1+t))^{-2} \int_0^t (1+t') F^2(t',u) dt'$$
$$\leqslant (\int_0^t F^2(t',u)dt')^{\frac{1}{2}} ((1+\log(1+t))^{-4} \int_0^t (1+t')^2 F^2(t',u) dt')^{\frac{1}{2}} \tag{5.147}$$

By (5.143) and (5.144),

$$\int_0^t F^2(t',u)dt' \leqslant C \int_0^u \mathcal{F}_0^t(u')du' + C\epsilon_0^2 \int_0^t (1+t')^{-2} \mathcal{E}_0^u(t') dt' \tag{5.148}$$

On the other hand, by (5.78) we have

$$\int_0^t (1+t')^2 F^2(t',u) dt' \leqslant C \int_0^u \mathcal{F}_1'^t(u') du' \tag{5.149}$$

So we have

$$f_0(t)g(t) \leqslant C(\int_0^u \mathcal{F}_0^t(u')du' \tag{5.150}$$
$$+ C\epsilon_0^2 \int_0^t (1+t')^{-2} \mathcal{E}_0^u(t')dt')^{\frac{1}{2}} ((1+\log(1+t))^{-4} \int_0^u \mathcal{F}_1'^t(u')du')^{\frac{1}{2}}$$

Moreover, since

$$|\frac{df_0}{dt}| \leqslant C(1+t)^{-2}(1+\log(1+t))^{-2} \tag{5.151}$$

by (5.146) and (5.149) we have

$$\int_0^t g(t') |\frac{df_0}{dt}| dt' \leqslant C \int_0^t (1+t')^{-2} (1+\log(1+t'))^{-2} (\int_0^u \mathcal{F}_1'^{t'}(u') du') dt' \tag{5.152}$$

We thus obtain

$$J_1 \leqslant C(\int_0^u \mathcal{F}_0^t(u')du' + C\epsilon_0^2 \int_0^t (1+t')^{-2}\mathcal{E}_0^u(t')dt')^{\frac{1}{2}} \cdot ((1+\log(1+t))^{-4} \int_0^u \mathcal{F}_1'^t(u')du')^{\frac{1}{2}} + C\int_0^t (1+t')^{-2}(1+\log(1+t'))^{-2}(\int_0^u \mathcal{F}_1'^{t'}(u')du')dt' \tag{5.153}$$

which together with (5.137) yields

$$\int_{W_u^t} |Q_{0,2}|d\mu_g \leqslant C\int_0^t (1+t')^{-1}(1+\log(1+t'))^{-2}\mathcal{E}_0^u(t')dt' + C(\int_0^u \mathcal{F}_0^t(u')du' + C\epsilon_0^2 \int_0^t (1+t')^{-2}\mathcal{E}_0^u(t')dt')^{\frac{1}{2}} \cdot ((1+\log(1+t))^{-4} \int_0^u \mathcal{F}_1'^t(u')du')^{\frac{1}{2}} + C\int_0^t (1+t')^{-2}(1+\log(1+t'))^{-2}(\int_0^u \mathcal{F}_1'^{t'}(u')du')dt' \tag{5.154}$$

We now turn to $Q_{0,6}$ and $Q_{0,7}$. The terms $Q_{0,3}, Q_{0,4}, Q_{0,5}$ shall be treated later.

From **B3**, **B4** and **A1**, **A3**, we have

$$|\hat{\tilde{\not\pi}}_0| \leqslant C(1+t)(1+\log(1+t))^{-6} \tag{5.155}$$

In view of (5.77) and (5.90) we can then estimate

$$\int_{W_u^t} |Q_{0,6}|d\mu_g \leqslant C\int_0^t (1+t')^{-1}(1+\log(1+t'))^{-6}\mathcal{E}_1'^u(t')dt' \leqslant C\int_0^t (1+t')^{-1}(1+\log(1+t'))^{-2}\bar{\mathcal{E}}_1'^u(t')dt' \tag{5.156}$$

Also, by **B1**, **B2**, **A1**, **A3**, we have

$$|\mathrm{tr}\ \tilde{\not\pi}_0| \leqslant C(1+t)^{-1}(1+\log(1+t))^4 \tag{5.157}$$

Writing $|L\psi| \leqslant |L\psi + \nu\psi| + |\nu\psi|$, we can then estimate

$$\int_{W_u^t} |Q_{0,7}|d\mu_g \leqslant C(J_2 + J_3) \tag{5.158}$$

where

$$J_2 = \int_0^t f_1(t') \int_{\Sigma_{t'}^u} |\nu\psi||\underline{L}\psi| \tag{5.159}$$

$$J_3 = \int_0^t f_1(t') \int_{\Sigma_{t'}^u} |L\psi + \nu\psi||\underline{L}\psi| \tag{5.160}$$

and

$$f_1(t) = (1+t)^{-1}(1+\log(1+t))^4 \tag{5.161}$$

We have

$$J_2 \leqslant I_0^{\frac{1}{2}} I_1^{\frac{1}{2}} \tag{5.162}$$

where

$$I_0 = \int_0^t (1+t')f_1(t')(\int_{\Sigma_{t'}^u} (\nu\psi)^2)dt' \tag{5.163}$$

$$I_1 = \int_0^t (1+t')^{-1}f_1(t')(\int_{\Sigma_{t'}^u} (\underline{L}\psi)^2)dt' \tag{5.164}$$

By **B1** and Lemma 5.1, we have, with

$$f_2(t) = (1+t)^{-1}f_1(t) = (1+t)^{-2}(1+\log(1+t))^4 \tag{5.165}$$

$$I_0 \leqslant C\epsilon_0 \int_0^t f_2(t')\mathcal{E}_0^u(t')dt' \tag{5.166}$$

Also, by (5.75),

$$I_1 \leqslant C \int_0^t f_2(t')\mathcal{E}_0^u(t')dt' \tag{5.167}$$

Thus we obtain

$$J_2 \leqslant C\epsilon_0^{\frac{1}{2}} \int_0^t f_2(t')\mathcal{E}_0^u(t')dt' \tag{5.168}$$

We have

$$J_3 \leqslant \frac{1}{2}(I_1 + I_2) \tag{5.169}$$

where

$$I_2 = \int_0^t (1+t')f_1(t')(\int_{\Sigma_{t'}^u} (L\psi + \nu\psi)^2)dt' \tag{5.170}$$

Recalling (5.139), we have

$$I_2 = \int_0^t f_2(t')\frac{dg(t')}{dt'}dt' \tag{5.171}$$

We can see that $I_2$ is similar to $J_1$, moreover, $f_2$ decays faster than $f_0$, thus $I_2$ has the same bound as $J_1$:

$$\begin{aligned} I_2 \leqslant & C(\int_0^u \mathcal{F}_0^t(u')du' + C\epsilon_0^2 \int_0^t (1+t')^{-2}\mathcal{E}_0^u(t')dt')^{\frac{1}{2}} \\ & \cdot((1+\log(1+t))^{-4} \int_0^u \mathcal{F}_1'^t(u')du')^{\frac{1}{2}} \\ & + C\int_0^t (1+t')^{-2}(1+\log(1+t'))^{-2}(\int_0^u \mathcal{F}_1'^{t'}(u')du')dt' \end{aligned} \tag{5.172}$$

We conclude that

$$\int_{W_t^t} |Q_{0,7}| d\mu_g \leqslant C \int_0^t f_2(t') \mathcal{E}_0^u(t') dt' \qquad (5.173)$$
$$+ C\Big(\int_0^u \mathcal{F}_0^t(u') du' + C\epsilon_0^2 \int_0^t (1+t')^{-2} \mathcal{E}_0^u(t') dt'\Big)^{\frac{1}{2}}$$
$$\cdot \Big((1+\log(1+t))^{-4} \int_0^u \mathcal{F}_1'^t(u') du'\Big)^{\frac{1}{2}}$$
$$+ C \int_0^t (1+t')^{-2}(1+\log(1+t'))^{-2}\Big(\int_0^u \mathcal{F}_1'^{t'}(u') du'\Big) dt'$$

We turn to $Q_1$. Since we are considering the case that $\rho$ vanishes, we have

$$Q_{1,0} = 0 \qquad (5.174)$$

Also, from (5.120), (5.123) and (5.126) we have

$$Q_{1,1} = Q_{1,4} = Q_{1,7} = 0 \qquad (5.175)$$

So we first consider $Q_{1,2}$. By $\mathbf{B1}, \mathbf{B10}, \mathbf{B13}, \mathbf{D1}, \mathbf{D3}, \mathbf{A1}$, we have

$$\mu^{-1} |\tilde{\pi}'_{1,\underline{LL}}| \leqslant C(1+t)(1+\log(1+t))^3 \qquad (5.176)$$

Hence,

$$\mu |Q_{1,2}| \leqslant C(1+t)(1+\log(1+t))^3 (L\psi)^2 \qquad (5.177)$$

Writing again: $(L\psi)^2 \leqslant 2(L\psi + \nu\psi)^2 + 2(\nu\psi)^2$, it then follows

$$\int_{W_u^t} |Q_{1,2}| d\mu_g \leqslant C(J_4 + J_5) \qquad (5.178)$$

where

$$J_4 = \int_0^t (1+t')(1+\log(1+t'))^3 \Big(\int_{\Sigma_{t'}^u} (\nu\psi)^2\Big) dt' \qquad (5.179)$$

$$J_5 = \int_0^t (1+t')(1+\log(1+t'))^3 \Big(\int_{\Sigma_{t'}^u} (L\psi + \nu\psi)^2\Big) dt' \qquad (5.180)$$

By **B1** and Lemma 5.1 we have

$$J_4 \leqslant C\epsilon_0^2 \int_0^t (1+t')^{-1}(1+\log(1+t'))^3 \mathcal{E}_0^u(t') dt' \leqslant C\epsilon_0^2 \bar{\mathcal{E}}_0^u(t)(1+\log(1+t))^4 \qquad (5.181)$$

From (5.139) we have

$$J_5 = \int_0^t f_3(t') \frac{dg(t')}{dt'} dt' \qquad (5.182)$$

where

$$f_3(t) = (1+t)^{-1}(1+\log(1+t))^3 \qquad (5.183)$$

Then we have

$$J_5 = f_3(t)g(t) - \int_0^t g(t')\frac{df_3(t')}{dt'}dt'$$
$$\leqslant g(t)(f_3(t) + \int_0^t |\frac{df_3(t')}{dt'}|dt') \leqslant C\int_0^u \mathcal{F}_1'^t(u')du' \tag{5.184}$$

We have used the fact that $g(t)$ is non-decreasing. In conclusion,

$$\int_{W_u^t} |Q_{1,2}|d\mu_g \leqslant C\int_0^u \mathcal{F}_1'^t(u')du' + C\epsilon_0^2\bar{\mathcal{E}}_0^u(t)(1+\log(1+t))^4 \tag{5.185}$$

Consider next $Q_{1,6}$. By **A1**, **B1**, **B3**, **D1**, we have

$$|\hat{\tilde{\not\pi}}_1'| \leqslant C(1+t)(1+\log(1+t))^{-2} \tag{5.186}$$

Hence, by (5.77),

$$\int_{W_u^t} |Q_{1,6}|d\mu_g \leqslant C\int_0^t (1+t')(1+\log(1+t'))^{-2}(\int_{\Sigma_{t'}^u} \mu|\not d\psi|^2)dt'$$
$$\leqslant C\int_0^t (1+t')^{-1}(1+\log(1+t'))^{-2}\mathcal{E}_1'^u(t')dt' \tag{5.187}$$

To estimate $Q_{1,8}$, we shall use **D5** and Lemma 5.1:

$$\int_{W_u^t} |Q_{1,8}|d\mu_g \leqslant \int_0^t (\int_0^u \sup_{S_{t',u'}}(\mu|\Box_{\tilde{g}}\omega|)\cdot(\int_{S_{t',u'}} \psi^2 d\mu_{\not g})du')dt'$$
$$\leqslant C\epsilon_0\int_0^t (\int_0^u \sup_{S_{t',u'}}(\mu|\Box_{\tilde{g}}\omega|)du')\mathcal{E}_0^u(t')dt' \tag{5.188}$$

So from **D5** we have

$$\int_{W_u^t} |Q_{1,8}|d\mu_g \leqslant C\epsilon_0(1+\log(1+t))^4\bar{\mathcal{E}}_0^u(t) \tag{5.189}$$

We now turn to the crucial terms, $Q_{0,3}$ and $Q_{1,3}$.

Consider first $Q_{1,3}$. We have

$$\tilde{\pi}_{1,L\underline{L}}' = -2\Omega\mu(\omega\nu^{-1})(\mu^{-1}L\mu + r_1) \tag{5.190}$$

where

$$r_1 = \omega^{-1}(L\omega - \nu\omega) - \nu^{-1}(L\nu + \nu^2) + L\log\Omega \tag{5.191}$$

Thus,

$$\int_{W_u^t} Q_{1,3}d\mu_g = \int_0^t (\int_{\Sigma_{t'}^u} \frac{\Omega}{2}(\omega\nu^{-1})(\mu^{-1}L\mu + r_1)\mu|\not d\psi|^2)dt' \tag{5.192}$$

Decomposing $\mu^{-1}L\mu$ into its positive and negative parts, we write

$$\int_{W_u^t} Q_{1,3}d\mu_g = \int_0^t \left(\int_{\Sigma_{t'}^u} \frac{\Omega}{2}(\omega\nu^{-1})(\mu^{-1}(L\mu)_+ + \mu^{-1}(L\mu)_- + r_1)\mu|\not{d}\psi|^2\right)dt' \quad (5.193)$$

Then by **C1** and (5.64) we have

$$\int_{W_u^t} Q_{1,3}d\mu_g \leqslant \int_0^t ((1+t')^{-1}(1+\log(1+t'))^{-1} + A(t') + \sup_{\Sigma_{t'}^u} |r_1|)\mathcal{E}_1'^u(t')dt' - K(t,u) \quad (5.194)$$

where $K(t,u)$ is the non-negative spacetime integral

$$K(t,u) = -\int_{W_u^t} \frac{\Omega}{2}\omega\nu^{-1}\mu^{-1}(L\mu)_-|\not{d}\psi|^2 d\mu_g \quad (5.195)$$

Moreover, by $\mathbf{B1}, \mathbf{D1}, \mathbf{B5}, \mathbf{B12}, \mathbf{D2}$ we have

$$\sup_{\Sigma_t^{\epsilon_0}} |r_1| \leqslant C(1+t)^{-1}(1+\log(1+t))^{-2} \quad (5.196)$$

Consider next $Q_{0,3}$. From (5.105) and (5.18) we have

$$\tilde{\pi}_{0,L\underline{L}} = -2\Omega\mu(\mu^{-1}(L\mu + \underline{L}\mu) + r_0) \quad (5.197)$$

where

$$\begin{aligned} r_0 &= \alpha^{-2}L\mu + 2L(\alpha^{-1}\kappa) + K_0 \log\Omega \\ &= 3L(\alpha^{-1}\kappa) + 2\alpha^{-2}\kappa L\alpha + (1+\alpha^{-1}\kappa)L\log\Omega + \underline{L}\log\Omega \end{aligned} \quad (5.198)$$

Thus,

$$\int_{W_u^t} Q_{0,3}d\mu_g = \int_0^t \left(\int_{\Sigma_{t'}^u} \frac{\Omega}{2}(\mu^{-1}(L\mu + \underline{L}\mu) + r_0)\mu|\not{d}\psi|^2\right)dt' \quad (5.199)$$

Decomposing $\mu^{-1}(L\mu+\underline{L}\mu)$ into its positive and negative parts, and then dropping the negative part, we obtain

$$\int_{W_u^t} Q_{0,3}d\mu_g \leqslant \int_0^t \left(\int_{\Sigma_{t'}^u} \frac{\Omega}{2}(\mu^{-1}(L\mu + \underline{L}\mu)_+ + |r_0|)\mu|\not{d}\psi|^2\right)dt' \quad (5.200)$$

Then by **C2** and (5.64) we have

$$\int_{W_u^t} Q_{0,3}d\mu_g \leqslant \int_0^t (1+t')^{-2}(B(t') + \sup_{\Sigma_{t'}^u} |r_0|)\mathcal{E}_1'^u(t')dt' \quad (5.201)$$

Moreover, by $\mathbf{B5}, \mathbf{B6}, \mathbf{B9}, \mathbf{B10}$ we have

$$\sup_{\Sigma_t^{\epsilon_0}} |r_0| \leqslant C(1+t)(1+\log(1+t))^{-6} \tag{5.202}$$

Finally, we consider $Q_{0,4}, Q_{0,5}$ and $Q_{1,5}$.

First, we have

$$\int_{W_u^t} |Q_{1,5}| d\mu_g \leqslant M_1 + R_1 \tag{5.203}$$

where

$$M_1 = \int_{W_u^t} \Omega(\omega\nu^{-1})|L\psi||\not{d}\psi||\zeta+\eta|\mu^{-1}d\mu_g \tag{5.204}$$

$$R_1 = \int_{W_u^t} \Omega|L\psi||\not{d}\psi||\not{d}(\omega\nu^{-1})|d\mu_g \tag{5.205}$$

First, we estimate $M_1$. We decompose

$$M_1 = M_1' + M_1'' \tag{5.206}$$

where

$$M_1' = \int_{\mathcal{U}\bigcap W_u^t} \Omega(\omega\nu^{-1})|L\psi||\not{d}\psi||\zeta+\eta|\mu^{-1}d\mu_g \tag{5.207}$$

$$M_1'' = \int_{\mathcal{U}^c\bigcap W_u^t} \Omega(\omega\nu^{-1})|L\psi||\not{d}\psi||\zeta+\eta|\mu^{-1}d\mu_g \tag{5.208}$$

The region $\mathcal{U}$ is defined by (5.14).

According to **C3** we have

$$-(L\mu)_- \geqslant C^{-1}(1+t')^{-1}(1+\log(1+t'))^{-1} \tag{5.209}$$

in $\mathcal{U}\bigcap W_u^t$.

Comparing with (5.195) we have

$$\begin{aligned} K(t,u) &\geqslant -\int_{\mathcal{U}\bigcap W_u^t} \frac{\Omega}{2}\omega\nu^{-1}\mu^{-1}(L\mu)_-|\not{d}\psi|^2 d\mu_g \\ &\geqslant \frac{1}{2C}\int_{\mathcal{U}\bigcap W_u^t} \Omega\omega\nu^{-1}(1+t')^{-1}(1+\log(1+t'))^{-1}\mu^{-1}|\not{d}\psi|^2 d\mu_g \end{aligned} \tag{5.210}$$

Thus we can estimate

$$M_1' \leqslant CK^{\frac{1}{2}}N_1^{\frac{1}{2}} \tag{5.211}$$

where

$$N_1 = \int_{W_u^t} \Omega\omega\nu^{-1}(1+t')(1+\log(1+t'))|\zeta+\eta|^2|L\psi|^2\mu^{-1}d\mu_g \tag{5.212}$$

By virtue of **B7**,

$$N_1 \leqslant \int_{W_u^t} \Omega\omega\nu^{-1}(1+t')^{-1}(1+\log(1+t'))^3|L\psi|^2\mu^{-1}d\mu_g \tag{5.213}$$

Thus

$$N_1 \leqslant C(N_{1,0} + N_{1,1}) \tag{5.214}$$

where

$$N_{1,0} = \int_{\mathcal{U}\bigcap W_u^t} \Omega\omega\nu^{-1}(1+t')^{-1}(1+\log(1+t'))^3|\nu\psi|^2\mu^{-1}d\mu_g \tag{5.215}$$

$$N_{1,1} = \int_{\mathcal{U}\bigcap W_u^t} \Omega\omega\nu^{-1}(1+t')^{-1}(1+\log(1+t'))^3|L\psi+\nu\psi|^2\mu^{-1}d\mu_g \tag{5.216}$$

Now by **B1**, **D1** and **A1**,

$$N_{1,0} \leqslant C\int_0^t (1+t')^{-1}(1+\log(1+t'))^3(\int_0^u (\int_{S_{t',u'}} \psi^2 d\mu_{\not g})du')dt' \tag{5.217}$$

hence by Lemma 5.1,

$$\begin{aligned} N_{1,0} &\leqslant C\epsilon_0^2 \int_0^t (1+t')^{-1}(1+\log(1+t'))^3\mathcal{E}_0^u(t')dt' \\ &\leqslant C\epsilon_0^2\bar{\mathcal{E}}_0^u(t)(1+\log(1+t))^4 \end{aligned} \tag{5.218}$$

Also, from (5.59),

$$N_{1,1} \leqslant C\int_0^u (\int_{C_{u'}^t} \Omega\omega\nu^{-1}(L\psi+\nu\psi)^2)du' = C\int_0^u \mathcal{F}_1'^t(u')du' \tag{5.219}$$

We thus obtain

$$N_1 \leqslant C\epsilon_0^2\bar{\mathcal{E}}_0^u(t)(1+\log(1+t))^4 + C\int_0^u \mathcal{F}_1'^t(u')du' \tag{5.220}$$

We conclude that

$$M_1' \leqslant CK(t,u)^{\frac{1}{2}}(\epsilon_0^2\bar{\mathcal{E}}_0^u(t)(1+\log(1+t))^4 + C\int_0^u \mathcal{F}_1'^t(u')du')^{\frac{1}{2}} \tag{5.221}$$

On the other hand, since in $\mathcal{U}^c\bigcap W_u^t$ we have $\mu \geqslant 1/4$,

$$\begin{aligned} M_1'' &\leqslant 2\int_{\mathcal{U}^c\bigcap W_u^t} \Omega\omega\nu^{-1}|L\psi||\not d\psi||\zeta+\eta|\mu^{-\frac{1}{2}}d\mu_g \\ &\leqslant 2\int_{W_u^t} \Omega\omega\nu^{-1}|L\psi||\not d\psi||\zeta+\eta|\mu^{-\frac{1}{2}}d\mu_g \\ &\leqslant 2(\int_{W_u^t}(1+t')^{-1}(1+\log(1+t'))^{-1}\Omega\omega\nu^{-1}|\not d\psi|^2d\mu_g)^{\frac{1}{2}}N_1^{\frac{1}{2}} \end{aligned} \tag{5.222}$$

where $N_1$ is given by (5.212). In view of (5.64),

$$M_1'' \leqslant C(\int_0^t (1+t')^{-1}(1+\log(1+t'))^{-1}\mathcal{E}_1'^u(t')dt')^{\frac{1}{2}}N_1^{\frac{1}{2}} \tag{5.223}$$

So by (5.220) we conclude

$$M_1'' \leqslant C\Big(\int_0^t (1+t')^{-1}(1+\log(1+t'))^{-1}\mathcal{E}_1'^u(t')dt'\Big)^{\frac{1}{2}} \cdot \Big(\epsilon_0^2\bar{\mathcal{E}}_0^u(t)(1+\log(1+t))^4 + C\int_0^u \mathcal{F}_1'^t(u')du'\Big)^{\frac{1}{2}} \tag{5.224}$$

In the following we estimate $R_1$ given by (5.205). Using **B14**, **D4**, **B1**, **D1**, we have

$$\omega^{-1}\nu|\not{d}(\omega\nu^{-1})| \leqslant C(1+t)^{-1}(1+\log(1+t))^{\frac{1}{2}} \tag{5.225}$$

so $\mu^{\frac{1}{2}}\omega^{-1}\nu|\not{d}(\omega\nu^{-1})|$ enjoys the same bound as $|\zeta+\eta|$, namely by $C(1+t)^{-1}(1+\log(1+t))$. It follows that $R_1$ is bounded in the same way as $M_1''$. So we conclude

$$\int_{W_u^t} |Q_{1,5}|d\mu_g \leqslant C\Big(\epsilon_0^2\bar{\mathcal{E}}_0^u(t)(1+\log(1+t))^4 + C\int_0^u \mathcal{F}_1'^t(u')du'\Big)^{\frac{1}{2}} \cdot \Big(K(t,u) + \int_0^t (1+t')^{-1}(1+\log(1+t'))^{-1}\mathcal{E}_1'^u(t')dt'\Big)^{\frac{1}{2}} \tag{5.226}$$

Next we consider $Q_{0,4}$. Obviously, we have

$$\int_{W_u^t} |Q_{0,4}|d\mu_g \leqslant M_0 \tag{5.227}$$

where

$$M_0 = \int_{W_u^t} \Omega|\underline{L}\psi||\not{d}\psi||\zeta+\eta|\mu^{-1}d\mu_g \tag{5.228}$$

Again, we decompose

$$M_0 = M_0' + M_0'' \tag{5.229}$$

where

$$M_0' = \int_{\mathcal{U}\bigcap W_u^t} \Omega|\underline{L}\psi||\not{d}\psi||\zeta+\eta|\mu^{-1}d\mu_g \tag{5.230}$$

$$M_0'' = \int_{\mathcal{U}^c\bigcap W_u^t} \Omega|\underline{L}\psi||\not{d}\psi||\zeta+\eta|\mu^{-1}d\mu_g \tag{5.231}$$

By **B7**, we estimate

$$\int_{\mathcal{U}\bigcap W_u^t} \Omega|\underline{L}\psi||\not{d}\psi||\zeta+\eta|\mu^{-1}d\mu_g \leqslant C\int_{\mathcal{U}\bigcap W_u^t} (1+t')^{-1}(1+\log(1+t'))|\underline{L}\psi||\not{d}\psi|d\mu_{\not{g}}du'dt' \tag{5.232}$$

By Holder Inequality, the right hand side of the above is bounded by

$$C\int_0^t (1+t')^{-1}(1+\log(1+t'))\sqrt{\mathcal{E}_0^u(t')}\sqrt{\int_{\mathcal{U}\bigcap\Sigma_{t'}^u} |\not{d}\psi|^2 dt'} \leqslant \Big(\int_0^t (1+t')^{-2}(1+\log(1+t'))^2\mathcal{E}_0^u(t')dt'\Big)^{1/2} \cdot \Big(\int_{\mathcal{U}\bigcap W_u^t} |\not{d}\psi|^2 d\mu_{\not{g}}du'dt'\Big)^{1/2} \tag{5.233}$$

Let us define

$$F(t,u) := \int_0^t (1+t')(1+\log(1+t'))^{-1}(\int_{\mathcal{U}\bigcap\Sigma_{t'}^u} |\not{d}\psi|^2)dt' \tag{5.234}$$

By (5.210) and **B1**, **D1**,

$$K(t,u) \geqslant \frac{1}{C}\int_{\mathcal{U}\bigcap W_u^t} \Omega(1+t')(1+\log(1+t'))^{-1}\mu^{-1}|\not{d}\psi|^2 d\mu_g \tag{5.235}$$

Let us define

$$\bar{K}(t,u) = \sup_{t'\in[0,t]} (1+\log(1+t'))^{-4}K(t',u)$$

Therefore

$$F(t,u) \leqslant CK(t,u) \leqslant C\bar{K}(t,u)(1+\log(1+t))^4 \tag{5.236}$$

and

$$\frac{dF}{dt}(t,u) = (1+t)(1+\log(1+t))^{-1}\int_{\mathcal{U}\bigcap\Sigma_t^u} |\not{d}\psi|^2 \tag{5.237}$$

What we need to estimate is the following integral:

$$I(t) := \int_0^t (\int_{\mathcal{U}\bigcap\Sigma_{t'}^u} |\not{d}\psi|^2)dt' \tag{5.238}$$

We have

$$\begin{aligned} I(t) &= \int_0^t (1+t')^{-1}(1+\log(1+t'))\frac{dF}{dt'}(t')dt' \\ &= (1+t)^{-1}(1+\log(1+t))F(t) \\ &+ \int_0^t F(t')((1+t')^{-2}(1+\log(1+t')) - (1+t')^{-2})dt' \end{aligned} \tag{5.239}$$

Substitute (5.236) we obtain

$$I(t) \leqslant C'\bar{K}(t,u) \tag{5.240}$$

Therefore

$$M_0' \leqslant C\bar{K}(t,u)^{1/2}(\int_0^t (1+t')^{-2}(1+\log(1+t'))^2\bar{\mathcal{E}}_0^u(t')dt')^{1/2} \tag{5.241}$$

To estimate $M_0''$ we note that since $\mu \geqslant \eta_0/4$ in $\mathcal{U}^c$,

$$\begin{aligned} M_0'' &\leqslant 2\int_{\mathcal{U}^c\bigcap W_u^t} \Omega|\underline{L}\psi||\not{d}\psi||\zeta+\eta|\mu^{-1/2}d\mu_g \\ &\leqslant 2\int_{W_u^t} \Omega|\underline{L}\psi||\not{d}\psi||\zeta+\eta|\mu^{-1/2}d\mu_g \end{aligned}$$

$$= 2\int_0^t \left(\int_{\Sigma_{t'}^u} \Omega|\underline{L}\psi||\not{d}\psi||\zeta + \eta|\mu^{1/2}\right)dt' \tag{5.242}$$

Hence

$$\begin{aligned} M_0'' &\leqslant 2\int_0^t \left(\int_{\Sigma_{t'}^u} \Omega\mu|\not{d}\psi|^2\right)^{1/2} \cdot \left(\int_{\Sigma_{t'}^u} \Omega|\underline{L}\psi|^2|\zeta + \eta|^2\right)^{1/2} dt' \\ &\leqslant C\int_0^t \frac{\mathcal{E}_1'^u(t')^{1/2}}{1+t'} \cdot \frac{(1+\log(1+t'))\mathcal{E}_0^u(t')^{1/2}}{1+t'} dt' \\ &\leqslant C\left(\int_0^t (1+t')^{-2}(1+\log(1+t'))^3 \bar{\mathcal{E}}_1'^u(t')dt'\right)^{1/2} \\ &\quad \cdot \left(\int_0^t (1+t')^{-2}(1+\log(1+t'))^3 \bar{\mathcal{E}}_0^u(t')dt'\right)^{1/2} \end{aligned} \tag{5.243}$$

From (5.227), (5.229), (5.241) and (5.243) we conclude that

$$\begin{aligned} \int_{W_u^t} |Q_{0,4}|d\mu_g &\leqslant C\bar{K}(t,u)^{\frac{1}{2}}\left(\int_0^t (1+t')^{-2}(1+\log(1+t'))^2 \bar{\mathcal{E}}_0^u(t')dt'\right)^{\frac{1}{2}} \\ &\quad + C\left(\int_0^t (1+t')^{-2}(1+\log(1+t'))^3 \bar{\mathcal{E}}_1'^u(t')dt'\right)^{1/2} \\ &\quad \cdot \left(\int_0^t (1+t')^{-2}(1+\log(1+t'))^3 \bar{\mathcal{E}}_0^u(t')dt'\right)^{1/2} \end{aligned} \tag{5.244}$$

We are left with $Q_{0,5}$. We have

$$\int_{W_u^t} |Q_{0,5}|d\mu_g \leqslant \tilde{M}_0 + R_0 \tag{5.245}$$

where

$$\begin{aligned} \tilde{M}_0 &= \int_{W_u^t} \Omega|L\psi||\not{d}\psi|(1+\eta^{-1}\kappa)|\zeta + \eta|\mu^{-1}d\mu_g \\ R_0 &= \int_{W_u^t} \Omega|L\psi||\not{d}\psi||\not{d}(\eta^{-1}\kappa)|d\mu_g \end{aligned} \tag{5.246}$$

Decomposing

$$\begin{aligned} \tilde{M}_0 &= \tilde{M}_0' + \tilde{M}_0'' \\ \tilde{M}_0' &= \int_{\mathcal{U}\bigcap W_u^t} \Omega|L\psi||\not{d}\psi|(1+\eta^{-1}\kappa)|\zeta + \eta|\mu^{-1}d\mu_g \\ \tilde{M}_0'' &= \int_{\mathcal{U}^c\bigcap W_u^t} \Omega|L\psi||\not{d}\psi|(1+\eta^{-1}\kappa)|\zeta + \eta|\mu^{-1}d\mu_g \end{aligned} \tag{5.247}$$

Similarly we have

$$\begin{aligned} &\int_{\mathcal{U}\bigcap W_u^t} \Omega|L\psi||\not{d}\psi|(1+\eta^{-1}\kappa)|\zeta + \eta|d\mu_g du'dt' \\ &\leqslant \int_0^t (1+t')^{-1}(1+\log(1+t'))^{3/2}\sqrt{\mathcal{E}_0^u(t')}\sqrt{\int_{\mathcal{U}\bigcap \Sigma_{t'}^u} |\not{d}\psi|^2 dt'} \end{aligned} \tag{5.248}$$

This is similar to (5.233). We then proceed as we estimate $M_0'$ to obtain

$$\tilde{M}_0' \leqslant C\bar{K}(t,u)^{1/2}\left(\int_0^u \mathcal{F}_0^t(u')du'\right)^{1/2} \tag{5.249}$$

Next we estimate $\tilde{M}_0''$. Since $\mu \geqslant 1/4$ in $\mathcal{U}^c \bigcap W_u^t$, we have, by **A2, A3** and **B7**,

$$\begin{aligned}\tilde{M}_0'' &\leqslant \frac{2}{\sqrt{\eta_0}}\int_{\mathcal{U}^c\bigcap W_u^t} \Omega|L\psi||\not{d}\psi||\zeta + \eta|(1+\eta^{-1}\kappa)\mu^{-1/2}d\mu_g \\ &\leqslant C\int_{W_u^t}(1+t')^{-1}(1+\log(1+t'))^{3/2}\Omega|L\psi||\not{d}\psi|(1+\eta^{-1}\kappa)^{1/2}\mu^{-1/2}d\mu_g\end{aligned} \tag{5.250}$$

Since the factor $(1+t')^{-1}(1+\log(1+t'))^{3/2}$ is bounded by a numerical constant we obtain

$$\tilde{M}_0'' \leqslant C\int_{W_u^t}\Omega((1+\eta^{-1}\kappa)(L\psi)^2 + \mu|\not{d}\psi|^2)\mu^{-1}d\mu_g \tag{5.251}$$

hence in view of (5.54) we get

$$\tilde{M}_0'' \leqslant C\int_0^u \mathcal{F}_0^t(u')du' \tag{5.252}$$

Finally we estimate $R_0$. By **B8** we have

$$R_0 \leqslant C\int_{W_u^t}(1+t')^{-1}(1+\log(1+t'))\Omega|L\psi||\not{d}\psi|(1+\eta^{-1}\kappa)^{1/2}\mu^{-1/2}d\mu_g \tag{5.253}$$

We have used the fact that $(1+\eta^{-1}\kappa)^{1/2}\mu^{-1/2} = ((1+\eta^{-2}\mu)/\mu)^{1/2} \geqslant \eta^{-1} \geqslant C^{-1}$.

Since the factor $(1+t')^{-1}(1+\log(1+t'))$ is bounded by a numerical constant we again obtain

$$R_0 \leqslant C\int_{W_u^t}\Omega((1+\eta^{-1}\kappa)(L\psi)^2 + \mu|\not{d}\psi|^2)\mu^{-1}d\mu_g \leqslant C\int_0^u \mathcal{F}_0^t(u')du' \tag{5.254}$$

So we conclude that

$$\int_{W_u^t}|Q_{0,5}|d\mu_g \leqslant C\bar{K}(t,u)^{1/2}\left(\int_0^u \mathcal{F}_0^t(u')du'\right)^{1/2} + C\int_0^u \mathcal{F}_0^t(u')du' \tag{5.255}$$

## 5.5 Treatment of the Integral Inequalities Depending on $t$ and $u$. Completion of the Proof

We now focus on the identity (5.73). By (5.174), (5.175), (5.185), (5.187), (5.189), (5.194), (5.196) and (5.226), the spacetime integral on right hand side of (5.73) is bounded from above by

$$\int_{W_u^t} Q_1 d\mu_g \leqslant CM(t,u) + L(t,u) + \int_0^t \tilde{A}(t')\mathcal{E}_1'^u(t')dt' \\ - K(t,u) + C(K(t,u)^{1/2} + L(t,u)^{1/2})M(t,u)^{1/2} \tag{5.256}$$

Here,

$$M(t,u) = \bar{\mathcal{E}}_0^u(t)(1+\log(1+t))^4 + \int_0^u \mathcal{F}_1'^t(u')du' \\ L(t,u) = \int_0^t (1+t')^{-1}(1+\log(1+t'))^{-1}\mathcal{E}_1'^u(t')dt' \\ \tilde{A}(t) = A(t) + C(1+t)^{-1}(1+\log(1+t))^{-2} \tag{5.257}$$

Now from **C1** we have

$$\int_0^t \tilde{A}(t')dt' \leqslant C \tag{5.258}$$

where $C$ is independent of $t$.

Using the inequalities

$$-K + CK^{1/2}M^{1/2} \leqslant -\frac{1}{2}K + CM, \quad CL^{1/2}M^{1/2} \leqslant \frac{1}{2}L + CM \tag{5.259}$$

we get the following:

$$\int_{W_u^t} Q_1 d\mu_g \leqslant -\frac{1}{2}K(t,u) + CM(t,u) + \frac{3}{2}L(t,u) + \int_0^t \tilde{A}(t')\mathcal{E}_1'^u(t')dt' \tag{5.260}$$

We have

$$L(t,u) \leqslant \int_0^t (1+t')^{-1}(1+\log(1+t'))^3\bar{\mathcal{E}}_1'^u(t')dt' \\ \leqslant \frac{1}{4}(1+\log(1+t))^4\bar{\mathcal{E}}_1'^u(t) \tag{5.261}$$

Also, since

$$\mathcal{F}_1'^t(u) \leqslant (1+\log(1+t))^4\bar{\mathcal{F}}_1'^t(u) \tag{5.262}$$

defining

$$V_1'(t,u) = \int_0^u \bar{\mathcal{F}}_1'^t(u')du' \tag{5.263}$$

we have

$$\int_0^u \mathcal{F}_1'^t(u')du' \leqslant (1+\log(1+t))^4 V_1'(t,u) \tag{5.264}$$

Note that $V_1'(t,u)$ is a non-decreasing function of $t$ at each $u$ as well as a non-decreasing function of $u$ at each $t$. Hence

$$M(t,u) \leqslant (1+\log(1+t))^4(\bar{\mathcal{E}}_0^u(t) + V_1'(t,u)) \tag{5.265}$$

and also,

$$\int_0^t \tilde{A}(t')\mathcal{E}_1'^u(t')dt' \leqslant (1+\log(1+t))^4 \int_0^t \tilde{A}(t')\bar{\mathcal{E}}_1'^u(t')dt' \tag{5.266}$$

From $\mathbf{A1}, \mathbf{B2}, \mathbf{D1}, \mathbf{D3}$ and Lemma 5.1, the spacelike hypersurface integrals on the right hand side of (5.73) are bounded by

$$\begin{aligned} &|\textstyle\int_{\Sigma_t^u} \frac{1}{2}\Omega(\underline{L}\omega + \underline{\nu}\omega)\psi^2| \leqslant C\bar{\mathcal{E}}_0^u(t)(1+\log(1+t))^4, \\ &|\textstyle\int_{\Sigma_0^u} \frac{1}{2}\Omega(\underline{L}\omega + \underline{\nu}\omega)\psi^2| \leqslant CD_0 \end{aligned} \tag{5.267}$$

Also, by (5.75) and (5.77) at $t=0$, the remaining term on the right hand side of (5.73) is bounded by

$$\mathcal{E}_1'^u(0) \leqslant CD_0 \tag{5.268}$$

In view of (5.260)—(5.268), the identity (5.73) implies

$$\begin{aligned} &\mathcal{E}_1'^u(t) + \mathcal{F}_1'^t(u) + \frac{1}{2}K(t,u) \\ &\leqslant (1+\log(1+t))^4\Big(\frac{3}{8}\bar{\mathcal{E}}_1'^u(t) + C(\bar{\mathcal{E}}_0^u(t) + V_1'(t,u)) \\ &+ \int_0^t \tilde{A}(t')\bar{\mathcal{E}}_1'^u(t')dt'\Big) + CD_0 \end{aligned} \tag{5.269}$$

Keeping only the first term on the left we have

$$\begin{aligned} &(1+\log(1+t))^{-4}\mathcal{E}_1'^u(t) \\ &\leqslant \frac{3}{8}\bar{\mathcal{E}}_1'^u(t) + C(\bar{\mathcal{E}}_0^u(t) + V_1'(t,u)) + \int_0^t \tilde{A}(t')\bar{\mathcal{E}}_1'^u(t')dt' + CD_0 \end{aligned} \tag{5.270}$$

The same holds with $t$ replaced by $t' \in [0,t]$. Since the right hand side is a non-decreasing function of $t$ at each $u$, we deduce

$$\bar{\mathcal{E}}_1'^u(t) \leqslant C(\bar{\mathcal{E}}_0^u(t) + V_1'(t,u)) + \int_0^t \tilde{A}(t')\bar{\mathcal{E}}_1'^u(t')dt' + CD_0 \tag{5.271}$$

Since $\bar{\mathcal{E}}_0^u(t) + V_1'(t,u)$ is a non-decreasing function of $t$ at each $u$ we can use Gronwall's inequality to obtain in view of (5.259),

$$\bar{\mathcal{E}}_1'^u(t) \leqslant C(\bar{\mathcal{E}}_0^u(t) + V_1'(t,u)) + CD_0 \tag{5.272}$$

hence also

$$\int_0^t \tilde{A}(t')\bar{\mathcal{E}}_1'^u(t')dt' \leqslant \bar{\mathcal{E}}_1'^u(t)\int_0^t \tilde{A}(t')dt' \leqslant C(\bar{\mathcal{E}}_0^u(t) + V_1'(t,u)) \tag{5.273}$$

Substituting (5.272) and (5.273) to (5.269) and keeping only the second term on the left, we get

$$(1+\log(1+t))^{-4}\mathcal{F}_1'^t(u) \leqslant C(\bar{\mathcal{E}}_0^u(t) + V_1'(t,u)) + CD_0 \tag{5.274}$$

The same holds with $t$ replaced by $t' \in [0,t]$. Since the right hand side is a non-decreasing function of $t$ at each $u$, we deduce

$$\bar{\mathcal{F}}_1'^t(u) \leqslant C(\bar{\mathcal{E}}_0^u(t) + V_1'(t,u)) + CD_0 \tag{5.275}$$

In view of the definition of $V_1'(t,u)$, this is

$$\bar{\mathcal{F}}_1'^t(u) \leqslant C\bar{\mathcal{E}}_0^u(t) + C\int_0^u \bar{\mathcal{F}}_1'^t(u')du' + CD_0 \tag{5.276}$$

In view of the fact that $\bar{\mathcal{E}}_0^u(t)$ is a non-decreasing function of $u$ at each $t$ and $[0,\epsilon_0]$ is a bounded interval, we can use Gronwall's inequality to get

$$\bar{\mathcal{F}}_1'^t(u) \leqslant C\bar{\mathcal{E}}_0^u(t) + CD_0 \tag{5.277}$$

hence also

$$V_1'(t,u) \leqslant C\epsilon_0\bar{\mathcal{E}}_0^u(t) + CD_0 \tag{5.278}$$

Substituting (5.278) into (5.272) we obtain

$$\bar{\mathcal{E}}_1'^u(t) \leqslant C\bar{\mathcal{E}}_0^u(t) + CD_0 \tag{5.279}$$

Substituting (5.273), (5.278) and (5.279) into (5.269) and keeping only the third term on the left we obtain

$$(1+\log(1+t))^{-4}K(t,u) \leqslant C\bar{\mathcal{E}}_0^u(t) + CD_0 \tag{5.280}$$

which implies

$$\bar{K}(t,u) \leqslant C\bar{\mathcal{E}}_0^u(t) + CD_0 \tag{5.281}$$

We now turn to the identity (5.74). By (5.129), (5.154), (5.156), (5.173), (5.201), (5.202), (5.244) and (5.255), the spacetime integral on the right hand side of (5.74) is bounded from above by, in view of (5.264),

$$\begin{aligned}
&\int_{W_u^t} Q_0 d\mu_g \\
&\leqslant \int_0^t (1+t')^{-2}(1+\log(1+t'))^4 B(t')\bar{\mathcal{E}}_1'^u(t')dt' \qquad (5.282)\\
&+C\int_0^t (1+t')^{-1}(1+\log(1+t'))^{-2}(\bar{\mathcal{E}}_1'^u(t') + \bar{\mathcal{E}}_0^u(t'))dt' + CV_0(t,u)\\
&+C(V_0(t,u) + \int_0^t (1+t')^{-1}(1+\log(1+t'))^{-2}\bar{\mathcal{E}}_0'^u(t')dt')^{1/2}(V_1'(t,u))^{1/2}\\
&+C\int_0^t (1+t')^{-2}(1+\log(1+t'))^2 V_1'(t',u)dt'\\
&+C\bar{K}(t,u)^{1/2}(\int_0^t (1+t')^{-1}(1+\log(1+t'))^{-2}\bar{\mathcal{E}}_0^u(t')dt')^{1/2}\\
&+C\bar{K}(t,u)^{1/2}(V_0(t,u))^{1/2}
\end{aligned}$$

Here we have defined

$$V_0(t,u) = \int_0^u \mathcal{F}_0^t(u')du' \tag{5.283}$$

We now substitute (5.278), (5.279) and (5.281) to the above. In doing this we estimate the fourth term on the right by

$$\begin{aligned} &C\bar{\mathcal{E}}_0^u(t)^{1/2}(V_0(t,u) + \int_0^t (1+t')^{-1}(1+\log(1+t'))^{-2}\bar{\mathcal{E}}_0^{\prime u}(t')dt')^{1/2} \\ &\leqslant \frac{\delta}{2}\bar{\mathcal{E}}_0^u(t) + \frac{C^2}{2\delta}(V_0(t,u) + \int_0^t (1+t')^{-1}(1+\log(1+t'))^{-2}\bar{\mathcal{E}}_0^{\prime u}(t')dt') \end{aligned} \tag{5.284}$$

and the sixth term by

$$\begin{aligned} &C\bar{\mathcal{E}}_0^u(t)^{1/2}(\int_0^t (1+t')^{-1}(1+\log(1+t'))^{-2}\bar{\mathcal{E}}_0^u(t')dt')^{1/2} \\ &\leqslant \frac{\delta}{2}\bar{\mathcal{E}}_0^u(t) + \frac{C^2}{2\delta}(\int_0^t (1+t')^{-1}(1+\log(1+t'))^{-2}\bar{\mathcal{E}}_0^u(t')dt') \end{aligned} \tag{5.285}$$

and the seventh term by

$$C\bar{\mathcal{E}}_0^u(t)^{1/2}(V_0(t,u))^{1/2} \leqslant \frac{\delta}{2}\bar{\mathcal{E}}_0^u(t) + \frac{C^2}{2\delta}V_0(t,u) \tag{5.286}$$

After these substitutions, we obtain

$$\int_{W_u^t} Q_0 d\mu_g \leqslant \frac{3\delta}{2}\bar{\mathcal{E}}_0^u(t) + C(1+\frac{1}{\delta})V_0(t,u) + C(1+\frac{1}{\delta})\int_0^t \tilde{B}(t')\bar{\mathcal{E}}_0^u(t')dt' + CD_0 \tag{5.287}$$

The constants $C$ are independent of $\delta$. Here

$$\tilde{B}(t) = (1+t)^{-2}(1+\log(1+t))^4 B(t) + C(1+t)^{-1}(1+\log(1+t))^{-2} \tag{5.288}$$

By **C2** we have

$$\int_0^t \tilde{B}(t')dt' \leqslant C \tag{5.289}$$

In view of (5.287) the identity (5.74) implies

$$\begin{aligned} &\mathcal{E}_0^u(t) + \mathcal{F}_0^t(u) \\ &\leqslant \mathcal{E}_0^u(0) + \frac{3\delta}{2}\bar{\mathcal{E}}_0^u(t) + C(1+\frac{1}{\delta})(V_0(t,u) + \int_0^t \tilde{B}(t')\bar{\mathcal{E}}_0^u(t')dt') \end{aligned} \tag{5.290}$$

We set $\delta = \frac{1}{3}$. Replacing $t$ by $t' \in [0,t]$ in the above, noting that the right hand side is a non-decreasing function of $t$ at each $u$, and keeping only the first term on the left, we obtain

$$\bar{\mathcal{E}}_0^u(t) \leqslant \mathcal{E}_0^u(0) + CV_0(t,u) + C\int_0^t \tilde{B}(t')\bar{\mathcal{E}}_0^u(t')dt' \tag{5.291}$$

Since $V_0(t,u)$ is a non-decreasing function of $t$ at each $u$, we can apply Gronwall's inequality to obtain

$$\bar{\mathcal{E}}_0^u(t) \leqslant C(\mathcal{E}_0^u(0) + V_0(t,u)) \tag{5.292}$$

hence also

$$\int_0^t \tilde{B}(t')\bar{\mathcal{E}}_0^u(t')dt' \leqslant \bar{\mathcal{E}}_0^u(t)\int_0^t \tilde{B}(t')dt' \leqslant C(\mathcal{E}_0^u(0) + V_0(t,u)) \tag{5.293}$$

Keeping only the second term on the left in (5.290) we obtain

$$\mathcal{F}_0^t(u) \leqslant C(\mathcal{E}_0^u(0) + V_0(t,u)) = C(\mathcal{E}_0^u(0) + \int_0^u \mathcal{F}_0^t(u')du') \tag{5.294}$$

In view that $\mathcal{E}_0^u(0)$ is a non-decreasing function of $u$ and $[0,\epsilon_0]$ is a bounded interval, we can use Gronwall's inequality to obtain

$$\mathcal{F}_0^t(u) \leqslant C\mathcal{E}_0^u(0) \tag{5.295}$$

hence also

$$V_0(t,u) \leqslant C\epsilon_0\mathcal{E}_0^u(0) \tag{5.296}$$

Substituting this into (5.292) we obtain

$$\bar{\mathcal{E}}_0^u(t) \leqslant C\mathcal{E}_0^u(0) \tag{5.297}$$

Substituting finally (5.297) to (5.277), (5.279), (5.281) we conclude that

$$\bar{\mathcal{E}}_1'^u(t), \bar{\mathcal{F}}_1'^t(u), \bar{K}(t,u) \leqslant C\mathcal{E}_0^u(0) \tag{5.298}$$

The proof of Theorem 5.1 is now complete. □

We conclude this chapter by noting certain key features of the above estimates.

(i) When we estimate $Q_{1,3}$, we do not just bound the contribution from $L\mu$ simply in absolute value. Instead, we split $L\mu$ into positive and negative parts. Because by **C3** and (5.195), we can use the negative part of $L\mu$ to estimate the tangential derivative of variations on the region $\mathcal{U}$, where $\mu$ is very small. In estimating $Q_{0,3}$, we also split the term $L\mu + \underline{L}\mu$ into positive part and negative part so that we can use the bootstrap assumption **C1** and **C2**.

(ii) Let us revisit the estimate (5.218) for $N_{1,0}$, which is original from $Q_{1,5}$. There is a growth of $(1+\log(1+t))^4$ which can not be improved on the right hand side, this is why we introduce the weight for $\mathcal{E}_1'^u(t)$ and $\mathcal{F}_1'^t(u)$, and also for $K(t,u)$ see (5.90), (5.91) and (5.235). Then we can complete the error estimates associated to $K_1$, since also this weight is enough to bound the contribution of $L(t,u)$ (see (5.257)), whose growth in time is like $\log(\log(1+t))$.

(iii) Finally, the treatment in Section 5.5 is not standard, because we have two variables $t, u$ in the integral inequalities with $u$ in a bounded interval. However, we can use Gronwall inequality with respect to one variable when the other is fixed to complete the estimates.

# 6

# Construction of Commutation Vectorfields

## 6.1 Commutation Vectorfields and Their Deformation Tensors

In this chapter, we shall construct the vectorfields $Y_i, i = 1, 2, 3, 4, 5$ used to define the higher order variations of the wave function $\phi$. That is,

$$\psi_n = Y_{i_1} \cdots Y_{i_{n-1}} \psi_1 \tag{6.1}$$

where $\psi_1$ is a first order variation, namely one of the functions (5.5). Here $i_1, \cdots, i_{n-1} = 1, 2, 3, 4, 5$. Since $\psi_1$ satisfies

$$\Box_{\tilde{g}} \psi_1 = 0 \tag{6.2}$$

$\psi_n$ satisfies

$$\Box_{\tilde{g}} \psi_n = \rho_n \tag{6.3}$$

where $\rho_n$ is obtained by commuting the vectorfields $Y_{i_1} \cdots Y_{i_{n-1}}$ with $\Box_{\tilde{g}}$. So we call $Y_i, i = 1, 2, 3, 4, 5$ the *commutation vectorfields*.

We require that, at each point, these commutation vectors span the tangent space to the spacetime manifold at the corresponding point. Thus, the set of all $\psi_n$ for fixed $n$ corresponding to a given $\psi_1$ contains all derivatives of $\psi_1$ of order $n-1$.

We take

$$Y_1 = T \tag{6.4}$$

Since this is transversal to $C_u$, we require $Y_i, i = 2, 3, 4, 5$ to be tangential to $C_u$. Moreover, for each $u \in [0, \epsilon_0]$ and each point $x \in C_u$, the set $\{Y_i(x) : i = 2, 3, 4, 5\}$ spans the tangent space to $C_u$ at $x$.

Next, we take

$$Y_2 = Q := (1 + t)L \tag{6.5}$$

In view of the fact that the range of $u$ on $W^*_{\epsilon_0}$ is the bounded interval $[0, \epsilon_0]$, $Q$ is an analogue in the acoustical spacetime of the dilation field $D$ of flat spacetime introduced in Chapter 1. Since $Y_2$ is transversal to the surfaces $S_{t,u}$, we require that $Y_i, i = 3,4,5$ to be tangential to $S_{t,u}$. Moreover, for each $t, u$ and each point $x \in S_{t,u}$, the set $\{Y_i(x) : i = 3,4,5\}$ spans the tangent plane to $S_{t,u}$.

We set

$$Y_{i+2} = R_i := \Pi \mathring{R}_i, i = 1,2,3 \tag{6.6}$$

Here $\mathring{R}_i$ are the generators of rotations about the three rectangular coordinate axes:

$$\mathring{R}_i = \epsilon_{ijk} x^j \frac{\partial}{\partial x^k} = \frac{1}{2} \epsilon_{ijk} \Omega_{jk} \tag{6.7}$$

where $\epsilon_{ijk}$ is the fully antisymmetric 3-dimensional symbol. In (6.6) $\Pi$ is the projection to the tangent plane to the surfaces $S_{t,u}$ with respect to the induced acoustical metric $\bar{g}$ on $\Sigma_t$, which, as we have seen, coincides with the Euclidean metric.

When we consider the commutator $[Y, \Box_{\tilde{g}}]$, the deformation tensor ${}^{(Y)}\tilde{\pi}$ appears. Since

$${}^{(Y)}\tilde{\pi} = \Omega^{(Y)}\pi + (Y\Omega)g \tag{6.8}$$

we must consider the expression for ${}^{(Y)}\pi$

$${}^{(Y)}\pi(Z_1, Z_2) = g(D_{Z_1}Y, Z_2) + g(D_{Z_2}Y, Z_1) \tag{6.9}$$

The components of the deformation tensors ${}^{(T)}\pi$, ${}^{(Q)}\pi$ in the null frame $(L, \underline{L}, X_1, X_2)$ can be directly computed from (3.99)—(3.106) and (3.130)—(3.137).

$${}^{(T)}\tilde{\pi}_{LL} = 0 \tag{6.10}$$

$${}^{(T)}\tilde{\pi}_{\underline{L}\underline{L}} = 4\Omega\mu T(\eta^{-1}\kappa) \tag{6.11}$$

$${}^{(T)}\tilde{\pi}_{L\underline{L}} = -2\Omega(T\mu + \mu T \log \Omega) \tag{6.12}$$

$${}^{(T)}\tilde{\pi}_{LA} = -\Omega(\zeta_A + \eta_A) \tag{6.13}$$

$${}^{(T)}\tilde{\pi}_{\underline{L}A} = -\Omega\alpha^{-1}\kappa(\zeta_A + \eta_A) \tag{6.14}$$

$${}^{(T)}\hat{\tilde{\not\pi}}_{AB} = \Omega(\hat{\underline{\chi}}_{AB} - \eta^{-1}\kappa\hat{\chi}_{AB}) \tag{6.15}$$

$$\mathrm{tr}\,{}^{(T)}\tilde{\not\pi} = 2\Omega(\underline{\nu} - \eta^{-1}\kappa\nu) \tag{6.16}$$

and

$${}^{(Q)}\tilde{\pi}_{LL} = 0 \tag{6.17}$$

$${}^{(Q)}\tilde{\pi}_{\underline{L}\underline{L}} = 4\Omega\mu(Q(\eta^{-1}\kappa) - \eta^{-1}\kappa) \tag{6.18}$$

$${}^{(Q)}\tilde{\pi}_{L\underline{L}} = -2\Omega(Q\mu + \mu Q \log \Omega + \mu) \tag{6.19}$$

$${}^{(Q)}\tilde{\pi}_{LA} = 0 \tag{6.20}$$

$${}^{(Q)}\tilde{\pi}_{\underline{L}A} = 2\Omega(1+t)(\zeta_A + \eta_A) \tag{6.21}$$

$${}^{(Q)}\hat{\tilde{\not\pi}}_{AB} = 2\Omega(1+t)\hat{\chi}_{AB} \tag{6.22}$$

$$\mathrm{tr}\,{}^{(Q)}\tilde{\not\pi} = 4\Omega(1+t)\nu \tag{6.23}$$

Here we denote by ${}^{(Y)}\tilde{\not\pi}$ the restriction of ${}^{(Y)}\tilde{\pi}$ to $S_{t,u}$, and by ${}^{(Y)}\hat{\tilde{\not\pi}}$ the trace-free part of ${}^{(Y)}\tilde{\not\pi}$.

Noting that the definition of $R_i, i = 1, 2, 3$ is intrinsic to each spacelike hypersurface $\Sigma_t$, we shall derive expressions for the components of ${}^{(R_i)}\pi$ in the frame $(L, T, X_1, X_2)$, taking advantage of the fact that $(T, X_1, X_2)$ is a frame field for each $\Sigma_t$. From (3.99)—(3.106), we have

$$ {}^{(R_i)}\pi_{LL} = 2g(D_L R_i, L) = -2g(R_i, D_L L) = 0 \tag{6.24} $$

This is because $R_i$ is tangential to $S_{t,u}$, while $L$ is orthogonal to $S_{t,u}$. For the same reason, $T$ is also orthogonal to $R_i$, hence

$$ {}^{(R_i)}\pi_{TT} = 2g(D_T R_i, T) = -2g(R_i, D_T T) \tag{6.25} $$

Since $R_i$ is tangential to $S_{t,u}$, we can expand

$$ R_i = R_i^A X_A \tag{6.26} $$

we obtain

$$ {}^{(R_i)}\pi_{TT} = 2\kappa R_i \kappa \tag{6.27} $$

Next,

$$ \begin{aligned} {}^{(R_i)}\pi_{LT} &= g(D_L R_i, T) + g(D_T R_i, L) \\ &= -g(R_i, D_L T) - g(R_i, D_T L) = -\eta_A R_i^A + \zeta_A R_i^A = -R_i \mu \end{aligned} \tag{6.28} $$

where we have used (3.55) in Chapter 3.

Next we have

$$ {}^{(R_i)}\pi_{LA} = g(D_L R_i, X_A) + g(D_{X_A} R_i, L) \tag{6.29} $$

Now,

$$ g(D_{X_A} R_i, L) = -g(R_i, D_{X_A} L) = -\chi_{AB} R_i^B \tag{6.30} $$

On the other hand, from (6.6) we have

$$ D_L R_i = (D_L \Pi)\mathring{R}_i + \Pi(D_L \mathring{R}_i) \tag{6.31} $$

Noting that for any vectorfield $Z$

$$ g(\Pi Z, X_A) = g(Z, X_A) \tag{6.32} $$

we obtain

$$ g(D_L R_i, X_A) = g((D_L \Pi)\mathring{R}_i, X_A) + g(D_L \mathring{R}_i, X_A) \tag{6.33} $$

Thus,

$$ {}^{(R_i)}\pi_{LA} = g((D_L \Pi)\mathring{R}_i, X_A) + g(D_L \mathring{R}_i, X_A) - \chi_{AB} R_i^B \tag{6.34} $$

Similarly, we have

$$ {}^{(R_i)}\pi_{TA} = g((D_T \Pi)\mathring{R}_i, X_A) + g(D_T \mathring{R}_i, X_A) - \kappa\theta_{AB} R_i^B \tag{6.35} $$

and

$$
\begin{aligned}
{}^{(\mathring{R}_i)}\pi_{AB} = &\, g((D_{X_A}\Pi)\mathring{R}_i, X_B) + g((D_{X_B}\Pi)\mathring{R}_i, X_A) \\
&+ g(D_{X_A}\mathring{R}_i, X_B) + g(D_{X_B}\mathring{R}_i, X_A)
\end{aligned}
\tag{6.36}
$$

Now we must compute $(D\Pi)\mathring{R}_i$ and $D\mathring{R}_i$. Since $\mathring{R}_i$ are tangential to $\Sigma_t$ we can expand

$$\mathring{R}_i = R_i^A X_A + \lambda_i \hat{T}, \quad \hat{T} = \kappa^{-1} T \tag{6.37}$$

for some functions $\lambda_i$. Since $\Pi T = 0$, for any vectorfield $Z$ we have

$$(D_Z \Pi)T = D_Z(\Pi T) - \Pi(D_Z T) = -\Pi(D_Z T) \tag{6.38}$$

Setting $Z$ equal to $L, T, X_A, A = 1, 2$, we obtain, using (3.99)—(3.106),

$$(D_L \Pi)T = \zeta^A X_A \tag{6.39}$$

$$(D_T \Pi)T = \kappa(\not{d}^A \kappa) X_A \tag{6.40}$$

$$(D_{X_A}\Pi)T = -\kappa \theta_A^B X_B \tag{6.41}$$

Since $\Pi X_A = X_A$, for any vectorfield $Z$ we have

$$(D_Z \Pi)X_A = D_Z(\Pi X_A) - \Pi(D_Z X_A) = D_Z X_A - \Pi(D_Z X_A) \tag{6.42}$$

Obviously, the right hand side of the above is orthogonal to $S_{t,u}$. So we get

$$g((D_L \Pi)\mathring{R}_i, X_A) = \kappa^{-1} \lambda_i \zeta_A \tag{6.43}$$

$$g((D_T \Pi)\mathring{R}_i, X_A) = \lambda_i \not{d}_A \kappa \tag{6.44}$$

$$g((D_{X_A}\Pi)\mathring{R}_i, X_B) = -\lambda_i \theta_{AB} \tag{6.45}$$

Next, we shall compute $D\mathring{R}_i$. In an arbitrary coordinate system we have

$$D_\mu \mathring{R}_i^\nu = \frac{\partial \mathring{R}_i^\nu}{\partial x^\mu} + \Gamma^\nu_{\mu\lambda} \mathring{R}_i^\lambda \tag{6.46}$$

where $\Gamma^\nu_{\mu\lambda}$ are the Christoffel symbols of the acoustical metric $g$ in the given coordinate system, and we have

$$\Gamma^\nu_{\mu\lambda} = (g^{-1})^{\nu\kappa} \Gamma_{\mu\lambda\kappa} \tag{6.47}$$

Recall from Chapter 3, that in a Galilean coordinate system we have

$$\Gamma_{000} = \frac{1}{2}\partial_0(-\eta^2 + |\mathbf{v}|^2) \tag{6.48}$$

$$\Gamma_{0i0} = \frac{1}{2}\partial_i(-\eta^2 + |\mathbf{v}|^2) \tag{6.49}$$

$$\Gamma_{ij0} = \partial_i \partial_j \phi \tag{6.50}$$

$$\Gamma_{00k} = \frac{1}{2}(2\partial_t \partial_k \phi - \partial_k(-\eta^2 + |\mathbf{v}|^2)) \tag{6.51}$$

$$\Gamma_{i0k} = \Gamma_{ijk} = 0 \tag{6.52}$$

Now we can compute $g(D_L \mathring{R}_i, X_A), g(D_T \mathring{R}_i, X_A)$ and $g(D_{X_A} \mathring{R}_i, X_B)$.

$$g(D_L \mathring{R}_i, X_A) = g_{\mu\nu} L^\lambda D_\lambda \mathring{R}_i^\mu X_A^\nu = g_{\mu\nu} L^\lambda (\frac{\partial \mathring{R}_i^\mu}{\partial x^\lambda} + \Gamma^\mu_{\lambda\gamma} \mathring{R}_i^\gamma) X_A^\nu \tag{6.53}$$

This is

$$\begin{aligned} L^\lambda (g_{\mu\nu} \frac{\partial \mathring{R}_i^\mu}{\partial x^\lambda} + \Gamma_{\lambda\mu\nu} \mathring{R}_i^\mu) X_A^\nu &= L^l \bar{g}_{mn} \frac{\partial \mathring{R}_i^m}{\partial x^l} X_A^n \\ &= L^l \delta_{mn} \epsilon_{ilm} X_A^n = L^l \epsilon_{ilm} X_A^m \end{aligned} \tag{6.54}$$

where we have used the expressions for Christoffel symbols and the fact that $\mathring{R}_i^0 = X_A^0 = 0$.

Similarly, we have

$$g(D_T \mathring{R}_i, X_A) = T^l \epsilon_{ilm} X_A^m \tag{6.55}$$

and

$$g(D_{X_A} \mathring{R}_i, X_B) = X_A^l \epsilon_{ilm} X_B^m \tag{6.56}$$

Substituting the above in (6.35)—(6.37) we obtain the following:

$$^{(R_i)}\pi_{LA} = -\chi_{AB} R_i^B + L^l \epsilon_{ilm} X_A^m + \kappa^{-1} \lambda_i \zeta_A \tag{6.57}$$

$$^{(R_i)}\pi_{TA} = -\kappa \theta_{AB} R_i^B + T^l \epsilon_{ilm} X_A^m + \lambda_i \not{d}_A \kappa \tag{6.58}$$

and noting that $\epsilon_{ilm}$ is antisymmetric in $l, m$, while $X_A^l X_B^m + X_A^m X_B^l$ is symmetric in $l, m$:

$$^{(R_i)}\pi_{AB} = -2\lambda_i \theta_{AB} \tag{6.59}$$

Since the coefficient of $\lambda_i$ in (6.57) is $\kappa^{-1}\zeta_A$, and recalling from Chapter 3 that $\kappa^{-1}\zeta_A$, $\theta_{AB}$ are regular as $\mu \to 0$, these expressions are regular as $\mu \to 0$.

We introduce the functions $y^i$ by setting

$$\hat{T}^i = -\frac{x^i}{1-u+t} + y^i \tag{6.60}$$

using this and the fact that

$$\epsilon_{ilm} x^l X_A^m = \sum_m \mathring{R}_i^m X_A^m = \not{g}_{AB} R_i^B \tag{6.61}$$

we rewrite the first two terms on the right hand side in (6.58) as

$$-\kappa \theta_{AB} R_i^B + T^l \epsilon_{ilm} X_A^m = -\kappa(\theta_{AB} + \frac{\not{g}_{AB}}{1-u+t}) R_i^B + \kappa \epsilon_{ilm} y^l X_A^m \tag{6.62}$$

Then (6.58) takes the form

$$^{(R_i)}\pi_{TA} = -\kappa(\theta_{AB} + \frac{\not{g}_{AB}}{1-u+t}) R_i^B + \kappa \epsilon_{ilm} y^l X_A^m + \lambda_i \not{d}_A \kappa \tag{6.63}$$

Also we define the function $z^i$ by setting

$$L^i = \frac{x^i}{1-u+t} + z^i \tag{6.64}$$

By (6.60) and (6.64) and recalling $L = \partial_0 - (\eta\hat{T}^i + \psi_i)\partial_i$, we get

$$z^i = -\eta y^i + \frac{(\eta-1)x^i}{1-u+t} - \psi_i \tag{6.65}$$

Using (6.62) and (6.65), we rewrite the first two terms of (6.57):

$$-\chi_{AB}R_i^B + L^l\epsilon_{ilm}X_A^m = -(\chi_{AB} - \frac{\not{g}_{AB}}{1-u+t})R_i^B + \epsilon_{ilm}z^l X_A^m \tag{6.66}$$

So (6.57) takes the form

$${}^{(R_i)}\pi_{LA} = -(\chi_{AB} - \frac{\not{g}_{AB}}{1-u+t})R_i^B + \epsilon_{ilm}z^l X_A^m + \lambda_i(\kappa^{-1}\zeta_A) \tag{6.67}$$

## 6.2 Preliminary Estimates for the Deformation Tensors

In the remainder of this chapter, we shall show how the deformation tensors ${}^{(Y)}\pi$ of the commutation vectorfields are to be controlled in terms of $\chi, \mu$ and $\psi_\mu$.

In the following, the assumptions **A1**, **A2**, **A3** are assumed to hold in $W^s_{\epsilon_0}$.

$$\mathbf{A1}: C^{-1} \leqslant \Omega \leqslant C$$
$$\mathbf{A2}: C^{-1} \leqslant \eta \leqslant C$$
$$\mathbf{A3}: \mu \leqslant C(1+\log(1+t))$$

where the constant $C$ is independent of $s$.

Recall that

$$\psi_\mu = \partial_\mu \phi \tag{6.68}$$

and we shall use the following bootstrap assumptions on $\psi_\mu$ and their first derivatives. In the following we denote by $\delta_0$ a positive constant, which is by definition less or equal to unity. This constant is used to keep track of the relative size of various quantities.

On the other hand, no smallness condition is assumed on $\epsilon_0$, other than the condition: $\epsilon_0 \leqslant \frac{1}{2}$.

There is a positive constant $C$ independent of $s$ such that in $W^s_{\epsilon_0}$,

$$\mathbf{E1}: |\psi_0 - h_0|, \quad |\psi_i| \leqslant C\delta_0(1+t)^{-1}$$
$$\mathbf{E2}: |T\psi_\mu| \leqslant C\delta_0(1+t)^{-1}, \quad |L\psi_\mu|, \quad |\not{d}\psi_\mu| \leqslant C\delta_0(1+t)^{-2}$$

The second of **E2** shall follow from $|Q\psi_\mu| \leqslant C\delta_0(1+t)^{-1}$, while the third shall follow from $|R_i\psi_\mu| \leqslant C\delta_0(1+t)^{-1}, i = 1,2,3$ in the actual estimates of Chapter 10, 11 and 12.

We also have the bootstrap assumptions on $\mu$ and $\chi$ in $W^s_{\epsilon_0}$:

$$\mathbf{F1}: |T\mu| \leqslant C\delta_0(1+\log(1+t)), \quad |\not{d}\mu| \leqslant C\delta_0(1+t)^{-1}(1+\log(1+t))$$

$$\mathbf{F2}: |\chi - \frac{\not{g}}{1-u+t}| \leqslant C\delta_0(1+t)^{-2}(1+\log(1+t))$$

Here the positive constant $C$ is also independent of $s$.

Under the assumptions $\mathbf{A}, \mathbf{E}, \mathbf{F}$, the estimates which we shall derive will hold in $W^s_{\epsilon_0}$ with constants which are independent of $s$.

First, recall that the enthalpy is $h = \psi_0 - \frac{1}{2}\sum_{i=1}^3 \psi_i^2$, so by $\mathbf{E}$ we have

$$|h| \leqslant C\delta_0(1+t)^{-1}, \quad |Th| \leqslant C\delta_0(1+t)^{-1}, \quad |Lh|, \quad |\not{d}h| \leqslant C\delta_0(1+t)^{-2} \tag{6.69}$$

Recall that $\hat{T}$ is the unit normal to the surface $S_{t,u}$ in $\Sigma_t$ with respect to the induced acoustical metric $\bar{g}$, which is the Euclidean metric, so

$$|\hat{T}| = 1 \tag{6.70}$$

It follows by $\mathbf{E1}$ that

$$|\psi_{\hat{T}}| \leqslant C\delta_0(1+t)^{-1} \tag{6.71}$$

Recalling that $L = \partial_0 - (\alpha\hat{T}^i + \psi_i)\partial_i$, from $\mathbf{E1}$ and (6.70) we have

$$|L^i| \leqslant C \tag{6.72}$$

Since $L^0 = 1$, we have

$$|\psi_L| \leqslant C\delta_0(1+t)^{-1} \tag{6.73}$$

Let us express $\bar{g}^{-1}$ in the frame $\hat{T}, X_1, X_2$:

$$\delta^{ij} = (\bar{g}^{-1})^{ij} = \hat{T}^i\hat{T}^j + (\not{g}^{-1})^{AB}X^i_A X^j_B \tag{6.74}$$

It follows from $\mathbf{E1}$ that

$$\begin{aligned}|\not{\psi}|^2 &= (\not{g}^{-1})^{AB}\not{\psi}_A\not{\psi}_B = (\not{g}^{-1})^{AB}X^i_A X^j_B\psi_i\psi_j \\ &\leqslant (\bar{g}^{-1})^{ij}\psi_i\psi_j = \sum_i(\psi_i)^2 \leqslant C\delta_0^2(1+t)^{-2}\end{aligned} \tag{6.75}$$

that is

$$|\not{\psi}| \leqslant C\delta_0(1+t)^{-1} \tag{6.76}$$

For a bilinear form $\omega$ in the tangent space to $S_{t,u}$ at a point, the magnitude $|\omega|$ is given by

$$|\omega|^2 = (\not{g}^{-1})^{AC}(\not{g}^{-1})^{BD}\omega_{AB}\omega_{CD} \tag{6.77}$$

Recall that $\not{k}_{AB} = X^i_A X^j_B k_{ij} = -\eta^{-1}X^i_A X^j_B\partial_i\psi_j$. So

$$\begin{aligned}|\not{k}|^2 &= (\not{g}^{-1})^{AC}(\not{g}^{-1})^{BD}\not{k}_{AB}\not{k}_{CD} \\ &= (\not{g}^{-1})^{AC}(\not{g}^{-1})^{BD}X^i_A X^j_B k_{ij}X^l_C X^m_D k_{lm} \\ &= \eta^{-2}(\not{g}^{-1})^{AC}(\not{g}^{-1})^{BD}X^i_A\not{d}_B\psi_i X^l_C\not{d}_D\psi_l\end{aligned}$$

$$\leqslant C\eta^{-2}(\bar{g}^{-1})^{ij}(\not{g}^{-1})^{BD}(\not{d}_B\psi_i)(\not{d}_D\psi_j)$$
$$= \eta^{-2}\sum_i(\not{g}^{-1})^{BD}(\not{d}_B\psi_i)(\not{d}_D\psi_i)$$
$$= \eta^{-2}\sum_i|\not{d}\psi_i|^2 \leqslant C\delta_0^2(1+t)^{-4} \tag{6.78}$$

That is

$$|\not{k}| \leqslant C\delta_0(1+t)^{-2} \tag{6.79}$$

By assumptions **A2** and **A3** this implies

$$\kappa|\not{k}| \leqslant C\delta_0(1+t)^{-2}(1+\log(1+t)) \tag{6.80}$$

For a linear form $v$ in the tangent space to $S_{t,u}$ at a point, the magnitude $|v|$ is given by

$$|v|^2 = (\not{g}^{-1})^{AB}v_Av_B \tag{6.81}$$

From the definition in Chapter 3, $\epsilon_A = \kappa^{-1}k(X_A, T)$, we have

$$|\epsilon|^2 = (\not{g}^{-1})^{AB}\epsilon_A\epsilon_B = (\not{g}^{-1})^{AB}k_{A\hat{T}}k_{B\hat{T}} = (\not{g}^{-1})^{AB}X_A^iX_B^j\hat{T}^l\hat{T}^m k_{il}k_{jm}$$
$$= \eta^{-2}(\not{g}^{-1})^{AB}\hat{T}^l\hat{T}^m\not{d}_A\psi_l\not{d}_B\psi_m \leqslant \eta^{-2}\sum_i|\not{d}\psi_i|^2 \leqslant C\delta_0^2(1+t)^{-4} \tag{6.82}$$

That is

$$|\epsilon| \leqslant C\delta_0(1+t)^{-2} \tag{6.83}$$

Next, we shall estimate $\eta$. Recall that

$$\eta = \left(\frac{\rho'(h)}{\rho(h)}\right)^{-\frac{1}{2}} \tag{6.84}$$

From the first inequality of (6.69), it then follows that

$$|\eta - 1| \leqslant C\delta_0(1+t)^{-1} \tag{6.85}$$

Differentiating (6.84) tangentially to $S_{t,u}$, we obtain $\not{d}\eta = ((\frac{\rho'(h)}{\rho(h)})^{-\frac{1}{2}})'\not{d}h$, hence by the last of (6.69),

$$|\not{d}\eta| \leqslant C\delta_0(1+t)^{-2} \tag{6.86}$$

Similarly, we have

$$|T\eta| \leqslant C\delta_0(1+t)^{-1}, \quad |L\eta| \leqslant C\delta_0(1+t)^{-2} \tag{6.87}$$

In view of the fact that $\kappa^{-1}\zeta_A = \eta\epsilon_A - \not{d}_A\alpha$, we then obtain

$$|\kappa^{-1}\zeta| \leqslant C\delta_0(1+t)^{-2} \tag{6.88}$$

hence

$$|\zeta| \leqslant C\delta_0(1+t)^{-2}(1+\log(1+t)) \tag{6.89}$$

We can also bound $L\mu$, given by (3.92) in Chapter 3:

$$L\mu = m + \mu e \tag{6.90}$$

where

$$m = \frac{1}{2}\frac{dH}{dh}Th, \quad e = \frac{1}{2\alpha^2}(\frac{\rho'}{\rho})'Lh + \alpha^{-1}\hat{T}^i(L\psi_i) \tag{6.91}$$

From (6.69) we have

$$|m| \leqslant C\delta_0(1+t)^{-1} \tag{6.92}$$

From (6.69), (6.70) and **E2** we have

$$|e| \leqslant C\delta_0(1+t)^{-2} \tag{6.93}$$

Hence

$$|L\mu| \leqslant C\delta_0(1+t)^{-1} \tag{6.94}$$

In view of $\kappa = \alpha^{-1}\mu$, and the second inequality of (6.87), we have

$$|L\kappa| \leqslant C\delta_0(1+t)^{-1} \tag{6.95}$$

Integrating this along the integral curves of $L$ yields

$$|\kappa - \kappa_0| \leqslant C\delta_0 \log(1+t) \tag{6.96}$$

Here the value of $\kappa_0$ at a point in $W^*_{\epsilon_0}$ is the value of $\kappa$ at the corresponding point on $\Sigma_0$ along the same integral curve of $L$. Combining (6.96) and the fact that on $\Sigma_0$ we have

$$\kappa_0 = 1$$

we obtain the following estimates in $W^s_{\epsilon_0}$:

$$|\kappa - 1| \leqslant C\delta_0(1 + \log(1+t)) \tag{6.97}$$

We now combine **F1** with (6.86) and the first inequality of (6.87) to estimate the first derivatives of $\kappa$ on $\Sigma_t$:

$$|T\kappa| \leqslant C\delta_0(1 + \log(1+t)), \quad |\not{d}\kappa| \leqslant C\delta_0(1+t)^{-1}(1 + \log(1+t)) \tag{6.98}$$

In view of the fact that $\eta = \zeta + \not{d}\mu$, **F1** and (6.89) yield

$$|\eta| \leqslant C\delta_0(1+t)^{-1}(1 + \log(1+t)) \tag{6.99}$$

Now we use **F2** to estimate

$$\nu = \frac{1}{2}(\mathrm{tr}\chi + \frac{d\log\Omega}{dh}Lh), \quad \underline{\nu} = \frac{1}{2}(\mathrm{tr}\underline{\chi} + \frac{d\log\Omega}{dh}\underline{L}h) \tag{6.100}$$

If $\delta_0$ is small enough, **F2** implies

$$\mathrm{tr}\chi \geqslant C^{-1}(1+t)^{-1} \tag{6.101}$$

This restriction on the size of $\delta_0$ is imposed from now on.

Recall that

$$\chi = \eta(\not k - \theta), \quad \underline{\chi} = \kappa(\not k + \theta) \tag{6.102}$$

Thus we have

$$\mathrm{tr}\underline{\chi} = 2\kappa \mathrm{tr}\not k - \eta^{-1}\kappa \mathrm{tr}\chi, \quad \hat{\underline{\chi}} = 2\kappa\hat{\not k} - \eta^{-1}\kappa\hat{\chi} \tag{6.103}$$

From (6.80), (6.96) and **F2** we have

$$|\mathrm{tr}\underline{\chi}| \leqslant C(1+t)^{-1}(1+\log(1+t)), \quad |\hat{\underline{\chi}}| \leqslant C\delta_0(1+t)^{-2}(1+\log(1+t))^2 \tag{6.104}$$

Also, using (6.79), (6.85) and **F2** we deduce

$$|\theta + \frac{\not\phi}{1-u+t}| \leqslant C\delta_0(1+t)^{-2}(1+\log(1+t)) \tag{6.105}$$

Finally, combining **F2**, (6.69) and (6.101), we obtain

$$\begin{gathered} C^{-1}(1+t)^{-1} \leqslant \nu \leqslant C(1+t)^{-1} \\ |\nu - (1-u+t)^{-1}| \leqslant C\delta_0(1+t)^{-2}(1+\log(1+t)) \end{gathered} \tag{6.106}$$

Also, from (6.69) and (6.104) we have

$$|\underline{\nu}| \leqslant C(1+t)^{-1}(1+\log(1+t)) \tag{6.107}$$

We can now give the bounds of the deformation tensor of $T$ and $Q$. In the following, we denote by ${}^{(Y)}\not\pi_L$ and ${}^{(Y)}\not\pi_{\underline{L}}$ the 1-forms on each surface $S_{t,u}$ with components:

$${}^{(Y)}\not\pi_L(X_A) = {}^{(Y)}\not\pi_{LA}, \quad {}^{(Y)}\not\pi_{\underline{L}}(X_A) = {}^{(Y)}\not\pi_{\underline{L}A} \tag{6.108}$$

and similarly for ${}^{(Y)}\tilde{\not\pi}_L$ and ${}^{(Y)}\tilde{\not\pi}_{\underline{L}}$.

First, we estimate the deformation tensor of $T$. By (6.11), (6.97), (6.98) and (6.87) we have

$$\mu^{-1}|{}^{(T)}\tilde{\pi}_{\underline{LL}}| \leqslant C\delta_0(1+\log(1+t)) \tag{6.109}$$

By (6.12), **F1**, **A** and (6.69) we have

$$|{}^{(T)}\tilde{\pi}_{L\underline{L}}| \leqslant C\delta_0(1+\log(1+t)) \tag{6.110}$$

By (6.13), (6.89) and (6.99) we have

$$|{}^{(T)}\tilde{\not\pi}_L| \leqslant C\delta_0(1+t)^{-1}(1+\log(1+t)) \tag{6.111}$$

By (6.14), (6.89), (6.97) and (6.99) we have

$$\mu^{-1}|{}^{(T)}\tilde{\not\pi}_{\underline{L}}| \leqslant C\delta_0(1+t)^{-1}(1+\log(1+t)) \tag{6.112}$$

By (6.15), (6.97), (6.104) and **F2** we have

$$|{}^{(T)}\hat{\tilde{\not\pi}}| \leqslant C\delta_0(1+t)^{-2}(1+\log(1+t))^2 \tag{6.113}$$

Finally by (6.16), (6.97), (6.106) and (6.107) we have

$$|\mathrm{tr}\,{}^{(T)}\tilde{\not\pi}| \leqslant C(1+t)^{-1}(1+\log(1+t)) \tag{6.114}$$

Next, we estimate the deformation tensor of $Q$. By (6.18), (6.87), (6.95) and (6.97) we have

$$\mu^{-1}|{}^{(Q)}\tilde{\pi}_{L\underline{L}}| \leqslant C(1+\log(1+t)) \tag{6.115}$$

By (6.19), (6.69), (6.94) we have

$$|{}^{(Q)}\tilde{\pi}_{L\underline{L}}| \leqslant C(1+\log(1+t)) \tag{6.116}$$

By (6.21), (6.89) and (6.99) we have

$$|{}^{(Q)}\tilde{\not\pi}_{\underline{L}}| \leqslant C\delta_0(1+\log(1+t)) \tag{6.117}$$

By (6.22) and **F2** we have

$$|{}^{(Q)}\hat{\tilde{\not\pi}}| \leqslant C\delta_0(1+t)^{-1}(1+\log(1+t)) \tag{6.118}$$

By (6.23) and (6.106) we have

$$|\mathrm{tr}\,{}^{(Q)}\tilde{\not\pi}| \leqslant C \tag{6.119}$$

We can obtain, in fact, a more precise estimate for $\mathrm{tr}\,{}^{(Q)}\tilde{\not\pi}$:

$$\mathrm{tr}\,{}^{(Q)}\tilde{\not\pi} - 4 = 4(\Omega-1)(1+t)\nu + 4(1+t)(\nu - \frac{1}{1+t}) \tag{6.120}$$

Recalling that $\Omega(0)=1$, i.e. at the constant state, the conformal factor $\Omega=1$, (6.69) implies

$$|\Omega-1| \leqslant C\delta_0(1+t)^{-1} \tag{6.121}$$

Then by (6.106), we get

$$|\mathrm{tr}\,{}^{(Q)}\tilde{\not\pi} - 4| \leqslant C(1+t)^{-1}(1+\log(1+t)) \tag{6.122}$$

where we have used the fact that the behavior of difference between $(1+t)^{-1}$ and $(1-u+t)^{-1}$ is like $(1+t)^{-2}$.

To bound the deformation tensor of $R_i$, we must get estimates for $\lambda_i$ defined by (6.37), as well as the functions $y_i$ and $z_i$ defined by (6.60) and (6.64).

First we shall get upper and lower bounds for the function

$$r = \sqrt{\sum_{i=1}^{3}(x^i)^2} \tag{6.123}$$

Since $r$ achieves its maximum value in $\Sigma_t \bigcap W^*_{\epsilon_0}$ at the outer boundary $\Sigma_t \bigcap C_0 = S_{t,0}$ where it is equal to $1+t$, we have

$$r \leqslant 1 + t \tag{6.124}$$

in $W^*_{\epsilon_0}$.

For the lower bound, we have

$$Tr = \sum_{i=1}^{3} \frac{x^i T^i}{r} = \kappa \sum_{i=1}^{3} \frac{x^i \hat{T}^i}{r} \tag{6.125}$$

In view of (6.70), we get

$$|Tr| \leqslant \kappa \tag{6.126}$$

then by (6.97), we have

$$|Tr| \leqslant 1 + C\delta_0(1 + \log(1+t)) \tag{6.127}$$

Integrating (6.127) along the integral curves of $T$ from $\Sigma_t \bigcap C_0 = S_{t,0}$ where $r = 1 + t$ to $S_{t,u}$, we get

$$r \geqslant 1 - u + t - C\delta_0 u(1 + \log(1+t)) \tag{6.128}$$

in $W^*_{\epsilon_0}$, $u \in [0, \epsilon_0]$, hence taking $\delta_0$ suitably small we get

$$r \geqslant C^{-1}(1+t) \tag{6.129}$$

We proceed to derive the estimates for $\lambda_i$, from (6.37) we have

$$\lambda_i = \bar{g}(\mathring{R}_i, \hat{T}) \tag{6.130}$$

First, we derive the estimates for $\mathring{R}_i$ and $R_i$:

$$|\mathring{R}_i| = \sqrt{r^2 - (x_i)^2} \leqslant r \leqslant 1 + t \tag{6.131}$$

From (6.37), we have

$$|R_i| \leqslant |\mathring{R}_i| \leqslant 1 + t \tag{6.132}$$

Next, we derive an ordinary differential equation for $\lambda_i$ along the integral curves of $L$. Note that the surfaces $S_{0,u}$ being spheres centered at the origin and we have $\hat{T} = -\frac{\partial}{\partial r}$ on $\Sigma_0$, hence

$$\lambda_i = 0 \quad \text{on} \quad \Sigma_0 \tag{6.133}$$

Now we have

$$L\lambda_i = \sum_m (L\mathring{R}_i^m)\hat{T}^m + \sum_m \mathring{R}_i^m L\hat{T}^m \tag{6.134}$$

Recall that $L = \partial_t - (\eta\hat{T}^i + \psi_i)\partial_i$ and $\mathring{R}_i^m = \epsilon_{ijm}x^j$.

$$\Rightarrow L\mathring{R}_i^m = -(\eta\hat{T}^j + \psi_j)\epsilon_{ijm} \tag{6.135}$$

While $L\hat{T}^m$ is given by (3.176):

$$L\hat{T}^m = p_L\hat{T}^m + q_L^m \tag{6.136}$$

Recall from Chapter 3, $p_L = 0$, and from (3.179) we have

$$q_L = q_L^A X_A = q_L^A X_A^m \partial_m \Rightarrow q_L^m = q_L^A X_A^m \tag{6.137}$$

we thus obtain

$$L\lambda_i = -\psi_j \hat{T}^m \epsilon_{ijm} + \epsilon_{ijm} q_L^A X_A^m x^j \tag{6.138}$$

where we have used the fact that $\epsilon_{ijm}\hat{T}^j\hat{T}^m = 0$.

Recalling from (3.182) that

$$q_L^A = -\kappa^{-1}\zeta^A \tag{6.139}$$

the estimate (6.88) gives

$$|q_L| \leqslant C\delta_0(1+t)^{-2} \tag{6.140}$$

The second term on the right in (6.138) is $\bar{g}(\mathring{R}_i, q_L)$.

By (6.131) and (6.140),

$$|\bar{g}(\mathring{R}_i, q_L)| \leqslant |\mathring{R}_i||q_L| \leqslant C\delta_0(1+t)^{-1} \tag{6.141}$$

using also **E1** we obtain

$$|L\lambda_i| \leqslant C\delta_0(1+t)^{-1} \tag{6.142}$$

Integrating this along the integral curve of $L$, we conclude that

$$|\lambda_i| \leqslant C\delta_0(1+\log(1+t)) \tag{6.143}$$

Next, we should derive the estimates of the derivatives of $\lambda_i$ tangent to the $\Sigma_t$. We have

$$T\lambda_i = \sum_m (T\mathring{R}_i^m)\hat{T}^m + \sum_m \mathring{R}_i^m T(\hat{T}^m) \tag{6.144}$$

From $\mathring{R}_i^m = \epsilon_{ijm}x^j$, we have

$$\begin{aligned} &\hat{T}\mathring{R}_i^m = \epsilon_{ijm}\hat{T}^j \\ &\Rightarrow \sum_m (T\mathring{R}_i^m)\hat{T}^m = \epsilon_{ijm}\hat{T}^j\hat{T}^m = 0 \end{aligned} \tag{6.145}$$

Hence

$$T\lambda_i = \sum_m \mathring{R}_i^m T(\hat{T}^m) \tag{6.146}$$

From (3.191), we know that

$$T(\hat{T}^m) = p_T\hat{T}^m + q_T^m \tag{6.147}$$

From (3.192), we have $p_T = 0$, and from (3.193) and (3.194), we have $q_T^A = -X^A(\kappa)$. So

$$T\lambda_i = \bar{g}(\mathring{R}_i, q_T) \Rightarrow |T\lambda_i| \leqslant |\mathring{R}_i||q_T| \tag{6.148}$$

and

$$|q_T|^2 = \not{g}_{AB} q_T^A q_T^B = |\not{d}\kappa|^2 \leqslant C\delta_0^2(1+t)^{-2}(1+\log(1+t))^2 \tag{6.149}$$

by the second inequality of (6.98). In view also of (6.131) we conclude that

$$|T\lambda_i| \leqslant C\delta_0(1+\log(1+t)) \tag{6.150}$$

Next, we have

$$\not{d}_A \mathring{R}_i^m = \epsilon_{ijm} \not{d}_A x^j = \epsilon_{ijm} X_A^j \tag{6.151}$$

$$\begin{aligned} \Rightarrow \not{d}_A \lambda_i &= \sum_m (\not{d}_A \mathring{R}_i^m)\hat{T}^m + \sum_m \mathring{R}_i^m \not{d}_A \hat{T}^m \\ &= \epsilon_{ijm} X_A^j \hat{T}^m + \sum_m \mathring{R}_i^m (\not{p}_A \hat{T}^m + \not{q}_A^m) \end{aligned} \tag{6.152}$$

where we have used (3.200). Also from (3.201) and (3.202) we have

$$\not{p}_A = 0, \quad \not{q}_A^m = \theta_A^B X_B^m \tag{6.153}$$

Therefore

$$\not{d}_A \lambda_i = \epsilon_{ijm} X_A^j \hat{T}^m + \bar{g}(\mathring{R}_i, \not{q}_A) \tag{6.154}$$

By (6.60), the first term on the right in (6.154) is

$$-\frac{\epsilon_{ijm} X_A^j x^m}{1-u+t} + \epsilon_{ijm} X_A^j y^m \tag{6.155}$$

On the other hand, the second term on the right in (6.154) is

$$\begin{aligned} \mathring{R}_i^m \theta_A^B X_B^m &= \epsilon_{ijm} x^j \theta_A^B X_B^m \\ &= -\frac{\epsilon_{ijm} x^j X_A^m}{1-u+t} + \epsilon_{ijm} x^j (\theta_A^B + \frac{\delta_A^B}{1-u+t}) X_B^m \end{aligned} \tag{6.156}$$

The first terms in (6.155) and (6.156) cancel each other, hence we obtain

$$\not{d}_A \lambda_i = \epsilon_{ijm}(X_A^j y^m + x^j(\theta_A^B + \frac{\delta_A^B}{1-u+t}) X_B^m) \tag{6.157}$$

In view of (6.105) and (6.124), the estimate

$$|y^i| \leqslant C\delta_0(1+t)^{-1}(1+\log(1+t))$$

would imply

$$|\not{d}\lambda_i| \leqslant C\delta_0(1+t)^{-1}(1+\log(1+t)) \tag{6.158}$$

To derive the above estimate for the functions $y^i$, and a similar estimate for the functions $x^i$, we shall need another bound for $r$.

Note that for any vector $V$ in Euclidean space $\mathbb{R}^3$, we have

$$\sum_i \bar{g}(\mathring{R}_i, V)^2 = r^2|\Sigma V|^2 \tag{6.159}$$

where $\Sigma$ is the Euclidean projection operator to the Euclidean spheres:

$$\Sigma^i_j = \delta^i_j - r^{-2}x^i x^j \tag{6.160}$$

We can check this with the identities

$$\sum_i \epsilon_{ijm}\epsilon_{ikn} = \delta_{jk}\delta_{mn} - \delta_{jn}\delta_{km}, \quad \sum_i \Sigma^i_m \Sigma^i_n = \Sigma^m_n \tag{6.161}$$

Taking in (6.159) $V$ to be $\hat{T}$, we obtain in view of (6.130)

$$|\Sigma\hat{T}|^2 = r^{-2}\sum_i (\lambda_i)^2 \tag{6.162}$$

Then by (6.129) and (6.143) we have

$$|\Sigma\hat{T}| \leqslant C\delta_0(1+t)^{-1}(1+\log(1+t)) \tag{6.163}$$

Consider $N$, the Euclidean outward unit normal to the Euclidean spheres:

$$N = \frac{x^i}{r}\partial_i \tag{6.164}$$

Then (6.125) takes the form

$$Tr = \kappa\bar{g}(N, \hat{T}) \tag{6.165}$$

We decompose

$$\begin{aligned} \hat{T} &= \bar{g}(N,\hat{T})N + \Sigma\hat{T} \\ \Rightarrow |\hat{T}|^2 &= |g(N,\hat{T})|^2 + |\Sigma\hat{T}|^2 \end{aligned} \tag{6.166}$$

Recalling that $|\hat{T}| = 1$, we have

$$1 - \bar{g}(N,\hat{T})^2 = |\Sigma\hat{T}|^2 \leqslant C\delta_0(1+t)^{-2}(1+\log(1+t))^2 \tag{6.167}$$

If $\delta_0$ is small enough, the right hand side above is no bigger than $1/2$. While on $S_{t,0}$, $\bar{g}(N,\hat{T}) = -1$, by continuity in $u$ this implies the angle between $\hat{T}$ and $N$ is greater than $\pi/2$. Hence

$$\bar{g}(\hat{T}, N) < 0$$

Therefore

$$0 \leqslant 1 + \bar{g}(N,\hat{T}) \leqslant C\delta_0(1+t)^{-2}(1+\log(1+t))^2 \tag{6.168}$$

Going back to (6.165),

$$Tr + 1 = \kappa(1 + \bar{g}(N,\hat{T})) - (\kappa - 1) \tag{6.169}$$

From (6.97) and (6.168) we have

$$|Tr + 1| \leqslant C\delta_0(1 + \log(1 + t)) \tag{6.170}$$

Since $r = 1 + t$ on $S_{t,0}$, we have, on $S_{t,u}$

$$r - 1 - t + u = \int_0^u (Tr + 1)du' \tag{6.171}$$

Using (6.170), we get

$$r \leqslant 1 + t - u(1 - C\delta_0(1 + \log(1 + t))) \tag{6.172}$$

If $C\delta_0(1 + \log(1 + t)) < 1$, (6.172) is stronger than (6.124), otherwise weaker. From (6.128) and (6.172), we get

$$|\frac{1}{r} - \frac{1}{1 - u + t}| \leqslant C\delta_0 u(1 + t)^{-2}(1 + \log(1 + t)) \tag{6.173}$$

Consider now the vectorfield

$$y' = \hat{T} + N = \Sigma\hat{T} + (1 + \bar{g}(N, \hat{T}))N \tag{6.174}$$

(6.163) and (6.168) imply

$$|y'| \leqslant C\delta_0(1 + t)^{-1}(1 + \log(1 + t)) \tag{6.175}$$

Recall (6.60) and $N^i = \frac{x^i}{r}$, we get

$$y'^i = -\frac{x^i}{1 - u + t} + y^i + \frac{x^i}{r} \Rightarrow y^i = y'^i - x^i(\frac{1}{r} - \frac{1}{1 - u + t}) \tag{6.176}$$

Using (6.124), (6.172) and (6.174) we have

$$|y^i| \leqslant C\delta_0(1 + t)^{-1}(1 + \log(1 + t)) \tag{6.177}$$

By this and (6.65), **E1**, we have

$$|z^i| \leqslant C\delta_0(1 + t)^{-1}(1 + \log(1 + t)) \tag{6.178}$$

We also used the fact that $|\alpha - 1| \leqslant C\delta_0(1 + t)^{-1}$.

We turn to estimate the deformation tensor of $R_i$. Since $\underline{L} = \alpha^{-1}\kappa L + 2T$, we get from (6.24) and (6.27)—(6.29):

$$^{(R_i)}\pi_{LL} = 0 \tag{6.179}$$

$$^{(R_i)}\pi_{L\underline{L}} = -2R_i\mu \tag{6.180}$$

$$^{(R_i)}\pi_{\underline{L}\underline{L}} = 4\mu R_i(\alpha^{-1}\kappa) \tag{6.181}$$

By **F1**, (6.85), (6.86), (6.97), (6.98) and (6.132), we have

$$|^{(R_i)}\pi_{L\underline{L}}| \leqslant C\delta_0(1 + \log(1 + t)) \tag{6.182}$$

$$\mu^{-1}|^{(R_i)}\pi_{\underline{L}\underline{L}}| \leqslant C\delta_0(1 + \log(1 + t)) \tag{6.183}$$

Next, we estimate the $S_{t,u}$ 1-form ${}^{(R_i)}\not\pi_L$, whose component ${}^{(R_i)}\pi_{LA}$ is given by (6.67). In view of **F2** and (6.132), the first term on the right of (6.67) is bounded by $C\delta_0(1+t)^{-1}(1+\log(1+t))$. The square magnitude of the second term is $(\not g^{-1})^{AB}\epsilon_{ilm}z^l X_A^m \epsilon_{irp}z^r X_B^p$. From (6.74), this is bounded by $\sum_m \epsilon_{ilm}z^l\epsilon_{irm}z^r = \sum_{l\neq i}(z^l)^2$. We have used the identity: $\sum_i \epsilon_{ijm}\epsilon_{ikn} = \delta_{jk}\delta_{mn} - \delta_{jn}\delta_{km}$. Then by (6.178), the second term in (6.67) is bounded by $C\delta_0(1+t)^{-1}(1+\log(1+t))$. From (6.88) and (6.143), the third term is bounded by $C\delta_0^2(1+t)^{-2}(1+\log(1+t))$.

$$\Rightarrow |{}^{(R_i)}\not\pi_L| \leqslant C\delta_0(1+t)^{-1}(1+\log(1+t)) \tag{6.184}$$

Next, we estimate ${}^{(R_i)}\pi_{TA}$, which is given by (6.63).

By (6.105), the first term is bounded by $C\delta_0(1+t)^{-1}(1+\log(1+t))^2$.

By (6.97) and (6.177) we know that the second term is bounded by $C\delta_0(1+t)^{-1}(1+\log(1+t))^2$

Lastly, using (6.98) and (6.143), the third term $\lambda_i \not d_A \kappa$ is bounded by $C\delta_0(1+t)^{-1}(1+\log(1+t))^2$.

So we get

$$|{}^{(R_i)}\not\pi_T| \leqslant C\delta_0(1+t)^{-1}(1+\log(1+t))^2 \tag{6.185}$$

Since $\underline{L} = \alpha^{-1}\kappa L + 2T$, we get

$$|{}^{(R_i)}\not\pi_{\underline{L}}| \leqslant C\delta_0(1+t)^{-1}(1+\log(1+t))^2 \tag{6.186}$$

Finally, we estimate ${}^{(R_i)}\not\pi_{AB}$, which is given by (6.59). Using **F2**, (6.80) and (6.143), this is bounded by $C\delta_0(1+t)^{-1}(1+\log(1+t))$. But we would like to bound its trace and trace-free part separately.

Since $\mathrm{tr}{}^{(R_i)}\not\pi = 2\lambda_i(\alpha^{-1}\mathrm{tr}\chi - \mathrm{tr}\not k)$, also by **F2**, (6.79) and (6.143),

$$|\mathrm{tr}{}^{(R_i)}\not\pi| \leqslant C\delta_0(1+t)^{-1}(1+\log(1+t)) \tag{6.187}$$

as well as

$$|{}^{(R_i)}\hat{\not\pi}| \leqslant C\delta_0(1+t)^{-2}(1+\log(1+t))^2 \tag{6.188}$$

Then we can substitute all these estimates to the expressions of ${}^{(R_i)}\tilde{\pi}$:

$${}^{(R_i)}\tilde{\pi} = \Omega{}^{(R_i)}\pi + \frac{d\Omega}{dh}(R_i h)g \tag{6.189}$$

By (6.69) and (6.132) we have $|R_i h| \leqslant C\delta_0(1+t)^{-1}$, then we get

$$\begin{aligned}
{}^{(R_i)}\tilde{\pi}_{LL} &= 0 && (6.190)\\
|{}^{(R_i)}\tilde{\pi}_{L\underline{L}}| &\leqslant C\delta_0(1+\log(1+t)) && (6.191)\\
\mu^{-1}|{}^{(R_i)}\tilde{\pi}_{\underline{L}\underline{L}}| &\leqslant C\delta_0(1+\log(1+t)) && (6.192)\\
|{}^{(R_i)}\tilde{\not\pi}_L| &\leqslant C\delta_0(1+t)^{-1}(1+\log(1+t)) && (6.193)\\
|{}^{(R_i)}\tilde{\not\pi}_{\underline{L}}| &\leqslant C\delta_0(1+t)^{-1}(1+\log(1+t))^2 && (6.194)\\
|\mathrm{tr}{}^{(R_i)}\tilde{\not\pi}| &\leqslant C\delta_0(1+t)^{-1}(1+\log(1+t)) && (6.195)\\
|{}^{(R_i)}\hat{\tilde{\not\pi}}| &\leqslant C\delta_0(1+t)^{-2}(1+\log(1+t))^2 && (6.196)
\end{aligned}$$

Finally, we shall estimate the functions ${}^{(R_i)}\delta$, defined by

$${}^{(R_i)}\delta = \frac{1}{2}\tilde{\mathrm{tr}}^{(Y)}\tilde{\pi} - \mu^{-1}Y\mu - 2\Omega^{-1}Y\Omega \tag{6.197}$$

which will play a role in Chapter 7. Here $\tilde{\mathrm{tr}}$ is the trace with respect to $\tilde{g}$. Since $\tilde{g}^{-1} = \Omega^{-1}g^{-1}$, we have

$$\tilde{\mathrm{tr}}^{(Y)}\tilde{\pi} = \Omega^{-1}\mathrm{tr}^{(Y)}\tilde{\pi} = \Omega^{-1}(-\mu^{-1(Y)}\tilde{\pi}_{L\underline{L}} + \mathrm{tr}\,{}^{(Y)}\tilde{\not\pi}) \tag{6.198}$$

Hence,

$${}^{(Y)}\delta = -\frac{1}{2}\Omega^{-1}\mu^{-1(Y)}\tilde{\pi}_{L\underline{L}} - \mu^{-1}Y\mu + \Omega^{-1}(\frac{1}{2}\mathrm{tr}\,{}^{(Y)}\tilde{\not\pi} - 2\frac{d\Omega}{dh}Yh) \tag{6.199}$$

From (6.12), we find

$${}^{(T)}\delta = \Omega^{-1}(\frac{1}{2}\mathrm{tr}\,{}^{(T)}\tilde{\not\pi} - \frac{d\Omega}{dh}Th) \tag{6.200}$$

From (6.19),

$${}^{(Q)}\delta = \Omega^{-1}(\frac{1}{2}\mathrm{tr}\,{}^{(Q)}\tilde{\not\pi} - \frac{d\Omega}{dh}Qh) + 1 \tag{6.201}$$

From (6.69) and (6.114), we have

$$|{}^{(T)}\delta| \leqslant C(1+t)^{-1}(1+\log(1+t)) \tag{6.202}$$

From (6.69) and (6.122), we have

$$|{}^{(Q)}\delta| \leqslant C \tag{6.203}$$

From (6.180) and (6.189), we have

$${}^{(R_i)}\tilde{\pi}_{L\underline{L}} = -2\Omega R_i\mu - 2\mu R_i\Omega \tag{6.204}$$

$$\Rightarrow {}^{(R_i)}\delta = \Omega^{-1}(\frac{1}{2}\mathrm{tr}\,{}^{(R_i)}\tilde{\not\pi} - \frac{d\Omega}{dh}R_ih) \tag{6.205}$$

By (6.69) and (6.195), we have

$$|{}^{(R_i)}\delta| \leqslant C\delta_0(1+t)^{-1}(1+\log(1+t)) \tag{6.206}$$

# 7

# Outline of the Derived Estimates of Each Order

## 7.1 The Inhomogeneous Wave Equations for the Higher Order Variations. The Recursion Formula for the Source Functions

**Proposition 7.1** Let $\psi$ be a solution of the inhomogeneous wave equation

$$\Box_{\tilde{g}}\psi = \rho \tag{7.1}$$

and let $Y$ be an arbitrary vectorfield. Then

$$\psi' = Y\psi \tag{7.2}$$

satisfies the inhomogeneous wave equation

$$\Box_{\tilde{g}}\psi' = \rho' \tag{7.3}$$

where the source function $\rho'$ is related to the source function $\rho$ by

$$\rho' = \tilde{\mathrm{div}}\,{}^{(Y)}\tilde{J} + Y\rho + \frac{1}{2}\tilde{\mathrm{tr}}\,{}^{(Y)}\tilde{\pi}\rho \tag{7.4}$$

Here

$${}^{(Y)}\tilde{J}^{\mu} = \frac{1}{2}((\tilde{g}^{-1})^{\mu\alpha}(\tilde{g}^{-1})^{\nu\beta} + (\tilde{g}^{-1})^{\nu\alpha}(\tilde{g}^{-1})^{\mu\beta} - (\tilde{g}^{-1})^{\mu\nu}(\tilde{g}^{-1})^{\alpha\beta})\,{}^{(Y)}\tilde{\pi}_{\alpha\beta}\partial_{\nu}\psi \tag{7.5}$$

is the commutation current associated to $\psi$ and $Y$.

*Proof*. Let $f_s$ be the local 1-parameter group generated by $Y$. We denote by $f_{s*}$ the corresponding pullback. We then have

$$\Box_{f_{s*}\tilde{g}}(f_{s*}\psi) = f_{s*}(\Box_{\tilde{g}}\psi) = f_{s*}\rho \tag{7.6}$$

Now, in an arbitrary system of local coordinates,

$$\Box_{\tilde{g}}\psi = \frac{1}{\sqrt{-\det\tilde{g}}}\partial_{\mu}((\tilde{g}^{-1})^{\mu\nu}\sqrt{-\det\tilde{g}}\,\partial_{\nu}\psi) \tag{7.7}$$

and

$$\square_{f_{s*}\tilde{g}}(f_{s*}\psi) = \frac{1}{\sqrt{-\det f_{s*}\tilde{g}}}\partial_\mu(((f_{s*}\tilde{g})^{-1})^{\mu\nu}\sqrt{-\det f_{s*}\tilde{g}}\partial_\nu(f_{s*}\psi)) \tag{7.8}$$

Let us differentiate (7.8) with respect to $s$ at $s = 0$. In view of the facts that

$$(\frac{d}{ds}f_{s*}\tilde{g})_{s=0} = \mathcal{L}_Y\tilde{g} = {}^{(Y)}\tilde{\pi} \tag{7.9}$$

$$(\frac{d}{ds}\sqrt{-\det f_{s*}\tilde{g}})_{s=0} = \frac{1}{2}\sqrt{-\det \tilde{g}}\tilde{\mathrm{tr}}^{(Y)}\tilde{\pi} \tag{7.10}$$

$$(\frac{d}{ds}((f_{s*}\tilde{g})^{-1})^{\mu\nu})_{s=0} = -(\tilde{g}^{-1})^{\mu\alpha}(\tilde{g}^{-1})^{\nu\beta(Y)}\tilde{\pi}_{\alpha\beta} \tag{7.11}$$

and

$$(\frac{d}{ds}f_{s*}\psi)_{s=0} = (\frac{d}{ds}\psi \circ f_s)_{s=0} = Y\psi \tag{7.12}$$

we obtain

$$\begin{aligned}
&(\frac{d}{ds}\square_{f_{s*}\tilde{g}}(f_{s*}\psi))_{s=0} \\
&= -\frac{1}{2}\frac{1}{\sqrt{-\det \tilde{g}}}\tilde{\mathrm{tr}}^{(Y)}\tilde{\pi}\partial_\mu((\tilde{g}^{-1})^{\mu\nu}\sqrt{-\det \tilde{g}}\partial_\nu\psi) \\
&+ \frac{1}{\sqrt{-\det \tilde{g}}}\partial_\mu(\sqrt{-\det \tilde{g}}(-(\tilde{g}^{-1})^{\mu\alpha}(\tilde{g}^{-1})^{\nu\beta(Y)}\tilde{\pi}_{\alpha\beta} \\
&+ \frac{1}{2}(\tilde{g}^{-1})^{\mu\nu}\tilde{\mathrm{tr}}^{(Y)}\tilde{\pi})\partial_\nu\psi + (\tilde{g}^{-1})^{\mu\nu}\sqrt{-\det \tilde{g}}\partial_\nu(Y\psi))
\end{aligned} \tag{7.13}$$

On the other hand, by (7.6), we have

$$(\frac{d}{ds}\square_{f_{s*}\tilde{g}}(f_{s*}\psi))_{s=0} = (\frac{d}{ds}f_{s*}\rho)_{s=0} = (\frac{d}{ds}\rho \circ f_s)_{s=0} = Y\rho \tag{7.14}$$

Comparing (7.13) and (7.14) and in view of the expression (7.7) with $Y\psi$ in the role of $\psi$, the proposition follows. □

Let $X$ be an arbitrary vectorfield. In an arbitrary system of local coordinates the divergence of $X$ with respect to the acoustical metric $g$ is expressed by

$$\mathrm{div}X = D_\mu X^\mu = \frac{1}{\sqrt{-\det g}}\frac{\partial}{\partial x^\mu}(\sqrt{-\det g}X^\mu) \tag{7.15}$$

while its divergence with respect to the conformal acoustical metric $\tilde{g}$ is expressed by

$$\tilde{\mathrm{div}}X = \tilde{D}_\mu X^\mu = \frac{1}{\sqrt{-\det \tilde{g}}}\frac{\partial}{\partial x^\mu}(\sqrt{-\det \tilde{g}}X^\mu) \tag{7.16}$$

Since

$$\tilde{g}_{\mu\nu} = \Omega g_{\mu\nu}, \quad \sqrt{-\det \tilde{g}} = \Omega^2\sqrt{-\det g} \tag{7.17}$$

we have

$$\tilde{D}_\mu X^\mu = \Omega^{-2} D_\mu (\Omega^2 X^\mu) \tag{7.18}$$

Applying this to the vectorfield ${}^{(Y)}\tilde{J}$, we obtain

$$\tilde{\mathrm{div}}^{(Y)}\tilde{J} = \Omega^{-2}\mathrm{div}^{(Y)}J \tag{7.19}$$

where

$${}^{(Y)}J = \Omega^{2(Y)}\tilde{J} \tag{7.20}$$

In an arbitrary local coordinates:

$${}^{(Y)}J^\mu = \frac{1}{2}((g^{-1})^{\mu\alpha}(g^{-1})^{\nu\beta} + (g^{-1})^{\nu\alpha}(g^{-1})^{\mu\beta} - (g^{-1})^{\mu\nu}(g^{-1})^{\alpha\beta})^{(Y)}\tilde{\pi}_{\alpha\beta}\partial_\nu\psi \tag{7.21}$$

Setting

$${}^{(Y)}\tilde{\pi}^{\mu\nu} = (g^{-1})^{\mu\alpha}(g^{-1})^{\nu\beta(Y)}\tilde{\pi}_{\alpha\beta} \tag{7.22}$$

we can write

$${}^{(Y)}J^\mu = ({}^{(Y)}\tilde{\pi}^{\mu\nu} - \frac{1}{2}(g^{-1})^{\mu\nu}\mathrm{tr}^{(Y)}\tilde{\pi})\partial_\nu\psi \tag{7.23}$$

We now consider the $n$th order variations $\psi_n$ of the wave function $\phi$, as defined in the beginning of Chapter 6, by applying to a first order variation $\psi_1$ a string of commutation vectorfields $Y_i, i = 1, 2, 3, 4, 5$ of length $n-1$. We shall use Proposition 7.1 to derive a recursion formula for the corresponding source functions $\rho_n$:

$$\Box_{\tilde{g}}\psi_n = \rho_n \tag{7.24}$$

First, we have

$$\rho_1 = 0 \tag{7.25}$$

Denoting by $Y$ one of the $Y_i$, $i - 1, 2, 3, 4, 5$, we have

$$\psi_n = Y\psi_{n-1} \tag{7.26}$$

Applying Proposition 7.1 to $\psi_{n-1}$ and $\psi_n$ we get

$$\rho_n = \tilde{\mathrm{div}}^{(Y)}\tilde{J}_{n-1} + Y\rho_{n-1} + \frac{1}{2}\tilde{\mathrm{tr}}^{(Y)}\tilde{\pi}\rho_{n-1} \tag{7.27}$$

Here,

$${}^{(Y)}\tilde{J}^\mu_{n-1} = \frac{1}{2}((\tilde{g}^{-1})^{\mu\alpha}(\tilde{g}^{-1})^{\nu\beta} + (\tilde{g}^{-1})^{\nu\alpha}(\tilde{g}^{-1})^{\mu\beta} - (\tilde{g}^{-1})^{\mu\nu}(\tilde{g}^{-1})^{\alpha\beta})^{(Y)}\tilde{\pi}_{\alpha\beta}\partial_\nu\psi_{n-1} \tag{7.28}$$

is the commutation current associated to $\psi_{n-1}$ and $Y$. Equation (7.27) is a recursion formula for the source functions $\rho_n$, but it is not quite in the form which can be used in our estimates. To obtain the appropriate form we consider instead the re-scaled sources:

$$\tilde{\rho}_n = \Omega^2\mu\rho_n \tag{7.29}$$

By direct computation and (7.27), (7.28) as well as (7.19)—(7.23), we have

$$\tilde{\rho}_n = {}^{(Y)}\sigma_{n-1} + Y\tilde{\rho}_{n-1} + {}^{(Y)}\delta\tilde{\rho}_{n-1} \tag{7.30}$$

Here,

$${}^{(Y)}\sigma_{n-1} = \mu \mathrm{div}^{(Y)}J_{n-1} \tag{7.31}$$

$${}^{(Y)}J^{\mu}_{n-1} = ({}^{(Y)}\tilde{\pi}^{\mu\nu} - \frac{1}{2}(g^{-1})^{\mu\nu}\mathrm{tr}^{(Y)}\tilde{\pi})\partial_\nu\psi_{n-1} \tag{7.32}$$

and ${}^{(Y)}\delta$ are defined in Chapter 6:

$${}^{(Y)}\delta = \frac{1}{2}\tilde{\mathrm{tr}}^{(Y)}\tilde{\pi} - \mu^{-1}Y\mu - 2\Omega^{-1}Y\Omega \tag{7.33}$$

Moreover, by (7.25),

$$\tilde{\rho}_1 = 0 \tag{7.34}$$

## 7.2 The First Term in $\tilde{\rho}_n$

Since $\psi_n$ is the solution of (7.24), we can apply Theorem 5.1 to $\psi_n$, provided that we can estimate

$$\int_{W^t_u} Q_{0,0,n} d\mu_g = -\int_{W^t_u} \Omega^2 \rho_n K_0\psi_n d\mu_g = -\int_{W^t_u} \tilde{\rho}_n (K_0\psi_n) dt'du'd\mu_{\not{g}} \tag{7.35}$$

and

$$\begin{aligned}\int_{W^t_u} Q_{1,0,n} d\mu_g &= -\int_{W^t_u} \Omega^2 \rho_n (K_1\psi_n + \omega\psi_n) d\mu_g \\ &= -\int_{W^t_u} \tilde{\rho}_n (K_1\psi_n + \omega\psi_n) dt'du'd\mu_{\not{g}}\end{aligned} \tag{7.36}$$

We have used here the fact that

$$d\mu_g = \mu dt du d\mu_{\not{g}} \tag{7.37}$$

First, we consider the contribution of the term ${}^{(Y)}\sigma_{n-1}$ in $\tilde{\rho}_n$.

Let $V$ be an arbitrary vectorfield defined in the spacetime domain $W_{\epsilon_0}$. We decompose

$$V = V^L L + V^{\underline{L}}\underline{L} + \not{V} \tag{7.38}$$

where $\not{V} = \Pi V = V^A X_A$ is tangent to $S_{t,u}$. We have

$$\begin{aligned} V^L &= -\frac{1}{2\mu} g(V, \underline{L}) \\ V^{\underline{L}} &= -\frac{1}{2\mu} g(V, L) \\ V^A &= (\not{g}^{-1})^{AB} g(V, X_B) \end{aligned} \tag{7.39}$$

The divergence of $V$ is then expressed by

$$\mathrm{div}V = (D_L V)^L + (D_{\underline{L}} V)^{\underline{L}} + (D_{X_A} V)^A \tag{7.40}$$

Replacing $V$ in (7.39) by $D_L V, D_{\underline{L}} V, D_{X_A} V$, we obtain

$$\begin{aligned}(D_L V)^L &= -\frac{1}{2\mu} g(D_L V, \underline{L}) \\ (D_{\underline{L}} V)^{\underline{L}} &= -\frac{1}{2\mu} g(D_{\underline{L}} V, L) \\ (D_{X_A} V)^A &= (\not{g}^{-1})^{AB} g(D_{X_A} V, X_B)\end{aligned} \tag{7.41}$$

Moreover, substituting (7.38) we obtain

$$\begin{aligned}g(D_{X_A} V, X_B) &= V^L g(D_{X_A} L, X_B) + V^{\underline{L}} g(D_{X_A} \underline{L}, X_B) + g(D_{X_A} \not{V}, X_B) \\ &= \chi_{AB} V^L + \underline{\chi}_{AB} V^{\underline{L}} + \not{g}(\not{D}_{X_A} \not{V}, X_B)\end{aligned} \tag{7.42}$$

hence

$$(D_{X_A} V)^A = \mathrm{tr}\chi V^L + \mathrm{tr}\underline{\chi} V^{\underline{L}} + \not{\mathrm{d}}\mathrm{iv}\not{V} \tag{7.43}$$

Thus we obtain

$$\mathrm{div}V = -\frac{1}{2\mu}(g(D_L V, \underline{L}) + g(D_{\underline{L}} V, L)) + \mathrm{tr}\chi V^L + \mathrm{tr}\underline{\chi} V^{\underline{L}} + \not{\mathrm{d}}\mathrm{iv}\not{V} \tag{7.44}$$

We can express the first term on the right-hand side in terms of

$$V_L = g(V, L), \quad V_{\underline{L}} = g(V, \underline{L}) \tag{7.45}$$

Using (3.130)—(3.137),

$$\begin{aligned}g(D_L V, \underline{L}) &= L(g(V, \underline{L})) - g(V, D_L \underline{L}) \\ &= L(V_{\underline{L}}) + g(V, 2\zeta^A X_A) = L(V_{\underline{L}}) + 2\zeta_A V^A\end{aligned} \tag{7.46}$$

$$\begin{aligned}g(D_{\underline{L}} V, L) &= \underline{L}(g(V, L)) - g(V, D_{\underline{L}} L) \\ &= \underline{L}(V_L) - g(V, -L(\eta^{-1}\kappa)L + 2\eta^A X_A) \\ &= \underline{L}(V_L) + L(\eta^{-1}\kappa)V_L - 2\eta_A V^A\end{aligned} \tag{7.47}$$

Substituting the equation

$$\eta_A - \zeta_A = \not{d}_A \mu \tag{7.48}$$

we have

$$-2\mu \mathrm{div}V = L(V_{\underline{L}}) + \underline{L}(V_L) - 2\not{\mathrm{d}}\mathrm{iv}(\mu\not{V}) + L(\eta^{-1}\kappa)V_L + \mathrm{tr}\chi V_{\underline{L}} + \mathrm{tr}\underline{\chi} V_L \tag{7.49}$$

We shall apply the above formula to ${}^{(Y)}J_{n-1}$, given by (7.32). We introduce the vectorfields ${}^{(Y)}\tilde{Z}$ and ${}^{(Y)}\tilde{\underline{Z}}$, associated to the commutation vectorfield $Y$, and tangent to $S_{t,u}$, by the condition that

$$g(\tilde{Z}, V) = {}^{(Y)}\tilde{\pi}(L, \Pi V), \quad g(\tilde{\underline{Z}}, V) = {}^{(Y)}\tilde{\pi}(\underline{L}, \Pi V) \tag{7.50}$$

for any vector $V \in TW_{\epsilon_0}$. In terms of the null frames we have

$$^{(Y)}\tilde{Z} = {}^{(Y)}\tilde{Z}^A X_A, \, {}^{(Y)}\tilde{\underline{Z}} = {}^{(Y)}\tilde{\underline{Z}}^A X_A \tag{7.51}$$

where

$$^{(Y)}\tilde{Z}^A = {}^{(Y)}\tilde{\pi}_{LB}(\not{g}^{-1})^{AB}, \quad {}^{(Y)}\tilde{\underline{Z}}^A = {}^{(Y)}\tilde{\pi}_{\underline{L}B}(\not{g}^{-1})^{AB} \tag{7.52}$$

Note that the $S_{t,u}$-tangential vectorfields ${}^{(Y)}\tilde{Z}$ and ${}^{(Y)}\tilde{\underline{Z}}$ correspond, through the induced metric $\not{g}$, to the 1-forms ${}^{(Y)}\not{\pi}_L$ and ${}^{(Y)}\not{\pi}_{\underline{L}}$, respectively. Taking into account these definitions as well as the fact that

$$\mathrm{tr}{}^{(Y)}\tilde{\pi} = -\frac{1}{\mu}{}^{(Y)}\tilde{\pi}_{L\underline{L}} + \mathrm{tr}{}^{(Y)}\tilde{\not{\pi}} \tag{7.53}$$

we deduce the following the components of ${}^{(Y)}J_{n-1}$:

$$^{(Y)}J_{n-1,L} = -\frac{1}{2}\mathrm{tr}{}^{(Y)}\tilde{\not{\pi}}(L\psi_{n-1}) + {}^{(Y)}\tilde{Z} \cdot \not{d}\psi_{n-1} \tag{7.54}$$

$$^{(Y)}J_{n-1,\underline{L}} = -\frac{1}{2}\mathrm{tr}{}^{(Y)}\tilde{\not{\pi}}(\underline{L}\psi_{n-1}) + {}^{(Y)}\tilde{\underline{Z}} \cdot \not{d}\psi_{n-1} - \frac{1}{2\mu}{}^{(Y)}\tilde{\pi}_{\underline{L}\underline{L}}(L\psi_{n-1}) \tag{7.55}$$

$$\begin{aligned}\mu{}^{(Y)}\not{J}^A_{n-1} = & -\frac{1}{2}{}^{(Y)}\tilde{Z}^A(\underline{L}\psi_{n-1}) - \frac{1}{2}{}^{(Y)}\tilde{\underline{Z}}^A(L\psi_{n-1}) \\ & + \frac{1}{2}({}^{(Y)}\tilde{\pi}_{L\underline{L}} - \mu\mathrm{tr}{}^{(Y)}\tilde{\not{\pi}})\not{d}^A\psi_{n-1} + \mu{}^{(Y)}\tilde{\not{\pi}}^A_B\not{d}^B\psi_{n-1}\end{aligned} \tag{7.56}$$

We note here that there is no $\frac{1}{\mu}{}^{(Y)}\tilde{\pi}_{L\underline{L}}$ in the expressions for ${}^{(Y)}J_{n-1,L}$, ${}^{(Y)}J_{n-1,\underline{L}}$, although such a term appears in (7.53). This is related to the fact that the operator $\Delta_g$ on the 2-dimensional Riemannian manifold $(M, g)$ is conformally covariant. Here in the role of such a 2-dimensional manifold we have the 2-dimensional distribution of time-like planes spanned by $L$ and $\underline{L}$ at each point. This distribution is not integrable, the obstruction to integrability being

$$\Pi[L, \underline{L}] = 2\Lambda, \quad \Lambda = -(\not{g}^{-1})^{AB}(\zeta_B + \eta_B)X_A \tag{7.57}$$

However, the conformal covariance is still reflected by the fact that the restriction of ${}^{(Y)}J_{n-1}$ to the plane spanned by $L$, $\underline{L}$ depends only on the trace-free, relative to this plane, part of the restriction of $\tilde{\pi}$ to the plane, therefore not on $\tilde{\pi}_{L\underline{L}}$. Similarly, one can easily check that ${}^{(Y)}\not{J}^A_{n-1}$ does not actually depend on $\mathrm{tr}{}^{(Y)}\tilde{\not{\pi}}$.

Applying (7.49) to ${}^{(Y)}J_{n-1}$ we obtain

$$\begin{aligned}{}^{(Y)}\sigma_{n-1} = & -\frac{1}{2}L({}^{(Y)}J_{n-1,\underline{L}}) - \frac{1}{2}\underline{L}({}^{(Y)}J_{n-1,L}) + \not{\mathrm{div}}(\mu{}^{(Y)}\not{J}_{n-1}) \\ & - \frac{1}{2}L(\eta^{-1}\kappa){}^{(Y)}J_{n-1,L} - \frac{1}{2}\mathrm{tr}\chi{}^{(Y)}J_{n-1,\underline{L}} - \frac{1}{2}\mathrm{tr}\underline{\chi}{}^{(Y)}J_{n-1,L}\end{aligned} \tag{7.58}$$

The right hand side of (7.58) contains the principal parts, the derivatives of products of components of ${}^{(Y)}\tilde{\pi}$ with derivatives of $\psi_{n-1}$, while the second line contains lower order terms of the form of triple products of connection coefficients of the null frame with components of ${}^{(Y)}\tilde{\pi}$ and derivatives of $\psi_{n-1}$. We decompose

$$ {}^{(Y)}\sigma_{n-1} = {}^{(Y)}\sigma_{1,n-1} + {}^{(Y)}\sigma_{2,n-1} + {}^{(Y)}\sigma_{3,n-1} \tag{7.59} $$

where ${}^{(Y)}\sigma_{1,n-1}$ contains the terms which are products of components of ${}^{(Y)}\tilde{\pi}$ with 2nd derivatives of $\psi_{n-1}$, ${}^{(Y)}\sigma_{2,n-1}$ contains the terms which are products of 1st derivatives of ${}^{(Y)}\tilde{\pi}$ with 1st derivatives of $\psi_{n-1}$, and ${}^{(Y)}\sigma_{3,n-1}$ contains the lower order terms.

Now, in view of (7.54)—(7.56), the first two terms on the right of (7.58) contain $L({}^{(Y)}\underline{\tilde{Z}}\cdot\not{d}\psi_{n-1})$ and $\underline{L}({}^{(Y)}\tilde{Z}\cdot\not{d}\psi_{n-1})$. In decomposing each of these into three parts as above, we need the following lemma.

**Lemma 7.1** Let $V$ be a vectorfield defined on the spacetime domain $W^*_{\epsilon_0}$ and tangent to $S_{t,u}$. Also, let $f$ be a function on $W^*_{\epsilon_0}$. We then have

$$ \begin{aligned} L(V\cdot\not{d}f) &= V\cdot\not{d}Lf + \not{\mathcal{L}}_L V\cdot\not{d}f \\ \underline{L}(V\cdot\not{d}f) &= V\cdot\not{d}\underline{L}f + \not{\mathcal{L}}_{\underline{L}} V\cdot\not{d}f - (V\cdot\not{d}(\eta^{-1}\kappa))Lf \end{aligned} $$

Here, we denote

$$ \not{\mathcal{L}}_L V = \Pi\mathcal{L}_L V = \Pi[L,V], \quad \not{\mathcal{L}}_{\underline{L}} V = \Pi\mathcal{L}_{\underline{L}} V = \Pi[\underline{L},V] \tag{7.60} $$

*Proof*. Since $V$ is tangent to $S_{t,u}$ we have

$$ V\cdot\not{d}f = V\cdot df \tag{7.61} $$

Hence,

$$ \begin{aligned} L(V\cdot\not{d}f) &= L(V\cdot df) = \mathcal{L}_L V\cdot df + V\cdot dLf = \mathcal{L}_L V\cdot df + V\cdot\not{d}Lf \\ \underline{L}(V\cdot\not{d}f) &= \underline{L}(V\cdot df) = \mathcal{L}_{\underline{L}} V\cdot df + V\cdot d\underline{L}f = \mathcal{L}_{\underline{L}} V\cdot df + V\cdot\not{d}\underline{L}f \end{aligned} \tag{7.62} $$

Since

$$ Lt = 1, \quad Lu = 0 $$

and

$$ S_{t,u} = \Sigma_t \bigcap C_u \tag{7.63} $$

and $V$ is $S_{t,u}$-tangential, thus

$$ V(u) = V(t) = 0 \tag{7.64} $$

it follows that

$$ \begin{aligned} (\mathcal{L}_L V)(t) &= L(Vt) - V(Lt) = 0 \\ (\mathcal{L}_L V)(u) &= L(Vu) - V(Lu) = 0 \end{aligned} $$

So we have

$$\mathcal{L}_L V = \not{\mathcal{L}}_L V \tag{7.65}$$

Concerning the term,

$$\mathcal{L}_{\underline{L}} V \tag{7.66}$$

we use the fact that

$$\underline{L} = \eta^{-1}\kappa L + 2T \tag{7.67}$$

Then we have

$$\mathcal{L}_{\underline{L}} V = (\eta^{-1}\kappa)\mathcal{L}_L V + 2\mathcal{L}_T V - (V \cdot \not{d}(\eta^{-1}\kappa))L \tag{7.68}$$

By an argument similar to that leading to (7.65), we can see that

$$\mathcal{L}_T V = \not{\mathcal{L}}_T V \tag{7.69}$$

Therefore

$$\not{\mathcal{L}}_{\underline{L}} V = (\eta^{-1}\kappa)\not{\mathcal{L}}_L V + 2\not{\mathcal{L}}_T V \tag{7.70}$$

So we obtain

$$\mathcal{L}_{\underline{L}} V = \not{\mathcal{L}}_{\underline{L}} V - (V(\eta^{-1}\kappa))L \tag{7.71}$$

In view of (7.62), (7.65) and (7.71), the lemma follows. □

Applying Lemma 7.1 to the case $V = {}^{(Y)}\tilde{\underline{Z}}, f = \psi_{n-1}$, we obtain

$$L({}^{(Y)}\tilde{\underline{Z}} \cdot \not{d}\psi_{n-1}) = {}^{(Y)}\tilde{\underline{Z}} \cdot \not{d}L\psi_{n-1} + \not{\mathcal{L}}_L {}^{(Y)}\tilde{\underline{Z}} \cdot \not{d}\psi_{n-1} \tag{7.72}$$

while applying Lemma 7.1 to the case $V = {}^{(Y)}\tilde{Z}, f = \psi_{n-1}$, yields

$$\underline{L}({}^{(Y)}\tilde{Z} \cdot \not{d}\psi_{n-1}) = {}^{(Y)}\tilde{Z} \cdot \not{d}\underline{L}\psi_{n-1} + \not{\mathcal{L}}_{\underline{L}} {}^{(Y)}\tilde{Z} \cdot \not{d}\psi_{n-1} - {}^{(Y)}\tilde{Z} \cdot \not{d}(\eta^{-1}\kappa)L\psi_{n-1} \tag{7.73}$$

Substituting (7.54)—(7.56), we get the following expressions in regard to the decomposition (7.59):

$$\begin{aligned}
{}^{(Y)}\sigma_{1,n-1} = &\frac{1}{2}\mathrm{tr}\,{}^{(Y)}\tilde{\not{\pi}}(L\underline{L}\psi_{n-1} + \nu\underline{L}\psi_{n-1}) \\
&+ \frac{1}{4}(\mu^{-1(Y)}\tilde{\pi}_{\underline{L}\underline{L}})L^2\psi_{n-1} \\
&- {}^{(Y)}\tilde{Z} \cdot \not{d}\underline{L}\psi_{n-1} - {}^{(Y)}\tilde{\underline{Z}} \cdot \not{d}L\psi_{n-1} \\
&+ \frac{1}{2}{}^{(Y)}\tilde{\pi}_{L\underline{L}}\not{\Delta}\psi_{n-1} + \mu\,{}^{(Y)}\hat{\tilde{\not{\pi}}} \cdot \not{D}^2\psi_{n-1}
\end{aligned} \tag{7.74}$$

$$\begin{aligned}
{}^{(Y)}\sigma_{2,n-1} = &\frac{1}{4}L(\mathrm{tr}\,{}^{(Y)}\tilde{\not{\pi}})\underline{L}\psi_{n-1} + \frac{1}{4}\underline{L}(\mathrm{tr}\,{}^{(Y)}\tilde{\not{\pi}})L\psi_{n-1} \\
&+ \frac{1}{4}L(\mu^{-1(Y)}\tilde{\pi}_{\underline{L}\underline{L}})L\psi_{n-1} \\
&- \frac{1}{2}\not{\mathcal{L}}_L {}^{(Y)}\tilde{\underline{Z}} \cdot \not{d}\psi_{n-1} - \frac{1}{2}\not{\mathcal{L}}_{\underline{L}} {}^{(Y)}\tilde{Z} \cdot \not{d}\psi_{n-1}
\end{aligned} \tag{7.75}$$

$$-\frac{1}{2}\mathrm{d\!/\!iv}\,{}^{(Y)}\tilde{Z}\underline{L}\psi_{n-1} - \frac{1}{2}\mathrm{d\!/\!iv}\,{}^{(Y)}\underline{\tilde{Z}}L\psi_{n-1}$$
$$+\frac{1}{2}\not{d}\,{}^{(Y)}\tilde{\pi}_{L\underline{L}}\cdot\not{d}\psi_{n-1} + \mathrm{d\!/\!iv}(\mu\,{}^{(Y)}\hat{\tilde{\not{\pi}}})\cdot\not{d}\psi_{n-1}$$

and

$$^{(Y)}\sigma_{3,n-1} = {}^{(Y)}\sigma^{L}_{3,n-1}L\psi_{n-1} + {}^{(Y)}\sigma^{\underline{L}}_{3,n-1}\underline{L}\psi_{n-1} + {}^{(Y)}\not{\sigma}_{3,n-1}\cdot\not{d}\psi_{n-1} \tag{7.76}$$

where

$$^{(Y)}\sigma^{L}_{3,n-1} = \frac{1}{4}\mathrm{tr}\underline{\chi}\,\mathrm{tr}\,{}^{(Y)}\tilde{\not{\pi}} + \frac{1}{4}\mathrm{tr}\chi(\mu^{-1}\,{}^{(Y)}\tilde{\pi}_{\underline{L}\underline{L}}) + \frac{1}{2}{}^{(Y)}\tilde{Z}\cdot\not{d}(\eta^{-1}\kappa) \tag{7.77}$$

$$^{(Y)}\sigma^{\underline{L}}_{3,n-1} = -\frac{1}{4}(L\log\Omega)\mathrm{tr}\,{}^{(Y)}\tilde{\not{\pi}} \tag{7.78}$$

$$^{(Y)}\not{\sigma}_{3,n-1} = -\frac{1}{2}(\mathrm{tr}\,{}^{(Y)}\tilde{\not{\pi}})\Lambda - \frac{1}{2}(\mathrm{tr}\underline{\chi} + L(\eta^{-1}\kappa))\,{}^{(Y)}\tilde{Z} - \frac{1}{2}\mathrm{tr}\chi\,{}^{(Y)}\underline{\tilde{Z}} \tag{7.79}$$

Note that we have incorporated the term $\frac{1}{2}\mathrm{tr}\,{}^{(Y)}\tilde{\not{\pi}}\nu\underline{L}\psi_{n-1}$ into the first term on the right in (7.74). This term comes from the contribution $\frac{1}{4}\mathrm{tr}\chi\mathrm{tr}\,{}^{(Y)}\tilde{\not{\pi}}\underline{L}\psi_{n-1}$ of the first term on the right in (7.55) to the second term in the second line in (7.58). This leaves as a remainder the coefficient (7.78) of $\underline{L}\psi_{n-1}$ in (7.76).

## 7.3 The Estimates of the Contribution of the First Term in $\tilde{\rho}_n$ to the Error Integrals

We shall begin our estimates of the contribution of ${}^{(Y)}\sigma_{n-1}$ to the error integrals (7.35), (7.36), with the estimates of the partial contribution of ${}^{(Y)}\sigma_{1,n-1}$. We shall use the following assumptions:

**G0**: There is a positive constant $C$ independent of $s$ such that in $W^s_{\epsilon_0}$, for all five commutation fields $Y$, we have

$$^{(Y)}\tilde{\pi}_{LL} = 0 \tag{7.80}$$
$$\mu^{-1}|{}^{(Y)}\tilde{\pi}_{\underline{L}\underline{L}}| \leqslant C(1+\log(1+t)) \tag{7.81}$$
$$|{}^{(Y)}\tilde{\pi}_{L\underline{L}}| \leqslant C(1+\log(1+t)) \tag{7.82}$$
$$|{}^{(Y)}\tilde{\not{\pi}}_{L}| \leqslant C(1+t)^{-1}(1+\log(1+t)) \tag{7.83}$$
$$|{}^{(Y)}\tilde{\not{\pi}}_{\underline{L}}| \leqslant C(1+\log(1+t)) \tag{7.84}$$
$$|{}^{(Y)}\hat{\tilde{\not{\pi}}}| \leqslant C(1+t)^{-1}(1+\log(1+t)) \tag{7.85}$$
$$|{}^{(Y)}\mathrm{tr}\tilde{\not{\pi}}| \leqslant C \tag{7.86}$$

This assumption comes from the results in Chapter 6.

We shall also use the assumptions concerning the set of rotation fields $\{R_i : i = 1, 2, 3\}$.

**H0**: There is a positive constant $C$ independent of $s$ such that for any function $f$ differentiable on each surface $S_{t,u}$ we have

$$|\not{d}f|^2 \leqslant C(1+t)^{-2}\sum_i (R_i f)^2 \tag{7.87}$$

**H1**: There is a positive constant $C$ independent of $s$ such that for any 1-form $\xi$ differentiable on each surface $S_{t,u}$ we have

$$|\not{D}\xi|^2 \leqslant C(1+t)^{-2}\sum_i |\not{\mathcal{L}}_{R_i}\xi|^2 \tag{7.88}$$

In particular, taking $\xi = \not{d}f$, where $f$ is any function twice differentiable on each $S_{t,u}$, we have

$$|\not{D}^2 f|^2 \leqslant C(1+t)^{-2}\sum_i |\not{d}R_i f|^2 \tag{7.89}$$

**H2**: There is a positive constant $C$ independent of $s$ such that for any differentiable traceless symmetric 2-covariant tensorfield $\vartheta$ on each surface $S_{t,u}$ we have

$$|\not{D}\vartheta|^2 \leqslant C(1+t)^{-2}\sum_i |\not{\mathcal{L}}_{R_i}\vartheta|^2 \tag{7.90}$$

Similarly as in Chapter 5 we define

$$\begin{aligned}
\mathcal{E}^u_{0,n}(t) &= \sum \int_{\Sigma^u_t} \frac{\Omega}{2}(\eta^{-1}\kappa(1+\eta^{-1}\kappa)(L\psi_n)^2 \\
&\quad + (\underline{L}\psi_n)^2 + (1+2\eta^{-1}\kappa)\mu|\not{d}\psi_n|^2)d\mu_{\not{g}}du' \\
\mathcal{F}^t_{0,n}(u) &= \sum \int_{C^t_u} \Omega((1+\eta^{-1}\kappa)(L\psi_n)^2 + \mu|\not{d}\psi_n|^2)d\mu_{\not{g}}dt' \\
\mathcal{E}'^u_{1,n}(t) &= \sum \int_{\Sigma^u_t} \frac{\Omega}{2}\omega\nu^{-1}(\eta^{-1}\kappa(L\psi_n+\nu\psi_n)^2 + \mu|\not{d}\psi_n|^2)d\mu_{\not{g}}du' \\
\mathcal{F}'^t_{1,n}(u) &= \sum \int_{C^t_u} \Omega\omega\nu^{-1}(L\psi_n+\nu\psi_n)^2 d\mu_{\not{g}}dt'
\end{aligned}$$

In each of the above the sum is over the set of $\psi_n$ of the form (7.26) as $Y$ ranges over the set $\{Y_i : i = 1,2,3,4,5\}$. Also as in Chapter 5, we have

$$\begin{aligned}
&C^{-1}\mathcal{E}^u_{0,n}(t) \\
&\leqslant \sum \int_0^u \left(\int_{S_{t,u'}} (\mu(1+\mu)((L\psi_n)^2 + |\not{d}\psi_n|^2) + (\underline{L}\psi_n)^2)d\mu_{\not{g}}\right)du' \leqslant C\mathcal{E}^u_{0,n}(t)
\end{aligned} \tag{7.91}$$

$$\begin{aligned}
&C^{-1}\mathcal{F}^t_{0,n}(u) \\
&\leqslant \sum \int_0^t \left(\int_{S_{t',u}} ((1+\mu)(L\psi_n)^2 + \mu|\not{d}\psi_n|^2)d\mu_{\not{g}}\right)dt' \leqslant C\mathcal{F}^t_{0,n}(u)
\end{aligned} \tag{7.92}$$

$$\begin{aligned}
&C^{-1}\mathcal{E}'^u_{1,n}(t) \\
&\leqslant (1+t)^2 \sum \int_0^u \left(\int_{S_{t,u'}} \mu((L\psi_n+\nu\psi_n)^2 + |\not{d}\psi_n|^2)d\mu_{\not{g}}\right)du' \leqslant C\mathcal{E}'^u_{1,n}(t)
\end{aligned} \tag{7.93}$$

$$C^{-1}\mathcal{F}'^t_{1,n}(u) \leqslant \sum \int_0^t (1+t')^2 (\int_{S_{t',u}} (L\psi_n + \nu\psi_n)^2 d\mu_{\not{g}}) dt' \leqslant C\mathcal{F}'^t_{1,n}(u) \tag{7.94}$$

Also, the sum has the same meaning as the above. And we can define the following quantities which are non-decreasing in $t$ at each $u$ as in Chapter 5:

$$\bar{\mathcal{E}}^u_{0,n}(t) = \sup_{t'\in[0,t]} \mathcal{E}^u_{0,n}(t') \tag{7.95}$$

$$\mathcal{F}^t_{0,n}(u) \tag{7.96}$$

$$\bar{\mathcal{E}}'^u_{1,n}(t) = \sup_{t'\in[0,t]} (1+\log(1+t'))^{-4}\mathcal{E}'^u_{1,n}(t') \tag{7.97}$$

$$\bar{\mathcal{F}}'^t_{1,n}(u) = \sup_{t'\in[0,t]} (1+\log(1+t'))^{-4}\mathcal{F}'^{t'}_{1,n}(u) \tag{7.98}$$

Obviously, $\bar{\mathcal{E}}^u_{0,n}(t)$, $\bar{\mathcal{E}}'^u_{1,n}(t)$ are also non-decreasing functions of $u$ at each $t$.

The contribution of ${}^{(Y)}\sigma_{n-1}$ to (7.35) and (7.36) are

$$-\int_{W^t_u} (K_0\psi_n)^{(Y)}\sigma_{n-1} dt' du' d\mu_{\not{g}} \tag{7.99}$$

and

$$-\int_{W^t_u} (K_1\psi_n + \omega\psi_n)^{(Y)}\sigma_{n-1} dt' du' d\mu_{\not{g}} \tag{7.100}$$

respectively.

We shall first consider the contribution from ${}^{(Y)}\sigma_{1,n-1}$ to each of these integrals and we begin with the contribution to the integral (7.99). We first consider the first term on the right-hand side of (7.74):

$$\frac{1}{2}\mathrm{tr}{}^{(Y)}\tilde{\not{\pi}}(L\underline{L}\psi_{n-1} + \nu\underline{L}\psi_{n-1}) \tag{7.101}$$

Now $\underline{L}$ can be expressed in terms of commutation vectorfields by

$$\underline{L} = \eta^{-1}\kappa(1+t)^{-1}Q + 2T \tag{7.102}$$

Thus we have

$$\begin{aligned} &\frac{1}{2}(L\underline{L}\psi_{n-1} + \nu\underline{L}\psi_{n-1}) \\ &= (LT\psi_{n-1} + \nu T\psi_{n-1}) + \frac{1}{2}\eta^{-1}\kappa(1+t)^{-1}(LQ\psi_{n-1} + \nu Q\psi_{n-1}) \\ &+ \frac{1}{2}(1+t)^{-1}(L(\eta^{-1}\kappa) - (1+t)^{-1}\eta^{-1}\kappa)Q\psi_{n-1} \end{aligned} \tag{7.103}$$

Here we estimate the contribution of the first term on the right in (7.103), which is the principal part. The contributions of the other two terms are easier to estimate because of the presence of the decay factor $(1+t)^{-1}$.

We thus first estimate

$$-\int_{W_u^t}(K_0\psi_n)\mathrm{tr}^{(Y)}\tilde{\not{\pi}}(LT\psi_{n-1}+\nu T\psi_{n-1})dt'du'd\mu_{\not{g}} \tag{7.104}$$

By the last of **G0**, this is bounded in absolute value by $C(M^L+M^{\underline{L}})$, where

$$M^L=\int_{W_u^t}(1+\eta^{-1}\kappa)|L\psi_n||LT\psi_{n-1}+\nu T\psi_{n-1}|dt'du'd\mu_{\not{g}} \tag{7.105}$$

$$M^{\underline{L}}=\int_{W_u^t}|\underline{L}\psi_n||LT\psi_{n-1}+\nu T\psi_{n-1}|dt'du'd\mu_{\not{g}} \tag{7.106}$$

Since $T$ is one of the commutation fields, we have, by (7.94) and (7.98), for all $u\in[0,\epsilon_0]$,

$$\int_{C_u^t}(1+t')^2(LT\psi_{n-1}+\nu T\psi_{n-1})^2d\mu_{\not{g}}dt'\leqslant C\bar{\mathcal{F}}'^t_{1,n}(u)(1+\log(1+t))^4 \tag{7.107}$$

Therefore

$$\begin{aligned}&\int_{W_u^t}(1+t')^2(LT\psi_{n-1}+\nu T\psi_{n-1})^2d\mu_{\not{g}}dt'du'\\ \leqslant\, &C\int_0^u\bar{\mathcal{F}}'^t_{1,n}(u')du'(1+\log(1+t))^4\end{aligned} \tag{7.108}$$

We estimate

$$M^{\underline{L}}\leqslant\int_0^t((1+t')^{-3/2}\mathcal{E}_0^u(t'))^{1/2}((1+t')^{3/2}\int_{\Sigma_{t'}^u}(LT\psi_{n-1}+\nu T\psi_{n-1})^2)^{1/2}dt' \tag{7.109}$$

Let us define

$$F(t,u):=\int_0^t(1+t')^2(\int_{\Sigma_{t'}^u}(LT\psi_{n-1}+\nu T\psi_{n-1})^2)dt' \tag{7.110}$$

Therefore

$$F(t,u)\leqslant C\int_0^u\bar{\mathcal{F}}'^t_{1,n}(u')du'(1+\log(1+t))^4 \tag{7.111}$$

Then

$$\frac{dF}{dt}(t,u)=(1+t)^2\int_{\Sigma_t^u}(LT\psi_{n-1}+\nu T\psi_{n-1})^2 \tag{7.112}$$

What we need to estimate is

$$I(t):=\int_0^t(1+t')^{3/2}(\int_{\Sigma_t^u}(LT\psi_{n-1}+\nu T\psi_{n-1})^2)dt' \tag{7.113}$$

We have

$$I(t) = \int_0^t (1+t')^{-1/2}\frac{dF}{dt'}(t')dt'$$
$$= (1+t)^{-1/2}F(t) + \frac{1}{2}\int_0^t (1+t')^{-3/2}F(t')dt' \quad (7.114)$$

Substitute (7.111) we obtain

$$I(t) \leqslant C\int_0^u \bar{\mathcal{F}}'^t_{1,n}(u')du'(1+t)^{-1/2}(1+\log(1+t))^4 \leqslant C'\int_0^u \bar{\mathcal{F}}'^t_{1,n}(u')du' \quad (7.115)$$

Therefore

$$M^{\underline{L}} \leqslant C(\int_0^u \bar{\mathcal{F}}'^t_{1,n}(u')du')^{1/2}(\int_0^t (1+t')^{-3/2}\mathcal{E}^u_{0,n}(t')dt')^{1/2} \quad (7.116)$$

We turn to $M^L$ given by (7.105):

$$\begin{aligned} M^L \leqslant &(\int_{W^t_u} \Omega(1+\eta^{-1}\kappa)(L\psi_n)^2 dt'du'd\mu_{\not g})^{1/2} \\ &\cdot(\int_{W^t_u} \Omega^{-1}(1+\eta^{-1}\kappa)(LT\psi_{n-1}+\nu T\psi_{n-1})^2 dt'du'd\mu_{\not g})^{1/2} \\ \leqslant &C(\int_0^u \mathcal{F}^t_{0,n}(u')du')^{1/2} \\ &\cdot(\int_{W^t_u} (1+\log(1+t'))(LT\psi_{n-1}+\nu T\psi_{n-1})^2 dt'du'd\mu_{\not g})^{1/2} \end{aligned} \quad (7.117)$$

where we have used assumption **A**. We can estimate the integral in the last factor in a similar way as we estimate $M^{\underline{L}}$:

$$\int_{W^t_u} (1+\log(1+t'))(LT\psi_{n-1}+\nu T\psi_{n-1})^2 dt'du'd\mu_{\not g} \leqslant C\int_0^u \bar{\mathcal{F}}'^t_{1,n}(u')du' \quad (7.118)$$

Therefore we obtain

$$M^L \leqslant C(\int_0^u \mathcal{F}^t_{0,n}(u')du')^{1/2}(\int_0^u \bar{\mathcal{F}'}^t_{1,n}(u')du')^{1/2} \quad (7.119)$$

This completes the estimate of the integral (7.104).

We then turn to estimate the contribution of the second term on the right in (7.74)

$$(1/4)(\mu^{-1(Y)}\tilde{\pi}_{\underline{L}L})L^2\psi_{n-1} \quad (7.120)$$

to the integral (7.99).

Writing

$$L^2\psi_{n-1} = L((1+t)^{-1}Q\psi_{n-1}) = (1+t)^{-1}LQ\psi_{n-1} - (1+t)^{-2}Q\psi_{n-1} \quad (7.121)$$

what we must estimate is

$$
\begin{aligned}
&-(1/4)\int_{W_u^t}(K_0\psi_n)(\mu^{-1(Y)}\tilde{\pi}_{\underline{L}L})(1+t')^{-1}(LQ\psi_{n-1})dt'du'd\mu_{\not g}\\
&+(1/4)\int_{W_u^t}(K_0\psi_n)(\mu^{-1(Y)}\tilde{\pi}_{\underline{L}L})(1+t')^{-2}(Q\psi_{n-1})dt'du'd\mu_{\not g}
\end{aligned}
\tag{7.122}
$$

By (7.81), the first integral is bounded in absolute value by

$$C(I_1^{\underline{L}}+I_1^L) \tag{7.123}$$

where

$$I_1^{\underline{L}}=\int_{W_u^t}(1+t')^{-1}|\underline{L}\psi_n|(1+\log(1+t'))|LQ\psi_{n-1}|dt'du'd\mu_{\not g} \tag{7.124}$$

$$I_1^L=\int_{W_u^t}(1+\eta^{-1}\kappa)|L\psi_n|(1+t')^{-1}(1+\log(1+t'))|LQ\psi_{n-1}|dt'du'd\mu_{\not g} \tag{7.125}$$

We estimate

$$
\begin{aligned}
I_1^{\underline{L}}\leqslant\ & C\Big(\int_{W_u^t}(\underline{L}\psi_n)^2(1+t')^{-2}(1+\log(1+t'))^2dt'du'd\mu_{\not g}\Big)^{1/2}\\
&\cdot\Big(\int_{W_u^t}(LQ\psi_{n-1})^2dt'du'd\mu_{\not g}\Big)^{1/2}\\
\leqslant\ & C\Big(\int_0^t(1+t')^{-2}(1+\log(1+t'))^2\mathcal{E}_{0,n}^u(t')dt'\Big)^{1/2}\Big(\int_0^u\mathcal{F}_{0,n}^t(u')du'\Big)^{1/2}
\end{aligned}
\tag{7.126}
$$

and

$$I_1^L\leqslant C\int_0^u\mathcal{F}_{0,n}^t(u')du' \tag{7.127}$$

where we have used the fact that for any $t\geqslant 0$, $(1+t)^{-1}(1+\log(1+t))$ is bounded.

Also, by (7.81), the second integral in (7.122) is bounded by

$$
\begin{aligned}
&C\int_{W_u^t}|\underline{L}\psi_n||Q\psi_{n-1}|(1+t')^{-2}(1+\log(1+t'))dt'du'd\mu_{\not g}\\
&+C\int_{W_u^t}(1+\eta^{-1}\kappa)|L\psi_n||Q\psi_{n-1}|(1+t')^{-2}(1+\log(1+t'))dt'du'd\mu_{\not g}\\
\leqslant\ & C\Big(\int_{W_u^t}(\underline{L}\psi_n)^2(1+t')^{-2}(1+\log(1+t'))dt'du'd\mu_{\not g}\Big)^{1/2}\\
&\cdot\Big(\int_{W_u^t}(Q\psi_{n-1})^2(1+t')^{-2}(1+\log(1+t'))dt'du'd\mu_{\not g}\Big)^{1/2}\\
&+C\Big(\int_{W_u^t}(1+\eta^{-1}\kappa)(L\psi_n)^2dt'du'd\mu_{\not g}\Big)^{1/2}\\
&\cdot\Big(\int_{W_u^t}(1+\eta^{-1}\kappa)(Q\psi_{n-1})^2(1+t')^{-4}(1+\log(1+t'))^2dt'du'd\mu_{\not g}\Big)^{1/2}
\end{aligned}
$$

$$
\begin{aligned}
&\leqslant C\Big(\int_0^t (1+t')^{-2}(1+\log(1+t'))\mathcal{E}^u_{0,n}(t')dt'\Big)^{1/2} \qquad (7.128)\\
&\cdot\Big(\epsilon_0^2\int_0^t (1+t')^{-2}(1+\log(1+t'))\mathcal{E}^u_{0,n}(t')dt'\Big)^{1/2}\\
&+C\Big(\int_0^u \mathcal{F}^t_{0,n}(u')du'\Big)^{1/2}\\
&\cdot\Big(\epsilon_0^2\int_0^t (1+t')^{-4}(1+\log(1+t'))^3\mathcal{E}^u_{0,n}(t')dt'\Big)^{1/2}
\end{aligned}
$$

Here we have made use of Lemma 5.1 with $\psi_n$ in the role of $\psi$, namely

$$\int_{S_{t,u}} \sum \psi_n^2 d\mu_{\not g} \leqslant C\epsilon_0\mathcal{E}^u_{0,n}(t) \qquad (7.129)$$

the sum being over the set of $\psi_n$ of the form (7.26) as $Y$ ranges over the set $\{Y_i : i = 1,2,3,4,5\}$.

Next, we shall estimate the contribution of the third and fourth terms on the right in (7.74):

$$-{}^{(Y)}\tilde{Z}\cdot \not d\underline{L}\psi_{n-1} - {}^{(Y)}\underline{\tilde{Z}}\cdot \not dL\psi_{n-1} \qquad (7.130)$$

to the integral (7.99).

Expressing $\underline{L}$ and $L$ in terms of commutation fields $T$, $Q$,

$$\underline{L} = \eta^{-1}\kappa(1+t)^{-1}Q + 2T, \quad L = (1+t)^{-1}Q \qquad (7.131)$$

(7.130) takes the form

$$
\begin{aligned}
&-2{}^{(Y)}\tilde{Z}\cdot\not dT\psi_{n-1} - (1+t)^{-1}(\eta^{-1}\kappa{}^{(Y)}\tilde{Z} + {}^{(Y)}\underline{\tilde{Z}})\cdot\not dQ\psi_{n-1}\\
&-(1+t)^{-1}{}^{(Y)}\tilde{Z}\cdot\not d(\eta^{-1}\kappa)Q\psi_{n-1} \qquad (7.132)
\end{aligned}
$$

The last term is a lower order term and its contribution is easily estimated.

Using (7.83), (7.84) and noting that $T\psi_{n-1}, Q\psi_{n-1}$ are among the $\psi_n$, the contribution of (7.132) to the error integral (7.99) can be bounded by

$$
\begin{aligned}
&\int_{W^t_u} |\underline{L}\psi_n||\not d\psi_n|(1+t')^{-1}(1+\log(1+t'))dt'du'd\mu_{\not g}\\
&+\int_{W^t_u} (1+\eta^{-1}\kappa)|L\psi_n||\not d\psi_n|(1+t')^{-1}(1+\log(1+t'))dt'du'd\mu_{\not g} \qquad (7.133)
\end{aligned}
$$

Here the first integral is bounded by

$$\Big(\int_{W^t_u}(1+t')^{-2}(1+\log(1+t'))^2(\underline{L}\psi_n)^2dt'du'd\mu_{\not g}\Big)^{1/2}\Big(\int_{W^t_u}|\not d\psi_n|^2dt'du'd\mu_{\not g}\Big)^{1/2} \qquad (7.134)$$

The integral in the first factor in (7.134) is bounded by

$$C\int_0^t (1+t')^{-2}(1+\log(1+t'))^2\mathcal{E}^u_{0,n}(t')dt' \qquad (7.135)$$

While the integral in the second factor decomposes into

$$\int_{\mathcal{U}\bigcap W_u^t} |\not{d}\psi_n|^2 dt' du' d\mu_{\not{g}} + \int_{\mathcal{U}^c\bigcap W_u^t} |\not{d}\psi_n|^2 dt' du' d\mu_{\not{g}} \tag{7.136}$$

where $\mathcal{U}$ is defined in Chapter 5:

$$\mathcal{U} = \{x \in W_{\epsilon_0}^* : \mu < 1/4\} \tag{7.137}$$

Since $4\mu \geqslant 1$ in $\mathcal{U}^c$,

$$\int_{\mathcal{U}^c\bigcap W_u^t} |\not{d}\psi_n|^2 dt' du' d\mu_{\not{g}} \leqslant 4 \int_{W_u^t} \mu |\not{d}\psi_n|^2 dt' du' d\mu_{\not{g}} \leqslant C \int_0^u \mathcal{F}_{0,n}^t(u') du' \tag{7.138}$$

To estimate the integral over $\mathcal{U}\bigcap W_u^t$ we use the spacetime integral $K_n(t,u)$ defined by (5.195), with $\psi_n$ in the role of $\psi$:

$$K_n(t,u) = -\int_{W_u^t} \frac{\Omega}{2} \omega \nu^{-1} \mu^{-1} (L\mu)_- |\not{d}\psi_n|^2 d\mu_{\not{g}} \tag{7.139}$$

According to **C3**, **B1**, **D1**, we have

$$K_n(t,u) \geqslant \frac{1}{C} \int_{\mathcal{U}\bigcap W_u^t} (1+t')(1+\log(1+t'))^{-1} |\not{d}\psi_n|^2 dt' du' d\mu_{\not{g}} \tag{7.140}$$

Recalling the definition

$$\bar{K}_n(t,u) = \sup_{t' \in [0,t]} (1+\log(1+t'))^{-4} K_n(t',u) \tag{7.141}$$

and noting that for $t' \in [t_{m-1}, t_m]$ it holds that

$$(1+t')^{-1}(1+\log(1+t')) \leqslant 2(1+t_m)^{-1}(1+\log(1+t_m)) \tag{7.142}$$

we then obtain

$$\int_{\mathcal{U}\bigcap W_u^{t_{m-1},t_m}} |\not{d}\psi_n|^2 dt' du' d\mu_{\not{g}} \leqslant C A'_m \bar{K}_n(t_m, u) \tag{7.143}$$

where

$$A'_m = (1+t_m)^{-1}(1+\log(1+t_m))^5 = (1+2^{m+r})^{-1}(1+\log(1+2^{m+r}))^5 \tag{7.144}$$

Hence, in view of the fact that $\sum_{m=0}^{\infty} A'_m$ is convergent and $\bar{K}_n(t,u)$ is a non-decreasing function of $t$, we conclude that

$$\begin{aligned} &\int_{\mathcal{U}\bigcap W_u^t} |\not{d}\psi_n|^2 dt' du' d\mu_{\not{g}} \\ &= \sum_{m=0}^{N} \int_{\mathcal{U}\bigcap W_u^{t_{m-1},t_m}} |\not{d}\psi_n|^2 dt' du' d\mu_{\not{g}} \leqslant C \bar{K}_n(t,u) \end{aligned} \tag{7.145}$$

Combining with (7.138) we get

$$\int_{W_u^t} |\not{d}\psi_n|^2 dt'du'd\mu_{\not{g}} \leqslant \bar{K}_n(t,u) + \int_0^u \mathcal{F}_{0,n}^t(u')du' \tag{7.146}$$

Thus, the first integral in (7.133) is bounded in absolute value by

$$C(\int_0^t (1+t')^{-2}(1+\log(1+t'))^2 \mathcal{E}_{0,n}^u(t')dt')^{1/2}(\bar{K}_n(t,u) + \int_0^u \mathcal{F}_{0,n}^t(u')du')^{1/2} \tag{7.147}$$

Also, in view of (7.92) and the estimate (7.146), the second integral in (7.133) is bounded in absolute value by

$$\begin{aligned} &(\int_{W_u^t} (1+t')^{-2}(1+\log(1+t'))^2(1+\alpha^{-1}\kappa)^2(L\psi_n)^2 dt'du'd\mu_{\not{g}})^{1/2} \\ &\cdot(\int_{W_u^t} |\not{d}\psi_n|^2 dt'du'd\mu_{\not{g}})^{1/2} \\ &\leqslant C(\int_0^u \mathcal{F}_{0,n}^t(u')du')^{1/2}(\bar{K}_n(t,u) + \int_0^u \mathcal{F}_{0,n}^t(u')du')^{1/2} \end{aligned} \tag{7.148}$$

Finally, we estimate the contribution of the last term in (7.74),

$$(1/2)^{(Y)}\tilde{\pi}_{L\underline{L}}\not{\Delta}\psi_{n-1} + \mu \,{}^{(Y)}\hat{\not{\tilde{\pi}}} \cdot \not{D}^2\psi_{n-1} \tag{7.149}$$

to the error integral (7.99). From (7.82) and (7.85), this contribution can be bounded by

$$C\int_{W_u^t} |K_0\psi_n||\not{D}^2\psi_{n-1}|(1+\log(1+t'))dt'du'd\mu_{\not{g}} \tag{7.150}$$

By **H1**,

$$|\not{D}^2\psi_{n-1}| \leqslant C(1+t)^{-1}\sum_i |\not{d}R_i\psi_{n-1}| \leqslant C(1+t)^{-1}\sum |\not{d}\psi_n| \tag{7.151}$$

since the $R_i\psi_{n-1}$ are among the $\psi_n$. So (7.160) can be bounded by

$$C\int_{W_u^t} |K_0\psi_n||\not{d}\psi_n|(1+t')^{-1}(1+\log(1+t'))dt'du'd\mu_{\not{g}} \tag{7.152}$$

which is the integral (7.133) bounded by (7.148).

We collect the above results in the following lemma:

**Lemma 7.2** We have

$$|\int_{W_u^t} (K_0\psi_n)^{(Y)}\sigma_{1,n-1}dt'du'd\mu_{\not{g}}| \tag{7.153}$$

$$\leqslant C(\int_0^t (1+t')^{-3/2}\mathcal{E}_{0,n}^u(t')dt' + \int_0^u \mathcal{F}_{0,n}^t(u')du') \tag{7.154}$$

$$+C((\int_0^u \bar{\mathcal{F}}'^t_{1,n}(u')du')^{1/2} + (\bar{K}_n(t.u))^{1/2}) \tag{7.155}$$

$$\cdot(\int_0^t (1+t')^{-3/2}\mathcal{E}^u_{0,n}(t')dt' + \int_0^u \mathcal{F}^t_{0,n}(u')du')^{1/2} \tag{7.156}$$

We now consider the contribution of ${}^{(Y)}\sigma_{1,n-1}$ to the integral (7.100). Since

$$K_1 = (\omega/\nu)L \tag{7.157}$$

(7.100) equals

$$-\int_{W^t_u} (\omega/\nu)(L\psi_n + \nu\psi_n){}^{(Y)}\sigma_{n-1}dt'du'd\mu_{\not{g}} \tag{7.158}$$

What we consider here is

$$-\int_{W^t_u} (\omega/\nu)(L\psi_n + \nu\psi_n){}^{(Y)}\sigma_{1,n-1}dt'du'd\mu_{\not{g}} \tag{7.159}$$

By **B1** and **D1** we have

$$C^{-1}(1+t)^2 \leqslant \omega/\nu \leqslant C(1+t)^2 \tag{7.160}$$

So (7.159) is bounded by

$$\int_{W^t_u} (1+t')^2|L\psi_n + \nu\psi_n||{}^{(Y)}\sigma_{1,n-1}|dt'du'd\mu_{\not{g}} \tag{7.161}$$

We shall estimate this integral.

We begin with the contribution of the first term on the right in (7.74). We expand this term as in (7.103). So the contribution of the principal part in (7.103) is

$$\int_{W^t_u} (1+t')^2|L\psi_n + \nu\psi_n||\mathrm{tr}{}^{(Y)}\tilde{\not{\pi}}||LT\psi_{n-1} + \nu T\psi_{n-1}|dt'du'd\mu_{\not{g}} \tag{7.162}$$

By (7.86) and the fact that $T\psi_{n-1}$ is one of the $\psi_n$, this is bounded by

$$C\int_0^u \mathcal{F}'^t_{1,n}(u')du' \tag{7.163}$$

To estimate the contribution of the second term on the right in (7.74), we write

$$\begin{aligned} L^2\psi_{n-1} &= (1+t)^{-1}LQ\psi_{n-1} - (1+t)^{-2}Q\psi_{n-1} \\ &= (1+t)^{-1}(LQ\psi_{n-1} + \nu Q\psi_{n-1}) - (1+t)^{-2}(1+(1+t)\nu)Q\psi_{n-1} \end{aligned} \tag{7.164}$$

So by (7.81) and the fact that $Q\psi_{n-1}$ is one of the $\psi_n$, the contribution in question can be bounded by

$$C\int_{W^t_u} (1+t')(1+\log(1+t'))|L\psi_n + \nu\psi_n|^2 dt'du'd\mu_{\not{g}}$$

$$+C\int_{W^t_u} (1+\log(1+t'))|L\psi_n + \nu\psi_n||\psi_n|dt'du'd\mu_{\not{g}}$$

$$\leqslant C\int_0^u \mathcal{F}'^{t}_{1,n}(u')du' + C(\int_0^u \mathcal{F}'^{t}_{1,n}(u')du')^{1/2} \tag{7.165}$$
$$\cdot(\epsilon_0^2\int_0^t (1+t')^{-2}(1+\log(1+t'))^2\mathcal{E}^u_{0,n}(t')dt')^{1/2}$$

To estimate the contribution of the third and fourth term on the right in (7.74), we shall use the expression (7.132). Using (7.83), (7.84) the contribution in question is bounded by

$$\int_{W^t_u}(1+t')(1+\log(1+t'))|L\psi_n+\nu\psi_n||\not{d}\psi_n|dt'du'd\mu_{\not{g}} \tag{7.166}$$

which is in turn bounded by

$$(\int_{W^t_u}(1+t')^2|L\psi_n+\nu\psi_n|^2dt'du'd\mu_{\not{g}})^{1/2}$$
$$\cdot(\int_{W^t_u}(1+\log(1+t'))^2|\not{d}\psi_n|^2dt'du'd\mu_{\not{g}})^{1/2} \tag{7.167}$$

Now, the integral in the first factor in (7.167) is bounded by

$$C\int_0^u \mathcal{F}'^{t}_{1,n}(u')du'^{1/2} \tag{7.168}$$

The integral in the second factor in (7.167) decomposes into

$$\int_{W^t_u\bigcap\mathcal{U}}(1+\log(1+t'))^2|\not{d}\psi_n|^2dt'du'd\mu_{\not{g}}$$
$$+\int_{W^t_u\bigcap\mathcal{U}^c}(1+\log(1+t'))^2|\not{d}\psi_n|^2dt'du'd\mu_{\not{g}} \tag{7.169}$$

In $\mathcal{U}^c$ we have $4\mu\geqslant 1$; we can thus estimate, in view of (7.93),

$$\int_{W^t_u\bigcap\mathcal{U}^c}(1+\log(1+t'))^2|\not{d}\psi_n|^2dt'du'd\mu_{\not{g}}$$
$$\leqslant 4\int_{W^t_u}(1+\log(1+t'))^2\mu|\not{d}\psi_n|^2dt'du'd\mu_{\not{g}} \tag{7.170}$$
$$\leqslant C\int_0^t(1+t')^{-2}(1+\log(1+t'))^2\mathcal{E}'^u_{1,n}(t')dt' \tag{7.171}$$

on the other hand, by (7.140),

$$\int_{\mathcal{U}\bigcap W^t_u}(1+\log(1+t'))^2|\not{d}\psi_n|^2dt'du'd\mu_{\not{g}}\leqslant CK_n(t,u) \tag{7.172}$$

So we obtain

$$\int_{W_u^t}(1+\log(1+t'))^2|\not{d}\psi_n|^2 dt'du'd\mu_{\not{g}}$$
$$\leqslant C(K_n(t,u)+\int_0^t (1+t')^{-2}(1+\log(1+t'))^2\mathcal{E}'^u_{1,n}(t')dt') \tag{7.173}$$

Thus (7.166) is bounded by

$$C(\int_0^u \mathcal{F}'^t_{1,n}(u')du')^{1/2}(K_n(t,u)+\int_0^t (1+t')^{-2}(1+\log(1+t'))^2\mathcal{E}'^u_{1,n}(t')dt')^{1/2} \tag{7.174}$$

Finally, we estimate the contribution of the last two terms on the right in (7.74). By (7.82), (7.85) and (7.151) this is bounded by (7.166), which has already been estimated.

We collect the above results in the following lemma:

**Lemma 7.3** We have

$$|\int_{W_u^t}(K_1\psi_n+\omega\psi_n)^{(Y)}\sigma_{1,n-1}dt'du'd\mu_{\not{g}}|$$
$$\leqslant C\int_0^u \mathcal{F}'^t_{1,n}(u')du' + C(\int_0^u \mathcal{F}'^t_{1,n}(u')du')^{1/2} \tag{7.175}$$
$$\cdot (K_n(t,u)+\int_0^t (1+t')^{-2}(1+\log(1+t'))^2(\mathcal{E}'^u_{1,n}(t')+\epsilon_0^2\mathcal{E}^u_{0,n}(t'))dt')^{1/2}$$

To estimate the contribution of ${}^{(Y)}\sigma_{2,n-1}$ to (7.99) and (7.100), we shall use the following assumptions:

**G1**: There is a positive constant $C$ independent of $s$ such that in $W^s_{\epsilon_0}$, for all five commutation vectorfields we have

$$Y({}^{(Y)}\tilde{\pi}_{LL})=0 \tag{7.176}$$
$$|Y(\mu^{-1(Y)}\tilde{\pi}_{\underline{L}\underline{L}})|\leqslant C(1+\log(1+t)) \tag{7.177}$$
$$|Y({}^{(Y)}\tilde{\pi}_{L\underline{L}})|\leqslant C(1+\log(1+t)) \tag{7.178}$$
$$|\not{\mathcal{L}}_Y({}^{(Y)}\tilde{\not{\pi}}_L)|\leqslant C(1+t)^{-1}(1+\log(1+t)) \tag{7.179}$$
$$|\not{\mathcal{L}}_Y({}^{(Y)}\tilde{\not{\pi}}_{\underline{L}})|\leqslant C(1+\log(1+t)) \tag{7.180}$$
$$|\not{\mathcal{L}}_Y({}^{(Y)}\hat{\tilde{\not{\pi}}})|\leqslant C(1+t)^{-1}(1+\log(1+t)) \tag{7.181}$$
$$|Y({}^{(Y)}\mathrm{tr}\tilde{\not{\pi}})|\leqslant C(1+t)^{-1}(1+\log(1+t)) \tag{7.182}$$

Here in each line $Y$ occurs twice. In each of these occurrences $Y$ is meant to range *independently* over the set of the five commutation fields $\{Y_i : i=1,2,3,4,5\}$.

The last of **G1** seems stronger than what corresponds to the last of **G0**, however, recall that by (6.114), (6.122) and (6.195) of Chapter 6, we have

$$|{}^{(Y_i)}\mathrm{tr}\tilde{\not{\pi}}|\leqslant C(1+t)^{-1}(1+\log(1+t)), i\neq 2$$

while for $Y_2 = Q$ we have

$$|\mathrm{tr}^{(Q)}\tilde{\not\pi} - 4| \leqslant C(1+t)^{-1}(1+\log(1+t))$$

So the last of **G1** is in fact in accordance with the other assumptions.

In the following we shall use the bounds

$$\begin{aligned}
&\int_{S_{t,u}} (\underline{L}\psi_{n-1})^2 d\mu_{\not g} \leqslant C\epsilon_0 \mathcal{E}^u_{0,n}(t) \\
&\int_{S_{t,u}} (L\psi_{n-1})^2 d\mu_{\not g} \leqslant C\epsilon_0 (1+t)^{-2} \mathcal{E}^u_{0,n}(t) \qquad (7.183)\\
&\int_{S_{t,u}} |\not d\psi_{n-1}|^2 d\mu_{\not g} \leqslant C\epsilon_0 (1+t)^{-2} \mathcal{E}^u_{0,n}(t)
\end{aligned}$$

These follow from (7.129) using the expressions of $L$ and $\underline{L}$ in terms of the commutation fields $T$, $Q$, as well as **H0**.

We consider now the error integral

$$-\int_{W^t_u} (K_0\psi_n)^{(Y)}\sigma_{2,n-1} dt' du' d\mu_{\not g} \qquad (7.184)$$

The first term in (7.75),

$$(1/4)L(\mathrm{tr}^{(Y)}\tilde{\not\pi})\underline{L}\psi_{n-1}$$

involves $L(\mathrm{tr}^{(Y)}\tilde{\not\pi})$. Writing $L = (1+t)^{-1}Q$, we have, from **G1**,

$$|L(\mathrm{tr}^{(Y)}\tilde{\not\pi})| \leqslant C(1+t)^{-2}(1+\log(1+t)) \qquad (7.185)$$

It follows that the contribution of the first term to the error integral (7.184) is bounded in absolute value by

$$C(I^{\underline{L}}_2 + I^{L}_2)$$

where

$$I^{\underline{L}}_2 = \int_{W^t_u} |\underline{L}\psi_n||\underline{L}\psi_{n-1}|(1+t')^{-2}(1+\log(1+t'))dt'du'd\mu_{\not g} \qquad (7.186)$$

$$I^{L}_2 = \int_{W^t_u} (1+\eta^{-1}\kappa)|L\psi_n||\underline{L}\psi_{n-1}|(1+t')^{-2}(1+\log(1+t'))dt'du'd\mu_{\not g} \qquad (7.187)$$

Using the first of (7.183) we can estimate

$$\begin{aligned}
I^{\underline{L}}_2 &\leqslant \int_0^t \Big(\int_0^u \int_{S_{t',u'}} (\underline{L}\psi_n)^2 d\mu_{\not g} du'\Big)^{1/2} \Big(\int_0^u \int_{S_{t',u'}} (\underline{L}\psi_{n-1})^2 d\mu_{\not g} du'\Big)^{1/2} \\
&\quad \cdot (1+t')^{-2}(1+\log(1+t'))dt' \\
&\leqslant C\int_0^t (\mathcal{E}^u_{0,n}(t'))^{1/2}(\epsilon_0^2\mathcal{E}^u_{0,n}(t'))^{1/2}(1+t')^{-2}(1+\log(1+t'))dt' \\
&= C\epsilon_0 \int_0^t (1+t')^{-2}(1+\log(1+t'))\mathcal{E}^u_{0,n}(t')dt' \qquad (7.188)
\end{aligned}$$

and (using **A**)

$$\begin{aligned}
I_2^{\not L} &\leqslant \Big(\int_{W_u^t}(1+\eta^{-1}\kappa)(L\psi_n)^2 dt'du'd\mu_{\not g}\Big)^{1/2} \\
&\cdot\Big(\int_{W_u^t}(1+\eta^{-1}\kappa)(\underline{L}\psi_{n-1})^2(1+t')^{-4}(1+\log(1+t'))^2 dt'du'd\mu_{\not g}\Big)^{1/2} \\
&\leqslant C\epsilon_0\Big(\int_0^u \mathcal{F}_{0,n}^t(u')du'\Big)^{1/2}\Big(\int_0^t (1+t')^{-4}(1+\log(1+t'))^3\mathcal{E}_{0,n}^u(t')dt'\Big)^{1/2}
\end{aligned} \tag{7.189}$$

The second term in (7.75),

$$(\frac{1}{4})\underline{L}(\mathrm{tr}{}^{(Y)}\tilde{\not\pi})L\psi_{n-1} \tag{7.190}$$

involves $\underline{L}(\mathrm{tr}{}^{(Y)}\tilde{\not\pi})$. Writing $\underline{L} = 2T + \alpha^{-1}\kappa(1+t)^{-1}Q$, we have, by **G1**,

$$|\underline{L}(\mathrm{tr}{}^{(Y)}\tilde{\not\pi})| \leqslant C(1+t)^{-1}(1+\log(1+t)) \tag{7.191}$$

So the contribution of the second term is bounded in absolute value by

$$C(I_3^{\underline{L}} + I_3^{L})$$

where

$$I_3^{\underline{L}} = \int_{W_u^t} |\underline{L}\psi_n||L\psi_{n-1}|(1+t')^{-1}(1+\log(1+t'))dt'du'd\mu_{\not g} \tag{7.192}$$

$$I_3^{L} = \int_{W_u^t}(1+\eta^{-1}\kappa)|L\psi_n||L\psi_{n-1}|(1+t')^{-1}(1+\log(1+t'))dt'du'd\mu_{\not g} \tag{7.193}$$

using the second of (7.183), we can estimate

$$\begin{aligned}
I_3^{\underline{L}} &\leqslant \int_0^t \Big(\int_0^u\int_{S_{t',u'}}(\underline{L}\psi_n)^2 d\mu_{\not g}du'\Big)^{1/2}\Big(\int_0^u\int_{S_{t',u'}}(L\psi_{n-1})^2 d\mu_{\not g}du'\Big)^{1/2} \\
&\cdot(1+t')^{-1}(1+\log(1+t'))dt' \\
&\leqslant C\int_0^t (\mathcal{E}_{0,n}^u(t'))^{1/2}(\epsilon_0^2(1+t')^{-2}\mathcal{E}_{0,n}^u(t'))^{1/2}(1+t')^{-1}(1+\log(1+t'))dt' \\
&= C\epsilon_0\int_0^t (1+t')^{-2}(1+\log(1+t'))\mathcal{E}_{0,n}^u(t')dt'
\end{aligned} \tag{7.194}$$

and (using **A**)

$$\begin{aligned}
I_3^{L} &\leqslant \Big(\int_{W_u^t}(1+\eta^{-1}\kappa)(L\psi_n)^2 dt'du'd\mu_{\not g}\Big)^{1/2} \\
&\cdot\Big(\int_{W_u^t}(1+\eta^{-1}\kappa)(L\psi_{n-1})^2(1+t')^{-2}(1+\log(1+t'))^2 dt'du'd\mu_{\not g}\Big)^{1/2} \\
&\leqslant C\epsilon_0\Big(\int_0^u \mathcal{F}_{0,n}^t(u')du'\Big)^{1/2}\Big(\int_0^t (1+t')^{-4}(1+\log(1+t'))^3\mathcal{E}_{0,n}^u(t')dt'\Big)^{1/2}
\end{aligned} \tag{7.195}$$

The third term in (7.75),

$$(\frac{1}{4})L(\mu^{-1(Y)}\tilde{\pi}_{\underline{L}L})L\psi_{n-1}$$

involves $L(\mu^{-1(Y)}\tilde{\pi}_{\underline{L}L})$. Writing $L = (1+t)^{-1}Q$, we have, from **G1**,

$$|L(\mu^{-1(Y)}\tilde{\pi}_{\underline{L}L})| \leqslant C(1+t)^{-1}(1+\log(1+t)) \tag{7.196}$$

It follows that the contribution in question can be also bounded in absolute value by

$$C(I_3^{\underline{L}} + I_3^{L})$$

The fourth and fifth terms in (7.75),

$$-\frac{1}{2}\not{\mathcal{L}}_L{}^{(Y)}\tilde{\underline{Z}}\cdot \not{d}\psi_{n-1} - \frac{1}{2}\not{\mathcal{L}}_{\underline{L}}{}^{(Y)}\tilde{Z}\cdot \not{d}\psi_{n-1}$$

involve $\not{\mathcal{L}}_L{}^{(Y)}\tilde{\underline{Z}}$ and $\not{\mathcal{L}}_{\underline{L}}{}^{(Y)}\tilde{Z}$. Now from (7.50)—(7.52), we have

$$^{(Y)}\tilde{\not{\pi}}_L = \not{g}\cdot {}^{(Y)}\tilde{Z}, \quad {}^{(Y)}\tilde{\not{\pi}}_{\underline{L}} = \not{g}\cdot {}^{(Y)}\tilde{\underline{Z}}$$

hence

$$\begin{aligned} \not{\mathcal{L}}_Y{}^{(Y)}\tilde{\not{\pi}}_L &= \not{g}\cdot\not{\mathcal{L}}_Y{}^{(Y)}\tilde{Z} + {}^{(Y)}\not{\pi}\cdot{}^{(Y)}\tilde{Z} \\ \not{\mathcal{L}}_Y{}^{(Y)}\tilde{\not{\pi}}_{\underline{L}} &= \not{g}\cdot\not{\mathcal{L}}_Y{}^{(Y)}\tilde{\underline{Z}} + {}^{(Y)}\not{\pi}\cdot{}^{(Y)}\tilde{\underline{Z}} \end{aligned} \tag{7.197}$$

Since

$$^{(Y)}\tilde{\pi} = \Omega({}^{(Y)}\pi + Y(\log\Omega)g)$$

and by **E1** and **E2** we have

$$|Yh| \leqslant C(1+t)^{-1} \tag{7.198}$$

which implies

$$|Y\log\Omega| \leqslant C(1+t)^{-1} \tag{7.199}$$

for all five commutation vectorfields $Y$, then by **G0** and **G1** we have

$$\begin{aligned} |\not{\mathcal{L}}_Y{}^{(Y)}\tilde{Z}| &\leqslant C(1+t)^{-1}(1+\log(1+t)) \\ |\not{\mathcal{L}}_Y{}^{(Y)}\tilde{\underline{Z}}| &\leqslant C(1+\log(1+t)) \end{aligned} \tag{7.200}$$

These imply, writing $\underline{L} = 2T + \eta^{-1}\kappa(1+t)^{-1}Q$, $L = (1+t)^{-1}Q$, that

$$|\not{\mathcal{L}}_{\underline{L}}{}^{(Y)}\tilde{Z}|, |\not{\mathcal{L}}_L{}^{(Y)}\tilde{\underline{Z}}| \leqslant C(1+t)^{-1}(1+\log(1+t)) \tag{7.201}$$

So the contribution of the fourth and fifth terms in (7.75) can be bounded in absolute value by

$$C(I_4^{\underline{L}} + I_4^{L})$$

where

$$I_4^{\underline{L}} = \int_{W_u^t} |\underline{L}\psi_n||\not{d}\psi_{n-1}|(1+t')^{-1}(1+\log(1+t'))dt'du'd\mu_{\not{g}} \tag{7.202}$$

$$I_4^{L} = \int_{W_u^t} (1+\eta^{-1}\kappa)|L\psi_n||\not{d}\psi_{n-1}|(1+t')^{-1}(1+\log(1+t'))dt'du'd\mu_{\not{g}} \tag{7.203}$$

Using the last inequality of (7.183), we can estimate

$$\begin{aligned} I_4^{\underline{L}} \leqslant & \int_0^t (\int_0^u \int_{S_{t',u'}} (\underline{L}\psi_n)^2 d\mu_{\not{g}} du')^{1/2} (\int_0^u \int_{S_{t',u'}} |\not{d}\psi_{n-1}|^2 d\mu_{\not{g}} du')^{1/2} \\ & \cdot (1+t')^{-1}(1+\log(1+t'))dt' \\ \leqslant & C \int_0^t (\mathcal{E}_{0,n}^u(t'))^{1/2} (\epsilon_0^2(1+t')^{-2}\mathcal{E}_{0,n}^u(t'))^{1/2} \\ & \cdot (1+t')^{-1}(1+\log(1+t'))dt' \\ = & C\epsilon_0 \int_0^t (1+t')^{-2}(1+\log(1+t'))\mathcal{E}_{0,n}^u(t')dt' \end{aligned} \tag{7.204}$$

and (using **A**)

$$\begin{aligned} I_4^{L} \leqslant & (\int_{W_u^t} (1+\eta^{-1}\kappa)(L\psi_n)^2 dt'du'd\mu_{\not{g}})^{1/2} \\ & (\int_{W_u^t} (1+\eta^{-1}\kappa)|\not{d}\psi_{n-1}|^2(1+t')^{-2}(1+\log(1+t'))^2 dt'du'd\mu_{\not{g}})^{1/2} \\ \leqslant & C\epsilon_0 (\int_0^u \mathcal{F}_{0,n}^t(u')du')^{1/2} (\int_0^t (1+t')^{-4}(1+\log(1+t'))^3 \mathcal{E}_{0,n}^u(t')dt')^{1/2} \end{aligned} \tag{7.205}$$

The sixth and seventh terms in (7.75) involve $\not{\mathrm{div}}^{(Y)}\tilde{Z}$ and $\not{\mathrm{div}}^{(Y)}\underline{\tilde{Z}}$. Now by **H1** applied to the 1-forms $\tilde{\not{\pi}}_L$, $\tilde{\not{\pi}}_{\underline{L}}$ on each $S_{t,u}$ yields

$$\begin{aligned} |\not{D}^{(Y)}\tilde{Z}|^2 = |\not{D}^{(Y)}\tilde{\not{\pi}}_L|^2 \leqslant C(1+t)^{-2}\sum_i |\mathcal{L}_{R_i}{}^{(Y)}\tilde{\not{\pi}}_L|^2 \\ |\not{D}^{(Y)}\underline{\tilde{Z}}|^2 = |\not{D}^{(Y)}\tilde{\not{\pi}}_{\underline{L}}|^2 \leqslant C(1+t)^{-2}\sum_i |\mathcal{L}_{R_i}{}^{(Y)}\tilde{\not{\pi}}_{\underline{L}}|^2 \end{aligned} \tag{7.206}$$

Then from **G1**

$$\begin{aligned} |\not{D}^{(Y)}\tilde{Z}| \leqslant C(1+t)^{-2}(1+\log(1+t)) \\ |\not{D}^{(Y)}\underline{\tilde{Z}}| \leqslant C(1+t)^{-1}(1+\log(1+t)) \end{aligned} \tag{7.207}$$

Thus, also

$$\begin{aligned} |\not{\mathrm{div}}^{(Y)}\tilde{Z}| \leqslant C(1+t)^{-2}(1+\log(1+t)) \\ |\not{\mathrm{div}}^{(Y)}\underline{\tilde{Z}}| \leqslant C(1+t)^{-1}(1+\log(1+t)) \end{aligned} \tag{7.208}$$

It then follows that the contribution of the sixth term in (7.75)

$$-(\frac{1}{2})\mathrm{d}\!\!/\mathrm{iv}\,{}^{(Y)}\tilde{Z}\underline{L}\psi_{n-1}$$

is bounded in absolute value by

$$C(I_{\underline{2}}^{L}+I_{\overline{2}}^{\underline{L}})$$

while the contribution of the seventh term in (7.75)

$$-(\frac{1}{2})\mathrm{d}\!\!/\mathrm{iv}\,{}^{(Y)}\underline{\tilde{Z}}L\psi_{n-1}$$

is bounded in absolute value by

$$C(I_{\underline{3}}^{L}+I_{\overline{3}}^{\underline{L}})$$

The eighth term in (7.75),

$$(\frac{1}{2})\not{d}\,{}^{(Y)}\tilde{\pi}_{L\underline{L}}\cdot\not{d}\psi_{n-1}$$

involves $\not{d}\,{}^{(Y)}\tilde{\pi}_{L\underline{L}}$. By **H0** and **G1**,

$$|\not{d}\,{}^{(Y)}\tilde{\pi}_{L\underline{L}}|\leqslant C(1+t)^{-1}\sqrt{\sum_{i}(R_i{}^{(Y)}\tilde{\pi}_{L\underline{L}})^2}\leqslant C(1+t)^{-1}(1+\log(1+t)) \tag{7.209}$$

It then follows that the contribution of this term is bounded by

$$C(I_{\underline{4}}^{L}+I_{\overline{4}}^{\underline{L}})$$

Finally, the last term in (7.75),

$$\mathrm{d}\!\!/\mathrm{iv}(\mu\,\overset{(Y)}{\hat{\not\pi}})\cdot\not{d}\psi_{n-1}$$

involves

$$\mathrm{d}\!\!/\mathrm{iv}(\mu\,\overset{(Y)}{\hat{\not\pi}})=\mu\,\mathrm{d}\!\!/\mathrm{iv}\,\overset{(Y)}{\hat{\not\pi}}+\not{d}\mu\cdot\overset{(Y)}{\hat{\not\pi}}$$

Now, **H2** applied to $\overset{(Y)}{\hat{\not\pi}}$ gives

$$|\not{D}\,\overset{(Y)}{\hat{\not\pi}}|^2\leqslant C(1+t)^{-2}\sum_{i}|\not{\mathcal{L}}_{R_i}\overset{(Y)}{\hat{\not\pi}}|^2 \tag{7.210}$$

Then **G1** implies

$$|\not{D}\,\overset{(Y)}{\hat{\not\pi}}|\leqslant C(1+t)^{-2}(1+\log(1+t)) \tag{7.211}$$

thus also

$$|\mathrm{d}\!\!/\mathrm{iv}\,\overset{(Y)}{\hat{\not\pi}}|\leqslant C(1+t)^{-2}(1+\log(1+t)) \tag{7.212}$$

Using then **A3**, **F1** and **G0** we obtain

$$|\mathrm{d}\!\!/\mathrm{iv}(\mu\,\overset{(Y)}{\hat{\not\pi}})|\leqslant C(1+t)^{-2}(1+\log(1+t))^2 \tag{7.213}$$

hence the contribution of this term is also bounded in absolute value by

$$C(I_4^{\underline{L}} + I_{\frac{L}{4}})$$

We collect the above results in the following lemma:

**Lemma 7.4** We have

$$|\int_{W_u^t} (K_0\psi_n)^{(Y)}\sigma_{2,n-1}dt'du'd\mu_{\not g}|$$

$$\leqslant C\epsilon_0(\int_0^t (1+t')^{-2}(1+\log(1+t'))\mathcal{E}_{0,n}^u(t')dt' + \int_0^u \mathcal{F}_{0,n}^t(u')du') \quad (7.214)$$

We now consider the error integral

$$-\int_{W_u^t} (K_1\psi_n + \omega\psi_n)^{(Y)}\sigma_{2,n-1}dt'du'd\mu_{\not g}$$

or

$$-\int_{W_u^t} (\omega/\nu)(L\psi_n + \nu\psi_n)^{(Y)}\sigma_{2,n-1}dt'du'd\mu_{\not g} \quad (7.215)$$

Recalling the bounds for $\omega$ and $\nu$, this is bounded by

$$C\int_{W_u^t} (1+t')^2|L\psi_n + \nu\psi_n||^{(Y)}\sigma_{2,n-1}|dt'du'd\mu_{\not g} \quad (7.216)$$

We shall estimate this integral.

In view of (7.195), the contribution of the first term in (7.75) to the integral (7.161) is bounded in absolute value by

$$C\int_0^u (\int_{C_{u'}} |L\psi_n + \nu\psi_n||\underline{L}\psi_{n-1}|(1+\log(1+t'))d\mu_{\not g}dt')du'$$

$$\leqslant C\int_0^u (\int_{C_{u'}} (1+t')^2(L\psi_n + \nu\psi_n)^2 d\mu_{\not g}dt')^{1/2}$$

$$\cdot(\int_{C_{u'}} (1+t')^{-2}(1+\log(1+t'))^2(\underline{L}\psi_{n-1})^2 d\mu_{\not g}dt')^{1/2}du'$$

$$\leqslant C\int_0^u (\mathcal{F}_{1,n}^{\prime t}(u'))^{1/2}(\int_0^t \epsilon_0\mathcal{E}_{0,n}^u(t')(1+t')^{-2}(1+\log(1+t'))^2dt')^{1/2}du'$$

$$\leqslant C\epsilon_0(\int_0^u \mathcal{F}_{1,n}^{\prime t}(u')du')^{1/2}(\int_0^t (1+t')^{-2}(1+\log(1+t'))^2\mathcal{E}_{0,n}^u(t')dt')^{1/2} \quad (7.217)$$

Here we have used the first of (7.183).

In view of (7.191), the contribution of the second term in (7.75) is bounded in absolute value by

$$C\int_0^u (\int_{C_{u'}} |L\psi_n + \nu\psi_n||L\psi_{n-1}|(1+t')(1+\log(1+t'))d\mu_{\not g}dt')du'$$

$$\leqslant C\int_0^u (\int_{C_{u'}} (1+t')^2(L\psi_n + \nu\psi_n)^2 d\mu_{\not g}dt')^{1/2}$$

$$
\begin{aligned}
&\cdot\Big(\int_{C_{u'}}(1+\log(1+t'))^2(L\psi_{n-1})^2 d\mu_{\not g}dt'\Big)^{1/2}du' \\
\leqslant & C\int_0^u(\mathcal{F}'^t_{1,n}(u'))^{1/2}\Big(\int_0^t \epsilon_0\mathcal{E}^u_{0,n}(t')(1+t')^{-2}(1+\log(1+t'))^2dt'\Big)^{1/2}du' \\
\leqslant & C\epsilon_0\Big(\int_0^u \mathcal{F}'^t_{1,n}(u')du'\Big)^{1/2}\Big(\int_0^t(1+t')^{-2}(1+\log(1+t'))^2\mathcal{E}^u_{0,n}(t')dt'\Big)^{1/2}
\end{aligned} \tag{7.218}
$$

Here we have used the second of (7.183).

In view of (7.196), the contribution of the third term in (7.75) is bounded in a same way as (7.218).

In view of (7.201), the contributions of the fourth and fifth terms in (7.75) is bounded in absolute value by

$$
\begin{aligned}
& C\int_0^u\Big(\int_{C_{u'}}|L\psi_n+\nu\psi_n||\not d\psi_{n-1}|(1+t')(1+\log(1+t'))d\mu_{\not g}dt'\Big)du' \\
\leqslant & C\int_0^u\Big(\int_{C_{u'}}(1+t')^2(L\psi_n+\nu\psi_n)^2d\mu_{\not g}dt'\Big)^{1/2} \\
&\cdot\Big(\int_{C_{u'}}(1+\log(1+t'))^2|\not d\psi_{n-1}|^2d\mu_{\not g}dt'\Big)^{1/2}du' \\
\leqslant & C\int_0^u(\mathcal{F}'^t_{1,n}(u'))^{1/2}\Big(\int_0^t \epsilon_0\mathcal{E}^u_{0,n}(t')(1+t')^{-2}(1+\log(1+t'))^2dt'\Big)^{1/2}du' \\
\leqslant & C\epsilon_0\Big(\int_0^u \mathcal{F}'^t_{1,n}(u')du'\Big)^{1/2}\Big(\int_0^t(1+t')^{-2}(1+\log(1+t'))^2\mathcal{E}^u_{0,n}(t')dt'\Big)^{1/2}
\end{aligned} \tag{7.219}
$$

Here we have used the last inequality of (7.183).

In view of the bounds (7.208), the contribution of the sixth term in (7.75) is bounded in the same way as (7.217), while that of the seventh is bounded in the same way as (7.228).

Finally, in view of (7.209) and (7.213), the contributions of the last two terms of (7.75) is bounded in the same way as (7.219).

We collect the above results in the following lemma:

**Lemma 7.5** We have

$$
\begin{aligned}
&\Big|\int_{W^t_u}(K_1\psi_n+\omega\psi_n)\,{}^{(Y)}\sigma_{2,n-1}dt'du'd\mu_{\not g}\Big| \\
\leqslant & C\epsilon_0\Big(\int_0^u \mathcal{F}'^t_{1,n}(u')du'\Big)^{1/2}\Big(\int_0^t(1+t')^{-2}(1+\log(1+t'))^2\mathcal{E}^u_{0,n}(t')dt'\Big)^{1/2}
\end{aligned} \tag{7.220}
$$

Finally, we consider the contribution of ${}^{(Y)}\sigma_{3,n-1}$ given by (7.77)—(7.79) to (7.99) and (7.100). From **E1**, **E2**, **F1**, **F2** and (6.86), (6.87), (6.89), (6.95), (6.98), (6.99) and (6.104), we have

$$
\begin{aligned}
&|L\log\Omega|\leqslant C(1+t)^{-2} \\
&|\mathrm{tr}\chi|\leqslant C(1+t)^{-1},\quad |\mathrm{tr}\underline{\chi}|\leqslant C(1+t)^{-1}(1+\log(1+t)) \\
&|\Lambda|\leqslant C(1+t)^{-1}(1+\log(1+t)) \\
&|L(\eta^{-1}\kappa)|\leqslant C(1+t)^{-1},\quad |\not d(\eta^{-1}\kappa)|\leqslant C(1+t)^{-1}(1+\log(1+t))
\end{aligned} \tag{7.221}
$$

Together with **G0** these imply

$$\begin{aligned}|^{(Y)}\sigma_{3,n-1}^{\underline{L}}| &\leqslant C(1+t)^{-2} \\ |^{(Y)}\sigma_{3,n-1}^{L}| &\leqslant (1+t)^{-1}(1+\log(1+t)) \\ |^{(Y)}\not{\sigma}_{3,n-1}| &\leqslant C(1+t)^{-1}(1+\log(1+t))\end{aligned} \tag{7.222}$$

Recalling (7.185), (7.191) and (7.201), we conclude that the contribution of ${}^{(Y)}\sigma_{3,n-1}$ to each of the error integrals (7.99), (7.100) can be bounded in the same way as the corresponding contribution of ${}^{(Y)}\sigma_{2,n-1}$.

Combining this with the previous conclusions we arrive at the following lemma:

**Lemma 7.6** We have

$$\begin{aligned}&|\int_{W_u^t}(K_0\psi_n)^{(Y)}\sigma_{n-1}dt'du'd\mu_{\not{g}}| \\ \leqslant &C(\int_0^t(1+t')^{-3/2}\mathcal{E}_{0,n}^u(t')dt'+\int_0^u\mathcal{F}_{0,n}^t(u')du') \\ &+C((\int_0^u\bar{\mathcal{F}}_{1,n}'^t(u')du')^{1/2}+(\bar{K}_n(t,u))^{1/2}) \\ &\cdot(\int_0^t(1+t')^{-3/2}\mathcal{E}_{0,n}^u(t')dt'+\int_0^u\mathcal{F}_{0,n}^t(u')du')^{1/2}\end{aligned}$$

and

$$\begin{aligned}&|\int_{W_u^t}(K_1\psi_n+\omega\psi_n)^{(Y)}\sigma_{n-1}dt'du'd\mu_{\not{g}}| \\ \leqslant &C\int_0^u\mathcal{F}_{1,n}'^t(u')du'+C(\int_0^u\mathcal{F}_{1,n}'(u')du')^{1/2}(K_n(t,u))^{1/2} \\ &+C(\int_0^u\mathcal{F}_{1,n}'^t(u')du')^{1/2} \\ &\cdot(\int_0^t(1+t')^{-2}(1+\log(1+t'))^2(\mathcal{E}_{1,n}'^u(t')+\epsilon_0^2\mathcal{E}_{0,n}^u(t')dt'))^{1/2}\end{aligned}$$

# 8

# Regularization of the Propagation Equation for $\not{d}\mathrm{tr}\chi$. Estimates for the Top Order Angular Derivatives of $\chi$

## 8.1 Preliminary

### 8.1.1 Regularization of The Propagation Equation

We denote

$$\tau_\mu = \partial_\mu h, \quad \omega_{\mu\nu} = \partial_\mu \psi_\nu = \partial_\nu \psi_\mu \tag{8.1}$$

**Proposition 8.1** The function $h$ satisfies the following inhomogeneous wave equation in the acoustical metric $g$:

$$\Box_g h = -\Omega^{-1}\frac{d\Omega}{dh}a - b \tag{8.2}$$

where $a$ and $b$ are the functions

$$a = (g^{-1})^{\mu\nu}\tau_\mu\tau_\nu, \quad b = \sum_i (g^{-1})^{\mu\nu}\omega_{\mu i}\omega_{\nu i} \tag{8.3}$$

*Proof*. From Chapter 1, we know that

$$\Box_{\tilde{g}}\psi_\alpha = 0 \tag{8.4}$$

It follows that $h$ being given by

$$h = \partial_t\phi - \frac{1}{2}\sum_i (\partial_i\phi)^2 \tag{8.5}$$

satisfies the equation

$$\Box_{\tilde{g}} h = -\sum_i (\tilde{g}^{-1})^{\mu\nu}\partial_\mu\psi_i\partial_\nu\psi_i \tag{8.6}$$

Since

$$\tilde{g}_{\mu\nu} = \Omega g_{\mu\nu} \tag{8.7}$$

then for an arbitrary function $f$, we have

$$\Box_{\tilde{g}} f = \Omega^{-1}\Box_g f + \Omega^{-2}(g^{-1})^{\mu\nu}\partial_\mu\Omega\partial_\nu f \tag{8.8}$$

and we have

$$\partial_\mu \Omega = \frac{d\Omega}{dh}\partial_\mu h \tag{8.9}$$

then the proposition follows. □

From Chapter 3 $\mathrm{tr}\chi$ satisfies the propagation equation

$$L\mathrm{tr}\chi = \mu^{-1}(L\mu)\mathrm{tr}\chi - |\chi|^2 - \mathrm{tr}\alpha \tag{8.10}$$

Since

$$S_{\mu\nu} = (g^{-1})^{\kappa\lambda}R_{\kappa\mu\lambda\nu} \tag{8.11}$$

is the Ricci tensor, we have

$$\mathrm{tr}\alpha = S(L, L) \tag{8.12}$$

We decompose

$$S(L, L) = S^{[P]}(L, L) + S^{[N]}(L, L) = \mathrm{tr}\alpha^{[P]} + \mathrm{tr}\alpha^{[N]} \tag{8.13}$$

From Chapter 4,

$$\mathrm{tr}\alpha^{[P]} = -\frac{1}{2}\frac{dH}{dh}(\not{g}^{-1})^{AB}v_{AB} \tag{8.14}$$

Since

$$-\mu^{-1}v_{L\underline{L}} + (\not{g}^{-1})^{AB}v_{AB} = \mathrm{tr}v = \Box_g h \tag{8.15}$$

we have

$$\mathrm{tr}\alpha^{[P]} = -\frac{1}{2}\frac{dH}{dh}(\mu^{-1}v_{L\underline{L}} + \Box_g h) \tag{8.16}$$

Also from Chapter 4,

$$\mathrm{tr}\alpha^{[N]} = \eta^{-2}\mathrm{tr}\alpha^{[A]} + \frac{1}{2}H_1\mathrm{tr}\alpha^{[C]} - \frac{1}{2}H_2\mathrm{tr}\alpha^{[B]} \tag{8.17}$$

where

$$H_1 = \eta^{-2}\frac{dH}{dh}, \quad H_2 = \frac{d^2H}{dh^2} + \frac{1}{2}\eta^{-2}(\frac{dH}{dh})^2 \tag{8.18}$$

and

$$\mathrm{tr}\alpha^{[A]} = w_{LL}\mathrm{tr}\not{\psi} - |\not{\psi}_L|^2 \tag{8.19}$$

$$\mathrm{tr}\alpha^{[B]} = |\not{\tau}|^2 \tag{8.20}$$

$$\mathrm{tr}\alpha^{[C]} = 2(\tau_L\mathrm{tr}\not{\psi} - \not{\tau}\cdot\not{\psi}_L) \tag{8.21}$$

Now,

$$v_{L\underline{L}} = D^2 h(L, \underline{L}) = L(\underline{L}h) - \tau(D_L \underline{L}) \tag{8.22}$$

and from Chapter 3,

$$D_L \underline{L} = -2\zeta^A X_A$$

Hence

$$v_{L\underline{L}} = L(\underline{L}h) + 2\zeta \cdot \not{\tau} \tag{8.23}$$

Let us define, noting that $\underline{L}h = \tau_{\underline{L}}$,

$$f = -\frac{1}{2\mu}\frac{dH}{dh}\tau_{\underline{L}} \tag{8.24}$$

We then have

$$S(L, L) = Lf + g \tag{8.25}$$

where

$$g = \frac{1}{2\mu}\frac{d^2H}{dh^2}\tau_L\tau_{\underline{L}} - \frac{1}{2\mu^2}(L\mu)\frac{dH}{dh}\tau_{\underline{L}} - \frac{1}{\mu}\frac{dH}{dh}\zeta \cdot \not{\tau} - \frac{1}{2}\frac{dH}{dh}\Box_g h + \mathrm{tr}\alpha^{[N]} \tag{8.26}$$

We now set

$$\check{f} = \mu f = -\frac{1}{2}\frac{dH}{dh}\tau_{\underline{L}} \tag{8.27}$$

which is regular as $\mu \to 0$. We then have

$$\mu S(L, L) = L\check{f} + \check{g} \tag{8.28}$$

where

$$\begin{aligned} \check{g} &= \mu g - (L\mu) f \\ &= \frac{1}{2}\frac{d^2H}{dh^2}\tau_L\tau_{\underline{L}} - \frac{dH}{dh}\zeta \cdot \not{\tau} - \frac{1}{2}\frac{dH}{dh}\mu\Box_g h + \mu \mathrm{tr}\alpha^{[N]} \end{aligned} \tag{8.29}$$

By Proposition 8.1,

$$\mu\Box_g h = -\Omega^{-1}\frac{d\Omega}{dh}\mu a - \mu b \tag{8.30}$$

where

$$\mu a = \mu(g^{-1})^{\mu\nu}\tau_\mu\tau_\nu = -\tau_L\tau_{\underline{L}} + \mu|\not{\tau}|^2 \tag{8.31}$$

and

$$\mu b = \sum_i \mu (g^{-1})^{\mu\nu} \partial_\mu \psi_i \partial_\nu \psi_i = \sum_i (-(L\psi_i)(\underline{L}\psi_i) + \mu |\not{d}\psi_i|^2) \tag{8.32}$$

We see that $\mu a$ and $\mu b$ are both regular as $\mu \to 0$, so is $\mu \Box_g h$. It follows that $\check{g}$ is regular as $\mu \to 0$. (8.10) takes the form

$$L(\mu \mathrm{tr}\chi + \check{f}) = 2(L\mu)\mathrm{tr}\chi - \frac{1}{2}\mu(\mathrm{tr}\chi)^2 - \mu|\hat{\chi}|^2 - \check{g} \tag{8.33}$$

Introduce the $S_{t,u}$ 1-form:

$$x_0 = \mu \not{d}\mathrm{tr}\chi + \not{d}\check{f} \tag{8.34}$$

We shall obtain a propagation equation for $x_0$, by differentiating (8.33) tangentially to $S_{t,u}$.

**Lemma 8.1** For an arbitrary function $\phi$, we have

$$\not{\mathcal{L}}_L(\not{d}\phi) = \not{d}(L\phi) \tag{8.35}$$

*Proof*. Since both sides of the equation are $S_{t,u}$ 1-forms, we need only evaluate both sides on $X_A$.

$$\begin{aligned} \not{\mathcal{L}}_L(\not{d}\phi)(X_A) &= \mathcal{L}_L(\not{d}\phi)(X_A) = L(X_A(\phi)) - (\not{d}\phi)(\mathcal{L}_L X_A) \\ &= L(X_A(\phi)) - [L, X_A]\phi = X_A(L\phi) \end{aligned}$$

While $\not{d}(L\phi)(X_A) = X_A(L\phi)$, so the lemma follows. □

Taking $\phi = \mu \mathrm{tr}\chi + \check{f}$ in Lemma 8.1, using (8.33) and (8.34), we get the propagation equation for $x_0$:

$$\not{\mathcal{L}}_L x_0 + (\mathrm{tr}\chi - 2\mu^{-1}(L\mu))x_0 = (\frac{1}{2}\mathrm{tr}\chi - 2\mu^{-1}(L\mu))\not{d}\check{f} - \mu\not{d}(|\hat{\chi}|^2) - g_0 \tag{8.36}$$

where

$$g_0 = \not{d}\check{g} - \frac{1}{2}\mathrm{tr}\chi\not{d}(\check{f} + 2L\mu) + (\not{d}\mu)(L\mathrm{tr}\chi + |\chi|^2) \tag{8.37}$$

We may substitute (3.24)

$$L\mathrm{tr}\chi + |\chi|^2 = \mu^{-1}(L\mu)\mathrm{tr}\chi - \mathrm{tr}\alpha \tag{8.38}$$

into the expression of $g_0$. From Chapter 4,

$$\mathrm{tr}\alpha = \mathrm{tr}\alpha^{[P]} + \mathrm{tr}\alpha^{[N]}$$

where $\mathrm{tr}\alpha^{[N]}$ is regular as $\mu \to 0$ while

$$\mathrm{tr}\alpha^{[P]} = -\frac{1}{2}\frac{dH}{dh}(\not{g}^{-1})^{AB} v_{AB} \tag{8.39}$$

where

$$v_{AB} = D^2 h(X_A, X_B) = \not{D}^2 h(X_A, X_B) - \eta^{-1}\not{k}_{AB} Lh - \mu^{-1}\chi_{AB} Th$$

hence

$$(\not{g}^{-1})^{AB}v_{AB} = \not{\Delta}h - \eta^{-1}\mathrm{tr}\not{k}Lh - \mu^{-1}\mathrm{tr}\chi Th \tag{8.40}$$

and

$$\mathrm{tr}\alpha^{[P]} = \mu^{-1}m\mathrm{tr}\chi - \frac{1}{2}\frac{dH}{dh}(\not{\Delta}h - \eta^{-1}\mathrm{tr}\not{k}Lh) \tag{8.41}$$

In view of the propagation equation for $\mu$

$$L\mu = m + \mu e \tag{8.42}$$

the right hand side of (8.38) is

$$e\mathrm{tr}\chi + \frac{1}{2}\frac{dH}{dh}(\not{\Delta}h - \eta^{-1}\mathrm{tr}\not{k}Lh) - \mathrm{tr}\alpha^{[N]} \tag{8.43}$$

which is regular as $\mu \to 0$.

Thus $g_0$ is of the order of the 2nd derivatives of $\psi_\mu$ and bounded as $\mu \to 0$.

### 8.1.2 Propagation Equations for Higher Order Angular Derivatives

To control the higher order angular derivatives of $\mathrm{tr}\chi$, we introduce the $S_{t,u}$ 1-forms

$${}^{(i_1\cdots i_l)}x_l = \mu\not{d}(R_{i_l}\cdots R_{i_1}\mathrm{tr}\chi) + \not{d}(R_{i_l}\cdots R_{i_1}\check{f}) \tag{8.44}$$

Thus, for a given positive integer $l$ we have the multi-indices $(i_1\cdots i_l)$ of length $l$, where each $i_k \in \{1,2,3\}$, $k = 1,\cdots,l$. To the multi-indices $(i_1\cdots i_l)$ there corresponds the string $(R_{i_l}\cdots R_{i_1})$ of rotation vectorfields. In deriving propagation equations for $x_l$, we shall use the following lemma.

**Lemma 8.2** Let $Y$ be an arbitrary $S_{t,u}$-tangential vectorfield in spacetime manifold. We have

$$[L,Y] = {}^{(Y)}Z \tag{8.45}$$

where ${}^{(Y)}Z$ is the $S_{t,u}$-tangential vectorfields, associated to $Y$, and defined by the condition that for any vector $V \in TW^*_{\epsilon_0}$,

$$g({}^{(Y)}Z, V) = {}^{(Y)}\pi(L, \Pi V) \tag{8.46}$$

In terms of $(L,T,X_1,X_2)$ or $(L,\underline{L},X_1,X_2)$,

$${}^{(Y)}Z = {}^{(Y)}Z^A X_A, \quad {}^{(Y)}Z^A = {}^{(Y)}\pi_{LB}(\not{g}^{-1})^{AB} \tag{8.47}$$

*Proof*. The vectorfield $[L,Y]$ satisfies

$$[L,Y]t = L(Yt) - Y(Lt) = 0$$
$$[L,Y]u = L(Yu) - Y(Lu) = 0$$

It follows that $[L, Y] = {}^{(Y)}Z$ is a vectorfield which is tangential to the surfaces $S_{t,u}$, consequently we expand

$$
{}^{(Y)}Z = {}^{(Y)}Z^B X_B \tag{8.48}
$$

Taking $g$-inner product with $X_A$ we obtain

$$
\begin{aligned}
\not{g}_{AB}{}^{(Y)}Z^B &= {}^{(Y)}Z_A = g({}^{(Y)}Z, X_A) \\
&= g([L, Y], X_A) = g(D_L Y, X_A) - g(D_Y L, X_A) \\
&= {}^{(Y)}\pi_{LA} - g(D_{X_A} Y, L) - g(D_Y L, X_A)
\end{aligned} \tag{8.49}
$$

Now since $g(L, X_A) = 0$ we have

$$
-g(D_Y L, X_A) = g(L, D_Y X_A) \tag{8.50}
$$

hence, substituting,

$$
\not{g}_{AB} Z^B = {}^{(Y)}\pi_{LA} + g(L, [Y, X_A]) = {}^{(Y)}\pi_{LA} \tag{8.51}
$$

We have used the fact that $[Y, X_A]$ is tangential to $S_{t,u}$, hence orthogonal to $L$. The lemma is proved. □

**Lemma 8.3** Let $Y$ be an arbitrary $S_{t,u}$-tangential vectorfield on the spacetime domain $W^*_{\epsilon_0}$ and let $\xi$ be an arbitrary $S_{t,u}$ 1-form on $W^*_{\epsilon_0}$. We have

$$
\not{\mathcal{L}}_L \not{\mathcal{L}}_Y \xi - \not{\mathcal{L}}_Y \not{\mathcal{L}}_L \xi = \not{\mathcal{L}}_{{}^{(Y)}Z} \xi \tag{8.52}
$$

*Proof.* Since both $L, Y$ are tangential to the hypersurfaces $C_u$ we can restrict ourselves to a given $C_u$. In defining $\not{\mathcal{L}}_Y \xi, \not{\mathcal{L}}_L \xi$ we are considering the extension of $\xi$ to $TC_u$ by the condition

$$
\xi(L) = 0 \tag{8.53}
$$

We have

$$
(\mathcal{L}_L \xi)(L) = L(\xi(L)) - \xi([L, L]) = 0 \tag{8.54}
$$

therefore

$$
\not{\mathcal{L}}_L \xi = \mathcal{L}_L \xi \tag{8.55}
$$

However

$$
(\mathcal{L}_Y \xi)(L) = Y(\xi(L)) - \xi([Y, L]) = \xi([L, Y]) = \xi({}^{(Y)}Z) \tag{8.56}
$$

Since $\not{\mathcal{L}}_Y \xi$ is defined by restricting $\mathcal{L}_Y \xi$ to $TS_{t,u}$ and then extending to $TC_u$ by the condition $(\not{\mathcal{L}}_Y \xi)(L) = 0$, it follows that on the manifold $C_u$,

$$
\not{\mathcal{L}}_Y \xi = \mathcal{L}_Y \xi - \xi({}^{(Y)}Z) dt \tag{8.57}
$$

in view of the fact that $dt(L) = Lt = 1, \not{d}t = 0$.

We have

$$
(\not{\mathcal{L}}_L \not{\mathcal{L}}_Y \xi - \not{\mathcal{L}}_Y \not{\mathcal{L}}_L \xi)(X_A) = (\mathcal{L}_L \not{\mathcal{L}}_Y \xi - \mathcal{L}_Y \not{\mathcal{L}}_L \xi)(X_A) \tag{8.58}
$$

Substituting (8.55) and (8.57), the right hand side of the above is

$$(\mathcal{L}_L\mathcal{L}_Y\xi - L(\xi(^{(Y)}Z))dt - \xi(^{(Y)}Z)\mathcal{L}_L(dt) - \mathcal{L}_Y\mathcal{L}_L\xi)(X_A) \tag{8.59}$$

Now, $dt(X_A) = 0$ and $\mathcal{L}_L(dt) = d(Lt) = 0$, therefore (8.58) becomes

$$(\mathcal{L}_L\mathcal{L}_Y\xi - \mathcal{L}_Y\mathcal{L}_L\xi)(X_A) = (\mathcal{L}_{[L,Y]}\xi)(X_A) \tag{8.60}$$

and by Lemma 8.2, this is $(\mathcal{L}_{^{(Y)}Z}\xi)(X_A)$. The lemma follows. □

In the sequel, the proofs of several propositions make use of a general elementary proposition on linear recursions, the proof of which is by a simple application of the principle of induction.

**Proposition 8.2** Let $\{y_n : n = 1, 2, \cdots\}$ be a given sequence in a space $X$ and $\{A_n : n = 1, 2, \cdots\}$ a given sequence of operators in $X$. Suppose that $\{x_n : n = 0, 1, 2, \cdots\}$ is a sequence in $X$ satisfying the recursion:

$$x_n = A_n x_{n-1} + y_n \tag{8.61}$$

Then for each $n = 1, 2, \cdots$ we have

$$x_n = A_n \cdots A_1 x_0 + \sum_{m=0}^{n-1} A_n \cdots A_{n-m+1} y_{n-m} \tag{8.62}$$

To present the propagation equation for $^{(i_1\cdots i_l)}x_l$, we introduce the functions

$$^{(i_1\cdots i_l)}\check{f}_l = R_{i_l}\cdots R_{i_1}\check{f} \tag{8.63}$$

$$^{(i_1\cdots i_l)}h_l = R_{i_l}\cdots R_{i_1}(|\hat{\chi}|^2) \tag{8.64}$$

**Proposition 8.3** For each nonnegative integer $l$ and each multi-index $(i_1\cdots i_l)$, the $S_{t,u}$ 1 form $^{(i_1\cdots i_l)}x_l$ satisfies the propagation equation

$$\begin{aligned}&\not{\mathcal{L}}_L{}^{(i_1\cdots i_l)}x_l + (\text{tr}\chi - 2\mu^{-1}(L\mu))^{(i_1\cdots i_l)}x_l\\ &= (\frac{1}{2}\text{tr}\chi - 2\mu^{-1}(L\mu))\not{d}^{(i_1\cdots i_l)}\check{f}_l - \mu\not{d}^{(i_1\cdots i_l)}h_l - {}^{(i_1\cdots i_l)}g_l\end{aligned}$$

where $^{(i_1\cdots i_l)}g_l$ is the $S_{t,u}$ 1-form given by

$$\begin{aligned}^{(i_1\cdots i_l)}g_l =& \not{\mathcal{L}}_{R_{i_l}}\cdots\not{\mathcal{L}}_{R_{i_1}}g_0 - \sum_{k=0}^{l-1}\not{\mathcal{L}}_{R_{i_l}}\cdots\not{\mathcal{L}}_{R_{i_{l-k+1}}}\not{\mathcal{L}}_{^{(R_{i_{l-k}})}Z}{}^{(i_1\cdots i_{l-k-1})}x_{l-k-1}\\ &+\sum_{k=0}^{l-1}\not{\mathcal{L}}_{R_{i_l}}\cdots\not{\mathcal{L}}_{R_{i_{l-k+1}}}{}^{(i_1\cdots i_{l-k})}y_{l-k}\end{aligned}$$

where $g_0$ is given by (8.37). Here for each $j = 1, \cdots, l$, $^{(i_1\cdots i_j)}y_j$ is the $S_{t,u}$ 1-form:

$$\begin{aligned}^{(i_1\cdots i_j)}y_j =& (R_{i_j}\mu)^{(i_1\cdots i_{j-1})}a_{j-1}\\ &+ (\mu R_{i_j}\text{tr}\chi - R_{i_j}L\mu + {}^{(R_{i_j})}Z\mu)\not{d}(R_{i_{j-1}}\cdots R_{i_1}\text{tr}\chi)\end{aligned}$$

$$+\frac{1}{2}(R_{i_j}\mathrm{tr}\chi)\not{d}^{(i_1\cdots i_{j-1})}\check{f}_{j-1}$$

where ${}^{(i_1\cdots i_{j-1})}a_{j-1}$ is the $S_{t,u}$ 1-form:

$${}^{(i_1\cdots i_{j-1})}a_{j-1} = \not{\mathcal{L}}_L\not{d}(R_{i_{j-1}}\cdots R_{i_1}\mathrm{tr}\chi) + \mathrm{tr}\chi\not{d}(R_{i_{j-1}}\cdots R_{i_1}\mathrm{tr}\chi) + \not{d}^{(i_1\cdots i_{j-1})}h_{j-1}$$

*Proof*. When $l = 0$, the propagation equation in the proposition is just Equation (8.36). Thus, by induction, assuming that the propagation equation holds with $l$ replaced by $l-1$, that is,

$$\begin{aligned}
&\not{\mathcal{L}}_L{}^{(i_1\cdots i_{l-1})}x_{l-1} + (\mathrm{tr}\chi - 2\mu^{-1}(L\mu))^{(i_1\cdots i_{l-1})}x_{l-1}\\
&=(\frac{1}{2}\mathrm{tr}\chi - 2\mu^{-1}(L\mu))\not{d}^{(i_1\cdots i_{l-1})}\check{f}_{l-1} - \mu\not{d}^{(i_1\cdots i_{l-1})}h_{l-1} - {}^{(i_1\cdots i_{l-1})}g_{l-1}
\end{aligned}\tag{8.65}$$

holds for some $S_{t,u}$ 1-form ${}^{(i_1\cdots i_{l-1})}g_{l-1}$, what we shall show is that a propagation equation of the form given by the proposition holds true for $l$, where ${}^{(i_1\cdots i_l)}g_l$ is an $S_{t,u}$ 1-form related to ${}^{(i_1\cdots i_{l-1})}g_{l-1}$ by a certain recursion relation. This recursion relation shall then determine ${}^{(i_1\cdots i_l)}g_l$, for each $l$, from the $S_{t,u}$ 1-form $g_0$, given by (8.37).

We begin by re-writing

$$2\mu^{-1}(L\mu)({}^{(i_1\cdots i_{l-1})}x_{l-1} - \not{d}^{(i_1\cdots i_{l-1})}\check{f}_{l-1})$$

as

$$2(L\mu)\not{d}(R_{i_{l-1}}\cdots R_{i_1}\mathrm{tr}\chi)$$

(see (8.44) and (8.63)) obtaining

$$\begin{aligned}
&\not{\mathcal{L}}_L{}^{(i_1\cdots i_{l-1})}x_{l-1} + \mathrm{tr}\chi^{(i_1\cdots i_{l-1})}x_{l-1}\\
&=2(L\mu)\not{d}(R_{i_{l-1}}\cdots R_{i_1}\mathrm{tr}\chi) + \frac{1}{2}\mathrm{tr}\chi\not{d}^{(i_1\cdots i_{l-1})}\check{f}_{l-1}\\
&\quad - \mu\not{d}^{(i_1\cdots i_{l-1})}h_{l-1} - {}^{(i_1\cdots i_{l-1})}g_{l-1}
\end{aligned}\tag{8.66}$$

We now apply $\not{\mathcal{L}}_{R_{i_l}}$ to this equation. By Lemma 8.1 applied to $R_{i_l}$ and to the functions

$$R_{i_{l-1}}\cdots R_{i_1}\mathrm{tr}\chi, \qquad {}^{(i_1\cdots i_{l-1})}\check{f}_{l-1}, \qquad {}^{(i_1\cdots i_{l-1})}h_{l-1}$$

we have

$$\begin{aligned}
\not{\mathcal{L}}_{R_{i_l}}\not{d}(R_{i_{l-1}}\cdots R_{i_1}\mathrm{tr}\chi) &= \not{d}(R_{i_l}\cdots R_{i_1}\mathrm{tr}\chi)\\
\not{\mathcal{L}}_{R_{i_l}}\not{d}({}^{(i_1\cdots i_{l-1})}\check{f}_{l-1}) &= \not{d}({}^{(i_1\cdots i_l)}\check{f}_l)\\
\not{\mathcal{L}}_{R_{i_l}}\not{d}({}^{(i_1\cdots i_{l-1})}h_{l-1}) &= \not{d}({}^{(i_1\cdots i_l)}h_l)
\end{aligned}\tag{8.67}$$

(see (8.63), (8.64)). We then obtain

$$\begin{aligned}
&\not{\mathcal{L}}_{R_{i_l}}\not{\mathcal{L}}_L{}^{(i_1\cdots i_{l-1})}x_{l-1}+\mathrm{tr}\chi\not{\mathcal{L}}_{R_{i_l}}{}^{(i_1\cdots i_{l-1})}x_{l-1}+(R_{i_l}\mathrm{tr}\chi)^{(i_1\cdots i_{l-1})}x_{l-1}\\
&=2(L\mu)\not{d}(R_{i_l}\cdots R_{i_1}\mathrm{tr}\chi)+2(R_{i_l}L\mu)\not{d}(R_{i_{l-1}}\cdots R_{i_1}\mathrm{tr}\chi) \qquad (8.68)\\
&+\frac{1}{2}\mathrm{tr}\chi\not{d}({}^{(i_1\cdots i_l)}\check{f}_l)+\frac{1}{2}(R_{i_l}\mathrm{tr}\chi)\not{d}({}^{(i_1\cdots i_{l-1})}\check{f}_{l-1})\\
&-\mu\not{d}({}^{(i_1\cdots i_l)}h_l)-(R_{i_l}\mu)\not{d}({}^{(i_1\cdots i_{l-1})}h_{l-1})-\not{\mathcal{L}}_{R_{i_l}}{}^{(i_1\cdots i_{l-1})}g_{l-1}
\end{aligned}$$

Next, we apply Lemma 8.3 to $Y=R_{i_l},\xi={}^{(i_1\cdots i_{l-1})}x_{l-1}$, to express

$$\not{\mathcal{L}}_{R_{i_l}}\not{\mathcal{L}}_L{}^{(i_1\cdots i_{l-1})}x_{l-1}=\not{\mathcal{L}}_L\not{\mathcal{L}}_{R_{i_l}}{}^{(i_1\cdots i_{l-1})}x_{l-1}-\not{\mathcal{L}}_{{}^{(R_{i_l})}Z}{}^{(i_1\cdots i_{l-1})}x_{l-1} \tag{8.69}$$

Now, from (8.63) we have

$${}^{(i_1\cdots i_{l-1})}x_{l-1}=\mu\not{d}(R_{i_{l-1}}\cdots R_{i_1}\mathrm{tr}\chi)+\not{d}(R_{i_{l-1}}\cdots R_{i_1}\check{f})$$

Applying $\not{\mathcal{L}}_{R_{i_l}}$ to this we obtain, using (8.63) and Lemma 8.1,

$$\not{\mathcal{L}}_{R_{i_l}}{}^{(i_1\cdots i_{l-1})}x_{l-1}={}^{(i_1\cdots i_l)}x_l+(R_{i_l}\mu)\not{d}(R_{i_{l-1}}\cdots R_{i_1}\mathrm{tr}\chi) \tag{8.70}$$

Applying $\not{\mathcal{L}}_L$ to (8.70) and expressing

$$LR_{i_l}\mu=R_{i_l}L\mu+{}^{(R_{i_l})}Z\mu \tag{8.71}$$

using the fact that by Lemma 8.2

$$[L,R_{i_l}]={}^{(R_{i_l})}Z \tag{8.72}$$

yields

$$\begin{aligned}
\not{\mathcal{L}}_L\not{\mathcal{L}}_{R_{i_l}}{}^{(i_1\cdots i_{l-1})}x_{l-1}=&\not{\mathcal{L}}_L{}^{(i_1\cdots i_l)}x_l+(R_{i_l}\mu)\not{\mathcal{L}}_L\not{d}(R_{i_{l-1}}\cdots R_{i_1}\mathrm{tr}\chi) \qquad (8.73)\\
&+(R_{i_l}L\mu+{}^{(R_{i_l})}Z\mu)\not{d}(R_{i_{l-1}}\cdots R_{i_1}\mathrm{tr}\chi)
\end{aligned}$$

Substituting (8.72) in (8.69) and the result in (8.68), and using (8.70) to re-write the second term on the left in (8.68), a propagation equation for ${}^{(i_1\cdots i_l)}x_l$ of the form given by the proposition results with ${}^{(i_1\cdots i_l)}g_l$ expressed in terms of ${}^{(i_1\cdots i_l)}g_{l-1}$ by the recursion formula

$${}^{(i_1\cdots i_l)}g_l=\not{\mathcal{L}}_{R_{i_l}}{}^{(i_1\cdots i_{l-1})}g_{l-1}-\not{\mathcal{L}}_{{}^{(R_{i_l})}Z}{}^{(i_1\cdots i_{l-1})}x_{l-1}+{}^{(i_1\cdots i_l)}y_l \tag{8.74}$$

Applying Proposition 8.2 to this recursion with the space of $S_{t,u}$ 1-forms in the role of the space $X$, and ${}^{(i_1\cdots i_l)}g_l$, ${}^{(i_1\cdots i_l)}y_l-\not{\mathcal{L}}_{{}^{(R_{i_l})}Z}{}^{(i_1\cdots i_{l-1})}x_{l-1}$, in the role of $x_n,y_n$, respectively, the proposition follows. □

We remark that the $S_{t,u}$ 1-form ${}^{(i_1\cdots i_{j-1})}a_{j-1}$ reduces for $j=1$ to

$$a_0=\not{\mathcal{L}}_L\not{d}\mathrm{tr}\chi+\mathrm{tr}\chi\not{d}\mathrm{tr}\chi+\not{d}h_0 \tag{8.75}$$

Define

$$L\mathrm{tr}\chi+|\chi|^2=f_0 \tag{8.76}$$

Applying $\not{d}$ to the above, we obtain, in view of Lemma 8.1 and the fact that

$$|\chi|^2 = \frac{1}{2}(\mathrm{tr}\chi)^2 + h_0,$$

simply

$$a_0 = \not{d} f_0 \tag{8.77}$$

For $j \geqslant 2$, we apply $\not{\mathcal{L}}_{R_{i_{j-1}}} \cdots \not{\mathcal{L}}_{R_{i_1}}$ to (8.75) to obtain, in view of (8.76), the following expression for ${}^{(i_1 \cdots i_{j-1})}a_{j-1}$:

$${}^{(i_1 \cdots i_{j-1})}a_{j-1} = \not{d}(R_{i_{j-1}} \cdots R_{i_1} f_0) + {}^{(i_1 \cdots i_{j-1})}b_{j-1} \tag{8.78}$$

where

$$\begin{aligned} &{}^{(i_1 \cdots i_{j-1})}b_{j-1} \\ &= [\not{\mathcal{L}}_L, \not{\mathcal{L}}_{R_{i_{j-1}}} \cdots \not{\mathcal{L}}_{R_{i_1}}]\not{d}\mathrm{tr}\chi \\ &\quad + \sum_{m=1}^{j-1} \sum_{k_1 < \cdots < k_m = 1}^{j-1} (R_{i_{k_m}} \cdots R_{i_{k_1}} \mathrm{tr}\chi)(\not{d} R_{i_{j-1}} \overset{> i_{k_m} \cdots i_{k_1} <}{\cdots} R_{i_1} \mathrm{tr}\chi) \end{aligned} \tag{8.79}$$

Here and in the sequel the symbol $>\ <$ signifies that the enclosed indices are absent. We can get an explicit expression for the first term on the right hand side of (8.79) by applying the following lemma, proved inductively, using Lemma 8.3.

**Lemma 8.4** For any positive integer $l$ and any $S_{t,u}$ 1-form $\xi$ we have

$$[\not{\mathcal{L}}_L, \not{\mathcal{L}}_{R_{i_l}} \cdots \not{\mathcal{L}}_{R_{i_1}}]\xi = \sum_{k=0}^{l-1} \not{\mathcal{L}}_{R_{i_l}} \cdots \not{\mathcal{L}}_{R_{i_{l-k+1}}} \not{\mathcal{L}}_{{}^{(R_{i_{l-k}})}Z} \not{\mathcal{L}}_{R_{i_{l-k+1}}} \cdots \not{\mathcal{L}}_{R_{i_1}} \xi$$

Applying this to $S_{t,u}$ 1-form $\not{d}\mathrm{tr}\chi$, we get

$$\begin{aligned} &[\not{\mathcal{L}}_L, \not{\mathcal{L}}_{R_{i_{j-1}}} \cdots \not{\mathcal{L}}_{R_{i_1}}]\not{d}\mathrm{tr}\chi \\ &= \sum_{m=0}^{j-2} \not{d}(R_{i_{j-1}} \cdots R_{i_{j-m}} {}^{(R_{i_{j-m-1}})}Z R_{i_{j-m-2}} \cdots R_{i_1} \mathrm{tr}\chi) \end{aligned} \tag{8.80}$$

Let us now investigate the order of the various terms in the propagation equation of Proposition 8.3. We agree to the convention that the order of $\psi_\alpha, \alpha = 0, 1, 2, 3$ is 0, so the order of enthalpy $h$ and the acoustical metric $g, \bar{g}, \not{g}$ is also 0. Since $\mu$, $\kappa$ are the components of the acoustical metric $g$ in another coordinate system, they are also of order 0. So $L^\mu$, $T^i$ and $\alpha$ are also of order 0, as well as $\psi_L$, $\psi_{\hat{T}}$ and $\not{\psi}$. The $\not{\psi}$, $\not{\psi}_L$, $\omega_{LL}$ and $\omega_{L\hat{T}}$ are of order 1, so are the $\kappa^{-1}\zeta$ and $\not{k}$. The $\chi$, the focus of this chapter is also of order 1. We are to set $l = n - 1$, in estimating the $n$th order angular derivatives of $\chi$ in terms of the $(n+1)$st order derivatives of $\psi_\alpha$. From the definition of $\check{f}$ (see (8.27)), we see that $\check{f}$ is of order 1. So $\not{d}^{(i_1 \cdots i_l)}\check{f}_l$ is of order $l+2$, its principal part contains $(l+2)$nd derivatives of $h$, and it contains $(l+1)$st order angular derivatives of $\mu$.

Next, we shall investigate the term ${}^{(i_1\cdots i_l)}g_l$. First, we must investigate the terms $\not{d}\check{g}$, $L\mathrm{tr}\chi + |\chi|^2$, together with $\not{d}\check{f}$, which express $g_0$. From (8.38), we see that $L\mathrm{tr}\chi + |\chi|^2$ is of order 2. Its principal part is $\frac{1}{2}\frac{dH}{dh}\not{\Delta}h$. The function $\mu$ is not involved in $L\mathrm{tr}\chi+|\chi|^2$. We then investigate $\not{d}\check{g}$. From the definition (8.29) we know $\not{d}\check{g}$ contains 1st angular derivatives of $\mu$, and 2nd derivatives of the $\psi_\alpha$. It follows that

$$\not{\mathcal{L}}_{R_{i_l}}\cdots\not{\mathcal{L}}_{R_{i_1}}g_0$$

the first term in the expression for ${}^{(i_1\cdots i_l)}g_l$, is of principal order $l+2$ and its principal parts are $l+2$ order derivatives of the $\psi_\alpha$. Also it contains $l+1$ order angular derivatives of $\mu$.

We now consider the last term in the expression for ${}^{(i_1\cdots i_l)}g_l$, namely,

$$\sum_{k=0}^{l-1}\not{\mathcal{L}}_{R_{i_l}}\cdots\not{\mathcal{L}}_{R_{i_{l-k+1}}}{}^{(i_1\cdots i_{l-k})}y_{l-k} \tag{8.81}$$

${}^{(i_1\cdots i_j)}y_j$ is the sum of

$${}^{(i_1\cdots i_{j-1})}a_{j-1},\quad \not{d}(R_{i_{j-1}}\cdots R_{i_1}\mathrm{tr}\chi),\quad \not{d}^{(i_1\cdots i_{j-1})}\check{f}_{j-1}$$

with coefficients

$$R_{i_j}\mu$$
$$\mu R_{i_j}\mathrm{tr}\chi - R_{i_j}L\mu + {}^{(R_{i_j})}Z\mu = \mu R_{i_j}(\mathrm{tr}\chi - e) - R_{i_j}m + {}^{(R_{i_j})}Z\mu - eR_{i_j}\mu$$
$$\frac{1}{2}(R_{i_j}\mathrm{tr}\chi)$$

of order 1, 2, 2, respectively. The first two of these coefficients involve the 1st angular derivatives of $\mu$. The order of $\not{d}(R_{i_{j-1}}\cdots R_{i_1}\mathrm{tr}\chi)$ and $\not{d}^{(i_1\cdots i_{j-1})}\check{f}_{j-1}$ is $j+1$. On the other hand, ${}^{(i_1\cdots i_{j-1})}a_{j-1}$ is given by (8.78). By the above discussion of $f_0$, the first term on the right of (8.78) is of order $j+2$, its principal parts being angular derivatives of $h$. This is the principal part in ${}^{(i_1\cdots i_{j-1})}a_{j-1}$. The second term on the right, ${}^{(i_1\cdots i_{j-1})}b_{j-1}$, given by (8.79) is of lower order, namely $j+1$. In fact only the first term in (8.79), given by (8.80), is of order $j+1$ while the second term, the double sum, is of order $j$. Also, ${}^{(i_1\cdots i_{j-1})}a_{j-1}$ does not involve the function $\mu$. So we conclude that ${}^{(i_1\cdots i_j)}y_j$ is of order $j+2$ and its principal parts are $j+2$ order angular derivatives of $h$. Then the order of (8.81) is $l+2$, its principal parts are $l+2$ order angular derivatives of $h$. Moreover, (8.81) contains $l$ order angular derivatives of $\mu$.

To complete the investigation of ${}^{(i_1\cdots i_l)}g_l$, we must consider

$$\sum_{k=0}^{l-1}\not{\mathcal{L}}_{R_{i_l}}\cdots\not{\mathcal{L}}_{R_{i_{l-k+1}}}\not{\mathcal{L}}_{{}^{(R_{i_{l-k}})}Z}{}^{(i_1\cdots i_{l-k-1})}x_{l-k-1} \tag{8.82}$$

which we can write as

$$\sum_{k=0}^{l-1} \not{\mathcal{L}}_{(R_{i_{l-k}})Z} \not{\mathcal{L}}_{R_{i_l}} \cdots \not{\mathcal{L}}_{R_{i_{l-k+1}}} {}^{(i_1 \cdots i_{l-k-1})}x_{l-k-1} \tag{8.83}$$

minus the commutator term

$$\sum_{k=1}^{l-1} [\not{\mathcal{L}}_{(R_{i_{l-k}})Z}, \not{\mathcal{L}}_{R_{i_l}} \cdots \not{\mathcal{L}}_{R_{i_{l-k+1}}}]^{(i_1 \cdots i_{l-k-1})}x_{l-k-1} \tag{8.84}$$

We shall show this commutator term is of lower order by using the following lemma.

**Lemma 8.5** For any $S_{t,u}$-tangential vectorfield $X$ and any $S_{t,u}$ 1-form $\xi$ and positive integer $l$ we have

$$[\not{\mathcal{L}}_X, \not{\mathcal{L}}_{R_{i_l}} \cdots \not{\mathcal{L}}_{R_{i_1}}]\xi = -\sum_{j=1}^{l} \sum_{k_1 < \cdots < k_j = 1}^{l} \not{\mathcal{L}}_{(i_{k_1} \cdots i_{k_j})Y} {}^{(i_1 \overset{>i_{k_1} \cdots i_{k_j}<}{\cdots} i_l)}\xi_{l-j}$$

Here ${}^{(i_{k_1} \cdots i_{k_j})}Y$ is the $S_{t,u}$-tangential vectorfield:

$${}^{(i_{k_1} \cdots i_{k_j})}Y = \not{\mathcal{L}}_{R_{i_{k_j}}} \cdots \not{\mathcal{L}}_{R_{i_{k_1}}} X$$

and ${}^{(i_1 \cdots i_m)}\xi_m$ is the $S_{t,u}$ 1-form:

$${}^{(i_1 \cdots i_m)}\xi_m = \not{\mathcal{L}}_{R_{i_m}} \cdots \not{\mathcal{L}}_{R_{i_1}} \xi$$

The proof is by induction, starting from the case $l = 1$, which is a standard fact in differential geometry, in view of the fact that everything restricts to $S_{t,u}$.

Applying Lemma 8.5 we see that (8.84) is minus:

$$\sum_{k=1}^{l-1} \sum_{j=1}^{k} \sum_{m_1 < \cdots < m_j = l-k+1}^{l} \not{\mathcal{L}}_{(i_{m_1} \cdots i_{m_j}; i_{l-k})Z} \not{\mathcal{L}}_{R_{i_l}} \overset{>i_{m_j} \cdots i_{m_1}<}{\cdots} \not{\mathcal{L}}_{R_{i_{l-k+1}}} {}^{(i_1 \cdots i_{l-k-1})}x_{l-k-1} \tag{8.85}$$

Here the indexes $i_{m_j}, \cdots, i_{m_1}$ are absent, and ${}^{(i_{m_1} \cdots i_{m_j}; i_{l-k})}Z$ is the $S_{t,u}$-tangential vectorfield:

$${}^{(i_{m_1} \cdots i_{m_j}; i_{l-k})}Z = \not{\mathcal{L}}_{R_{i_{m_j}}} \cdots \not{\mathcal{L}}_{R_{i_{m_1}}} {}^{(R_{i_{l-k}})}Z \tag{8.86}$$

From (6.57), we know that ${}^{(R_i)}Z$ is of order 1 and its principal acoustical parts involves $\chi$. Hence ${}^{(i_{m_1} \cdots i_{m_j}; i_{l-k})}Z$ is of order $j+1$ and its principal acoustical parts is a $j$th order angular derivative of $\chi$. In (8.85), the 1-form

$$\not{\mathcal{L}}_{R_{i_l}} \overset{>i_{m_j} \cdots i_{m_1}<}{\cdots} \not{\mathcal{L}}_{R_{i_{l-k+1}}} {}^{(i_1 \cdots i_{l-k-1})}x_{l-k-1} \tag{8.87}$$

is of order $l-j+1$ as it is an angular derivative of order $k-j$ of ${}^{(i_1 \cdots i_{l-k-1})}x_{l-k-1}$. If $X$ is an $S_{t,u}$-tangential vectorfield of order $m$ and $\xi$ an $S_{t,u}$ 1-form of order $n$, then the order of $\not{\mathcal{L}}_X\xi$ is $\max\{m, n\} + 1$. This is evident from the formula

$$(\not{\mathcal{L}}_X \xi)(Y) = X(\xi(Y)) - \xi(\not{\mathcal{L}}_X Y)$$

Thus, the order of the terms in the triple sum in (8.85) is

$$\max\{j+1, l-j+1\} + 1$$

and the two extreme terms corresponding to $j = 1$ and $j = l - 1$ are of highest order, namely $l + 1$. So (8.85) is of order $l + 1$, thus a lower order term.

We now turn to the principal part (8.83). In view of (8.44) and the fact that for any function $\phi$,

$$\not{\mathcal{L}}_{R_{i_l}} \cdots \not{\mathcal{L}}_{R_{i_1}} \not{d}\phi = \not{d}(R_{i_l} \cdots R_{i_1} \phi)$$

we obtain

$$\begin{aligned} &\not{\mathcal{L}}_{R_{i_l}} \cdots \not{\mathcal{L}}_{R_{i_{l-k+1}}} {}^{(i_1 \cdots i_{l-k-1})}x_{l-k-1} \\ &= {}^{(i_l \cdots i_{l-k+1} i_{l-k-1} \cdots i_1)}x_{l-1} \\ &\quad + \sum_{j=1}^{k} \sum_{m_1 < \cdots < m_j = l-k+1}^{l} (R_{i_{m_j}} \cdots R_{i_{m_1}} \mu) \not{d}(R_{i_l} \overset{> i_{m_j} \cdots i_{m_1} i_{l-k} <}{\cdots} R_{i_1} \mathrm{tr}\chi) \end{aligned} \tag{8.88}$$

Here the indexes $i_{m_j} \cdots i_{m_1} i_{l-k}$ are absent. The first term is of order $l + 1$, while the double sum is of order $l$. The contribution of the first term in (8.88) to (8.83) is then the principal part of (8.82). This contribution is

$$\sum_{k=0}^{l-1} \not{\mathcal{L}}_{(R_{i_{l-k}})} {}_Z{}^{(i_1 \cdots i_{l-k-1} i_{l-k+1} \cdots i_l)}x_{l-1} \tag{8.89}$$

We conclude that the principal acoustical parts of ${}^{(i_1 \cdots i_l)}g_l$ are all the terms of the sum (8.89). Moreover (8.82) does not involve the principal order of function $\mu$.

We shall estimate each term in (8.89) pointwise in terms of

$$\max_j |{}^{(i_1 \cdots i_{l-k-1} i_{l-k+1} \cdots i_l j)}x_l|$$

**Lemma 8.6** Let $\phi$ and $\rho$ be arbitrary functions and $\mu$ a nonnegative function. Denote by $\xi$ the $S_{t,u}$ 1-form:

$$\xi = \mu \not{d}\phi + \not{d}\rho$$

and by $\eta_j$ the $S_{t,u}$ 1-forms:

$$\eta_j = \mu \not{d}(R_j \phi) + \not{d}(R_j \rho), \quad j = 1, 2, 3$$

Then for any $S_{t,u}$-tangential vectorfield $X$ we have the pointwise estimate

$$\begin{aligned} |\not{\mathcal{L}}_X \xi| \leqslant\, & C(1+t)^{-1}(|X|(\max_j |\eta_j| + 2\max_j |\not{d}(R_j \rho)| + (1+t)|\not{d}\mu||\not{d}\phi|) \\ & + (\max_j |\not{\mathcal{L}}_{R_j} X|)(\mu|\not{d}\phi| + |\not{d}\rho|)) \end{aligned}$$

*Proof*. We have

$$\not{\mathcal{L}}_X\xi = \mu\not{d}(X\phi) + (X\mu)\not{d}\phi + \not{d}(X\rho) \tag{8.90}$$

hence

$$|\not{\mathcal{L}}_X\xi| \leqslant \mu|\not{d}(X\phi)| + |X||\not{d}\mu||\not{d}\phi| + |\not{d}(X\rho)| \tag{8.91}$$

Now, by **H0**,

$$|\not{d}(X\phi)| \leqslant C(1+t)^{-1}\max_j|R_j(X\phi)| \tag{8.92}$$

and we have

$$R_j(X\phi) = X(R_j\phi) + [R_j, X]\phi = X\cdot\not{d}(R_j\phi) + (\not{\mathcal{L}}_{R_j}X)\cdot\not{d}\phi \tag{8.93}$$

hence

$$\max_j|R_j(X\phi)| \leqslant |X|\max_j|\not{d}(R_j\phi)| + (\max_j|\not{\mathcal{L}}_{R_j}X|)|\not{d}\phi| \tag{8.94}$$

Then we get

$$|\not{d}(X\phi)| \leqslant C(1+t)^{-1}(|X|\max_j|\not{d}(R_j\phi)| + (\max_j|\not{\mathcal{L}}_{R_j}X|)|\not{d}\phi|) \tag{8.95}$$

By the definition of $\eta_j$,

$$\begin{aligned}\mu\max_j|\not{d}(R_j\phi)| &= \max_j|\mu\not{d}(R_j\phi)| \\ &= \max_j|\eta_j - \not{d}(R_j\rho)| \leqslant \max_j|\eta_j| + \max_j|\not{d}(R_j\rho)|\end{aligned} \tag{8.96}$$

Moreover, from (8.95) with $\rho$ in the role of $\phi$ we have

$$|\not{d}(X\rho)| \leqslant C(1+t)^{-1}(|X|\max_j|\not{d}(R_j\rho)| + (\max_j|\not{\mathcal{L}}_{R_j}X|)|\not{d}\rho|) \tag{8.97}$$

In view of (8.95)—(8.97), the lemma follows through (8.91). □

We now take $\phi = R_{i_l}\cdots R_{i_{l-k+1}}R_{i_{l-k-1}}\cdots R_{i_1}\mathrm{tr}\chi$, and $\rho = R_{i_l}\cdots R_{i_{l-k+1}}R_{i_{l-k-1}}\cdots R_{i_1}\check{f} = {}^{(i_1\cdots i_{l-k-1}i_{l-k+1}\cdots i_l)}\check{f}_{l-1}$. Then $\xi$ is the $S_{t,u}$ 1-form ${}^{(i_1\cdots i_{l-k-1}i_{l-k+1}\cdots i_l)}x_{l-1}$ and $\eta_j$ the $S_{t,u}$ 1-form ${}^{(i_1\cdots i_{l-k-1}i_{l-k+1}\cdots i_l j)}x_l$. Moreover, we take $X$ to be the $S_{t,u}$-tangential vectorfield ${}^{(R_{i_{l-k}})}Z$. Noting that

$$\mu|\not{d}\phi| + |\not{d}\rho| \leqslant |\xi| + 2|\not{d}\rho|,$$

we obtain

$$\begin{aligned}&|\not{\mathcal{L}}_{{}^{(R_{i_{l-k}})}Z}\,{}^{(i_1\cdots i_{l-k-1}i_{l-k+1}\cdots i_l)}x_{l-1}| \\ &\leqslant C(1+t)^{-1}(|{}^{(R_{i_{l-k}})}Z|(\max_j|{}^{(i_1\cdots i_{l-k-1}i_{l-k+1}\cdots i_l j)}x_l| \\ &\quad + 2\max_j|\not{d}({}^{(i_1\cdots i_{l-k-1}i_{l-k+1}\cdots i_l j)}\check{f}_l)| \\ &\quad + (1+t)|\not{d}\mu||\not{d}(R_{i_l}\cdots R_{i_{l-k+1}}R_{i_{l-k-1}}\cdots R_{i_1}\mathrm{tr}\chi)|) \\ &\quad + (\max_j|\not{\mathcal{L}}_{R_j}{}^{(R_{i_{l-k}})}Z|)(|{}^{(i_1\cdots i_{l-k-1}i_{l-k+1}\cdots i_l)}x_{l-1}| \\ &\quad + 2|\not{d}({}^{(i_1\cdots i_{l-k-1}i_{l-k+1}\cdots i_l)}\check{f}_{l-1})|))\end{aligned} \tag{8.98}$$

### 8.1.3 Elliptic Theory on $S_{t,u}$

Returning to the propagation equation for ${}^{(i_1\cdots i_l)}x_l$ given by Proposition 8.3, the term $\mu\not{d}({}^{(i_1\cdots i_l)}h_l)$ remains to be considered. Since ${}^{(i_1\cdots i_l)}h_l$ given by (8.64) is $l$th order angular derivative of $|\hat{\chi}|^2$, so the term to be considered is a principal acoustical part. Moreover, it involves $(l+1)$st order angular derivatives of $\hat{\chi}$, not $\mathrm{tr}\chi$, which is what the propagation equation for ${}^{(i_1\cdots i_l)}x_l$ allows us to control.

The term in question comes from the right-hand side of propagation equation (8.36) for $x_0 = \mu\not{d}\mathrm{tr}\chi + \not{d}\check{f}$, from the term

$$\mu\not{d}(|\hat{\chi}|^2) = 2\mu\hat{\chi}\cdot\not{D}\hat{\chi}$$

Now, the propagation equation (8.36) must be considered in conjunction with the Codazzi equation (4.30):

$$\not{\mathrm{div}}\hat{\chi} = \frac{1}{2}\not{d}\mathrm{tr}\chi + i \tag{8.99}$$

where

$$i = \beta^*_A - \mu^{-1}(\zeta_B\hat{\chi}^B_A - \zeta_A\mathrm{tr}\chi) \tag{8.100}$$

which is regular as $\mu \to 0$. Equation (8.99) allows us to estimate, on $S_{t,u}$, $\hat{\chi}$ and $\not{D}\hat{\chi}$ in terms of $\not{d}\mathrm{tr}\chi$ and the derivatives of $\psi_\mu$ of up to 1st order. The propagation equation (8.36) allows us to estimate, along each generator, $\not{d}\mathrm{tr}\chi$ in terms of $\not{d}(|\hat{\chi}|^2)$, that is in terms of $\hat{\chi}$ and $\not{D}\hat{\chi}$, and the derivatives of $\psi_\mu$ of up to the 2nd order. Thus, considering on $C_u$ the propagation equation (8.36) in conjunction with the Codazzi equation (8.99) we can estimate $\chi$ and its 1st order angular derivatives in terms of $\psi_\mu$ and their derivatives of up to the 2nd order, avoiding in this way loss of differentiability. The idea coupling the propagation equations for $\chi$ along the generators of $C_u$ with the Codazzi equations at each $S_{t,u}$ was also used in [6], where the Ricci curvature tensor vanishes.

To get a higher order version of (8.99), we shall use the following lemma:

**Lemma 8.7** Let $(M, g)$ be a 2-dimensional Riemannian manifold, let $X$ be an arbitrary vectorfield on $M$ and let $\theta$ be a trace-free symmetric 2-covariant tensorfield on $(M, g)$ satisfying the equation

$$\mathrm{div}_g\theta = f \tag{8.101}$$

for some 1-form $f$ on $M$. Then $\hat{\mathcal{L}}_X\theta$, the trace-free part of $\mathcal{L}_X\theta$, satisfies the equation

$$\mathrm{div}_g\hat{\mathcal{L}}_X\theta = {}^{(X)}\dot{f} \tag{8.102}$$

where ${}^{(X)}\dot{f}$ is the 1-form given, in an arbitrary local frame, by

$$\begin{aligned}{}^{(X)}\dot{f}_a =& (\mathcal{L}_X f)_a + \frac{1}{2}\mathrm{tr}{}^{(X)}\pi f_a \\ &+ \frac{1}{2}{}^{(X)}\hat{\pi}^{bc}(\nabla_b\theta_{ac} + \nabla_c\theta_{ab} - \nabla_a\theta_{bc}) + (\mathrm{div}_g{}^{(X)}\hat{\pi})^b\theta_{ab}\end{aligned}$$

Here, ${}^{(X)}\pi = \mathcal{L}_X g$.

*Proof*. Let $\phi_t$ be the local 1-parameter group generated by $X$ and let $\phi_{t*}$ be the corresponding pullback. We then have

$$\mathrm{div}_{\phi_{t*}g}(\phi_{t*}\theta) = \phi_{t*}(\mathrm{div}_g\theta) = \phi_{t*}f \tag{8.103}$$

Now in an arbitrary local coordinates

$$\begin{aligned}(\mathrm{div}_g\theta)_a &= (g^{-1})^{bc}\nabla_c\theta_{ab} \\ &= (g^{-1})^{bc}(\frac{\partial\theta_{ab}}{\partial x^c} - \Gamma^d_{ca}\theta_{db} - \Gamma^d_{cb}\theta_{ad})\end{aligned} \tag{8.104}$$

where $\Gamma^c_{ab}$ are the Christoffel symbols of the metric $g$. Similarly,

$$(\mathrm{div}_{\phi_{t*}g}(\phi_{t*}\theta))_a = ((\phi_{t*}g)^{-1})^{bc}(\frac{\partial(\phi_{t*}\theta)_{ab}}{\partial x^c} - \overset{\phi_{t*}g}{\Gamma^d_{ca}}(\phi_{t*}\theta)_{db} - \overset{\phi_{t*}g}{\Gamma^d_{cb}}(\phi_{t*}\theta)_{ad}) \tag{8.105}$$

where $\overset{\phi_{t*}g}{\Gamma^c_{ab}}$ is the Christoffel symbols of the metric $\phi_{t*}g$. We have

$$\Gamma^c_{ab} = \frac{1}{2}(g^{-1})^{cd}(\frac{\partial g_{bd}}{\partial x^a} + \frac{\partial g_{ad}}{\partial x^b} - \frac{\partial g_{ab}}{\partial x^d}) \tag{8.106}$$

and

$$\overset{\phi_{t*}g}{\Gamma^c_{ab}} = \frac{1}{2}((\phi_{t*}g)^{-1})^{cd}(\frac{\partial(\phi_{t*}g)_{bd}}{\partial x^a} + \frac{\partial(\phi_{t*}g)_{ad}}{\partial x^b} - \frac{\partial(\phi_{t*}g)_{ab}}{\partial x^d}) \tag{8.107}$$

Since, by definition,

$$(\frac{d}{dt}\phi_{t*}g)_{t=0} = \mathcal{L}_X g = {}^{(X)}\pi \tag{8.108}$$

differentiating (8.107) with respect to $t$ at $t = 0$, we obtain, in view of (8.106), the following formula:

$$(\frac{d}{dt}\overset{\phi_{t*}g}{\Gamma^c_{ab}})_{t=0} = \frac{1}{2}(g^{-1})^{cd}(\nabla_a{}^{(X)}\pi_{bd} + \nabla_b{}^{(X)}\pi_{ad} - \nabla_d{}^{(X)}\pi_{ab}) \tag{8.109}$$

Differentiating then (8.105) with respect to $t$ at $t = 0$ and using (8.109) and (8.108) and the fact, by definition,

$$(\frac{d}{dt}\phi_{t*}\theta)_{t=0} = \mathcal{L}_X\theta \tag{8.110}$$

we obtain

$$\begin{aligned}&(\frac{d}{dt}(\mathrm{div}_{\phi_{t*}g}(\phi_{t*}\theta))_a)_{t=0} \\ &= (\mathrm{div}_g(\mathcal{L}_X\theta))_a - {}^{(X)}\pi^{bc}\nabla_c\theta_{ab} \\ &\quad - \frac{1}{2}((\nabla_a{}^{(X)}\pi^{bc})\theta_{bc} + (2\nabla_c{}^{(X)}\pi^{bc} - \nabla^b\mathrm{tr}{}^{(X)}\pi)\theta_{ab})\end{aligned} \tag{8.111}$$

From (8.103) and the definition

$$(\frac{d}{dt}\phi_{t*}f)_{t=0} = \mathcal{L}_X f \tag{8.112}$$

we conclude

$$\mathrm{div}_g(\mathcal{L}_X\theta) = {}^{(X)}f' \tag{8.113}$$

where

$$\begin{aligned} {}^{(X)}f'_a = &(\mathcal{L}_X f)_a + {}^{(X)}\pi^{bc}\nabla_c\theta_{ab} \\ &+ \frac{1}{2}((\nabla_a{}^{(X)}\pi^{bc})\theta_{bc} + (2\nabla_c{}^{(X)}\pi^{bc} - \nabla^b\mathrm{tr}^{(X)}\pi)\theta_{ab}) \end{aligned} \tag{8.114}$$

In fact the above hold for any $n$-dimensional manifold $M$, and irrespective of the trace-free nature of $\theta$. In the present case where $M$ is 2-dimensional, decomposing ${}^{(X)}\pi$ into its trace-free part and its trace,

$${}^{(X)}\pi = {}^{(X)}\hat{\pi} + \frac{1}{2}g\mathrm{tr}^{(X)}\pi \tag{8.115}$$

we may write (8.114) as

$$\begin{aligned} {}^{(X)}f'_a = &(\mathcal{L}_X f)_a + \frac{1}{2}\mathrm{tr}^{(X)}\pi f_a + {}^{(X)}\hat{\pi}^{bc}\nabla_c\theta_{ab} \\ &+ \frac{1}{2}(\nabla_a{}^{(X)}\hat{\pi}^{bc})\theta_{bc} + (\nabla_c{}^{(X)}\hat{\pi}^{bc})\theta_{ab} \end{aligned} \tag{8.116}$$

Now,

$$\mathrm{tr}(\mathcal{L}_X\theta) = (g^{-1})^{ab}(\mathcal{L}_X\theta)_{ab} = -(\mathcal{L}_X g^{-1})^{ab}\theta_{ab} = {}^{(X)}\pi^{ab}\theta_{ab} = {}^{(X)}\hat{\pi}\cdot\theta \tag{8.117}$$

where we have used the fact that $\theta$ is trace-free. Therefore, $\hat{\mathcal{L}}_X\theta$ is given by

$$\hat{\mathcal{L}}_X\theta = \mathcal{L}_X\theta - \frac{1}{2}g({}^{(X)}\hat{\pi}\cdot\theta) \tag{8.118}$$

and noting that for any function $\psi$ we have

$$\mathrm{div}_g(g\psi) = d\psi$$

we obtain

$$\mathrm{div}_g(\hat{\mathcal{L}}_X\theta) = \mathrm{div}_g(\mathcal{L}_X\theta) - \frac{1}{2}d({}^{(X)}\hat{\pi}\cdot\theta) \tag{8.119}$$

Hence, defining

$${}^{(X)}\dot{f} = {}^{(X)}f' - \frac{1}{2}d({}^{(X)}\hat{\pi}\cdot\theta) \tag{8.120}$$

we conclude that $\mathrm{div}_g\hat{\mathcal{L}}_X\theta = {}^{(X)}\dot{f}$ and ${}^{(X)}\dot{f}$ is as given in the lemma. □

**Proposition 8.4** For each non-negative integer $l$ and each multi-index $(i_1, \cdots, i_l)$, the trace-free symmetric 2-covariant tensorfield

$${}^{(i_1\cdots i_l)}\hat{\chi}_l = \hat{\mathcal{L}}_{R_{i_l}}\cdots\hat{\mathcal{L}}_{R_{i_1}}\hat{\chi}$$

on $S_{t,u}$ satisfies the elliptic system

$$\not{d}\mathrm{iv}({}^{(i_1\cdots i_l)}\hat{\chi}_l) = \frac{1}{2}\not{d}(R_{i_l}\cdots R_{i_1}\mathrm{tr}\chi) + {}^{(i_1\cdots i_l)}i_l$$

where the $S_{t,u}$ 1-form ${}^{(i_1\cdots i_l)}i_l$ is given by

$$\begin{aligned}{}^{(i_1\cdots i_l)}i_l = &(\not{\mathcal{L}}_{R_{i_l}} + \frac{1}{2}\mathrm{tr}\,{}^{(R_{i_l})}\not{\pi})\cdots(\not{\mathcal{L}}_{R_{i_1}} + \frac{1}{2}\mathrm{tr}\,{}^{(R_{i_1})}\not{\pi})i \\ &+ \sum_{k=0}^{l-1}(\not{\mathcal{L}}_{R_{i_l}} + \frac{1}{2}\mathrm{tr}\,{}^{(R_{i_l})}\not{\pi})\cdots(\not{\mathcal{L}}_{R_{i_{l-k+1}}} + \frac{1}{2}\mathrm{tr}\,{}^{(R_{i_{l-k+1}})}\not{\pi}){}^{(i_1\cdots i_{l-k})}q_{l-k}\end{aligned}$$

Here $i$ is the $S_{t,u}$ 1-form given by (8.100) and for each $j = 1, \cdots, l$, ${}^{(i_1\cdots i_j)}q_j$ is the $S_{t,u}$ 1-form given by

$$\begin{aligned}{}^{(i_1\cdots i_j)}q_{j,A} = &\frac{1}{4}\mathrm{tr}\,{}^{(R_{i_j})}\not{\pi}\,\not{d}(R_{i_{j-1}}\cdots R_{i_1}\mathrm{tr}\chi) \\ &+ \frac{1}{2}{}^{(R_{i_j})}\hat{\not{\pi}}^{BC}(\not{D}_B({}^{(i_1\cdots i_{j-1})}\hat{\chi}_{j-1})_{AC} + \not{D}_C({}^{(i_1\cdots i_{j-1})}\hat{\chi}_{j-1})_{AB} \\ &- \not{D}_A({}^{(i_1\cdots i_{j-1})}\hat{\chi}_{j-1})_{BC}) + (\not{d}\mathrm{iv}\,{}^{(R_{i_j})}\hat{\not{\pi}})^B({}^{(i_1\cdots i_{j-1})}\hat{\chi}_{j-1})_{AB}\end{aligned}$$

*Proof*. The proposition reduces for $l = 0$, since $i = i_0$, to (8.99). By induction, we assume the elliptic system holds with $l$ replaced by $l - 1$:

$$\not{d}\mathrm{iv}({}^{(i_1\cdots i_{l-1})}\hat{\chi}_{l-1}) = \frac{1}{2}\not{d}(R_{i_{l-1}}\cdots R_{i_1}\mathrm{tr}\chi) + {}^{(i_1\cdots i_{l-1})}i_{l-1}$$

Taking $(M, g)$ to be $(S_{t,u}, \not{g})$, $\theta$ to be ${}^{(i_1\cdots i_{l-1})}\hat{\chi}_{l-1}$, $X$ to be $R_{i_l}$, and $f$ to be

$$\frac{1}{2}\not{d}(R_{i_{l-1}}\cdots R_{i_1}\mathrm{tr}\chi) + {}^{(i_1\cdots i_{l-1})}i_{l-1}$$

in Lemma 8.7, and taking into account the fact that

$$\not{\mathcal{L}}_{R_{i_l}}\not{d}(R_{i_{l-1}}\cdots R_{i_1}\mathrm{tr}\chi) = \not{d}(R_{i_l}\cdots R_{i_1}\mathrm{tr}\chi)$$

we get a recursion formula for ${}^{(i_1\cdots i_l)}i_l$:

$${}^{(i_1\cdots i_l)}i_l = \not{\mathcal{L}}_{R_{i_l}}{}^{(i_1\cdots i_{l-1})}i_{l-1} + \frac{1}{2}\mathrm{tr}\,{}^{(R_{i_l})}\not{\pi}\,{}^{(i_1\cdots i_{l-1})}i_{l-1} + {}^{(i_1\cdots i_l)}q_l \tag{8.121}$$

Applying Proposition 8.2 to this recursion with the space of $S_{t,u}$ 1-forms in the role of the space $X$, the operators

$$\not{\mathcal{L}}_{R_{i_l}} + \frac{1}{2}\mathrm{tr}\,{}^{(R_{i_l})}\not{\pi}$$

in the role of the operators $A_n$ and ${}^{(i_1\cdots i_l)}i_l$, ${}^{(i_1\cdots i_l)}q_l$, in the role of $x_n$, $y_n$, respectively, the proposition is proved. □

Let us investigate the order of the terms in ${}^{(i_1\cdots i_l)}i_l$. We begin with $i$, given by (8.100). From the expression (4.30), (4.32) and (4.33), we know that there is no 2nd order term in $i$, so the first term in the expression for ${}^{(i_1\cdots i_l)}i_l$ contains no principal part. Moreover, the second term is of order $l+1$. This can be easily got from the fact that ${}^{(i_1\cdots i_j)}q_j$ is of order $j+1$. Also ${}^{(i_1\cdots i_l)}i_l$ does not involve $\mu$.

*Remark.* From the propagation equation in Proposition 8.3, we can control $\not{d}(R_{i_l}\cdots R_{i_1}\mathrm{tr}\chi)$ in terms of $2\mu\hat{\chi}\not{D}(\hat{\not{\mathcal{L}}}_{R_{i_l}}\cdots\hat{\not{\mathcal{L}}}_{R_{i_1}}\hat{\chi})$ and the derivatives of $\psi_\mu$ up to $(l+2)$nd order. While using the elliptic system in Proposition 8.4, we can control $\not{D}(\hat{\not{\mathcal{L}}}_{R_{i_l}}\cdots\hat{\not{\mathcal{L}}}_{R_{i_1}}\hat{\chi})$ in terms of $\not{d}(R_{i_l}\cdots R_{i_1}\mathrm{tr}\chi)$ and the derivatives of $\psi_\mu$ up to $(l+1)$st order.

Since the elliptic system is to be considered in conjunction with the propagation equation of Proposition 8.3, the right hand side of the elliptic system for $\hat{\not{\mathcal{L}}}_{R_{i_l}}\cdots\hat{\not{\mathcal{L}}}_{R_{i_1}}\hat{\chi}$ is thus to be expressed in terms of ${}^{(i_1\cdots i_l)}x_l$. From (8.44) we have

$$\mu\not{d}(R_{i_l}\cdots R_{i_1}\mathrm{tr}\chi) = {}^{(i_1\cdots i_l)}x_l - \not{d}^{(i_1\cdots i_l)}\check{f}_l$$

Then we can only control the product of $\mu$ with the right-hand side of the elliptic system in terms of ${}^{(i_1\cdots i_l)}x_l$. This shall lead to $\mu$-weighted $L^2$ estimates on $S_{t,u}$.

**Lemma 8.8** Let $(M,g)$ be a compact 2-dimensional Riemannian manifold, and let $\theta$ be a trace-free symmetric 2-covariant tensorfield on $(M,g)$ satisfying the equation

$$\mathrm{div}_g\theta = f \tag{8.122}$$

for some 1-form $f$ on $M$. Let also $\mu$ be an arbitrary non-negative function on $M$. Then the following estimate holds on $M$:

$$\int_M \mu^2(\frac{1}{2}|\nabla\theta|^2 + 2K|\theta|^2)d\mu_g \leqslant 3\int_M \mu^2|f|^2 d\mu_g + 3\int_M |d\mu|^2|\theta|^2 d\mu_g$$

where $K$ is the Gauss curvature of $(M,g)$.

*Proof.* Consider the 3-covariant tensorfield given in an arbitrary local frame by

$$\omega_{abc} = \nabla_a\theta_{bc} - \nabla_b\theta_{ac} \tag{8.123}$$

For any vectorfield $X$, $\omega_{abc}X^c$ defines a 2-form on $M$. Since $\mathrm{dim}M = 2$, this 2-form must be a function $\phi$, times the volume form $\epsilon$ of $(M,g)$. Since $\phi$ depends linearly on $X$, there must be a 1-form $e$ on $M$ such that $e\cdot X = \phi$. So we can write

$$\omega_{abc} = \epsilon_{ab}e_c \tag{8.124}$$

Contracting the right hand side of (8.124) with $(g^{-1})^{ac}$ we get

$$(g^{-1})^{ac}\epsilon_{ab}e_c = e^a\epsilon_{ab} = {}^*e_b$$

(the Hodge dual of $e$). Contracting the left hand side of (8.124) with $(g^{-1})^{ac}$ and taking into account the fact that $(g^{-1})^{ac}\theta_{ac} = 0$ we obtain

$$(g^{-1})^{ac}\omega_{abc} = (\mathrm{div}\theta)_b = f_b$$

We conclude that

$$^*e = f, \quad e = -^*f \tag{8.125}$$

and (8.124) becomes

$$\omega_{abc} = -\epsilon_{ab}{}^* f_c \tag{8.126}$$

In view of the fact that

$$\epsilon_{ac}\epsilon_b{}^c = g_{ab}$$

It follows that

$$\frac{1}{2}\omega_{abc}\omega^{abc} = \frac{1}{2}\epsilon_{ab}{}^* f_c \epsilon^{ab*} f^c = |^* f|^2 = |f|^2 \tag{8.127}$$

On the other hand, from the definition (8.123) we have

$$\begin{aligned}\frac{1}{2}\omega_{abc}\omega^{abc} &= \frac{1}{2}(\nabla_a\theta_{bc} - \nabla_b\theta_{ac})(\nabla^a\theta^{bc} - \nabla^b\theta^{ac}) \\ &= |\nabla\theta|^2 - \nabla_a\theta_{bc}\nabla^b\theta^{ac}\end{aligned} \tag{8.128}$$

We write

$$\nabla_a\theta_{bc}\nabla^b\theta^{ac} = \nabla^b(\theta^{ac}\nabla_a\theta_{bc}) - \theta^{ac}\nabla^b\nabla_a\theta_{bc} \tag{8.129}$$

Using a well known fact in differential geometry, we have

$$\begin{aligned}\nabla^b\nabla_a\theta_{bc} - \nabla_a\nabla^b\theta_{bc} &= R_b{}^{db}{}_a\theta_{dc} + R_c{}^{db}{}_a\theta_{bd} \\ &= S^d{}_a\theta_{dc} + R_c{}^{db}{}_a\theta_{bd}\end{aligned} \tag{8.130}$$

where $R_{abcd}$ is the curvature tensor and $S_{ab} = (g^{-1})^{cd}R_{cadb}$ the Ricci tensor of $(M, g)$. Since $M$ is 2-dimensional, we have

$$R_{abcd} = K(g_{ac}g_{bd} - g_{ad}g_{bc}), \quad S_{ab} = Kg_{ab}$$

Hence, the right hand side of (8.130) is

$$K\delta^d_a\theta_{dc} + K(\delta^b_c\delta^d_a - g_{ca}(g^{-1})^{db})\theta_{bd} = 2K\theta_{ac}$$

in view of the fact that $\mathrm{tr}\theta = 0$. Thus, (8.130) reduces to

$$\nabla^b\nabla_a\theta_{bc} - \nabla_a\nabla^b\theta_{bc} = 2K\theta_{ac} \tag{8.131}$$

Substituting in (8.129) we obtain

$$\nabla_a\theta_{bc}\nabla^b\theta^{ac} = \nabla^b(\theta^{ac}\nabla_a\theta_{bc}) - \theta^{ac}\nabla_a\nabla^b\theta_{bc} - 2K|\theta|^2 \tag{8.132}$$

In reference to the second term on the right, we write

$$\theta^{ac}\nabla_a\nabla^b\theta_{bc} = \nabla_a(\theta^{ac}\nabla^b\theta_{bc}) - (\nabla_a\theta^{ac})(\nabla^b\theta_{bc}) = \nabla_a(\theta^{ac}\nabla^b\theta_{bc}) - |f|^2$$

Substituting in (8.132) then yields

$$\nabla_a\theta_{bc}\nabla^b\theta^{ac} = \mathrm{div}_g J - 2K|\theta|^2 + |f|^2 \tag{8.133}$$

where $J$ is the vectorfield

$$J^a = \theta^b{}_c \nabla_b \theta^{ac} - \theta^a{}_c \nabla_b \theta^{bc} = \theta^b{}_c \nabla_b \theta^{ac} - \theta^a{}_c f^c \tag{8.134}$$

Substituting (8.133) in (8.128) we get

$$\frac{1}{2}\omega_{abc}\omega^{abc} = |\nabla\theta|^2 + 2K|\theta|^2 - |f|^2 - \mathrm{div}_g J \tag{8.135}$$

Comparing (8.135) with (8.127) we conclude that

$$|\nabla\theta|^2 + 2K|\theta|^2 = 2|f|^2 + \mathrm{div}_g J \tag{8.136}$$

Multiplying this equation by $\mu^2$ and integrating over $M$, we obtain

$$\begin{aligned}&\int_M \mu^2(|\nabla\theta|^2 + 2K|\theta|^2)d\mu_g\\ =&2\int_M \mu^2|f|^2 d\mu_g + \int_M \mu^2 \mathrm{div}_g J d\mu_g\end{aligned} \tag{8.137}$$

Writing

$$\mu^2 \mathrm{div}_g J = \mathrm{div}_g(\mu^2 J) - 2\mu(J\cdot d\mu)$$

we obtain, since $M$ is compact,

$$\int_M \mu^2 \mathrm{div}_g J d\mu_g = -2\int_M \mu(J\cdot d\mu)d\mu_g \tag{8.138}$$

From (8.134), we can estimate

$$|J| \leqslant |\theta|(|\nabla\theta| + |f|) \tag{8.139}$$

therefore

$$\begin{aligned}-2\int_M \mu(J\cdot d\mu)d\mu_g &\leqslant 2\int_M \mu|J||d\mu|d\mu_g\\ &\leqslant 2\int_M |d\mu||\theta|(\mu|\nabla\theta| + \mu|f|)d\mu_g\end{aligned}$$

Applying the inequalities

$$2|d\mu||\theta|\mu|\nabla\theta| \leqslant \frac{1}{2}\mu^2|\nabla\theta|^2 + 2|d\mu|^2|\theta|^2$$
$$2|d\mu||\theta|\mu|f| \leqslant \mu^2|f|^2 + |d\mu|^2|\theta|^2$$

we then obtain

$$\begin{aligned}&-2\int_M \mu(J\cdot d\mu)d\mu_g\\ \leqslant&\frac{1}{2}\int_M \mu^2|\nabla\theta|^2 d\mu_g + \int_M \mu^2|f|^2 d\mu_g + 3\int_M |d\mu|^2|\theta|^2 d\mu_g\end{aligned} \tag{8.140}$$

So the lemma follows. □

Next, we shall derive an estimate for $K$, the Gauss curvature of $S_{t,u}$. Recall the bootstrap assumptions $\mathbf{A}, \mathbf{E}, \mathbf{F}$ of Chapter 6.

**Lemma 8.9** Under the bootstrap assumptions $\mathbf{A}, \mathbf{E}, \mathbf{F}$, the following estimate holds:

$$|K - r^{-2}| \leqslant C\delta_0(1+t)^{-3}(1+\log(1+t))$$

In particular, taking $\delta_0$ suitably small, we have

$$K \geqslant C^{-1}(1+t)^{-2}$$

*Proof.* From Equations (3.27), (3.31) and (3.36) in Chapter 3, we have the following expression for $K$:

$$K = \frac{1}{2}\eta^{-2}((\mathrm{tr}\chi)^2 - |\chi|^2) - \eta^{-1}(\mathrm{tr}\not{k}\,\mathrm{tr}\chi - \chi \cdot \not{k}) + \rho$$

From (3.5), (3.33) and (4.5), we know that $K$ is regular as $\mu \longrightarrow 0$. Also from $\mathbf{E}$ we know that

$$|\rho| \leqslant C\delta_0^2(1+t)^{-4} \tag{8.141}$$

From $\mathbf{A2}, \mathbf{E}$ and $\mathbf{F2}$ we have

$$|\eta^{-1}(\mathrm{tr}\not{k}\,\mathrm{tr}\chi - \chi \cdot \not{k})| \leqslant C\delta_0(1+t)^{-3} \tag{8.142}$$

Finally, we consider the first term in the expression of $K$. Assumption $\mathbf{F2}$ implies

$$|\frac{1}{2}((\mathrm{tr}\chi)^2 - |\chi|^2) - (\frac{1}{r})^2| \leqslant C\delta_0 \frac{1+\log(1+t)}{(1+t)^3} \tag{8.143}$$

In fact, we can decompose

$$|\chi|^2 = |\hat{\chi}|^2 + \frac{1}{2}(\mathrm{tr}\chi)^2$$

and

$$|\frac{1}{4}(\mathrm{tr}\chi)^2 - (\frac{1}{r})^2| \leqslant |\frac{1}{2}\mathrm{tr}\chi - \frac{1}{r}||\frac{1}{2}\mathrm{tr}\chi + \frac{1}{r}|$$

From (6.124), (6.129) and $\mathbf{F2}$ we have

$$|\frac{1}{4}(\mathrm{tr}\chi)^2 - (\frac{1}{r})^2| \leqslant C\delta_0(1+t)^{-3}(1+\log(1+t))$$

Also from $\mathbf{F2}$, we have

$$|\hat{\chi}|^2 \leqslant C\delta_0(1+t)^{-4}(1+\log(1+t))^2$$

So (8.143) results.

Then from $\mathbf{E}$, we have

$$|\eta^{-2} - 1| \leqslant C\delta_0(1+t)^{-1} \tag{8.144}$$

so from (8.143) and (6.124), (6.129), we have

$$|\frac{1}{2}\eta^{-2}((\mathrm{tr}\chi)^2 - |\chi|^2) - \frac{1}{r^2}| \leqslant C\delta_0(1+t)^{-3}(1+\log(1+t)) \tag{8.145}$$

The lemma then follows. □

### 8.1.4 Preliminary Estimates for the Solutions of the Propagation Equations

We now apply Lemma 8.8 to Proposition 8.4, taking $(M, g)$ to be $(S_{t,u}, \not{g})$, the trace-free symmetric 2-covariant tensorfields $\theta$ to be ${}^{(i_1 \cdots i_l)}\hat{\chi}_l$, and the 1-form $f$ to be

$$\frac{1}{2}\not{d}(R_{i_l} \cdots R_{i_1}\mathrm{tr}\chi) + {}^{(i_1 \cdots i_l)}i_l$$

In view of the fact that by Lemma 8.9, taking $\delta_0$ suitably small, we have $K \geqslant 0$, we obtain

$$\begin{aligned} &\|\mu \not{D}^{(i_1 \cdots i_l)}\hat{\chi}_l\|_{L^2(S_{t,u})} \\ &\leqslant C\|\mu\not{d}(R_{i_l} \cdots R_{i_1}\mathrm{tr}\chi)\|_{L^2(S_{t,u})} + C\|\mu^{(i_1 \cdots i_l)}i_l\|_{L^2(S_{t,u})} \\ &\quad + C\|\not{d}\mu\|_{L^\infty(S_{t,u})}\|^{(i_1 \cdots i_l)}\hat{\chi}_l\|_{L^2(S_{t,u})} \end{aligned} \tag{8.146}$$

Now, by **F1** we have

$$|\not{d}\mu| \leqslant C\delta_0(1+t)^{-1}(1+\log(1+t)) \tag{8.147}$$

Also,

$$\mu\not{d}(R_{i_l} \cdots R_{i_1}\mathrm{tr}\chi) = {}^{(i_1 \cdots i_l)}x_l - \not{d}^{(i_1 \cdots i_l)}\check{f}_l \tag{8.148}$$

Substituting (8.147) and (8.148) in (8.146) yields

$$\begin{aligned} \|\mu \not{D}^{(i_1 \cdots i_l)}\hat{\chi}_l\|_{L^2(S_{t,u})} \leqslant & C\|^{(i_1 \cdots i_l)}x_l\|_{L^2(S_{t,u})} \\ & + C\|\not{d}^{(i_1 \cdots i_l)}\check{f}_l\|_{L^2(S_{t,u})} + C\|\mu^{(i_1 \cdots i_l)}i_l\|_{L^2(S_{t,u})} \\ & + C\delta_0(1+t)^{-1}(1+\log(1+t))\|^{(i_1 \cdots i_l)}\hat{\chi}_l\|_{L^2(S_{t,u})} \end{aligned} \tag{8.149}$$

Let us return to the propagation equation of Proposition 8.3. Setting

$${}^{(i_1 \cdots i_l)}\tilde{g}_l = \mu\not{d}^{(i_1 \cdots i_l)}h_l + {}^{(i_1 \cdots i_l)}g_l \tag{8.150}$$

This propagation equation takes the form

$$\begin{aligned} &\not{\mathcal{L}}_L{}^{(i_1 \cdots i_l)}x_l + (\mathrm{tr}\chi - 2\mu^{-1}(L\mu))^{(i_1 \cdots i_l)}x_l \\ &= (\frac{1}{2}\mathrm{tr}\chi - 2\mu^{-1}(L\mu))\not{d}^{(i_1 \cdots i_l)}\check{f}_l - {}^{(i_1 \cdots i_l)}\tilde{g}_l \end{aligned} \tag{8.151}$$

Recall that $S_{0,0}$ is the unit sphere $S^2$ in $\mathbb{R}^3$ ($\Sigma_0$ is identified with $\mathbb{R}^3$). We define a diffeomorphism $\Phi_{t,u}$ of $S^2$ onto $S_{t,u}$ as follows. First, at $t = 0$, $\Phi_{0,u}$ is the diffeomorphism of $S^2$ onto $S_{0,u}$ defined by the flow of $T$ on $\Sigma_0$. Then $\Phi_{t,u} = \Phi_t \circ \Phi_{0,u}$ where $\Phi_t$ is the flow of $L$ on $W^*_{\epsilon_0}$. That is, $\Phi_{t|S_{0,u}}$ is the diffeomorphism of $S_{0,u}$ onto $S_{t,u}$ defined by the generators of $C_u$. Given an arbitrary $S_{t,u}$ 1-form $\xi$, we may pull it back by $\Phi_{t,u}$ to $S^2$ and consider the 1-form $\xi(t,u) = \Phi^*_{t,u}\xi$ as a 1-form on $S^2$ depending on the parameters $t$ and $u$. If $\xi = \not{\mathcal{L}}_L\zeta$, where $\zeta$ is an arbitrary $S_{t,u}$ 1-form, then $\xi(t,u) = \frac{\partial\zeta(t,u)}{\partial t}$. If $\xi = \not{d}\phi$ where $\phi$ is a function, then $\xi(t,u) = \not{d}\phi(t,u)$, where $\phi(t,u) = \phi \circ \Phi_{t,u}$ is the corresponding function on $S^2$ and $\not{d}\phi(t,u)$ denotes its differential on $S^2$. Also, we may pull the induced metric $\not{g}$ on $S_{t,u}$ back to $S^2$ and consider $\not{g}(t,u) = \Phi^*_{t,u}\not{g}$ as a metric on $S^2$ depending on $t$ and $u$. The foregoing correspond simply to the description in terms of acoustical coordinates $(t,u,\vartheta^1,\vartheta^2)$, where $(\vartheta^1,\vartheta^2)$ are arbitrary local coordinates on $S^2$ extended to $\Sigma_0$ so that the lines corresponding to constant values of $(\vartheta^1,\vartheta^2)$ are orthogonal to the $S_{0,u}$ (hence $\Xi = 0$ on $\Sigma_0$).

In view of above, we may think of (8.151) as an equation for the 1-form $^{(i_1\cdots i_l)}x_l(t,u)$ on $S^2$ depending on $(t,u)$:

$$\begin{aligned}&\frac{\partial}{\partial t}{}^{(i_1\cdots i_l)}x_l + (\mathrm{tr}\chi - 2\mu^{-1}(L\mu))^{(i_1\cdots i_l)}x_l\\ &= (\frac{1}{2}\mathrm{tr}\chi - 2\mu^{-1}(L\mu))\not{d}^{(i_1\cdots i_l)}\check{f}_l - {}^{(i_1\cdots i_l)}\tilde{g}_l\end{aligned} \tag{8.152}$$

where $^{(i_1\cdots i_l)}x_l = {}^{(i_1\cdots i_l)}x_l(t,u)$ is a 1-form on $S^2$ depending on $(t,u)$.

In the following we denote by $(\ ,\ )$ the pointwise inner product of tensorfields on $S^2$, depending on $t$ and $u$, with respect to the metric $\not{g}(t,u)$. For 1-forms $\xi(t,u)$ and $\zeta(t,u)$, we have

$$(\xi(t,u),\zeta(t,u)) = (\not{g}^{-1})^{AB}(t,u)\xi_A(t,u)\zeta_B(t,u) \tag{8.153}$$

We also denote by $|\ \ |$ the pointwise magnitude of tensorfields on $S^2$, depending on $t$ and $u$, with respect to $\not{g}(t,u)$. Thus,

$$|\xi(t,u)| = \sqrt{(\xi(t,u),\xi(t,u))} \tag{8.154}$$

Now by the definition of $\chi$ we have

$$\begin{aligned}&\frac{\partial \not{g}_{AB}(t,u)}{\partial t} = 2\chi_{AB}(t,u), \quad \frac{\partial(\not{g}^{-1})^{AB}(t,u)}{\partial t} = -2\chi^{AB}(t,u);\\ &\chi^{AB}(t,u) = (\not{g}^{-1})^{AC}(t,u)(\not{g}^{-1})^{BD}(t,u)\chi_{CD}(t,u)\end{aligned} \tag{8.155}$$

It follows from (8.153) and (8.155) that

$$\frac{\partial}{\partial t}(\xi,\zeta) = (\frac{\partial\xi}{\partial t},\zeta) + (\xi,\frac{\partial\zeta}{\partial t}) - 2\xi\cdot\chi\cdot\zeta \tag{8.156}$$

Here,

$$\xi\cdot\chi\cdot\zeta = \xi_A\chi^{AB}\zeta_B \tag{8.157}$$

In particular, taking $\zeta = \xi$ we obtain, in view of (8.154),

$$|\xi|\frac{\partial}{\partial t}|\xi| = \frac{1}{2}\frac{\partial}{\partial t}(\xi,\xi) = (\frac{\partial\xi}{\partial t},\xi) - \xi\cdot\chi\cdot\xi \tag{8.158}$$

Decomposing

$$\chi_{AB} = \hat{\chi}_{AB} + \frac{1}{2}\not{g}_{AB}\mathrm{tr}\chi, \quad \chi^{AB} = \hat{\chi}^{AB} + \frac{1}{2}(\not{g}^{-1})^{AB}\mathrm{tr}\chi \tag{8.159}$$

then we have

$$\xi \cdot \chi \cdot \zeta = \xi \cdot \hat{\chi} \cdot \zeta + \frac{1}{2}\mathrm{tr}\chi(\xi, \zeta) \tag{8.160}$$

hence (8.158) takes the form

$$|\xi|\frac{\partial}{\partial t}|\xi| = (\frac{\partial \xi}{\partial t}, \xi) - \xi \cdot \hat{\chi} \cdot \xi - \frac{1}{2}\mathrm{tr}\chi|\xi|^2 \tag{8.161}$$

We shall apply (8.161) taking $\xi = {}^{(i_1 \cdots i_l)}x_l$. From (8.152) we have

$$\begin{aligned}
&(\frac{\partial {}^{(i_1\cdots i_l)}x_l}{\partial t}, {}^{(i_1\cdots i_l)}x_l) - \frac{1}{2}\mathrm{tr}\chi|{}^{(i_1\cdots i_l)}x_l|^2 \\
&= -(-2\mu^{-1}\frac{\partial \mu}{\partial t} + \frac{3}{2}\mathrm{tr}\chi)|{}^{(i_1\cdots i_l)}x_l|^2 \\
&+ (-2\mu^{-1}\frac{\partial \mu}{\partial t} + \frac{1}{2}\mathrm{tr}\chi)({}^{(i_1\cdots i_l)}x_l, \not{d}^{(i_1\cdots i_l)}\check{f}_l) \\
&- ({}^{(i_1\cdots i_l)}x_l, {}^{(i_1\cdots i_l)}\tilde{g}_l)
\end{aligned} \tag{8.162}$$

We shall use the following assumption:
In $W^s_{\epsilon_0}$ we have

$$\mathbf{AS}: -2\mu^{-1}\frac{\partial \mu}{\partial t} + \frac{1}{2}\mathrm{tr}\chi \geqslant 0$$

From this assumption and (8.162) we have

$$\begin{aligned}
&(\frac{\partial {}^{(i_1\cdots i_l)}x_l}{\partial t}, {}^{(i_1\cdots i_l)}x_l) - \frac{1}{2}\mathrm{tr}\chi|{}^{(i_1\cdots i_l)}x_l|^2 \\
&\leqslant -(-2\mu^{-1}\frac{\partial \mu}{\partial t} + \frac{3}{2}\mathrm{tr}\chi)|{}^{(i_1\cdots i_l)}x_l|^2 \\
&+ (-2\mu^{-1}\frac{\partial \mu}{\partial t} + \frac{1}{2}\mathrm{tr}\chi)|{}^{(i_1\cdots i_l)}x_l||\not{d}^{(i_1\cdots i_l)}\check{f}_l| \\
&+ |{}^{(i_1\cdots i_l)}x_l||{}^{(i_1\cdots i_l)}\tilde{g}_l|
\end{aligned} \tag{8.163}$$

In reference to the second term on the right in (8.161) with ${}^{(i_1\cdots i_l)}x_l$ in the role of $\xi$, we have

$$|{}^{(i_1\cdots i_l)}x_l \cdot \hat{\chi} \cdot {}^{(i_1\cdots i_l)}x_l| \leqslant |\hat{\chi}||{}^{(i_1\cdots i_l)}x_l|^2 \tag{8.164}$$

Substituting (8.164) in (8.163) we obtain, in view of (8.161),

$$\begin{aligned}
\frac{\partial |{}^{(i_1\cdots i_l)}x_l|}{\partial t} \leqslant &-(-2\mu^{-1}\frac{\partial \mu}{\partial t} + \frac{3}{2}\mathrm{tr}\chi - |\hat{\chi}|)|{}^{(i_1\cdots i_l)}x_l| \\
&+ (-2\mu^{-1}\frac{\partial \mu}{\partial t} + \frac{1}{2}\mathrm{tr}\chi)|\not{d}^{(i_1\cdots i_l)}\check{f}_l| + |{}^{(i_1\cdots i_l)}\tilde{g}_l|
\end{aligned} \tag{8.165}$$

The integrating factor here is

$$\exp(\int_0^t (-2\mu^{-1}\frac{\partial \mu}{\partial t} + \frac{3}{2}\mathrm{tr}\chi - |\hat{\chi}|)(t', u)dt') = (\frac{\mu(t,u)}{\mu(0,u)})^{-2}(A(t,u))^{3/2}e^{-S(t,u)} \tag{8.166}$$

Here

$$A(t,u) = \exp\left(\int_0^t \mathrm{tr}\chi(t',u)dt'\right), \quad S(t,u) = \int_0^t |\hat{\chi}|(t',u)dt' \tag{8.167}$$

Since by (8.155),

$$\mathrm{tr}\chi = \frac{1}{\sqrt{\det \not{g}}}\frac{\partial}{\partial t}\sqrt{\det \not{g}} \tag{8.168}$$

we have

$$A(t,u) = \frac{\sqrt{\det \not{g}(t,u)}}{\sqrt{\det \not{g}(0,u)}} \tag{8.169}$$

that is, $A(t,u)$ is the ratio of the area elements of $S_{t,u}$ and $S_{0,u}$ at corresponding points along the same generator of $C_u$. Integrating (8.165) and using (8.166) we get

$$|{}^{(i_1\cdots i_l)}x_l(t,u)| \leqslant {}^{(i_1\cdots i_l)}F_l(t,u) + {}^{(i_1\cdots i_l)}G_l(t,u) \tag{8.170}$$

where

$$\begin{aligned} &{}^{(i_1\cdots i_l)}F_l(t,u) \\ &= e^{S(t,u)}(A(t,u))^{-3/2}(\mu(t,u))^2((\mu(0,u))^{-2}|{}^{(i_1\cdots i_l)}x_l(0,u)| \\ &+ \int_0^t (\mu(t',u))^{-2}(A(t',u))^{3/2}e^{-S(t',u)}(-2\mu^{-1}\frac{\partial\mu}{\partial t} + \frac{1}{2}\mathrm{tr}\chi)(t',u)|\not{d}^{(i_1\cdots i_l)}\check{f}_l(t',u)|) \end{aligned} \tag{8.171}$$

and

$$\begin{aligned} {}^{(i_1\cdots i_l)}G_l(t,u) &= e^{S(t,u)}(A(t,u))^{-3/2}(\mu(t,u))^2 \\ &\cdot \int_0^t (\mu(t',u))^{-2}(A(t',u))^{3/2}e^{-S(t',u)}|{}^{(i_1\cdots i_l)}\tilde{g}_l(t',u)|dt' \end{aligned} \tag{8.172}$$

We begin our estimates with $A(t,u)$ and $S(t,u)$. By **F2** and (8.167) we have

$$|\log A(t,u) - \int_0^t \frac{2}{1-u+t'}dt'| \leqslant C\delta_0 \int_0^t (1+t')^{-2}(1+\log(1+t'))dt' \leqslant C\delta_0$$

Since

$$\int_0^t \frac{2}{1-u+t'}dt' = 2\log\left(\frac{1-u+t}{1-u}\right)$$

it follows that

$$e^{-C\delta_0}\left(\frac{1-u+t}{1-u}\right)^2 \leqslant A(t,u) \leqslant e^{C\delta_0}\left(\frac{1-u+t}{1-u}\right)^2 \tag{8.173}$$

Also by **F2** and (8.167) we have

$$S(t,u) \leqslant C\delta_0 \tag{8.174}$$

the integral

$$\int_0^\infty (1+t')^{-2}(1+\log(1+t'))dt'$$

being convergent.

In view of (8.173) and (8.174) we obtain from (8.171) and (8.172)

$$\begin{aligned}&{}^{(i_1\cdots i_l)}F_l(t,u)\\ &\leqslant e^{C\delta_0}(1-u+t)^{-3}({}^{(i_1\cdots i_l)}M_l^0(t,u)+{}^{(i_1\cdots i_l)}M_l^1(t,u)+{}^{(i_1\cdots i_l)}M_l^2(t,u))\end{aligned} \tag{8.175}$$

where

$$\begin{aligned}&{}^{(i_1\cdots i_l)}M_l^0(t,u)\\ &=(\frac{\mu(t,u)}{\mu(0,u)})^2(1-u)^3|{}^{(i_1\cdots i_l)}x_l(0,u)|\end{aligned} \tag{8.176}$$

$$\begin{aligned}&{}^{(i_1\cdots i_l)}M_l^1(t,u)\\ &=\int_0^t(\frac{\mu(t,u)}{\mu(t',u)})^2(1-u+t')^3(-2\mu^{-1}(\frac{\partial\mu}{\partial t})_-(t',u))\cdot|\not{d}^{(i_1\cdots i_l)}\check{f}_l(t',u)|dt'\end{aligned} \tag{8.177}$$

$$\begin{aligned}&{}^{(i_1\cdots i_l)}M_l^2(t,u)\\ &=\int_0^t(\frac{\mu(t,u)}{\mu(t',u)})^2(1-u+t')^3(\frac{1}{2}\mathrm{tr}\chi(t',u))|\not{d}^{(i_1\cdots i_l)}\check{f}_l(t',u)|dt'\end{aligned} \tag{8.178}$$

Also

$$\begin{aligned}&{}^{(i_1\cdots i_l)}G_l(t,u)\\ &\leqslant e^{C\delta_0}(1-u+t)^{-3}\cdot\int_0^t(\frac{\mu(t,u)}{\mu(t',u)})^2(1-u+t')^3|{}^{(i_1\cdots i_l)}\tilde{g}_l(t',u)|dt'\end{aligned} \tag{8.179}$$

To proceed we must analyze the behavior of $\mu$. We need the following assumption:

$$\not{\mathbf{E}}\mathbf{3}_0:|\not{\Delta}\psi_0|\leqslant C\delta_0(1+t)^{-3}$$

## 8.2 Crucial Lemmas Concerning the Behavior of $\mu$

**Lemma 8.10** Under the assumptions $\mathbf{E1},\mathbf{E2},\not{\mathbf{E}}\mathbf{3}_0,\mathbf{F2}$ and $\mathbf{A}$, the following hold. Let us denote

$$P_s(u,\vartheta)=(1+s)(\underline{L}\psi_0)(s,u,\vartheta)$$

We then have, for all $t\in[0,s]$,

$$(\underline{L}\psi_0)(t,u,\vartheta)=\frac{P_s(u,\vartheta)}{1+t}+R_s(t,u,\vartheta)$$

and

$$|R_s(t,u,\vartheta)| \leqslant C\delta_0 \frac{1+\log(1+t)}{(1+t)^2}\frac{s-t}{1+s}$$

where $C$ is a constant independent of $s$.

*Proof*. The function $\psi_0$ satisfies the equation

$$\Box_{\tilde{g}}\psi_0 = 0$$

From the expression for $\Box_{\tilde{g}}$ in Chapter 3, this reads

$$L(\underline{L}\psi_0) + \nu(\underline{L}\psi_0) = \rho_0 \tag{8.180}$$

where

$$\rho_0 = \mu\not{\Delta}\psi_0 - \underline{\nu}L\psi_0 - 2\zeta \cdot \not{d}\psi_0 + \mu\frac{d\log\Omega}{dh}\not{d}h \cdot \not{d}\psi_0 \tag{8.181}$$

Now, by $\not{\mathbf{E}}\mathbf{3}_0$ and **A3**, the first term on the right hand side of (8.181) is bounded in absolute value by

$$C\delta_0(1+t)^{-3}(1+\log(1+t))$$

By (6.107) and **E2**, the second term on the right hand side of (8.181) is bounded in absolute value by

$$C\delta_0(1+t)^{-3}(1+\log(1+t))$$

By (6.89) and **E2**, the third term on the right hand side of (8.181) is bounded in absolute value by

$$C\delta_0(1+t)^{-4}(1+\log(1+t))$$

Finally, by **A3** and **E**, the last term on the right of (8.181) is bounded in absolute value by

$$C\delta_0(1+t)^{-4}(1+\log(1+t))$$

We conclude

$$|\rho_0| \leqslant C\delta_0(1+t)^{-3}(1+\log(1+t)) \tag{8.182}$$

Consider on $S^2$ the function

$$\tilde{A}(t,u) = A(t,u)\frac{\Omega(t,u)}{\Omega(0,u)} \tag{8.183}$$

depending on the parameters $t$ and $u$, where $A(t,u)$ is defined by (8.169). We have, using (8.167) and the definition of $\nu$,

$$\frac{\partial\tilde{A}(t,u)}{\partial t} = 2\nu(t,u)\tilde{A}(t,u), \quad \tilde{A}(0,u) = 1 \tag{8.184}$$

Thus, setting, in terms of the pullbacks by $\Phi_{t,u}$ to $S^2$,

$$\tau_0(t,u) = (\tilde{A}(t,u))^{1/2}(\underline{L}\psi_0)(t,u), \quad \tilde{\rho}_0(t,u) = (\tilde{A}(t,u))^{1/2}\rho_0(t,u) \tag{8.185}$$

(8.180) becomes

$$\frac{\partial \tau_0}{\partial t} = \tilde{\rho}_0 \tag{8.186}$$

Integrating (8.184) we obtain

$$\tilde{A}(t,u) = e^{2N(t,u)}, \quad N(t,u) = \int_0^t \nu(t',u)dt' \tag{8.187}$$

Writing

$$\nu(t,u) = \frac{1}{1-u+t} + \hat{\nu} \tag{8.188}$$

we have

$$N(t,u) = \log(\frac{1-u+t}{1-u}) + \hat{N}(t,u), \quad \hat{N}(t,u) = \int_0^t \hat{\nu}(t',u)dt' \tag{8.189}$$

hence

$$\tilde{A}(t,u) = (\frac{1-u+t}{1-u})^2 e^{2\hat{N}(t,u)} \tag{8.190}$$

Now, by **E** and **F2**,

$$|\hat{\nu}(t,u)| \leqslant C\delta_0(1+t)^{-2}(1+\log(1+t)) \tag{8.191}$$

It follows that

$$|\hat{N}(s,u) - \hat{N}(t,u)| \leqslant \int_t^s |\hat{\nu}(t',u)|dt' \leqslant C\delta_0 \int_t^s (1+t')^{-2}(1+\log(1+t'))dt' \tag{8.192}$$

The last integral is

$$\begin{aligned} &\int_t^s (1+t')^{-2}(1+\log(1+t'))dt' \\ &= 2(\frac{1}{1+t} - \frac{1}{1+s}) + (\frac{\log(1+t)}{1+t} - \frac{\log(1+s)}{1+s}) \\ &\leqslant 2\frac{1+\log(1+t)}{1+t}\frac{s-t}{1+s} \end{aligned} \tag{8.193}$$

Thus, we obtain

$$|\hat{N}(s,u) - \hat{N}(t,u)| \leqslant C\delta_0 \frac{1+\log(1+t)}{1+t}\frac{s-t}{1+s} \tag{8.194}$$

This implies in particular, replacing $s, t$ by $t, 0$ respectively and noting that $N(0,u) = 0$,

$$|\hat{N}(t,u)| \leqslant C\delta_0 \tag{8.195}$$

We now return to (8.186). Integrating on $[t, s]$ we obtain

$$\tau_0(t, u) = \tau_0(s, u) - \int_t^s \tilde{\rho}_0(t', u)dt' \tag{8.196}$$

By (8.182), (8.185), (8.190) and (8.195) we have

$$|\tilde{\rho}_0(t, u)| \leqslant C\delta_0(1 + t)^{-2}(1 + \log(1 + t)) \tag{8.197}$$

Hence, in view of (8.193),

$$|\int_t^s \tilde{\rho}_0(t', u)dt'| \leqslant C\delta_0 \frac{1 + \log(1 + t)}{1 + t} \frac{s - t}{1 + s} \tag{8.198}$$

From (8.185) and (8.190) we have

$$(\underline{L}\psi_0)(t, u) = (\tilde{A}(t, u))^{-1/2}\tau_0(t, u) = (\frac{1 - u}{1 - u + t})e^{-\hat{N}(t,u)}\tau_0(t, u) \tag{8.199}$$

In particular, this holds at $t = s$. On the other hand, according to the definition of $P_s(u)$ in the statement of Lemma 8.10,

$$(\underline{L}\psi_0)(s, u) = \frac{P_s(u)}{1 + s}$$

it follows that

$$\tau_0(s, u) = \frac{P_s(u)}{1 - u} \frac{1 - u + s}{1 + s} e^{\hat{N}(s,u)} \tag{8.200}$$

Substituting (8.200) in (8.196) and the result in (8.199) we get

$$\begin{aligned}(\underline{L}\psi_0)(t, u) = &\frac{P_s(u)}{1 + s} \frac{1 - u + s}{1 - u + t} e^{\hat{N}(s,u) - \hat{N}(t,u)} \\ &- \frac{1 - u}{1 - u + t} e^{-\hat{N}(t,u)} \int_t^s \tilde{\rho}_0(t', u)dt'\end{aligned} \tag{8.201}$$

Setting

$$R_s(t, u) = P_s(u)B_s(t, u) - \frac{1 - u}{1 - u + t} e^{-\hat{N}(t,u)} \int_t^s \tilde{\rho}_0(t', u)dt' \tag{8.202}$$

where

$$B_s(t, u) = \frac{1}{1 + s} \frac{1 - u + s}{1 - u + t} e^{\hat{N}(s,u) - \hat{N}(t,u)} - \frac{1}{1 + t} \tag{8.203}$$

We obtain

$$(\underline{L}\psi_0)(t, u) = \frac{P_s(u)}{1 + t} + R_s(t, u) \tag{8.204}$$

Moreover, (8.203) can be written as

$$B_s(t, u) = \frac{1}{1 + t}(e^{\hat{N}(s,u) - \hat{N}(t,u)} - 1 - \frac{-u}{1 - u + t} \frac{s - t}{1 + s} e^{\hat{N}(s,u) - \hat{N}(t,u)}) \tag{8.205}$$

Since from (8.194),

$$\begin{aligned} e^{|\hat{N}(s,u)-\hat{N}(t,u)|} &\leqslant e^{C\delta_0}, \\ |e^{\hat{N}(s,u)-\hat{N}(t,u)} - 1| &\leqslant e^{C\delta_0}|\hat{N}(s,u) - \hat{N}(t,u)| \\ &\leqslant C\delta_0 \frac{1+\log(1+t)}{1+t}\frac{s-t}{1+s} \end{aligned}$$

it follows that

$$|B_s(t,u)| \leqslant C\frac{1+\log(1+t)}{(1+t)^2}\frac{s-t}{1+s}$$

Since by **E2**, we also have

$$|P_s(u)| \leqslant C\delta_0 \tag{8.206}$$

it follows from (8.202), in view of (8.198), that

$$|R_s(t,u)| \leqslant C\delta_0\frac{1+\log(1+t)}{(1+t)^2}\frac{s-t}{1+s} \tag{8.207}$$

and the lemma is proved. □

We now introduce the additional bootstrap assumptions:
There is a positive constant $C$ independent of $s$ such that in $W^s_{\epsilon_0}$,

$$\begin{aligned} \mathbf{E_{LT}3} &: |LT\psi_\mu| \leqslant C\delta_0(1+t)^{-2} \\ \mathbf{E_{LL}3} &: |LL\psi_\mu| \leqslant C\delta_0(1+t)^{-3} \end{aligned}$$

**Proposition 8.5** Let the assumptions of Lemma 8.10 hold. Let the additional bootstrap assumptions $\mathbf{E_{LT}3}, \mathbf{E_{LL}3}$ hold as well. Denoting

$$E_s(u,\vartheta) = (1+s)(\frac{\partial\mu}{\partial t})(s,u,\vartheta)$$

we then have, for all $t \in [0,s]$,

$$(\frac{\partial\mu}{\partial t})(t,u,\vartheta) = \frac{E_s(u,\vartheta)}{1+t} + Q_{1,s}(t,u,\vartheta)$$

and

$$|Q_{1,s}(t,u,\vartheta)| \leqslant C\delta_0\frac{1+\log(1+t)}{(1+t)^2}\frac{s-t}{1+s}$$

*Proof*. We begin with the equation for $\mu$ (see (3.92)). We have

$$m = \frac{1}{2}\frac{dH}{dh}(Th) \tag{8.208}$$

Setting

$$m_0 = \frac{1}{2}\ell(T\psi_0), \quad m_1 = m - m_0 \tag{8.209}$$

where $\ell$ is the constant

$$\ell = \frac{dH}{dh}(0) \tag{8.210}$$

we have

$$m_1 = \frac{1}{2}(\frac{dH}{dh} - \ell)(Th) + \frac{1}{2}\ell((Th) - (T\psi_0)) \tag{8.211}$$

From **E1**,

$$|h| \leqslant C\delta_0(1+t)^{-1} \tag{8.212}$$

which implies

$$|\frac{dH}{dh} - \ell| \leqslant C\delta_0(1+t)^{-1} \tag{8.213}$$

Moreover, since

$$Th = T\psi_0 - \sum_i \psi_i T\psi_i$$

**E** implies

$$|m_1| \leqslant C\delta_0(1+t)^{-2} \tag{8.214}$$

Also using $\mathbf{E_{LT}3}$ we deduce

$$|Lm_1| \leqslant C\delta_0(1+t)^{-3} \tag{8.215}$$

Now we set

$$m_{0,1} = \frac{1}{4}\ell(\underline{L}\psi_0) \tag{8.216}$$

Then, since $\underline{L} = 2T + \alpha^{-2}\mu L$, we have

$$m_{0,1} = m_0 + \frac{1}{4}\ell\alpha^{-2}\mu(L\psi_0) \tag{8.217}$$

Thus, setting also

$$e_1 = e - \frac{1}{4}\ell\alpha^{-2}(L\psi_0), \quad n_1 = m_1 + \mu e_1 \tag{8.218}$$

we get

$$L\mu = m_{0,1} + n_1 \tag{8.219}$$

From (8.214), **A** and **E**, we have

$$|e_1| \leqslant C\delta_0(1+t)^{-2}, \quad |n_1| \leqslant C\delta_0(1+t)^{-2}(1+\log(1+t)) \tag{8.220}$$

Moreover, by $\mathbf{E_{LL}3}$,

$$|Le_1| \leqslant C\delta_0(1+t)^{-3} \tag{8.221}$$

which, together with (8.215) and the fact that

$$|L\mu| \leqslant C\delta_0(1+t)^{-1} \tag{8.222}$$

yields

$$|Ln_1| \leqslant C\delta_0(1+t)^{-3}(1+\log(1+t)) \tag{8.223}$$

By virtue of Lemma 8.10 we have

$$m_{0,1}(t,u) = \frac{1}{4}\ell\frac{P_s(u)}{1+t} + \frac{1}{4}\ell R_s(t,u) \tag{8.224}$$

We define the function $E_s(u)$ on $S^2$, depending on the parameters $u$ and $s$:

$$E_s(u) = \frac{1}{4}\ell P_s(u) + (1+s)n_1(s,u) \tag{8.225}$$

Then by the second of (8.220),

$$|E_s(u) - \frac{1}{4}\ell P_s(u)| \leqslant C\delta_0(1+s)^{-1}(1+\log(1+s)) \tag{8.226}$$

From (8.219), (8.224) and the fact that $R_s(s,u) = 0$, we have

$$(1+s)\frac{\partial\mu}{\partial t}(s,u) = E_s(u) \tag{8.227}$$

Moreover we have

$$\begin{aligned} \frac{\partial\mu}{\partial t}(t,u) &= \frac{E_s(u)}{1+t} + Q_{1,s}(t,u) \\ Q_{1,s}(t,u) &= \frac{1}{4}\ell R_s(t,u) - n_1(s,u)\frac{s-t}{1+t} - (n_1(s,u) - n_1(t,u)) \end{aligned} \tag{8.228}$$

From Lemma 8.10, the first term on the right of above satisfies the required bound. From (8.220) and the fact that $f(t) = (1+t)^{-1}(1+\log(1+t))$ is decreasing in $t$, the second term on the right of above also satisfies the required bound. For the last term, we have, by (8.223),

$$|n_1(s,u) - n_1(t,u)| \leqslant \int_t^s |\frac{\partial n_1}{\partial t}(t',u)|dt' \leqslant C\delta_0 \int_t^s (1+t')^{-3}(1+\log(1+t'))dt' \tag{8.229}$$

and

$$\begin{aligned} &\int_t^s (1+t')^{-3}(1+\log(1+t'))dt' \\ &= \frac{3}{4}(\frac{1}{(1+t)^2} - \frac{1}{(1+s)^2}) + \frac{1}{2}(\frac{\log(1+t)}{(1+t)^2} - \frac{\log(1+s)}{(1+s)^2}) \\ &\leqslant \frac{3}{4}\frac{1+\log(1+t)}{(1+t)^2}\frac{s-t}{1+s} \end{aligned} \tag{8.230}$$

We conclude that indeed

$$|Q_{1,s}(t,u)| \leqslant C\delta_0\frac{1+\log(1+t)}{(1+t)^2}\frac{s-t}{1+s} \tag{8.231}$$

and the proposition is proved. □

From Proposition 8.5, we have

$$\mu(t,u) = \mu(0,u) + E_s(u)\log(1+t) + \int_0^t Q_{1,s}(t',u)dt' \tag{8.232}$$

Recall that $\mu = \eta\kappa$. Assumption **E1** restricted to $\Sigma_0^{\epsilon_0}$ implies

$$|\eta(0,u) - 1| \leqslant C\delta_0 \tag{8.233}$$

In view of the fact that

$$\kappa = 1 \quad \text{on} \quad \Sigma_0 \tag{8.234}$$

we obtain

$$|\mu(0,u) - 1| \leqslant C\delta_0 \tag{8.235}$$

According the bound for $Q_{1,s}(t,u)$ in Proposition 8.5,

$$|Q_{1,s}(t,u)| \leqslant C\delta_0(1+t)^{-2}(1+\log(1+t)) \tag{8.236}$$

Let us define on $S^2$ the function

$$\mu_{1,s}(u) = \int_0^s Q_{1,s}(t,u)dt \tag{8.237}$$

Substituting (8.236) we obtain

$$|\mu_{1,s}(u)| \leqslant C\delta_0 \tag{8.238}$$

We then define on $S^2$ the function

$$\mu_{[1],s}(u) = \mu(0,u) + \mu_{1,s}(u) \tag{8.239}$$

From (8.235) and (8.238) we have

$$|\mu_{[1],s}(u) - 1| \leqslant C\delta_0 \tag{8.240}$$

Taking $\delta_0$ suitably small, this implies

$$\inf_{u\in[0,\epsilon_0]} \inf_{S^2} \mu_{[1],s}(u) \geqslant \frac{1}{2} \tag{8.241}$$

In view of (8.237) and (8.239), (8.232) can be written as

$$\mu(t,u) = \mu_{[1],s}(u) + E_s(u)\log(1+t) + Q_{0,s}(t,u) \tag{8.242}$$

where

$$Q_{0,s}(t,u) = -\int_t^s Q_{1,s}(t',u)dt' \tag{8.243}$$

We have

$$|Q_{0,s}(t,u)| \leqslant \int_t^s |Q_{1,s}(t',u)|dt' \leqslant C\delta_0 \int_t^s (1+t')^{-2}(1+\log(1+t'))dt' \tag{8.244}$$

$$\leqslant 2C\delta_0 \frac{1+\log(1+t)}{1+t} \frac{s-t}{1+s} \tag{8.245}$$

(see (8.193)).

Finally, setting

$$\begin{aligned} \hat{\mu}_s(t,u) &= \frac{\mu(t,u)}{\mu_{[1],s}(u)}, \quad \hat{E}_s(u) = \frac{E_s(u)}{\mu_{[1],s}(u)} \\ \hat{Q}_{0,s}(t,u) &= \frac{Q_{0,s}(t,u)}{\mu_{[1],s}(u)}, \quad \hat{Q}_{1,s}(t,u) = \frac{Q_{1,s}(t,u)}{\mu_{[1],s}(u)} \end{aligned} \tag{8.246}$$

We collect the above results in the following proposition:

**Proposition 8.6** Let the assumptions of Proposition 8.5 hold. Then, there is a function $\mu_{[1],s}(u,\vartheta)$ on $S^2$ depending on the parameters $u$ and $s$, satisfying

$$|\mu_{[1],s}(u,\vartheta) - 1| \leqslant C\delta_0, \quad \inf_{u\in[0,\epsilon_0]} \inf_{\vartheta\in S^2} \mu_{[1],s}(u,\vartheta) \geqslant \frac{1}{2}$$

such that setting

$$\hat{\mu}_s(t,u,\vartheta) = \frac{\mu(t,u,\vartheta)}{\mu_{[1],s}(u,\vartheta)}$$

we have

$$\begin{aligned} \hat{\mu}_s(t,u,\vartheta) &= 1 + \hat{E}_s(u,\vartheta)\log(1+t) + \hat{Q}_{0,s}(t,u,\vartheta) \\ \frac{\partial\hat{\mu}_s}{\partial t}(t,u,\vartheta) &= \frac{\hat{E}_s(u,\vartheta)}{1+t} + \hat{Q}_{1,s}(t,u,\vartheta) \end{aligned}$$

where

$$\hat{Q}_{0,s}(t,u,\vartheta) = -\int_t^s \hat{Q}_{1,s}(t',u,\vartheta)dt'$$

and the functions $\hat{Q}_{0,s}, \hat{Q}_{1,s}$ satisfy the bounds

$$\begin{aligned} |\hat{Q}_{0,s}(t,u,\vartheta)| &\leqslant C\delta_0 \frac{1+\log(1+t)}{1+t} \frac{s-t}{1+s} \\ |\hat{Q}_{1,s}(t,u,\vartheta)| &\leqslant C\delta_0 \frac{1+\log(1+t)}{(1+t)^2} \frac{s-t}{1+s} \end{aligned}$$

Let us define

$$\mu_m(t) = \min_{\Sigma_t^{\epsilon_0}} \mu = \min_{(u,\vartheta)\in[0,\epsilon_0]\times S^2} \mu(t,u,\vartheta) \tag{8.247}$$

We set

$$\bar{\mu}_m(t) = \min\{\mu_m(t), 1\} \tag{8.248}$$

Let us also define

$$M(t) = \max_{\Sigma_t^{\epsilon_0}}(-\mu^{-1}(L\mu)_-) = \max_{(u,\vartheta)\in[0,\epsilon_0]\times S^2}(-\mu^{-1}(\frac{\partial\mu}{\partial t})_-(t,u,\vartheta)) \tag{8.249}$$

The following lemma plays a crucial role in the sequel.

**Lemma 8.11** Let the assumptions of Proposition 8.6 hold on $W^s_{\epsilon_0}$. Then for any constant $a \geqslant 4$, there is a positive constant $C$ independent of $s$ and $a$ such that for all $t \in [0, s]$ we have

$$I_a(t) := \int_0^t \bar{\mu}_m^{-a}(t')M(t')dt' \leqslant Ca^{-1}\bar{\mu}_m^{-a}(t)$$

provided $\delta_0$ is suitably small depending on $a$.

*Proof*. We can express $M(t)$ as

$$M(t) = \max_{(u,\vartheta)\in[0,\epsilon_0]\times S^2}\left(-\frac{1}{\hat{\mu}_s}\left(\frac{\partial\hat{\mu}_s}{\partial t}\right)_-(t,u,\vartheta)\right) \tag{8.250}$$

Let us denote

$$\hat{E}_{s,m} = \min_{(u,\vartheta)\in[0,\epsilon_0]\times S^2} \hat{E}_s(u,\vartheta) \tag{8.251}$$

Case 1) $\hat{E}_{s,m} \geqslant 0$. In this case,

$$\hat{E}_s(u,\vartheta) \geqslant 0, \quad \forall (u,\vartheta) \in [0,\epsilon_0] \times S^2 \tag{8.252}$$

From Proposition 8.6 we then have

$$\frac{\partial\hat{\mu}_s}{\partial t}(t,u,\vartheta) \geqslant \hat{Q}_{1,s}(t,u,\vartheta)$$

hence

$$-\left(\frac{\partial\hat{\mu}_s}{\partial t}\right)_-(t,u,\vartheta) \leqslant -(\hat{Q}_{1,s})_-(t,u,\vartheta) \leqslant C\delta_0\frac{1+\log(1+t)}{(1+t)^2} \tag{8.253}$$

Also

$$\hat{\mu}_s(t,u,\vartheta) \geqslant 1 + \hat{Q}_{0,s}(t,u,\vartheta) \geqslant 1 - C\delta_0 \tag{8.254}$$

Now (8.253) and (8.254) together imply

$$M(t) \leqslant C\delta_0\frac{1+\log(1+t)}{(1+t)^2} \tag{8.255}$$

provided $\delta_0$ is suitably small.

On the other hand, we have

$$\mu(t,u,\vartheta) = \mu_{[1],s}(u,\vartheta)\hat{\mu}_s(t,u,\vartheta)$$

From Proposition 8.6,

$$|\mu_{[1],s}(u,\vartheta) - 1| \leqslant C\delta_0 \tag{8.256}$$

The lower bound of (8.254) then implies

$$\mu_m(t) \geqslant 1 - C\delta_0 \tag{8.257}$$

hence also

$$\bar{\mu}_m(t) \geqslant 1 - C\delta_0 \tag{8.258}$$

Substituting (8.255) and (8.258) into $I_a(t)$ we obtain

$$I_a(t) \leqslant \int_0^t (1 - C\delta_0)^{-a} C\delta_0 \frac{1 + \log(1 + t')}{(1 + t')^2} dt' \leqslant C'\delta_0 \tag{8.259}$$

provided that

$$\delta_0 a \leqslant \frac{1}{C} \tag{8.260}$$

This is because

$$(1 - C\delta_0)^{-a} \leqslant (1 - a^{-1})^{-a} \to e, \quad \text{as} \quad a \to \infty$$

On the other hand, we have, in any case, by definition,

$$\bar{\mu}_m(t) \leqslant 1 \Rightarrow \bar{\mu}_m^{-a}(t) \geqslant 1$$

Therefore, the lemma holds in Case 1).

Case 2) $\hat{E}_{s,m} < 0$. Let us set in this case

$$\hat{E}_{s,m} = -\delta_1, \quad \delta_1 > 0 \tag{8.261}$$

From the definition of $E_s(u)$ and (8.222) we have

$$\delta_1 \leqslant C\delta_0 \tag{8.262}$$

In the following we denote

$$b(t, s) = \frac{1 + \log(1 + t)}{1 + t} \frac{s - t}{1 + s}, \quad c(t) = \frac{1 + \log(1 + t)}{(1 + t)^2} \tag{8.263}$$

We also denote

$$\hat{\mu}_{s,m}(t) = \min_{(u,\vartheta)\in[0,\epsilon_0]\times S^2} \hat{\mu}_s(t, u, \vartheta) \tag{8.264}$$

From Proposition 8.6 we have, in the present case,

$$\hat{\mu}_s(t, u, \vartheta) \geqslant 1 - \delta_1 \log(1 + t) - C\delta_0 b(t, s) \tag{8.265}$$

for all $(u, \vartheta) \in [0, \epsilon_0] \times S^2$, hence

$$\hat{\mu}_{s,m}(t) \geqslant 1 - \delta_1 \log(1 + t) - C\delta_0 b(t, s) \tag{8.266}$$

On the other hand, if

$$\hat{E}_s(u_m, \vartheta_m) = \hat{E}_{s,m}$$

then

$$\begin{aligned}\hat{\mu}_{s,m}(t) &\leqslant \hat{\mu}_s(t,u_m,\vartheta_m)\\ &= 1-\delta_1\log(1+t)+\hat{Q}_{0,s}(t,u_m,\vartheta_m)\end{aligned}$$

hence

$$\hat{\mu}_{s,m}(t) \leqslant 1-\delta_1\log(1+t)+C\delta_0 b(t,s) \tag{8.267}$$

Moreover, from Proposition 8.6 we have

$$-(\frac{\partial\hat{\mu}_s}{\partial t})_-(t,u,\vartheta) \leqslant -\frac{(\hat{E}_s)_-(u,\vartheta)}{1+t} - (\hat{Q}_{1,s})_-(t,u,\vartheta)$$

hence

$$-(\frac{\partial\hat{\mu}_s}{\partial t})_-(t,u,\vartheta) \leqslant \frac{\delta_1}{1+t}+C\delta_0 c(t) \tag{8.268}$$

Consider now the integral $I_a(t)$. Let us set

$$t_1 = e^{\frac{1}{2a\delta_1}}-1 \tag{8.269}$$

We have two subcases to consider.

Subcase 2a) $t \leqslant t_1$. Since $t' \leqslant t$, where $t'$ is the variable of the integration in the integral $I_a(t)$, in this subcase we have

$$1-\delta_1\log(1+t') \geqslant 1-\frac{1}{2a} \tag{8.270}$$

hence, by (8.266),

$$\hat{\mu}_{s,m}(t') \geqslant 1-\frac{1}{a} \tag{8.271}$$

provided that

$$\delta_0 a \leqslant \frac{1}{2C} \tag{8.272}$$

where we have used the fact that $b(t,s) \leqslant 1$. In view of (8.256), (8.271) implies

$$\mu_m(t') \geqslant 1-\frac{2}{a}$$

hence also

$$\bar{\mu}_m(t') \geqslant 1-\frac{2}{a} \tag{8.273}$$

Now by (8.268) and (8.271) we have

$$M(t') \leqslant (1-\frac{1}{a})^{-1}(\frac{\delta_1}{1+t'}+C\delta_0 c(t')) \tag{8.274}$$

Using (8.273) and (8.274) we obtain

$$I_a(t) \leqslant (1-\frac{1}{a})^{-1}(1-\frac{2}{a})^{-a}\int_0^t(\frac{\delta_1}{1+t'}+C\delta_0 c(t'))dt' \tag{8.275}$$

Here,

$$\int_0^t \frac{\delta_1}{1+t'} dt' = \delta_1 \log(1+t) \leqslant \delta_1 \log(1+t_1) = \frac{1}{2a} \tag{8.276}$$

while,

$$\int_0^t C\delta_0 c(t') dt' \leqslant C'\delta_0 \leqslant \frac{1}{2a} \tag{8.277}$$

provided $\delta_0$ is suitably small. In view of the fact that the coefficient in (8.275):

$$(1-\frac{1}{a})^{-1}(1-\frac{2}{a})^{-a} \to e^2, \quad \text{as} \quad a \to \infty$$

it follows that for $t \leqslant t_1$,

$$I_a(t) \leqslant \frac{C}{a} \tag{8.278}$$

Since $\bar{\mu}_m(t) \leqslant 1$, the lemma holds in subcase 2a).

Subcase 2b) $t > t_1$. We write

$$I_a(t) = I_a(t_1) + \int_{t_1}^t \bar{\mu}_m^{-a}(t') M(t') dt' \tag{8.279}$$

By (8.278) with $t$ replaced by $t_1$,

$$I_a(t) \leqslant \frac{C}{a} + \int_{t_1}^t \bar{\mu}_m^{-a}(t') M(t') dt' \tag{8.280}$$

Consider now the upper bound (8.267) evaluated at $t = s$. Since $b(s,s) = 0$, we have

$$\hat{\mu}_{s,m}(s) \leqslant 1 - \delta_1 \log(1+s) \tag{8.281}$$

On the other hand, $\mu > 0$ in $W^s_{\epsilon_0}$. It follows that

$$s \leqslant t_*, \quad \text{where} \quad t_* = e^{\frac{1}{\delta_1}} - 1 \tag{8.282}$$

Let us introduce the variable

$$\tau' = \log(1+t') \tag{8.283}$$

and correspondingly write

$$\tau = \log(1+t), \quad \sigma = \log(1+s) \tag{8.284}$$

and

$$\tau_1 = \log(1+t_1) = \frac{1}{2a\delta_1}, \quad \tau_* = \log(1+t_*) = \frac{1}{\delta_1} \tag{8.285}$$

Then in (8.280) we have

$$\tau_1 \leqslant \tau' \leqslant \tau \leqslant \sigma \leqslant \tau_* \tag{8.286}$$

Also, we express

$$b(t,s) = \frac{1+\tau}{e^\tau}\frac{e^\sigma - e^\tau}{e^\sigma} \tag{8.287}$$

Since

$$0 \leqslant \frac{e^\sigma - e^\tau}{e^\sigma} = 1 - e^{-(\sigma-\tau)} \leqslant \sigma - \tau$$

it follows that

$$0 \leqslant b(t,s) \leqslant \frac{1+\tau}{e^\tau}(\sigma - \tau) \tag{8.288}$$

Substituting (8.288) in (8.266) with $t$ replaced by $t'$ we obtain

$$\begin{aligned} \hat{\mu}_{s,m}(t') &\geqslant 1 - \delta_1\tau' - C\delta_0\frac{1+\tau'}{e^{\tau'}}(\sigma - \tau') \\ &= 1 - \delta_1\sigma + \delta_1(1 - \frac{C\delta_0}{\delta_1}\frac{1+\tau'}{e^{\tau'}})(\sigma - \tau') \end{aligned} \tag{8.289}$$

We have

$$\frac{1+\tau'}{e^{\tau'}} \leqslant \frac{1+\tau_1}{e^{\tau_1}} = (1 + \frac{1}{2a\delta_1})e^{-\frac{1}{2a\delta_1}} \tag{8.290}$$

Hence, under the smallness condition

$$C\delta_0 \leqslant \frac{1}{2a}$$

on $\delta_0$,

$$\frac{C\delta_0}{\delta_1}\frac{1+\tau'}{e^{\tau'}} \leqslant \frac{1}{2a\delta_1}(1 + \frac{1}{2a\delta_1})e^{-\frac{1}{2a\delta_1}} \tag{8.291}$$

In view of the fact that

$$\lim_{x\to 0_+}\frac{1}{x}(1 + \frac{1}{x})e^{-\frac{1}{x}} = 0$$

there is a positive constant $K(a)$, depending on $a$, such that

$$\frac{1}{x}(1 + \frac{1}{x})e^{-\frac{1}{x}} \leqslant \frac{1}{a}, \quad \forall x \leqslant \frac{1}{K(a)}$$

Here we set

$$x = 2a\delta_1 \leqslant 2aC\delta_0$$

So the smallness condition

$$\delta_0 \leqslant \frac{1}{2CaK(a)} \tag{8.292}$$

on $\delta_0$ implies

$$\frac{1}{2a\delta_1}(1 + \frac{1}{2a\delta_1})e^{-\frac{1}{2a\delta_1}} \leqslant \frac{1}{a} \tag{8.293}$$

It then follows from (8.291) that

$$1 - \frac{C\delta_0}{\delta_1}\frac{1+\tau'}{e^{\tau'}} \geqslant 1 - \frac{1}{a} \tag{8.294}$$

In reference to (8.289), taking into account the fact that

$$1 - \delta_1 \sigma \geqslant 1 - \delta_1 \tau_* = 0 \tag{8.295}$$

we obtain

$$\begin{aligned} &1 - \delta_1 \sigma + \delta_1 (1 - \frac{C\delta_0}{\delta_1} \frac{1+\tau'}{e^{\tau'}})(\sigma - \tau') \\ &\geqslant 1 - \delta_1 \sigma + (1 - \frac{1}{a})\delta_1(\sigma - \tau') \\ &\geqslant (1 - \frac{1}{a})(1 - \delta_1 \sigma) + (1 - \frac{1}{a})\delta_1(\sigma - \tau') = (1 - \frac{1}{a})(1 - \delta_1 \tau') \end{aligned} \tag{8.296}$$

From (8.289) we then conclude that

$$\hat{\mu}_s(t', u, \vartheta) \geqslant \hat{\mu}_{s,m}(t') \geqslant (1 - \frac{1}{a})(1 - \delta_1 \tau'), \quad \forall (u, \vartheta) \in [0, \epsilon_0] \times S^2, \forall t' \geqslant t_1 \tag{8.297}$$

We now consider the bound (8.268). Since

$$(1 + t')c(t') = \frac{1 + \tau'}{e^{\tau'}} \tag{8.298}$$

the bound (8.268) with $t$ replaced by $t'$ reads

$$-(1 + t')(\frac{\partial \hat{\mu}_s}{\partial t})_-(t', u, \vartheta) \leqslant \delta_1 (1 + \frac{C\delta_0}{\delta_1} \frac{1 + \tau'}{e^{\tau'}}) \tag{8.299}$$

If (8.292) holds, this implies

$$-(1 + t')(\frac{\partial \hat{\mu}_s}{\partial t})_-(t', u, \vartheta) \leqslant \delta_1 (1 + \frac{1}{a}), \quad \forall (u, \vartheta) \in [0, \epsilon_0] \times S^2, \forall t' \geqslant t_1 \tag{8.300}$$

The bounds (8.297) and (8.300) together yield

$$(1 + t')M(t') \leqslant \frac{1 + \frac{1}{a}}{1 - \frac{1}{a}} \frac{\delta_1}{1 - \delta_1 \tau'} \tag{8.301}$$

By (8.256), taking $C\delta_0 \leqslant a^{-1}$, and (8.297),

$$\mu_m(t') \geqslant (1 - \frac{2}{a})(1 - \delta_1 \tau') \tag{8.302}$$

hence also

$$\bar{\mu}_m(t') \geqslant (1 - \frac{2}{a})(1 - \delta_1 \tau') \tag{8.303}$$

We now consider the integral in (8.280). In terms of the variable $\tau'$ we have

$$\int_{t_1}^{t} \bar{\mu}_m^{-a}(t') M(t') dt' = \int_{\tau_1}^{\tau} \bar{\mu}_m^{-a}(t')(1 + t')M(t') d\tau'$$

thus (8.301) and (8.303) yields

$$\int_{t_1}^{t} \bar{\mu}_m^{-a}(t')M(t')dt' \leqslant \frac{1+\frac{1}{a}}{1-\frac{1}{a}} \frac{1}{(1-\frac{2}{a})^a} \int_{\tau_1}^{\tau} (1-\delta_1\tau')^{-a-1}\delta_1 d\tau' \tag{8.304}$$

Since

$$\int_{\tau_1}^{\tau} (1-\delta_1\tau')^{-a-1}\delta_1 d\tau' = \frac{1}{a}((1-\delta_1\tau)^{-a} - (1-\delta_1\tau_1)^{-a}) \leqslant \frac{1}{a}(1-\delta_1\tau)^{-a}$$

while the coefficient

$$\frac{1+\frac{1}{a}}{1-\frac{1}{a}} \frac{1}{(1-\frac{2}{a})^a} \to e^2, \quad \text{as} \quad a \to \infty$$

we conclude that

$$\int_{t_1}^{t} \bar{\mu}_m^{-a}(t')M(t')dt' \leqslant \frac{C}{a}(1-\delta_1\tau)^{-a} \tag{8.305}$$

On the other hand, by (8.267) and (8.288),

$$\begin{aligned} \hat{\mu}_{s,m}(t) &\leqslant 1 - \delta_1\tau + C\delta_0 \frac{1+\tau}{e^\tau}(\sigma - \tau) \\ &= 1 - \delta_1\sigma + \delta_1(1 + \frac{C\delta_0}{\delta_1}\frac{1+\tau}{e^\tau})(\sigma - \tau) \end{aligned} \tag{8.306}$$

Now by (8.291) with $\tau'$ replaced by $\tau$ and (8.293),

$$\frac{C\delta_0}{\delta_1}\frac{1+\tau}{e^\tau} \leqslant \frac{1}{a} \tag{8.307}$$

provided (8.292) holds. Hence,

$$1 + \frac{C\delta_0}{\delta_1}\frac{1+\tau}{e^\tau} \leqslant 1 + \frac{1}{a} \tag{8.308}$$

Then,

$$\begin{aligned} &1 - \delta_1\sigma + \delta_1(1 + \frac{C\delta_0}{\delta_1}\frac{1+\tau}{e^\tau})(\sigma - \tau) \\ \leqslant &1 - \delta_1\sigma + (1 + \frac{1}{a})\delta_1(\sigma - \tau) \\ \leqslant &(1 + \frac{1}{a})(1 - \delta_1\sigma) + (1 + \frac{1}{a})\delta_1(\sigma - \tau) \\ = &(1 + \frac{1}{a})(1 - \delta_1\tau) \end{aligned} \tag{8.309}$$

Thus we obtain

$$\hat{\mu}_{s,m}(t) \leqslant (1 + \frac{1}{a})(1 - \delta_1\tau) \tag{8.310}$$

which together with (8.256) and a smallness condition on $\delta_0$ of the form (8.260) yields

$$\mu_m(t) \leqslant (1+\frac{2}{a})(1-\delta_1\tau) \tag{8.311}$$

hence also

$$\bar{\mu}_m(t) \leqslant (1+\frac{2}{a})(1-\delta_1\tau) \tag{8.312}$$

Therefore

$$\bar{\mu}_m^{-a}(t) \geqslant \frac{1}{(1+\frac{2}{a})^a}(1-\delta_1\tau)^{-a} \tag{8.313}$$

Since

$$(1+\frac{2}{a})^a \to e^2, \quad \text{as} \quad a \to \infty$$

we conclude that

$$\bar{\mu}_m^{-a}(t) \geqslant \frac{1}{C}(1-\delta_1\tau)^{-a} \tag{8.314}$$

Comparing with (8.305), then yields

$$\int_{t_1}^{t} \bar{\mu}_m^{-a}(t')M(t')dt' \leqslant \frac{C}{a}\bar{\mu}_m^{-a}(t) \tag{8.315}$$

In view of above, (8.280) and the fact that $\bar{\mu}_m(t) \leqslant 1$, the lemma follows as well in subcase 2b).

The proof of the lemma is now complete. □

**Corollary 1** Under the assumptions of Lemma 8.11, there is a positive constant $C$ independent of $s$ such that the following holds. If at some $(u,\vartheta) \in [0,\epsilon_0] \times S^2$ we have $\hat{E}_s(u,\vartheta) < 0$, then we have

$$\frac{\hat{\mu}_s(t,u,\vartheta)}{\hat{\mu}_s(t',u,\vartheta)} \leqslant C$$

for all $t' \in [0,t]$ and all $t \in [0,s]$.

*Proof*. Suppose at $(u,\vartheta)$, we have

$$\hat{E}_s(u,\vartheta) = -\delta_1, \delta_1 > 0,$$

From (8.266) and (8.267),

$$1-\delta_1\log(1+t) - C\delta_0 b(t,s) \leqslant \hat{\mu}_s(t,u,\vartheta) \leqslant 1-\delta_1\log(1+t) + C\delta_0 b(t,s) \tag{8.316}$$

Fixing $a=4$, and setting, in accordance with (8.269),

$$t_1 = e^{\frac{1}{8\delta_1}} - 1 \tag{8.317}$$

we have to consider the two subcases according as to whether $t \leqslant t_1$ or $t > t_1$. If $t \leqslant t_1$, from (8.271),

$$\hat{\mu}_s(t',u,\vartheta) \geqslant \frac{1}{2} \tag{8.318}$$

while by the upper bound in (8.316),

$$\hat{\mu}_s(t,u,\vartheta) \leqslant \frac{5}{4} \tag{8.319}$$

So the corollary holds when $t \leqslant t_1$.

If $t > t_1$, from (8.310) we have

$$\hat{\mu}_s(t,u,\vartheta) \leqslant \frac{5}{4}(1-\delta_1\tau) \tag{8.320}$$

If then $t' \leqslant t_1$, the corollary holds by (8.318). While if $t' > t_1$, from (8.297) we have

$$\hat{\mu}_s(t',u,\vartheta) \geqslant \frac{3}{4}(1-\delta_1\tau') \tag{8.321}$$

Since $(1-\delta_1\tau) \leqslant (1-\delta_1\tau')$, the corollary follows. □

**Corollary 2** Under the assumptions of Lemma 8.11, there is a positive constant $C$ independent of $s$ and $a$ such that

$$\bar{\mu}_m^{-a}(t') \leqslant C\bar{\mu}_m^{-a}(t)$$

holds for all $t' \in [0,t]$ and all $t \in [0,s]$.

*Proof.* Case 1) $\hat{E}_{s,m} \geqslant 0$. In this case, from (8.258),

$$\bar{\mu}_m(t') \geqslant 1 - C\delta_0 \tag{8.322}$$

On the other hand, by definition,

$$\bar{\mu}_m(t) \leqslant 1$$

It follows that

$$\frac{\bar{\mu}_m^{-a}(t')}{\bar{\mu}_m^{-a}(t)} \leqslant (1-C\delta_0)^{-a} \tag{8.323}$$

So the corollary holds provided that

$$C\delta_0 \leqslant \frac{1}{a}$$

Case 2) $\hat{E}_{s,m} < 0$. In this case we define $\delta_1$ as in (8.261) and $t_1$ as in (8.269) and we consider the subcases 2a) $t' \leqslant t_1$ and 2b) $t' > t_1$ separately.

In Subcase 2a), (8.273) holds, thus we have

$$\frac{\bar{\mu}_m^{-a}(t')}{\bar{\mu}_m^{-a}(t)} \leqslant (1-\frac{2}{a})^{-a} \tag{8.324}$$

which is bounded by a constant independent of $a$ (for $a \geqslant 4$).

In Subcase 2b), (8.303) and (8.312) hold, hence

$$\frac{\bar{\mu}_m^{-a}(t')}{\bar{\mu}_m^{-a}(t)} \leqslant (\frac{1-\frac{2}{a}}{1+\frac{2}{a}})^{-a}(\frac{1-\delta_1\tau}{1-\delta_1\tau'})^{a} \leqslant (\frac{1-\frac{2}{a}}{1+\frac{2}{a}})^{-a} \tag{8.325}$$

(for $\tau' \leqslant \tau$) which is bounded by a constant independent of $a$ (for $a \geqslant 4$), as required. □

**Corollary 3** The assumptions of Lemma 8.11 imply the assumption **AS**.

*Proof*. We consider a given $(u,\vartheta) \in [0,\epsilon_0] \times S^2$ as in the proof of Corollary 1. Again, we have the two cases $\hat{E}_s(u,\vartheta) \geqslant 0$ and $\hat{E}_s(u,\vartheta) < 0$ to consider.

In the first case, recalling (8.254), we have

$$\hat{\mu}_s(t,u,\vartheta) \geqslant 1 - C\delta_0$$

On the other hand, from Proposition 8.6 we have

$$\frac{\partial \hat{\mu}_s}{\partial t}(t,u,\vartheta) = \frac{\hat{E}_s(u,\vartheta)}{1+t} + \hat{Q}_{1,s}(t,u,\vartheta) \leqslant C\delta_0(1+t)^{-1}$$

where we have used (8.206), (8.226) and (8.231). Therefore

$$-2\mu^{-1}\frac{\partial \mu}{\partial t} = -2\hat{\mu}_s^{-1}\frac{\partial \hat{\mu}_s}{\partial t} \geqslant -C\delta_0(1+t)^{-1}$$

On the other hand, **F2** implies

$$\mathrm{tr}\chi \geqslant C^{-1}(1+t)^{-1}$$

So **AS** follows in the first case if $\delta_0$ is suitably small.

In the second case we set as in the proof of Corollary 1, $\hat{E}_s(u,\vartheta) = -\delta_1, \delta_1 > 0, a = 4$ and define $t_1$ as (8.317). Then we have to consider the subcases $t \leqslant t_1$ and $t > t_1$.

In the first subcase we have (see (8.318))

$$\hat{\mu}_s(t,u,\vartheta) \geqslant \frac{3}{4}$$

from Proposition 8.6

$$\frac{\partial \hat{\mu}_s}{\partial t}(t,u,\vartheta) \leqslant \hat{Q}_{1,s}(t,u,\vartheta) \leqslant C\delta_0(1+t)^{-2}(1+\log(1+t))$$

Therefore

$$-2\mu^{-1}\frac{\partial \mu}{\partial t} = -2\hat{\mu}_s^{-1}\frac{\partial \hat{\mu}_s}{\partial t} \geqslant -C\delta_0(1+t)^{-2}(1+\log(1+t))$$

So **AS** follows in the first subcase if $\delta_0$ is suitably small.

In the second subcase we have from Proposition 8.6,

$$\begin{aligned}(1+t)\frac{\partial \hat{\mu}_s}{\partial t}(t,u,\vartheta) &= -\delta_1 + (1+t)\hat{Q}_{1,s}(t,u,\vartheta) \\ &\leqslant -\delta_1\Big(1 - \frac{C\delta_0}{\delta_1}\frac{1+\tau}{e^{\tau}}\Big) \leqslant -\frac{3\delta_1}{4}\end{aligned}$$

if $\delta_0$ is suitably small, according to (8.292) with $a = 4$. It follows that in the second subcase

$$-2\mu^{-1}\frac{\partial \mu}{\partial t} > 0$$

So **AS** again holds. □

## 8.3 The Actual Estimates for the Solutions of the Propagation Equations

We now return to (8.175). We would like to obtain an estimate for the $L^2$ norm of ${}^{(i_1\cdots i_l)}F_l(t)$ on $[0,\epsilon_0]\times S^2$. In general, if $\phi$ is a function defined on $W^*_{\epsilon_0}$ and we consider $\phi(t,u)$, the corresponding function on $S^2$ depending on $t$ and $u$, then the $L^2$ norm of $\phi$ on $\Sigma_t^{\epsilon_0}$ is

$$\begin{aligned}\|\phi\|_{L^2(\Sigma_t^{\epsilon_0})} &= \sqrt{\int_{\Sigma_t^{\epsilon_0}} \phi^2 d\mu_{\not{g}} du} \\ &= \sqrt{\int_{[0,\epsilon_0]\times S^2} (\phi(t,u))^2 d\mu_{\not{g}(t,u)} du}\end{aligned} \tag{8.326}$$

While the $L^2$ norm of $\phi(t,.)$ on $[0,\epsilon_0]\times S^2$ is

$$\|\phi(t)\|_{L^2([0,\epsilon_0]\times S^2)} = \sqrt{\int_{[0,\epsilon_0]\times S^2} (\phi(t,u))^2 d\mu_{\not{g}(0,0)} du} \tag{8.327}$$

$\not{g}(0,0)$ being the standard metric on $S^2$. We have

$$d\mu_{\not{g}(t,u)} = \frac{\sqrt{\det \not{g}(t,u)}}{\sqrt{\det \not{g}(0,0)}} d\mu_{\not{g}(0,0)} = A(t,u)\frac{\sqrt{\det \not{g}(0,u)}}{\sqrt{\det \not{g}(0,0)}} d\mu_{\not{g}(0,0)} \tag{8.328}$$

Now, by the definition of $\theta$, we have,

$$\kappa \text{tr}\theta = \frac{1}{\sqrt{\det \not{g}}}\frac{\partial}{\partial u}\sqrt{\det \not{g}} \tag{8.329}$$

on $\Sigma_0^{\epsilon_0}$, because $\Xi$ vanishes there. Using **F2**, (6.85) and the relation

$$\chi = \eta(\not{k} - \theta)$$

we have

$$|\kappa \text{tr}\theta + \frac{2}{1-u}| \leqslant C\delta_0 \tag{8.330}$$

on $\Sigma_0^{\epsilon_0}$.

Integrating (8.330) we get

$$e^{-C\delta_0}(1-u)^2 \leqslant \frac{\sqrt{\det \not{g}(0,u)}}{\sqrt{\det \not{g}(0,0)}} \leqslant e^{C\delta_0}(1-u)^2 \tag{8.331}$$

Combining with (8.173), we obtain from (8.328)

$$e^{-C\delta_0}(1-u+t)^2 d\mu_{\not{g}(0,0)} \leqslant d\mu_{\not{g}(t,u)} \leqslant e^{C\delta_0}(1-u+t)^2 d\mu_{\not{g}(0,0)} \tag{8.332}$$

In view of (8.326) and (8.327) it follows that

$$C^{-1}(1+t)\|\phi(t)\|_{L^2([0,\epsilon_0]\times S^2)} \leqslant \|\phi\|_{L^2(\Sigma_t^{\epsilon_0})} \leqslant C(1+t)\|\phi(t)\|_{L^2([0,\epsilon_0]\times S^2)} \quad (8.333)$$

Moreover, if $\xi$ is an $S_{t,u}$ 1-form defined on $W^*_{\epsilon_0}$ and we consider $\xi(t,u)$, the corresponding 1-form on $S^2$ depending on $t$ and $u$, then setting $\phi = |\xi|$, $\phi(t,u) = |\xi(t,u)|$, we have

$$\|\phi\|_{L^2(\Sigma_t^{\epsilon_0})} = \|\xi\|_{L^2(\Sigma_t^{\epsilon_0})}, \|\phi(t)\|_{L^2([0,\epsilon_0]\times S^2)} = \||\xi(t)|\|_{L^2([0,\epsilon_0]\times S^2)} \quad (8.334)$$

hence, by (8.333),

$$C^{-1}(1+t)\||\xi(t)|\|_{L^2([0,\epsilon_0]\times S^2)} \leqslant \|\xi\|_{L^2(\Sigma_t^{\epsilon_0})} \leqslant C(1+t)\||\xi(t)|\|_{L^2([0,\epsilon_0]\times S^2)} \quad (8.335)$$

We now consider the terms ${}^{(i_1\cdots i_l)}M_l^0(t,u)$, ${}^{(i_1\cdots i_l)}M_l^1(t,u)$, ${}^{(i_1\cdots i_l)}M_l^2(t,u)$ on the right hand side of (8.175), given by (8.176) (8.177) and (8.178) respectively.

First, by **A3**, (8.335) and (8.235),

$$\|{}^{(i_1\cdots i_l)}M_l^0(t,u)\|_{L^2([0,\epsilon_0]\times S^2)} \leqslant C(1+\log(1+t))^2\|{}^{(i_1\cdots i_l)}x_l(0)\|_{L^2(\Sigma_0^{\epsilon_0})} \quad (8.336)$$

We turn to ${}^{(i_1\cdots i_l)}M_l^1(t,u)$. We partition $[0,\epsilon_0]\times S^2$ into

$$\mathcal{V}_{s-} = \{(u,\vartheta)\in[0,\epsilon_0]\times S^2 : \hat{E}_s(u,\vartheta) < 0\} \quad (8.337)$$

and

$$\mathcal{V}_{s+} = \{(u,\vartheta)\in[0,\epsilon_0]\times S^2 : \hat{E}_s(u,\vartheta) \geqslant 0\} \quad (8.338)$$

We then have

$$\begin{aligned}&\|{}^{(i_1\cdots i_l)}M_l^1(t)\|^2_{L^2([0,\epsilon_0]\times S^2)}\\ &= \|{}^{(i_1\cdots i_l)}M_l^1(t)\|^2_{L^2(\mathcal{V}_{s-})} + \|{}^{(i_1\cdots i_l)}M_l^1(t)\|^2_{L^2(\mathcal{V}_{s+})}\end{aligned} \quad (8.339)$$

By Minkowski's inequality, we have

$$\|{}^{(i_1\cdots i_l)}M_l^1(t)\|^2_{L^2(\mathcal{V}_{s-})} \leqslant \int_0^t \|{}^{(i_1\cdots i_l)}N_l^1(t,t')\|_{L^2(\mathcal{V}_{s-})}dt' \quad (8.340)$$

where

$$\begin{aligned}&{}^{(i_1\cdots i_l)}N_l^1(t,t',u)\\ &= \Big(\frac{\mu(t,u)}{\mu(t',u)}\Big)^2(1-u+t')^3\Big(-2\mu^{-1}\Big(\frac{\partial\mu}{\partial t}\Big)_-(t',u)\Big)|\not{d}^{(i_1\cdots i_l)}\check{f}_l(t',u)|\end{aligned} \quad (8.341)$$

We have

$$\begin{aligned}&\|{}^{(i_1\cdots i_l)}N_l^1(t,t')\|_{L^2(\mathcal{V}_{s-})}\\ &\leqslant(1+t')^3\Big(\max_{\mathcal{V}_{s-}}\frac{\mu(t)}{\mu(t')}\Big)^2\max_{\mathcal{V}_{s-}}\Big(-2\mu^{-1}\Big(\frac{\partial\mu}{\partial t}\Big)_-(t')\Big)\||\not{d}^{(i_1\cdots i_l)}\check{f}_l(t')|\|_{L^2(\mathcal{V}_{s-})}\end{aligned} \quad (8.342)$$

Now, by Corollary 1 to Lemma 8.11 we have

$$\max_{\mathcal{V}_{s-}} \frac{\mu(t)}{\mu(t')} \leqslant C \tag{8.343}$$

while by (8.249),

$$\max_{\mathcal{V}_{s-}} (-2\mu^{-1}(\frac{\partial \mu}{\partial t})_-(t')) \leqslant 2M(t') \tag{8.344}$$

Substituting in (8.342),

$$\|{}^{(i_1\cdots i_l)}N_l^1(t,t')\|_{L^2(\mathcal{V}_{s-})} \leqslant C(1+t')^3 M(t')\|\not{d}^{(i_1\cdots i_l)}\check{f}_l(t')\|_{L^2([0,\epsilon_0]\times S^2)} \tag{8.345}$$

We define

$${}^{(i_1\cdots i_l)}P_l(t) = (1+t)^2\|\not{d}^{(i_1\cdots i_l)}\check{f}_l(t')\|_{L^2([0,\epsilon_0]\times S^2)} \tag{8.346}$$

Suppose that, for non-negative quantities ${}^{(i_1\cdots i_l)}P_l^{(0)}$, ${}^{(i_1\cdots i_l)}P_l^{(1)}$, we have

$${}^{(i_1\cdots i_l)}P_l(t) \leqslant {}^{(i_1\cdots i_l)}P_l^{(0)}(t) + {}^{(i_1\cdots i_l)}P_l^{(1)}(t) \tag{8.347}$$

Defining then the non-decreasing non-negative quantities ${}^{(i_1\cdots i_l)}\bar{P}_{l,a}^{(0)}$, ${}^{(i_1\cdots i_l)}\bar{P}_{l,a}^{(1)}$, by

$${}^{(i_1\cdots i_l)}\bar{P}_{l,a}^{(0)} = \sup_{t'\in[0,t]} (\bar{\mu}_m^a(t')^{(i_1\cdots i_l)}P_l^{(0)}(t')) \tag{8.348}$$

$${}^{(i_1\cdots i_l)}\bar{P}_{l,a}^{(1)} = \sup_{t'\in[0,t]} ((1+t')^{1/2}\bar{\mu}_m^a(t')^{(i_1\cdots i_l)}P_l^{(1)}(t')) \tag{8.349}$$

Then for $t' \in [0,t]$ we have

$${}^{(i_1\cdots i_l)}P_l(t') \leqslant \bar{\mu}_m^{-a}(t')({}^{(i_1\cdots i_l)}\bar{P}_{l,a}^{(0)}(t) + (1+t')^{-1/2\,(i_1\cdots i_l)}\bar{P}_{l,a}^{(1)}(t)) \tag{8.350}$$

Substituting in (8.345) we obtain

$$\begin{aligned}&\|{}^{(i_1\cdots i_l)}N_l^1(t,t')\|_{L^2(\mathcal{V}_{s-})}\\ &\leqslant C((1+t)^{(i_1\cdots i_l)}\bar{P}_{l,a}^{(0)}(t) + (1+t)^{1/2\,(i_1\cdots i_l)}\bar{P}_{l,a}^{(1)}(t))\bar{\mu}_m^{-a}(t')M(t')\end{aligned} \tag{8.351}$$

for all $t' \in [0,t]$. Substituting this in (8.340) we obtain

$$\begin{aligned}&\|{}^{(i_1\cdots i_l)}M_l^1(t)\|_{L^2(\mathcal{V}_{s-})}\\ &\leqslant C((1+t)^{(i_1\cdots i_l)}\bar{P}_{l,a}^{(0)}(t) + (1+t)^{1/2\,(i_1\cdots i_l)}\bar{P}_{l,a}^{(1)}(t))I_a(t)\end{aligned} \tag{8.352}$$

We then apply Lemma 8.11 to conclude that

$$\begin{aligned}&\|{}^{(i_1\cdots i_l)}M_l^1(t)\|_{L^2(\mathcal{V}_{s-})}\\ &\leqslant Ca^{-1}((1+t)^{(i_1\cdots i_l)}\bar{P}_{l,a}^{(0)}(t) + (1+t)^{1/2\,(i_1\cdots i_l)}\bar{P}_{l,a}^{(1)}(t))\bar{\mu}_m^{-a}(t)\end{aligned} \tag{8.353}$$

In analogy with (8.340) and (8.342), with $\mathcal{V}_{s+}$ in the role of $\mathcal{V}_{s-}$, we have

$$\|^{(i_1\cdots i_l)}M_l^1(t)\|^2_{L^2(\mathcal{V}_{s+})} \leqslant \int_0^t \|^{(i_1\cdots i_l)}N_l^1(t,t')\|_{L^2(\mathcal{V}_{s+})}dt' \tag{8.354}$$

and

$$\begin{aligned}&\|^{(i_1\cdots i_l)}N_l^1(t,t')\|_{L^2(\mathcal{V}_{s+})}\\ &\leqslant (1+t')^3(\max_{\mathcal{V}_{s+}}\frac{\mu(t)}{\mu(t')})^2\max_{\mathcal{V}_{s+}}(-2\mu^{-1}(\frac{\partial\mu}{\partial t})_-(t'))\|\not{d}^{(i_1\cdots i_l)}\check{f}_l(t')\|_{L^2(\mathcal{V}_{s+})}\end{aligned} \tag{8.355}$$

From (8.255) we have

$$\max_{\mathcal{V}_{s+}}(-2\mu^{-1}(\frac{\partial\mu}{\partial t})_-(t')) \leqslant C\delta_0(1+t')^{-2}(1+\log(1+t')) \tag{8.356}$$

and from (8.254) we have

$$\max_{\mathcal{V}_{s+}}\frac{\mu(t)}{\mu(t')} \leqslant C(1+\log(1+t)) \tag{8.357}$$

Substituting (8.356) and (8.357) as well as (8.346) in (8.354) and (8.355) we obtain

$$\begin{aligned}&\|^{(i_1\cdots i_l)}M_l^1(t)\|_{L^2(\mathcal{V}_{s+})}\\ &\leqslant C\delta_0(1+\log(1+t))^2\int_0^t\frac{1+\log(1+t')}{1+t'}{}^{(i_1\cdots i_l)}P_l(t')dt'\end{aligned} \tag{8.358}$$

hence, by (8.350),

$$\begin{aligned}&\|^{(i_1\cdots i_l)}M_l^1(t)\|_{L^2(\mathcal{V}_{s+})}\\ &\leqslant C\delta_0(1+\log(1+t))^2({}^{(i_1\cdots i_l)}\bar{P}^{(0)}_{l,a}(t)+{}^{(i_1\cdots i_l)}\bar{P}^{(1)}_{l,a}(t))\\ &\quad\cdot\int_0^t(1+t')^{-1}(1+\log(1+t'))\bar{\mu}_m^{-a}(t')dt'\end{aligned} \tag{8.359}$$

To estimate the integral in (8.359), we follow the proof of Corollary 2 to Lemma 8.11. We have two cases $\hat{E}_{s,m} \geqslant 0$ and $\hat{E}_{s,m} < 0$ to consider. In the first case (8.322) holds, hence if

$$\delta_0 a \leqslant \frac{1}{C}$$

we have

$$\bar{\mu}_m(t')^{-a} \leqslant (1-\frac{1}{a})^{-a} \leqslant C \tag{8.360}$$

and then

$$\int_0^t\frac{1+\log(1+t')}{1+t'}\bar{\mu}_m^{-a}(t')dt' \leqslant C(1+\log(1+t))^2 \tag{8.361}$$

In the second case we define as before, $\delta_1 > 0$ by $\hat{E}_{s,m} = -\delta_1$, and $t_1$ according to (8.269). Then for $t' \leqslant t_1$ (8.273) holds, so an estimate of the form (8.361) holds for all $t \leqslant t_1$.

If $t' > t_1$, then (8.303) holds, under the smallness condition (8.292) on $\delta_0$. Thus we have

$$\begin{aligned}&\int_{t_1}^{t} \frac{1+\log(1+t')}{1+t'}\bar{\mu}_m^{-a}(t')dt' \\ \leqslant & C\int_{\tau_1}^{\tau}(1+\tau')(1-\delta_1\tau')^{-a}d\tau' \\ \leqslant & C(1+\tau)\int_{\tau_1}^{\tau}(1-\delta_1\tau')^{-a}d\tau' \leqslant C(1+\tau)\cdot\frac{(1-\delta_1\tau)^{1-a}}{\delta_1(a-1)} \\ = & 2C(1-\frac{1}{a})^{-1}\tau_1(1+\tau)(1-\delta_1\tau)^{1-a} \end{aligned} \tag{8.362}$$

The last equality comes from the first of (8.285). By the upper bound (8.312) we conclude that the last is bounded by a constant multiple of

$$(1+\log(1+t))^2\bar{\mu}_m^{1-a}(t)$$

We conclude that in general,

$$\int_0^t \frac{1+\log(1+t')}{1+t'}\bar{\mu}_m^{-a}(t')dt' \leqslant C(1+\log(1+t))^2\bar{\mu}_m^{1-a}(t) \tag{8.363}$$

which, substituted in (8.359), yields

$$\begin{aligned}&\|{}^{(i_1\cdots i_l)}M_l^1(t)\|_{L^2(\mathcal{V}_{s+})} \\ \leqslant & C\delta_0(1+\log(1+t))^4({}^{(i_1\cdots i_l)}\bar{P}_{l,a}^{(0)}(t) + {}^{(i_1\cdots i_l)}\bar{P}_{l,a}^{(1)}(t))\bar{\mu}_m^{1-a}(t)\end{aligned} \tag{8.364}$$

Combining (8.364) and (8.353), and taking account into the fact that $C\delta_0 \leqslant \frac{1}{a}$, we obtain

$$\begin{aligned}&\|{}^{(i_1\cdots i_l)}M_l^1(t)\|_{L^2([0,\epsilon_0]\times S^2)} \\ \leqslant & Ca^{-1}((1+t){}^{(i_1\cdots i_l)}\bar{P}_{l,a}^{(0)}(t) + (1+t)^{1/2}{}^{(i_1\cdots i_l)}\bar{P}_{l,a}^{(1)}(t))\bar{\mu}_m^{-a}(t)\end{aligned} \tag{8.365}$$

We turn to ${}^{(i_1\cdots i_l)}M_l^2(t,u)$. We have

$$\begin{aligned}&\|{}^{(i_1\cdots i_l)}M_l^2(t)\|^2_{L^2([0,\epsilon_0]\times S^2)} \\ = & \|{}^{(i_1\cdots i_l)}M_l^2(t)\|^2_{L^2(\mathcal{V}_{s-})} + \|{}^{(i_1\cdots i_l)}M_l^2(t)\|^2_{L^2(\mathcal{V}_{s+})}\end{aligned} \tag{8.366}$$

and also

$$\|{}^{(i_1\cdots i_l)}M_l^2(t)\|_{L^2(\mathcal{V}_{s-})} \leqslant \int_0^t \|{}^{(i_1\cdots i_l)}N_l^2(t,t')\|_{L^2(\mathcal{V}_{s-})}dt' \tag{8.367}$$

$$\|{}^{(i_1\cdots i_l)}M_l^2(t)\|_{L^2(\mathcal{V}_{s+})} \leqslant \int_0^t \|{}^{(i_1\cdots i_l)}N_l^2(t,t')\|_{L^2(\mathcal{V}_{s+})}dt' \tag{8.368}$$

where

$$^{(i_1\cdots i_l)}N_l^2(t,t',u)=(\frac{\mu(t,u)}{\mu(t',u)})^2(1-u+t')^3(\frac{1}{2}\mathrm{tr}\chi(t',u))|\not{d}^{(i_1\cdots i_l)}\check{f}_l(t',u)| \quad (8.369)$$

We have

$$\begin{aligned}&\|^{(i_1\cdots i_l)}N_l^2(t,t')\|_{L^2(\mathcal{V}_{s-})}\\&\leqslant(1+t')^3(\max_{\mathcal{V}_{s-}}\frac{\mu(t)}{\mu(t')})^2\max_{\mathcal{V}_{s-}}(\frac{1}{2}\mathrm{tr}\chi(t'))\||\not{d}^{(i_1\cdots i_l)}\check{f}_l(t')|\|_{L^2(\mathcal{V}_{s-})}\end{aligned} \quad (8.370)$$

Now, by **F2**,

$$\max_{[0,\epsilon_0]\times S^2}(\frac{1}{2}\mathrm{tr}\chi(t'))\leqslant C(1+t')^{-1} \quad (8.371)$$

This together with (8.343) imply

$$\|^{(i_1\cdots i_l)}N_l^2(t,t')\|_{L^2(\mathcal{V}_{s-})}\leqslant C(1+t')^2\||\not{d}^{(i_1\cdots i_l)}\check{f}_l(t')|\|_{L^2([0,\epsilon_0]\times S^2)} \quad (8.372)$$

Then from (8.346), (8.350) and Corollary 2 of Lemma 8.11, we have

$$\begin{aligned}&\|^{(i_1\cdots i_l)}N_l^2(t,t')\|_{L^2(\mathcal{V}_{s-})}\leqslant C^{(i_1\cdots i_l)}P_l(t')\\&\leqslant C(^{(i_1\cdots i_l)}\bar{P}_{l,a}^{(0)}(t)+(1+t')^{-1/2(i_1\cdots i_l)}\bar{P}_{l,a}^{(1)}(t))\bar{\mu}_m^{-a}(t')\\&\leqslant C'(^{(i_1\cdots i_l)}\bar{P}_{l,a}^{(0)}(t)+(1+t')^{-1/2(i_1\cdots i_l)}\bar{P}_{l,a}^{(1)}(t))\bar{\mu}_m^{-a}(t)\end{aligned} \quad (8.373)$$

Substituting (8.373) in (8.367) we get

$$\begin{aligned}&\|^{(i_1\cdots i_l)}M_l^2(t)\|_{L^2(\mathcal{V}_{s-})}\\&\leqslant C((1+t)^{(i_1\cdots i_l)}\bar{P}_{l,a}^{(0)}(t)+(1+t)^{1/2(i_1\cdots i_l)}\bar{P}_{l,a}^{(1)}(t))\bar{\mu}_m^{-a}(t)\end{aligned} \quad (8.374)$$

In analogy with (8.370) we have

$$\begin{aligned}&\|^{(i_1\cdots i_l)}N_l^2(t,t')\|_{L^2(\mathcal{V}_{s+})}\\&\leqslant(1+t')^3(\max_{\mathcal{V}_{s+}}\frac{\mu(t)}{\mu(t')})^2\max_{\mathcal{V}_{s+}}(\frac{1}{2}\mathrm{tr}\chi(t'))\||\not{d}^{(i_1\cdots i_l)}\check{f}_l(t')|\|_{L^2(\mathcal{V}_{s+})}\end{aligned} \quad (8.375)$$

To estimate

$$\max_{\mathcal{V}_{s+}}\frac{\mu(t)}{\mu(t')}=\max_{(u,\vartheta)\in\mathcal{V}_{s+}}\frac{\hat{\mu}_s(t,u,\vartheta)}{\hat{\mu}_s(t',u,\vartheta)}$$

we appeal to Proposition 8.6 to obtain

$$\max_{\mathcal{V}_{s+}}\frac{\mu(t)}{\mu(t')}\leqslant\max_{(u,\vartheta)\in\mathcal{V}_{s+}}\frac{1+C\delta_0+\hat{E}_s(u,\vartheta)\log(1+t)}{1-C\delta_0+\hat{E}_s(u,\vartheta)\log(1+t')} \quad (8.376)$$

We have two cases to distinguish according as to whether $t'<\sqrt{t}$ or $t'\geqslant\sqrt{t}$. In the second case we have $1+t'\geqslant 1+\sqrt{t}\geqslant\sqrt{1+t}$, hence

$$\hat{E}_s(u,\vartheta)\log(1+t')\geqslant\frac{1}{2}\hat{E}_s(u,\vartheta)\log(1+t),\quad\forall(u,\vartheta)\in\mathcal{V}_{s+}$$

Then from (8.376) we obtain

$$\max_{\mathcal{V}_{s+}} \frac{\mu(t)}{\mu(t')} \leqslant 2, \quad t' \geqslant \sqrt{t} \tag{8.377}$$

provided that

$$C\delta_0 \leqslant \frac{1}{3}$$

On the other hand, in the first case, we have, trivially

$$\max_{\mathcal{V}_{s+}} \frac{\mu(t)}{\mu(t')} \leqslant C(1 + \log(1+t)), \quad t' < \sqrt{t} \tag{8.378}$$

In view of (8.371) and the definition (8.346) we obtain

$$\begin{aligned}
&\|^{(i_1\cdots i_l)}N_l^2(t,t')\|_{L^2(\mathcal{V}_{s+})} \leqslant C(1+\log(1+t))^{2(i_1\cdots i_l)}P_l(t'), t' < \sqrt{t}\\
&\|^{(i_1\cdots i_l)}N_l^2(t,t')\|_{L^2(\mathcal{V}_{s+})} \leqslant C^{(i_1\cdots i_l)}P_l(t'), t' \geqslant \sqrt{t}
\end{aligned} \tag{8.379}$$

Substituting (8.379) in (8.368) yields

$$\begin{aligned}
&\|^{(i_1\cdots i_l)}M_l^2(t)\|_{L^2(\mathcal{V}_{s+})}\\
&\leqslant \int_0^{\sqrt{t}} \|^{(i_1\cdots i_l)}N_l^2(t,t')\|_{L^2(\mathcal{V}_{s+})}dt' + \int_{\sqrt{t}}^t \|^{(i_1\cdots i_l)}N_l^2(t,t')\|_{L^2(\mathcal{V}_{s+})}dt'\\
&\leqslant C(1+\log(1+t))^2 \int_0^{\sqrt{t}} {}^{(i_1\cdots i_l)}P_l(t')dt' + C\int_{\sqrt{t}}^t {}^{(i_1\cdots i_l)}P_l(t')dt'\\
&\leqslant C(1+\log(1+t))^2 \int_0^{\sqrt{t}} \bar{\mu}_m^{-a}(t')({}^{(i_1\cdots i_l)}\bar{P}_{l,a}^{(0)}(t) + (1+t')^{-1/2(i_1\cdots i_l)}\bar{P}_{l,a}^{(1)}(t))dt'\\
&\quad + C\int_{\sqrt{t}}^t \bar{\mu}_m^{-a}(t')({}^{(i_1\cdots i_l)}\bar{P}_{l,a}^{(0)}(t) + (1+t')^{-1/2(i_1\cdots i_l)}\bar{P}_{l,a}^{(1)}(t))dt'\\
&\leqslant C(1+\log(1+t))^2((1+t)^{1/2(i_1\cdots i_l)}\bar{P}_{l,a}^{(0)}(t) + (1+t)^{1/4(i_1\cdots i_l)}\bar{P}_{l,a}^{(1)}(t))\bar{\mu}_m^{-a}(t)\\
&\quad + C((1+t)^{(i_1\cdots i_l)}\bar{P}_{l,a}^{(0)}(t) + (1+t)^{1/2(i_1\cdots i_l)}\bar{P}_{l,a}^{(1)}(t))\bar{\mu}_m^{-a}(t)\\
&\leqslant C'((1+t)^{(i_1\cdots i_l)}\bar{P}_{l,a}^{(0)}(t) + (1+t)^{1/2(i_1\cdots i_l)}\bar{P}_{l,a}^{(1)}(t))\bar{\mu}_m^{-a}(t)
\end{aligned} \tag{8.380}$$

Here we have used (8.350) and Corollary 2 of Lemma 8.11.

Combining (8.374) and (8.380) we obtain

$$\begin{aligned}
&\|^{(i_1\cdots i_l)}M_l^2(t)\|_{L^2([0,\epsilon_0]\times S^2)}\\
&\leqslant C((1+t)^{(i_1\cdots i_l)}\bar{P}_{l,a}^{(0)}(t) + (1+t)^{1/2(i_1\cdots i_l)}\bar{P}_{l,a}^{(1)}(t))\bar{\mu}_m^{-a}(t)
\end{aligned} \tag{8.381}$$

Finally, (8.336), (8.365) and (8.381) yield, through (8.175), the following estimate:

$$\begin{aligned}&\|^{(i_1\cdots i_l)}F_l(t)\|_{L^2([0,\epsilon_0]\times S^2)}\\&\leqslant C(1+t)^{-3}(1+\log(1+t))^2\|^{(i_1\cdots i_l)}x_l(0)\|_{L^2(\Sigma_0^{\epsilon_0})}\\&+C(1+t)^{-2}({}^{(i_1\cdots i_l)}\bar{P}^{(0)}_{l,a}(t)+(1+t)^{-1/2}{}^{(i_1\cdots i_l)}\bar{P}^{(1)}_{l,a}(t))\bar{\mu}_m^{-a}(t)\end{aligned}\tag{8.382}$$

We now consider the estimate (8.179) for ${}^{(i_1\cdots i_l)}G_l(t,u)$. By (8.343) and (8.357) we have, in general,

$$\max_{[0,\epsilon_0]\times S^2}\frac{\mu(t)}{\mu(t')}\leqslant C(1+\log(1+t))\tag{8.383}$$

Hence (8.179) implies

$${}^{(i_1\cdots i_l)}G_l(t,u)\leqslant C(1+t)^{-3}(1+\log(1+t))^2\int_0^t(1+t')^3|{}^{(i_1\cdots i_l)}\tilde{g}_l(t',u)|dt'\tag{8.384}$$

It follows by Minkowski's inequality that

$$\begin{aligned}&\|^{(i_1\cdots i_l)}G_l(t)\|_{L^2([0,\epsilon_0]\times S^2)}\\&\leqslant C(1+t)^{-3}(1+\log(1+t))^2\int_0^t(1+t')^3\||^{(i_1\cdots i_l)}\tilde{g}_l(t')|\|_{L^2([0,\epsilon_0]\times S^2)}dt'\end{aligned}\tag{8.385}$$

Now the $S_{t,u}$ 1-form ${}^{(i_1\cdots i_l)}\tilde{g}_l$ is given by (8.150). Here we must distinguish the principal acoustical parts. In reference to (8.64), we define the function

$${}^{(i_1\cdots i_l)}\dot{h}_l={}^{(i_1\cdots i_l)}h_l-2\hat{\chi}\cdot{}^{(i_1\cdots i_l)}\hat{\chi}_l\tag{8.386}$$

then $\not{d}^{(i_1\cdots i_l)}\dot{h}_l$ does not contain principal acoustical parts. Also, according to the discussion following Lemma 8.4, the principal acoustical part of ${}^{(i_1\cdots i_l)}g_l$ consists of the sum (8.89). Therefore the $S_{t,u}$ 1-form

$${}^{(i_1\cdots i_l)}\dot{g}_l={}^{(i_1\cdots i_l)}g_l+\sum_{k=0}^{l-1}\not{\mathcal{L}}_{{}^{(R_{i_{l-k}})}Z}{}^{(i_1\cdots i_{l-k-1}i_{l-k+1}\cdots i_l)}x_{l-1}\tag{8.387}$$

does not contain principal acoustical parts. We conclude that, writing

$${}^{(i_1\cdots i_l)}\tilde{g}_l=2\mu\hat{\chi}\cdot\not{D}^{(i_1\cdots i_l)}\hat{\chi}_l-\sum_{k=0}^{l-1}\not{\mathcal{L}}_{{}^{(R_{i_{l-k}})}Z}{}^{(i_1\cdots i_{l-k-1}i_{l-k+1}\cdots i_l)}x_{l-1}+{}^{(i_1\cdots i_l)}\ddot{g}_l\tag{8.388}$$

where

$${}^{(i_1\cdots i_l)}\ddot{g}_l={}^{(i_1\cdots i_l)}\dot{g}_l+\mu(2\not{D}\hat{\chi}\cdot{}^{(i_1\cdots i_l)}\hat{\chi}_l+\not{d}^{(i_1\cdots i_l)}\dot{h}_l)\tag{8.389}$$

does not contain principal acoustical parts.

Let us define

$$X_l(t)=\max_{i_1\cdots i_l}\||^{(i_1\cdots i_l)}x_l(t)|\|_{L^2([0,\epsilon_0]\times S^2)}\tag{8.390}$$

We shall estimate $\||^{(i_1\cdots i_l)}\tilde{g}_l(t)\|_{L^2([0,\epsilon_0]\times S^2)}$ in terms of $X_l(t)$. Consider first the second term on the right of (8.388). By (8.98) we have

$$\begin{aligned}
&|\sum_{k=0}^{l-1}\not{\mathcal{L}}_{{}^{(R_{i_{l-k}})}Z}{}^{(i_1\cdots i_{l-k-1}i_{l-k+1}\cdots i_l)}x_{l-1}|\\
&\leqslant C(1+t)^{-1}\Big(\sum_{k=0}^{l-1}|^{(R_{i_{l-k}})}Z|\sum_j|^{(i_1\cdots i_{l-k-1}i_{l-k+1}\cdots i_l j)}x_l|+{}^{(i_1\cdots i_l)}z_l\Big)
\end{aligned}\tag{8.391}$$

where

$$\begin{aligned}
{}^{(i_1\cdots i_l)}z_l=&\sum_{k=0}^{l-1}|^{(R_{i_{l-k}})}Z|\Big(\sum_j|\not{d}({}^{(i_1\cdots i_{l-k-1}i_{l-k+1}\cdots i_l j)}\check{f}_l)|\\
&+(1+t)|\not{d}\mu||\not{d}(R_{i_l}\cdots R_{i_{l-k+1}}R_{i_{l-k-1}}\cdots R_{i_1}\mathrm{tr}\chi)|\Big)\\
&\cdot\sum_{k=0}^{l-1}(\max_j|\not{\mathcal{L}}_{R_j}{}^{(R_{i_{l-k}})}Z|)\Big(|^{(i_1\cdots i_{l-k-1}i_{l-k+1}\cdots i_l)}x_{l-1}|\\
&+|\not{d}({}^{(i_1\cdots i_{l-k-1}i_{l-k+1}\cdots i_l)}\check{f}_{l-1})|\Big)
\end{aligned}\tag{8.392}$$

From (6.184), we have

$$|^{(R_i)}Z|\leqslant C\delta_0(1+t)^{-1}(1+\log(1+t))\tag{8.393}$$

Substituting this in (8.391), we deduce

$$\begin{aligned}
&\||\sum_{k=0}^{l-1}\not{\mathcal{L}}_{{}^{(R_{i_{l-k}})}Z}{}^{(i_1\cdots i_{l-k-1}i_{l-k+1}\cdots i_l)}x_{l-1}|\|_{L^2([0,\epsilon_0]\times S^2)}\\
&\leqslant Cl\delta_0(1+t)^{-2}(1+\log(1+t))X_l(t)\\
&\quad+C(1+t)^{-1}\|^{(i_1\cdots i_l)}z_l(t)\|_{L^2([0,\epsilon_0]\times S^2)}
\end{aligned}\tag{8.394}$$

Consider next the first term on the right in (8.388). By the 2nd of **F2** we have

$$\begin{aligned}
&\|2\mu|\hat{\chi}\cdot\not{D}^{(i_1\cdots i_l)}\hat{\chi}_l|(t)\|_{L^2([0,\epsilon_0]\times S^2)}\\
&\leqslant C\delta_0(1+t)^{-2}(1+\log(1+t))\|\mu|\not{D}^{(i_1\cdots i_l)}\hat{\chi}_l|(t)\|_{L^2([0,\epsilon_0]\times S^2)}
\end{aligned}\tag{8.395}$$

Using the elliptic estimate (8.149) and the inequality

$$C^{-1}(1+t)\||\xi(t,u)|\|_{L^2(S^2)}\leqslant\|\xi\|_{L^2(S_{t,u})}\leqslant C(1+t)\||\xi(t,u)|\|_{L^2(S^2)}\tag{8.396}$$

which holds for any covariant $S_{t,u}$ tensorfield $\xi$ of any rank, $\xi(t,u)$ being the corresponding tensorfield on $S^2$, we obtain

$$\begin{aligned}
&\|\mu|\not{D}^{(i_1\cdots i_l)}\hat{\chi}_l|(t,u)\|_{L^2(S^2)}\\
&\leqslant C\||^{(i_1\cdots i_l)}x_l(t,u)|\|_{L^2(S^2)}\\
&\quad+C\||\not{d}^{(i_1\cdots i_l)}\check{f}_l(t,u)|\|_{L^2(S^2)}+C\|\mu|^{(i_1\cdots i_l)}i_l|(t,u)\|_{L^2(S^2)}\\
&\quad+C\delta_0(1+t)^{-1}(1+\log(1+t))\||^{(i_1\cdots i_l)}\hat{\chi}_l(t,u)|\|_{L^2(S^2)}
\end{aligned}\tag{8.397}$$

Taking the $L^2$ norms on $[0,\epsilon_0]$ then yields

$$\begin{aligned}
&\|\mu|\not{D}^{(i_1\cdots i_l)}\hat{\chi}_l|(t)\|_{L^2([0,\epsilon_0]\times S^2)} \\
&\leqslant C\||^{(i_1\cdots i_l)}x_l(t)|\|_{L^2([0,\epsilon_0]\times S^2)} \\
&+C\||\not{d}^{(i_1\cdots i_l)}\check{f}_l(t)|\|_{L^2([0,\epsilon_0]\times S^2)}+C\|\mu|^{(i_1\cdots i_l)}i_l|(t)\|_{L^2([0,\epsilon_0]\times S^2)} \\
&+C\delta_0(1+t)^{-1}(1+\log(1+t))\||^{(i_1\cdots i_l)}\hat{\chi}_l(t)|\|_{L^2([0,\epsilon_0]\times S^2)}
\end{aligned} \tag{8.398}$$

In view of (8.394), (8.395) and (8.398), we obtain from (8.388)

$$\begin{aligned}
&\||^{(i_1\cdots i_l)}\tilde{g}_l(t)|\|_{L^2([0,\epsilon_0]\times S^2)} \\
&\leqslant C(l+1)\delta_0(1+t)^{-2}(1+\log(1+t))X_l(t)+{}^{(i_1\cdots i_l)}Q_l(t)
\end{aligned} \tag{8.399}$$

where

$$\begin{aligned}
{}^{(i_1\cdots i_l)}Q_l(t)=&\,C\delta_0^2(1+t)^{-3}(1+\log(1+t))^2\||^{(i_1\cdots i_l)}\hat{\chi}_l(t)|\|_{L^2([0,\epsilon_0]\times S^2)} \\
&C\delta_0(1+t)^{-2}(1+\log(1+t))\||\not{d}^{(i_1\cdots i_l)}\check{f}_l(t)|\|_{L^2([0,\epsilon_0]\times S^2)} \\
&+C\delta_0(1+t)^{-2}(1+\log(1+t))\|\mu|^{(i_1\cdots i_l)}i_l|(t)\|_{L^2([0,\epsilon_0]\times S^2)} \\
&+C(1+t)^{-1}\|^{(i_1\cdots i_l)}z_l(t)\|_{L^2([0,\epsilon_0]\times S^2)} \\
&+\||^{(i_1\cdots i_l)}\ddot{g}_l(t)|\|_{L^2([0,\epsilon_0]\times S^2)}
\end{aligned} \tag{8.400}$$

Returning to (8.170) and taking $L^2$ norms on $[0,\epsilon_0]\times S^2$ we obtain

$$\begin{aligned}
&\||^{(i_1\cdots i_l)}x_l(t)|\|_{L^2([0,\epsilon_0]\times S^2)} \\
&\leqslant \|^{(i_1\cdots i_l)}F_l(t)\|_{L^2([0,\epsilon_0]\times S^2)}+\|^{(i_1\cdots i_l)}G_l(t)\|_{L^2([0,\epsilon_0]\times S^2)}
\end{aligned} \tag{8.401}$$

Substituting (8.382), (8.385) and (8.399) then yields

$$\begin{aligned}
&\||^{(i_1\cdots i_l)}x_l(t)|\|_{L^2([0,\epsilon_0]\times S^2)} \\
&\leqslant {}^{(i_1\cdots i_l)}B_l(t)+C(l+1)\delta_0(1+t)^{-3}(1+\log(1+t))^2 \\
&\cdot\int_0^t(1+t')(1+\log(1+t'))X_l(t')dt'
\end{aligned} \tag{8.402}$$

where

$$\begin{aligned}
{}^{(i_1\cdots i_l)}B_l(t)=&\,C(1+t)^{-3}(1+\log(1+t))^2\|^{(i_1\cdots i_l)}x_l(0)\|_{L^2(\Sigma_0^{\epsilon_0})} \\
&+C(1+t)^{-2}({}^{(i_1\cdots i_l)}\bar{P}^{(0)}_{l,a}(t)+(1+t)^{-1/2}{}^{(i_1\cdots i_l)}\bar{P}^{(1)}_{l,a}(t))\bar{\mu}_m^{-a}(t) \\
&+C(1+t)^{-3}(1+\log(1+t))^2\int_0^t(1+t')^{3}{}^{(i_1\cdots i_l)}Q_l(t')dt'
\end{aligned} \tag{8.403}$$

Taking in (8.402) the maximum over $i_1,\cdots,i_l$, recalling the definition (8.390), we obtain the integral inequality

$$X_l(t) \leqslant B_l(t) + C(l+1)\delta_0(1+t)^{-3}(1+\log(1+t))^2 \cdot \int_0^t (1+t')(1+\log(1+t'))X_l(t')dt' \tag{8.404}$$

where

$$B_l(t) = \max_{i_1 \cdots i_l} {}^{(i_1\cdots i_l)}B_l(t) \tag{8.405}$$

Setting

$$Y_l(t) = \int_0^t (1+t')(1+\log(1+t'))X_l(t')dt' \tag{8.406}$$

then (8.404) becomes

$$\frac{dY_l(t)}{dt} \leqslant (1+t)(1+\log(1+t))B_l(t) + C(l+1)\delta_0(1+t)^{-2}(1+\log(1+t))^3 Y_l(t) \tag{8.407}$$

which implies

$$Y_l(t) \leqslant e^{C_l(t)} \int_0^t e^{-C_l(t')}(1+t')(1+\log(1+t'))B_l(t')dt' \tag{8.408}$$

where

$$C_l(t) = C(l+1)\delta_0 \int_0^t \frac{(1+\log(1+t'))^3}{(1+t')^2} dt' \tag{8.409}$$

Obviously,

$$0 \leqslant C_l(t) \leqslant C'(l+1)\delta_0 \tag{8.410}$$

Hence, if

$$\delta_0 \leqslant \frac{\log 2}{C'(l+1)} \tag{8.411}$$

(8.408) implies

$$Y_l(t) \leqslant 2\int_0^t (1+t')(1+\log(1+t'))B_l(t')dt' \tag{8.412}$$

Substituting this in (8.402), recalling (8.406), we finally conclude that

$$\begin{aligned} &\||{}^{(i_1\cdots i_l)}x_l(t)\||_{L^2([0,\epsilon_0]\times S^2)} \\ &\leqslant {}^{(i_1\cdots i_l)}B_l(t) + 2C(l+1)\delta_0(1+t)^{-3}(1+\log(1+t))^2 \\ &\quad \cdot \int_0^t (1+t')(1+\log(1+t'))B_l(t')dt' \end{aligned} \tag{8.413}$$

# 9

# Regularization of the Propagation Equation for $\not{\Delta}\mu$. Estimates for the Top Order Spatial Derivatives of $\mu$

## 9.1 Regularization of the Propagation Equation

In this chapter, we shall concentrate on the propagation equation (3.92) in Chapter 3:

$$L\mu = m + \mu e \tag{9.1}$$

where

$$m = \frac{1}{2}\frac{dH}{dh}Th \tag{9.2}$$

and

$$e = \frac{1}{2\eta^2}(\frac{\rho}{\rho'})'Lh + \frac{1}{\eta}\hat{T}^i(L\psi_i) \tag{9.3}$$

Since

$$\frac{\rho}{\rho'} = \frac{\rho}{d\rho/dh} = \frac{dp/dh}{d\rho/dh} = \eta^2$$

therefore (9.3) takes the form

$$e = \frac{1}{\eta}\frac{d\eta}{dh}Lh + \frac{1}{\eta}\hat{T}^i(L\psi_i) \tag{9.4}$$

**Lemma 9.1** The following commutation formula holds:

$$[L, \not{\Delta}]\phi + \text{tr}\,\chi\not{\Delta}\phi = -2\hat{\chi}\cdot\hat{\not{D}}^2\phi - 2\not{\text{d}}\text{iv}\hat{\chi}\cdot\not{d}\phi$$

and

$$[T, \not{\Delta}]\phi + \kappa\text{tr}\,\theta\not{\Delta}\phi = -2\kappa\hat{\theta}\cdot\hat{\not{D}}^2\phi - 2\not{\text{d}}\text{iv}(\kappa\hat{\theta})\cdot\not{d}\phi$$

Here, $\phi$ is an arbitrary function defined on $W^*_{\epsilon_0}$.

*Proof*. Let us work in acoustical coordinates $(t, u, \vartheta^1, \vartheta^2)$. We have

$$\not{\Delta}\phi = (\not{g}^{-1})^{AB}\not{D}_A(\not{d}_B\phi) = (\not{g}^{-1})^{AB}(\frac{\partial^2\phi}{\partial\vartheta^A\partial\vartheta^B} - \not{\Gamma}^C_{AB}\frac{\partial\phi}{\partial\vartheta^C}) \tag{9.5}$$

Since in acoustical coordinates $L = \frac{\partial}{\partial t}$, we obtain

$$L\not{\Delta}\phi = (\not{g}^{-1})^{AB}\left(\frac{\partial^2}{\partial\vartheta^A\partial\vartheta^B}\left(\frac{\partial\phi}{\partial t}\right) - \not{\Gamma}^C_{AB}\frac{\partial}{\partial\vartheta^C}\left(\frac{\partial\phi}{\partial t}\right)\right) + \frac{\partial}{\partial t}(\not{g}^{-1})^{AB}\not{D}_A(\not{d}_B\phi) - (\not{g}^{-1})^{AB}\frac{\partial\not{\Gamma}^C_{AB}}{\partial t}\frac{\partial\phi}{\partial\vartheta^C} \tag{9.6}$$

We have

$$\frac{\partial}{\partial t}(\not{g}^{-1})^{AB} = -2\chi^{AB} \tag{9.7}$$

and

$$\frac{\partial\not{\Gamma}^C_{AB}}{\partial t} = \not{D}_A\chi^C_B + \not{D}_B\chi^C_A - \not{D}^C\chi_{AB} \tag{9.8}$$

Here, we used the fact that for a non-degenerate matrix $M$, we have

$$\frac{dM^{-1}}{dt} = -M^{-1}\frac{dM}{dt}M^{-1}$$

and the way we use to derive (8.109).

Hence

$$(\not{g}^{-1})^{AB}\frac{\partial\not{\Gamma}^C_{AB}}{\partial t} = 2\not{\mathrm{d}}\mathrm{iv}\chi^C - \not{d}^C\mathrm{tr}\chi = 2\not{\mathrm{d}}\mathrm{iv}\hat{\chi}^C \tag{9.9}$$

Substituting (9.6) and (9.8) in (9.5), we get

$$L\not{\Delta}\phi = \not{\Delta}L\phi - 2\chi^{AB}\not{D}_A(\not{d}_B\phi) - 2(\not{\mathrm{d}}\mathrm{iv}\hat{\chi})^A\not{d}_A\phi \tag{9.10}$$

Writing then

$$2\chi^{AB}\not{D}_A(\not{d}_B\phi) = \mathrm{tr}\chi\not{\Delta}\phi + 2\hat{\chi}^{AB}\not{D}_A(\not{d}_B\phi)$$

the first formula follows. The second formula can be proved in a similar way. □

Now we apply the second identity by taking $\phi = h$ to obtain

$$\not{\Delta}Th = T\not{\Delta}h + c_T \tag{9.11}$$

where

$$c_T = \kappa\mathrm{tr}\,\theta\not{\Delta}h + 2\kappa\hat{\theta}\cdot\hat{\not{D}}^2 h + 2\not{\mathrm{d}}\mathrm{iv}(\kappa\hat{\theta})\cdot\not{d}h \tag{9.12}$$

$c_T$ is of order 2 and regular as $\mu \to 0$. The last term is a principal (i.e. order 2) acoustical part.

We now appeal to the wave equation for $h$ given in Chapter 8

$$\mu\Box_g h = -\Omega^{-1}\frac{d\Omega}{dh}\breve{a} - \breve{b} \tag{9.13}$$

where

$$\check{a} = \mu a, \quad \check{b} = \mu b \tag{9.14}$$

are regular as $\mu \to 0$. Since for any function $\phi$ we have

$$\mu \Box_g \phi = \mu \not{\Delta} \phi - L\underline{L}\phi - \frac{1}{2}\mathrm{tr}\,\chi \underline{L}\phi - \frac{1}{2}\mathrm{tr}\,\underline{\chi} L\phi - 2\zeta \cdot \not{d}\phi \tag{9.15}$$

Then (9.13) reads

$$\mu \not{\Delta} h = L\underline{L}h + \frac{1}{2}\mathrm{tr}\chi \underline{L}h + \frac{1}{2}\mathrm{tr}\underline{\chi} Lh + j \tag{9.16}$$

where

$$j = 2\zeta \cdot \not{d}h - \Omega^{-1}\frac{d\Omega}{dh}\check{a} - \check{b} \tag{9.17}$$

is of order 1 and regular as $\mu \to 0$. Applying $T$ to (9.16), in view of the fact that

$$[L, T] = \Lambda \tag{9.18}$$

we have

$$T(\mu \not{\Delta} h) = L(T\underline{L}h) - \Lambda \cdot \not{d}\underline{L}h + \frac{1}{2}T(\mathrm{tr}\,\chi \underline{L}h) + \frac{1}{2}T(\mathrm{tr}\,\underline{\chi} Lh) + Tj \tag{9.19}$$

On the other hand, from (9.11) we have

$$\mu \not{\Delta} Th = T(\mu \not{\Delta} h) - (T\mu)\not{\Delta} h + \mu c_T \tag{9.20}$$

Combining, we obtain

$$\mu \not{\Delta} Th = L(T\underline{L}h) + \frac{1}{2}T(\mathrm{tr}\,\chi \underline{L}h) + \frac{1}{2}T(\mathrm{tr}\,\underline{\chi} Lh) + j' \tag{9.21}$$

where

$$j' = Tj - \Lambda \cdot \not{d}\underline{L}h + \mu c_T - (T\mu)\not{\Delta} h \tag{9.22}$$

$j'$ is of order 2 and regular as $\mu \to 0$. The only principal acoustical part in $j'$ is contributed by the last term in $c_T$ noted as above.

We have

$$\not{\Delta} m = \frac{1}{2}\frac{dH}{dh}\not{\Delta} Th + w_0 \tag{9.23}$$

where

$$w_0 = \frac{d^2H}{dh^2}\not{d}h \cdot \not{d}Th + \frac{1}{2}\left(\frac{d^2H}{dh^2}\not{\Delta} h + \frac{d^3H}{dh^3}|\not{d}h|^2\right)Th \tag{9.24}$$

is of order 2 and regular as $\mu \to 0$. Then by (9.21),

$$\mu\not{\Delta}m = Lf_0' - \frac{1}{2}\frac{d^2H}{dh^2}(Lh)(T\underline{L}h) + \frac{1}{4}\frac{dH}{dh}((Lh)T\mathrm{tr}\underline{\chi} + (\underline{L}h)T\mathrm{tr}\chi) + \frac{1}{4}\frac{dH}{dh}(\mathrm{tr}\underline{\chi}(TLh) + \mathrm{tr}\chi(T\underline{L}h)) + \frac{1}{2}\frac{dH}{dh}j' + \mu w_0 \tag{9.25}$$

where

$$f_0' = \frac{1}{2}\frac{dH}{dh}T\underline{L}h \tag{9.26}$$

here we note that the 3rd term on the right contains principal acoustical parts. In view of the equation for $\not{\mathcal{L}}_T\chi$ in Chapter 3, the principal acoustical part of $T\mathrm{tr}\chi$ is $\not{\Delta}\mu$. Also, since

$$\underline{\chi} = \eta^{-1}\kappa(-\chi + 2\not{k}) \tag{9.27}$$

the principal acoustical part of $T\mathrm{tr}\underline{\chi}$ is $-\eta^{-1}\kappa\not{\Delta}\mu$. It follows that the principal acoustical part of the 3rd term on the right in (9.25) is

$$\frac{1}{4}\frac{dH}{dh}(-\eta^{-1}\kappa(Lh) + (\underline{L}h))\not{\Delta}\mu = \frac{1}{2}\frac{dH}{dh}(Th)\not{\Delta}\mu = m\not{\Delta}\mu \tag{9.28}$$

We can therefore write

$$\mu\not{\Delta}m = Lf_0' + \frac{1}{2}\mathrm{tr}\chi f_0' + m\not{\Delta}\mu + w_0' \tag{9.29}$$

where

$$w_0' = -\frac{1}{2}\frac{d^2H}{dh^2}(Lh)(T\underline{L}h) + \frac{1}{4}\frac{dH}{dh}(T\mathrm{tr}\chi - \not{\Delta}\mu)\underline{L}h + \frac{1}{4}\frac{dH}{dh}(T\mathrm{tr}\underline{\chi} + \eta^{-1}\kappa\not{\Delta}\mu)Lh + \frac{1}{4}\frac{dH}{dh}\mathrm{tr}\underline{\chi}TLh + \frac{1}{2}\frac{dH}{dh}j' + \mu w_0 \tag{9.30}$$

contains principal acoustical parts only through $j'$.

Let us turn to $e$. We have

$$\not{\Delta}e = \frac{1}{\eta}\frac{d\eta}{dh}\not{\Delta}Lh + \frac{1}{\eta}\hat{T}^i\not{\Delta}L\psi_i + w_1 \tag{9.31}$$

where

$$w_1 = 2\frac{d}{dh}(\frac{1}{\eta}\frac{d\eta}{dh})\not{d}h\cdot\not{d}Lh + 2(\frac{d}{dh}(\frac{1}{\eta})\hat{T}^i\not{d}h\cdot\not{d}L\psi_i + \frac{1}{\eta}\not{d}\hat{T}^i\cdot\not{d}L\psi_i) + (\frac{d}{dh}(\frac{1}{\eta}\frac{d\eta}{dh})\not{\Delta}h + \frac{d^2}{dh^2}(\frac{1}{\eta}\frac{d\eta}{dh})|\not{d}h|^2)Lh + ((\frac{d}{dh}(\frac{1}{\eta})\not{\Delta}h + \frac{d^2}{dh^2}(\frac{1}{\eta})|\not{d}h|^2)\hat{T}^i + 2\frac{d}{dh}(\frac{1}{\eta})\not{d}h\cdot\not{d}\hat{T}^i + \frac{1}{\eta}\not{\Delta}\hat{T}^i)L\psi_i \tag{9.32}$$

is of order 2 and regular as $\mu \to 0$. The only principal acoustical part in $w_1$ is the term

$$\frac{1}{\eta} \not{\Delta} \hat{T}^i L\psi_i \tag{9.33}$$

contained in the last term on the right in (9.32).

Applying Lemma 9.1, taking $\phi = h, \psi_i$ we then obtain

$$\not{\Delta} e = \frac{1}{\eta}\frac{d\eta}{dh} L\not{\Delta} h + \frac{1}{\eta}\hat{T}^i L\not{\Delta}\psi_i + c_L + w_1 \tag{9.34}$$

where

$$c_L = \frac{2}{\eta}\frac{d\eta}{dh}(\chi \cdot \not{D}^2 h + \not{d}\mathrm{iv}\hat{\chi} \cdot \not{d}h) + \frac{2}{\eta}\hat{T}^i(\chi \cdot \not{D}^2\psi_i + \not{d}\mathrm{iv}\hat{\chi} \cdot \not{d}\psi_i) \tag{9.35}$$

contains principal acoustical parts in the terms involving $\not{d}\mathrm{iv}\hat{\chi}$. Defining

$$f_1' = \frac{1}{\eta}\frac{d\eta}{dh}\not{\Delta} h + \frac{1}{\eta}\hat{T}^i \not{\Delta}\psi_i \tag{9.36}$$

we then have

$$\mu^2 \not{\Delta} e = L(\mu^2 f_1') + \mu w_1' \tag{9.37}$$

where

$$\begin{aligned} w_1' = &- 2(L\mu) f_1' - \mu\Big(\frac{d}{dh}\big(\frac{1}{\eta}\frac{d\eta}{dh}\big)(Lh)\not{\Delta} h + \big(\frac{d}{dh}\big(\frac{1}{\eta}\big)(Lh)\hat{T}^i + \frac{1}{\eta}L\hat{T}^i\big)\not{\Delta}\psi_i\Big) \\ &+ \mu(c_L + w_1) \end{aligned} \tag{9.38}$$

is of order 2 and regular as $\mu \to 0$.

We commute $\not{\Delta}$ with Equation (9.1) and multiply it by $\mu$ to obtain

$$\begin{aligned} &L(\mu\not{\Delta}\mu) + (\mu\mathrm{tr}\chi - L\mu)\not{\Delta}\mu \\ &= - 2\mu\hat{\chi} \cdot \hat{\not{D}}^2 \mu - \mu\not{d}\mu \cdot (\not{d}\mathrm{tr}\chi + 2i - 2\not{d}e) + \mu\not{\Delta} m + \mu^2\not{\Delta} e + e\mu\not{\Delta}\mu \end{aligned} \tag{9.39}$$

Substituting (9.29) in (9.39), the term $m\not{\Delta}\mu$ in (9.29) combines with the last term on the right in (9.39) to $(L\mu)\not{\Delta}\mu$. This is brought to the left hand side of (9.39), making the coefficient of $\not{\Delta}\mu$ on the left equal to $\mu\mathrm{tr}\chi - 2L\mu$. Also, defining

$$\check{f}' = f_0' + \mu^2 f_1' \tag{9.40}$$

and

$$x' = \mu\not{\Delta}\mu - \check{f}' \tag{9.41}$$

the term $L\check{f}'$ is brought to the left hand side of (9.39). Then left hand side becomes

$$Lx' + (\mathrm{tr}\chi - 2\mu^{-1}(L\mu))(x' + \check{f}') \tag{9.42}$$

and the resulting right hand side is

$$\frac{1}{2}\mathrm{tr}\chi \check{f}_0' - 2\mu\hat{\chi} \cdot \hat{\not{D}}^2 \mu - \mu \not{d}\mu \cdot (\not{d}\mathrm{tr}\chi + 2i - 2\not{d}e) + w_0' + \mu w_1' \tag{9.43}$$

Defining then

$$\check{g}' = -\frac{\mu^2}{2}\mathrm{tr}\chi \check{f}_1' - \mu\not{d}\mu \cdot (\not{d}\mathrm{tr}\chi + 2i - 2\not{d}e) + w_0' + \mu w_1' \tag{9.44}$$

the resulting equation takes the form

$$\begin{aligned} &Lx' + (\mathrm{tr}\chi - 2\mu^{-1}(L\mu))x' \\ &= -(\frac{1}{2}\mathrm{tr}\chi - 2\mu^{-1}(L\mu))\check{f}' - 2\mu\hat{\chi} \cdot \hat{\not{D}}^2 \mu + \check{g}' \end{aligned} \tag{9.45}$$

The functions $\check{f}'$, $\check{g}'$ are of order 2 and regular as $\mu \to 0$. The function $\check{f}'$ does not contain principal acoustical parts. From the above discussion, the principal acoustical part of $\check{g}'$ is, besides the term $-\mu\not{d}\mu \cdot \not{d}\mathrm{tr}\chi$, contained in the last two terms in (9.43). In fact, the principal acoustical part of $w_0'$ is contained in the term

$$\frac{1}{2}\mu\frac{dH}{dh}c_T \tag{9.46}$$

contributed by the term $\frac{1}{2}\frac{dH}{dh}j'$ in (9.30). Since

$$\theta = \not{k} - \eta^{-1}\chi \tag{9.47}$$

taking into account the Codazzi equation

$$\not{\mathrm{div}}\hat{\chi} = \frac{1}{2}\not{d}\mathrm{tr}\chi + i \tag{9.48}$$

this principal acoustical part is

$$-\frac{\mu^2}{\eta^2}\frac{dH}{dh}\not{d}h \cdot \not{d}\mathrm{tr}\chi \tag{9.49}$$

The principal acoustical part of $\mu w_1'$ is contained in the term

$$\mu^2(c_L + w_1) \tag{9.50}$$

Taking into account the Codazzi equation, the principal acoustical part of $\mu^2 c_L$ is

$$\frac{\mu^2}{\eta}(\frac{d\eta}{dh}\not{d}h + \hat{T}^i\not{d}\psi_i) \cdot \not{d}\mathrm{tr}\chi \tag{9.51}$$

Finally, the principal acoustical part of $\mu^2 w_1$ is contained in

$$\frac{\mu^2}{\eta}\not{\Delta}\hat{T}^i L\psi_i \tag{9.52}$$

By the results in Chapter 3, the principal acoustical part of $\not{\Delta}\hat{T}^i$ is

$$-\eta^{-1}\not{d}x^i \cdot \not{d}\mathrm{tr}\chi \tag{9.53}$$

Now,

$$\frac{\partial x^i}{\partial \vartheta^A} L\psi_i = X^i_A L\psi_i = L^\mu X_A \psi_\mu \tag{9.54}$$

hence

$$\begin{aligned}(\not{d}x^i)L\psi_i &= L^\mu \not{d}\psi_\mu = \not{d}\psi_t + L^i \not{d}\psi_i \\ &= \not{d}\psi_t - \psi^i \not{d}\psi_i - \eta \hat{T}^i \not{d}\psi_i = \not{d}h - \eta\hat{T}^i \not{d}\psi_i\end{aligned} \tag{9.55}$$

Therefore the principal acoustical part of (9.52) is

$$\frac{\mu^2}{\eta}(-\frac{1}{\eta}\not{d}h + \hat{T}^i \not{d}\psi_i) \cdot \not{d}\mathrm{tr}\chi \tag{9.56}$$

Combining the above results and noting that

$$-\frac{1}{2}\frac{dH}{dh} = 1 + \eta\frac{d\eta}{dh} \tag{9.57}$$

we conclude that the principal acoustical part of $\breve{g}'$ is

$$\xi \cdot (\mu \not{d}\mathrm{tr}\chi) \tag{9.58}$$

where

$$\xi = -\not{d}\mu + \frac{2\mu}{\eta}\hat{T}^i \not{d}\psi_i + \frac{2\mu}{\eta}\frac{d\eta}{dh}\not{d}h \tag{9.59}$$

## 9.2 Propagation Equations for the Higher Order Spatial Derivatives

We now return to the propagation equation (9.45). To control the higher order spatial derivatives of $\mu$, we introduce the function

$$^{(i_1 \cdots i_l)}x'_{m,l} = \mu R_{i_l} \cdots R_{i_1} T^m \not{\Delta}\mu - R_{i_l} \cdots R_{i_1} T^m \not{f}' \tag{9.60}$$

We have

$$x'_{0,0} = x' \tag{9.61}$$

To derive a propagation equation for $x'_{m,l}$, we shall use the following lemma.

**Lemma 9.2** Let $\alpha$ and $\beta$ be $S_{t,u}$ trace-free symmetric 2-covariant tensorfields defined in the spacetime domain $W^*_{\epsilon_0}$. Then we have

$$T(\alpha \cdot \beta) = (\hat{\not{\mathcal{L}}}_T \alpha) \cdot \beta + \alpha \cdot (\hat{\not{\mathcal{L}}}_T \beta) - \mathrm{tr}^{(T)}\not{\pi}(\alpha \cdot \beta)$$

*Proof*. Since $T$ is tangential to $\Sigma_t$, we can confine ourselves to $\Sigma_t$. To define $\hat{\not{\mathcal{L}}}_T\alpha$ and $\hat{\not{\mathcal{L}}}_T\beta$, we extend $\alpha$ and $\beta$ to $T\Sigma_t$ as

$$\alpha(V,T) = \beta(V,T) = 0, \quad \forall V \in T\Sigma_t \tag{9.62}$$

Since

$$\bar{g}^{-1} = \hat{T} \otimes \hat{T} + (\not{g}^{-1})^{AB} X_A \otimes X_B \tag{9.63}$$

we have

$$\alpha \cdot \beta = (\not{g}^{-1})^{AC}(\not{g}^{-1})^{BD}\alpha_{AB}\beta_{CD} = (\bar{g}^{-1})^{ac}(\bar{g}^{-1})^{bd}\alpha_{ab}\beta_{cd} \tag{9.64}$$

Then,

$$T(\alpha \cdot \beta) = -{}^{(T)}\bar{\pi}^{ab}\gamma_{ab} + (\mathcal{L}_T\alpha)_{ab}\beta^{ab} + \alpha^{ab}(\mathcal{L}_T\beta)_{ab} \tag{9.65}$$

Here,

$${}^{(T)}\bar{\pi}_{ab} = (\mathcal{L}_T\bar{g})_{ab}, \quad {}^{(T)}\bar{\pi}^{ab} = (\bar{g}^{-1})^{ac}(\bar{g}^{-1})^{bd(T)}\bar{\pi}_{cd} \tag{9.66}$$

Here, $\gamma$ is a symmetric 2-covariant tensorfield on $\Sigma_t$ given by

$$\gamma_{ab} = (\bar{g}^{-1})^{cd}(\alpha_{ac}\beta_{bd} + \beta_{ac}\alpha_{bd}) \tag{9.67}$$

We have

$$\gamma_{ab}T^b = 0$$

so, we can view $\gamma$ as an $S_{t,u}$ symmetric 2-covariant tensorfield. Now for any two symmetric trace-free 2-dimensional matrices $A$ and $B$ we have

$$AB + BA - \operatorname{tr}(AB)I = 0$$

In an orthonormal frame relative to $\not{g}$ the components of $\alpha$ and $\beta$ form $A$ and $B$. The components of $\gamma$ form $AB + BA$. It follows that

$$\gamma_{AB} = (\alpha \cdot \beta)\not{g}_{AB} \tag{9.68}$$

Hence, we have

$${}^{(T)}\bar{\pi}^{AB}\gamma_{AB} = {}^{(T)}\not{\pi}^{AB}\gamma_{AB} = \operatorname{tr}{}^{(T)}\not{\pi}(\alpha \cdot \beta) \tag{9.69}$$

Since $\alpha$ and $\beta$ are trace-free, the lemma follows. □

Similarly, we have

**Lemma 9.3** Let $\alpha$ and $\beta$ be $S_{t,u}$ trace-free symmetric 2-covariant tensorfields defined in the spacetime domain $W^*_{\epsilon_0}$. Then we have

$$R_i(\alpha \cdot \beta) = (\hat{\not{\mathcal{L}}}_{R_i}\alpha) \cdot \beta + \alpha \cdot (\hat{\not{\mathcal{L}}}_{R_i}\beta) - \operatorname{tr}{}^{(R_i)}\not{\pi}(\alpha \cdot \beta)$$

To simplify the derivation of the propagation equation for $x'_{m,l}$, we introduce the functions

$${}^{(i_1\cdots i_l)}\check{f}'_{m,l} = R_{i_l}\cdots R_{i_1}T^m\check{f}' \tag{9.70}$$

$$\hat{\not{\mu}}_2 = \hat{\not{D}}^2 \mu \tag{9.71}$$

and

$$^{(i_1\cdots i_l)}\hat{\not{\mu}}_{2,m,l} = \hat{\not{\mathcal{L}}}_{R_{i_l}} \cdots \hat{\not{\mathcal{L}}}_{R_{i_1}} (\hat{\not{\mathcal{L}}}_T)^m \hat{\not{\mu}}_2 \tag{9.72}$$

**Proposition 9.1** For each non-negative integer $m$ the function $x'_{m,0}$ satisfies the propagation equation

$$Lx'_{m,0} + (\mathrm{tr}\chi - 2\mu^{-1}(L\mu))x'_{m,0} = -(\frac{1}{2}\mathrm{tr}\chi - 2\mu^{-1}(L\mu))\check{f}'_{m,0} - 2\mu\hat{\chi} \cdot \hat{\not{\mu}}_{2,m,0} + g'_{m,0}$$

where $g'_{m,0}$ is given by

$$g'_{m,0} = T^m \check{g}' + \sum_{k=0}^{m-1} T^k \Lambda x'_{m-k-1,0} + \sum_{k=0}^{m-1} T^k y'_{m-k,0}$$

Here, for each $j = 1, \cdots, m$, $y'_{j,0}$ is the function

$$\begin{aligned} y'_{j,0} = & -(T\mu)a'_{j-1,0} + (TL\mu - \mu T\mathrm{tr}\chi - \Lambda\mu)T^{j-1}\not{\Delta}\mu \\ & + \frac{1}{2}(T\mathrm{tr}\chi)\check{f}'_{j-1,0} - 2\mu(\hat{\not{\mathcal{L}}}_T\hat{\chi} - \mathrm{tr}^{(T)}\not{\pi}\hat{\chi}) \cdot \hat{\not{\mu}}_{2,j-1,0} \end{aligned}$$

where

$$a'_{j-1,0} = LT^{j-1}\not{\Delta}\mu + \mathrm{tr}\chi T^{j-1}\not{\Delta}\mu + 2\hat{\chi} \cdot \hat{\not{\mu}}_{2,j-1,0}$$

*Proof*. The proposition reduces for $m = 0$ to the propagation equation (9.45). By induction on $m$, assuming that the propagation equation holds with $m$ replaced by $m - 1$:

$$\begin{aligned} & Lx'_{m-1,0} + (\mathrm{tr}\chi - 2\mu^{-1}(L\mu))x'_{m-1,0} \\ = & -(\frac{1}{2}\mathrm{tr}\chi - 2\mu^{-1}(L\mu))\check{f}'_{m-1,0} - 2\mu\hat{\chi} \cdot \hat{\not{\mu}}_{2,m-1,0} + g'_{m-1,0} \end{aligned} \tag{9.73}$$

for some $g'_{m-1,0}$, we shall show that a propagation equation of the form given by the proposition holds for $m$, where $g'_{m,0}$ is related to $g'_{m-1,0}$ by a certain recursion relation. Rewrite the term

$$2\mu^{-1}(L\mu)(x'_{m-1,0} + \check{f}'_{m-1,0})$$

in (9.73) as

$$2(L\mu)T^{m-1}\not{\Delta}\mu$$

(see (9.60) and (9.70)) obtaining the equation

$$\begin{aligned} & Lx'_{m-1,0} + \mathrm{tr}\chi x'_{m-1,0} \\ = & \, 2(L\mu)T^{m-1}\not{\Delta}\mu - \frac{1}{2}\mathrm{tr}\chi\check{f}'_{m-1,0} - 2\mu\hat{\chi} \cdot \hat{\not{\mu}}_{2,m-1,0} + g'_{m-1,0} \end{aligned} \tag{9.74}$$

We apply $T$ to this equation. Since

$$TT^{m-1}\Delta\mu = T^m\Delta\mu$$
$$T\check{f}'_{m-1,0} = \check{f}'_{m,0}$$

and by Lemma 9.2,

$$T(\hat{\chi}\cdot\hat{\not\mu}_{2,m-1,0}) = (\hat{\not{\mathcal{L}}}_T\hat{\chi})\cdot\hat{\not\mu}_{2,m-1,0} + \hat{\chi}\cdot\hat{\not\mu}_{2,m,0} - \mathrm{tr}^{(T)}\not\pi(\hat{\chi}\cdot\hat{\not\mu}_{2,m-1,0}) \tag{9.75}$$

(see definition (9.72)), we obtain

$$\begin{aligned} &TLx'_{m-1,0} + \mathrm{tr}\chi Tx'_{m-1,0} + (T\mathrm{tr}\chi)x'_{m-1,0} \\ &= 2(L\mu)T^m\Delta\mu + 2(TL\mu)T^{m-1}\Delta\mu \\ &- \frac{1}{2}\mathrm{tr}\chi\check{f}'_{m,0} - \frac{1}{2}(T\mathrm{tr}\chi)\check{f}'_{m-1,0} \\ &- 2\mu\hat{\chi}\cdot\hat{\not\mu}_{2,m,0} - 2\mu(\hat{\not{\mathcal{L}}}_T\hat{\chi})\cdot\hat{\not\mu}_{2,m-1,0} \\ &+ 2(\mu\mathrm{tr}^{(T)}\not\pi - (T\mu))(\hat{\chi}\cdot\hat{\not\mu}_{2,m-1,0}) + Tg'_{m-1,0} \end{aligned} \tag{9.76}$$

Using the fact

$$[L,T] = \Lambda = -(\not g^{-1})^{AB}(\zeta_B + \eta_B)X_A \tag{9.77}$$

we express

$$TLx'_{m-1,0} = LTx'_{m-1,0} - \Lambda x'_{m-1,0} \tag{9.78}$$

From the definition (9.60) with $l = 0$ and $m$ replaced by $m-1$:

$$x'_{m-1,0} = \mu T^{m-1}\Delta\mu - T^{m-1}\check{f}'$$

we get

$$Tx'_{m-1,0} = x'_{m,0} + (T\mu)T^{m-1}\Delta\mu \tag{9.79}$$

Applying $L$ to (9.79) and expressing

$$LT\mu = TL\mu + \Lambda\mu$$

yields

$$LTx'_{m-1,0} = Lx'_{m,0} + (T\mu)LT^{m-1}\Delta\mu + (TL\mu + \Lambda\mu)T^{m-1}\Delta\mu \tag{9.80}$$

Substituting (9.80) to (9.78) and the result in (9.76), substituting also (9.79) on the left, we obtain a propagation equation of the required form with

$$g'_{m,0} = Tg'_{m-1,0} + \Lambda x'_{m-1,0} + y'_{m,0} \tag{9.81}$$

Applying then Proposition 8.2, the proposition follows. □

**Proposition 9.2** For each pair of non-negative integers $(m, l)$ and each multi-index $(i_1 \cdots i_l)$, the function ${}^{(i_1\cdots i_l)}x'_{m,l}$ satisfies the propagation equation

$$L^{(i_1\cdots i_l)}x'_{m,l} + (\mathrm{tr}\chi - 2\mu^{-1}(L\mu))^{(i_1\cdots i_l)}x'_{m,l}$$
$$= -(\frac{1}{2}\mathrm{tr}\chi - 2\mu^{-1}(L\mu))^{(i_1\cdots i_l)}\check{f}'_{m,l} - 2\mu\hat{\chi}\cdot{}^{(i_1\cdots i_l)}\hat{\not{\mu}}_{2,m,l} + {}^{(i_1\cdots i_l)}g'_{m,l}$$

where

$$^{(i_1\cdots i_l)}g'_{m,l} = R_{i_l}\cdots R_{i_1}g'_{m,0} + \sum_{k=0}^{l-1} R_{i_l}\cdots R_{i_{l-k+1}}{}^{(R_{i_{l-k}})}Z^{(i_1\cdots i_{l-k-1})}x'_{m,l-k-1}$$
$$+ \sum_{k=0}^{l-1} R_{i_l}\cdots R_{i_{l-k+1}}{}^{(i_1\cdots i_{l-k})}y'_{m,l-k}$$

where $g'_{m,0}$ is given by Proposition 9.1, and for each $j = 1, \cdots, l$,

$$^{(i_1\cdots i_j)}y'_{m,j} = -(R_{i_j}\mu)^{(i_1\cdots i_{j-1})}a'_{m,j-1}$$
$$+ (R_{i_j}L\mu - \mu R_{i_j}\mathrm{tr}\chi - {}^{(R_{i_j})}Z\mu)(R_{i_{j-1}}\cdots R_{i_1}T^m \not{\Delta}\mu)$$
$$+ \frac{1}{2}(R_{i_j}\mathrm{tr}\chi)^{(i_1\cdots i_{j-1})}\check{f}'_{m,j-1}$$
$$- 2\mu(\hat{\not{\mathcal{L}}}_{R_{i_j}}\hat{\chi} - \mathrm{tr}^{(R_{i_j})}\not{\pi}\hat{\chi})\cdot{}^{(i_1\cdots i_{j-1})}\hat{\not{\mu}}_{2,m,j-1}$$

where

$$^{(i_1\cdots i_{j-1})}a'_{m,j-1} = L(R_{i_{j-1}}\cdots R_{i_1}T^m \not{\Delta}\mu)$$
$$+ \mathrm{tr}\chi(R_{i_{j-1}}\cdots R_{i_1}T^m \not{\Delta}\mu) + 2\hat{\chi}\cdot{}^{(i_1\cdots i_{j-1})}\hat{\not{\mu}}_{2,m,j-1}$$

*Proof*. The proposition reduces for $l = 0$ to Proposition 9.1. By induction on $l$, assuming the proposition holds with $l$ replaced by $l - 1$:

$$L^{(i_1\cdots i_l)}x'_{m,l-1} + (\mathrm{tr}\chi - 2\mu^{-1}(L\mu))^{(i_1\cdots i_{l-1})}x'_{m,l-1}$$
$$= -(\frac{1}{2}\mathrm{tr}\chi - 2\mu^{-1}(L\mu))^{(i_1\cdots i_{l-1})}\check{f}'_{m,l-1} \tag{9.82}$$
$$- 2\mu\hat{\chi}\cdot{}^{(i_1\cdots i_{l-1})}\hat{\not{\mu}}_{2,m,l-1} + {}^{(i_1\cdots i_{l-1})}g'_{m,l-1}$$

for some ${}^{(i_1\cdots i_{l-1})}g'_{m,l-1}$, we shall show that a propagation equation of the form given by the proposition holds for $l$, where ${}^{(i_1\cdots i_l)}g'_{m,l}$ is related to ${}^{(i_1\cdots i_{l-1})}g'_{m,l-1}$ by a certain recursion relation. Rewrite the term

$$2\mu^{-1}(L\mu)({}^{(i_1\cdots i_{l-1})}x'_{m,l-1} + {}^{(i_1\cdots i_{l-1})}\check{f}'_{m,l-1})$$

in (9.82) as

$$2(L\mu)R_{i_{l-1}}\cdots R_{i_1}T^m \not{\Delta}\mu$$

to obtain the equation

$$\begin{aligned}
&L^{(i_1\cdots i_l)}x'_{m,l-1} + \mathrm{tr}\chi^{(i_1\cdots i_{l-1})}x'_{m,l-1} \\
&= 2(L\mu)R_{i_{l-1}}\cdots R_{i_1}T^m\not{\Delta}\mu - \frac{1}{2}\mathrm{tr}\chi^{(i_1\cdots i_{l-1})}\check{f}'_{m,l-1} \\
&\quad - 2\mu\hat{\chi}\cdot{}^{(i_1\cdots i_{l-1})}\hat{\not{\mu}}_{2,m,l-1} + {}^{(i_1\cdots i_{l-1})}g'_{m,l-1}
\end{aligned} \tag{9.83}$$

We now apply $R_{i_l}$ to this equation. Since

$$R_{i_l}{}^{(i_1\cdots i_{l-1})}\check{f}'_{m,l-1} = {}^{(i_1\cdots i_l)}\check{f}'_{m,l}$$

and by Lemma 9.3

$$\begin{aligned}
&R_{i_l}(\hat{\chi}\cdot\hat{\not{\mu}}_{2,m,l-1}) \\
&= (\hat{\not{\mathcal{L}}}_{R_{i_l}}\hat{\chi})\cdot{}^{(i_1\cdots i_{l-1})}\hat{\not{\mu}}_{2,m,l-1} + \hat{\chi}^{(i_1\cdots i_l)}\hat{\not{\mu}}_{2,m,l} - \mathrm{tr}^{(R_{i_l})}\not{\pi}(\hat{\chi}\cdot{}^{(i_1\cdots i_{l-1})}\hat{\not{\mu}}_{2,m,l-1})
\end{aligned} \tag{9.84}$$

we obtain

$$\begin{aligned}
&R_{i_l}L^{(i_1\cdots i_{l-1})}x'_{m,l-1} + \mathrm{tr}\chi R_{i_l}{}^{(i_1\cdots i_{l-1})}x'_{m,l-1} + (R_{i_l}\mathrm{tr}\chi)^{(i_1\cdots i_{l-1})}x'_{m,l-1} \\
&= 2(L\mu)R_{i_l}\cdots R_{i_1}T^m\not{\Delta}\mu + 2(R_{i_l}L\mu)R_{i_{l-1}}\cdots R_{i_1}T^m\not{\Delta}\mu \\
&\quad - \frac{1}{2}\mathrm{tr}\chi^{(i_1\cdots i_l)}\check{f}'_{m,l} - \frac{1}{2}(R_{i_l}\mathrm{tr}\chi)^{(i_1\cdots i_{l-1})}\check{f}'_{m,l-1} \\
&\quad - 2\mu\hat{\chi}\cdot{}^{(i_1\cdots i_l)}\hat{\not{\mu}}_{2,m,l} - 2\mu(\hat{\not{\mathcal{L}}}_{R_{i_l}}\hat{\chi})\cdot{}^{(i_1\cdots i_{l-1})}\hat{\not{\mu}}_{2,m,l-1} \\
&\quad + 2(\mu\mathrm{tr}^{(R_{i_l})}\not{\pi} - (R_{i_l}\mu))(\hat{\chi}\cdot{}^{(i_1\cdots i_{l-1})}\hat{\not{\mu}}_{2,m,l-1}) \\
&\quad + R_{i_l}{}^{(i_1\cdots i_{l-1})}g'_{m,l-1}
\end{aligned} \tag{9.85}$$

By Lemma 8.2 we have

$$[L, R_{i_l}] = {}^{(R_{i_l})}Z \tag{9.86}$$

hence

$$R_{i_l}L^{(i_1\cdots i_{l-1})}x'_{m,l-1} = LR_{i_l}{}^{(i_1\cdots i_{l-1})}x'_{m,l-1} - {}^{(R_{i_l})}Z^{(i_1\cdots i_{l-1})}x'_{m,l-1} \tag{9.87}$$

From the definition of ${}^{(i_1\cdots i_{l-1})}x'_{m,l-1}$ we obtain

$$R_{i_l}{}^{(i_1\cdots i_{l-1})}x'_{m,l-1} = {}^{(i_1\cdots i_l)}x'_{m,l} + (R_{i_l}\mu)(R_{i_{l-1}}\cdots R_{i_1}T^m\not{\Delta}\mu) \tag{9.88}$$

Applying $L$ to (9.88) and expressing

$$LR_{i_l}\mu = R_{i_l}L\mu + {}^{(R_{i_l})}Z$$

yields

$$\begin{aligned}
LR_{i_l}{}^{(i_1\cdots i_{l-1})}x'_{m,l-1} &= L^{(i_1\cdots i_l)}x'_{m,l} + (R_{i_l}\mu)L(R_{i_{l-1}}\cdots R_{i_1}T^m\not{\Delta}\mu) \\
&\quad + (R_{i_l}L\mu + {}^{(R_{i_l})}Z\mu)(R_{i_{l-1}}\cdots R_{i_1}T^m\not{\Delta}\mu)
\end{aligned} \tag{9.89}$$

We substitute (9.89) in (9.87) and the result in (9.85), substituting also (9.88) on the left, a propagation equation of the required form results, with

$$^{(i_1\cdots i_l)}g'_{m,l} = R_{i_l}\,^{(i_1\cdots i_{l-1})}g'_{m,l-1} + {}^{(R_{i_l})}Z\,^{(i_1\cdots i_{l-1})}x'_{m,l-1} + {}^{(i_1\cdots i_l)}y'_{m,l} \tag{9.90}$$

Applying then Proposition 8.2, the proposition follows. □

The function $^{(i_1\cdots i_{j-1})}a'_{m,j-1}$ defined in Proposition 9.2 reduces for $m=0, j=1$, to the function

$$a'_{0,0} = L\not{\Delta}\mu + \mathrm{tr}\chi\not{\Delta}\mu + 2\hat{\chi}\cdot\hat{\not{\mu}}_2 \tag{9.91}$$

According to equation

$$L(\not{\Delta}\mu) + (\mathrm{tr}\chi - e)\not{\mu} = -2\hat{\chi}\cdot\hat{\not{D}}^2\mu - \not{d}\mu\cdot(\not{d}\mathrm{tr}\chi + 2i - 2\not{d}e) + \not{\Delta}m + \mu\not{\Delta}e$$

which is derived by Lemma 9.1, we have

$$a'_{0,0} = e\not{\Delta}\mu - 2\not{d}\mu\cdot \not{\mathrm{div}}\hat{\chi} + \not{\Delta}m + \mu\not{\Delta}e + 2\not{d}\mu\cdot\not{d}e \tag{9.92}$$

**Lemma 9.4** For each non-negative integer $m$ we have

$$a'_{m,0} = T^m a'_{0,0} + b'_{m,0}$$

where

$$b'_{m,0} = \sum_{k=0}^{m-1} T^k \Lambda T^{m-k-1}\not{\Delta}\mu + \sum_{k=0}^{m-1} T^k c'_{m-k}$$

and for each $j = 1,\cdots,m$,

$$c'_j = -(T\mathrm{tr}\chi)T^{j-1}\not{\Delta}\mu - 2(\hat{\not{\mathcal{L}}}_T\hat{\chi} - \mathrm{tr}^{(T)}\not{\pi}\hat{\chi})\cdot\hat{\not{\mu}}_{2,j-1,0}$$

Moreover, for each pair of non-negative integers $(m,l)$ and each multi-index $(i_1\cdots i_l)$ we have

$$^{(i_1\cdots i_l)}a'_{m,l} = R_{i_l}\cdots R_{i_1}T^m a'_{0,0} + {}^{(i_1\cdots i_l)}b'_{m,l}$$

where

$$\begin{aligned}^{(i_1\cdots i_l)}b'_{m,l} =& R_{i_l}\cdots R_{i_1}b'_{m,0} + \sum_{k=0}^{l-1} R_{i_l}\cdots R_{i_{l-k+1}}\,^{(R_{i_{l-k}})}Z R_{i_{l-k-1}}\cdots R_{i_1}T^m\not{\Delta}\mu\\ &+\sum_{k=0}^{l-1} R_{i_l}\cdots R_{i_{l-k+1}}\,^{(i_1\cdots i_{l-k})}c''_{m,l-k}\end{aligned}$$

and for each $j=1,\cdots,l$ and each multi-index $(i_1\cdots i_j)$,

$$\begin{aligned}&^{(i_1\cdots i_j)}c''_{m,j}\\ &= -(R_{i_j}\mathrm{tr}\chi)(R_{i_{j-1}}\cdots R_{i_1}T^m\not{\Delta}\mu) - 2(\hat{\not{\mathcal{L}}}_{R_{i_j}}\hat{\chi} - \mathrm{tr}^{(R_{i_j})}\not{\pi}\hat{\chi})\cdot\,^{(i_1\cdots i_{j-1})}\hat{\not{\mu}}_{2,m,j-1}\end{aligned}$$

*Proof*. To prove the first part, we note that it holds trivially for $m=0$ with

$$b'_{0,0} = 0 \tag{9.93}$$

Thus, by induction on $m$, assume that

$$a'_{m-1,0} = T^{m-1}a'_{0,0} + b'_{m-1,0} \tag{9.94}$$

holds for some $b'_{m-1,0}$. Applying $T$ to this we obtain

$$Ta'_{m-1,0} = T^m a'_{0,0} + Tb'_{m-1,0} \tag{9.95}$$

On the other hand, from the definition,

$$a'_{m-1,0} = LT^{m-1}\Delta\mu + \mathrm{tr}\chi T^{m-1}\Delta\mu + 2\hat{\chi}\cdot\hat{\not{\mu}}_{2,m-1,0} \tag{9.96}$$

Applying $T$ to this and using Lemma 9.2 yields

$$\begin{aligned} Ta'_{m-1,0} =& TLT^{m-1}\Delta\mu + \mathrm{tr}\chi T^m\Delta\mu + 2\hat{\chi}\cdot\hat{\not{\mu}}_{2,m,0} \\ &+ (T\mathrm{tr}\chi)T^{m-1}\Delta\mu + 2(\hat{\not{\mathcal{L}}}_T\hat{\chi} - \mathrm{tr}^{(T)}\not{\pi}\hat{\chi})\cdot\hat{\not{\mu}}_{2,m-1,0} \end{aligned} \tag{9.97}$$

Writing

$$TLT^{m-1}\Delta\mu = LT^m\Delta\mu - \Lambda T^{m-1}\Delta\mu$$

(9.97), in view of the definition of $a'_{m,0}$, reads

$$Ta'_{m-1,0} = a'_{m,0} - \Lambda T^{m-1}\Delta\mu - c'_m \tag{9.98}$$

where $c'_m$ is as in the statement of the lemma. Equating the two expressions for $Ta'_{m-1,0}$, that is, (9.95) with (9.98), we obtain the formula for $a'_{m,0}$, if we set

$$b'_{m,0} = Tb'_{m-1,0} + \Lambda T^{m-1}\Delta\mu + c'_m \tag{9.99}$$

using then Proposition 8.2, the result follows.

To prove the second part, we apply induction on $l$. Assume

$$a'_{m,l-1} = R_{i_{l-1}}\cdots R_{i_1}T^m a'_{0,0} + {}^{(i_1\cdots i_{l-1})}b'_{m,l-1} \tag{9.100}$$

Applying $R_{i_l}$ to this we obtain

$$R_{i_l}a'_{m,l-1} = R_{i_l}\cdots R_{i_1}T^m a'_{0,0} + R_{i_l}{}^{(i_1\cdots i_{l-1})}b'_{m,l-1} \tag{9.101}$$

On the other hand, from the definition in Proposition 9.2,

$$\begin{aligned} {}^{(i_1\cdots i_{l-1})}a'_{m,l-1} =& L(R_{i_{l-1}}\cdots R_{i_1}T^m\Delta\mu) \\ &+ \mathrm{tr}\chi(R_{i_{l-1}}\cdots R_{i_1}T^m\Delta\mu) + 2\hat{\chi}\cdot{}^{(i_1\cdots i_{l-1})}\hat{\not{\mu}}_{2,m,l-1} \end{aligned} \tag{9.102}$$

Applying $R_{i_l}$ to this and using Lemma 9.3,

$$
\begin{aligned}
R_{i_l}{}^{(i_1\cdots i_{l-1})}a'_{m,l-1} = & R_{i_l}L(R_{i_{l-1}}\cdots R_{i_1}T^m \not{\Delta}\mu) + \mathrm{tr}\chi(R_{i_l}\cdots R_{i_1}T^m \not{\Delta}\mu) \\
& + 2\hat{\chi}\cdot {}^{(i_1\cdots i_l)}\hat{\not{\mu}}_{2,m,l} + (R_{i_l}\mathrm{tr}\chi)(R_{i_{l-1}}\cdots R_{i_1}T^m \not{\Delta}\mu) \\
& + 2(\hat{\not{\mathcal{L}}}_{R_{i_l}}\hat{\chi} - \mathrm{tr}^{(R_{i_l})}\not{\pi}\hat{\chi})\cdot {}^{(i_1\cdots i_{l-1})}\hat{\not{\mu}}_{2,m,l-1}
\end{aligned} \tag{9.103}
$$

Writing

$$
R_{i_l}L(R_{i_{l-1}}\cdots R_{i_1}T^m \not{\Delta}\mu) = L(R_{i_l}\cdots R_{i_1}T^m \not{\Delta}\mu) - {}^{(R_{i_l})}ZR_{i_{l-1}}\cdots R_{i_1}T^m \not{\Delta}\mu
$$

(9.103) becomes, in view of the definition of $a'_{m,l}$,

$$
R_{i_l}{}^{(i_1\cdots i_{l-1})}a'_{m,l-1} = {}^{(i_1\cdots i_l)}a'_{m,l} - {}^{(R_{i_l})}ZR_{i_{l-1}}\cdots R_{i_1}T^m \not{\Delta}\mu - c''_{m,l} \tag{9.104}
$$

Equating the two expressions for $R_{i_l}{}^{(i_1\cdots i_{l-1})}a'_{m,l-1}$, that is (9.101) with (9.104), we obtain the formula for ${}^{(i_1\cdots i_l)}a'_{m,l}$, if we set

$$
{}^{(i_1\cdots i_l)}b'_{m,l} = R_{i_l}{}^{(i_1\cdots i_{l-1})}b'_{m,l-1} + {}^{(R_{i_l})}ZR_{i_{l-1}}\cdots R_{i_1}T^m \not{\Delta}\mu + c''_{m,l} \tag{9.105}
$$

Using again Proposition 8.2, the result follows. □

Next, we shall investigate the order of various terms in the propagation equation of Proposition 9.2. In obtaining estimates for the $(n+1)$st order spatial derivatives of $\mu$, of which at least two are angular, we are to set $m+l=n-1$ in Proposition 9.2. The principal parts shall be of order $m+l+2=n+1$, and the principal acoustical parts shall be the principal parts in the spatial derivatives of $\chi$ and $\mu$.

From (9.26), (9.36) and (9.40), we know that $\check{f}'$ is of order 2, and contains no acoustical part of order 2. It follows that ${}^{(i_1\cdots i_l)}\check{f}_{m,l}$ is of order $m+l+2$, but contains no principal acoustical part.

We turn to ${}^{(i_1\cdots i_l)}g'_{m,l}$. From the expression of ${}^{(i_1\cdots i_l)}g'_{m,l}$, we must investigate $g'_{m,0}$. From (9.44), (9.36), (9.38), (9.35), (9.32), (9.30) and (9.24) we know that $\check{g}'$ is of order 2, and its principal acoustical part is given by (9.58), consisting of 1st angular derivatives of $\mathrm{tr}\chi$ multiplied by $\mu$. It follows that

$$
T^m\check{g}' \tag{9.106}
$$

the first term in $g'_{m,0}$ is of order $m+2$ and its principal acoustical part consists of

$$
\xi\cdot(\mu \not{d} T^m \mathrm{tr}\chi) \tag{9.107}
$$

or, in view of Equations (3.124)—(3.125),

$$
\begin{aligned}
& \xi\cdot(\mu \not{d}\,\mathrm{tr}\chi), m=0 \\
& \xi\cdot(\mu \not{d} T^{m-1}\not{\Delta}\mu), m\geqslant 1
\end{aligned} \tag{9.108}
$$

to principal acoustical parts. In view of (9.60) and (8.34), (9.108) can be expressed to principal acoustical parts in terms of

$$
\begin{aligned}
& \xi\cdot x_0, m=0 \\
& \xi\cdot \not{d}x'_{m-1,0}, m\geqslant 1
\end{aligned} \tag{9.109}
$$

The second term in $g'_{m,0}$ can be written as

$$\sum_{k=0}^{m-1} \Lambda T^k x'_{m-k-1,0} - \sum_{k=0}^{m-1} [\Lambda, T^k] x'_{m-k-1,0} \tag{9.110}$$

The commutator sum is of lower order $m+1$, while

$$\begin{aligned} T^k x'_{m-k-1,0} &= T^k(\mu T^{m-k-1} \not\Delta\mu - T^{m-k-1} \check{f}') \\ &= x'_{m-1,0} + \sum_{j=1}^{k} C_k^j (T^j\mu)(T^{m-j-1} \not\Delta\mu) \end{aligned} \tag{9.111}$$

The contribution of the second term in (9.111) to the first term of (9.110) is of lower order $m+1$, while the contribution of the first term is

$$m\Lambda x'_{m-1,0} \tag{9.112}$$

This is the principal part of the second term in $g'_{m,0}$.

The third term in $g'_{m,0}$ is

$$\sum_{k=0}^{m-1} T^k y'_{m-k,0} \tag{9.113}$$

Consider the expression for $y'_{j,0}$. Here, all terms except the first term

$$-(T\mu)a'_{j-1,0} = -(T\mu)(T^{j-1}a'_{0,0} + b'_{j-1,0}) \tag{9.114}$$

are of lower order $j+1$. The principal part in $a'_{0,0}$ is

$$\not\Delta m + \mu \not\Delta e$$

It follows that $T^{j-1}a'_{0,0}$ is of principal order $j+2$ but contains no principal acoustical part, its principal part being

$$T^{j-1} \not\Delta m + \mu T^{j-1} \not\Delta e$$

From Lemma 9.4,

$$b'_{j-1,0} = \sum_{k=0}^{j-2} T^k \Lambda T^{j-k-2} \not\Delta\mu + \sum_{k=0}^{j-2} T^k c'_{j-k-1} \tag{9.115}$$

The first sum in (9.115) is of order $j+1$, while from the expression of $c'_j$, the second sum in (9.115) is of order $j$. We conclude that the principal part of the third term in $g'_{m,0}$ is

$$-m(T\mu)T^{m-1}a'_{0,0} \tag{9.116}$$

This is of principal order $m+2$, but contains no principal acoustical part. This completes the investigation of $g'_{m,0}$.

We conclude from above that the contributions of (9.109), (9.112) and (9.116) to

$$R_{i_l} \cdots R_{i_1} g'_{m,0} \tag{9.117}$$

are all of principal order $l + m + 2$. The contribution of (9.109) to (9.117) is

$$\begin{aligned} &\xi \cdot {}^{(i_1 \cdots i_l)} x_l, m = 0 \\ &\xi \cdot \not{d}^{(i_1 \cdots i_l)} x'_{m-1,l}, m \geqslant 1 \end{aligned} \tag{9.118}$$

where $\xi$ is given by (9.59). By **H0**,

$$\begin{aligned} |\xi \cdot {}^{(i_1 \cdots i_l)} x_l| &\leqslant |\xi| |{}^{(i_1 \cdots i_l)} x_l| \\ |\xi \cdot \not{d}^{(i_1 \cdots i_l)} x'_{m-1,l}| &\leqslant C(1+t)^{-1} |\xi| \max_j |{}^{(i_1 \cdots i_l j)} x'_{m-1,l+1}| \end{aligned} \tag{9.119}$$

The contribution of (9.112) is

$$m \Lambda^{(i_1 \cdots i_l)} x'_{m-1,l} \tag{9.120}$$

From **H0**, this is bounded by

$$C m (1+t)^{-1} |\Lambda| \max_j |{}^{(i_1 \cdots i_l j)} x'_{m-1,l+1}| \tag{9.121}$$

Finally, the contribution of (9.116) is

$$-m(T\mu) R_{i_l} \cdots R_{i_1} T^{m-1} a'_{0,0} \tag{9.122}$$

This does not contain principal acoustical part. This completes the investigation of the first term of ${}^{(i_1 \cdots i_l)} g'_{m,l}$.

We turn to the second term of ${}^{(i_1 \cdots i_l)} g'_{m,l}$. This is

$$\sum_{k=0}^{l-1} R_{i_l} \cdots R_{i_{l-k+1}} {}^{(R_{i_{l-k}})} Z^{(i_1 \cdots i_{l-k-1})} x'_{m,l-k-1} \tag{9.123}$$

which we write as

$$\sum_{k=0}^{l-1} {}^{(R_{i_{l-k}})} Z R_{i_l} \cdots R_{i_{l-k+1}} {}^{(i_1 \cdots i_{l-k-1})} x'_{m,l-k-1} \tag{9.124}$$

minus

$$\sum_{k=0}^{l-1} [{}^{(R_{i_{l-k}})} Z, R_{i_l} \cdots R_{i_{l-k+1}}] {}^{(i_1 \cdots i_{l-k-1})} x'_{m,l-k-1} \tag{9.125}$$

Obviously, the commutator term is of lower order $l + m + 1$, while the principal part of (9.123) is

$$\sum_{k=0}^{l-1} {}^{(R_{i_{l-k}})} Z^{(i_1 \cdots i_{l-k-1} i_{l-k+1} \cdots i_l)} x'_{m,l-1} \tag{9.126}$$

This is a principal acoustical part. By **H0**, it can be bounded by

$$C(1+t)^{-1}\left(\sum_{k=1}^{l}|^{(R_{i_k})}Z|\right)\max_j|^{(i_1\cdots i_{k-1}i_{k+1}\cdots i_l j)}x'_{m,l}| \tag{9.127}$$

Finally, we consider the third term of $^{(i_1\cdots i_l)}g'_{m,l}$. This is

$$\sum_{k=0}^{l-1}R_{i_l}\cdots R_{i_{l-k+1}}{}^{(i_1\cdots i_{l-k})}y'_{m,l-k} \tag{9.128}$$

Note that only the first term in $^{(i_1\cdots i_j)}y'_{m,j}$ is of principal order $m+l+2$:

$$-(R_{i_j}\mu)^{(i_1\cdots i_{j-1})}a'_{m,j-1}=-(R_{i_j}\mu)(R_{i_{j-1}}\cdots R_{i_1}T^m a'_{0,0}+{}^{(i_1\cdots i_{j-1})}b'_{m,j-1}) \tag{9.129}$$

Since $a'_{0,0}$ is of order 3, but contains no acoustical part of order 3, it follows that $R_{i_{j-1}}\cdots R_{i_1}T^m a'_{0,0}$ is of principal order $m+j+2$ but contains no principal acoustical part. $^{(i_1\cdots i_{j-1})}b'_{m,j-1}$ is given by Lemma 9.4:

$$\begin{aligned}{}^{(i_1\cdots i_{j-1})}b'_{m,j-1}=&R_{i_{j-1}}\cdots R_{i_1}b'_{m,0}\\&+\sum_{k=0}^{j-2}R_{i_{j-1}}\cdots R_{i_{j-k}}{}^{(R_{i_{j-k-1}})}ZR_{i_{j-k-2}}\cdots R_{i_1}T^m\not\Delta\mu\\&+\sum_{k=0}^{j-2}R_{i_{j-1}}\cdots R_{i_{j-k}}{}^{(i_1\cdots i_{j-k-1})}c''_{m,j-k-1}\end{aligned} \tag{9.130}$$

By the discussions after (9.115), we know that the first term on the right of (9.130) is of order $m+j+1$. Obviously, the second term is also of order $m+j+1$. While $^{(i_1\cdots i_j)}c''_{m,j}$ given in Lemma 9.4 is of order $m+j+1$, hence, the third term on the right of (9.130) is of order $m+j$. Thus $^{(i_1\cdots i_{j-1})}b'_{m,j-1}$ is of lower order $m+j+1$. We conclude that the principal part of (9.128) is

$$-\sum_{k=0}^{l-1}(R_{i_{l-k}}\mu)R_{i_l}\cdots R_{i_{l-k+1}}R_{i_{l-k-1}}\cdots R_{i_1}T^m a'_{0,0} \tag{9.131}$$

This is of principal order $m+l+2$, but contains no principal acoustical part. This completes the investigation of $^{(i_1\cdots i_l)}g'_{m,l}$.

## 9.3 Elliptic Theory on $S_{t,u}$

Returning to the propagation equation for $^{(i_1\cdots i_l)}x'_{m,l}$ of Proposition 9.2, the term $2\mu\hat{\chi}\cdot{}^{(i_1\cdots i_l)}\hat{\not\mu}_{2,m,l}$ remains to be considered. Of course, $^{(i_1\cdots i_l)}\hat{\not\mu}_{2,m,l}$ is a principal acoustical part, it involves the $(m+l)$th order spatial derivatives of $\hat{\not\mu}_2$, not $\not\Delta\mu$, which is what the propagation equation for ${}^{(i_1\cdots i_l)}x'_{m,l}$ allows us to control. The term in question comes from (9.45), namely, from the term

$$2\mu\hat{\chi}\cdot\hat{\not{D}}^2\mu = 2\mu\hat{\chi}\cdot\hat{\not{\mu}}_2$$

Now, the propagation equation must be considered in conjunction with the definition (9.41):

$$\mu\not{\Delta}\mu = x' + \check{f}' \tag{9.132}$$

which is an elliptic equation for $\mu$ on each surface $S_{t,u}$. Then we can use this equation to estimate on $S_{t,u}$, $\not{d}\mu$ as well as $\hat{\not{\mu}}_2$ in terms of $x'$ and the 2nd derivatives of $\psi_\mu$. On the other hand, (9.45) allows us to estimate $x'$ along each generator of $C_u$ in terms of $\hat{\not{\mu}}_2$ and 2nd derivatives of $\psi_\mu$. Thus we can estimate $\not{d}\mu$ and $\not{D}^2\mu$ in terms of 2nd derivatives of $\psi_\mu$.

Similarly, the propagation equation for ${}^{(i_1\cdots i_l)}x'_{m,l}$ must be considered in conjunction with an elliptic equation for

$${}^{(i_1\cdots i_l)}\mu_{m,l} = R_{i_l}\cdots R_{i_1}T^m\mu \tag{9.133}$$

then we can estimate $\not{D}^2\,{}^{(i_1\cdots i_l)}\mu_{m,l}$ in terms of $(m+l+2)$nd derivatives of $\psi_\mu$. The required elliptic equation will be derived from (9.60):

$$\mu R_{i_l}\cdots R_{i_1}T^m\not{\Delta}\mu = {}^{(i_1\cdots i_l)}x'_{m,l} + {}^{(i_1\cdots i_l)}\check{f}'_{m,l} \tag{9.134}$$

with the help of the following lemmas.

**Lemma 9.5** Let $(M,g)$ be a 2-dimensional Riemannian manifold, let $X$ be an arbitrary vectorfield and $f$ an arbitrary function on $M$. Then the following commutation formulas hold:

$$X(\Delta_g f) - \Delta_g(Xf) = -{}^{(X)}\pi^{ab}(\nabla^2 f)_{ab} - \mathrm{tr}^{(X)}\pi_1^b d_b f$$

and

$$(\hat{\mathcal{L}}_X\hat{\nabla}^2 f)_{ab} - (\hat{\nabla}^2(Xf))_{ab} = -\frac{1}{2}{}^{(X)}\hat{\pi}_{ab}\Delta_g f - {}^{(X)}\hat{\pi}^c_{1,ab}d_c f$$

Here,

$$\begin{aligned}
\mathrm{tr}^{(X)}\pi_1^b &= \nabla^{a(X)}\pi_a^b - \frac{1}{2}\nabla^b\mathrm{tr}^{(X)}\pi \\
{}^{(X)}\hat{\pi}^c_{1,ab} &= \frac{1}{2}(\nabla_a{}^{(X)}\hat{\pi}_b^c + \nabla_b{}^{(X)}\hat{\pi}_a^c - \nabla^{c(X)}\hat{\pi}_{ab} - g_{ab}\nabla^{d(X)}\hat{\pi}_d^c) \\
&\quad + \frac{1}{4}(\delta_a^c d_b\mathrm{tr}^{(X)}\pi + \delta_b^c d_a\mathrm{tr}^{(X)}\pi - g_{ab}d^c\mathrm{tr}^{(X)}\pi)
\end{aligned}$$

where $\mathcal{L}_X g = {}^{(X)}\pi$.

*Proof*. Let $\phi_t$ be the local 1-parameter group generated by $X$ and let $\phi_{t*}$ be the corresponding pull back. Consider an arbitrary 1-form $\alpha$ on $M$. We then have

$$\phi_{t*}(\overset{g}{\nabla}\alpha) = \overset{\phi_{t*}g}{\nabla}(\phi_{t*}\alpha) \tag{9.135}$$

Now, in an arbitrary coordinates we have

$$(\overset{g}{\nabla}\alpha)_{ab} = \frac{\partial\alpha_b}{\partial x^a} - \overset{g}{\Gamma}{}^c_{ab}\alpha_c \tag{9.136}$$

where $\overset{g}{\Gamma}{}^c_{ab}$ is the Christoffel symbol of metric $g$ in these coordinates. Similarly, in the same coordinates,

$$(\overset{\phi_{t*}g}{\nabla}(\phi_{t*}\alpha))_{ab} = \frac{\partial(\phi_{t*}\alpha)_b}{\partial x^a} - \overset{\phi_{t*}g}{\Gamma}{}^c_{ab}(\phi_{t*}\alpha)_c \tag{9.137}$$

Differentiating (9.137) with respect to $t$ at $t = 0$ we obtain

$$\begin{aligned}&(\frac{d}{dt}(\overset{\phi_{t*}g}{\nabla}(\phi_{t*}\alpha))_{ab})_{t=0}\\ &= \frac{\partial}{\partial x^a}(\frac{d}{dt}(\phi_{t*}\alpha)_b)_{t=0} - \overset{g}{\Gamma}{}^c_{ab}(\frac{d}{dt}(\phi_{t*}\alpha)_c)_{t=0} - (\frac{d}{dt}\overset{\phi_{t*}g}{\Gamma}{}^c_{ab})_{t=0}\alpha_c\end{aligned} \tag{9.138}$$

Now, from the definition,

$$(\frac{d}{dt}(\phi_{t*}\alpha))_{t=0} = \mathcal{L}_X\alpha$$

By (9.135), the left hand-side of (9.138) is the $ab$-component of

$$(\frac{d}{dt}(\phi_{t*}(\overset{g}{\nabla}\alpha)))_{t=0} = \mathcal{L}_X(\nabla\alpha)$$

Recalling (8.109), we have

$$(\frac{d}{dt}\overset{\phi_{t*}g}{\Gamma}{}^c_{ab})_{t=0} = {}^{(X)}\pi^c_{1,ab} \tag{9.139}$$

where

$${}^{(X)}\pi^c_{1,ab} = \frac{1}{2}(\nabla_a{}^{(X)}\pi^c_b + \nabla_b{}^{(X)}\pi^c_a - \nabla^{c(X)}\pi_{ab}) \tag{9.140}$$

we conclude that

$$(\mathcal{L}_X(\nabla\alpha))_{ab} = (\nabla(\mathcal{L}_X\alpha))_{ab} - {}^{(X)}\pi^c_{1,ab}\alpha_c \tag{9.141}$$

In particular, in the case $\alpha = df$ for some function $f$, since

$$\mathcal{L}_X(df) = d(Xf)$$

we obtain

$$(\mathcal{L}_X(\nabla^2 f))_{ab} = (\nabla^2(Xf))_{ab} - {}^{(X)}\pi^c_{1,ab}d_c f \tag{9.142}$$

for any function $f$ on $M$. Since

$$\Delta_g f = (g^{-1})^{ab}(\nabla^2 f)_{ab}$$

it follows that

$$X(\Delta_g f) = \mathcal{L}_X(\Delta_g f) = -{}^{(X)}\pi^{ab}(\nabla^2 f)_{ab} + (g^{-1})^{ab}(\mathcal{L}_X(\nabla^2 f))_{ab} \tag{9.143}$$

Since

$$(g^{-1})^{ab}(\nabla^2(Xf))_{ab} = \Delta_g(Xf)$$

then by (9.142) and (9.143), the first part of the lemma follows. In fact, the first part holds for an arbitrary $n$-dimensional manifold $M$.

For the second part, we consider the case of 2-dimensional manifold $M$. The trace-free part of $\nabla^2 f$ is

$$\hat{\nabla}^2 f = \nabla^2 f - \frac{1}{2} g \Delta_g f$$

hence

$$\mathcal{L}_X(\hat{\nabla}^2 f) = \mathcal{L}_X(\nabla^2 f) - \frac{1}{2}{}^{(X)}\pi \Delta_g f - \frac{1}{2} g X(\Delta_g f) \tag{9.144}$$

Substituting from (9.142) and the first part of the lemma we then obtain

$$\begin{aligned}(\mathcal{L}_X(\hat{\nabla}^2 f))_{ab} =& (\nabla^2(Xf))_{ab} - {}^{(X)}\pi^c_{1,ab} d_c f - \frac{1}{2}{}^{(X)}\pi_{ab}\Delta_g f \\ & - \frac{1}{2} g_{ab}(\Delta_g(Xf) - {}^{(X)}\pi^{cd}(\nabla^2 f)_{cd} - \mathrm{tr}^{(X)}\pi^c_1 d_c f)\end{aligned} \tag{9.145}$$

Taking the trace-free part on both sides of above, then we get the second part of the lemma. □

**Lemma 9.6** Let $f$ be an arbitrary function defined on a given hypersurface $\Sigma_t$. Then the following commutation formulas hold:

$$T(\not{\Delta} f) - \not{\Delta}(Tf) = -{}^{(T)}\not{\pi}^{AB}(\not{D}^2 f)_{AB} - \mathrm{tr}^{(T)}\not{\pi}_1^B \not{d}_B f$$

and

$$(\hat{\not{\mathcal{L}}}_T \hat{\not{D}}^2 f)_{AB} - (\hat{\not{D}}^2(Tf))_{AB} = -\frac{1}{2}{}^{(T)}\hat{\not{\pi}}_{AB}\not{\Delta} f - {}^{(T)}\hat{\not{\pi}}^C_{1,AB}\not{d}_C f$$

Here,

$${}^{(T)}\not{\pi}^C_{1,AB} = \frac{1}{2}(\not{D}_A{}^{(T)}\not{\pi}^C_B + \not{D}_B{}^{(T)}\not{\pi}^C_A - \not{D}^{C(T)}\not{\pi}_{AB})$$

is the Lie derivative with respect to $T$ of the induced connection on $S_{t,u}$, and the trace in the lemma is with respect to the induced metric on $S_{t,u}$.

*Proof*. Since $T$ is tangential to $\Sigma_t$, we can confine attention to $\Sigma_t$. Also, we can choose acoustical coordinates $(\vartheta_1, \vartheta_2)$ so that $\Xi = 0$ on the given $\Sigma_t$. So on this $\Sigma_t$,

$$T = \frac{\partial}{\partial u} \tag{9.146}$$

Consider an arbitrary $S_{t,u}$ 1-form $\alpha$ defined on $\Sigma_t$. In terms of $(u, \vartheta_1, \vartheta_2)$,

$$(\not D\alpha)_{AB} = \frac{\partial \alpha_B}{\partial \vartheta^A} - \not\Gamma^C_{AB}\alpha_C \tag{9.147}$$

Differentiating (9.147) with respect to $u$ we obtain

$$\frac{\partial}{\partial u}(\not D\alpha)_{AB} = \frac{\partial}{\partial \vartheta^A}(\frac{\partial \alpha_B}{\partial u}) - \not\Gamma^C_{AB}\frac{\partial \alpha_C}{\partial u} - \frac{\partial \not\Gamma^C_{AB}}{\partial u}\alpha_C \tag{9.148}$$

In view of (9.146),

$$\frac{\partial \alpha_A}{\partial u} = (\not\mathcal{L}_T\alpha)_A, \quad \frac{\partial}{\partial u}(\not D\alpha)_{AB} = (\not\mathcal{L}_T(\not D\alpha))_{AB} \tag{9.149}$$

Moreover, from Chapter 3 we have, in view of (9.146),

$$\frac{\partial \not g_{AB}}{\partial u} = {}^{(T)}\not\pi_{AB} = 2\kappa\theta_{AB}, \quad \frac{\partial}{\partial u}(\not g^{-1})^{AB} = -{}^{(T)}\not\pi^{AB} = -2\kappa\theta^{AB} \tag{9.150}$$

hence

$$\begin{aligned}
\frac{\partial \not\Gamma^C_{AB}}{\partial u} &= \frac{1}{2}\frac{\partial}{\partial u}((\not g^{-1})^{CD}(\frac{\partial \not g_{BD}}{\partial \vartheta_A} + \frac{\partial \not g_{AD}}{\partial \vartheta_B} - \frac{\partial \not g_{AB}}{\partial \vartheta^D})) \\
&= -{}^{(T)}\not\pi^{CD}\not\Gamma^E_{AB}\not g_{DE} + \frac{1}{2}(\not g^{-1})^{CD}(\frac{\partial^{(T)}\not\pi_{BD}}{\partial \vartheta^A} + \frac{\partial^{(T)}\not\pi_{AD}}{\partial \vartheta^B} - \frac{\partial^{(T)}\not\pi_{AB}}{\partial \vartheta^D}) \\
&= \frac{1}{2}(\not g^{-1})^{CD}(\not D_A{}^{(T)}\not\pi_{BD} + \not D_B{}^{(T)}\not\pi_{AD} - \not D_D{}^{(T)}\not\pi_{AB})
\end{aligned}$$

that is

$$\frac{\partial \not\Gamma^C_{AB}}{\partial u} = {}^{(T)}\not\pi^C_{1,AB} \tag{9.151}$$

where

$${}^{(T)}\not\pi^C_{1,AB} = \frac{1}{2}(\not D_A{}^{(T)}\not\pi^C_B + \not D_B{}^{(T)}\not\pi^C_A - \not D^{C(T)}\not\pi_{AB}) \tag{9.152}$$

Substituting (9.151) in (9.148), and in view of (9.149), we then obtain

$$(\not\mathcal{L}_T(\not D\alpha))_{AB} = (\not D(\not\mathcal{L}_T\alpha))_{AB} - {}^{(T)}\not\pi^C_{1,AB}\alpha_C \tag{9.153}$$

This holds for any $S_{t,u}$ 1-form $\alpha$ defined on $\Sigma_t$. In particular, it holds in the case $\alpha = \not d f$ for some function $f$ defined on $\Sigma_t$. In this case, we have,

$$\alpha_A = \frac{\partial f}{\partial \vartheta^A}, \quad \frac{\partial \alpha_A}{\partial u} = \frac{\partial}{\partial \vartheta^A}(\frac{\partial f}{\partial u})$$

i.e.

$$\not\mathcal{L}_T(\not d f) = \not d(Tf) \tag{9.154}$$

So by (9.153) we have

$$(\not{\mathcal{L}}_T(\not{D}^2 f))_{AB} = (\not{D}^2(Tf))_{AB} - {}^{(T)}\not{\pi}^C_{1,AB}\not{d}_C f \tag{9.155}$$

for any function $f$ on $\Sigma_t$. Since

$$\not{\Delta} f = (\not{g}^{-1})^{AB}(\not{D}^2 f)_{AB}$$

it follows by (9.150) that

$$T(\not{\Delta} f) = \not{\mathcal{L}}_T(\not{\Delta} f) = -{}^{(T)}\not{\pi}^{AB}(\not{D}^2 f)_{AB} + (\not{g}^{-1})^{AB}(\not{\mathcal{L}}_T(\not{D}^2 f))_{AB} \tag{9.156}$$

Since

$$(\not{g}^{-1})^{AB}(\not{D}^2(Tf))_{AB} = \not{\Delta}(Tf)$$

the first part of the lemma follows.

To prove the second part, we note that the trace-free part of $\not{D}^2 f$ is

$$\hat{\not{D}}^2 f = \not{D}^2 f - \frac{1}{2}\not{g}\not{\Delta} f$$

hence

$$\not{\mathcal{L}}_T(\hat{\not{D}}^2 f) = \not{\mathcal{L}}_T(\not{D}^2 f) - \frac{1}{2}{}^{(T)}\not{\pi}\not{\Delta} f - \frac{1}{2}\not{g}T(\not{\Delta} f) \tag{9.157}$$

So using (9.155) and the first part of the lemma we obtain

$$\begin{aligned}(\not{\mathcal{L}}_T(\hat{\not{D}}^2 f))_{AB} =& (\not{D}^2(Tf))_{AB} - {}^{(T)}\not{\pi}^C_{1,AB}\not{d}_C f - \frac{1}{2}{}^{(T)}\not{\pi}_{AB}\not{\Delta} f \\ &- \frac{1}{2}\not{g}_{AB}(\not{\Delta}(Tf) - {}^{(T)}\not{\pi}^{CD}(\not{D}^2 f)_{CD} - \mathrm{tr}^{(T)}\not{\pi}^C_1\not{d}_C f)\end{aligned} \tag{9.158}$$

Then taking the trace-free part on the above, we get the second part of the lemma. □

**Proposition 9.3** For each pair of non-negative integers $(m, l)$ and each multi-index $(i_1 \cdots i_l)$ we have

$$\not{\Delta}^{(i_1\cdots i_l)}\mu_{m,l} - R_{i_l}\cdots R_{i_1}T^m\not{\Delta}\mu = {}^{(i_1\cdots i_l)}d_{m,l}$$

where

$$\begin{aligned}{}^{(i_1\cdots i_l)}d_{m,l} =& R_{i_l}\cdots R_{i_1}d_{m,0} + \sum_{k=0}^{l-1} R_{i_l}\cdots R_{i_{l-k+1}}({}^{(R_{i_{l-k}})}\not{\pi}\cdot\not{D}^{2(i_1\cdots i_{l-k-1})}\mu_{m,l-k-1}) \\ &+ \sum_{k=0}^{l-1} R_{i_l}\cdots R_{i_{l-k+1}}(\mathrm{tr}^{(R_{i_{l-k}})}\not{\pi}_1\cdot\not{d}^{(i_1\cdots i_{l-k-1})}\mu_{m,l-k-1})\end{aligned}$$

and

$$d_{m,0} = \sum_{k=0}^{m-1} T^k({}^{(T)}\not{\pi} \cdot \not{D}^2 \mu_{m-k-1,0}) + \sum_{k=0}^{m-1} T^k(\mathrm{tr}{}^{(T)}\not{\pi}_1 \cdot \not{d}\mu_{m-k-1,0})$$

*Proof.* We first consider the case $l = 0$. If also $m = 0$ we have trivially

$$d_{0,0} = 0 \tag{9.159}$$

We shall derive a recursion formula for $d_{m,0}$. Consider then

$$d_{m-1,0} = \not{\Delta}\mu_{m-1,0} - T^{m-1}\not{\Delta}\mu$$

Applying $T$ to this we obtain

$$Td_{m-1,0} = T(\not{\Delta}\mu_{m-1,0}) - T^m \not{\Delta}\mu \tag{9.160}$$

From the first part of Lemma 9.6,

$$T(\not{\Delta}\mu_{m-1,0}) = \not{\Delta}\mu_{m,0} - {}^{(T)}\not{\pi} \cdot \not{D}^2 \mu_{m-1,0} - \mathrm{tr}{}^{(T)}\not{\pi}_1 \cdot \not{d}\mu_{m-1,0} \tag{9.161}$$

Substituting (9.161) in (9.160) and noting that

$$\not{\Delta}\mu_{m,0} - T^m \not{\Delta}\mu = d_{m,0}$$

we obtain the recursion relation

$$d_{m,0} = Td_{m-1,0} + {}^{(T)}\not{\pi} \cdot \not{D}^2 \mu_{m-1,0} + \mathrm{tr}{}^{(T)}\not{\pi}_1 \cdot \not{d}\mu_{m-1,0} \tag{9.162}$$

Applying Proposition 8.2, we get the expression for $d_{m,0}$.

Next, we consider the case $l \geqslant 1$. We have

$$^{(i_1 \cdots i_{l-1})}d_{m,l-1} = \not{\Delta}^{(i_1 \cdots i_{l-1})}\mu_{m,l-1} - R_{i_{l-1}} \cdots R_{i_1} T^m \not{\Delta}\mu$$

Applying $R_{i_l}$ to this we obtain

$$R_{i_l}{}^{(i_1 \cdots i_{l-1})}d_{m,l-1} = R_{i_l}\not{\Delta}^{(i_1 \cdots i_{l-1})}\mu_{m,l-1} - R_{i_l} \cdots R_{i_1} T^m \not{\Delta}\mu \tag{9.163}$$

From the first part of Lemma 9.5 we have

$$\begin{aligned} &R_{i_l}\not{\Delta}^{(i_1 \cdots i_{l-1})}\mu_{m,l-1} \\ &= \not{\Delta}^{(i_1 \cdots i_l)}\mu_{m,l} - {}^{(R_{i_l})}\not{\pi} \cdot \not{D}^{2(i_1 \cdots i_{l-1})}\mu_{m,l-1} - \mathrm{tr}{}^{(R_{i_l})}\not{\pi}_1 \cdot \not{d}^{(i_1 \cdots i_{l-1})}\mu_{m,l-1} \end{aligned} \tag{9.164}$$

Substituting (9.164) in (9.163) and noting that

$$\not{\Delta}^{(i_1 \cdots i_l)}\mu_{m,l} - R_{i_l} \cdots R_{i_1} T^m \not{\Delta}\mu = {}^{(i_1 \cdots i_l)}d_{m,l}$$

we obtain the recursion relation

$$\begin{aligned} &^{(i_1 \cdots i_l)}d_{m,l} \\ &= R_{i_l}{}^{(i_1 \cdots i_{l-1})}d_{m,l-1} + {}^{(R_{i_l})}\not{\pi} \cdot \not{D}^{2(i_1 \cdots i_{l-1})}\mu_{m,l-1} \\ &\quad + \mathrm{tr}{}^{(R_{i_l})}\not{\pi}_1 \cdot \not{d}^{(i_1 \cdots i_{l-1})}\mu_{m,l-1} \end{aligned} \tag{9.165}$$

Applying again Proposition 8.2, the result follows. □

**Proposition 9.4** For each pair of non-negative integers $(m, l)$ and each multi-index $(i_1 \cdots i_l)$ we have

$$\hat{\not{D}}^2 {}^{(i_1\cdots i_l)}\mu_{m,l} - {}^{(i_1\cdots i_l)}\hat{\not{\mu}}_{2,m,l} = {}^{(i_1\cdots i_l)}e_{m,l}$$

where

$$\begin{aligned}{}^{(i_1\cdots i_l)}e_{m,l} = {} & \hat{\not{\mathcal{L}}}_{R_{i_l}} \cdots \hat{\not{\mathcal{L}}}_{R_{i_1}} e_{m,0} \\ & + \frac{1}{2}\sum_{k=0}^{l-1} \hat{\not{\mathcal{L}}}_{R_{i_l}} \cdots \hat{\not{\mathcal{L}}}_{R_{i_{l-k+1}}} \left( {}^{(R_{i_{l-k}})}\hat{\not{\pi}} \not{\Delta}^{(i_1\cdots i_{l-k-1})}\mu_{m,l-k-1} \right) \\ & + \sum_{k=0}^{l-1} \hat{\not{\mathcal{L}}}_{R_{i_l}} \cdots \hat{\not{\mathcal{L}}}_{R_{i_{l-k+1}}} \left( {}^{(R_{i_{l-k}})}\hat{\not{\pi}}_1 \cdot \not{d}^{(i_1\cdots i_{l-k-1})}\mu_{m,l-k-1} \right)\end{aligned}$$

and

$$e_{m,0} = \frac{1}{2}\sum_{k=0}^{m-1} \hat{\not{\mathcal{L}}}_T^k \left( {}^{(T)}\hat{\not{\pi}} \not{\Delta}\mu_{m-k-1,0} \right) + \sum_{k=0}^{m-1} \hat{\not{\mathcal{L}}}_T^k \left( {}^{(T)}\hat{\not{\pi}}_1 \cdot \not{d}\mu_{m-k-1,0} \right)$$

*Proof.* We first consider the case $l = 0$. If also $m = 0$, from the definition, we have

$$e_{0,0} = 0 \tag{9.166}$$

Consider then

$$e_{m-1,0} = \hat{\not{D}}^2 \mu_{m-1,0} - \hat{\not{\mu}}_{2,m-1,0}$$

Apply $\hat{\not{\mathcal{L}}}_T$ to this to obtain

$$\hat{\not{\mathcal{L}}}_T e_{m-1,0} = \hat{\not{\mathcal{L}}}_T \hat{\not{D}}^2 \mu_{m-1,0} - \hat{\not{\mu}}_{2,m,0} \tag{9.167}$$

From the second part of Lemma 9.6,

$$\hat{\not{\mathcal{L}}}_T \hat{\not{D}}^2 \mu_{m-1,0} = \hat{\not{D}}^2 \mu_{m,0} - \frac{1}{2}{}^{(T)}\hat{\not{\pi}} \not{\Delta}\mu_{m-1,0} - {}^{(T)}\hat{\not{\pi}}_1 \cdot \not{d}\mu_{m-1,0} \tag{9.168}$$

Substituting (9.168) in (9.167) and noting that

$$\hat{\not{D}}^2 \mu_{m,0} - \hat{\not{\mu}}_{2,m,0} = e_{m,0}$$

we obtain

$$e_{m,0} = \hat{\not{\mathcal{L}}}_T e_{m-1,0} + \frac{1}{2}{}^{(T)}\hat{\not{\pi}} \not{\Delta}\mu_{m-1,0} + {}^{(T)}\hat{\not{\pi}}_1 \cdot \not{d}\mu_{m-1,0} \tag{9.169}$$

Applying Proposition 8.2, the expression for $e_{m,0}$ follows.

Next, we consider the case $l \geqslant 1$. Consider

$$^{(i_1\cdots i_{l-1})}e_{m,l-1} = \hat{\not{D}}^{2\,(i_1\cdots i_{l-1})}\mu_{m,l-1} - {}^{(i_1\cdots i_{l-1})}\hat{\not{\mu}}_{2,m,l-1}$$

Applying $\hat{\not{\mathcal{L}}}_{R_{i_l}}$ to this we obtain

$$\hat{\not{\mathcal{L}}}_{R_{i_l}}{}^{(i_1\cdots i_{l-1})}e_{m,l-1} = \hat{\not{\mathcal{L}}}_{R_{i_l}}\hat{\not{D}}^{2\,(i_1\cdots i_{l-1})}\mu_{m,l-1} - {}^{(i_1\cdots i_l)}\hat{\not{\mu}}_{2,m,l} \tag{9.170}$$

From the second part of Lemma 9.5,

$$\begin{aligned}&\hat{\not{\mathcal{L}}}_{R_{i_l}}\hat{\not{D}}^{2\,(i_1\cdots i_{l-1})}\mu_{m,l-1}\\ &= \hat{\not{D}}^{2\,(i_1\cdots i_l)}\mu_{m,l} - \frac{1}{2}{}^{(R_{i_l})}\hat{\not{\pi}}\not{\Delta}^{(i_1\cdots i_{l-1})}\mu_{m,l-1} - {}^{(R_{i_l})}\hat{\not{\pi}}_1 \cdot \not{d}^{(i_1\cdots i_{l-1})}\mu_{m,l-1}\end{aligned} \tag{9.171}$$

Substituting (9.171) in (9.170) and noting that

$$\hat{\not{D}}^{2\,(i_1\cdots i_l)}\mu_{m,l} - {}^{(i_1\cdots i_l)}\hat{\not{\mu}}_{2,m,l} = {}^{(i_1\cdots i_l)}e_{m,l}$$

we obtain

$$\begin{aligned}{}^{(i_1\cdots i_l)}e_{m,l} =& \hat{\not{\mathcal{L}}}_{R_{i_l}}{}^{(i_1\cdots i_{l-1})}e_{m,l-1}\\ &+ \frac{1}{2}{}^{(R_{i_l})}\hat{\not{\pi}}\not{\Delta}^{(i_1\cdots i_{l-1})}\mu_{m,l-1} + {}^{(R_{i_l})}\hat{\not{\pi}}_1 \cdot \not{d}^{(i_1\cdots i_{l-1})}\mu_{m,l-1}\end{aligned} \tag{9.172}$$

Thus, applying again Proposition 8.2, the result follows. □

Let us investigate the order of ${}^{(i_1\cdots i_l)}d_{m,l}$. First, we should investigate $d_{m,0}$. Here, the principal parts are of order $m+1$ and are contributed by $\not{\mathcal{L}}_T^k(\not{D}^2\mu_{m-k-1})$ in the first sum, a 2nd angular derivative of $T^{m-1}\mu$ to principal parts, and by $\not{\mathcal{L}}_T^{m-1}\mathrm{tr}^{(T)}\not{\pi}_1$ in the second sum, a 1st angular derivative of $\not{\mathcal{L}}_T^{m-1}\theta$ to principal parts. The principal acoustical part of the latter is a first angular derivative of $\chi$ if $m=1$, a 3rd angular derivative of $T^{m-2}\mu$ if $m \geqslant 2$ (by (3.124)—(3.125)). It follows that the principal parts in the first term of $d_{m,l}$

$$R_{i_l}\cdots R_{i_1}d_{m,0}$$

are of order $m+l+1$, the principal acoustical parts being $(l+1)$st angular derivatives of $\chi$ and $(l+2)$nd angular derivatives of $\mu$ if $m=1$, $(l+3)$rd angular derivatives of $T^{m-2}\mu$ and $(l+2)$nd angular derivatives of $T^{m-1}\mu$ if $m \geqslant 2$.

The principal parts in the second term of ${}^{(i_1\cdots i_l)}d_{m,l}$

$$\sum_{k=0}^{l-1} R_{i_l}\cdots R_{i_{l-k+1}}({}^{(R_{i_{l-k}})}\not{\pi}\cdot\not{D}^2\mu_{m,l-k-1})$$

are also of order $m+l+1$, being $(l+1)$st angular derivatives of $T^m\mu$, $l \geqslant 1$. Finally, the principal part in the third term of ${}^{(i_1\cdots i_l)}d_{m,l}$:

$$\sum_{k=0}^{l-1} R_{i_l}\cdots R_{i_{l-k+1}}(\mathrm{tr}^{(R_{i_{l-k}})}\not{\pi}_1\cdot\not{d}^{(i_1\cdots i_{l-k-1})}\mu_{m,l-k-1})$$

is

$$(\not{\mathcal{L}}_{R_{i_l}} \cdots \not{\mathcal{L}}_{R_{i_2}} \mathrm{tr}^{(R_{i_1})}\not{\pi}_1) \cdot \not{d}\mu_{m,0}$$

From the expression for ${}^{(R_i)}\pi_{AB}$, this term involves the $l$th angular derivatives of $\chi$ and is thus of order $l+1$, which coincides with the order of the principal parts in the first two terms of the expression for ${}^{(i_1\cdots i_l)}d_{m,l}$ if $m = 0$. We conclude that ${}^{(i_1\cdots i_l)}d_{m,l}$ is of order $m+l+1$—one less than principal—and its leading acoustical parts are $l$th angular derivatives of $\chi$ and (for $l \geqslant 1$) $(l+1)$st angular derivatives of $\mu$ if $m = 0$, $(l+1)$st angular derivatives of $\chi$ and $(l+2)$nd angular derivatives of $\mu$ and (for $l \geqslant 1$) $(l+1)$st angular derivatives of $T\mu$ if $m = 1$, $(l+3)$rd angular derivatives of $T^{m-2}\mu$ and $(l+2)$nd angular derivatives of $T^{m-1}\mu$ and (for $l \geqslant 1$) $(l+1)$st angular derivatives of $T^m\mu$ if $m \geqslant 2$.

Let us also investigate the order of ${}^{(i_1\cdots i_l)}e_{m,l}$. First, we should investigate $e_{m,0}$. Here, the principal parts are of order $m+1$ and are contributed by $T^k(\not{\Delta}\mu_{m-k-1,0})$ in the first sum, a 2nd angular derivative of $T^{m-1}\mu$ to principal parts and by $\hat{\not{\mathcal{L}}}_T^{m-1}{}^{(T)}\hat{\not{\pi}}_1$ in the second sum, a 1st angular derivative of $\not{\mathcal{L}}_T^{m-1}\theta$ to principal parts. The principal acoustical part of the latter is a 1st angular derivative of $\chi$ if $m = 1$, a 3rd angular derivative of $T^{m-2}\mu$ if $m \geqslant 2$ (by (3.124)—(3.125)). It follows that the principal parts in the first term of ${}^{(i_1\cdots i_l)}e_{m,l}$

$$\hat{\not{\mathcal{L}}}_{R_{i_l}} \cdots \hat{\not{\mathcal{L}}}_{R_{i_1}} e_{m,0}$$

are of order $m+l+1$, the leading acoustical parts being $(l+1)$st angular derivatives of $\chi$ and $(l+2)$nd angular derivatives of $\mu$ if $m = 1$, $(l+3)$rd angular derivatives of $T^{m-2}\mu$ and $(l+2)$nd angular derivatives of $T^{m-1}\mu$ if $m \geqslant 2$. The principal parts in the second term of ${}^{(i_1\cdots i_l)}e_{m,l}$

$$\frac{1}{2}\sum_{k=0}^{l-1} \hat{\not{\mathcal{L}}}_{R_{i_l}} \cdots \hat{\not{\mathcal{L}}}_{R_{i_{l-k+1}}}({}^{(R_{i_{l-k}})}\hat{\not{\pi}}\not{\Delta}^{(i_1\cdots i_{l-k-1})}\mu_{m,l-k-1})$$

are also of order $m+l+1$, being $(l+1)$st angular derivatives of $T^m\mu, l \geqslant 1$. Finally, the principal part in the third term of ${}^{(i_1\cdots i_l)}e_{m,l}$:

$$\sum_{k=0}^{l-1} \hat{\not{\mathcal{L}}}_{R_{i_l}} \cdots \hat{\not{\mathcal{L}}}_{R_{i_{l-k+1}}}({}^{(R_{i_{l-k}})}\hat{\not{\pi}}_1 \cdot \not{d}^{(i_1\cdots i_{l-k-1})}\mu_{m,l-k-1})$$

is

$$(\hat{\not{\mathcal{L}}}_{R_{i_l}} \cdots \hat{\not{\mathcal{L}}}_{R_{i_2}} {}^{(R_{i_1})}\hat{\not{\pi}}_1) \cdot \not{d}\mu_{m,0}$$

This term involves the $l$th order angular derivatives of $\chi$ and is thus of order $l+1$, which coincides with the order of the principal parts in the first two terms of ${}^{(i_1\cdots i_l)}e_{m,l}$ if $m = 0$. We conclude that ${}^{(i_1\cdots i_l)}e_{m,l}$ is of order $m+l+1$—one less than principal—and its leading acoustical parts are $l$th angular derivatives of $\chi$ and (for $l \geqslant 1$) $(l+1)$st angular derivatives of $\mu$ if $m = 0$, $(l+1)$st angular derivatives of $\chi$ and $(l+2)$nd angular derivative of $\mu$ and (for $l \geqslant 1$) $(l+1)$st angular derivatives of $T\mu$ if $m=1$, $(l+3)$rd angular derivatives of $T^{m-2}\mu$ and $(l+2)$nd angular derivatives of $T^{m-1}\mu$ and (for $l \geqslant 1$) $(l+1)$st angular derivatives of $T^m\mu$ if $m \geqslant 2$.

By Equation (9.134) and Proposition 9.3, ${}^{(i_1\cdots i_l)}\mu_{m,l}$ satisfies on each $S_{t,u}$ the elliptic equation

$$\not{\Delta}^{(i_1\cdots i_l)}\mu_{m,l} = \mu^{-1}({}^{(i_1\cdots i_l)}x'_{m,l} + {}^{(i_1\cdots i_l)}\check{f}'_{m,l}) + {}^{(i_1\cdots i_l)}d_{m,l} \tag{9.173}$$

We shall need the following $\mu$-weighted $L^2$ estimate:

**Lemma 9.7** Let $(M,g)$ be a compact 2-dimensional Riemannian manifold, and let $\phi$ be a function on $(M,g)$ satisfying the equation

$$\Delta_g\phi = \rho$$

for some function $\rho$ on $M$. Let also $\mu$ be an arbitrary non-negative function on $M$. Then the following estimate holds on $M$:

$$\int_M \mu^2(\frac{1}{2}|\nabla^2\phi|^2 + K|d\phi|^2)d\mu_g \leqslant 2\int_M \mu^2\rho^2 d\mu_g + 3\int_M |d\mu|^2|d\phi|^2 d\mu_g$$

where $K$ is the Gauss curvature of $(M,g)$.

*Proof*. Consider the 1-form

$$\psi = d\phi \tag{9.174}$$

We have

$$\mathrm{div}_g\psi = \Delta_g\phi = \rho \tag{9.175}$$

So we get

$$\mu d(\mathrm{div}_g\psi) = d(\mu\rho) - \rho d\mu \tag{9.176}$$

Now, in arbitrary local coordinate $d(\mathrm{div}_g\psi)$ is

$$\nabla_a(\nabla^b\psi_b)$$

and we have

$$\nabla_a(\nabla^b\psi_b) = \nabla^b(\nabla_a\psi_b) - S_a{}^b\psi_b \tag{9.177}$$

where $S_a{}^b = S_{ac}(g^{-1})^{bc}$ and $S_{ac}$ are the components of the Ricci curvature of $(M,g)$. Since $\mathrm{dim}M = 2$, we have

$$S_a{}^b = K\delta_a^b$$

Noting also that

$$\nabla_a\psi_b = \nabla_b\psi_a$$

(9.177) reduces to

$$\nabla_a(\nabla^b\psi_b) = \nabla^b(\nabla_b\psi_a) - K\psi_a \tag{9.178}$$

Substituting in (9.176) we obtain

$$\mu\nabla^b(\nabla_b\psi_a) - \mu K\psi_a = d_a(\mu\rho) - \rho d_a\mu \tag{9.179}$$

We multiply this equation by $-\mu\psi^a$ and integrate over $M$. We have, integrating by parts,

$$-\int_M \mu^2\psi^a\nabla^b(\nabla_b\psi_a)d\mu_g = \int_M \mu^2|\nabla\psi|^2 d\mu_g + \int_M 2\mu I^a d_a\mu d\mu_g \tag{9.180}$$

where

$$I^a = \psi^b\nabla_b\psi^a \tag{9.181}$$

Moreover, we have, also integrating by parts,

$$-\int_M \mu\psi^a d_a(\mu\rho)d\mu_g = \int_M \mu\rho\,\mathrm{div}_g(\mu\psi)d\mu_g = \int_M (\mu^2\rho^2 + \mu\rho\psi^a d_a\mu)d\mu_g \tag{9.182}$$

In view of (9.180) and (9.182), we obtain

$$\int_M \mu^2(|\nabla\psi|^2 + K|\psi|^2)d\mu_g + \int_M 2\mu I^a d_a\mu d\mu_g = \int_M (\mu^2\rho^2 + 2\mu\rho\psi^a d_a\mu)d\mu_g \tag{9.183}$$

Now, we have

$$2\mu|I^a d_a\mu| \leqslant 2\mu|I||d\mu|$$

and

$$|I| \leqslant |\nabla\psi||\psi|$$

hence we can estimate

$$2\mu|I^a d_a\mu| \leqslant 2\mu|d\mu||\psi||\nabla\psi| \leqslant \frac{1}{2}\mu^2|\nabla\psi|^2 + 2|d\mu|^2|\psi|^2 \tag{9.184}$$

Also we can estimate

$$2\mu|\rho\psi^a d_a\mu| \leqslant \mu^2\rho^2 + |d\mu|^2|\psi|^2 \tag{9.185}$$

In view of (9.184)—(9.185), the identity (9.183) implies

$$\int_M \mu^2(\frac{1}{2}|\nabla\psi|^2 + K|\psi|^2)d\mu_g \leqslant 2\int_M \mu^2\rho^2 d\mu_g + 3\int_M |d\mu|^2|\psi|^2 d\mu_g \tag{9.186}$$

The lemma thus follows. □

We now apply Lemma 9.7 to (9.173), taking $(M, g)$ to be $(S_{t,u}, \not{g})$, $\phi$ to be ${}^{(i_1\cdots i_l)}\mu_{m,l}$, and the function $\rho$ to be

$$\mu^{-1}({}^{(i_1\cdots i_l)}x'_{m,l} + {}^{(i_1\cdots i_l)}\check{f}'_{m,l}) + {}^{(i_1\cdots i_l)}d_{m,l}$$

In view of Lemma 8.9, taking $\delta_0$ suitably small, we have $K \geqslant 0$, thus we obtain

$$\begin{aligned}
&\|\mu \not{D}^{2\,(i_1\cdots i_l)}\mu_{m,l}\|_{L^2(S_{t,u})} \\
&\leqslant C\|^{(i_1\cdots i_l)}x'_{m,l}\|_{L^2(S_{t,u})} + C\|^{(i_1\cdots i_l)}\check{f}'_{m,l}\|_{L^2(S_{t,u})} \\
&\quad + C\|\mu^{(i_1\cdots i_l)}d_{m,l}\|_{L^2(S_{t,u})} + C\|\not{d}\mu\|_{L^\infty(S_{t,u})}\|\not{d}^{(i_1\cdots i_l)}\mu_{m,l}\|_{L^2(S_{t,u})}
\end{aligned} \tag{9.187}$$

By **F1**,

$$|\not{d}\mu| \leqslant C\delta_0(1+t)^{-1}(1+\log(1+t)) \tag{9.188}$$

Substituting (9.188) in (9.187) yields

$$\begin{aligned}
&\|\mu \not{D}^{2\,(i_1\cdots i_l)}\mu_{m,l}\|_{L^2(S_{t,u})} \\
&\leqslant C\|^{(i_1\cdots i_l)}x'_{m,l}\|_{L^2(S_{t,u})} + C\|^{(i_1\cdots i_l)}\check{f}'_{m,l}\|_{L^2(S_{t,u})} + C\|\mu^{(i_1\cdots i_l)}d_{m,l}\|_{L^2(S_{t,u})} \\
&\quad + C\delta_0(1+t)^{-1}(1+\log(1+t))\|\not{d}^{(i_1\cdots i_l)}\mu_{m,l}\|_{L^2(S_{t,u})}
\end{aligned} \tag{9.189}$$

In view of Proposition 9.4, we then obtain

$$\begin{aligned}
&\|\mu^{(i_1\cdots i_l)}\hat{\not{\mu}}_{2,m,l}\|_{L^2(S_{t,u})} \\
&\leqslant C\|^{(i_1\cdots i_l)}x'_{m,l}\|_{L^2(S_{t,u})} + C\|^{(i_1\cdots i_l)}\check{f}'_{m,l}\|_{L^2(S_{t,u})} \\
&\quad + C\|\mu^{(i_1\cdots i_l)}d_{m,l}\|_{L^2(S_{t,u})} + C\|\mu^{(i_1\cdots i_l)}e_{m,l}\|_{L^2(S_{t,u})} \\
&\quad + C\delta_0(1+t)^{-1}(1+\log(1+t))\|\not{d}^{(i_1\cdots i_l)}\mu_{m,l}\|_{L^2(S_{t,u})}
\end{aligned} \tag{9.190}$$

## 9.4 The Estimates for the Solutions of the Propagation Equations

We now return to the propagation equation of Proposition 9.2. Setting

$$^{(i_1\cdots i_l)}\tilde{g}'_{m,l} = -2\mu\hat{\chi}\cdot{}^{(i_1\cdots i_l)}\hat{\not{\mu}}_{2,m,l} + {}^{(i_1\cdots i_l)}g'_{m,l} \tag{9.191}$$

the propagation equation takes the form

$$\begin{aligned}
&L^{(i_1\cdots i_l)}x'_{m,l} + (\mathrm{tr}\chi - 2\mu^{-1}(L\mu))^{(i_1\cdots i_l)}x'_{m,l} \\
&= -\left(\frac{1}{2}\mathrm{tr}\chi - 2\mu^{-1}(L\mu)\right)^{(i_1\cdots i_l)}\check{f}'_{m,l} + {}^{(i_1\cdots i_l)}\tilde{g}'_{m,l}
\end{aligned} \tag{9.192}$$

Defining as in Chapter 8 the diffeomorphisms $\Phi_{t,u}$ of $S^2$ onto $S_{t,u}$, if $\omega$ is any $S_{t,u}$ $r$-covariant tensorfield defined in $W^*_{\epsilon_0}$, we consider the pullback $\omega(t,u) = \Phi^*_{t,u}\omega$, a $r$-covariant tensorfield on $S^2$ depending on the parameters $t$ and $u$. If in place of $\omega$ we have $\not{\mathcal{L}}_L\omega$, the corresponding tensorfield on $S^2$ is $\frac{\partial\omega(t,u)}{\partial t}$. The propagation equation (9.192) can then be viewed as an equation for the function $^{(i_1\cdots i_l)}x'(t,u)$ on $S^2$ depending on $t$ and $u$:

$$\frac{\partial}{\partial t}{}^{(i_1\cdots i_l)}x'_{m,l} + (\mathrm{tr}\chi - 2\mu^{-1}\frac{\partial\mu}{\partial t}){}^{(i_1\cdots i_l)}x'_{m,l}$$
$$= -(\frac{1}{2}\mathrm{tr}\chi - 2\mu^{-1}\frac{\partial\mu}{\partial t}){}^{(i_1\cdots i_l)}\check{f}'_{m,l} + {}^{(i_1\cdots i_l)}\tilde{g}'_{m,l} \tag{9.193}$$

Consider a function $\phi(t,u)$ on $S^2$ depending on the parameters $t$ and $u$. We have

$$|\phi|\frac{\partial}{\partial t}|\phi| = \frac{1}{2}\frac{\partial}{\partial t}\phi^2 = \phi\frac{\partial\phi}{\partial t} \tag{9.194}$$

Then from the propagation equation (9.193) we have

$${}^{(i_1\cdots i_l)}x'_{m,l}\frac{\partial}{\partial t}{}^{(i_1\cdots i_l)}x'_{m,l} + (\mathrm{tr}\chi - 2\mu^{-1}\frac{\partial\mu}{\partial t})({}^{(i_1\cdots i_l)}x'_{m,l})^2$$
$$= -(\frac{1}{2}\mathrm{tr}\chi - 2\mu^{-1}\frac{\partial\mu}{\partial t}){}^{(i_1\cdots i_l)}x'_{m,l}\,{}^{(i_1\cdots i_l)}\check{f}'_{m,l} + {}^{(i_1\cdots i_l)}x'_{m,l}\,{}^{(i_1\cdots i_l)}\tilde{g}'_{m,l} \tag{9.195}$$

Since

$$|{}^{(i_1\cdots i_l)}x'_{m,l}\,{}^{(i_1\cdots i_l)}\check{f}'_{m,l}| \leqslant |{}^{(i_1\cdots i_l)}x'_{m,l}||{}^{(i_1\cdots i_l)}\check{f}'_{m,l}|$$

and

$$|{}^{(i_1\cdots i_l)}x'_{m,l}\,{}^{(i_1\cdots i_l)}\tilde{g}'_{m,l}| \leqslant |{}^{(i_1\cdots i_l)}x'_{m,l}||{}^{(i_1\cdots i_l)}\tilde{g}'_{m,l}|$$

the assumption **AS** implies through (9.195)

$${}^{(i_1\cdots i_l)}x'_{m,l}\frac{\partial}{\partial t}{}^{(i_1\cdots i_l)}x'_{m,l} + (\mathrm{tr}\chi - 2\mu^{-1}\frac{\partial\mu}{\partial t})|{}^{(i_1\cdots i_l)}x'_{m,l}|^2$$
$$\leqslant(\frac{1}{2}\mathrm{tr}\chi - 2\mu^{-1}\frac{\partial\mu}{\partial t})|{}^{(i_1\cdots i_l)}x'_{m,l}||{}^{(i_1\cdots i_l)}\check{f}'_{m,l}| + |{}^{(i_1\cdots i_l)}x'_{m,l}||{}^{(i_1\cdots i_l)}\tilde{g}'_{m,l}| \tag{9.196}$$

Applying (9.194) taking $\phi = {}^{(i_1\cdots i_l)}x'_{m,l}$, we get

$$\frac{\partial}{\partial t}|{}^{(i_1\cdots i_l)}x'_{m,l}| + (\mathrm{tr}\chi - 2\mu^{-1}\frac{\partial\mu}{\partial t})|{}^{(i_1\cdots i_l)}x'_{m,l}|$$
$$\leqslant(\frac{1}{2}\mathrm{tr}\chi - 2\mu^{-1}\frac{\partial\mu}{\partial t})|{}^{(i_1\cdots i_l)}\check{f}'_{m,l}| + |{}^{(i_1\cdots i_l)}\tilde{g}'_{m,l}| \tag{9.197}$$

The integrating factor here is

$$\exp(\int_0^t(\mathrm{tr}\chi - 2\mu^{-1}\frac{\partial\mu}{\partial t})(t',u)dt') = (\frac{\mu(t,u)}{\mu(0,u)})^{-2}A(t,u) \tag{9.198}$$

where $A(t,u)$ is defined by (8.167) and is given by (8.169). From (9.197) we deduce

$$|{}^{(i_1\cdots i_l)}x'_{m,l}| \leqslant {}^{(i_1\cdots i_l)}F'_{m,l}(t,u) + {}^{(i_1\cdots i_l)}G'_{m,l}(t,u) \tag{9.199}$$

where

$${}^{(i_1\cdots i_l)}F'_{m,l}(t,u)$$
$$=(A(t,u))^{-1}(\mu(t,u))^2((\mu(0,u))^{-2}|{}^{(i_1\cdots i_l)}x'_{m,l}(0,u)| \tag{9.200}$$

$$+\int_0^t (\mu(t',u))^{-2}A(t',u)(\frac{1}{2}\mathrm{tr}\chi - 2\mu^{-1}\frac{\partial\mu}{\partial t})(t',u)|^{(i_1\cdots i_l)}\check{f}'_{m,l}(t',u)|dt')$$

and

$$\begin{aligned}&{}^{(i_1\cdots i_l)}G'_{m,l}(t,u)\\&=(A(t,u))^{-1}(\mu(t,u))^2\cdot\int_0^t(\mu(t',u))^{-2}A(t',u)|^{(i_1\cdots i_l)}\tilde{g}'_{m,l}(t',u)|dt' \qquad (9.201)\end{aligned}$$

Using the bound (8.173), we obtain from (9.200) and (9.201)

$$\begin{aligned}{}^{(i_1\cdots i_l)}F'_{m,l}(t,u) \leqslant e^{C\delta_0}(1-u+t)^{-2}({}^{(i_1\cdots i_l)}M'^0_{m,l}(t,u) \qquad (9.202)\\ + {}^{(i_1\cdots i_l)}M'^1_{m,l}(t,u) + {}^{(i_1\cdots i_l)}M'^2_{m,l}(t,u))\end{aligned}$$

where

$$\begin{aligned}&{}^{(i_1\cdots i_l)}M'^0_{m,l}(t,u)\\&=(\frac{\mu(t,u)}{\mu(0,u)})^2(1-u)^2|^{(i_1\cdots i_l)}x'_{m,l}(0,u)| \qquad (9.203)\\&{}^{(i_1\cdots i_l)}M'^1_{m,l}(t,u)\\&=\int_0^t(\frac{\mu(t,u)}{\mu(t',u)})^2(1-u+t')^2\cdot(-2\mu^{-1}(\frac{\partial\mu}{\partial t})_-(t',u))\cdot|^{(i_1\cdots i_l)}\check{f}'_{m,l}(t',u)|dt' \quad (9.204)\\&{}^{(i_1\cdots i_l)}M'^2_{m,l}(t,u)\\&=\frac{1}{2}\int_0^t(\frac{\mu(t,u)}{\mu(t',u)})^2(1-u+t')^2\mathrm{tr}\chi(t',u)|^{(i_1\cdots i_l)}\check{f}'_{m,l}(t',u)|dt' \qquad (9.205)\end{aligned}$$

Also

$$\begin{aligned}&{}^{(i_1\cdots i_l)}G'_{m,l}(t,u)\\&\leqslant e^{C\delta_0}(1-u+t)^{-2}\cdot\int_0^t(\frac{\mu(t,u)}{\mu(t',u)})^2(1-u+t')^2|^{(i_1\cdots i_l)}\tilde{g}'_{m,l}(t',u)|dt' \qquad (9.206)\end{aligned}$$

We shall estimate the $L^2$ norm of ${}^{(i_1\cdots i_l)}F'_{m,l}(t)$ on $[0,\epsilon_0]\times S^2$.

First, by **A3** and (8.333) at $t=0$,

$$\|^{(i_1\cdots i_l)}M'^0_{m,l}(t)\|_{L^2([0,\epsilon_0]\times S^2)} \leqslant C(1+\log(1+t))^2\|^{(i_1\cdots i_l)}x'_{m,l}(0)\|_{L^2(\Sigma_0^{\epsilon_0})} \quad (9.207)$$

We turn to ${}^{(i_1\cdots i_l)}M'^1_{m,l}$. Partitioning $[0,\epsilon_0]\times S^2$ into the sets $\mathcal{V}_{s-}$, $\mathcal{V}_{s+}$ defined by (8.337) and (8.338), respectively, we have

$$\begin{aligned}&\|^{(i_1\cdots i_l)}M'^1_{m,l}(t)\|^2_{L^2([0,\epsilon_0]\times S^2)}\\&=\|^{(i_1\cdots i_l)}M'^1_{m,l}(t)\|^2_{L^2(\mathcal{V}_{s-})}+\|^{(i_1\cdots i_l)}M'^1_{m,l}(t)\|^2_{L^2(\mathcal{V}_{s+})} \qquad (9.208)\end{aligned}$$

By Minkowski's inequality,

$$\|^{(i_1\cdots i_l)}M'^{1}_{m,l}(t)\|_{L^2(\mathcal{V}_{s-})}\leqslant\int_0^t\|^{(i_1\cdots i_l)}N'^{1}_{m,l}(t,t')\|_{L^2(\mathcal{V}_{s-})}dt' \tag{9.209}$$

where

$$\begin{aligned}&^{(i_1\cdots i_l)}N'^{1}_{m,l}(t,t',u)\\&=(\frac{\mu(t,u)}{\mu(t',u)})^2(1-u+t')^2(-2\mu^{-1}(\frac{\partial\mu}{\partial t})_-(t',u))|^{(i_1\cdots i_l)}\check{f}'_{m,l}(t',u)|\end{aligned} \tag{9.210}$$

We have

$$\begin{aligned}&\|^{(i_1\cdots i_l)}N'^{1}_{m,l}(t,t')\|_{L^2(\mathcal{V}_{s-})}\\&\leqslant(1+t')^2(\max_{\mathcal{V}_{s-}}\frac{\mu(t)}{\mu(t')})^2\max_{\mathcal{V}_{s-}}(-2\mu^{-1}(\frac{\partial\mu}{\partial t})_-(t'))\|^{(i_1\cdots i_l)}\check{f}'_{m,l}(t')\|_{L^2(\mathcal{V}_{s-})}\end{aligned} \tag{9.211}$$

In view of (8.343) and (8.344) (see definition (8.249)),

$$\|^{(i_1\cdots i_l)}N'^{1}_{m,l}(t,t')\|_{L^2(\mathcal{V}_{s-})}\leqslant C(1+t')^2M(t')\|^{(i_1\cdots i_l)}\check{f}'_{m,l}(t')\|_{L^2([0,\epsilon_0]\times S^2)} \tag{9.212}$$

Let us define

$$^{(i_1\cdots i_l)}P'_{m,l}(t)=(1+t)\|^{(i_1\cdots i_l)}\check{f}'_{m,l}(t)\|_{L^2([0,\epsilon_0]\times S^2)} \tag{9.213}$$

Suppose that, for non-negative quantities $^{(i_1\cdots i_l)}P'^{(0)}_{m,l}$, $^{(i_1\cdots i_l)}P'^{(1)}_{m,l}$, we have

$$^{(i_1\cdots i_l)}P'_{m,l}(t)\leqslant{}^{(i_1\cdots i_l)}P'^{(0)}_{m,l}(t)+{}^{(i_1\cdots i_l)}P'^{(1)}_{m,l}(t) \tag{9.214}$$

We define the non-decreasing non-negative quantities $^{(i_1\cdots i_l)}\bar{P}'^{(0)}_{m,l,a}$, $^{(i_1\cdots i_l)}\bar{P}'^{(1)}_{m,l,a}$ by

$$^{(i_1\cdots i_l)}\bar{P}'^{(0)}_{m,l,a}(t)=\sup_{t'\in[0,t]}(\bar{\mu}^a_m(t')^{(i_1\cdots i_l)}P'^{(0)}_{m,l}(t')) \tag{9.215}$$

$$^{(i_1\cdots i_l)}\bar{P}'^{(1)}_{m,l,a}(t)=\sup_{t'\in[0,t]}((1+t')^{1/2}\bar{\mu}^a_m(t')^{(i_1\cdots i_l)}P'^{(1)}_{m,l}(t')) \tag{9.216}$$

Then for $t'\in[0,t]$ we have

$$^{(i_1\cdots i_l)}P'_{m,l}(t')\leqslant\bar{\mu}_m^{-a}(t')(^{(i_1\cdots i_l)}\bar{P}'^{(0)}_{m,l,a}(t)+(1+t')^{-1/2(i_1\cdots i_l)}\bar{P}'^{(1)}_{m,l,a}(t)) \tag{9.217}$$

Substituting in (9.212) and in view of (9.213) we have

$$\begin{aligned}&\|^{(i_1\cdots i_l)}N'^{1}_{m,l}(t,t')\|_{L^2(\mathcal{V}_{s-})}\\&\leqslant C((1+t)^{(i_1\cdots i_l)}\bar{P}'^{(0)}_{m,l,a}(t)+(1+t)^{1/2(i_1\cdots i_l)}\bar{P}'^{(1)}_{m,l,a}(t))\bar{\mu}_m^{-a}(t')M(t')\end{aligned} \tag{9.218}$$

for all $t'\in[0,t]$. Substituting this in (9.209) we obtain

$$\begin{aligned}&\|^{(i_1\cdots i_l)}M'^{1}_{m,l}(t)\|_{L^2(\mathcal{V}_{s-})}\\&\leqslant C((1+t)^{(i_1\cdots i_l)}\bar{P}'^{(0)}_{m,l,a}(t)+(1+t)^{1/2(i_1\cdots i_l)}\bar{P}'^{(1)}_{m,l,a}(t))I_a(t)\end{aligned} \tag{9.219}$$

Applying Lemma 8.11 we conclude that

$$\begin{aligned}
&\|{}^{(i_1\cdots i_l)}M'^1_{m,l}(t)\|_{L^2(\mathcal{V}_{s-})}\\
&\leqslant Ca^{-1}((1+t)^{(i_1\cdots i_l)}\bar{P}'^{(0)}_{m,l,a}(t)+(1+t)^{1/2(i_1\cdots i_l)}\bar{P}'^{(1)}_{m,l,a}(t))\bar{\mu}_m^{-a}(t)
\end{aligned}\tag{9.220}$$

In analogy with (9.209) and (9.211), with $\mathcal{V}_{s+}$ in the role of $\mathcal{V}_{s-}$, we have

$$\|{}^{(i_1\cdots i_l)}M'^1_{m,l}(t)\|_{L^2(\mathcal{V}_{s+})}\leqslant\int_0^t\|{}^{(i_1\cdots i_l)}N'^1_{m,l}(t,t')\|_{L^2(\mathcal{V}_{s+})}dt'\tag{9.221}$$

and

$$\begin{aligned}
&\|{}^{(i_1\cdots i_l)}N'^1_{m,l}(t,t')\|_{L^2(\mathcal{V}_{s+})}\\
&\leqslant(1+t')^2(\max_{\mathcal{V}_{s+}}\frac{\mu(t)}{\mu(t')})^2\max_{\mathcal{V}_{s+}}(-2\mu^{-1}(\frac{\partial\mu}{\partial t})_-(t'))\|{}^{(i_1\cdots i_l)}\check{f}'_{m,l}(t')\|_{L^2(\mathcal{V}_{s+})}
\end{aligned}\tag{9.222}$$

Substituting (8.356), (8.357) and (9.213) in (9.222), and the result in (9.221), we obtain

$$\begin{aligned}
&\|{}^{(i_1\cdots i_l)}M'^1_{m,l}(t)\|_{L^2(\mathcal{V}_{s+})}\\
&\leqslant C\delta_0(1+\log(1+t))^2\int_0^t\frac{1+\log(1+t')}{1+t'}{}^{(i_1\cdots i_l)}P'_{m,l}(t')dt'
\end{aligned}\tag{9.223}$$

hence, by (9.217) we have

$$\begin{aligned}
&\|{}^{(i_1\cdots i_l)}M'^1_{m,l}(t)\|_{L^2(\mathcal{V}_{s+})}\\
&\leqslant C\delta_0(1+\log(1+t))^2({}^{(i_1\cdots i_l)}\bar{P}'^{(0)}_{m,l,a}(t)+{}^{(i_1\cdots i_l)}\bar{P}'^{(1)}_{m,l,a}(t))\\
&\quad\cdot\int_0^t\frac{1+\log(1+t')}{1+t'}\bar{\mu}_m^{-a}(t')dt'
\end{aligned}\tag{9.224}$$

In view of (8.363), we conclude

$$\begin{aligned}
&\|{}^{(i_1\cdots i_l)}M'^1_{m,l}(t)\|_{L^2(\mathcal{V}_{s+})}\\
&\leqslant C\delta_0(1+\log(1+t))^4({}^{(i_1\cdots i_l)}\bar{P}'^{(0)}_{m,l,a}(t)+{}^{(i_1\cdots i_l)}\bar{P}'^{(1)}_{m,l,a}(t))\bar{\mu}_m^{1-a}(t)
\end{aligned}\tag{9.225}$$

Combining finally (9.220) and (9.225) and taking into account the assumption $a\delta_0\leqslant C^{-1}$, we obtain

$$\begin{aligned}
&\|{}^{(i_1\cdots i_l)}M'^1_{m,l}(t)\|_{L^2([0,\epsilon_0]\times S^2)}\\
&\leqslant Ca^{-1}((1+t)^{(i_1\cdots i_l)}\bar{P}'^{(0)}_{m,l,a}(t)+(1+t)^{1/2(i_1\cdots i_l)}\bar{P}'^{(1)}_{m,l,a}(t))\bar{\mu}_m^{-a}(t)
\end{aligned}\tag{9.226}$$

Here, we also have used the definition (8.248) of Chapter 8.

We turn to ${}^{(i_1\cdots i_l)}M'^2_{m,l}(t,u)$. We have

$$\begin{aligned}
&\|{}^{(i_1\cdots i_l)}M'^2_{m,l}(t)\|^2_{L^2([0,\epsilon_0]\times S^2)}\\
&=\|{}^{(i_1\cdots i_l)}M'^2_{m,l}(t)\|^2_{L^2(\mathcal{V}_{s-})}+\|{}^{(i_1\cdots i_l)}M'^2_{m,l}(t)\|^2_{L^2(\mathcal{V}_{s+})}
\end{aligned}\tag{9.227}$$

and

$$\|^{(i_1\cdots i_l)}M'^2_{m,l}(t)\|_{L^2(\mathcal{V}_{s-})} \leqslant \int_0^t \|^{(i_1\cdots i_l)}N'^2_{m,l}(t,t')\|_{L^2(\mathcal{V}_{s-})}dt' \tag{9.228}$$

$$\|^{(i_1\cdots i_l)}M'^2_{m,l}(t)\|_{L^2(\mathcal{V}_{s+})} \leqslant \int_0^t \|^{(i_1\cdots i_l)}N'^2_{m,l}(t,t')\|_{L^2(\mathcal{V}_{s+})}dt' \tag{9.229}$$

where

$$^{(i_1\cdots i_l)}N'^2_{m,l}(t,t',u) = \frac{1}{2}(\frac{\mu(t,u)}{\mu(t',u)})^2(1-u+t')^2\mathrm{tr}\chi(t',u)|^{(i_1\cdots i_l)}\check{f}'_{m,l}(t',u)| \tag{9.230}$$

We have

$$\begin{aligned}&\|^{(i_1\cdots i_l)}N'^2_{m,l}(t,t')\|_{L^2(\mathcal{V}_{s-})}\\ \leqslant&\frac{1}{2}(1+t')^2(\max_{\mathcal{V}_{s-}}\frac{\mu(t)}{\mu(t')})^2\max_{\mathcal{V}_{s-}}(\mathrm{tr}\chi(t'))\|^{(i_1\cdots i_l)}\check{f}'_{m,l}(t')\|_{L^2(\mathcal{V}_{s-})}\end{aligned} \tag{9.231}$$

By (8.343) and (8.371),

$$\|^{(i_1\cdots i_l)}N'^2_{m,l}(t,t')\|_{L^2(\mathcal{V}_{s-})} \leqslant C(1+t')\|^{(i_1\cdots i_l)}\check{f}'_{m,l}(t')\|_{L^2([0,\epsilon_0]\times S^2)} \tag{9.232}$$

Then by (9.213) and (9.217),

$$\begin{aligned}&\|^{(i_1\cdots i_l)}N'^2_{m,l}(t,t')\|_{L^2(\mathcal{V}_{s-})} \leqslant C^{(i_1\cdots i_l)}P'_{m,l}(t')\\ \leqslant&C(^{(i_1\cdots i_l)}\bar{P}'^{(0)}_{m,l,a}(t)+(1+t')^{-1/2(i_1\cdots i_l)}\bar{P}'^{(1)}_{m,l,a}(t))\bar{\mu}_m^{-a}(t')\\ \leqslant&C'(^{(i_1\cdots i_l)}\bar{P}'^{(0)}_{m,l,a}(t)+(1+t')^{-1/2(i_1\cdots i_l)}\bar{P}'^{(1)}_{m,l,a}(t))\bar{\mu}_m^{-a}(t)\end{aligned} \tag{9.233}$$

where we have used Corollary 2 to Lemma 8.11. Substituting (9.233) in (9.228) we get

$$\begin{aligned}&\|^{(i_1\cdots i_l)}M'^2_{m,l}(t)\|_{L^2(\mathcal{V}_{s-})}\\ \leqslant&C((1+t)^{(i_1\cdots i_l)}\bar{P}'^{(0)}_{m,l,a}(t)+(1+t)^{1/2(i_1\cdots i_l)}\bar{P}'^{(1)}_{m,l,a}(t))\bar{\mu}_m^{-a}(t)\end{aligned} \tag{9.234}$$

In analogy with (9.231) we have

$$\begin{aligned}&\|^{(i_1\cdots i_l)}N'^2_{m,l}(t,t')\|_{L^2(\mathcal{V}_{s+})}\\ \leqslant&\frac{1}{2}(1+t')^2(\max_{\mathcal{V}_{s+}}\frac{\mu(t)}{\mu(t')})^2\max_{\mathcal{V}_{s+}}(\mathrm{tr}\chi(t'))\|^{(i_1\cdots i_l)}\check{f}'_{m,l}(t')\|_{L^2(\mathcal{V}_{s+})}\end{aligned} \tag{9.235}$$

Here, we have two cases to distinguish according as to whether $t'$ is $< \sqrt{t}$ or $\geqslant \sqrt{t}$. In the first case (8.378) holds, and in the second case (8.377) holds. In view of these, recalling (8.371), we obtain

$$\begin{aligned}&\|^{(i_1\cdots i_l)}N'^2_{m,l}(t,t')\|_{L^2(\mathcal{V}_{s+})} \leqslant C(1+\log(1+t))^{2(i_1\cdots i_l)}P'_{m,l}(t'), t'\leqslant\sqrt{t}\\ &\|^{(i_1\cdots i_l)}N'^2_{m,l}(t,t')\|_{L^2(\mathcal{V}_{s+})} \leqslant C^{(i_1\cdots i_l)}P'_{m,l}(t'), t'\geqslant\sqrt{t}\end{aligned} \tag{9.236}$$

Substituting (9.236) in (9.229), we then have

$$\begin{aligned}
&\|^{(i_1\cdots i_l)}M'^2_{m,l}(t)\|_{L^2(\mathcal{V}_{s+})}\\
&\leqslant \int_0^{\sqrt{t}} \|^{(i_1\cdots i_l)}N'^2_{m,l}(t,t')\|_{L^2(\mathcal{V}_{s+})}dt' + \int_{\sqrt{t}}^t \|^{(i_1\cdots i_l)}N'^2_{m,l}(t,t')\|_{L^2(\mathcal{V}_{s+})}dt'\\
&\leqslant C(1+\log(1+t))^2 \int_0^{\sqrt{t}} {}^{(i_1\cdots i_l)}P'_{m,l}(t')dt' + C\int_{\sqrt{t}}^t {}^{(i_1\cdots i_l)}P'_{m,l}(t')dt'\\
&\leqslant C(1+\log(1+t))^2 \int_0^{\sqrt{t}} \bar{\mu}_m^{-a}(t')({}^{(i_1\cdots i_l)}\bar{P}'^{(0)}_{m,l,a}(t) + (1+t')^{-1/2}\,{}^{(i_1\cdots i_l)}\bar{P}'^{(1)}_{m,l,a}(t))dt'\\
&\quad + C\int_{\sqrt{t}}^t \bar{\mu}_m^{-a}(t')({}^{(i_1\cdots i_l)}\bar{P}'^{(0)}_{m,l,a}(t) + (1+t')^{-1/2}\,{}^{(i_1\cdots i_l)}\bar{P}'^{(1)}_{m,l,a}(t))dt'\\
&\leqslant C(1+\log(1+t))^2((1+t)^{1/2}\,{}^{(i_1\cdots i_l)}\bar{P}'^{(0)}_{m,l,a}(t) + (1+t)^{1/4}\,{}^{(i_1\cdots i_l)}\bar{P}'^{(1)}_{m,l,a}(t))\bar{\mu}_m^{-a}(t)\\
&\quad + C((1+t)\,{}^{(i_1\cdots i_l)}\bar{P}'^{(0)}_{m,l,a}(t) + (1+t)^{1/2}\,{}^{(i_1\cdots i_l)}\bar{P}'^{(1)}_{m,l,a}(t))\bar{\mu}_m^{-a}(t)\\
&\leqslant C'((1+t)\,{}^{(i_1\cdots i_l)}\bar{P}'^{(0)}_{m,l,a}(t) + (1+t)^{1/2}\,{}^{(i_1\cdots i_l)}\bar{P}'^{(1)}_{m,l,a}(t))\bar{\mu}_m^{-a}(t) \qquad (9.237)
\end{aligned}$$

where we have used Corollary 2 to Lemma 8.11 and (9.217).

Combining (9.234) and (9.237) we obtain

$$\begin{aligned}
&\|^{(i_1\cdots i_l)}M'^2_{m,l}(t)\|_{L^2([0,\epsilon_0]\times S^2)}\\
&\leqslant C((1+t)\,{}^{(i_1\cdots i_l)}\bar{P}'^{(0)}_{m,l,a}(t) + (1+t)^{1/2}\,{}^{(i_1\cdots i_l)}\bar{P}'^{(1)}_{m,l,a}(t))\bar{\mu}_m^{-a}(t) \qquad (9.238)
\end{aligned}$$

The estimates (9.202), (9.207), (9.226) and (9.238) yield

$$\begin{aligned}
&\|^{(i_1\cdots i_l)}F'_{m,l}(t)\|_{L^2([0,\epsilon_0]\times S^2)}\\
&\leqslant C(1+t)^{-2}(1+\log(1+t))^2\|^{(i_1\cdots i_l)}x'_{m,l}(0)\|_{L^2(\Sigma_0^{\epsilon_0})} \qquad (9.239)\\
&\quad + C(1+t)^{-1}({}^{(i_1\cdots i_l)}\bar{P}'^{(0)}_{m,l,a}(t) + (1+t)^{-1/2}\,{}^{(i_1\cdots i_l)}\bar{P}'^{(1)}_{m,l,a}(t))\bar{\mu}_m^{-a}(t)
\end{aligned}$$

We now consider the estimate (9.206). From (8.383),

$$^{(i_1\cdots i_l)}G'_{m,l}(t,u) \leqslant C(1+t)^{-2}(1+\log(1+t))^2 \int_0^t (1+t')^2|^{(i_1\cdots i_l)}\tilde{g}'_{m,l}(t',u)|dt' \qquad (9.240)$$

It follows that

$$\begin{aligned}
\|^{(i_1\cdots i_l)}G'_{m,l}(t)\|_{L^2([0,\epsilon_0]\times S^2)} \leqslant &C(1+t)^{-2}(1+\log(1+t))^2 \qquad (9.241)\\
&\cdot\int_0^t (1+t')^2\|^{(i_1\cdots i_l)}\tilde{g}'_{m,l}(t')\|_{L^2([0,\epsilon_0]\times S^2)}dt'
\end{aligned}$$

Now ${}^{(i_1\cdots i_l)}\tilde{g}'_{m,l}$ is given by (9.191). According to the discussion following Lemma 9.4, the principal part of ${}^{(i_1\cdots i_l)}g'_{m,l}$ consists of (9.118), (9.120) and (9.126). Define the function ${}^{(i_1\cdots i_l)}\dot{g}'_{m,l}$ by

$$\begin{aligned} {}^{(i_1\cdots i_l)}\dot{g}'_{0,l} = {}& {}^{(i_1\cdots i_l)}g'_{0,l} - \xi \cdot {}^{(i_1\cdots i_l)}x_l \\ & - \sum_{k=0}^{l-1} {}^{(R_{i_{l-k}})}Z^{(i_1\cdots i_{l-k-1}i_{l-k+1}\cdots i_l)}x'_{0,l-1}, m = 0 \end{aligned} \tag{9.242}$$

and

$$\begin{aligned} {}^{(i_1\cdots i_l)}\dot{g}'_{m,l} = {}& {}^{(i_1\cdots i_l)}g'_{m,l} - \xi \cdot \not{d}^{(i_1\cdots i_l)}x'_{m-1,l} \\ & - m\Lambda^{(i_1\cdots i_l)}x'_{m-1,l} - \sum_{k=0}^{l-1} {}^{(R_{i_{l-k}})}Z^{(i_1\cdots i_{l-k-1}i_{l-k+1}\cdots i_l)}x'_{m,l-1}, m \geqslant 1 \end{aligned} \tag{9.243}$$

Then ${}^{(i_1\cdots i_l)}\dot{g}'_{m,l}$ does not contain principal acoustical parts and we have

$$\begin{aligned} {}^{(i_1\cdots i_l)}\tilde{g}'_{0,l} = {}& -2\mu\hat{\chi} \cdot {}^{(i_1\cdots i_l)}\hat{\not{\mu}}_{2,0,l} + \xi \cdot {}^{(i_1\cdots i_l)}x_l \\ & + \sum_{k=0}^{l-1} {}^{(R_{i_{l-k}})}Z^{(i_1\cdots i_{l-k-1}i_{l-k+1}\cdots i_l)}x'_{0,l-1} + {}^{(i_1\cdots i_l)}\dot{g}'_{0,l}, m = 0 \end{aligned} \tag{9.244}$$

and

$$\begin{aligned} {}^{(i_1\cdots i_l)}\tilde{g}'_{m,l} = {}& -2\mu\hat{\chi} \cdot {}^{(i_1\cdots i_l)}\hat{\not{\mu}}_{2,m,l} + \xi \cdot \not{d}^{(i_1\cdots i_l)}x'_{m-1,l} + m\Lambda^{(i_1\cdots i_l)}x'_{m-1,l} \\ & + \sum_{k=0}^{l-1} {}^{(R_{i_{l-k}})}Z^{(i_1\cdots i_{l-k-1}i_{l-k+1}\cdots i_l)}x'_{m,l-1} + {}^{(i_1\cdots i_l)}\dot{g}'_{m,l}, m \geqslant 1 \end{aligned} \tag{9.245}$$

Let us define

$$X'_{m,l} = \max_{i_1\cdots i_l} \|{}^{(i_1\cdots i_l)}x'_{m,l}(t)\|_{L^2([0,\epsilon_0]\times S^2)} \tag{9.246}$$

We shall estimate $\|\tilde{g}'_{m,l}(t)\|_{L^2([0,\epsilon_0]\times S^2)}$ in terms of $X'_{0,l}(t)$ and $X_l(t)$, defined by (8.390), for $m = 0$, and in terms of $X'_{m,l}(t)$ and $X'_{m-1,l+1}(t)$, for $m \geqslant 1$. Consider first the second term on the right of (9.244) and (9.245). The $S_{t,u}$ 1-form $\xi$ is given by (9.59). By the bootstrap assumption **A, E** and **F** we have

$$|\xi| \leqslant C\delta_0(1+t)^{-1}(1+\log(1+t)) \tag{9.247}$$

Thus, by the pointwise estimates (9.119),

$$\begin{aligned} \|\xi \cdot {}^{(i_1\cdots i_l)}x_l\|_{L^2([0,\epsilon_0]\times S^2)} &\leqslant C\delta_0(1+t)^{-1}(1+\log(1+t))X_l(t) \\ \|\xi \cdot \not{d}^{(i_1\cdots i_l)}x'_{m-1,l}\|_{L^2([0,\epsilon_0]\times S^2)} &\leqslant C\delta_0(1+t)^{-2}(1+\log(1+t))X'_{m-1,l+1}(t) \end{aligned} \tag{9.248}$$

Consider next the third term on the right of (9.245). From (6.89) and (6.99) we have

$$|\Lambda| \leqslant C\delta_0(1+t)^{-1}(1+\log(1+t)) \tag{9.249}$$

Thus, by the pointwise estimate (9.121),

$$\|m\Lambda^{(i_1\cdots i_l)}x'_{m-1,l}\|_{L^2([0,\epsilon_0]\times S^2)} \leqslant Cm\delta_0(1+t)^{-2}(1+\log(1+t))X'_{m-1,l+1}(t) \tag{9.250}$$

Consider next the term next to the last term on the right in (9.244) and (9.245). By the pointwise estimates (9.127) and (8.393),

$$\begin{aligned}&\Big\|\sum_{k=0}^{l-1}{}^{(R_{i_{l-k}})}Z^{(i_1\cdots i_{l-k-1}i_{l-k+1}\cdots i_l)}x'_{m,l-1}\Big\|_{L^2([0,\epsilon_0]\times S^2)}\\&\leqslant Cl\delta_0(1+t)^{-2}(1+\log(1+t))X'_{m,l}(t)\end{aligned}\tag{9.251}$$

Finally, we consider the first term on the right in (9.244) and (9.245). By **F2** we have

$$\begin{aligned}&\|2\mu\hat{\chi}\cdot{}^{(i_1\cdots i_l)}\hat{\not\mu}_{2,m,l}\|_{L^2([0,\epsilon_0]\times S^2)}\\&\leqslant C\delta_0(1+t)^{-2}(1+\log(1+t))\|\mu|^{(i_1\cdots i_l)}\hat{\not\mu}_{2,m,l}|(t)\|_{L^2([0,\epsilon_0]\times S^2)}\end{aligned}\tag{9.252}$$

We now appeal to (9.190). In view of (8.396), (9.190) is equivalent to

$$\begin{aligned}&\|\mu|^{(i_1\cdots i_l)}\hat{\not\mu}_{2,m,l}|(t,u)\|_{L^2(S^2)}\\&\leqslant C\|^{(i_1\cdots i_l)}x'_{m,l}(t,u)\|_{L^2(S^2)}+C\|^{(i_1\cdots i_l)}\check{f}'_{m,l}(t,u)\|_{L^2(S^2)}\\&\quad+C\|\mu^{(i_1\cdots i_l)}d_{m,l}(t,u)\|_{L^2(S^2)}+C\|\mu|^{(i_1\cdots i_l)}e_{m,l}|(t,u)\|_{L^2(S^2)}\\&\quad+C\delta_0(1+t)^{-1}(1+\log(1+t))\||\not d^{(i_1\cdots i_l)}\mu_{m,l}|(t,u)\|_{L^2(S^2)}\end{aligned}\tag{9.253}$$

Taking $L^2$ norms on $[0,\epsilon_0]$ then yields

$$\begin{aligned}&\|\mu|^{(i_1\cdots i_l)}\hat{\not\mu}_{2,m,l}|(t)\|_{L^2([0,\epsilon_0]\times S^2)}\\&\leqslant C\|^{(i_1\cdots i_l)}x'_{m,l}(t)\|_{L^2([0,\epsilon_0]\times S^2)}+C\|^{(i_1\cdots i_l)}\check{f}'_{m,l}(t)\|_{L^2([0,\epsilon_0]\times S^2)}\\&\quad+C\|\mu^{(i_1\cdots i_l)}d_{m,l}(t)\|_{L^2([0,\epsilon_0]\times S^2)}+C\|\mu|^{(i_1\cdots i_l)}e_{m,l}|(t)\|_{L^2([0,\epsilon_0]\times S^2)}\\&\quad+C\delta_0(1+t)^{-1}(1+\log(1+t))\||\not d^{(i_1\cdots i_l)}\mu_{m,l}|(t)\|_{L^2([0,\epsilon_0]\times S^2)}\end{aligned}\tag{9.254}$$

In view of (9.248), (9.250), (9.251), (9.252) and (9.254), we obtain from (9.244) and (9.245)

$$\begin{aligned}&\|^{(i_1\cdots i_l)}\tilde{g}_{0,l}(t)\|_{L^2([0,\epsilon_0]\times S^2)}\\&\leqslant C(l+1)\delta_0(1+t)^{-2}(1+\log(1+t))X'_{0,l}(t)\\&\quad+C\delta_0(1+t)^{-1}(1+\log(1+t))X_l(t)+{}^{(i_1\cdots i_l)}Q'_{0,l}(t),\ m=0\\&\|^{(i_1\cdots i_l)}\tilde{g}'_{m,l}(t)\|_{L^2([0,\epsilon_0]\times S^2)}\\&\leqslant C(l+1)\delta_0(1+t)^{-2}(1+\log(1+t))X'_{m,l}(t)\\&\quad+C(m+1)\delta_0(1+t)^{-2}(1+\log(1+t))X'_{m-1,l+1}(t)\\&\quad+{}^{(i_1\cdots i_l)}Q'_{m,l}(t),\ m\geqslant 1\end{aligned}\tag{9.255}$$

where

$$
\begin{aligned}
&{}^{(i_1\cdots i_l)}Q'_{m,l}(t) \\
&= C\delta_0(1+t)^{-3}(1+\log(1+t))^2\||\not{d}^{(i_1\cdots i_l)}\mu_{m,l}|(t)\|_{L^2([0,\epsilon_0]\times S^2)} \\
&\quad + C\delta_0(1+t)^{-2}(1+\log(1+t))\|{}^{(i_1\cdots i_l)}\check{f}'_{m,l}(t)\|_{L^2([0,\epsilon_0]\times S^2)} \\
&\quad + C\delta_0(1+t)^{-2}(1+\log(1+t))\|\mu^{(i_1\cdots i_l)}d_{m,l}(t)\|_{L^2([0,\epsilon_0]\times S^2)} \\
&\quad + C\delta_0(1+t)^{-2}(1+\log(1+t))\||\mu|^{(i_1\cdots i_l)}e_{m,l}|(t)\|_{L^2([0,\epsilon_0]\times S^2)} \\
&\quad + \|{}^{(i_1\cdots i_l)}\dot{g}'_{m,l}(t)\|_{L^2([0,\epsilon_0]\times S^2)}
\end{aligned}
\tag{9.256}
$$

We now return to (9.193). Taking $L^2$ norms on $[0,\epsilon_0]\times S^2$ yields

$$
\begin{aligned}
&\|{}^{(i_1\cdots i_l)}x'_{m,l}(t)\|_{L^2([0,\epsilon_0]\times S^2)} \\
&\leqslant \|{}^{(i_1\cdots i_l)}F'_{m,l}(t)\|_{L^2([0,\epsilon_0]\times S^2)} + \|{}^{(i_1\cdots i_l)}G'_{m,l}(t)\|_{L^2([0,\epsilon_0]\times S^2)}
\end{aligned}
\tag{9.257}
$$

Substituting (9.239), (9.241) and (9.255) then yields

$$
\begin{aligned}
&\|{}^{(i_1\cdots i_l)}x'_{0,l}(t)\|_{L^2([0,\epsilon_0]\times S^2)} \\
&\leqslant {}^{(i_1\cdots i_l)}B'_{0,l}(t) + C\delta_0(1+t)^{-2}(1+\log(1+t))^2\int_0^t(1+t')(1+\log(1+t'))X_l(t')dt' \\
&\quad + C(l+1)\delta_0(1+t)^{-2}(1+\log(1+t))^2\int_0^t(1+\log(1+t'))X'_{0,l}(t')dt',\, m=0 \\
&\|{}^{(i_1\cdots i_l)}x'_{m,l}(t)\|_{L^2([0,\epsilon_0]\times S^2)} \\
&\leqslant {}^{(i_1\cdots i_l)}B'_{m,l}(t) + C(m+1)\delta_0(1+t)^{-2}(1+\log(1+t))^2\int_0^t(1+\log(1+t'))X'_{m-1,l+1}(t')dt' \\
&\quad + C(l+1)\delta_0(1+t)^{-2}(1+\log(1+t))^2\int_0^t(1+\log(1+t'))X'_{m,l}(t')dt',\, m\geqslant 1
\end{aligned}
\tag{9.258}
$$

where

$$
\begin{aligned}
{}^{(i_1\cdots i_l)}B'_{m,l}(t) &= C(1+t)^{-2}(1+\log(1+t))^2\|{}^{(i_1\cdots i_l)}x'_{m,l}(0)\|_{L^2(\Sigma_0^{\epsilon_0})} \\
&\quad + C(1+t)^{-1}({}^{(i_1\cdots i_l)}\bar{P}'^{(0)}_{m,l,a}(t) + (1+t)^{-1/2}{}^{(i_1\cdots i_l)}\bar{P}'^{(1)}_{m,l,a}(t))\bar{\mu}_m^{-a}(t) \\
&\quad + C(1+t)^{-2}(1+\log(1+t))^2\int_0^t(1+t')^2\,{}^{(i_1\cdots i_l)}Q'_{m,l}(t')dt'
\end{aligned}
\tag{9.259}
$$

Taking in (9.258) the maximum over $i_1\cdots i_l$ and recalling the definition (9.246) we obtain

$$
\begin{aligned}
X'_{0,l} &\leqslant B'_{0,l}(t) + C\delta_0(1+t)^{-2}(1+\log(1+t))^2 \\
&\quad \cdot\int_0^t(1+t')(1+\log(1+t'))X_l(t')dt' \\
&\quad + C(l+1)\delta_0(1+t)^{-2}(1+\log(1+t))^2 \\
&\quad \cdot\int_0^t(1+\log(1+t'))X'_{0,l}(t')dt',\, m=0
\end{aligned}
$$

$$X'_{m,l} \leqslant B'_{m,l}(t) + C(m+1)\delta_0(1+t)^{-2}(1+\log(1+t))^2 \cdot \int_0^t (1+\log(1+t'))X'_{m-1,l+1}(t')dt' + C(l+1)\delta_0(1+t)^{-2}(1+\log(1+t))^2 \cdot \int_0^t (1+\log(1+t'))X'_{m,l}(t')dt', m \geqslant 1 \tag{9.260}$$

where

$$B'_{m,l}(t) = \max_{i_1\cdots i_l} {}^{(i_1\cdots i_l)}B'_{m,l}(t) \tag{9.261}$$

Setting

$$l = n-1-m$$

(9.260) is a set of ordinary integral inequalities for $X'_{m,n-1-m}, m = 0,\cdots,n-1$. For $m=0$, we have an integral inequality for $X'_{0,n-1}$ containing on the right hand side the quantity $X_{n-1}$ which has been estimated in Chapter 8. For $m \geqslant 1$, we have an integral inequality for $X'_{m,n-1-m}$ containing on the right-hand side the quantity $X'_{m-1,n-1-(m-1)}$. Thus, the integral inequalities, considered successively in the order of increasing $m$, depend only on quantities already estimated. Setting

$$Y'_{l,m}(t) = \int_0^t (1+\log(1+t'))X'_{m,l}(t')dt' \tag{9.262}$$

and recalling (8.406), since

$$\frac{dY'_{m,l}(t)}{dt} = (1+\log(1+t))X'_{m,l}(t)$$

the $Y'_{m,l}$ satisfy

$$\begin{aligned}\frac{dY'_{0,l}(t)}{dt} &\leqslant (1+\log(1+t))B'_{0,l}(t) + C\delta_0(1+t)^{-2}(1+\log(1+t))^3 Y_l(t)\\ &\quad + C(l+1)\delta_0(1+t)^{-2}(1+\log(1+t))^3 Y'_{0,l}(t), m=0\\ \frac{dY'_{m,l}(t)}{dt} &\leqslant (1+\log(1+t))B'_{m,l}(t)\\ &\quad + C(m+1)\delta_0(1+t)^{-2}(1+\log(1+t))^3 Y'_{m-1,l+1}(t)\\ &\quad + C(l+1)\delta_0(1+t)^{-2}(1+\log(1+t))^3 Y'_{m,l}(t), m \geqslant 1\end{aligned} \tag{9.263}$$

The integrating factor here is

$$e^{-C_l(t)}$$

where $C_l(t)$ is of the form (8.409). We thus obtain

$$Y'_{0,l}(t) \leqslant e^{C_l(t)} \int_0^t e^{-C_l(t')}(1+\log(1+t'))B'_{0,l}(t')dt'$$

$$C\delta_0 \int_0^t (1+t')^{-2}(1+\log(1+t'))^3 Y_l(t')dt', m=0$$

$$Y'_{m,l}(t) \leqslant e^{C_l(t)} \int_0^t e^{-C_l(t')}(1+\log(1+t'))B'_{m,l}(t')dt'$$
$$+C(m+1)\delta_0 \int_0^t (1+t')^{-2}(1+\log(1+t'))^3 Y'_{m-1,l+1}(t')dt', m \geqslant 1 \tag{9.264}$$

Since the integral

$$\int_0^\infty \frac{(1+\log(1+t'))^3}{(1+t')^2}dt'$$

is convergent, if $\delta_0$ satisfies a smallness condition of the form (8.411), taking into account the fact that $Y_l, Y'_{m-1,l+1}$ are non-decreasing functions of $t$, (9.264) implies

$$Y'_{0,l}(t) \leqslant 2\int_0^t (1+\log(1+t))B'_{0,l}(t')dt' + C\delta_0 Y_l(t), m=0$$
$$Y'_{m,l}(t) \leqslant 2\int_0^t (1+\log(1+t'))B'_{m,l}(t')dt' + C(m+1)\delta_0 Y'_{m-1,l+1}(t), m \geqslant 1 \tag{9.265}$$

Setting $l = n-1-m, m = 0, 1, \cdots, n-1$, we then have

$$Y'_{0,n-1}(t) \leqslant 2\int_0^t (1+\log(1+t'))B'_{0,n-1}(t')dt' + C\delta_0 Y_{n-1}(t) \tag{9.266}$$

and, for $m = 1, \cdots, n-1$, using a modified version of Proposition 8.2 appropriate to inequalities, we deduce

$$Y'_{m,n-1-m}(t) \leqslant (m+1)!(C\delta_0)^m Y'_{0,n-1}(t) \tag{9.267}$$
$$+2\sum_{k=0}^{m-1} \frac{(m+1)!}{(m+1-k)!}(C\delta_0)^k \int_0^t (1+\log(1+t'))B'_{m-k,n-1-(m-k)}(t')dt'$$

Substituting the bound (8.412), these imply that

$$Y'_{m,n-1-m}(t) \leqslant 2\int_0^t (1+t')(1+\log(1+t'))B_{n-1}(t')dt' \tag{9.268}$$
$$+2\sum_{k=0}^{m} \int_0^t (1+\log(1+t'))B'_{k,n-1-k}(t')dt', m = 0, \cdots, n-1$$

provided $\delta_0$ is suitably small depending on $n$. The bounds (8.412) and (9.268) in turn imply

$$Y_{n-1}(t) + nY'_{0,n-1}(t)$$
$$\leqslant 2(n+1)\int_0^t (1+t')(1+\log(1+t'))B_{n-1}(t')dt'$$
$$+2n\int_0^t (1+\log(1+t'))B'_{0,n-1}(t')dt', m=0$$
$$(m+1)Y'_{m-1,n-1-(m-1)}(t) + (n-m)Y'_{m,n-1-m}(t)$$

$$\leqslant 2(n+1)\int_0^t (1+t')(1+\log(1+t'))B_{n-1}(t')dt' \tag{9.269}$$
$$+2(n+1)\sum_{k=0}^{m-1}\int_0^t (1+\log(1+t'))B'_{k,n-1-k}(t')dt'$$
$$+2(n-m)\int_0^t (1+\log(1+t'))B'_{m,n-1-m}(t')dt', m=1,\cdots,n-1$$

Recalling then (9.258) and setting $l=n-1-m$, we conclude that

$$\begin{aligned}
&\|^{(i_1\cdots i_l)}x'_{0,n-1}(t)\|_{L^2([0,\epsilon_0]\times S^2)}\\
\leqslant\, &^{(i_1\cdots i_{n-1})}B'_{0,n-1}(t)\\
&+C\delta_0(1+t)^{-2}(1+\log(1+t))^2(Y_{n-1}(t)+nY'_{0,n-1}(t)), m=0\\
&\|^{(i_1\cdots i_{n-1-m})}x'_{m,n-1-m}(t)\|_{L^2([0,\epsilon_0]\times S^2)}\\
\leqslant\, &^{(i_1\cdots i_{n-1-m})}B'_{m,n-1-m}(t)C\delta_0(1+t)^{-2}(1+\log(1+t))^2\\
&\cdot((m+1)Y'_{m-1,n-1-(m-1)}(t)+(n-m)Y'_{m,n-1-m}(t)), m\geqslant 1
\end{aligned} \tag{9.270}$$

So finally we obtain the following estimates by substituting (9.269) in (9.270):

$$\begin{aligned}
&\|^{(i_1\cdots i_l)}x'_{0,n-1}(t)\|_{L^2([0,\epsilon_0]\times S^2)}\\
\leqslant\, &^{(i_1\cdots i_{n-1})}B'_{0,n-1}(t)+C\delta_0(1+t)^{-2}(1+\log(1+t))^2\\
&\cdot(2(n+1)\int_0^t (1+t')(1+\log(1+t'))B_{n-1}(t')dt'\\
&+2n\int_0^t (1+\log(1+t'))B'_{0,n-1}(t')dt'), m=0
\end{aligned} \tag{9.271}$$

and

$$\begin{aligned}
&\|^{(i_1\cdots i_{n-1-m})}x'_{m,n-1-m}(t)\|_{L^2([0,\epsilon_0]\times S^2)}\\
\leqslant\, &^{(i_1\cdots i_{n-1-m})}B'_{m,n-1-m}(t)+C\delta_0(1+t)^{-2}(1+\log(1+t))^2\\
&\cdot(2(n+1)\int_0^t (1+t')(1+\log(1+t'))B_{n-1}(t')dt'\\
&+2(n+1)\sum_{k=0}^{m-1}\int_0^t (1+\log(1+t'))B'_{k,n-1-k}(t')dt'\\
&+2(n-m)\int_0^t (1+\log(1+t'))B'_{m,n-1-m}(t')dt'), m\geqslant 1
\end{aligned} \tag{9.272}$$

# 10

# Control of the Angular Derivatives of the First Derivatives of the $x^i$. Assumptions and Estimates in Regard to $\chi$

## 10.1 Preliminary

One of the aims of this chapter is concerned with derivation of estimates for angular derivatives of the spatial rectangular coordinates $x^i$ and $\hat{T}x^i, i = 1, 2, 3$ with respect to $R_j, j = 1, 2, 3$. The estimates are based on bootstrap assumptions in regard to $\chi$. These estimates are then used to obtain estimates for angular derivatives of the deformation tensors of the commutation fields.

Before we proceed, we must discuss the general definition of the operator $\not{\mathcal{L}}_X$, with $X$ an arbitrary $S_{t,u}$-tangential vectorfield, as well as the operator $\not{\mathcal{L}}_L$, $\not{\mathcal{L}}_T$, as applied to an arbitrary $T^q_p\ S_{t,u}$ tensorfield. Consider first the case of an $S_{t,u}$ 1-form $\xi$. Let $X$ be an $S_{t,u}$-tangential vectorfield. Then $\not{\mathcal{L}}_X\xi$ is a notion intrinsic to $S_{t,u}$. To define $\not{\mathcal{L}}_L\xi$ however, we must consider the extension of $\xi$ to a given $C_u$ as an $S_{t,u}$ 1-form, that is, by requiring $\xi(L) = 0$. We then define $\not{\mathcal{L}}_L\xi$ to be the restriction to $TS_{t,u}$ of the usual Lie derivative $\mathcal{L}_L\xi$, a notion intrinsic to $C_u$. Note that in any case $(\mathcal{L}_L\xi)(L) = 0$. We can define $\not{\mathcal{L}}_T\xi$ in a similar way by replacing $C_u$ by $\Sigma_t$.

The case of any $p$-covariant $S_{t,u}$ tensorfield in the role of $\xi$, is formally identical to the case of an $S_{t,u}$ 1-form. Consider next the case of an $S_{t,u}$-tangential vectorfield $Y$, and $X$ is another $S_{t,u}$-tangential vectorfield. Then $\not{\mathcal{L}}_XY$ is defined to be simply $\mathcal{L}_XY = [X, Y]$. From Lemma 8.2, $\mathcal{L}_LY = [L, Y] = {}^{(Y)}Z$ is an $S_{t,u}$-tangential vectorfield, so $\not{\mathcal{L}}_LY$ is defined to be $\mathcal{L}_LY$. Similarly (see Lemma 10.18), $\not{\mathcal{L}}_TY$ is defined to be $\mathcal{L}_TY$.

The case of any $q$-contravariant $S_{t,u}$ tensorfield is formally identical. And also, the general case of an arbitrary $T^q_p\ S_{t,u}$ tensorfield $\vartheta$ is similar. Note that the above definitions of $\not{\mathcal{L}}_X\vartheta$, $\not{\mathcal{L}}_L\vartheta$, $\not{\mathcal{L}}_T\vartheta$ do not depend on the acoustical metric $g$. So we have

$$\not{\mathcal{L}}_X(\vartheta \otimes \varphi) = (\not{\mathcal{L}}_X\vartheta) \otimes \varphi + \vartheta \otimes (\not{\mathcal{L}}_X\varphi) \tag{10.1}$$

with $X$ an arbitrary $S_{t,u}$-tangential vectorfield. And also

$$\not{\mathcal{L}}_L(\vartheta \otimes \varphi) = (\not{\mathcal{L}}_L\vartheta) \otimes \varphi + \vartheta \otimes (\not{\mathcal{L}}_L\varphi) \tag{10.2}$$

$$\not{\mathcal{L}}_T(\vartheta \otimes \varphi) = (\not{\mathcal{L}}_T\vartheta) \otimes \varphi + \vartheta \otimes (\not{\mathcal{L}}_T\varphi) \tag{10.3}$$

We also note that the same results hold with $Q$ in the role of $L$.

From the above discussion, we have, for $X, Y \in TS_{t,u}$,

$$\begin{aligned}(\mathcal{L}_L g)(X,Y) &= L(g(X,Y)) - g([L,X],Y) - g(X,[L,Y]) \\ &= L(\not{g}(X,Y)) - \not{g}([L,X],Y) - \not{g}(X,[L,Y]) \\ &= (\not{\mathcal{L}}_L\not{g})(X,Y)\end{aligned}$$

That is

$$^{(L)}\not{\pi} = \not{\mathcal{L}}_L\not{g} = 2\chi \tag{10.4}$$

Similarly, we have

$$^{(T)}\not{\pi} = \not{\mathcal{L}}_T\not{g} \tag{10.5}$$

and

$$\begin{aligned}(\mathcal{L}_T\bar{g})(X,Y) &= T(\bar{g}(X,Y)) - \bar{g}([T,X],Y) - \bar{g}(X,[T,Y]) \\ &= T(\not{g}(X,Y)) - \not{g}([T,X],Y) - \not{g}(X,[T,Y]) \\ &= (\not{\mathcal{L}}_T\not{g})(X,Y)\end{aligned}$$

thus,

$$\not{\mathcal{L}}_T\not{g} = 2\kappa\theta \tag{10.6}$$

Given now a positive integer $l$ let us denote by $\not{\mathbf{E}}_l$ the bootstrap assumption that there is a constant $C$ independent of $s$ such that for all $t \in [0,s]$:

$$\not{\mathbf{E}}_l : \max_{i_1\cdots i_l}\max_{\alpha}\|R_{i_l}\cdots R_{i_1}\psi_\alpha\|_{L^\infty(\Sigma_t^{\epsilon_0})} \leqslant C\delta_0(1+t)^{-1}$$

Also, for convenience of notation, let us denote by $\not{\mathbf{E}}_0$ the basic bootstrap assumption **E1** of Chapter 6:

$$\not{\mathbf{E}}_0 : \max_{\alpha}|\psi_\alpha| \leqslant C\delta_0(1+t)^{-1}$$

Given a positive integer $l$ we then denote by $\not{\mathbf{E}}_{[l]}$ the bootstrap assumption:

$$\not{\mathbf{E}}_{[l]} : \not{\mathbf{E}}_0 \quad \text{and} \cdots \text{and} \quad \not{\mathbf{E}}_l$$

Given a positive integer $n$ we denote

$$n_* = \begin{cases}\dfrac{n}{2}, & \text{if} \quad n \quad \text{is even} \\ \dfrac{n-1}{2}, & \text{if} \quad n \quad \text{is odd}\end{cases} \tag{10.7}$$

**Lemma 10.1** Let $G$ be a smooth function of the variations $(\psi_\alpha, \alpha = 0, 1, 2, 3)$. Suppose that the bootstrap assumption $\mathbf{E}_{[l_*]}$ holds for some positive integer $l$. Then there are constants $C$, $C_l$ independent of $s$ such that the following estimate holds:

$$\|R_{i_l} \cdots R_{i_1} G\|_{L^2(\Sigma_t^{\epsilon_0})}$$

$$\leqslant C \sum_\alpha \|R_{i_l} \cdots R_{i_1} \psi_\alpha\|_{L^2(\Sigma_t^{\epsilon_0})} + C_l \delta_0 (1+t)^{-1} \sum_{k=1}^{l-1} \max_{i_1 \cdots i_k} \max_\alpha \|R_{i_k} \cdots R_{i_1} \psi_\alpha\|_{L^2(\Sigma_t^{\epsilon_0})}$$

*Proof*. It is a direct calculation. □

Consider now the functions $\hat{T}^j$. By (3.200)—(3.202), we have

$$R_i \hat{T}^j = R_i^A \not{d}_A = \not{d}_i \cdot \not{d} x^j \tag{10.8}$$

where

$$\not{d}_i = m \cdot R_i \tag{10.9}$$

Here, $m$ is given by

$$m = m_\flat \cdot \not{g}^{-1} \tag{10.10}$$

where

$$m_\flat = \theta = -\alpha^{-1} \chi + \not{k} \tag{10.11}$$

Recall the definition of functions $y^i$ in Chapter 6:

$$\hat{T}^j = -\frac{x^j}{1-u+t} + y^j \tag{10.12}$$

Since $R_i x^j = R_i \cdot \not{d} x^j$, we have

$$R_i y^j = \not{d}'_i \cdot \not{d} x^j \tag{10.13}$$

where

$$\not{d}'_i = \not{d}_i + \frac{R_i}{1-u+t} \tag{10.14}$$

By (10.9)—(10.11) we have

$$\not{d}'_i = m' \cdot R_i \tag{10.15}$$

where

$$m' = m'_\flat \cdot \not{g}^{-1} \tag{10.16}$$

and

$$m'_{b} = m_b + \frac{\not{s}}{1-u+t} \tag{10.17}$$

is the 2-covariant $S_{t,u}$ tensorfield:

$$m'_{b} = -\frac{(\alpha^{-1}-1)\not{s}}{1-u+t} - \alpha^{-1}(\chi - \frac{\not{s}}{1-u+t}) + \not{k} \tag{10.18}$$

We consider next the angular derivatives of the rectangular coordinate functions $x^j$. According to (6.6) we have

$$(R_i)^j = \Pi^j_k (\mathring{R}_i)^k \tag{10.19}$$

where $\Pi^j_k$ are the components of $\Pi$, the $\bar{g}$-orthogonal projection to $S_{t,u}$ on $\Sigma_t$:

$$\Pi^j_k = \delta^j_k - \bar{g}_{kl}\hat{T}^l\hat{T}^j = \delta^j_k - \hat{T}^k\hat{T}^j \tag{10.20}$$

and

$$(\mathring{R}_i)^j = \epsilon_{imj}x^m \tag{10.21}$$

Recall (6.130):

$$\bar{g}(\mathring{R}_i, \hat{T}) = \lambda_i = \epsilon_{imj}x^m y^j \tag{10.22}$$

we then obtain

$$R_i = \mathring{R}_i - \lambda_i \hat{T}^j \partial_j \tag{10.23}$$

Define, for each non-negative integer $k$, the functions

$${}^{(k)}x^j_{i_1\cdots i_k} = \mathring{R}_{i_k}\cdots\mathring{R}_{i_1}x^j \tag{10.24}$$

These are linear functions of the rectangular coordinates, therefore the

$${}^{(k)}c^j_{l;i_1\cdots i_k} = \frac{\partial\, {}^{(k)}x^j_{i_1\cdots i_k}}{\partial x^l} \tag{10.25}$$

are constants. Define for each non-negative integer $k$ the functions ${}^{(k)}\delta^j_{i_1\cdots i_j}$ by the equation

$$R_{i_k}\cdots R_{i_1}x^j = {}^{(k)}x^j_{i_1\cdots i_k} - {}^{(k)}\delta^j_{i_1\cdots i_k} \tag{10.26}$$

In particular,

$${}^{(0)}\delta^j = 0 \tag{10.27}$$

and by (10.23),

$${}^{(1)}\delta^j_i = \lambda_i \hat{T}^j \tag{10.28}$$

Replacing $k$ by $k-1$ in (10.26) and applying $R_{i_k}$ we obtain

$$R_{i_k}R_{i_{k-1}}\cdots R_{i_1}x^j = R_{i_k}{}^{(k-1)}x^j_{i_1\cdots i_{k-1}} - R_{i_k}{}^{(k-1)}\delta^j_{i_1\cdots i_{k-1}} \tag{10.29}$$

By (10.23)—(10.25),

$$R_{i_k}{}^{(k-1)}x^j_{i_1\cdots i_{k-1}} = {}^{(k)}x^j_{i_1\cdots i_k} - {}^{(k-1)}c^j_{l;i_1\cdots i_{k-1}}\lambda_{i_k}\hat{T}^l \tag{10.30}$$

Substituting this in (10.29) and comparing with (10.26), we get

$${}^{(k)}\delta^j_{i_1\cdots i_k} = R_{i_k}{}^{(k-1)}\delta^j_{i_1\cdots i_{k-1}} + {}^{(k-1)}c^j_{l;i_1\cdots i_{k-1}}\lambda_{i_k}\hat{T}^l \tag{10.31}$$

Now we can apply Proposition 8.2 with $A_n$ in the role of $R_{i_k}$, $x_n$ in the role of ${}^{(k)}\delta^j_{i_1\cdots i_k}$ and $y_n$ in the role of

$${}^{(k-1)}c^j_{l;i_1\cdots i_{k-1}}\lambda_{i_k}\hat{T}^l$$

Since by (10.27) $x_0 = 0$, we obtain

$${}^{(k)}\delta^j_{i_1\cdots i_k} = \sum_{m=1}^{k} {}^{(m-1)}c^j_{l;i_1\cdots i_{m-1}} R_{i_k}\cdots R_{i_{k-m+2}}(\lambda_{i_m}\hat{T}^l) \tag{10.32}$$

Since the ${}^{(k)}x$ are linear functions of the rectangular coordinates, we have, from (10.25),

$${}^{(k)}x^j_{i_1\cdots i_k} = {}^{(k)}c^j_{l;i_1\cdots i_k}x^l \tag{10.33}$$

From (10.21) and (10.24) we obtain

$${}^{(0)}x^j = x^j, \quad {}^{(1)}x^j_i = \epsilon_{ilj}x^l \tag{10.34}$$

Moreover,

$${}^{(2)}x^j_{ik} = \delta_{jk}x^i - \delta_{ik}x^j \tag{10.35}$$

For $k \geqslant 2$ we have from (10.24) and (10.35)

$$\begin{aligned} {}^{(k)}x^j_{i_1\cdots i_k} &= \mathring{R}_{i_k}\cdots\mathring{R}_{i_3}{}^{(2)}x^j_{i_1i_2} \\ &= \mathring{R}_{i_k}\cdots\mathring{R}_{i_3}(\delta_{ji_2}x^{i_1} - \delta_{i_1i_2}x^j) \\ &= \delta_{ji_2}{}^{(k-2)}x^{i_1}_{i_3\cdots i_k} - \delta_{i_1i_2}{}^{(k-2)}x^j_{i_3\cdots i_k} \end{aligned} \tag{10.36}$$

It follows from (10.36) that for $k \geqslant 2$,

$$\max_{j;i_1\cdots i_k}|{}^{(k)}x^j_{i_1\cdots i_k}| = \max_{j;i_1\cdots i_{k-2}}|{}^{(k-2)}x^j_{i_1\cdots i_{k-2}}| \tag{10.37}$$

while from (10.34),

$$\max_{j;i}|{}^{(1)}x^j_i| = \max_j|{}^{(0)}x^j| = \max_j|x^j| \tag{10.38}$$

Then we have

$$\max_{j;i_1\cdots i_k} |{}^{(k)}x^j_{i_1\cdots i_k}| = \max_j |x^j| \tag{10.39}$$

which by (6.131) implies

$$\max_{j;i_1\cdots i_k} \|{}^{(k)}x^j_{i_1\cdots i_k}\|_{L^\infty(\Sigma_t^{\epsilon_0})} \leqslant 1+t \tag{10.40}$$

**Lemma 10.2** Let

$$\max_{j;i_1\cdots i_k} \|R_{i_k}\cdots R_{i_1}y^j\|_{L^\infty(\Sigma_t^{\epsilon_0})} \leqslant C_l\delta_0(1+t)^{-1}(1+\log(1+t))$$

hold for $k=0,\cdots,l$. Then if $\delta_0$ is suitably small (perhaps depending on $l$), we have

$$\max_{i;i_1\cdots i_k} \|R_{i_k}\cdots R_{i_1}\lambda_i\|_{L^\infty(\Sigma_t^{\epsilon_0})} \leqslant C_l\delta_0(1+\log(1+t))$$

and

$$\max_{j;ii_1\cdots i_k} \|{}^{(k+1)}\delta^j_{ii_1\cdots i_k}\|_{L^\infty(\Sigma_t^{\epsilon_0})} \leqslant C_l\delta_0(1+\log(1+t))$$

for all $k=0,\cdots,l$.

*Proof*. Noting (6.143) and (10.27), the lemma holds when $l=0$. Then by (10.12), (10.22), (10.26), (10.40) and (10.32), we can use an induction argument. □

**Lemma 10.3** Let

$$\max_{j;i_1\cdots i_k} \|R_{i_k}\cdots R_{i_1}y^j\|_{L^\infty(\Sigma_t^{\epsilon_0})} \leqslant C_l\delta_0(1+t)^{-1}(1+\log(1+t))$$

hold for $k=0,\cdots,l_*$. Then if $\delta_0$ is suitably small (perhaps depending on $l$) we have

$$\max_{i;i_1\cdots i_k} \|R_{i_k}\cdots R_{i_1}\lambda_i\|_{L^2(\Sigma_t^{\epsilon_0})} \leqslant C_l(1+t)\mathcal{Y}_{[l]}$$

and

$$\max_{j;ii_1\cdots i_k} \|{}^{(k+1)}\delta^j_{ii_1\cdots i_k}\|_{L^2(\Sigma_t^{\epsilon_0})} \leqslant C_l(1+t)\mathcal{Y}_{[l]}$$

for all $k=0,\cdots,l$. Here

$$\mathcal{Y}_k = \max_{j;i_1\cdots i_k} \|R_{i_k}\cdots R_{i_1}y^j\|_{L^2(\Sigma_t^{\epsilon_0})}, \quad \mathcal{Y}_{[l]} = \sum_{k=0}^{l}\mathcal{Y}_k$$

We also define

$$\mathcal{W}_0 = \max_\alpha \|\psi_\alpha\|_{L^2(\Sigma_t^{\epsilon_0})}$$

and for $k \neq 0$,

$$\mathcal{W}_k = \max_{\alpha; i_1 \cdots i_k} \| R_{i_k} \cdots R_{i_1} \psi_\alpha \|_{L^2(\Sigma_t^{\epsilon_0})}$$

and

$$\mathcal{W}_{[l]} = \sum_{k=0}^{l} \mathcal{W}_k$$

*Proof*. By (10.22), (10.32) and (10.40), the lemma holds for $l = 0$. We then use an induction argument. □

**Lemma 10.4** For every non-negative integer $k$ we have

$$R_{i_k} \cdots R_{i_1} R_i y^j = {}^{(k)}\not{q}'_{ii_1 \cdots i_k} \cdot \not{d} x^j + {}^{(k)}\not{r}'_{ii_1 \cdots i_k}$$

Here,

$${}^{(k)}\not{q}'_{ii_1 \cdots i_k} = \not{\mathcal{L}}_{R_{i_k}} \cdots \not{\mathcal{L}}_{R_{i_1}} \not{q}'_i$$

and

$${}^{(k)}\not{r}^j_{ii_1 \cdots i_k} = \sum_{m=0}^{k-1} R_{i_k} \cdots R_{i_{k-m+1}} \left( {}^{(k-m-1)}\not{q}'_{ii_1 \cdots i_{k-m-1}} \cdot \not{d}(R_{i_{k-m}} x^j) \right)$$

*Proof*. Since

$${}^{(0)}\not{q}'_i = \not{q}'_i, \quad {}^{(0)}\not{r}^j_i = 0$$

the lemma reduces for $k = 0$ to Equation (10.13). Then arguing by induction, and using Proposition 8.2, the lemma follows. □

Now it is evident from (10.15) and the formulas for the coefficients ${}^{(k)}\not{q}'_{ii_1 \cdots i_k}$ and ${}^{(k)}\not{r}^j_{ii_1 \cdots i_k}$ given in the statement of Lemma 10.4, that ${}^{(k)}\not{q}'_{ii_1 \cdots i_k}$ contains the angular derivatives of the $R_i$ of order up to $k$ and ${}^{(k)}\not{r}^j_{ii_1 \cdots i_k}$ contains the angular derivatives of the $R_i$ of order up to $k-1$, where by "the angular derivatives of the $R_i$ of order up to $l$" we mean the $S_{t,u}$-tangential vectorfields $\not{\mathcal{L}}_{R_{i_m}} \cdots \not{\mathcal{L}}_{R_{i_1}} R_i$ for $m = 0, \cdots, l$. We shall thus derive the expressions below for these vectorfields. The basic formula is the formula for

$$\not{\mathcal{L}}_{R_i} R_j = [R_i, R_j]$$

given by the following lemma. Let $w_i$ be

$$w_i = (w_i)_b \cdot \not{g}^{-1} \tag{10.41}$$

where

$$(w_i)_b = \epsilon_{ijm} y^j \bar{g}_{mn} \not{d} x^n = \epsilon_{ijm} y^j \not{d} x^m \tag{10.42}$$

Let us then define

$$\tilde{\not{d}}'_i = \not{d}'_i - w_i \tag{10.43}$$

**Lemma 10.5** We have

$$[R_i, R_j] = -\epsilon_{ijk} R_k + \lambda_i \tilde{\not{d}}'_j - \lambda_j \tilde{\not{d}}'_i$$

*Proof.* We express

$$R_i = (R_i)^m \partial_m$$

According to (10.23),

$$(R_i)^m = (\mathring{R}_i)^m - \lambda_i \hat{T}^m$$

Thus, we have

$$\begin{aligned}[R_i, R_j] &= (R_i((R_j)^m) - R_j((R_i)^m))\partial_m \\ &= (R_i((\mathring{R}_j)^m - \lambda_j \hat{T}^m) - R_j((\mathring{R}_i)^m - \lambda_i \hat{T}^m))\partial_m \end{aligned} \tag{10.44}$$

Now,

$$\begin{aligned} R_i((\mathring{R}_j)^m) &= R_i(\epsilon_{jkm} x^k) = \epsilon_{jkm}(R_i)^k \\ &= -\epsilon_{jmk}((\mathring{R}_i)^k - \lambda_i \hat{T}^k) = -\epsilon_{jmk}\epsilon_{ilk} x^l + \epsilon_{jmk}\lambda_i \hat{T}^k \end{aligned}$$

Since,

$$\epsilon_{jmk}\epsilon_{ilk} + \epsilon_{mik}\epsilon_{jlk} + \epsilon_{ijk}\epsilon_{mlk} = 0$$

we then obtain

$$R_i((\mathring{R}_j)^m) - R_j((\mathring{R}_i)^m) = -\epsilon_{ijk}(\mathring{R}_k)^m + (\epsilon_{ikm}\lambda_j - \epsilon_{jkm}\lambda_i)\hat{T}^k \tag{10.45}$$

Substituting (10.45) in (10.44) yields

$$\begin{aligned}[R_i, R_j] = &- \epsilon_{ijk} \mathring{R}_k - \lambda_i v_j + \lambda_j v_i \\ &- (R_i(\lambda_j) - R_j(\lambda_i))\hat{T} - (\lambda_j R_i(\hat{T}^m) - \lambda_i R_j(\hat{T}^m))\partial_m \end{aligned} \tag{10.46}$$

where

$$v_i = \epsilon_{ikm} \hat{T}^k \partial_m \tag{10.47}$$

By (10.12), (10.13),

$$R_i(\hat{T}^m)\partial_m = -(1 - u + t)^{-1} R_i + \not{d}'_i \tag{10.48}$$

Substituting (10.48) in (10.46), the latter becomes

$$[R_i, R_j] = -\epsilon_{ijk}\mathring{R}_k - (R_i(\lambda_j) - R_j(\lambda_i))\hat{T} + \lambda_i(\not{d}'_j - v'_j) - \lambda_j(\not{d}'_i - v'_i) \tag{10.49}$$

where

$$v'_i = v_i + (1-u+t)^{-1}R_i \tag{10.50}$$

Since $[R_i, R_j]$ is $S_{t,u}$-tangential, we may apply the projection operator $\Pi$ to the right hand side of (10.49). This annihilates the second term, and since $\Pi\mathring{R}_i = R_i$, we obtain

$$[R_i, R_j] = -\epsilon_{ijk}R_k + \lambda_i(\not{d}'_j - \Pi v'_j) - \lambda_j(\not{d}'_i - \Pi v'_i) \tag{10.51}$$

In view of (10.43), we must show that

$$\Pi v'_i = w_i \tag{10.52}$$

We have, in view of (10.50),

$$\Pi v'_i = \Pi\tilde{v}_i \tag{10.53}$$

where

$$\tilde{v}_i = v_i + (1-u+t)^{-1}\mathring{R}_i \tag{10.54}$$

In view of (10.47), (10.12) and (10.21), we have

$$\tilde{v}_i = \epsilon_{ikm}y^k\partial_m \tag{10.55}$$

We then have, in terms of the frame $(X_A, A=1,2)$,

$$\Pi\tilde{v}_i = \bar{g}(\tilde{v}_i, X_B)(\not{g}^{-1})^{BA}X_A \tag{10.56}$$

and

$$\bar{g}(\tilde{v}_i, X_B) = \bar{g}_{mn}(\tilde{v}_i)^m X_B^n = \epsilon_{ikm}y^k\bar{g}_{mn}\not{d}_B x^n = \epsilon_{ikm}y^k\not{d}_B x^m \tag{10.57}$$

Comparing (10.57) with (10.42) we see that

$$\bar{g}(\tilde{v}_i, X_B) = (w_i)_b(X_B) \tag{10.58}$$

Substituting (10.58) in (10.56) we conclude that

$$\Pi\tilde{v}_i = w_i \tag{10.59}$$

Therefore by (10.53), the lemma follows. □

Since by (10.15), (10.43),

$$\tilde{\not{d}}'_i = m'\cdot R_i - w_i \tag{10.60}$$

defining the $S_{t,u}$ tensorfields $\mu_i$:

$$\mu_i = \lambda_i m' \tag{10.61}$$

and the $S_{t,u}$-tangential vectorfields $\nu_{ij}$:

$$\nu_{ij} = -\lambda_i w_j + \lambda_j w_i \tag{10.62}$$

the formula of Lemma 10.5 takes the form

$$[R_i, R_j] = -\epsilon_{ijk} R_k + \mu_i \cdot R_j - \mu_j \cdot R_i + \nu_{ij} \tag{10.63}$$

**Lemma 10.6** For every non-negative integer $k$ we have

$$\not{\mathcal{L}}_{R_{i_k}} \cdots \not{\mathcal{L}}_{R_{i_1}} R_j = {}^{(k)}\alpha^m_{j;i_1\cdots i_k} R_m + {}^{(k)}\beta_{j;i_1\cdots i_k} \cdot R_m + {}^{(k)}\gamma_{j;i_1\cdots i_k}$$

The coefficients ${}^{(k)}\alpha^m_{j;i_1\cdots i_k}$ are constants, given by

$${}^{(k)}\alpha^m_{j;i_1\cdots i_k} = (-1)^k \epsilon_{i_k n_k m} \epsilon_{i_{k-1} n_{k-1} n_k} \cdots \epsilon_{i_1 n_1 n_2} \delta^{n_1}_j$$

The coefficients ${}^{(k)}\beta_{j;i_1\cdots i_k} = ({}^{(k)}\beta^m_{i_1\cdots i_k}, m = 1,2,3)$ are triplets of $T^1_1$-type $S_{t,u}$ tensorfields given by

$${}^{(k)}\beta_{j;i_1\cdots i_k} = \sum_{n=0}^{k-1} A_{i_k} \cdots A_{i_{k-n+1}} {}^{(k-n)}\rho_{j;i_1\cdots i_k}$$

Here $A_i$ are the operators acting on triplets $z = (z^m, m = 1,2,3)$ of $T^1_1$-type $S_{t,u}$ tensorfields given by

$$(A_i z)^m = \not{\mathcal{L}}_{R_i} z^m - \sum_n \epsilon_{inm} z^n + z^m \cdot \mu_i - (\sum_n z^n \cdot \mu_n)\delta^m_i$$

and the ${}^{(k)}\rho_{j;i_1\cdots i_k} = ({}^{(k)}\rho^m_{j;i_1\cdots i_k}, m = 1,2,3)$ are the triplets of $T^1_1$-type $S_{t,u}$ tensorfields, given by

$${}^{(k)}\rho^m_{j;i_1\cdots i_k} = {}^{(k-1)}\alpha^m_{j;i_1\cdots i_k} \mu_{i_k} - (\sum_n {}^{(k-1)}\alpha^n_{j;i_1\cdots i_k} \mu_n)\delta^m_{i_k}$$

Finally, the coefficients ${}^{(k)}\gamma_{j;i_1\cdots i_k}$ are $S_{t,u}$-tangential vectorfields, given by

$$\begin{aligned}{}^{(k)}\gamma_{j;i_1\cdots i_k} = &\sum_{n=0}^{k-1} {}^{(k-1-n)}\alpha^m_{j;i_1\cdots i_{k-n-1}} \not{\mathcal{L}}_{R_{i_k}} \cdots \not{\mathcal{L}}_{R_{i_{k-n+1}}} \nu_{i_{k-n}m} \\ &+ \sum_{n=0}^{k-1} \not{\mathcal{L}}_{R_{i_k}} \cdots \not{\mathcal{L}}_{R_{i_{k-n+1}}} ({}^{(k-n-1)}\beta^m_{j;i_1\cdots i_{k-n-1}} \cdot \nu_{i_{k-n}m})\end{aligned}$$

*Proof.* With

$${}^{(0)}\alpha^m_j = \delta^m_j, \quad {}^{(0)}\beta^m_j = 0, \quad {}^{(0)}\gamma_j = 0 \tag{10.64}$$

the lemma is trivial for $k = 0$. Moreover, one can easily check that the lemma reduces for $k = 1$ to (10.63). Then arguing by induction, using Proposition 8.2 and noting (10.64), the lemma follows. □

To analyze the operators $A_i$, we define the multiplication operators $M_i$. These are 3-dimensional matrices, the entries $(M_i)^m_n, m, n = 1,2,3$ of which are $T^1_1$-type of $S_{t,u}$ tensorfields given by

$$(M_i)_n^m = -\epsilon_{inm} I + \delta_n^m \mu_i - \delta_i^m \mu_n \tag{10.65}$$

Here $I$ is the $T_1^1$-type of $S_{t,u}$ tensorfield which is the identity as a transformation in the tangent space of $S_{t,u}$ at each point. These matrices act on triplets $z = (z^m, m = 1, 2, 3)$ of $T_1^1$-type of $S_{t,u}$ tensorfields on the right:

$$(z \cdot M_i)^m = \sum_n z^n \cdot (M_i)_n^m \tag{10.66}$$

so that with matrix multiplication defined according to

$$(M_i \cdot M_j)_n^m = \sum_k (M_i)_n^k \cdot (M_j)_k^m \tag{10.67}$$

we have

$$(z \cdot M_i) \cdot M_j = z \cdot (M_i \cdot M_j) \tag{10.68}$$

The operators $A_i$ are defined in terms of the matrices $M_i$ by

$$A_i z = \not{\mathcal{L}}_{R_i} z + z \cdot M_i \tag{10.69}$$

We may also define the action of the operators $A_i$ on matrices $K$ by

$$A_i K = \not{\mathcal{L}}_{R_i} K + K \cdot M_i \tag{10.70}$$

Note that if we also denote by $I$ the identity matrix, then according to (10.70) we have

$$A_i I = M_i \tag{10.71}$$

From (10.69) and (10.70), we readily deduce the following formula by induction on $k$:

$$A_{i_k} \cdots A_{i_1} z = \sum_{partitions} (\not{\mathcal{L}}_R)^{s_1} z \cdot (M)^{s_2} \tag{10.72}$$

Here the sum is over all ordered partitions $\{s_1, s_2\}$ of the set $\{1, \cdots, k\}$ into two ordered subsets $s_1, s_2$. Also, we denote

$$(M)^{s_2} = A_{i_{n_p}} \cdots A_{i_{n_1}} I, \quad \text{if} \quad s_2 = \{n_1, \cdots, n_p\}, p = |s_2| \tag{10.73}$$

Introducing the matrices $I_i$ and $N_i$ by

$$(I_i)_n^m = -\epsilon_{inm} I \tag{10.74}$$

$$(N_i)_n^m = \delta_n^m \mu_i - \delta_i^m \mu_n \tag{10.75}$$

we may write

$$M_i = I_i + N_i \tag{10.76}$$

Let us introduce for any positive integer $k$ and multi-index $(i_1, \cdots, i_k)$ the matrices ${}^{(k)}I_{i_1 \cdots i_k}$ and ${}^{(k)}B_{i_1 \cdots i_k}$ by

$$ {}^{(k)}I_{i_1 \cdots i_k} = I_{i_1} \cdots I_{i_k} \tag{10.77} $$

and

$$ {}^{(k)}B_{i_1 \cdots i_k} = A_{i_k} \cdots A_{i_1} I - {}^{(k)}I_{i_1 \cdots i_k} \tag{10.78} $$

We also set ${}^{(0)}I = I$, ${}^{(0)}B = 0$. It is then readily seen that

$$ ({}^{(k)}I_{i_1 \cdots i_k})^m_n = {}^{(k)}\alpha^m_{n;i_1 \cdots i_k} I \tag{10.79} $$

Obviously, the ${}^{(k)}B_{i_1 \cdots i_k}$ contain terms of degree 1 up to $k$ in the $N_i$, and the terms of degree $q$ contain up to $k-q$ angular derivatives. Thus the ${}^{(k)}B_{i_1 \cdots i_k}$ contain the angular derivatives of the $N_i$ of order up to $k-1$. Also, (10.73) can be written as

$$ (M)^{s_2} = {}^{(p)}I_{i_{n_1} \cdots i_{n_p}} + {}^{(p)}B_{i_{n_1} \cdots i_{n_p}} \tag{10.80} $$

Finally, by the way in which we derive (10.39), we have

$$ \max_{m,j;i_1 \cdots i_k} |{}^{(k)}\alpha^m_{j;i_1 \cdots i_k}| = \max_{m,j} |{}^{(0)}\alpha^m_j| = 1 \tag{10.81} $$

which implies

$$ \max_{n,m;i_1 \cdots i_k} |({}^{(k)}I_{i_1 \cdots i_k})^m_n| = 1 \tag{10.82} $$

## 10.2 Estimates for $y^i$

Given a non-negative integer $l$ let us denote by $\not{\mathbf{E}}^Q_l$ the bootstrap assumption that there is a constant $C$ independent of $s$ such that for all $t \in [0, s]$,

$$ \not{\mathbf{E}}^Q_l : \max_{i_1 \cdots i_l} \max_{\alpha} \|R_{i_l} \cdots R_{i_1} Q \psi_\alpha\|_{L^\infty(\Sigma_t^{\epsilon_0})} \leqslant C\delta_0 (1+t)^{-1} $$

We then denote by $\not{\mathbf{E}}^Q_{[l]}$ the bootstrap assumption:

$$ \not{\mathbf{E}}^Q_{[l]} : \not{\mathbf{E}}^Q_0 \quad \text{and} \cdots \text{and} \quad \not{\mathbf{E}}^Q_l $$

Given a positive integer $l$, denote by $\not{\mathbf{X}}_l$ the bootstrap assumption that there is a constant $C$ independent of $s$ such that for all $t \in [0, s]$,

$$ \not{\mathbf{X}}_l : \max_{i_1 \cdots i_l} \|\not{\mathcal{L}}_{R_{i_l}} \cdots \not{\mathcal{L}}_{R_{i_1}} \chi\|_{L^\infty(\Sigma_t^{\epsilon_0})} \leqslant C\delta_0 (1+t)^{-2} (1 + \log(1+t)) $$

Let us also denote by $\not{\mathbf{X}}_0$ the bootstrap assumption:

$$ \not{\mathbf{X}}_0 : |\chi - \frac{\not{g}}{1-u+t}| \leqslant C\delta_0 (1+t)^{-2} (1 + \log(1+t)) $$

This coincides with **F2** in Chapter 6. We then denote

$$ \not{\mathbf{X}}_{[l]} : \not{\mathbf{X}}_0 \quad \text{and} \cdots \text{and} \quad \not{\mathbf{X}}_l $$

### 10.2.1 $L^\infty$ Estimates for $R_{i_k} \cdots .R_{i_1} y^j$

**Proposition 10.1** Let hypothesis **H0** and the estimate (6.177) hold. Let also the bootstrap assumptions $\not{\mathbf{E}}_{[l]}$, $\not{\mathbf{E}}^Q_{[l-1]}$ and $\not{\mathbf{X}}_{[l-1]}$ hold, for some positive integer $l$. Then if $\delta_0$ is suitably small (depending on $l$) we have

$$\max_{j;i_1 \cdots i_k} \|R_{i_k} \cdots R_{i_1} y^j\|_{L^\infty(\Sigma_t^{\epsilon_0})} \leqslant C_l \delta_0 (1+t)^{-1}(1+\log(1+t))$$

for all $k = 0, \cdots, l$.

*Proof.* We shall argue by induction. The proposition reduces for $l = 0$ to (6.177), which is assumed here. Let then the proposition hold with $l$ replaced by $l-1$. We then have

$$\max_{j;i_1 \cdots i_k} \|R_{i_k} \cdots R_{i_1} y^j\|_{L^\infty(\Sigma_t^{\epsilon_0})} \leqslant C_{l-1} \delta_0 (1+t)^{-1}(1+\log(1+t)) \tag{10.83}$$

for all $k = 0, \cdots, l-1$.

Here, we shall use the expressions in Lemma 10.4. First, note that by Lemma 10.2,

$$\max_{j;i_1 \cdots i_k} \|{}^{(k)}\delta^j_{i_1 \cdots i_k}\|_{L^\infty(\Sigma_t^{\epsilon_0})} \leqslant C_{l-1} \delta_0 (1+\log(1+t)) \tag{10.84}$$

for all $k = 0, \cdots, l$.

Then by (10.26) and (10.40) we have

$$\max_{j;i_1 \cdots i_k} \|R_{i_k} \cdots R_{i_1} x^j\|_{L^\infty(\Sigma_t^{\epsilon_0})} \leqslant C(1+t) \tag{10.85}$$

for all $k = 0, \cdots, l$.

So what we need to do is to estimate $\not{\mathcal{L}}_{R_{i_k}} \cdots \not{\mathcal{L}}_{R_{i_1}} \not{d}'_i$ for $0 \leqslant k \leqslant l-1$. First, by $\not{\mathbf{E}}_{[l]}$ and **H0**, we have

$$\max_{i_1 \cdots i_k} \|R_{i_k} \cdots R_{i_1}(\alpha^{-1})\|_{L^\infty(\Sigma_t^{\epsilon_0})} \leqslant C\delta_0(1+t)^{-1} \tag{10.86}$$

$$\max_{i_1 \cdots i_k} \|\not{\mathcal{L}}_{R_{i_k}} \cdots \not{\mathcal{L}}_{R_{i_1}} \not{k}\|_{L^\infty(\Sigma_t^{\epsilon_0})} \leqslant C\delta_0(1+t)^{-2} \tag{10.87}$$

for $1 \leqslant k \leqslant l-1$.

Then by (6.59) and $\not{\mathbf{X}}_{[l-1]}$, we have

$$\max_{j;i_1 \cdots i_k} \|\not{\mathcal{L}}_{R_{i_k}} \cdots \not{\mathcal{L}}_{R_{i_1}} {}^{(R_j)}\not{\pi}\|_{L^\infty(\Sigma_t^{\epsilon_0})} \leqslant C_l \delta_0 (1+t)^{-1}(1+\log(1+t)) \tag{10.88}$$

for $0 \leqslant k \leqslant l-1$. Here we also used the induction assumption so that we have

$$\max_{i;i_1 \cdots i_k} \|R_{i_k} \cdots R_{i_1} \lambda_i\|_{L^\infty(\Sigma_t^{\epsilon_0})} \leqslant C_{l-1} \delta_0 (1+\log(1+t)) \tag{10.89}$$

So by $\not{\mathbf{X}}_{[l-1]}$, and (10.16), (10.18), we have

$$\max_{i_1 \cdots i_k} \|\not{\mathcal{L}}_{R_{i_k}} \cdots \not{\mathcal{L}}_{R_{i_1}} m'\|_{L^\infty(\Sigma_t^{\epsilon_0})} \leqslant C\delta_0(1+t)^{-2}(1+\log(1+t)) \tag{10.90}$$

for $0 \leqslant k \leqslant l-1$.

Second, by Lemma 10.6, we easily obtain

$$\|\not{\mathcal{L}}_{R_{i_k}} \cdots \not{\mathcal{L}}_{R_{i_1}} R_j\|_{L^\infty(\Sigma_t^{\epsilon_0})} \leqslant C(1+t) \tag{10.91}$$

for $0 \leqslant k \leqslant l$.

In fact, from (10.89) and (10.90) we have

$$\max_{i;i_1\cdots i_k} \|\not{\mathcal{L}}_{R_{i_k}} \cdots \not{\mathcal{L}}_{R_{i_1}} \mu_i\|_{L^\infty(\Sigma_t^{\epsilon_0})} \leqslant C_l\delta_0^2(1+t)^{-2}(1+\log(1+t))^2 \tag{10.92}$$

for all $k = 0, \cdots, l-1$.

Then by (10.75),

$$\max_{i;i_1\cdots i_k} \|\not{\mathcal{L}}_{R_{i_k}} \cdots \not{\mathcal{L}}_{R_{i_1}} N_i\|_{L^\infty(\Sigma_t^{\epsilon_0})} \leqslant C_l\delta_0^2(1+t)^{-2}(1+\log(1+t))^2 \tag{10.93}$$

for all $k = 0, \cdots, l-1$.

By the discussion preceding (10.80),

$$\max_{i_1\cdots i_k} \|{}^{(k)}B_{i_1\cdots i_k}\|_{L^\infty(\Sigma_t^{\epsilon_0})} \leqslant C_l\delta_0^2(1+t)^{-2}(1+\log(1+t))^2 \tag{10.94}$$

By (10.80), (10.82) and (10.94) we have

$$\|(M)^{s_2}\|_{L^\infty(\Sigma_t^{\epsilon_0})} \leqslant C \quad \text{for} \quad |s_2| \leqslant l \tag{10.95}$$

(provided that $\delta_0$ is suitably small depending on $l$). It then follows from (10.72),

$$\max_{i_1\cdots i_k} \|A_{i_k} \cdots A_{i_1} z\|_{L^\infty(\Sigma_t^{\epsilon_0})} \leqslant C_k \sum_{m=0}^{k} \max_{i_1\cdots i_m} \|\not{\mathcal{L}}_{R_{i_m}} \cdots \not{\mathcal{L}}_{R_{i_1}} z\|_{L^\infty(\Sigma_t^{\epsilon_0})} \tag{10.96}$$

for all $k = 0, \cdots, l$.

Then by Lemma 10.6 and (10.92),

$$\begin{aligned} &\max_{i_1\cdots i_{m+n}} \|\not{\mathcal{L}}_{R_{i_{m+n}}} \cdots \not{\mathcal{L}}_{R_{i_{m+1}}} {}^{(m)}\rho_{j;i_1\cdots i_m}\|_{L^\infty(\Sigma_t^{\epsilon_0})} \\ &\leqslant C_l\delta_0^2(1+t)^{-2}(1+\log(1+t))^2 \end{aligned} \tag{10.97}$$

for all $m$ and all $n \leqslant l-1$.

Then by (10.96) we obtain

$$\|A_{i_{m+n}} \cdots A_{i_{m+1}} {}^{(m)}\rho_{j;i_1\cdots i_m}\|_{L^\infty(\Sigma_t^{\epsilon_0})} \leqslant C_l\delta_0^2(1+t)^{-2}(1+\log(1+t))^2 \tag{10.98}$$

for all $m$ and all $n \leqslant l-1$.

So finally we have

$$\begin{aligned} &\max_{j;i_1\cdots i_{m+n}} \|\not{\mathcal{L}}_{R_{i_{m+n}}} \cdots \not{\mathcal{L}}_{R_{i_{m+1}}} {}^{(m)}\beta_{j;i_1\cdots i_m}\|_{L^\infty(\Sigma_t^{\epsilon_0})} \\ &\leqslant C_l\delta_0^2(1+t)^{-2}(1+\log(1+t))^2 \end{aligned} \tag{10.99}$$

for all $m + n \leqslant l$.

We still have to consider ${}^{(k)}\gamma_{j;i_1\cdots i_k}$. By the expression in Lemma 10.6 and the induction assumption, we easily deduce

$$\max_{j;i_1\cdots i_k} \|{}^{(k)}\gamma_{j;i_1\cdots i_k}\|_{L^\infty(\Sigma_t^{\epsilon_0})} \leqslant C_l\delta_0^2(1+t)^{-1}(1+\log(1+t))^2 \tag{10.100}$$

for all $k=0,\cdots,l$.

We thus arrive at (10.91). Then from (10.91), (10.90) and (10.15) we have

$$\max_{i;i_1\cdots i_k} \|\not{\mathcal{L}}_{R_{i_k}}\cdots\not{\mathcal{L}}_{R_{i_1}}\not{\ell}'_i\|_{L^\infty(\Sigma_t^{\epsilon_0})} \leqslant C_l\delta_0(1+t)^{-1}(1+\log(1+t)) \tag{10.101}$$

for all $k=0,\cdots,l-1$.

The proposition then follows. □

There are some immediate corollaries of this proposition. First by Lemma 10.2, we have

**Corollary 10.1.a** Under the assumptions of Proposition 10.1 we have

$$\max_{i;i_1\cdots i_k} \|R_{i_k}\cdots R_{i_1}\lambda_i\|_{L^\infty(\Sigma_t^{\epsilon_0})} \leqslant C_l\delta_0(1+\log(1+t))$$

and

$$\max_{j;ii_1\cdots i_k} \|{}^{(k+1)}\delta_{ii_1\cdots i_k}\|_{L^\infty(\Sigma_t^{\epsilon_0})} \leqslant C_l\delta_0(1+\log(1+t))$$

for all $k=0,\cdots,l$.

The second of above implies

$$\max_{j;i_1\cdots i_k} \|R_{i_k}\cdots R_{i_1}x^j\|_{L^\infty(\Sigma_t^{\epsilon_0})} \leqslant C(1+t)$$

for all $k=0,\cdots,l+1$, if $\delta_0$ is suitably small (depending on $l$)

Next, from the proof of Proposition 10.1 we have

**Corollary 10.1.b** Under the assumptions of Proposition 10.1 we have

$$\max_{j;i_1\cdots i_k} \|R_{i_k}\cdots R_{i_1}\hat{T}^j\|_{L^\infty(\Sigma_t^{\epsilon_0})} \leqslant C \quad \text{for all} \quad k=0,\cdots,l$$

and

$$\max_{i_1\cdots i_k} \|R_{i_k}\cdots R_{i_1}\psi_{\hat{T}}\|_{L^\infty(\Sigma_t^{\epsilon_0})} \leqslant C_l\delta_0(1+t)^{-1} \quad \text{for all} \quad k=0,\cdots,l$$

Moreover,

$$\max_{i_1\cdots i_k} \|\not{\mathcal{L}}_{R_{i_k}}\cdots\not{\mathcal{L}}_{R_{i_1}}\not{\psi}\|_{L^\infty(\Sigma_t^{\epsilon_0})} \leqslant C_l\delta_0(1+t)^{-1} \quad \text{for all} \quad k=0,\cdots,l$$

and

$$\max_{i_1\cdots i_k} \|\not{\mathcal{L}}_{R_{i_k}}\cdots\not{\mathcal{L}}_{R_{i_1}}\not{\psi}\|_{L^\infty(\Sigma_t^{\epsilon_0})} \leqslant C_l\delta_0(1+t)^{-1} \quad \text{for all} \quad k=0,\cdots,l-1$$

**Corollary 10.1.c** Under the assumptions of Proposition 10.1 we have

$$\max_{i_1\cdots i_k} \|\not{\mathcal{L}}_{R_{i_k}}\cdots\not{\mathcal{L}}_{R_{i_1}}\not{k}\|_{L^\infty(\Sigma_t^{\epsilon_0})} \leqslant C_l\delta_0(1+t)^{-2} \quad \text{for all} \quad k=0,\cdots,l-1$$

**Corollary 10.1.d** Under the assumptions of Proposition 10.1 we have

$$\max_{i_1\cdots i_k} \|\not{\mathcal{L}}_{R_{i_k}}\cdots\not{\mathcal{L}}_{R_{i_1}}{}^{(R_i)}\not{\pi}\|_{L^\infty(\Sigma_t^{\epsilon_0})} \leqslant C_l\delta_0(1+t)^{-1}(1+\log(1+t))$$
$$\text{for all} \quad k=0,\cdots,l-1$$

**Corollary 10.1.e** Under the assumptions of Proposition 10.1 we have, if $\delta_0$ is suitably small (depending on $l$),

$$\max_{j;i_1\cdots i_k} \|\not{\mathcal{L}}_{R_{i_k}}\cdots\not{\mathcal{L}}_{R_{i_1}}R_j\|_{L^\infty(\Sigma_t^{\epsilon_0})} \leqslant C_l\delta_0(1+t) \quad \text{for all} \quad k=0,\cdots,l$$

More precisely, we have

$$\max_{m,j;i_1\cdots i_k} |{}^{(k)}\alpha^m_{j;i_1\cdots i_k}| = 1 \quad \text{for all} \quad k$$
$$\max_{m,j;i_1\cdots i_k} \|{}^{(k)}\beta^m_{j;i_1\cdots i_k}\|_{L^\infty(\Sigma_t^{\epsilon_0})} \leqslant C_l\delta_0^2(1+t)^{-2}(1+\log(1+t))^2$$
$$\text{for all} \quad k=0,\cdots,l \ ({}^{(0)}\beta^m_j=0)$$
$$\max_{j;i_1\cdots i_k} \|{}^{(k)}\gamma_{j;i_1\cdots i_k}\|_{L^\infty(\Sigma_t^{\epsilon_0})} \leqslant C_l\delta_0^2(1+t)^{-1}(1+\log(1+t))^2$$
$$\text{for all} \quad k=0,\cdots,l \ ({}^{(0)}\gamma_j=0)$$

Finally, we have

**Corollary 10.1.f** Under the assumptions of Proposition 10.1, we have

$$\max_{i;i_1\cdots i_k} \|{}^{(k)}\not{q}'_{ii_1\cdots i_k}\|_{L^\infty(\Sigma_t^{\epsilon_0})} \leqslant C_l(1+t)^{-1}(1+\log(1+t)) \quad \text{for all} \quad k=0,\cdots,l-1$$

and

$$\max_{i;i_1\cdots i_k} \|{}^{(k)}\not{r}^j_{ii_1\cdots i_k}\|_{L^\infty(\Sigma_t^{\epsilon_0})} \leqslant C_l(1+t)^{-1}(1+\log(1+t)) \quad \text{for all} \quad k=0,\cdots,l-1$$

### 10.2.2 $L^2$ Estimates for $R_{i_k}\cdots R_{i_1}y^j$

We define

$$\mathcal{A}_0 = \|\chi - \frac{\not{g}}{1-u+t}\|_{L^2(\Sigma_t^{\epsilon_0})}$$

and for $k\neq 0$,

$$\mathcal{A}_k = \max_{i_1\cdots i_k}\|\not{\mathcal{L}}_{R_{i_k}}\cdots\not{\mathcal{L}}_{R_{i_1}}\chi\|_{L^2(\Sigma_t^{\epsilon_0})}$$

and

$$\mathcal{A}_{[l]} = \sum_{k=0}^{l} \mathcal{A}_k$$

Also, we define

$$\mathcal{W}_k^Q = \max_{\alpha;i_1\cdots i_k} \|R_{i_k}\cdots R_{i_1} Q\psi_\alpha\|_{L^2(\Sigma_t^{\epsilon_0})}, \quad \mathcal{W}_{[l]}^Q = \sum_{k=0}^{l} \mathcal{W}_k^Q$$

**Proposition 10.2** Let the hypothesis **H0** and the estimate (6.177) hold. Let $l$ be a positive integer and let the bootstrap assumptions $\mathbf{E}_{[l_*]}$, $\mathbf{E}_{[l_*-1]}^Q$ and $\mathbf{X}_{[l_*-1]}$ hold. Then if $\delta_0$ is suitably small (depending on $l$) we have

$$\mathcal{Y}_k \leqslant C_l(\mathcal{Y}_0 + (1+t)\mathcal{A}_{[l-1]} + \mathcal{W}_{[l]})$$

for all $k = 0, \cdots, l$.

*Proof*. The proof is again by induction. The proposition extends trivially to the case $l = 0$. Let then the proposition hold with $l$ replaced by $1, \cdots, l-1$ and consider the case $l$. The assumptions of Proposition 10.2 coincide with those of Proposition 10.1 with $l$ replaced by $l_*$. Therefore the conclusion of Proposition 10.1 holds with $l$ replaced by $l_*$. Therefore, we have the $L^\infty$-bounds for $\not{\mathcal{L}}_{R_{i_k}} \cdots \not{\mathcal{L}}_{R_{i_1}} \not{q}'_i$ where $k = 0, \cdots, l_*-1$ and for $\not{d}((R)^{s_2} x^j)$ where $|s_2| \leqslant l_*$. Also, all the corollaries of Proposition 10.1 hold with $l$ replaced by $l_*$. As before, we shall use the expression in Lemma 10.4 and we must get an $L^2$ estimate for $\not{\mathcal{L}}_{R_{i_k}} \cdots \not{\mathcal{L}}_{R_{i_1}} \not{q}'_i$ where $k = 0, \cdots, l-1$ and for $\not{d}((R)^{s_2} x^j)$ where $|s_2| \leqslant l-1$. By (10.18), we have

$$\|(\not{\mathcal{L}}_R)^s m'_b\|_{L^2(\Sigma_t^{\epsilon_0})} \leqslant C(\delta_0(1+t)^{-1}\mathcal{W}_{[l]} + \mathcal{A}_{[l-1]}) + C_l(1+t)^{-1}\mathcal{Y}_{[l-1]} \quad (10.102)$$

for all $|s| \leqslant l-1$. Here we have used Lemma 10.3, which yields

$$\max_{i;i_1\cdots i_k} \|R_{i_1}\cdots R_{i_k}\lambda_i\|_{L^2(\Sigma_t^{\epsilon_0})} \leqslant C_{l-1}(1+t)\mathcal{Y}_{[l-1]} \quad (10.103)$$

for all $k = 0, \cdots, l-1$.

This is because the hypothesis of Lemma 10.3 with $l$ replaced by $l-1$ are a fortiori satisfied.

Also we have

$$\begin{aligned}&\|(\not{\mathcal{L}}_R)^{s(R)}\not{\chi}\|_{L^2(\Sigma_t^{\epsilon_0})}\\ &\leqslant C_l\mathcal{Y}_{[l-2]} + C_l\delta_0(1+t)^{-1}(1+\log(1+t))\mathcal{W}_{[l-1]} \\ &\quad + C_l\delta_0(1+\log(1+t))\mathcal{A}_{[l-2]}\end{aligned} \quad (10.104)$$

for all $s = 0, \cdots, l-2$. Here we also used the induction hypothesis for $\lambda_i$.

Finally, we have to deal with the case that more than half of the derivatives hit $R_i$ in the expression (10.15). Here we must use Lemma 10.6. The first term on the right of this expression is obviously bounded in $L^\infty$, as well as the second factor in the second term. So first we consider the term ${}^{(k)}\beta^m_{j;i_1\cdots i_k}$ for all $k = 0, \cdots, l-1$.

Let us now consider expression (10.72) for an arbitrary triplet $z$ of $T_1^1$-type $S_{t,u}$ tensorfields and $k = 0, \cdots, l-1$. We have two cases to consider in regard to the terms in the sum in (10.72).

Case 1: $|s_2| \leqslant l_*$ and Case 2: $|s_1| \leqslant l_* - 1$.

In Case 1 by (10.95) with $l_*$ in the role of $l$, we have

$$\begin{aligned}\|(\not{\mathcal{L}}_R)^{s_1} z \cdot (M)^{s_2}\|_{L^2(\Sigma_t^{\epsilon_0})} &\leqslant \|(\not{\mathcal{L}}_R)^{s_1} z\|_{L^2(\Sigma_t^{\epsilon_0})} \|(M)^{s_2}\|_{L^\infty(\Sigma_t^{\epsilon_0})} \\ &\leqslant C \sum_{m=0}^{l-1} \max_{i_1 \cdots i_m} \|\not{\mathcal{L}}_{R_{i_m}} \cdots \not{\mathcal{L}}_{R_{i_1}} z\|_{L^2(\Sigma_t^{\epsilon_0})} \end{aligned} \tag{10.105}$$

In Case 2 we express $(M)^{s_2}$ as in (10.80). We estimate the contribution of the first term in (10.80) by placing this term in $L^\infty$ using (10.82):

$$\begin{aligned}\|(\not{\mathcal{L}}_R)^{s_1} z \cdot {}^{(p)}I_{i_{n_1} \cdots i_{n_p}}\|_{L^2(\Sigma_t^{\epsilon_0})} &\leqslant \|(\not{\mathcal{L}}_R)^{s_1} z\|_{L^2(\Sigma_t^{\epsilon_0})} \|{}^{(p)}I_{i_{n_1} \cdots i_{n_p}}\|_{L^\infty(\Sigma_t^{\epsilon_0})} \\ &\leqslant C \sum_{m=0}^{l-1} \max_{i_1 \cdots i_m} \|\not{\mathcal{L}}_{R_{i_m}} \cdots \not{\mathcal{L}}_{R_{i_1}} z\|_{L^2(\Sigma_t^{\epsilon_0})} \end{aligned} \tag{10.106}$$

Concerning the contribution of the second term in (10.80), we need an $L^2$ estimate for this term. By the discussion above (10.80), we need an estimate for

$$\max_{i; i_1 \cdots i_k} \|\not{\mathcal{L}}_{R_{i_k}} \cdots \not{\mathcal{L}}_{R_{i_1}} N_i\|_{L^2(\Sigma_t^{\epsilon_0})}$$

for all $k = 0, \cdots, l-2$.

Then by (10.75), we need an estimate for

$$\max_{i; i_1 \cdots i_k} \|\not{\mathcal{L}}_{R_{i_k}} \cdots \not{\mathcal{L}}_{R_{i_1}} \mu_i\|_{L^2(\Sigma_t^{\epsilon_0})}$$

for all $k = 0, \cdots, l-2$.

To obtain this estimate, we use the definition (10.61). By (10.103) (with $l$ replaced by $l-1$), we have an $L^2$ estimate for $\lambda_i$. For an $L^2$ estimate of $m'$, we use (10.104) and (10.102) as well as (10.88) and (10.90), to obtain

$$\begin{aligned}&\max_{i_1 \cdots i_k} \|\not{\mathcal{L}}_{R_{i_k}} \cdots \not{\mathcal{L}}_{R_{i_1}} m'\|_{L^2(\Sigma_t^{\epsilon_0})} \\ &\leqslant C(\delta_0 (1+t)^{-1} \mathcal{W}_{[l-1]} + \mathcal{A}_{[l-2]}) + C_l (1+t)^{-1} \mathcal{Y}_{[l-2]} \end{aligned} \tag{10.107}$$

for $k = 0, \cdots, l-2$.

Combining this and (10.103) with (10.90), (10.89) (with $l$ replaced by $l_*$), we get

$$\begin{aligned}&\max_{i; i_1 \cdots i_k} \|\not{\mathcal{L}}_{R_{i_k}} \cdots \not{\mathcal{L}}_{R_{i_1}} \mu_i\|_{L^2(\Sigma_t^{\epsilon_0})} \\ &\leqslant C_l \delta_0 (1 + \log(1+t))(\delta_0 (1+t)^{-1} \mathcal{W}_{[l-1]} + \mathcal{A}_{[l-2]}) \\ &\quad + C_l (1+t)^{-1} (1 + \log(1+t)) \mathcal{Y}_{[l-2]} \end{aligned} \tag{10.108}$$

for $k = 0, \cdots, l-2$.

Therefore we have

$$\begin{aligned}&\max_{i;i_1\cdots i_k}\|\not{\mathcal{L}}_{R_{i_k}}\cdots\not{\mathcal{L}}_{R_{i_1}}N_i\|_{L^2(\Sigma_t^{\epsilon_0})}\\&\leqslant C_l\delta_0(1+\log(1+t))(\delta_0(1+t)^{-1}\mathcal{W}_{[l-1]}+\mathcal{A}_{[l-2]})\\&\quad+C_l(1+t)^{-1}(1+\log(1+t))\mathcal{Y}_{[l-2]}\end{aligned}\tag{10.109}$$

So the contribution of the second term in (10.80) in Case 2 is bounded by

$$\begin{aligned}&\|(\not{\mathcal{L}}_R)^{s_1}z\cdot{}^{(p)}B_{i_{n_1}\cdots i_{n_p}}\|_{L^2(\Sigma_t^{\epsilon_0})}\\&\leqslant\|(\not{\mathcal{L}}_R)^{s_1}z\|_{L^\infty(\Sigma_t^{\epsilon_0})}\|{}^{(p)}B_{i_{n_1}\cdots i_{n_p}}\|_{L^2(\Sigma_t^{\epsilon_0})}\\&\leqslant C_l\delta_0(1+t)^{-1}(1+\log(1+t))\max_{m\leqslant l_*-1}\max_{i_1\cdots i_m}\|\not{\mathcal{L}}_{R_{i_m}}\cdots\not{\mathcal{L}}_{R_{i_1}}z\|_{L^\infty(\Sigma_t^{\epsilon_0})}\\&\quad\cdot(\mathcal{Y}_{[l-2]}+(1+t)\mathcal{A}_{[l-2]}+\mathcal{W}_{[l-1]})\end{aligned}\tag{10.110}$$

We thus conclude that for any triplet $z$ of $T^1_1$-type of $S_{t,u}$ tensorfields

$$\begin{aligned}&\max_{i_1\cdots i_k}\|A_{i_k}\cdots A_{i_1}z\|_{L^2(\Sigma_t^{\epsilon_0})}\\&\leqslant C_l\sum_{m=0}^{l-1}\max_{i_1\cdots i_m}\|\not{\mathcal{L}}_{R_{i_m}}\cdots\not{\mathcal{L}}_{R_{i_1}}z\|_{L^2(\Sigma_t^{\epsilon_0})}\\&\quad+C_l\delta_0(1+t)^{-1}(1+\log(1+t))\max_{m\leqslant l_*-1}\max_{i_1\cdots i_m}\|\not{\mathcal{L}}_{R_{i_m}}\cdots\not{\mathcal{L}}_{R_{i_1}}z\|_{L^\infty(\Sigma_t^{\epsilon_0})}\\&\quad\cdot(\mathcal{Y}_{[l-2]}+(1+t)\mathcal{A}_{[l-2]}+\mathcal{W}_{[l-1]})\end{aligned}\tag{10.111}$$

for all $k=0,\cdots,l-1$.

From the expression for the triplets ${}^{(k)}\rho_{j;i_1\cdots i_k}$ of Lemma 10.6 and the bounds (10.81) and (10.108) with $l-1$ replaced by $l$, we obtain

$$\begin{aligned}&\max_{i_1\cdots i_{m+n}}\|\not{\mathcal{L}}_{R_{i_{m+n}}}\cdots\not{\mathcal{L}}_{R_{i_{m+1}}}{}^{(m)}\rho_{j;i_1\cdots i_m}\|_{L^2(\Sigma_t^{\epsilon_0})}\\&\leqslant C_l\delta_0(1+t)^{-1}(1+\log(1+t))(\mathcal{Y}_{[l-1]}+(1+t)\mathcal{A}_{[l-1]}+\mathcal{W}_{[l]})\end{aligned}\tag{10.112}$$

for all $m$ and all $n\leqslant l-1$.

We then apply (10.111) to the triplet ${}^{(m)}\rho_{j;i_1\cdots i_m}$ using $L^2$ estimate (10.112) as well as $L^\infty$ bound (10.97) with $l_*$ in the role of $l$. This yields

$$\begin{aligned}&\|A_{i_{m+n}}\cdots A_{i_{m+1}}{}^{(m)}\rho_{j;i_1\cdots i_m}\|_{L^2(\Sigma_t^{\epsilon_0})}\\&\leqslant C_l\delta_0(1+t)^{-1}(1+\log(1+t))(\mathcal{Y}_{[l-1]}+(1+t)\mathcal{A}_{[l-1]}+\mathcal{W}_{[l]})\end{aligned}\tag{10.113}$$

for all $m$ and all $n\leqslant l-1$.

It then follows that

$$\begin{aligned}&\max_{j;i_1\cdots i_k}\|{}^{(k)}\beta_{j;i_1\cdots i_k}\|_{L^2(\Sigma_t^{\epsilon_0})}\\&\leqslant C_l\delta_0(1+t)^{-1}(1+\log(1+t))(\mathcal{Y}_{[l-1]}+(1+t)\mathcal{A}_{[l-1]}+\mathcal{W}_{[l]})\end{aligned}\tag{10.114}$$

for all $k=0,\cdots,l$.

More generally, we have

$$\max_{j;i_1\cdots i_{m+n}} \|\not{\mathcal{L}}_{R_{i_{m+n}}}\cdots\not{\mathcal{L}}_{R_{i_{m+1}}}{}^{(m)}\beta_{j;i_1\cdots i_m}\|_{L^2(\Sigma_t^{\epsilon_0})}$$
$$\leqslant C_l\delta_0(1+t)^{-1}(1+\log(1+t))(\mathcal{Y}_{[l-1]}+(1+t)\mathcal{A}_{[l-1]}+\mathcal{W}_{[l]}) \tag{10.115}$$

for all $m$ and all $n$ such that $m+n\leqslant l$.

We turn to consider the expression for the coefficients ${}^{(k)}\gamma_{j;i_1\cdots i_k}$ given in the statement of Lemma 10.6. Since we already have (10.115), what we need is an estimate for $\|\not{\mathcal{L}}_{R_{i_k}}\cdots\not{\mathcal{L}}_{R_{i_1}}\nu_{ij}\|_{L^2(\Sigma_t^{\epsilon_0})}$ for all $k=0,\cdots,l-2$. In view of (10.62), what is needed is an estimate for $\|\not{\mathcal{L}}_{R_{i_k}}\cdots\not{\mathcal{L}}_{R_{i_1}}w_i\|_{L^2(\Sigma_t^{\epsilon_0})}$, therefore an estimate for $\|\not{\mathcal{L}}_{R_{i_k}}\cdots\not{\mathcal{L}}_{R_{i_1}}(w_i)_b\|_{L^2(\Sigma_t^{\epsilon_0})}$for all $k=0,\cdots,l-2$. Concerning (10.42), the $L^2$ estimate for derivatives of $y^i$ is just the induction hypothesis. While since we have

$$R_q(R)^{s_2}x^j={}^{(p+1)}x^j_{i_{n_1}\cdots i_{n_p}q}-{}^{(p+1)}\delta^j_{i_{n_1}\cdots i_{n_p}q} \tag{10.116}$$

where $s_2=\{n_1,\cdots,n_p\}$, $p=|s_2|$. So by **H0**,

$$|\not{d}(R)^{s_2}x^j|\leqslant C(1+t)^{-1}\sum_{q=1}^{3}(|{}^{(p+1)}x^j_{i_{n_1}\cdots i_{n_p}q}|+|{}^{(p+1)}\delta^j_{i_{n_1}\cdots i_{n_p}q}|) \tag{10.117}$$

By (10.40), we can estimate the first term on the right of (10.117) in $L^\infty$ norm. By virtue of Proposition 10.1 with $l$ replaced by $l_*$, the hypothesis of Lemma 10.3 is satisfied, and of course, the hypothesis of Lemma 10.3 with $l$ replaced by $l-1$ is a fortiori satisfied. Therefore we have

$$\max_{j;i_1\cdots i_k}\|{}^{(k+1)}\delta^j_{ii_1\cdots i_k}\|_{L^2(\Sigma_t^{\epsilon_0})}\leqslant C_{l-1}(1+t)\mathcal{Y}_{[l-1]} \tag{10.118}$$

for all $k=0,\cdots,l-1$. So we can estimate the second term in $L^2$ norm. Combining (10.118), (10.85) with induction hypothesis and the conclusion of Proposition 10.1, we obtain

$$\max_{i;i_1\cdots i_k}\|\not{\mathcal{L}}_{R_{i_k}}\cdots\not{\mathcal{L}}_{R_{i_1}}(w_i)_b\|_{L^2(\Sigma_t^{\epsilon_0})}\leqslant C_l\mathcal{Y}_{[l-1]} \tag{10.119}$$

for all $k=0,\cdots,l-1$. Then combining this with (10.104), (10.88) and $L^\infty$ bounds for $\not{\mathcal{L}}_{R_{i_k}}\cdots\not{\mathcal{L}}_{R_{i_1}}(w_i)_b$, which is the same as the conclusion of Proposition 10.1, we obtain

$$\max_{i;i_1\cdots i_k}\|\not{\mathcal{L}}_{R_{i_k}}\cdots\not{\mathcal{L}}_{R_{i_1}}w_i\|_{L^2(\Sigma_t^{\epsilon_0})}$$
$$\leqslant C_l(\mathcal{Y}_{[l-1]}+\delta_0(1+t)^{-2}(1+\log(1+t))^2((1+t)\mathcal{A}_{[l-2]}+\mathcal{W}_{[l-1]})) \tag{10.120}$$

for all $k=0,\cdots,l-1$. Then we can easily obtain by (10.62)

$$\max_{i,j;i_1\cdots i_k}\|\not{\mathcal{L}}_{R_{i_k}}\cdots\not{\mathcal{L}}_{R_{i_1}}\nu_{ij}\|_{L^2(\Sigma_t^{\epsilon_0})}$$
$$\leqslant C_l\delta_0(1+\log(1+t))(\mathcal{Y}_{[l-1]}+\delta_0(1+t)^{-2}(1+\log(1+t))^2((1+t)\mathcal{A}_{[l-2]}+\mathcal{W}_{[l-1]})) \tag{10.121}$$

Combining this with (10.114), we obtain that the last two terms on the right in the expression of Lemma 10.6 are bounded in $L^2$ by

$$C_l\delta_0(1+\log(1+t))(\mathcal{Y}_{[l-1]}+(1+t)\mathcal{A}_{[l-2]}+\mathcal{W}_{[l-1]}) \qquad (10.122)$$

Recall that the first term is estimated in $L^\infty$ by $1+t$. Combining (10.122) with (10.107), Corollary 10.1.e and (10.90), we obtain

$$\|\not{\mathcal{L}}_{R_{i_k}}\cdots\not{\mathcal{L}}_{R_{i_1}}\not{f}'_i\|_{L^2(\Sigma_t^{\epsilon_0})} \leqslant C_l\mathcal{Y}_{[l-1]}+C_l\delta_0((1+t)\mathcal{A}_{[l-1]}+\mathcal{W}_{[l]}) \qquad (10.123)$$

for all $k=0,\cdots,l-1$.

Next, we must consider the $L^2$ estimate for $\not{d}((R)^{s_2}x^j)$ for $|s_2|\leqslant l-1$. Recalling the discussion from (10.116) to (10.118), the $L^2$ contribution is bounded by (10.118). Combining this with Corollary 10.1.f and (10.85) as well as **H0**, we obtain

$$\|{}^{(k)}\not{f}^j_{ii_1\cdots i_k}\|_{L^2(\Sigma_t^{\epsilon_0})} \leqslant C_l\mathcal{Y}_{[l-1]}+C_l\delta_0((1+t)\mathcal{A}_{[l-2]}+\mathcal{W}_{[l-1]}) \qquad (10.124)$$

for all $k=0,\cdots,l-1$.

Thus we finally obtain

$$\|R_{i_k}\cdots R_{i_1}y^j\|_{L^2(\Sigma_t^{\epsilon_0})} \leqslant C_l\mathcal{Y}_{[l-1]}+C_l\delta_0((1+t)\mathcal{A}_{[l-1]}+\mathcal{W}_{[l]}) \qquad (10.125)$$

for all $k=0,\cdots,l$.

Recall the induction hypothesis,

$$\|R_{i_k}\cdots R_{i_1}y^j\|_{L^2(\Sigma_t^{\epsilon_0})} \leqslant C_l(\mathcal{Y}_0+(1+t)\mathcal{A}_{[l-2]}+\mathcal{W}_{[l-1]}) \qquad (10.126)$$

for all $k=0,\cdots,l-1$.

Substituting this in (10.125), the proposition is proved. □

As before, we have a lot of corollaries. First, in conjunction with Lemma 10.3 we have

**Corollary 10.2.a** Under the assumptions of Proposition 10.2 we have

$$\max_{i;i_1\cdots i_k}\|R_{i_k}\cdots R_{i_1}\lambda_i\|_{L^2(\Sigma_t^{\epsilon_0})} \leqslant C_l(1+t)(\mathcal{Y}_0+(1+t)\mathcal{A}_{[l-1]}+\mathcal{W}_{[l-1]}) \quad (10.127)$$

for all $k=0,\cdots,l$, and

$$\max_{j;ii_1\cdots i_k}\|{}^{(k+1)}\delta^j_{ii_1\cdots i_k}\|_{L^2(\Sigma_t^{\epsilon_0})} \leqslant C_l(1+t)(\mathcal{Y}_0+(1+t)\mathcal{A}_{[l-1]}+\mathcal{W}_{[l-1]}) \quad (10.128)$$

for all $k=0,\cdots,l$.

Next, it is not difficult to obtain from the proof of Proposition 10.2 the following

**Corollary 10.2.b** Under the assumptions of Proposition 10.2 we have

$$\max_{i_1\cdots i_k}\|R_{i_k}\cdots R_{i_1}\psi_{\hat{T}}\|_{L^2(\Sigma_t^{\epsilon_0})} \leqslant C_l(\mathcal{W}_{[l]}+\delta_0(1+t)^{-1}(\mathcal{Y}_0+(1+t)\mathcal{A}_{[l-1]}))$$

for all $k=0,\cdots,l$,

$$\max_{i_1\cdots i_k} \|\not{\mathcal{L}}_{R_{i_k}}\cdots\not{\mathcal{L}}_{R_{i_1}}\not{\psi}\|_{L^2(\Sigma_t^{\epsilon_0})} \leqslant C_l(\mathcal{W}_{[l]}+\delta_0(1+t)^{-1}(\mathcal{Y}_0+(1+t)\mathcal{A}_{[l-1]}))$$

for all $k=0,\cdots,l$, and

$$\max_{i_1\cdots i_k} \|\not{\mathcal{L}}_{R_{i_k}}\cdots\not{\mathcal{L}}_{R_{i_1}}\not{\psi}\|_{L^2(\Sigma_t^{\epsilon_0})} \leqslant C_l(1+t)^{-1}(\mathcal{W}_{[l]}+\delta_0(1+t)^{-1}(\mathcal{Y}_0+(1+t)\mathcal{A}_{[l-2]}))$$

for all $k=0,\cdots,l-1$.

**Corollary 10.2.c** Under the assumptions of Proposition 10.2 we have

$$\max_{i_1\cdots i_k} \|\not{\mathcal{L}}_{R_{i_k}}\cdots\not{\mathcal{L}}_{R_{i_1}}\not{k}\|_{L^2(\Sigma_t^{\epsilon_0})} \leqslant C_l(1+t)^{-1}(\mathcal{W}_{[l]}+\delta_0(1+t)^{-1}(\mathcal{Y}_0+(1+t)\mathcal{A}_{[l-2]}))$$

for all $k=0,\cdots,l-1$.

**Corollary 10.2.d** Under the assumption of Proposition 10.2 we have

$$\max_{i;i_1\cdots i_k} \|\not{\mathcal{L}}_{R_{i_k}}\cdots\not{\mathcal{L}}_{R_{i_1}}{}^{(R_i)}\not{\pi}\|_{L^2(\Sigma_t^{\epsilon_0})}$$
$$\leqslant C_l(\delta_0(1+t)^{-1}(1+\log(1+t))((1+t)\mathcal{A}_{[l-1]}+\mathcal{W}_{[l]})+\mathcal{Y}_0+(1+t)\mathcal{A}_{[l-2]}+\mathcal{W}_{[l-1]})$$

for all $k=0,\cdots,l-1$.

**Corollary 10.2.e** Under the assumptions of Proposition 10.2, the coefficients of the expression for $\not{\mathcal{L}}_{R_{i_k}}\cdots\not{\mathcal{L}}_{R_{i_1}}R_j$ of Lemma 10.6 satisfy

$$\max_{j;i_1\cdots i_k} \|{}^{(k)}\beta_{j;i_1\cdots i_k}\|_{L^2(\Sigma_t^{\epsilon_0})} \leqslant C_l\delta_0(1+t)^{-1}(1+\log(1+t))(\mathcal{Y}_0+(1+t)\mathcal{A}_{[l-1]}+\mathcal{W}_{[l]})$$

for all $k=1,\cdots,l$,

$$\max_{j;i_1\cdots i_k} \|{}^{(k)}\gamma_{j;i_1\cdots i_k}\|_{L^2(\Sigma_t^{\epsilon_0})} \leqslant C_l\delta_0(1+\log(1+t))(\mathcal{Y}_0+(1+t)\mathcal{A}_{[l-2]}+\mathcal{W}_{[l-1]})$$

for all $k=1,\cdots,l$.

Finally we have

**Corollary 10.2.f** Under the assumptions of Proposition 10.2, the coefficients of the expression of $R_{i_k}\cdots R_{i_1}R_j y^j$ of Lemma 10.4 satisfy

$$\max_{ii_1\cdots i_k} \|{}^{(k)}\not{q}'_{ii_1\cdots i_k}\|_{L^2(\Sigma_t^{\epsilon_0})} \leqslant C_l(\mathcal{Y}_0+(1+t)\mathcal{A}_{[l-1]}+\mathcal{W}_{[l]})$$

for all $k=0,\cdots,l-1$, and

$$\max_{ii_1\cdots i_k} \|{}^{(k)}\not{r}^j_{ii_1\cdots i_k}\|_{L^2(\Sigma_t^{\epsilon_0})} \leqslant C_l(\mathcal{Y}_0+(1+t)\mathcal{A}_{[l-2]}+\mathcal{W}_{[l-1]})$$

for all $k=0,\cdots,l-1$.

We shall now estimate the angular derivatives of the functions $L^i$. After this we shall proceed with the estimates for the $S_{t,u}$ 1-form $\kappa^{-1}\zeta$ and the $S_{t,u}$-tangential vectorfields ${}^{(R_i)}Z$. According to Equations (6.64) and (6.65),

$$L^j = \frac{x^j}{1-u+t} + z^j \tag{10.129}$$

where

$$z^j = -\alpha y^j + \frac{(\alpha - 1)x^j}{1-u+t} - \not\psi^j \tag{10.130}$$

Let us define

$$\not\psi_L = L^\mu \not d \psi_\mu \tag{10.131}$$

and

$$\omega_{L\hat{T}} = \hat{T}^i(L\psi_i) \tag{10.132}$$

As a corollary of Proposition 10.1 we have, in connection with the first two statements of Corollary 10.1.b, the following

**Corollary 10.1.g** Under the assumptions of Proposition 10.1 we have

$$\max_{j;i_1\cdots i_k} \|R_{i_k}\cdots R_{i_1} z^j\|_{L^\infty(\Sigma_t^{\epsilon_0})} \leqslant C\delta_0(1+t)^{-1}(1+\log(1+t))$$

for all $k = 0, \cdots, l$.

$$\max_{j;i_1\cdots i_k} \|R_{i_k}\cdots R_{i_1} L^j\|_{L^\infty(\Sigma_t^{\epsilon_0})} \leqslant C$$

for all $k = 0, \cdots, l$. Moreover,

$$\max_{i_1\cdots i_k} \|R_{i_k}\cdots R_{i_1} \not\psi_L\|_{L^\infty(\Sigma_t^{\epsilon_0})} \leqslant C_l\delta_0(1+t)^{-1}$$

for all $k - 1, \cdots, l$. And

$$\max_{i_1\cdots i_k} \|\not\mathcal{L}_{R_{i_k}}\cdots \not\mathcal{L}_{R_{i_1}} \not\psi_L\|_{L^\infty(\Sigma_t^{\epsilon_0})} \leqslant C_l\delta_0(1+t)^{-2}$$

for all $k = 0, \cdots, l-1$.

$$\max_{i_1\cdots i_k} \|\not\mathcal{L}_{R_{i_k}}\cdots \not\mathcal{L}_{R_{i_1}} \omega_{L\hat{T}}\|_{L^\infty(\Sigma_t^{\epsilon_0})} \leqslant C_l\delta_0(1+t)^{-2}$$

for all $k = 0, \cdots, l-1$.

Also, as a corollary of Proposition 10.2 we have, in connection with the first statement of Corollary 10.2.b

**Corollary 10.2.g** Under the assumptions of Proposition 10.2 we have

$$\max_{j;i_1\cdots i_k} \|R_{i_k}\cdots R_{i_1} z^j\|_{L^2(\Sigma_t^{\epsilon_0})} \leqslant C_l(\mathcal{Y}_0 + (1+t)\mathcal{A}_{[l-1]} + \mathcal{W}_{[l]})$$

for all $k = 0, \cdots, l$. Moreover,

$$\max_{i_1 \cdots i_k} \|R_{i_k} \cdots R_{i_1} \psi_L\|_{L^2(\Sigma_t^{\epsilon_0})} \leqslant C_l(\mathcal{W}_{[l]} + \delta_0(1+t)^{-1}(\mathcal{Y}_0 + (1+t)\mathcal{A}_{[l-1]}))$$

for all $k = 1, \cdots, l$. And

$$\max_{i_1 \cdots i_k} \|\not{\mathcal{L}}_{R_{i_k}} \cdots \not{\mathcal{L}}_{R_{i_1}} \not{\psi}_L\|_{L^2(\Sigma_t^{\epsilon_0})}$$
$$\leqslant C_l(1+t)^{-1}(\mathcal{W}_{[l]} + \delta_0(1+t)^{-1}(\mathcal{Y}_0 + (1+t)\mathcal{A}_{[l-2]}))$$

for all $k = 0, \cdots, l-1$.

$$\max_{i_1 \cdots i_l} \|\not{\mathcal{L}}_{R_{i_k}} \cdots \not{\mathcal{L}}_{R_{i_1}} \omega_{L\hat{T}}\|_{L^2(\Sigma_t^{\epsilon_0})}$$
$$\leqslant C_l(1+t)^{-1}(\mathcal{W}^Q_{[l-1]} + \delta_0(1+t)^{-1}(\mathcal{Y}_0 + (1+t)\mathcal{A}_{[l-2]} + \mathcal{W}_{[l-1]}))$$

for all $k = 0, \cdots, l-1$.

We turn to the $S_{t,u}$ 1-form $\kappa^{-1}\zeta$. This is given by (3.53):

$$\kappa^{-1}\zeta = \alpha\varepsilon - \not{d}\alpha \tag{10.133}$$

We first consider, under the assumptions of Proposition 10.1, $L^\infty$ estimates on $\Sigma_t^{\epsilon_0}$ of the angular derivatives of $\kappa^{-1}\zeta$ up to order $l-1$. Using Corollary 10.1.b, we readily deduce

**Corollary 10.1.h** Under the assumptions of Proposition 10.1 we have

$$\max_{i_1 \cdots i_k} \|\not{\mathcal{L}}_{R_{i_k}} \cdots \not{\mathcal{L}}_{R_{i_1}} (\kappa^{-1}\zeta)\|_{L^\infty(\Sigma_t^{\epsilon_0})} \leqslant C_l\delta_0(1+t)^{-2}$$

for all $k = 0, \cdots, l-1$.

We consider next, under the assumptions of Proposition 10.2, $L^2$ estimates on $\Sigma_t^{\epsilon_0}$ of the angular derivatives of $\kappa^{-1}\zeta$ up to order $l-1$. By Corollary 10.2.g and (10.133), it is easy to see that

**Corollary 10.2.h** Under the assumptions of Proposition 10.2 we have

$$\max_{i_1 \cdots i_k} \|\not{\mathcal{L}}_{R_{i_k}} \cdots \not{\mathcal{L}}_{R_{i_1}} (\kappa^{-1}\zeta)\|_{L^2(\Sigma_t^{\epsilon_0})}$$
$$\leqslant C_l(1+t)^{-1}(\mathcal{W}_{[l]} + \delta_0(1+t)^{-1}(\mathcal{Y}_0 + (1+t)\mathcal{A}_{[l-2]}))$$

for all $k = 0, \cdots, l-1$.

We turn to the $S_{t,u}$-tangential vectorfields ${}^{(R_i)}Z$. These are related to the $S_{t,u}$ 1-form ${}^{(R_i)}\not{\pi}_L$ by

$${}^{(R_i)}Z = {}^{(R_i)}\not{\pi}_L \cdot \not{g}^{-1} \tag{10.134}$$

According to (6.67),

$${}^{(R_i)}\not{\pi}_L = -(\chi - \frac{\not{g}}{1-u+t}) \cdot R_i + \epsilon_{ijm} z^j \not{d}x^m + \lambda_i(\kappa^{-1}\zeta) \tag{10.135}$$

Using the corollaries of Proposition 10.1 we readily deduce

**Corollary 10.1.i** Under the assumptions of Proposition 10.1 we have

$$\max_{i;i_1\cdots i_k} \|\not{\mathcal{L}}_{R_{i_k}}\cdots\not{\mathcal{L}}_{R_{i_1}}{}^{(R_i)}\not{\pi}_L\|_{L^\infty(\Sigma_t^{\epsilon_0})} \leqslant C_l\delta_0(1+t)^{-1}(1+\log(1+t))$$

for all $k=0,\cdots,l-1$, and

$$\max_{i;i_1\cdots i_k} \|\not{\mathcal{L}}_{R_{i_k}}\cdots\not{\mathcal{L}}_{R_{i_1}}{}^{(R_i)}Z\|_{L^\infty(\Sigma_t^{\epsilon_0})} \leqslant C_l\delta_0(1+t)^{-1}(1+\log(1+t))$$

for all $k=0,\cdots,l-1$.

Also, using the corollaries of Proposition 10.1 with $l_*$ in the role of $l$, as well as the corollaries of Proposition 10.2 we can easily deduce

**Corollary 10.2.i** Under the assumptions of Proposition 10.2 we have

$$\max_{i;i_1\cdots i_k} \|\not{\mathcal{L}}_{R_{i_k}}\cdots\not{\mathcal{L}}_{R_{i_1}}{}^{(R_i)}\not{\pi}_L\|_{L^2(\Sigma_t^{\epsilon_0})}$$
$$\leqslant C_l(\mathcal{Y}_0+(1+t)\mathcal{A}_{[l-1]}+\delta_0(1+t)^{-1}(1+\log(1+t))\mathcal{W}_{[l]}+\mathcal{W}_{[l-1]})$$

for all $k=0,\cdots,l-1$, and

$$\max_{i;i_1\cdots i_k} \|\not{\mathcal{L}}_{R_{i_k}}\cdots\not{\mathcal{L}}_{R_{i_1}}{}^{(R_i)}Z\|_{L^2(\Sigma_t^{\epsilon_0})}$$
$$\leqslant C_l(\mathcal{Y}_0+(1+t)\mathcal{A}_{[l-1]}+\delta_0(1+t)^{-1}(1+\log(1+t))\mathcal{W}_{[l]}+\mathcal{W}_{[l-1]})$$

for all $k=0,\cdots,l-1$.

## 10.3 Bounds for the quantities $Q_l$ and $P_l$

The object of this section is to obtain appropriate bounds for the quantities ${}^{(i_1\cdots i_l)}Q_l$ and ${}^{(i_1\cdots i_l)}P_l$, which, through ${}^{(i_1\cdots i_l)}B_l$, enter the final estimates of Chapter 8 for the 1-forms ${}^{(i_1\cdots i_l)}x_l$.

### 10.3.1 Estimates for $Q_l$

We begin with the $S_{t,u}$ 1-form $i$ which is given by (8.100):

$$\begin{aligned} i &= \beta_A^* - \alpha^{-1}((\chi - \frac{\not{g}}{1-u+t})\cdot\not{g}^{-1} - (\mathrm{tr}\chi - \frac{2}{1-u+t})I)\cdot(\kappa^{-1}\zeta) \\ &:= i_1 + i_2 \end{aligned} \tag{10.136}$$

By (4.32)—(4.33), the first term on the right of above is given by

$$\begin{aligned} i_1 = \beta_A^* &= \eta^{-2}(\not{g}^{-1})^{BC}(\frac{1}{2}X_A(-\eta^2+|\mathbf{v}|^2)\not{\phi}(X_B,X_C) \\ &-\frac{1}{2}X_B(-\eta^2+|\mathbf{v}|^2)\not{\phi}(X_A,X_C)) \\ &-\eta^{-2}(\not{g}^{-1})^{BC}(\not{\phi}(X_A,X_C)L^jX_B(\psi_j) \\ &-L^jX_A(\psi_j)\not{\phi}(X_C,X_B)) \end{aligned} \tag{10.137}$$

It is obvious that, under the assumptions of Proposition 10.1, with $l$ replaced by $l+1$, we have

$$\max_{i_1\cdots i_k} \|\not{\mathcal{L}}_{R_{i_k}} \cdots \not{\mathcal{L}}_{R_{i_1}} i_1\|_{L^\infty(\Sigma_t^{\epsilon_0})} \leqslant C_l\delta_0^2(1+t)^{-4} \tag{10.138}$$

for all $k=0,\cdots,l$.

Also, it is easy to see that

$$\max_{i_1\cdots i_k} \|\not{\mathcal{L}}_{R_{i_k}} \cdots \not{\mathcal{L}}_{R_{i_1}} i_2\|_{L^\infty(\Sigma_t^{\epsilon_0})} \leqslant C_l\delta_0^2(1+t)^{-4}(1+\log(1+t)) \tag{10.139}$$

for all $k=0,\cdots,l$. We thus obtain

**Lemma 10.7** Let **H0** and (6.177) hold. Let also the bootstrap assumptions $\not{\mathbf{E}}_{[l+1]}$, $\not{\mathbf{E}}^Q_{[l]}$ and $\mathbf{X}_{[l]}$ hold, for some positive integer $l$. Then if $\delta_0$ is suitably small (depending on $l$) we have

$$\max_{i_1\cdots i_k} \|\not{\mathcal{L}}_{R_{i_k}} \cdots \not{\mathcal{L}}_{R_{i_1}} i\|_{L^\infty(\Sigma_t^{\epsilon_0})} \leqslant C_l\delta_0^2(1+t)^{-4}(1+\log(1+t))$$

for all $k=0,\cdots,l$.

By direct calculation and the corollaries of Proposition 10.2, with $l$ replaced by $l+1$, it is not difficult to see that

$$\begin{aligned}&\|\not{\mathcal{L}}_{R_{i_k}} \cdots \not{\mathcal{L}}_{R_{i_1}} i_1\|_{L^2(\Sigma_t^{\epsilon_0})}\\ &\leqslant C_l\delta_0(1+t)^{-3}(\mathcal{W}_{[l+1]}+\delta_0(1+t)^{-1}(\mathcal{Y}_0+(1+t)\mathcal{A}_{[l-1]}))\end{aligned} \tag{10.140}$$

for all $k=0,\cdots,l$.

Similarly, we have

$$\begin{aligned}&\|\not{\mathcal{L}}_{R_{i_k}} \cdots \not{\mathcal{L}}_{R_{i_1}} i_2\|_{L^2(\Sigma_t^{\epsilon_0})}\\ &\leqslant C_l\delta_0(1+t)^{-2}\mathcal{A}_{[l]}+\delta_0(1+t)^{-3}(1+\log(1+t))(\mathcal{W}_{[l+1]}+\delta_0(1+t)^{-1}\mathcal{Y}_0)\end{aligned} \tag{10.141}$$

for all $k=0,\cdots,l$. We thus obtain

**Lemma 10.8** Let **H0** and (6.177) hold. Let $l$ be a positive integer and let the bootstrap assumptions $\not{\mathbf{E}}_{l_*+1}$, $\not{\mathbf{E}}^Q_{[l_*]}$ and $\mathbf{X}_{[l_*]}$ hold. Then we have

$$\begin{aligned}&\|\not{\mathcal{L}}_{R_{i_k}} \cdots \not{\mathcal{L}}_{R_{i_1}} i\|_{L^2(\Sigma_t^{\epsilon_0})}\\ &\leqslant C_l\delta_0(1+t)^{-2}\mathcal{A}_{[l]}+\delta_0(1+t)^{-3}(1+\log(1+t))(\mathcal{W}_{[l+1]}+\delta_0(1+t)^{-1}\mathcal{Y}_0)\end{aligned} \tag{10.142}$$

for all $k=0,\cdots,l$.

We proceed to estimate, in $L^2(\Sigma_t^{\epsilon_0})$, the $S_{t,u}$ 1-form ${}^{(i_1\cdots i_l)}i_l$ given by Proposition 8.4,

$$\begin{aligned}{}^{(i_1\cdots i_l)}i_l=&(\not{\mathcal{L}}_{R_{i_l}}+\frac{1}{2}\mathrm{tr}^{(R_{i_l})}\not{\pi})\cdots(\not{\mathcal{L}}_{R_{i_1}}+\frac{1}{2}\mathrm{tr}^{(R_{i_1})}\not{\pi})i\\ &+\sum_{k=0}^{l-1}(\not{\mathcal{L}}_{R_{i_l}}+\frac{1}{2}\mathrm{tr}^{(R_{i_l})}\not{\pi})\cdots(\not{\mathcal{L}}_{R_{i_{l-k+1}}}+\frac{1}{2}\mathrm{tr}^{(R_{i_{l-k+1}})}\not{\pi})^{(i_1\cdots i_{l-k})}q_{l-k}\end{aligned} \tag{10.143}$$

under the assumptions of Lemma 10.8.

We begin with the first term on the right of (10.143). We have

$$(\not{\mathcal{L}}_{R_{i_l}} + \frac{1}{2}\mathrm{tr}^{(R_{i_l})}\not{\pi}) \cdots (\not{\mathcal{L}}_{R_{i_1}} + \frac{1}{2}\mathrm{tr}^{(R_{i_l})}\not{\pi}) i$$

$$= \not{\mathcal{L}}_{R_{i_l}} \cdots \not{\mathcal{L}}_{R_{i_1}} i + \sum_{j=1}^{l} \sum_{k_1 < \cdots < k_j = 1}^{l} \not{\mathcal{L}}_{R_{i_l}} \cdots \overset{>\not{\mathcal{L}}_{R_{i_{kj}}}<}{(\frac{1}{2}\mathrm{tr}^{(R_{i_{kj}})}\not{\pi})} \cdots \overset{>\not{\mathcal{L}}_{R_{i_{k1}}}<}{(\frac{1}{2}\mathrm{tr}^{(R_{i_{k1}})}\not{\pi})} \cdots \not{\mathcal{L}}_{R_{i_1}} i \tag{10.144}$$

We have two cases to consider:

Case 1: At most $l_*$ angular derivatives fall on one of the factors

$$\frac{1}{2}\mathrm{tr}^{(R_{i_{k1}})}\not{\pi}, \cdots, \frac{1}{2}\mathrm{tr}^{(R_{i_{kj}})}\not{\pi}$$

and

Case 2: one of these factors receives more than $l_*$ angular derivatives. Then all the other factors receive at most $l_*$ angular derivatives.

In Case 1 we just apply Corollary 10.1.d with $l_* + 1$ in the role of $l$, then we know that the second term on the right in (10.144) is bounded by

$$\begin{aligned} &C_l(\delta_0(1+t)^{-1}(1+\log(1+t)))^j(\delta_0(1+t)^{-2}\mathcal{A}_{[l-j]} \\ &+ \delta_0(1+t)^{-3}(1+\log(1+t))(\mathcal{W}_{[l-j+1]} + \delta_0(1+t)^{-1}\mathcal{Y}_0)) \end{aligned}$$

In Case 2 the number of angular derivatives falling on $i$ is at most $l_*$. So by Lemma 10.7 and Corollary 10.2.d, we have an $L^2(\Sigma_t^{\epsilon_0})$ bound for Case 2, by

$$\begin{aligned} &C_l(\delta_0(1+t)^{-1}(1+\log(1+t)))^{j-1} \cdot \delta_0(1+t)^{-4}(1+\log(1+t)) \\ &\cdot (\mathcal{Y}_0 + (1+t)\mathcal{A}_{[l-j]} + \mathcal{W}_{[l-j+1]}) \end{aligned}$$

Here, we have used the following facts that under the assumptions of Proposition 10.2, we have, by Corollary 10.2.d,

$$\max_{i;i_1\cdots i_k} \|\not{\mathcal{L}}_{R_{i_k}} \cdots \not{\mathcal{L}}_{R_{i_1}} {}^{(R_i)}\not{\pi}\|_{L^2(\Sigma_t^{\epsilon_0})} \leqslant C_l(\mathcal{Y}_0 + (1+t)\mathcal{A}_{[l-1]} + \mathcal{W}_{[l]}) \tag{10.145}$$

Combining the results, we obtain a bound for the second term on the right in (10.144) by

$$\begin{aligned} &C_l(\delta_0(1+t)^{-1}(1+\log(1+t)))^{j-1}(\delta_0(1+t)^{-2}\mathcal{A}_{[l-j]} \\ &+ \delta_0(1+t)^{-3}(1+\log(1+t))(\mathcal{W}_{[l-j+1]} + \delta_0(1+t)^{-1}\mathcal{Y}_0)) \end{aligned} \tag{10.146}$$

Next we consider the second term on the right in (10.143). First, we define the operator $\check{D}$:

$$(\check{\not{D}}\vartheta)_{ABC} = \frac{1}{2}(\not{D}_A\vartheta_{BC} + \not{D}_B\vartheta_{AC} - \not{D}_C\vartheta_{AB}) \tag{10.147}$$

where $\vartheta$ is a symmetric 2-covariant $S_{t,u}$ tensorfield.

So the second term in the expression for ${}^{(i_1\cdots i_j)}q_j$ of Proposition 8.4 is simply

$$(\check{\not{D}}^{(i_1\cdots i_{j-1})}\hat{\chi}_{j-1})\cdot(\not{g}^{-1}\cdot{}^{(R_j)}\hat{\not{\pi}}\cdot\not{g}^{-1})$$

Also, the third term in the same expression is

$${}^{(i_1\cdots i_{j-1})}\hat{\chi}_{j-1}\cdot(\not{\mathrm{div}}\,{}^{(R_j)}\hat{\not{\pi}}\cdot\not{g}^{-1})$$

While by a direct calculation, we have

$$\not{\mathrm{div}}\,{}^{(X)}\hat{\not{\pi}}\cdot\not{g}^{-1}=\mathrm{tr}\,{}^{(X)}\not{\pi}_1 \tag{10.148}$$

Here the definition of ${}^{(X)}\not{\pi}_1$ is (9.140):

$${}^{(X)}\not{\pi}^{C}_{1,AB}=\frac{1}{2}(\not{D}_A{}^{(X)}\not{\pi}^C_B+\not{D}_B{}^{(X)}\not{\pi}^C_A-\not{D}^{C}\,{}^{(X)}\not{\pi}_{AB})$$

We conclude that

$$\begin{aligned}{}^{(i_1\cdots i_j)}q_j=&\frac{1}{4}\mathrm{tr}^{(R_j)}\not{\pi}\not{d}(R_{i_{j-1}}\cdots R_{i_1}\mathrm{tr}\chi)\\&+(\check{\not{D}}^{(i_1\cdots i_{j-1})}\hat{\chi}_{j-1})\cdot(\not{g}^{-1}\cdot{}^{(R_j)}\hat{\not{\pi}}\cdot\not{g}^{-1})\\&+{}^{(i_1\cdots i_{j-1})}\hat{\chi}_{j-1}\cdot\mathrm{tr}^{(R_{i_j})}\not{\pi}_1\end{aligned} \tag{10.149}$$

To estimate the contribution of the above terms, we need some lemmas, which can be proved in a straightforward manner.

**Lemma 10.9** Let $h$ be an arbitrary smooth function on $S_{t,u}$ and $\vartheta$ be an arbitrary symmetric 2-covariant $S_{t,u}$ tensorfield. The following commutation formulas hold for any non-negative integer $k$:

$$\begin{aligned}{}^{(i_1\cdots i_k)}c_k:=&\not{\mathcal{L}}_{R_{i_k}}\cdots\not{\mathcal{L}}_{R_{i_1}}\not{D}^2h-\not{D}^2(R_{i_k}\cdots R_{i_1}h)\\=&-\sum_{m=0}^{k-1}\not{\mathcal{L}}_{R_{i_k}}\cdots\not{\mathcal{L}}_{R_{i_{k-m+1}}}({}^{(R_{i_{k-m}})}\not{\pi}_1\cdot\not{d}(R_{i_{k-m-1}}\cdots R_{i_1}h))\end{aligned}$$

and

$$\begin{aligned}&\not{\mathcal{L}}_{R_{i_k}}\cdots\not{\mathcal{L}}_{R_{i_1}}\check{\not{D}}\vartheta-\check{\not{D}}\not{\mathcal{L}}_{R_{i_k}}\cdots\not{\mathcal{L}}_{R_{i_1}}\vartheta\\=&-\sum_{m=0}^{k-1}\not{\mathcal{L}}_{R_{i_k}}\cdots\not{\mathcal{L}}_{R_{i_{k-m+1}}}({}^{(R_{i_{k-m}})}\not{\pi}_1\cdot\not{\mathcal{L}}_{R_{i_{k-m-1}}}\cdots\not{\mathcal{L}}_{R_{i_1}}\vartheta)\end{aligned}$$

The following assumption, which is another version of **H2**, is used in the proof of the following lemmas.

**H2$'$**: There is a constant $C$ (independent of $s$) such that for any 2-covariant symmetric $S_{t,u}$ tensorfield $\vartheta$ we have, pointwise,

$$|\not{D}\vartheta|^2 \leqslant C(1+t)^{-2}\left(\sum_j |\not{\mathcal{L}}_{R_j}\vartheta|^2 + |\vartheta|^2\right)$$

**Lemma 10.10** Let the hypotheses **H0**, **H1**, **H2′** and (6.177) hold. Let also the bootstrap assumptions $\not{\mathbf{E}}_{[l+1]}$, $\not{\mathbf{E}}^Q_{[l]}$ and $\not{\mathbf{X}}_{[l]}$ hold, for some positive integer $l$. Then if $\delta_0$ is suitably small (depending on $l$) we have

$$\max_{j;i_1\cdots i_k} \|\not{\mathcal{L}}_{R_{i_k}}\cdots\not{\mathcal{L}}_{R_{i_1}}{}^{(R_j)}\not{\pi}_1\|_{L^\infty(\Sigma_t^{\epsilon_0})} \leqslant C_l\delta_0(1+t)^{-2}(1+\log(1+t))$$

for all $k = 0, \cdots, l-1$.

**Lemma 10.11** Let the hypotheses **H0**, **H1**, **H2′** and (6.177) hold. Let also the bootstrap assumptions $\not{\mathbf{E}}_{[l_*+1]}$, $\not{\mathbf{E}}^Q_{[l_*]}$ and $\not{\mathbf{X}}_{[l_*]}$ hold, for some positive integer $l$. Then if $\delta_0$ is suitably small (depending on $l$) we have

$$\max_{j;i_1\cdots i_k} \|\not{\mathcal{L}}_{R_{i_k}}\cdots\not{\mathcal{L}}_{R_{i_1}}{}^{(R_j)}\not{\pi}_1\|_{L^2(\Sigma_t^{\epsilon_0})} \leqslant C_l(1+t)^{-1}(\mathcal{Y}_0 + (1+t)\mathcal{A}_{[l]} + \mathcal{W}_{[l+1]})$$

for all $k = 0, \cdots, l-1$.

Using these lemmas, it is not difficult to obtain the following $L^2$ estimate for the second term in (10.143):

$$\begin{aligned}&\max_{i_1\cdots i_l} \|\sum_{k=0}^{l-1}(\not{\mathcal{L}}_{R_{i_l}} + \frac{1}{2}\mathrm{tr}^{(R_{i_l})}\not{\pi})\cdots(\not{\mathcal{L}}_{R_{i_{l-k+1}}} + \frac{1}{2}\mathrm{tr}^{(R_{i_{l-k+1}})}\not{\pi})^{(i_1\cdots i_{l-k})}q_{l-k}\|_{L^2(\Sigma_t^{\epsilon_0})} \\ &\leqslant C_l\delta_0(1+t)^{-3}(1+\log(1+t))(\mathcal{Y}_0 + (1+t)\mathcal{A}_{[l]} + \mathcal{W}_{[l+1]})\end{aligned} \tag{10.150}$$

Combining this with (10.145) and (10.142) we obtain

**Proposition 10.3** Under the assumptions of Lemma 10.14 we have

$$\max_{i_1\cdots i_l} \|{}^{(i_1\cdots i_l)}i_l\|_{L^2(\Sigma_t^{\epsilon_0})} \leqslant C_l\delta_0(1+t)^{-3}(1+\log(1+t))(\mathcal{Y}_0 + (1+t)\mathcal{A}_{[l]} + \mathcal{W}_{[l+1]})$$

provided that $\delta_0$ is suitably small (depending on $l$).

We turn to the $S_{t,u}$ 1-form ${}^{(i_1\cdots i_l)}\dot{g}_l$:

$$ {}^{(i_1\cdots i_l)}\dot{g}_l = \not{\mathcal{L}}_{R_{i_l}}\cdots\not{\mathcal{L}}_{R_{i_1}}g_0 - {}^{(i_1\cdots i_l)}w_l + \sum_{k=0}^{l-1}\not{\mathcal{L}}_{R_{i_l}}\cdots\not{\mathcal{L}}_{R_{i_{l-k+1}}}{}^{(i_1\cdots i_{l-k})}y_{l-k} \tag{10.151}$$

where ${}^{(i_1\cdots i_j)}y_j$ is given in Proposition 8.3, and ${}^{(i_1\cdots i_l)}w_l$ is

$$\begin{aligned}{}^{(i_1\cdots i_l)}w_l = &\sum_{k=0}^{l-1}\not{\mathcal{L}}_{R_{i_l}}\cdots\not{\mathcal{L}}_{R_{i_{l-k+1}}}\not{\mathcal{L}}_{{}^{(R_{i_{l-k}})}Z}{}^{(i_1\cdots i_{l-k-1})}x_{l-k-1} \\ &- \sum_{k=0}^{l-1}\not{\mathcal{L}}_{{}^{(R_{i_{l-k}})}Z}{}^{(i_1\overset{>i_{l-k}<}{\cdots}i_l)}x_{l-1}\end{aligned} \tag{10.152}$$

We now introduce bootstrap assumptions in regard to the angular derivatives of the function $\mu$. Given a positive integer $l$ let us denote by $\mathbf{M}_l$ the bootstrap assumption that there is a constant $C$ independent of $s$ such that for all $t \in [0, s]$:

$$\mathbf{M}_l : \max_{i_1 \cdots i_l} \|R_{i_l} \cdots R_{i_1} \mu\|_{L^\infty(\Sigma_t^{\epsilon_0})} \leqslant C\delta_0(1 + \log(1 + t))$$

Let us also denote by $\mathbf{M}_0$ the bootstrap assumption

$$\mathbf{M}_0 : |\mu - 1| \leqslant C\delta_0(1 + \log(1 + t))$$

We then denote by $\mathbf{M}_{[l]}$ the bootstrap assumption

$$\mathbf{M}_{[l]} : \mathbf{M}_0 \quad \text{and} \cdots \text{and} \quad \mathbf{M}_l$$

We also introduce the quantities

$$\mathcal{B}_0 = \|\mu - 1\|_{L^2(\Sigma_t^{\epsilon_0})}$$

and for $k \neq 0$:

$$\mathcal{B}_k = \max_{i_1 \cdots i_k} \|R_{i_k} \cdots R_{i_1} \mu\|_{L^2(\Sigma_t^{\epsilon_0})} \quad \text{and} \quad \mathcal{B}_{[l]} = \sum_{k=0}^{l} \mathcal{B}_k$$

Then by the expression (8.27) and (8.29) we obtain

**Lemma 10.12** Let hypothesis **H0** and the estimate (6.177) hold. Let also the bootstrap assumptions $\mathbf{E}^T_{[l_*+1]}$, $\mathbf{E}^Q_{[l_*+1]}$, $\mathbf{E}_{[l_*+2]}$, as well as $\mathbf{X}_{[l_*]}$ and $\mathbf{M}_{[l_*]}$ hold, for some positive integer $l$. We then have

$$\begin{aligned} &\max_{i_1 \cdots i_l} \|\not{d} R_{i_l} \cdots R_{i_1} \check{g}\|_{L^2(\Sigma_t^{\epsilon_0})} \\ &\leqslant C_l \delta_0 (1+t)^{-3} (\mathcal{W}^T_{[l+1]} + \mathcal{W}^Q_{[l+1]} + (1+t)^{-1}(1 + \log(1+t))\mathcal{W}_{[l+2]} \\ &\quad + \delta_0 (1+t)^{-1}(1 + \log(1+t))(\mathcal{Y}_0 + (1+t)\mathcal{A}_{[l]}) + \delta_0 (1+t)^{-2} \mathcal{B}_{[l+1]}) \end{aligned}$$

provided that $\delta_0$ is suitably small (depending on $l$). The factor $1 + \log(1+t)$ comes from the product of $\mu$ and a regular term.

Here, the bootstrap assumption $\mathbf{E}^T_{[l]}$ is defined as follows:

$$\mathbf{E}^T_{[l]} : \mathbf{E}^T_0 \quad \text{and} \cdots \text{and} \quad \mathbf{E}^T_l$$

where $\mathbf{E}^T_l$ is the assumption

$$\mathbf{E}^T_l : \max_{i_1 \cdots i_l} \max_{\alpha} \|R_{i_l} \cdots R_{i_1} T\psi_\alpha\|_{L^\infty(\Sigma_t^{\epsilon_0})} \leqslant C\delta_0(1+t)^{-1}$$

We now return to Equation (8.37) for the $S_{t,u}$ 1-form $g_0$, to estimate the remaining terms. By the equation

$$L\mu = m + \mu e$$

and (8.27), we have

$$\begin{aligned}
&\max_{i_1\cdots i_l} \|\not{\mathcal{L}}_{R_{i_l}} \cdots \not{\mathcal{L}}_{R_{i_1}} (\mathrm{tr}\chi \not{d}(\check{f} + 2L\mu))\|_{L^2(\Sigma_t^{\epsilon_0})} \\
&\leqslant C_l(1+t)^{-3}((1+\log(1+t))(\mathcal{W}^Q_{[l+1]} + \delta_0(1+t)^{-1}(\mathcal{W}_{[l+1]} + \mathcal{Y}_0 \\
&+ (1+t)\mathcal{A}_{[l]})) + \delta_0(1+t)^{-1}\mathcal{B}_{[l+1]})
\end{aligned} \tag{10.153}$$

under the assumptions of Lemma 10.12 except for the assumption $\not{\mathbf{E}}^T_{[l_*+1]}$, due to the cancellation between $m$ and $\check{f}$. Also due to this cancellation, the coefficient of $\mathcal{A}_{[l]}$ has the decay factor $(1+t)^{-3}(1+\log(1+t))$.

Next, we consider the last term in (8.37), which is, by (8.38),

$$(\not{d}\mu)(\mu^{-1}(L\mu)\mathrm{tr}\chi - \mathrm{tr}\alpha)$$

then under the assumptions of Lemma 10.12 except for $\not{\mathbf{E}}^T_{[l_*+1]}$, we have

$$\begin{aligned}
&\max_{i_1\cdots i_l} \|\not{\mathcal{L}}_{R_{i_l}} \cdots \not{\mathcal{L}}_{R_{i_1}} ((\not{d}\mu)(\mu^{-1}(L\mu)\mathrm{tr}\chi - \mathrm{tr}\alpha))\|_{L^2(\Sigma_t^{\epsilon_0})} \\
&\leqslant C_l(1+t)^{-3}((1+\log(1+t))(\mathcal{W}^Q_{[l+1]} + \delta_0(1+t)^{-1}(\mathcal{W}_{[l+2]} + \mathcal{Y}_0 \\
&+ (1+t)\mathcal{A}_{[l]})) + \delta_0(1+t)^{-1}\mathcal{B}_{[l+1]})
\end{aligned} \tag{10.154}$$

The calculation of this can be found in Chapter 12.

Combining the above results, we have

**Lemma 10.13** Let the hypotheses **H0**, **H1**, **H2**$'$ and the estimate (6.177) hold. Let also the bootstrap assumptions $\not{\mathbf{E}}_{[l_*+2]}$, $\not{\mathbf{E}}^Q_{[l_*+1]}$ and $\not{\mathbf{X}}_{[l_*]}$, $\not{\mathbf{M}}_{[l_*+1]}$ hold, for some positive integer $l$. Then if $\delta_0$ is suitably small (depending on $l$) we have

$$\begin{aligned}
&\max_{i_1\cdots i_l} \|\not{\mathcal{L}}_{R_{i_l}} \cdots \not{\mathcal{L}}_{R_{i_1}} y_0\|_{L^2(\Sigma_t^{\epsilon_0})} \\
&\leqslant C_l(1+t)^{-3}((1+\log(1+t))(\mathcal{W}^Q_{[l+1]} + \delta_0(1+t)^{-1}\mathcal{W}_{[l+2]}) + \delta_0\mathcal{W}^T_{[l+1]} \\
&+ \delta_0(1+t)^{-1}(1+\log(1+t))(\mathcal{Y}_0 + (1+t)\mathcal{A}_{[l]}) + \delta_0(1+t)^{-1}\mathcal{B}_{[l+1]})
\end{aligned}$$

We turn to the third term in the formula (10.151) for ${}^{(i_1\cdots i_l)}\dot{g}_l$:

$$\sum_{k=0}^{l-1} \not{\mathcal{L}}_{R_{i_l}} \cdots \not{\mathcal{L}}_{R_{i_{l-k+1}}} {}^{(i_1\cdots i_{l-k})}y_{l-k} \tag{10.155}$$

where the $S_{t,u}$ 1-form ${}^{(i_1\cdots i_j)}y_j$, $j = 1, \cdots, l$ is given in the statement of Proposition 8.3:

$$\begin{aligned}
{}^{(i_1\cdots i_j)}y_j =& (R_{i_j}\mu)^{(i_1\cdots i_{j-1})}a_{j-1} \\
&+ (\mu R_{i_j}(\mathrm{tr}\chi - e) - R_{i_j}m + {}^{(R_{i_j})}Z\mu - eR_{i_j}\mu) \cdot \not{d}(R_{i_{j-1}} \cdots R_{i_1}\mathrm{tr}\chi) \\
&+ \frac{1}{2}(R_{i_j}\mathrm{tr}\chi)\not{d}^{(i_1\cdots i_{j-1})}\check{f}_{j-1}
\end{aligned} \tag{10.156}$$

for $j = 1, \cdots, l$.

The $S_{t,u}$ 1-form ${}^{(i_1 \cdots i_{j-1})}a_{j-1}$ is given by (8.78):

$$ {}^{(i_1 \cdots i_{j-1})}a_{j-1} = \not{d}(R_{i_{j-1}} \cdots R_{i_1} f_0) + {}^{(i_1 \cdots i_{j-1})}b_{j-1} \tag{10.157} $$

where

$$ f_0 = L\mathrm{tr}\chi + |\chi|^2 $$

which we have dealt with in deriving (10.153), and

$$ \begin{aligned} {}^{(i_1 \cdots i_{j-1})}b_{j-1} = & \sum_{m=0}^{j-2} \not{d}(R_{i_{j-1}} \cdots R_{i_{j-m}} {}^{(R_{i_{j-m-1}})}Z R_{i_{j-m-2}} \cdots R_{i_1} \mathrm{tr}\chi) \\ & + \sum_{m=1}^{j-1} \sum_{k_1 < \cdots < k_m = 1}^{j-1} (R_{i_{k_m}} \cdots R_{i_{k_1}} \mathrm{tr}\chi)(\not{d} R_{i_{j-1}} \overset{> i_{k_m} \cdots i_{k_1} <}{\cdots} R_{i_1} \mathrm{tr}\chi) \end{aligned} \tag{10.158} $$

for $j \geqslant 2$; $b_0 = 0$.

To estimate the sum (10.154) in $L^2(\Sigma_t^{\epsilon_0})$, we must estimate

$$ \max_{i_1 \cdots i_{j+k}} \| \not{\mathcal{L}}_{R_{i_{j+k}}} \cdots \not{\mathcal{L}}_{R_{i_{j+1}}} {}^{(i_1 \cdots i_j)}y_j \|_{L^2(\Sigma_t^{\epsilon_0})}, j + k = l \tag{10.159} $$

To estimate this requires estimating:

$$ \max_{i_1 \cdots i_{j-1+k}} \| \not{\mathcal{L}}_{R_{i_{j-1+k}}} \cdots \not{\mathcal{L}}_{R_{i_j}} {}^{(i_1 \cdots i_{j-1})}a_{j-1} \|_{L^2(\Sigma_t^{\epsilon_0})}, j + k \leqslant l \tag{10.160} $$

as well as

$$ \max_{i_1 \cdots i_{j-1+k}} \| \not{\mathcal{L}}_{R_{i_{j-1+k}}} \cdots \not{\mathcal{L}}_{R_{i_j}} {}^{(i_1 \cdots i_{j-1})}a_{j-1} \|_{L^\infty(\Sigma_t^{\epsilon_0})}, j + k \leqslant l_* \tag{10.161} $$

Since we have dealt with the first term on the right in (10.156) in deriving (10.153), we just need to estimate

$$ \max_{i_1 \cdots i_{j-1+k}} \| \not{\mathcal{L}}_{R_{i_{j-1+k}}} \cdots \not{\mathcal{L}}_{R_{i_j}} {}^{(i_1 \cdots i_{j-1})}b_{j-1} \|_{L^2(\Sigma_t^{\epsilon_0})}, j + k \leqslant l \tag{10.162} $$

as well as

$$ \max_{i_1 \cdots i_{j-1+k}} \| \not{\mathcal{L}}_{R_{i_{j-1+k}}} \cdots \not{\mathcal{L}}_{R_{i_j}} {}^{(i_1 \cdots i_{j-1})}b_{j-1} \|_{L^\infty(\Sigma_t^{\epsilon_0})}, j + k \leqslant l_* \tag{10.163} $$

These quantities enjoy the bounds

$$ \begin{aligned} & \max_{i_1 \cdots i_{j-1+k}} \| \not{\mathcal{L}}_{R_{i_{j-1+k}}} \cdots \not{\mathcal{L}}_{R_{i_j}} {}^{(i_1 \cdots i_{j-1})}b_{j-1} \|_{L^\infty(\Sigma_t^{\epsilon_0})} \\ & \leqslant C_l \delta_0^2 (1+t)^{-5} (1 + \log(1+t))^2 \end{aligned} \tag{10.164} $$

for $j + k \leqslant l_*$, and

$$\max_{i_1\cdots i_{j-1+k}} \|\not{\mathcal{L}}_{R_{i_{j-1+k}}}\cdots\not{\mathcal{L}}_{R_{i_j}}{}^{(i_1\cdots i_{j-1})}b_{j-1}\|_{L^2(\Sigma_t^{\epsilon_0})}$$
$$\leqslant C_l\delta_0(1+t)^{-4}(1+\log(1+t))((1+t)\mathcal{A}_{[l]}+\mathcal{Y}_0+\mathcal{W}_{[l]}) \tag{10.165}$$

for $j+k\leqslant l$ under the assumptions of Proposition 10.1 with $l_*+1$ in the role of $l$.

The above estimates together with those for $f_0$ yield the following lemma.

**Lemma 10.14** Let the hypotheses **H0**, **H1**, **H2**$'$, and the estimate (6.177) hold. Let also the bootstrap assumptions $\not{\mathbf{E}}_{[l_*+2]}$, $\not{\mathbf{E}}^Q_{[l_*+1]}$ and $\not{\mathbf{X}}_{[l_*]}$ hold, for some positive integer $l$. Then provided that $\delta_0$ is suitably small (depending on $l$), the $S_{t,u}$ 1-form ${}^{(i_1\cdots i_{j-1})}a_{j-1}$ satisfies the following estimates:

$$\max_{i_1\cdots i_{j-1+k}} \|\not{\mathcal{L}}_{R_{i_{j-1+k}}}\cdots\not{\mathcal{L}}_{R_{i_j}}{}^{(i_1\cdots i_{j-1})}a_{j-1}\|_{L^\infty(\Sigma_t^{\epsilon_0})}\leqslant C_l\delta_0(1+t)^{-4}$$

for $j+k\leqslant l_*$, and

$$\max_{i_1\cdots i_{j-1+k}} \|\not{\mathcal{L}}_{R_{i_{j-1+k}}}\cdots\not{\mathcal{L}}_{R_{i_j}}{}^{(i_1\cdots i_{j-1})}a_{j-1}\|_{L^2(\Sigma_t^{\epsilon_0})}$$
$$\leqslant C_l(1+t)^{-3}(\mathcal{W}_{[l+2]}+\delta_0(1+t)^{-1}\mathcal{W}^Q_{[l+1]}+\mathcal{W}^Q_{[l]}$$
$$+\delta_0(1+t)^{-1}(1+\log(1+t))(\mathcal{Y}_0+(1+t)\mathcal{A}_{[l]}))$$

for $j+k\leqslant l$.

Using Lemma 10.14 and the bootstrap assumption $\not{\mathbf{M}}_{[l_*+1]}$ we readily obtain

$$\|\not{\mathcal{L}}_{R_{i_{j+k}}}\cdots\not{\mathcal{L}}_{R_{i_{j+1}}}((R_{i_j}\mu)^{(i_1\cdots i_{j-1})}a_{j-1})\|_{L^2(\Sigma_t^{\epsilon_0})}$$
$$\leqslant C_l\delta_0(1+t)^{-3}(1+\log(1+t))(\mathcal{W}_{[l+2]}+\delta_0(1+t)^{-1}\mathcal{W}^Q_{[l+1]}+\mathcal{W}^Q_{[l]} \tag{10.166}$$
$$+\delta_0(1+t)^{-1}(1+\log(1+t))(\mathcal{Y}_0+(1+t)\mathcal{A}_{[l]}))+C_l\delta_0(1+t)^{-4}\mathcal{B}_{[l]}$$

for $j+k=l$.

We consider next the second term on the right in (10.156). We have the following estimates:

$$\|\not{\mathcal{L}}_{R_{i_{j+1}}}\cdots\not{\mathcal{L}}_{R_{i_{j+k}}}((\mu R_{i_j}(\mathrm{tr}\chi-e)-R_{i_j}m+{}^{(R_{i_j})}Z\mu-eR_{i_j}\mu)$$
$$\cdot\not{d}(R_{i_{j-1}}\cdots R_{i_1}\mathrm{tr}\chi))\|_{L^2(\Sigma_t^{\epsilon_0})}$$
$$\leqslant C_l\delta_0(1+t)^{-3}((1+t)\mathcal{A}_{[l]}+\mathcal{W}^T_{[l]}+\mathcal{Y}_0+\mathcal{W}_{[l]}$$
$$+(1+\log(1+t))\mathcal{W}^Q_{[l]}+\delta_0(1+t)^{-1}(1+\log(1+t))\mathcal{W}_{[l+1]})$$

Here, we also need the assumption $\not{\mathbf{X}}_{[(l+1)_*]}$.

We turn to the last term on the right in (10.156). Recall that

$$\check{f}=\mu f=-\frac{1}{2}\frac{dH}{dh}\tau_{\underline{L}} \tag{10.167}$$

Then by (8.24) and $\not{\mathbf{M}}_{[l_*]}$ we obtain

$$\max_{i_1\cdots i_k} \|R_{i_k}\cdots R_{i_1}\check{f}\|_{L^\infty(\Sigma_t^{\epsilon_0})} \leqslant C_l\delta_0(1+t)^{-1}, k\leqslant l_* \tag{10.168}$$

and also

$$\max_{i_1\cdots i_k} \|R_{i_k}\cdots R_{i_1}\check{f}\|_{L^2(\Sigma_t^{\epsilon_0})} \leqslant C_l(\mathcal{W}^T_{[l]} + \delta_0(1+t)^{-1}(\mathcal{W}_{[l]} + (1+t)^{-1}\mathcal{B}_{[l]}) \tag{10.169}$$
$$+ (1+t)^{-1}(1+\log(1+t))\mathcal{W}^Q_{[l]})$$

for $k\leqslant l$.

So with the assumption $\mathbf{X}_{[(l+1)_*]}$, we have

$$\|\not{\mathcal{L}}_{R_{i_{j+k}}}\cdots\not{\mathcal{L}}_{R_{i_{j+1}}}((R_{i_j}\mathrm{tr}\chi)(\not{d}R_{i_{j-1}}\cdots R_{i_1}\check{f}))\|_{L^2(\Sigma_t^{\epsilon_0})}$$
$$\leqslant C_l\delta_0(1+t)^{-3}(\mathcal{Y}_0 + (1+t)\mathcal{A}_{[l]} + \mathcal{W}_{[l]} + (1+\log(1+t))(\mathcal{W}^T_{[l]} \tag{10.170}$$
$$+ (1+t)^{-1}(1+\log(1+t))\mathcal{W}^Q_{[l]} + \delta_0(1+t)^{-2}\mathcal{B}_{[l]}))$$

for $j+k=l$.

Combining the results we obtain the following lemma:

**Lemma 10.15** Let the hypotheses **H0**, **H1**, **H2′**, and the estimate (6.177) hold. Let also the bootstrap assumptions $\mathbf{E}_{[l_*+2]}$, $\mathbf{E}^Q_{[l_*+1]}$, $\mathbf{E}^T_{[l_*]}$, $\mathbf{X}_{[(l+1)_*]}$ and $\mathbf{M}_{[l_*+1]}$ hold, for some positive integer $l$. Then provided that $\delta_0$ is suitably small (depending on $l$), we have

$$\max_{i_1\cdots i_l} \|\sum_{k=0}^{l-1}\not{\mathcal{L}}_{R_{i_l}}\cdots\not{\mathcal{L}}_{R_{i_{l-k+1}}}{}^{(i_1\cdots i_{l-k})}y_{l-k}\|_{L^2(\Sigma_t^{\epsilon_0})}$$
$$\leqslant C_l\delta_0(1+t)^{-3}((1+\log(1+t))\times(\mathcal{W}_{[l+2]} + \delta_0(1+t)^{-1}\mathcal{W}^Q_{[l+1]}$$
$$+ \mathcal{W}^T_{[l]} + (1+t)^{-1}(1+\log(1+t))\mathcal{W}^Q_{[l]}) + \mathcal{W}_{[l]}$$
$$+ \mathcal{Y}_0 + (1+t)\mathcal{A}_{[l]} + (1+t)^{-1}\mathcal{B}_{[l]})$$

To estimate the $S_{t,u}$ 1-form ${}^{(i_1\cdots i_l)}\dot{g}_l$, given by (10.151), it remains for us to consider the $S_{t,u}$ 1-form ${}^{(i_1\cdots i_l)}w_l$, defined by (10.152). A treatment similar to the above yields

**Lemma 10.16** Let hypotheses **H0** and (6.177) hold. Let also the bootstrap assumptions $\mathbf{E}_{[l_*+2]}$, $\mathbf{E}^Q_{[l_*+1]}$, $\mathbf{E}^T_{[l_*+1]}$, $\mathbf{X}_{[(l+1)_*]}$ and $\mathbf{M}_{[l_*+1]}$ hold, for some positive integer $l$. Then, provided that $\delta_0$ is suitably small (depending on $l$), we have

$$\max_{i_1\cdots i_l}\|{}^{(i_1\cdots i_l)}w_l\|_{L^2(\Sigma_t^{\epsilon_0})}$$
$$\leqslant C_l\delta_0(1+t)^{-3}(\mathcal{Y}_0 + (1+t)\mathcal{A}_{[l]} + \mathcal{W}_{[l+1]} + (1+\log(1+t))\mathcal{W}^T_{[l]}$$
$$+ (1+t)^{-1}(1+\log(1+t))^2(\mathcal{W}^Q_{[l]} + \delta_0(1+t)^{-1}\mathcal{B}_{[l]}))$$

Combining Lemma 10.13, Lemma 10.15 and Lemma 10.16, we arrive at the following proposition:

**Proposition 10.4** Let the hypotheses **H0**, **H1**, **H2′** and (6.177) hold. Let also the bootstrap assumptions $\mathbf{E}_{[l_*+2]}$, $\mathbf{E}^Q_{[l_*+1]}$ and $\mathbf{E}^T_{[l_*+1]}$, as well as $\mathbf{X}_{[(l+1)_*]}$ and

$\mathbf{M}_{[l_*+1]}$ hold, for some positive integer $l$. Then if $\delta_0$ is suitably small (depending on $l$) we have

$$\max_{i_1\cdots i_l} \|^{(i_1\cdots i_l)}\dot{g}_l\|_{L^2(\Sigma_t^{\epsilon_0})}$$
$$\leqslant C_l(1+t)^{-3}((1+\log(1+t))(\mathcal{W}^Q+\delta_0(\mathcal{W}^T_{[l+1]}+\mathcal{W}_{[l+2]}))$$
$$+\delta_0(\mathcal{Y}_0+(1+t)\mathcal{A}_{[l]}+(1+t)^{-1}\mathcal{B}_{[l+1]}))$$

Let us now consider the additional terms

$$2\mu \not{D}\hat{\chi}\cdot {}^{(i_1\cdots i_l)}\hat{\chi}_l + \mu \not{d}^{(i_1\cdots i_l)}\dot{h}_l \tag{10.171}$$

Using the same methods we establish

**Lemma 10.17** Let **H0** and (6.177) hold. Let also the bootstrap assumptions $\mathbf{E}_{[l_*+1]}$, $\mathbf{E}^Q_{[l_*]}$ and $\mathbf{X}_{[(l+1)_*]}$ hold for some positive integer $l$. Then we have

$$\max_{i_1\cdots i_l} \|\mu \not{d}^{(i_1\cdots i_l)}\dot{h}_l\|_{L^2(\Sigma_t^{\epsilon_0})}$$
$$\leqslant C_l\delta_0(1+t)^{-4}(1+\log(1+t))^2(\mathcal{Y}_0+(1+t)\mathcal{A}_{[l]}+\mathcal{W}_{[l+1]})$$

provided that $\delta_0$ is suitably small (depending on $l$).

Finally the first term in (10.171) can be estimated directly:

$$\max_{i_1\cdots i_l} \|\mu \not{D}\hat{\chi}\cdot {}^{(i_1\cdots i_l)}\hat{\chi}_l\|_{L^2(\Sigma_t^{\epsilon_0})}$$
$$\leqslant C_l\delta_0(1+t)^{-4}(1+\log(1+t))^2(\mathcal{Y}_0+(1+t)\mathcal{A}_{[l]}+\mathcal{W}_{[l]}) \tag{10.172}$$

Proposition 10.4 together with Lemma 10.17 and (10.170) gives us an appropriate bound for $\|^{(i_1\cdots i_l)}\ddot{g}_l\|_{L^2(\Sigma_t^{\epsilon_0})}$. To get the estimate for $^{(i_1\cdots i_l)}Q_l(t)$, it remains for us to bound the quantities $\|\not{d}^{(i_1\cdots i_l)}\check{f}_l\|_{L^2(\Sigma_t^{\epsilon_0})}$ and $\|^{(i_1\cdots i_l)}z_l\|_{L^2(\Sigma_t^{\epsilon_0})}$. From (10.168) we have

$$\max_{i_1\cdots i_l} \|\not{d}^{(i_1\cdots i_l)}\check{f}_l\|_{L^2(\Sigma_t^{\epsilon_0})}$$
$$\leqslant C_l(1+t)^{-1}(\mathcal{W}^T_{[l+1]}+\delta_0(1+t)^{-1}(\mathcal{W}_{[l+1]}+(1+t)^{-1}\mathcal{B}_{[l+1]}) \tag{10.173}$$
$$+(1+t)^{-1}(1+\log(1+t))\mathcal{W}^Q_{[l+1]})$$

The non-negative function $^{(i_1\cdots i_l)}z_l$ is given by (8.392). Using the same methods we deduce

$$\max_{i_1\cdots i_l} \|^{(i_1\cdots i_l)}z_l\|_{L^2(\Sigma_t^{\epsilon_0})}$$
$$\leqslant C_l\delta_0(1+t)^{-2}(1+\log(1+t)) \tag{10.174}$$
$$\cdot(\mathcal{W}^T_{[l+1]}+(1+t)^{-1}(1+\log(1+t))(\mathcal{W}^Q_{[l+1]}+\delta_0(1+t)^{-1}\mathcal{W}_{[l+2]})$$
$$+(1+t)^{-1}(1+\log(1+t))(\mathcal{Y}_0+(1+t)\mathcal{A}_{[l]})+\delta_0(1+t)^{-2}\mathcal{B}_{[l+1]})$$

We now can write down the estimate for ${}^{(i_1\cdots i_l)}Q_l$:

**Proposition 10.5** Let the hypotheses **H0**, **H1**, **H2′** and the estimate (6.177) hold. Let also the bootstrap assumptions $\not{\mathbf{E}}_{[l_*+2]}$, $\not{\mathbf{E}}^Q_{[l_*+1]}$ and $\not{\mathbf{E}}^T_{[l_*+1]}$, as well as $\not{\mathbf{X}}_{[(l+1)_*]}$ and $\not{\mathbf{M}}_{[l_*+1]}$ hold, for some positive integer $l$. Then we have

$$\max_{i_1\cdots i_l} {}^{(i_1\cdots i_l)}Q_l \leqslant C_l(1+t)^{-4}((1+\log(1+t))(\mathcal{W}^Q_{[l+1]}+\delta_0(\mathcal{W}^T_{[l+1]}+\mathcal{W}_{[l+2]}))$$
$$+\delta_0(\mathcal{Y}_0+(1+t)\mathcal{A}_{[l]}+(1+t)^{-1}\mathcal{B}_{[l+1]}))$$

provided that $\delta_0$ is suitably small (depending on $l$).

### 10.3.2 Estimates for $P_l$

We now consider the principal part in the defining expression (8.403) for ${}^{(i_1\cdots i_l)}B_l$, namely, the term

$$C(1+t)^{-2}({}^{(i_1\cdots i_l)}\bar{P}^{(0)}_{l,a}(t)+(1+t)^{-1/2}{}^{(i_1\cdots i_l)}\bar{P}^{(1)}_{l,a}(t))\bar{\mu}_m^{-a}(t)$$

Here, the quantities ${}^{(i_1\cdots i_l)}\bar{P}^{(0)}_{l,a}$, ${}^{(i_1\cdots i_l)}\bar{P}^{(1)}_{l,a}$ are defined by (8.348) and (8.349). Recall the definition

$${}^{(i_1\cdots i_l)}P_l(t)=(1+t)^2\|\|\not{d}^{(i_1\cdots i_l)}\check{f}_l(t)\|\|_{L^2([0,\epsilon_0]\times S^2)} \tag{10.175}$$

Here we need a more precise estimate for $\|\not{d}^{(i_1\cdots i_l)}\check{f}_l\|_{L^2(\Sigma_t^{\epsilon_0})}$.

By (8.333) and **H0** we have

$${}^{(i_1\cdots i_l)}P_l \leqslant C\sum_j \|R_j{}^{(i_1\cdots i_l)}\check{f}_l\|_{L^2(\Sigma_t^{\epsilon_0})}=C\sum_j\|R_jR_{i_l}\cdots R_{i_1}\check{f}\|_{L^2(\Sigma_t^{\epsilon_0})} \tag{10.176}$$

Here, we must recall the definition of $\check{f}$ in Chapter 8:

$$\check{f}=\mu f=-\frac{1}{2}\frac{dH}{dh}\tau_{\underline{L}} \tag{10.177}$$

We can use the definition of the acoustical metric and (8.24) to write the above as follows:

$$\check{f}=-2m-\frac{1}{2}\alpha^{-1}\kappa\frac{dH}{dh}Lh:=-2m-\mu\bar{e} \tag{10.178}$$

where we have used the fact that

$$\alpha=\eta,\quad \partial_t-\psi_i\partial_i=B=L+\alpha\kappa^{-1}T$$

We can express

$$m=m^\alpha_T(T\psi_\alpha) \tag{10.179}$$

where

$$m^0_T=\frac{1}{2}\frac{dH}{dh},\quad m^i_T=-\frac{1}{2}\frac{dH}{dh}\psi_i \tag{10.180}$$

and

$$\bar{e} = \bar{e}_L^{\alpha}(L\psi_\alpha) \tag{10.181}$$

where

$$\bar{e}_L^0 = \frac{1}{2}\alpha^{-2}\frac{dH}{dh}, \quad \bar{e}_L^i = -\frac{1}{2}\alpha^{-2}\frac{dH}{dh}\psi_i \tag{10.182}$$

Then

$$\begin{aligned} & R_j R_{i_l} \cdots R_{i_1} \check{f} \\ = & -2m_T^{\alpha}(R_j R_{i_l} \cdots R_{i_1} T\psi_\alpha) - \mu \bar{e}_L^{\alpha}(R_j R_{i_l} \cdots R_{i_1} L\psi_\alpha) + {}^{(i_1\cdots i_l j)}n_{l+1} \end{aligned} \tag{10.183}$$

Here, ${}^{(i_1\cdots i_l j)}n_{l+1}$ is the lower order term:

$$\begin{aligned} {}^{(i_1\cdots i_l j)}n_{l+1} = & -\sum_{|s_1|\geqslant 1}((R)^{s_1}\mu)((R)^{s_2}\bar{e}) \\ & -2\sum_{|s_1|\geqslant 1}((R)^{s_1}m_T^{\alpha})((R)^{s_2}T\psi_\alpha) \\ & -\sum_{|s_1|\geqslant 1}\mu((R)^{s_1}\bar{e}_L^{\alpha})((R)^{s_2}L\psi_\alpha) \end{aligned} \tag{10.184}$$

We have

$$|m_T^0 - \frac{1}{2}\ell| \leqslant C\delta_0(1+t)^{-1}, \quad |m_T^i| \leqslant C\delta_0(1+t)^{-1} \tag{10.185}$$

and also

$$|\bar{e}_L^0| \sim |\ell| \tag{10.186}$$

From (10.181)—(10.183) and (10.185)—(10.186) we have

$$\begin{aligned} & \|R_j R_{i_l} \cdots R_{i_1} \check{f}\|_{L^2(\Sigma_t^{\epsilon_0})} \\ \leqslant & (|\ell| + C\delta_0(1+t)^{-1})\sum_{\alpha}\|R_j R_{i_l} \cdots R_{i_1} T\psi_\alpha\|_{L^2(\Sigma_t^{\epsilon_0})} \\ & + C\sum_{\alpha}\|\mu R_j R_{i_l} \cdots R_{i_1} L\psi_\alpha\|_{L^2(\Sigma_t^{\epsilon_0})} + \|{}^{(i_1\cdots i_l j)}n_{l+1}\|_{L^2(\Sigma_t^{\epsilon_0})} \end{aligned} \tag{10.187}$$

where $C$ is a constant which is *independent* of $l$. Under the assumptions of Proposition 10.5 we readily deduce

$$\begin{aligned} & \max_{i_1\cdots i_l j}\|{}^{(i_1\cdots i_l j)}n_{l+1}\|_{L^2(\Sigma_t^{\epsilon_0})} \\ \leqslant & C_l\delta_0(1+t)^{-1}(\mathcal{W}_{[l]}^T + (1+\log(1+t))\mathcal{W}_{[l]}^Q + \mathcal{W}_{[l+1]} + (1+t)^{-1}\mathcal{B}_{[l+1]}) \end{aligned} \tag{10.188}$$

To proceed, we need the following lemma, which can be proved in a similar way as Lemma 8.2.

**Lemma 10.18** Let $Y$ be an arbitrary $S_{t,u}$-tangential vectorfield on the spacetime domain $W_{\epsilon_0}^*$. We have

$$[T, Y] = {}^{(Y)}\Theta$$

where ${}^{(Y)}\Theta$ is an $S_{t,u}$-tangential vectorfield, associated to $Y$, and defined by the condition that for any vector $V \in TW^*_{\epsilon_0}$:

$$g({}^{(Y)}\Theta, V) = {}^{(Y)}\pi(T, \Pi V)$$

In terms of the $(L, T, X_1, X_2)$ frame,

$${}^{(Y)}\Theta = {}^{(Y)}\Theta^A X_A, \quad {}^{(Y)}\Theta^A = {}^{(Y)}\pi_{TB}(\not{g}^{-1})^{AB}$$

We can express, in rectangular coordinates,

$$[T, R_i] = (T(R_i)^j - R_i T^j)\partial_j \tag{10.189}$$

Since $T = \kappa\hat{T}$, we have

$$R_i T^j = (R_i\kappa)\hat{T}^j + \kappa R_i \hat{T}^j$$

Using (10.8) we have

$$R_i T^j = (R_i\kappa)\hat{T}^j + \kappa \not{q}_i \cdot \not{d}x^j \tag{10.190}$$

Projecting to $S_{t,u}$:

$$\Pi(R_i \hat{T}^j \partial_j) = \kappa \not{q}_i \tag{10.191}$$

On the other hand, from (10.21), (10.23),

$$(R_i)^j = \epsilon_{imj} x^m - \lambda_i \hat{T}^j$$

hence

$$T(R_i)^j = \epsilon_{imj} T^m - (T\lambda_i)\hat{T}^j - \lambda_i T(\hat{T}^j) \tag{10.192}$$

According to (3.191),

$$T(\hat{T}^j) = q_T^j \tag{10.193}$$

Here $q_T$ is an $S_{t,u}$-tangential vectorfield given by

$$q_T = (q_T)_\flat \cdot \not{g}^{-1} \tag{10.194}$$

where

$$(q_T)_\flat = -\not{d}\kappa \tag{10.195}$$

Substituting (10.193) in (10.192) we obtain

$$T(R_i)^j \partial_j = \kappa v_i - (T\lambda_i)\hat{T} - \lambda_i q_T \tag{10.196}$$

where $v_i$ is defined by (10.47). Projecting to $S_{t,u}$:

$$\Pi(T(R_i)^j\partial_j) = \kappa\Pi v_i - \lambda_i q_T \tag{10.197}$$

So we obtain

$$[T,R_i] = \kappa\Pi v_i - \lambda_i q_T - \kappa\not{q}_i \tag{10.198}$$

Defining the vectorfields $v_i'$ as in (10.50), Equation (10.52) holds, hence

$$\Pi v_i = w_i - (1-u+t)^{-1}R_i \tag{10.199}$$

Therefore by (10.43) and (10.14), we obtain

$$[T,R_i] = -\kappa\tilde{\not{q}}' - \lambda_i q_T \tag{10.200}$$

We have thus proved the following lemma:

**Lemma 10.19** We have

$$[T,R_i] = {}^{(R_i)}\Theta = -\kappa\tilde{\not{q}}' - \lambda_i q_T$$

Now, under the assumptions of Proposition 10.1 with $l_* + 1$ in the role of $l$ augmented by assumption $\mathbf{M}_{[l_*+1]}$, we have, by (10.194) and (10.195),

$$\max_{i_1\cdots i_k} \|\not{\mathcal{L}}_{R_{i_k}}\cdots\not{\mathcal{L}}_{R_{i_1}} q_T\|_{L^\infty(\Sigma_t^{\epsilon_0})} \leqslant C_l\delta_0(1+t)^{-1}(1+\log(1+t)) \tag{10.201}$$

for $k \leqslant l_*$. Also by Corollary 10.2.d, with $l$ replaced by $l+1$, we have

$$\begin{aligned}&\max_{i_1\cdots i_k} \|\not{\mathcal{L}}_{R_{i_k}}\cdots\not{\mathcal{L}}_{R_{i_1}} q_T\|_{L^2(\Sigma_t^{\epsilon_0})}\\ &\leqslant C_l((1+t)^{-1}\mathcal{B}_{[l+1]} + \delta_0(1+t)^{-1}(1+\log(1+t))(\mathcal{Y}_0 + (1+t)\mathcal{A}_{[l-1]} + \mathcal{W}_{[l]}))\end{aligned} \tag{10.202}$$

By (10.101) and (10.41)—(10.42), Proposition 10.1, with $l_* + 1$ in the role of $l$ we have

$$\max_{i;i_1\cdots i_k} \|\not{\mathcal{L}}_{R_{i_k}}\cdots\not{\mathcal{L}}_{R_{i_1}} \tilde{\not{q}}_i'\|_{L^\infty(\Sigma_t^{\epsilon_0})} \leqslant C_l\delta_0(1+t)^{-1}(1+\log(1+t)) \tag{10.203}$$

for $k \leqslant l_*$, and by (10.120) and (10.123) with $l+1$ in the role of $l$

$$\max_{i;i_1\cdots i_k} \|\not{\mathcal{L}}_{R_{i_k}}\cdots\not{\mathcal{L}}_{R_{i_1}} \tilde{\not{q}}_i'\|_{L^2(\Sigma_t^{\epsilon_0})} \leqslant C_l(\mathcal{Y}_0 + (1+t)\mathcal{A}_{[l]} + \mathcal{W}_{[l+1]}) \tag{10.204}$$

for $k \leqslant l$.

The above estimates together with Corollary 10.1.a with $l_*$ in the role of $l$ and Corollary 10.2.a yield the following lemma.

**Lemma 10.20** Under the assumptions of Proposition 10.1 with $l_* + 1$ in the role of $l$ and $\mathbf{M}_{[l_*+1]}$, we have

$$\max_{i;i_1\cdots i_k} \|\not{\mathcal{L}}_{R_{i_k}} \cdots \not{\mathcal{L}}_{R_{i_1}} {}^{(R_i)}\Theta\|_{L^\infty(\Sigma_t^{\epsilon_0})} \leqslant C_l \delta_0 (1+t)^{-1}(1+\log(1+t))^2$$

for $k \leqslant l_*$, and

$$\max_{i;i_1\cdots i_k} \|\not{\mathcal{L}}_{R_{i_k}} \cdots \not{\mathcal{L}}_{R_{i_1}} {}^{(R_i)}\Theta\|_{L^2(\Sigma_t^{\epsilon_0})}$$
$$\leqslant C_l(1+\log(1+t))(\mathcal{Y}_0 + (1+t)\mathcal{A}_{[l]} + \delta_0(1+t)^{-1}\mathcal{B}_{[l+1]} + \mathcal{W}_{[l+1]})$$

for $k \leqslant l$.

Provided that $\delta_0$ is suitably small (depending on $l$).

Consider now the commutator

$$TR_{i_l}\cdots R_{i_1} - R_{i_l}\cdots R_{i_1}T = {}^{(i_1\cdots i_l)}A_l^T \tag{10.205}$$

By Lemma 10.19,

$${}^{(i_1\cdots i_l)}A_l^T = R_{i_l}\,{}^{(i_1\cdots i_{l-1})}A_{l-1}^T + {}^{(R_{i_l})}\Theta R_{i_{l-1}}\cdots R_{i_1} \tag{10.206}$$

Applying Proposition 8.2 to this recursion yields

$${}^{(i_1\cdots i_l)}A_l^T = \sum_{k=0}^{l-1} R_{i_l}\cdots R_{i_{l-k+1}}\,{}^{(R_{i_{l-k}})}\Theta R_{i_{l-k-1}}\cdots R_{i_1} \tag{10.207}$$

So we have

$$\begin{aligned} R_{i_{l+1}}R_{i_l}\cdots R_{i_1}T\psi_\alpha =& TR_{i_{l+1}}R_{i_l}\cdots R_{i_1}\psi_\alpha \\ &- \sum_{k=0}^{l} R_{i_{l+1}}\cdots R_{i_{l-k+2}}\,{}^{(R_{i_{l-k+1}})}\Theta R_{i_{l-k}}\cdots R_{i_1}\psi_\alpha \end{aligned} \tag{10.208}$$

We shall now use the following lemma, which is proved in a similar way as Lemma 8.5.

**Lemma 10.21** For any $S_{t,u}$-tangential vectorfield $X$, any function $\phi$ and positive integer $l$ we have

$$[X, R_{i_l}\cdots R_{i_1}]\phi = -\sum_{j=1}^{l}\sum_{k_1<\cdots<k_j=1}^{l} {}^{(i_{k_1}\cdots i_{k_j})}YR_{i_l} \overset{>i_{k_j}\cdots i_{k_1}<}{\cdots} R_{i_1}\phi$$

where ${}^{(i_{k_1}\cdots i_{k_j})}Y$ is the $S_{t,u}$-tangential vectorfield

$${}^{(i_{k_1}\cdots i_{k_j})}Y = \not{\mathcal{L}}_{R_{i_{k_j}}}\cdots\not{\mathcal{L}}_{R_{i_{k_1}}}X$$

Consider then an arbitrary term of the sum on the right in (10.207), corresponding to some $k \in \{0, \cdots, l\}$. Setting

$$X = {}^{(R_{i_{l-k+1}})}\Theta, \quad \phi = R_{i_{l-k}}\cdots R_{i_1}\psi_\alpha \tag{10.209}$$

we write this term in the form, using Lemma 10.21,

$$\begin{aligned}&R_{i_{l+1}}\cdots R_{i_{l-k+2}}X\phi\\&=X\cdot\not{d}(R_{i_{l+1}}\cdots R_{i_{l-k+2}}\phi)+\sum_{|s_1|>0}(\not{\mathcal{L}}_R)^{s_1}X\cdot\not{d}(R)^{s_2}\phi\end{aligned}\tag{10.210}$$

By Lemma 10.20, the first term on the right in (10.209) is bounded in $L^2(\Sigma_t^{\epsilon_0})$ by

$$C_l\delta_0(1+t)^{-2}(1+\log(1+t))^2\mathcal{W}_{[l+1]}\tag{10.211}$$

A term in the sum in (10.210) is the product of an angular derivative of $X$ of order $|s_1|$ with an angular derivative of $\psi_\alpha$ of order $|s_2|+l-k+1$, and we have $|s_1|+|s_2|=k$. Thus the sum in (10.210) contains angular derivatives of $X$ of order at most $k\leqslant l$ and angular derivatives of $\psi_\alpha$ of order at most $l$ (for, $|s_1|\geqslant 1$).

We have the following two cases.

Case 1: $|s_1|\leqslant l_*$ and Case 2: $|s_1|\geqslant l_*+1$.

In Case 1 we use the first part of Lemma 10.20 to place the first factor in $L^\infty$, obtaining an $L^2$ bound by

$$C_l\delta_0(1+t)^{-2}(1+\log(1+t))^2\mathcal{W}_{[l]}\tag{10.212}$$

In Case 2 we have $|s_2|\leqslant k-l_*-1$, hence

$$|s_2|+l-k+1\leqslant l-l_*\leqslant l_*+1$$

and the bootstrap assumption $\not{\mathbf{E}}_{[l_*+1]}$ allows us to place the second factor in $L^\infty$. Using also the second part of Lemma 10.20 in placing the first factor in $L^2$, we obtain an $L^2$ bound by

$$C_l(1+t)^{-2}(1+\log(1+t))(\mathcal{Y}_0+(1+t)\mathcal{A}_{[l]}+\delta_0(1+t)^{-1}\mathcal{B}_{[l+1]}+\mathcal{W}_{[l+1]})\tag{10.213}$$

Combining the above results (10.211)—(10.213) we conclude that

$$\begin{aligned}&\|\sum_{k=0}^{l}R_{i_{l+1}}\cdots R_{i_{l-k+2}}{}^{(R_{i_{l-k+1}})}\Theta R_{i_{l-k}}\cdots R_{i_1}\psi_\alpha\|_{L^2(\Sigma_t^{\epsilon_0})}\\&\leqslant C_l\delta_0(1+t)^{-2}(1+\log(1+t))(\mathcal{Y}_0+(1+t)\mathcal{A}_{[l]}+\delta_0(1+t)^{-1}\mathcal{B}_{[l+1]}\\&\quad+(1+\log(1+t))\mathcal{W}_{[l+1]})\end{aligned}\tag{10.214}$$

for $k=0,\cdots,l$.

A commutation relation similar to (10.208) with ${}^{(R_i)}Z$ in the role of ${}^{(R_i)}\Theta$ holds for $L$ in the role of $T$, that is

$$\begin{aligned}R_{i_{l+1}}R_{i_l}\cdots R_{i_1}L\psi_\alpha=&LR_{i_{l+1}}R_{i_l}\cdots R_{i_1}\psi_\alpha\\&-\sum_{k=0}^{l}R_{i_{l+1}}\cdots R_{i_{l-k+2}}{}^{(R_{i_{l-k+1}})}ZR_{i_{l-k}}\cdots R_{i_1}\psi_\alpha\end{aligned}\tag{10.215}$$

Using Corollary 10.1.i with $l_* + 1$ in the role of $l$ and Corollary 10.2.i with $l+1$ in the role of $l$, in place of the estimates of Lemma 10.20, we deduce that

$$\|\mu \sum_{k=0}^{l} R_{i_{l+1}} \cdots R_{i_{l-k+2}} {}^{(R_{i_{l-k+1}})}Z R_{i_{l-k}} \cdots R_{i_1}\psi_\alpha\|_{L^2(\Sigma_t^{\epsilon_0})}$$
$$\leqslant C_l\delta_0(1+t)^{-2}(1+\log(1+t))(\mathcal{Y}_0+(1+t)\mathcal{A}_{[l]} + (1+\log(1+t))\mathcal{W}_{[l+1]}) \quad (10.216)$$

for $k = 0, \cdots, l$.

Now we have, from Chapter 5, the following hold for any variation $\psi$:

$$\begin{aligned} \|T\psi\|_{L^2(\Sigma_t^{\epsilon_0})} &\leqslant C\sqrt{\mathcal{E}_0[\psi]} \\ \|\sqrt{\mu}(L\psi+\nu\psi)\|_{L^2(\Sigma_t^{\epsilon_0})} &\leqslant C(1+t)^{-1}\sqrt{\mathcal{E}_1'[\psi]} \end{aligned} \quad (10.217)$$

while

$$\|\mu\nu R_j R_{i_l} \cdots R_{i_1}\psi_\alpha\|_{L^2(\Sigma_t^{\epsilon_0})} \leqslant C(1+t)^{-1}(1+\log(1+t))\mathcal{W}_{[l+1]} \quad (10.218)$$

Thus, up to the estimates for the commutators, the first term on the right in (10.187) is bounded by

$$(|\ell| + C\delta_0(1+t)^{-1})\sqrt{\sum_\alpha \mathcal{E}_0[R_j R_{i_l} \cdots R_{i_1}\psi_\alpha]} \quad (10.219)$$

and the second term on the right in (10.187) is bounded by

$$\begin{aligned} &C(1+t)^{-1}(1+\log(1+t))^{1/2}\sqrt{\sum_\alpha \mathcal{E}_1'[R_j R_{i_l} \cdots R_{i_1}\psi_\alpha]} \\ &+C(1+t)^{-1}(1+\log(1+t))\mathcal{W}_{[l+1]} \end{aligned} \quad (10.220)$$

Actually, taking into account the commutator estimates (10.214), (10.216), we have the following proposition.

**Proposition 10.6** Under the assumptions of Proposition 10.5, we have

$${}^{(i_1\cdots i_l)}P_l \leqslant {}^{(i_1\cdots i_l)}P_l^{(0)} + {}^{(i_1\cdots i_l)}P_l^{(1)}$$

where

$${}^{(i_1\cdots i_l)}P_l^{(0)} = |\ell|\sqrt{\sum_{j,\alpha} \mathcal{E}_0[R_j R_{i_l} \cdots R_{i_1}\psi_\alpha]}$$

and

$$\begin{aligned} {}^{(i_1\cdots i_l)}P_l^{(1)} =& C\delta_0(1+t)^{-1}\sqrt{\sum_{j,\alpha} \mathcal{E}_0[R_j R_{i_l} \cdots R_{i_1}\psi_\alpha]} \\ &+ C(1+t)^{-1}(1+\log(1+t))^{1/2}\sqrt{\sum_{j,\alpha} \mathcal{E}_1'[R_j R_{i_l} \cdots R_{i_1}\psi_\alpha]} \\ &+ C(1+t)^{-1}(1+\log(1+t))\mathcal{W}_{[l+1]} \\ &+ C_l\delta_0(1+t)^{-1}(\mathcal{W}_{[l]}^T + (1+\log(1+t))\mathcal{W}_{[l]}^Q \\ &+ (1+t)^{-1}((1+\log(1+t))(\mathcal{Y}_0 + (1+t)\mathcal{A}_{[l]}) + \mathcal{B}_{[l+1]})) \end{aligned}$$

# 11

# Control of the Spatial Derivatives of the First Derivatives of the $x^i$. Assumptions and Estimates in Regard to $\mu$

## 11.1 Estimates for $T\hat{T}^i$

### 11.1.1 Basic Lemmas

The first part of the present chapter is concerned with the derivation of estimates for the spatial derivatives of the $\hat{T}x^i = \hat{T}^i$, of which at least one is a $T$-derivative. By spatial derivatives we mean the derivatives with respect to $T$ and the rotation fields $R_j, j = 1, 2, 3$. Combining with the estimates of Chapter 10 we then obtain estimates for all spatial derivatives of the first derivatives of the $x^i$ with respect to all five commutation fields. The derivation of the estimates of this chapter is based on bootstrap assumptions in regard to $\mu$, in addition to the bootstrap assumptions of Chapter 10 in regard to $\chi$. The estimates are then used to obtain estimates for the spatial derivatives of the deformation tensors of the commutation fields.

We begin by defining certain norms. In the following $\xi$ is an arbitrary $S_{t,u}$ tensorfield defined on $\Sigma_t^{\epsilon_0}$. We first define $L^\infty$ and $L^2$ norms which consider only angular derivatives. Given a non-negative integer $l$ we define

$$\|\xi\|_{\infty,[l],\Sigma_t^{\epsilon_0}} = \max_{0\leqslant n\leqslant l}\max_{i_1\cdots i_n}\|\not{\mathcal{L}}_{R_{i_n}}\cdots\not{\mathcal{L}}_{R_{i_1}}\xi\|_{L^\infty(\Sigma_t^{\epsilon_0})} \tag{11.1}$$

$$\|\xi\|_{2,[l],\Sigma_t^{\epsilon_0}} = \sum_{n=0}^{l}\max_{i_1\cdots i_n}\|\not{\mathcal{L}}_{R_{i_n}}\cdots\not{\mathcal{L}}_{R_{i_1}}\xi\|_{L^2(\Sigma_t^{\epsilon_0})} \tag{11.2}$$

We then define $L^\infty$ and $L^2$ norms which take into consideration all spatial derivatives.

Given a pair of non-negative integers $k, l$ with $k \leqslant l$, we define

$$\|\xi\|_{\infty,[k,l],\Sigma_t^{\epsilon_0}} = \max_{0\leqslant m\leqslant k}\|(\not{\mathcal{L}}_T)^m\xi\|_{\infty,[l-m],\Sigma_t^{\epsilon_0}} \tag{11.3}$$

$$\|\xi\|_{2,[k,l],\Sigma_t^{\epsilon_0}} = \sum_{m=0}^{k}\|(\not{\mathcal{L}}_T)^m\xi\|_{2,[l-m],\Sigma_t^{\epsilon_0}} \tag{11.4}$$

Also, we shall denote

$$\|\xi\|_{\infty,\{l\},\Sigma_t^{\epsilon_0}} = \|\xi\|_{\infty,[l,l],\Sigma_t^{\epsilon_0}}, \quad \|\xi\|_{2,\{l\},\Sigma_t^{\epsilon_0}} = \|\xi\|_{2,[l,l],\Sigma_t^{\epsilon_0}} \tag{11.5}$$

According to the above definitions $\|\xi\|_{\infty,[k,l],\Sigma_t^{\epsilon_0}}$ is the maximum of the quantities:

$$\max_{i_1\cdots i_n} \|\not{\mathcal{L}}_{R_{i_n}} \cdots \not{\mathcal{L}}_{R_{i_1}} (\not{\mathcal{L}}_T)^m \xi\|_{L^\infty(\Sigma_t^{\epsilon_0})} \tag{11.6}$$

over

$$\{(m,n) : 0 \leqslant n \leqslant l-m, 0 \leqslant m \leqslant k\} \tag{11.7}$$

Similarly for $\|\xi\|_{2,[k,l],\Sigma_t^{\epsilon_0}}$.

Also, $\|\xi\|_{\infty,\{l\},\Sigma_t^{\epsilon_0}}$ is the maximum of the quantities (11.6) over

$$\{(m,n) : m,n \geqslant 0, m+n \leqslant l\} \tag{11.8}$$

Similarly for $\|\xi\|_{2,\{l\},\Sigma_t^{\epsilon_0}}$.

We now introduce some bootstrap assumptions. In the following, $C$ is a constant independent of $s$, and $t \in [0,s]$:

$$\mathbf{E}_{m,n} : \max_{\alpha;i_1\cdots i_n} \|R_{i_n} \cdots R_{i_1}(T)^m \psi_\alpha\|_{L^\infty(\Sigma_t^{\epsilon_0})} \leqslant C\delta_0(1+t)^{-1}$$

In particular,

$$\mathbf{E}_{0,0} : \max_\alpha \|\psi_\alpha\|_{L^\infty(\Sigma_t^{\epsilon_0})} \leqslant C\delta_0(1+t)^{-1}$$

We denote by $\mathbf{E}_{\{l\}}$ the conjunction of assumptions $\mathbf{E}_{m,n}$ corresponding to (11.8). The constant $C$ depends on $l$ only. Note that $\mathbf{E}_{0,n}$ coincide with $\not{\mathbf{E}}_n$ and that $\mathbf{E}_{\{l\}}$ contain $\not{\mathbf{E}}_{[l]}$.

Given non-negative integers $m,n$ not both 0, we denote by $\mathcal{W}_{m,n}$ the quantity

$$\mathcal{W}_{m,n} = \max_{\alpha;i_1\cdots i_n} \|R_{i_n} \cdots R_{i_1}(T)^m \psi_\alpha\|_{L^2(\Sigma_t^{\epsilon_0})} \tag{11.9}$$

In particular we denote by $\mathcal{W}_{0,0}$ the quantity

$$\mathcal{W}_{0,0} = \max_\alpha \|\psi_\alpha\|_{L^2(\Sigma_t^{\epsilon_0})} \tag{11.10}$$

We then denote by $\mathcal{W}_{\{l\}}$ the sum of the quantities $\mathcal{W}_{m,n}$ corresponding to (11.8). Note that $\mathcal{W}_{0,n}$ coincide with $\mathcal{W}_n$, and that $\mathcal{W}_{\{l\}}$ dominate $\mathcal{W}_{[l]}$. Moreover, we have

$$\max_\alpha \|\psi_\alpha\|_{2,\{l\},\Sigma_t^{\epsilon_0}} \leqslant \mathcal{W}_{\{l\}} \tag{11.11}$$

The following lemma, which can be proved by a direct calculation, extends Lemma 10.1 to all spatial derivatives.

**Lemma 11.1** Let $G$ be a smooth function of the $(\psi_\alpha, \alpha = 0,1,2,3)$ defined in a neighborhood of $(0,0,0,0)$, and let $G_0$ be the constant:

$$G_0 = G(0,0,0,0)$$

Suppose that the bootstrap assumption $\mathbf{E}_{\{l_*\}}$ holds for some positive integer $l$. Then, if $\delta_0$ is suitably small (depending on $l$), we have

$$\|G - G_0\|_{2,\{l\},\Sigma_t^{\epsilon_0}} \leqslant C\mathcal{W}_{\{l\}}$$

where $C$ is a constant which is independent of $l$.

We continue with a product lemma for the norms (11.3) and (11.4), which shall be applied repeatedly in the sequel.

**Lemma 11.2** **a.** Let $\xi_1, \cdots, \xi_N$ be arbitrary $S_{t,u}$ tensorfields, defined on $\Sigma_t^{\epsilon_0}$ and let us denote by

$$\xi_1 \cdots \xi_N$$

an arbitrary tensor product with contractions. Let also $k, l$ be non-negative integers, $k \leqslant l$. We then have

$$\|\xi_1 \cdots \xi_N\|_{2,[k,l],\Sigma_t^{\epsilon_0}} \leqslant C_l \sum_{i=1}^{N} \Big(\prod_{j \neq i} \|\xi_j\|_{\infty,[k,l_*],\Sigma_t^{\epsilon_0}}\Big) \|\xi_i\|_{2,[k,l],\Sigma_t^{\epsilon_0}}$$

**b.** Let $\xi_1, \cdots, \xi_N$ be as above, and $\vartheta$ another arbitrary $S_{t,u}$ tensorfield defined on $\Sigma_t^{\epsilon_0}$. Let $k, l$ be positive integers, $k \leqslant l$. We then have

$$\begin{aligned}
&\|\xi_1 \cdots \xi_N \cdot \vartheta\|_{2,[k-1,l-1],\Sigma_t^{\epsilon_0}} \\
&\leqslant C_l \|\vartheta\|_{\infty,[k-1,l_*-1],\Sigma_t^{\epsilon_0}} \sum_{i=1}^{N} \Big(\prod_{j \neq i} \|\xi_j\|_{\infty,[k-1,l_*],\Sigma_t^{\epsilon_0}}\Big) \|\xi_i\|_{2,[k-1,l-1],\Sigma_t^{\epsilon_0}} \\
&\quad + C_l \Big(\prod_{i=1}^{N} \|\xi_i\|_{\infty,[k-1,l_*],\Sigma_t^{\epsilon_0}}\Big) \|\vartheta\|_{2,[k-1,l-1],\Sigma_t^{\epsilon_0}}
\end{aligned}$$

*Proof*. It is a direct calculation. □

Part **a** shall be applied to $S_{t,u}$ tensorfields $\xi_1, \cdots, \xi_N$ of order at most 1, and part **b** shall be applied when the $\xi_1, \cdots, \xi_N$ are of order at most 1, but the $S_{t,u}$ tensorfield $\vartheta$ is of order 2.

As in Chapter 10 a primary role was played by the estimates, given by Propositions 10.1 and 10.2, for the angular derivatives of the rectangular components $\hat{T}^i$, or equivalently of the functions $y^i$, a primary role shall be played in the present chapter by the estimates for the spartial derivatives of the $\hat{T}^i$ of which at least one is a $T$-derivative. From (10.193) we have

$$T\hat{T}^i = q_T \cdot \not{d}x^i \tag{11.12}$$

where

$$q_T = (q_T)_\flat \cdot \not{g}^{-1} \tag{11.13}$$

and

$$(q_T)_\flat = -\not{d}\kappa \tag{11.14}$$

The higher order $T$-derivatives of the $\hat{T}^i$ shall be expressed recursively through the following lemma.

**Lemma 11.3** For any non-negative integer $m$ we have

$$(T)^{m+1}\hat{T}^i = p_{T,m}\hat{T}^i + q_{T,m} \cdot \not{d}x^i + \sum_{n=0}^{m-1} r_{T,m}^n \cdot \not{d}(T)^n\hat{T}^i$$

Here the $p_{T,m}$ are the functions and the $q_{T,m}$ are $S_{t,u}$-tangential vectorfields determined by the recursion relations

$$\begin{aligned} p_{T,m} &= Tp_{T,m-1} + \not{d}\kappa \cdot q_{T,m-1} \\ q_{T,m} &= \not{\mathcal{L}}_T q_{T,m-1} + q_T p_{T,m-1} \end{aligned}$$

and the initial conditions

$$p_{T,0} = 0, \quad q_{T,0} = q_T$$

Also, the $r_{T,m}^n$ are $S_{t,u}$-tangential vectorfields determined by the recursion relations:

$$\begin{aligned} r_{T,m}^0 &= \not{\mathcal{L}}_T r_{T,m-1}^0 + \kappa q_{T,m-1} \\ r_{T,m}^n &= \not{\mathcal{L}}_T r_{T,m-1}^n + r_{T,m-1}^{n-1} \quad \text{for} \quad n \in \{1, \cdots, m-2\} \\ r_{T,m}^{m-1} &= r_{T,m-1}^{m-2} \end{aligned}$$

and the initial condition

$$r_{T,0}^0 = 0$$

*Proof.* The proof is a direct calculation. □

Let us now investigate the recursions of the above lemma. First we have the recursion satisfied by

$$\begin{bmatrix} p_{T,m} \\ q_{T,m} \end{bmatrix}$$

Here we considering columns

$$\begin{bmatrix} \phi \\ X \end{bmatrix}$$

where $\phi$ is a function and $X$ is an $S_{t,u}$-tangential vectorfield. We define the operator $\mathbf{A}$ acting on such columns by

$$\mathbf{A} = \not{\mathcal{L}}_T + \mathbf{B} \tag{11.15}$$

where, naturally,

$$\not{\mathcal{L}}_T \begin{bmatrix} \phi \\ X \end{bmatrix} = \begin{bmatrix} T\phi \\ \not{\mathcal{L}}_T X \end{bmatrix}$$

and **B** is the multiplication operator:

$$\mathbf{B} \begin{bmatrix} \phi \\ X \end{bmatrix} = \begin{bmatrix} 0 & \not{d}\kappa \\ q_T & 0 \end{bmatrix} \cdot \begin{bmatrix} \phi \\ X \end{bmatrix} = \begin{bmatrix} \not{d}\kappa \cdot X \\ q_T \phi \end{bmatrix} \tag{11.16}$$

In terms of the operator **A** the recursion in question takes the form, simply,

$$\begin{bmatrix} p_{T,m} \\ q_{T,m} \end{bmatrix} = \mathbf{A} \begin{bmatrix} p_{T,m-1} \\ q_{T,m-1} \end{bmatrix} \tag{11.17}$$

and the solution satisfying the initial condition

$$\begin{bmatrix} p_{T,0} \\ q_{T,0} \end{bmatrix} = \begin{bmatrix} 0 \\ q_T \end{bmatrix} \tag{11.18}$$

is

$$\begin{bmatrix} p_{T,m} \\ q_{T,m} \end{bmatrix} = \mathbf{A}^m \begin{bmatrix} 0 \\ q_T \end{bmatrix} \tag{11.19}$$

We turn to the 2-dimensional recursion satisfied by the $S_{t,u}$-tangential vectorfields $r^n_{T,m}$. The first recursion is

$$r^0_{T,m} = \not{\mathcal{L}}_T r^0_{T,m-1} + \kappa q_{T,m-1} \tag{11.20}$$

by the initial condition

$$r^0_{T,0} = 0 \tag{11.21}$$

we obtain

$$r^0_{T,m} = \sum_{j=0}^{m-1} (\not{\mathcal{L}}_T)^j (\kappa q_{T,m-1-j}) \tag{11.22}$$

Next, we consider the third recursion relation. This is

$$r^{m-1}_{T,m} = r^{m-2}_{T,m-1} \tag{11.23}$$

It follows that

$$r^{m-1}_{T,m} = r^0_{T,1} = \kappa q_T \tag{11.24}$$

Finally, we consider the second recursion relation, which takes the form (setting $k = m - n$)

$$r^{m-k}_{T,m} = \not{\mathcal{L}}_T r^{m-k}_{T,m-1} + r^{m-1-k}_{T,m-1} \tag{11.25}$$

for $k = 2, \cdots, m-1$.

Given any positive integer $k$ let us denote by $\mathcal{N}_k$ the set of positive integers:

$$\mathcal{N}_k = \{m \geqslant k\} \tag{11.26}$$

On $\mathcal{N}_k$ we define the function $x_k$ with values in the space $\mathcal{X}$ of $S_{t,u}$-tangential vectorfields by

$$x_k(m) = r_{T,m}^{m-k} \tag{11.27}$$

Let $\mathbf{L}_k$ be the linear map taking $\mathcal{X}$-valued functions defined on $\mathcal{N}_{k-1}$ into $\mathcal{X}$-valued functions defined on $\mathcal{N}_k$ by

$$(\mathbf{L}_k f)(m) = \not{\mathcal{L}}_T(f(m-1)), \quad \forall m \in \mathcal{N}_k \tag{11.28}$$

Then (11.25) takes the form

$$x_k(m) - x_k(m-1) = (\mathbf{L}_k x_{k-1})(m), \quad \forall m \in \mathcal{N}_{k+1}, k \geqslant 2 \tag{11.29}$$

Setting

$$y_k = \mathbf{L}_k x_{k-1} \tag{11.30}$$

we can write (11.29) as

$$x_k(m) - x_k(m-1) = y_k(m), \quad \forall m \in \mathcal{N}_{k+1}, k \geqslant 2 \tag{11.31}$$

Let us sum (11.31) over $m \in \{k+1, \cdots, j\}$. We obtain

$$\sum_{m=k+1}^{j} (x_k(m) - x_k(m-1)) = x_k(j) - x_k(k)$$

i.e.

$$x_k(m) = x_k(k) + \sum_{j=k+1}^{m} y_k(j), \quad \forall m \in \mathcal{N}_k, k \geqslant 2 \tag{11.32}$$

Now, from (11.27) with $m = k$,

$$x_k(k) = c_k = r_{T,k}^0 \tag{11.33}$$

is given by (11.22), while from (11.30) and (11.28) the sum on the right is

$$\sum_{j=k+1}^{m} y_k(j) = \sum_{j=k+1}^{m} (\mathbf{L}_k x_{k-1})(j) = \sum_{j=k+1}^{m} \not{\mathcal{L}}_T(x_{k-1}(j-1)) \tag{11.34}$$

Let us define a linear operator $\mathbf{M}_k$ acting on $\mathcal{X}$-valued functions $f$ defined on $\mathcal{N}_k$ by

$$(\mathbf{M}_k f)(m) = \sum_{j=k}^{m-1} (\not{L}_T f)(j) \tag{11.35}$$

In terms of the operator $\mathbf{M}_k$ and the constants $c_k$, (11.32) becomes the following recursion in $k$ for the $\mathcal{X}$-valued functions $x_k$:

$$x_k = c_k + \mathbf{M}_k x_{k-1}, \quad k \geqslant 2 \tag{11.36}$$

Here we consider $c_k$ as constant functions defined on $\mathcal{N}_k$. Now, by (11.27) and (11.24) for $k = 1$ we have

$$x_1(m) = r_{T,m}^{m-1} = \kappa q_T, \quad \forall m \in \mathcal{N}_1 \tag{11.37}$$

hence, by (11.33) for $k = 1$,

$$x_1 = c_1 \tag{11.38}$$

Thus (11.36) holds also for $k = 1$ if we set $x_0 = 0$:

$$\begin{aligned} x_k &= c_k + \mathbf{M}_k x_{k-1}, \quad k \geqslant 1 \\ x_0 &= 0 \end{aligned} \tag{11.39}$$

To this recursion Proposition 8.2 applies, yielding

$$x_k = \sum_{i=0}^{k-1} \mathbf{M}_k \cdots \mathbf{M}_{k-i+1} c_{k-i} \tag{11.40}$$

Now by (11.35),

$$(\mathbf{M}_k c)(m) = (m-k)\not{L}_T c, \quad \forall m \in \mathcal{N}_k \tag{11.41}$$

Let us define, for each non-negative integer $l$, the function $N_l$ on the set of non-negative integers, with the value in the same set, recursively by

$$\begin{aligned} N_0(n) &= 1 \quad \text{for all non-negative integers} \quad n \\ N_l(n) &= \sum_{m=1}^{n} N_{l-1}(m) \end{aligned} \tag{11.42}$$

for all non-negative integers $n$ and all positive integers $l$.

In particular, we have

$$N_l(0) = 0 \quad \text{for all positive integers} \quad l \tag{11.43}$$

In fact, it is readily seen that

$$N_l(n) = \frac{n \cdots (n+l-1)}{l!} \quad \begin{aligned} &\text{for all non-negative integers} \quad n \\ &\text{and all positive integers} \quad l \end{aligned} \tag{11.44}$$

We can then show, by induction on $j$ , that for all non-negative integers $j$ we have

$$(\mathbf{M}_{k-i+j}\cdots\mathbf{M}_{k-i+1}c_{k-i})(m) = N_j(m-k+i-j)(\not{\mathcal{L}}_T)^j c_{k-i} \tag{11.45}$$

Setting $j = i$ in the above we obtain

$$(\mathbf{M}_k\cdots\mathbf{M}_{k-i+1}c_{k-i})(m) = N_i(m-k)(\not{\mathcal{L}}_T)^i c_{k-i}, \quad \forall m \in \mathcal{N}_k \tag{11.46}$$

Substituting in (11.40) then yields

$$x_k(m) = \sum_{i=0}^{k-1} N_i(m-k)(\not{\mathcal{L}}_T)^i c_{k-i}, \quad \forall m \in \mathcal{N}_k, \forall k \geqslant 1 \tag{11.47}$$

We summarize the above results in the following lemma.

**Lemma 11.4** Let $\mathbf{A}$ be the operator

$$\mathbf{A} = \not{\mathcal{L}}_T + \mathbf{B} \quad \text{acting on columns} \quad \begin{bmatrix} \phi \\ X \end{bmatrix}$$

where $\phi$ is a function and $X$ an $S_{t,u}$-tangential vectorfield, $\mathbf{B}$ being the multiplication operator:

$$\mathbf{B} = \begin{bmatrix} 0 & \not{d}\kappa \\ q_T & 0 \end{bmatrix}$$

Then functions $p_{T,m}$ and $S_{t,u}$-tangential vectorfields $q_{T,m}$ are given by

$$\begin{bmatrix} p_{T,m} \\ q_{T,m} \end{bmatrix} = \mathbf{A}^m \begin{bmatrix} p_T \\ q_T \end{bmatrix}$$

Moreover, the $S_{t,u}$-tangential vectorfields $r^n_{T,m-1}$, $n \in \{0, \cdots, m-1\}$, are given by

$$r^n_{T,m} = \sum_{i=0}^{m-1-n} N_i(n) \sum_{j=i}^{m-1-n} (\not{\mathcal{L}}_T)^j (\kappa q_{T,m-1-n-j})$$

where $N_i(n)$ are the non-negative integers:

$$N_i(n) = \begin{cases} 1, & \text{if} \quad i = 0 \\ \dfrac{n\cdots(n+i-1)}{i!}, & \text{if} \quad i \geqslant 1 \end{cases}$$

The $S_{t,u}$ 1-form $(q_T)_\flat$ can be directly estimated, however to estimate the $S_{t,u}$-tangential vectorfield $q_T$, which enters all the above formulas, we must estimate the $T$-derivatives of $\not{g}^{-1}$. Now,

$$\not{\mathcal{L}}_T \not{g}^{-1} = -\not{g}^{-1} \cdot {}^{(T)}\not{\pi} \cdot \not{g}^{-1}, \quad {}^{(T)}\not{\pi} = \not{\mathcal{L}}_T \not{g} \tag{11.48}$$

Thus estimating $(\not{\mathcal{L}}_T)^{m+1}\not{g}^{-1}$ reduces to estimating $(\not{\mathcal{L}}_T)^{m\,(T)}\not{\pi}$. Now from (3.27) we have

$$ {}^{(T)}\not{\pi} = 2\kappa\theta = -2\alpha^{-1}\kappa\chi + 2\kappa\not{k} \tag{11.49} $$

Let us define

$$ \chi' = \chi - \frac{\not{g}}{1-u+t} \tag{11.50} $$

and

$$ {}^{(T)}\not{\pi}' = -2\alpha^{-1}\kappa\chi' + 2\kappa\not{k} \tag{11.51} $$

We then have

$$ {}^{(T)}\not{\pi} = {}^{(T)}\not{\pi}' + \lambda(1-u+t)^{-1}\not{g} \tag{11.52} $$

where $\lambda$ is the function

$$ \lambda = -2(\alpha^{-1}\kappa) \tag{11.53} $$

We shall be able to estimate in a more direct manner the $T$-derivatives of ${}^{(T)}\not{\pi}'$. From these estimates for the $T$-derivatives of ${}^{(T)}\not{\pi}$, we shall use the auxiliary symmetric 2-covariant $S_{t,u}$ tensorfields ${}^{(m;T)}\not{\pi}$ defined below. Their definition requires the polynomials $p_m(x)$, which we shall presently define.

For each non-negative integer $m$ we define

$$ p_m(x) = \begin{cases} 1, & \text{for} \quad m = 0 \\ (m+x)\cdots(1+x), & \text{for} \quad m \geqslant 1 \end{cases} \tag{11.54} $$

We can see inductively, that

$$ m!(1 + x\sum_{n=0}^{m-1}\frac{p_n(x)}{(n+1)!}) = p_m(x) \tag{11.55} $$

We define, for any non-negative integer $m$,

$$ {}^{(m;T)}\not{\pi} = (\not{\mathcal{L}}_T)^{m\,(T)}\not{\pi} - \lambda p_m(\lambda)(1-u+t)^{-m-1}\not{g} \tag{11.56} $$

Note that

$$ {}^{(0;T)}\not{\pi} = {}^{(T)}\not{\pi}' \tag{11.57} $$

Considering the fact, from (11.53), that $\lambda = -2$ in the constant state, we define

$$ \lambda' = \lambda + 2 \tag{11.58} $$

**Lemma 11.5** Let $l$ be a positive integer and $m$ a non-negative integer, $m \leqslant l$. Suppose that

$$\max_i \|{}^{(R_i)}\not{\pi}\|_{\infty,[l-1],\Sigma_t^{\epsilon_0}} \leqslant C_l\delta_0(1+t)^{-1}(1+\log(1+t))$$

and

$$\|\lambda'\|_{\infty,[m,l],\Sigma_t^{\epsilon_0}} \leqslant C_l\delta_0(1+\log(1+t))$$

Then if $\delta_0$ is suitably small (depending on $l$) we have

$$\max_{0\leqslant k\leqslant m} \|{}^{(k;T)}\not{\pi}\|_{\infty,[l-k],\Sigma_t^{\epsilon_0}} \leqslant C_l\|{}^{(T)}\not{\pi}'\|_{\infty,[m,l],\Sigma_t^{\epsilon_0}} + C_l\delta_0(1+t)^{-1}(1+\log(1+t))$$

*Proof*. We apply $(\not{\mathcal{L}}_T)^k$ to (11.52), for $k \in \{0,\cdots,m\}$. A direct calculation gives

$$\begin{aligned}{}^{(k;T)}\not{\pi} = (\not{\mathcal{L}}_T)^{k\,(T)}\not{\pi}' &+ \sum_{k_1+k_3=k,k_3\geqslant 1}\frac{k!}{k_3!}(1-u+t)^{-k_1-1}\lambda^{(k_3-1;T)}\not{\pi} \\ &+ \sum_{k_1+k_2=k,k_2\geqslant 1}\frac{k!}{k_2!}(1-u+t)^{-k_1-1}((T)^{k_2}\lambda)\not{g} \\ &+ \sum_{k_1+k_2+k_3=k,k_2,k_3\geqslant 1}\frac{k!}{k_2!k_3!}(1-u+t)^{-k_1-k_3-1}((T)^{k_2}\lambda)\lambda p_{k_3-1}(\lambda)\not{g} \\ &+ \sum_{k_1+k_2+k_3=k,k_2,k_3\geqslant 1}\frac{k!}{k_2!k_3!}(1-u+t)^{-k_1-1}((T)^{k_2}\lambda)^{(k_3-1;T)}\not{\pi}\end{aligned} \tag{11.59}$$

We apply $\not{\mathcal{L}}_{R_{i_n}}\cdots\not{\mathcal{L}}_{R_{i_1}}$ to (11.59) with $n \in \{0,\cdots,l-k\}$. The first term on the right is bounded by

$$\|(\not{\mathcal{L}}_T)^{k\,(T)}\not{\pi}'\|_{\infty,[l-k],\Sigma_t^{\epsilon_0}} \tag{11.60}$$

We decompose the first sum on the right according to

$$\lambda = -2 + \lambda' \tag{11.61}$$

Setting

$$k' = k_3 - 1 \tag{11.62}$$

the contribution of the term $-2$ is

$$2\sum_{k'=0}^{k-1}\frac{k!}{(k'+1)!}(1-u+t)^{k'-k}\|{}^{(k';T)}\not{\pi}\|_{\infty,[l-k],\Sigma_t^{\epsilon_0}} \tag{11.63}$$

While the term $\lambda'$ is bounded by

$$C_l\delta_0(1+\log(1+t))\sum_{k'=0}^{k-1}\frac{k!}{(k'+1)!}(1-u+t)^{k'-k}\|{}^{(k';T)}\not{\pi}\|_{\infty,[l-k],\Sigma_t^{\epsilon_0}} \tag{11.64}$$

The last sum is bounded by

$$C_l\delta_0(1+\log(1+t))$$
$$\times\sum_{k'=0}^{k-2}\sum_{j=0}^{k-2-k'}\frac{k!}{(j+1)!(k'+1)!}(1-u+t)^{k'-k+1+j}\|^{(k';T)}\not\chi\|_{\infty,[l-k],\Sigma_t^{\epsilon_0}} \tag{11.65}$$

where we have set

$$k_2 = 1+j \tag{11.66}$$

Next, we consider the second sum, which is

$$\sum_{j=0}^{k-1}\frac{k!}{(j+1)!}(1-u+t)^{j-k}((T)^{j+1}\lambda')\not g \tag{11.67}$$

By the first assumption,

$$\|((T)^{j+1}\lambda')\not g\|_{\infty,[l-k],\Sigma_t^{\epsilon_0}} \leqslant C\|(T)^{j+1}\lambda'\|_{\infty,[l-k],\Sigma_t^{\epsilon_0}} \tag{11.68}$$

for $k \geqslant 1$, provided that $\delta_0$ is suitably small (depending on $l$). Then by the second assumption we know that the second sum is bounded by

$$C_l\delta_0(1+t)^{-1}(1+\log(1+t)) \tag{11.69}$$

Finally, we have to deal with the third sum. We express

$$\lambda p_{k'}(\lambda) = -2p_{k'}(-2) + q_{k'+1}(\lambda') \tag{11.70}$$

where $q_{k'+1}(\lambda')$ is a polynomial of degree $k'+1$ in $\lambda'$ with vanishing constant term. By the second assumption of the lemma, the constant term is bounded by

$$C_l\delta_0(1+t)^{-2}(1+\log(1+t)) \tag{11.71}$$

while the contribution of the polynomial $q_{k'+1}(\lambda')$ is bounded by

$$C_l\delta_0(1+\log(1+t))\sum_{k'=0}^{k-2}\Big(\sum_{i=1}^{k'+1}(\delta_0(1+\log(1+t)))^i\Big)(1+t)^{-2-k'} \tag{11.72}$$

A simple calculation yields that this is bounded by

$$C_l\delta_0^2(1+t)^{-2}(1+\log(1+t))^2 \tag{11.73}$$

provided that $\delta_0$ is suitably small.

Combining the results (11.60), (11.63), (11.64), (11.65), (11.69), (11.71) and (11.73), we conclude that

$$\|^{(k;T)}\not\chi\|_{\infty,[l-k],\Sigma_t^{\epsilon_0}} \leqslant \|(\not\mathcal{L}_T)^{k(T)}\not\chi'\|_{\infty,[l-k],\Sigma_t^{\epsilon_0}} \tag{11.74}$$
$$+ C_l(1+t)^{-1}(1+\log(1+t))\sum_{k'=0}^{k-1}\|^{(k';T)}\not\chi\|_{\infty,[l-k],\Sigma_t^{\epsilon_0}}$$
$$+ C_l\delta_0(1+t)^{-1}(1+\log(1+t))$$

Since

$$\sum_{k'=0}^{k-1} \|{}^{(k';T)}\not{\chi}\|_{\infty,[l-k],\Sigma_t^{\epsilon_0}} \leqslant \sum_{k'=0}^{k-1} \|{}^{(k';T)}\not{\chi}\|_{\infty,[l-1-k'],\Sigma_t^{\epsilon_0}} \leqslant \sum_{k'=0}^{k-1} \|{}^{(k';T)}\not{\chi}\|_{\infty,[l-k'],\Sigma_t^{\epsilon_0}} \tag{11.75}$$

(11.74) reduce to

$$x_k \leqslant c_k + a \sum_{k'=0}^{k-1} x_{k'}, \quad \text{for} \quad k = 0, \cdots, m \tag{11.76}$$

where

$$\begin{aligned} x_k &= \|{}^{(k;T)}\not{\chi}\|_{\infty,[l-k],\Sigma_t^{\epsilon_0}} \\ c_k &= \|(\not{\mathcal{L}}_T)^{k(T)}\not{\chi}'\|_{\infty,[l-k],\Sigma_t^{\epsilon_0}} + C_l\delta_0(1+t)^{-1}(1+\log(1+t)) \\ a &= C_l(1+t)^{-1}(1+\log(1+t)) \end{aligned} \tag{11.77}$$

It follows by direct calculation that

$$x_k \leqslant c_k + a \sum_{k'=0}^{k-1} (1+a)^{k-1-k'} c_{k'}, \quad \text{for} \quad k = 0, \cdots, m \tag{11.78}$$

hence

$$\max_{0 \leqslant k \leqslant m} x_k \leqslant \max_{0 \leqslant k \leqslant m} c_k (1 + a \sum_{k'=0}^{m-1} (1+a)^{m-1-k'}) = (1+a)^m \max_{0 \leqslant k \leqslant m} c_k \tag{11.79}$$

Substituting (11.77) and recalling that $(1+t)^{-1}(1+\log(1+t)) \leqslant 1$ we obtain

$$\max_{0 \leqslant k \leqslant m} \|{}^{(k;T)}\not{\chi}\|_{\infty,[l-k],\Sigma_t^{\epsilon_0}} \leqslant C_l(\|{}^{(T)}\not{\chi}'\|_{\infty,[m,l],\Sigma_t^{\epsilon_0}} + C_l\delta_0(1+t)^{-1}(1+\log(1+t)))$$

□

**Lemma 11.6** Let $l$ be a positive integer and $m$ a non-negative integer, $m \leqslant l$. Suppose that

$$\begin{aligned} \max_i \|{}^{(R_i)}\not{\chi}\|_{\infty,[l_*-1],\Sigma_t^{\epsilon_0}} &\leqslant C_l\delta_0(1+t)^{-1}(1+\log(1+t)) \\ \|\lambda'\|_{\infty,[m,l_*],\Sigma_t^{\epsilon_0}} &\leqslant C_l\delta_0(1+\log(1+t)) \end{aligned}$$

and

$$\|{}^{(T)}\not{\chi}'\|_{\infty,[m-1,l_*-1],\Sigma_t^{\epsilon_0}} \leqslant C_l\delta_0(1+t)^{-1}(1+\log(1+t))$$

Then, if $\delta_0$ is suitably small (depending on $l$) we have

$$\begin{aligned} \sum_{k=0}^{m} \|{}^{(k;T)}\not{\chi}\|_{2,[l-k],\Sigma_t^{\epsilon_0}} \leqslant{}& C_l\|{}^{(T)}\not{\chi}'\|_{2,[m,l],\Sigma_t^{\epsilon_0}} + C_l(1+t)^{-1}(\|\lambda'\|_{2,[m,l],\Sigma_t^{\epsilon_0}} \\ &+ \delta_0(1+\log(1+t)) \max_i \|{}^{(R_i)}\not{\chi}\|_{2,[l-2],\Sigma_t^{\epsilon_0}}) \end{aligned}$$

*Proof*. We use (11.59). The proof is quite similar to that of Lemma 11.5. As in Chapter 10, when we estimate a product in $L^2$, we put the lower order factors in $L^\infty$, and the highest order factor in $L^2$. □

We use Lemmas 11.5 and 11.6 to derive $L^\infty$ and $L^2$ estimates for the $T$-derivatives of ${}^{(T)}\not\pi$ in terms of estimates for the $T$-derivatives of ${}^{(T)}\not\pi'$.

**Lemma 11.7** Under the assumptions of Lemma 11.5 we have

$$\begin{aligned}
&\|{}^{(T)}\not\pi + 2(1-u+t)^{-1}\not g\|_{\infty,[l],\Sigma_t^{\epsilon_0}}\\
&\leqslant C_l\|{}^{(T)}\not\pi'\|_{\infty,[0,l],\Sigma_t^{\epsilon_0}} + C_l\delta_0(1+t)^{-1}(1+\log(1+t))\\
&\|\not{\mathcal{L}}_T{}^{(T)}\not\pi - 2(1-u+t)^{-2}\not g\|_{\infty,[l-1],\Sigma_t^{\epsilon_0}}\\
&\leqslant C_l\|{}^{(T)}\not\pi'\|_{\infty,[1,l],\Sigma_t^{\epsilon_0}} + C_l\delta_0(1+t)^{-1}(1+\log(1+t))
\end{aligned}$$

and

$$\begin{aligned}
&\max_{2\leqslant k\leqslant m}\|(\not{\mathcal{L}}_T)^{k\,(T)}\not\pi\|_{\infty,[l-k],\Sigma_t^{\epsilon_0}}\\
&\leqslant C_l\|{}^{(T)}\not\pi'\|_{\infty,[m,l],\Sigma_t^{\epsilon_0}} + C_l\delta_0(1+t)^{-1}(1+\log(1+t))
\end{aligned}$$

*Proof*. We consider formula (11.56) with $m$ replaced by $k\in\{0,\cdots,m\}$:

$$(\not{\mathcal{L}}_T)^{k\,(T)}\not\pi = {}^{(k;T)}\not\pi + \lambda p_k(\lambda)(1-u+t)^{-k-1}\not g \tag{11.80}$$

We express $\lambda p_k(\lambda)$ as in (11.70) and we bring the contribution of the constant term $-2p_k(-2)$ on the left. We obtain

$$\begin{aligned}
{}^{(T)}\not\pi + 2(1-u+t)^{-1}\not g &= {}^{(0;T)}\not\pi + q_1(\lambda')(1-u+t)^{-1}\not g\\
\not{\mathcal{L}}_T{}^{(T)}\not\pi - 2(1-u+t)^{-2}\not g &= {}^{(1;T)}\not\pi + q_2(\lambda')(1-u+t)^{-2}\not g\\
(\not{\mathcal{L}}_T)^{k\,(T)}\not\pi &= {}^{(k;T)}\not\pi + q_{k+1}(\lambda')(1-u+t)^{-k-1}\not g
\end{aligned} \tag{11.81}$$

for $k\geqslant 2$.

We first take the $\|\quad\|_{\infty,[l],\Sigma_t^{\epsilon_0}}$-norm of each side of the first of (11.81). Since

$$\begin{aligned}
&\|q_1(\lambda')\not g\|_{\infty,[l],\Sigma_t^{\epsilon_0}}\\
&\leqslant C\|\lambda'\|_{\infty,[l],\Sigma_t^{\epsilon_0}} + C_l\|\lambda'\|_{\infty,[l-1],\Sigma_t^{\epsilon_0}}\max_j\|{}^{(R_i)}\not\pi\|_{\infty,[l-1],\Sigma_t^{\epsilon_0}}\\
&\leqslant C_l\delta_0(1+\log(1+t))
\end{aligned} \tag{11.82}$$

(provided that $\delta_0$ is suitably small), the first statement of the lemma results by substituting the estimate for $\|{}^{(0;T)}\not\pi\|_{\infty,[l],\Sigma_t^{\epsilon_0}}$ of Lemma 11.5 with $m=0$. The next two cases in (11.81) can be dealt in a similar way, then the lemma is proved. □

**Lemma 11.8** Let the assumptions of Lemma 11.6 hold. Let us define

$$\mathcal{T}_{[0,l]} = \|{}^{(T)}\not\pi + 2(1-u+t)^{-1}\not g\|_{2,[l],\Sigma_t^{\epsilon_0}}$$

$$\mathcal{T}_{[1,l]} = \|{}^{(T)}\not\pi + 2(1-u+t)^{-1}\not g\|_{2,[l],\Sigma_t^{\epsilon_0}} + \|\not\mathcal{L}_T{}^{(T)}\not\pi - 2(1-u+t)^{-2}\not g\|_{2,[l-1],\Sigma_t^{\epsilon_0}}$$

and for $m \geqslant 2$,

$$\mathcal{T}_{[m,l]} = \|{}^{(T)}\not\pi + 2(1-u+t)^{-1}\not g\|_{2,[l],\Sigma_t^{\epsilon_0}} + \|\not\mathcal{L}_T{}^{(T)}\not\pi - 2(1-u+t)^{-2}\not g\|_{2,[l-1],\Sigma_t^{\epsilon_0}} + \sum_{k=2}^{m} \|(\not\mathcal{L}_T)^{k\,(T)}\not\pi\|_{2,[l-k],\Sigma_t^{\epsilon_0}}$$

We then have

$$\mathcal{T}_{[m,l]} \leqslant C_l \|{}^{(T)}\not\pi'\|_{2,[m,l],\Sigma_t^{\epsilon_0}} + C_l(1+t)^{-1}(\|\lambda'\|_{2,[m,l],\Sigma_t^{\epsilon_0}} + \delta_0(1+\log(1+t)) \max_i \|{}^{(R_i)}\not\pi\|_{2,[l-1],\Sigma_t^{\epsilon_0}})$$

*Proof*. Again, the proof is similar to that of Lemma 11.7 and Lemma 11.6. □

In establishing the primary propositions of the present chapter, which concern estimates for the spatial derivatives of the $\hat{T}^i$ of which at least one is a $T$-derivative, we shall have to estimate, for given $S_{t,u}$ 1-forms $\xi$, the commutators

$$^{(i_1\cdots i_n)}c_{m,n}[\xi] = \not\mathcal{L}_{R_{i_n}} \cdots \not\mathcal{L}_{R_{i_1}} (\not\mathcal{L}_T)^m \not D\xi - \not D \not\mathcal{L}_{R_{i_n}} \cdots \not\mathcal{L}_{R_{i_1}} (\not\mathcal{L}_T)^m \xi \tag{11.83}$$

**Lemma 11.9** The commutators $^{(i_1\cdots i_n)}c_{m,n}[\xi]$ are given by

$$^{(i_1\cdots i_n)}c_{m,n}[\xi] = \not\mathcal{L}_{R_{i_n}} \cdots \not\mathcal{L}_{R_{i_1}} c_{m,0}[\xi] - \sum_{j=0}^{n-1} \not\mathcal{L}_{R_{i_n}} \cdots \not\mathcal{L}_{R_{i_{n-j+1}}} ({}^{(R_{i_{n-j}})}\not\pi_1 \cdot \not\mathcal{L}_{R_{i_{n-j-1}}} \cdots \not\mathcal{L}_{R_{i_1}} (\not\mathcal{L}_T)^m \xi)$$

where $c_{m,0}[\xi]$ are given by

$$c_{m,0}[\xi] = -\sum_{i=0}^{m-1} (\not\mathcal{L}_T)^i ({}^{(T)}\not\pi_1 \cdot (\not\mathcal{L}_T)^{m-i-1}\xi)$$

Here ${}^{(T)}\not\pi_1$ is defined by (9.151) and (9.152).

*Proof*. The proof is a direct calculation which relies on Proposition 8.2. □

We must now obtain estimates for ${}^{(T)}\not\pi_1$. In terms of the operator $\check{\not D}$ defined by (10.147), we can express

$${}^{(T)}\not\pi_1 = {}^{(T)}(\not\pi_1)_b \cdot \not g^{-1}, \quad {}^{(T)}(\not\pi_1)_b = \check{\not D}\,{}^{(T)}\not\pi \tag{11.84}$$

We shall need the following analogue of Lemma 10.9, which can be also proved in a straightforward manner.

**Lemma 11.10** Let $\vartheta$ be an arbitrary symmetric 2-covariant $S_{t,u}$ tensorfield. The following commutation formula holds for any non-negative integer $m$:

$$(\not{\mathcal{L}}_T)^m \check{\not{D}}\vartheta - \check{\not{D}}(\not{\mathcal{L}}_T)^m\vartheta = -\sum_{k=0}^{m-1}(\not{\mathcal{L}}_T)^k({}^{(T)}\not{\pi}_1 \cdot (\not{\mathcal{L}}_T)^{m-k-1}\vartheta)$$

We proceed to

**Lemma 11.11** Let $l$ be a positive integer and $m$ a non-negative integer $m \leqslant l-1$. Let also hypothesis **H2**$'$ hold. Suppose that

$$\begin{aligned}
\|\lambda'\|_{\infty,[m,l],\Sigma_t^{\epsilon_0}} &\leqslant C_l\delta_0(1+\log(1+t)) \\
\|{}^{(T)}\not{\pi}'\|_{\infty,[m,l],\Sigma_t^{\epsilon_0}} &\leqslant C_l\delta_0(1+t)^{-1}(1+\log(1+t)) \\
\max_i \|{}^{(R_i)}\not{\pi}\|_{\infty,[l-1],\Sigma_t^{\epsilon_0}} &\leqslant C_l\delta_0(1+t)^{-1}(1+\log(1+t))
\end{aligned}$$

and, if $l \geqslant 2$,

$$\max_i \|{}^{(R_i)}\not{\pi}_1\|_{\infty,[l-2],\Sigma_t^{\epsilon_0}} \leqslant C_l\delta_0(1+t)^{-2}(1+\log(1+t))$$

Then, if $\delta_0$ is suitably small (depending on $l$) we have

$$\|{}^{(T)}\not{\pi}_1\|_{\infty,[m,l-1],\Sigma_t^{\epsilon_0}} \leqslant C_l\delta_0(1+t)^{-2}(1+\log(1+t))$$

*Proof*. To prove the lemma, we just need to estimate

$$\|(\not{\mathcal{L}}_T)^{k}\,{}^{(T)}\not{\pi}_1\|_{\infty,[l-1-k],\Sigma_t^{\epsilon_0}} \tag{11.85}$$

To do this, we proceed inductively on $k$. Since under the assumptions of the present lemma, Lemma 11.7 holds, substituting the bound for

$$\|{}^{(T)}\not{\pi}'\|_{\infty,[m,l],\Sigma_t^{\epsilon_0}}$$

yields

$$\begin{aligned}
\|{}^{(T)}\not{\pi} + 2(1-u+t)^{-1}\not{g}\|_{\infty,[l],\Sigma_t^{\epsilon_0}} &\leqslant C_l\delta_0(1+t)^{-1}(1+\log(1+t)) \\
\|\not{\mathcal{L}}_T{}^{(T)}\not{\pi} - 2(1-u+t)^{-2}\not{g}\|_{\infty,[l-1],\Sigma_t^{\epsilon_0}} &\leqslant C_l\delta_0(1+t)^{-1}(1+\log(1+t)) \\
\max_{2\leqslant k\leqslant m}\|(\not{\mathcal{L}}_T)^{k}\,{}^{(T)}\not{\pi}\|_{\infty,[l-k],\Sigma_t^{\epsilon_0}} &\leqslant C_l\delta_0(1+t)^{-1}(1+\log(1+t))
\end{aligned} \tag{11.86}$$

We can write

$${}^{(T)}(\not{\pi}_1)_\flat = \check{\not{D}}({}^{(T)}\not{\pi} + 2(1-u+t)^{-1}\not{g}) \tag{11.87}$$

Then by Lemma 10.9 and a direct calculation, we obtain the conclusion when $k = 0$.

Next, we apply Lemma 11.9 to obtain

$$(\not{\mathcal{L}}_T)^{k(T)}(\not\pi_1)_b = \check{\not{D}}(\not{\mathcal{L}}_T)^{k(T)}\not\pi + d'_k \tag{11.88}$$

The estimate for $k = 1$ is also straightforward, we just use Lemma 11.9 to obtain an estimate for $d'_1$.

Then we assume, for $k \in \{2, \cdots, m\}$, that

$$\|(\not{\mathcal{L}}_T)^{k'(T)}\not\pi_1\|_{\infty,[l-1-k'],\Sigma_t^{\epsilon_0}} \leqslant C_l\delta_0(1+t)^{-2}(1+\log(1+t)) \tag{11.89}$$

holds for all $k' \leqslant k-1$.

To obtain the estimates in the case $k' = k$, we again use Lemma 11.9, and then use the induction hypothesis to estimate $d'_k$. Then the lemma follows. □

To proceed with $L^2$ estimates for ${}^{(T)}\not\pi_1$, we shall need the following lemma. This lemma shall also be used in the sequel.

**Lemma 11.12** Let $k, l$ be positive integers $k \leqslant l$ and let

$$\|\lambda'\|_{\infty,[k-1,l_*-1],\Sigma_t^{\epsilon_0}} \leqslant C_l\delta_0(1+\log(1+t))$$
$$\max_i \|{}^{(R_i)}\not\pi\|_{\infty,[l_*-1],\Sigma_t^{\epsilon_0}} \leqslant C_l\delta_0(1+t)^{-1}(1+\log(1+t))$$
$$\|{}^{(T)}\not\pi'\|_{\infty,[k-1,l_*-1],\Sigma_t^{\epsilon_0}} \leqslant C_l\delta_0(1+t)^{-1}(1+\log(1+t))$$

Then provided that $\delta_0$ is suitably small (depending on $l$), we have, for any $S_{t,u}$ tensorfield $\xi$,

$$\|\xi\cdot\not g\|_{2,[k-1,l-1],\Sigma_t^{\epsilon_0}}, \|\xi\cdot\not g^{-1}\|_{2,[k-1,l-1],\Sigma_t^{\epsilon_0}}$$
$$\leqslant C_l(\|\xi\|_{2,[k-1,l-1],\Sigma_t^{\epsilon_0}} + \|\xi\|_{\infty,[k-1,l_*-1],\Sigma_t^{\epsilon_0}}(\max_i \|{}^{(R_i)}\not\pi\|_{2,[l-2],\Sigma_t^{\epsilon_0}} + \mathcal{T}_{[k-2,l-2]}))$$

where the $\mathcal{T}_{[k,l]}$ are the quantities defined in the statement of Lemma 11.8 (the term $\mathcal{T}_{[k-2,l-2]}$ is present only for $k \geqslant 2$).

*Proof*. Under the assumptions of the present lemma, the conclusions of Lemma 11.7 hold with $(k-1, l_*-1)$ in the role of $(m, l)$. Then substituting in these the bound for

$$\|{}^{(T)}\not\pi'\|_{\infty,[k-1,l_*-1],\Sigma_t^{\epsilon_0}}$$

yields

$$\|{}^{(T)}\not\pi + 2(1-u+t)^{-1}\not g\|_{\infty,[l_*-1],\Sigma_t^{\epsilon_0}} \leqslant C_l\delta_0(1+t)^{-1}(1+\log(1+t))$$
$$\|\not{\mathcal{L}}_T{}^{(T)}\not\pi - 2(1-u+t)^{-2}\not g\|_{\infty,[l_*-2],\Sigma_t^{\epsilon_0}} \leqslant C_l\delta_0(1+t)^{-1}(1+\log(1+t)) \tag{11.90}$$
$$\max_{2\leqslant i\leqslant k-1} \|(\not{\mathcal{L}}_T)^{i(T)}\not\pi\|_{\infty,[l_*-1-i],\Sigma_t^{\epsilon_0}} \leqslant C_l\delta_0(1+t)^{-1}(1+\log(1+t))$$

We shall estimate

$$\|(\not{\mathcal{L}}_T)^{j-1}(\xi\cdot\not g)\|_{2,[l-j],\Sigma_t^{\epsilon_0}} \quad \text{for} \quad j = 1, \cdots, k \tag{11.91}$$

The case $j = 1$ is easy:

$$\|\xi \cdot \not{g}\|_{2,[l-1],\Sigma_t^{\epsilon_0}} \leqslant C_l(\|\xi\|_{2,[l-1],\Sigma_t^{\epsilon_0}} + \|\xi\|_{\infty,[l_*-1],\Sigma_t^{\epsilon_0}} \max_i \|{}^{(R_i)}\not{\pi}\|_{2,[l-2],\Sigma_t^{\epsilon_0}}) \tag{11.92}$$

For $j \geqslant 2$ we express

$$\begin{aligned}(\not{\mathcal{L}}_T)^{j-1}(\xi \cdot \not{g}) =& \xi_{j-1} \cdot \not{g} + (j-1)((\not{\mathcal{L}}_T)^{j-2}\xi) \cdot ({}^{(T)}\not{\pi} + 2(1-u+t)^{-1}\not{g}) \\ &+ \frac{(j-1)(j-2)}{2}((\not{\mathcal{L}}_T)^{j-3}\xi) \cdot (\not{\mathcal{L}}_T{}^{(T)}\not{\pi} - 2(1-u+t)^{-2}\not{g}) \\ &+ \sum_{i=2}^{j-2} \frac{(j-1)!}{(i+1)!(j-2-i)!}((\not{\mathcal{L}}_T)^{j-2-i}\xi) \cdot ((\not{\mathcal{L}}_T)^{i(T)}\not{\pi})\end{aligned} \tag{11.93}$$

Here

$$\begin{aligned}\xi_{j-1} =& (\not{\mathcal{L}}_T)^{j-1}\xi - 2(j-1)(1-u+t)^{-1}(\not{\mathcal{L}}_T)^{j-2}\xi \\ &+ (j-1)(j-2)(1-u+t)^{-2}(\not{\mathcal{L}}_T)^{j-3}\xi\end{aligned} \tag{11.94}$$

Only the first two terms in (11.93) and (11.94) are present for $j = 2$. Also, the sum in (11.93) is present only for $j \geqslant 4$. What follows is just a direct calculation. The estimate for $\|\xi \cdot \not{g}^{-1}\|_{2,[k-1,l-1],\Sigma_t^{\epsilon_0}}$ follows in a similar manner. □

We proceed with $L^2$ estimates for ${}^{(T)}\not{\pi}_1$.

**Lemma 11.13** Let $l$ be a positive integer and $m$ a non-negative integer $m \leqslant l-1$. Let also hypothesis **H2′** hold. Suppose that

$$\begin{aligned}\|\lambda'\|_{\infty,[m,l_*],\Sigma_t^{\epsilon_0}} &\leqslant C_l\delta_0(1+\log(1+t)) \\ \max_i \|{}^{(R_i)}\not{\pi}\|_{\infty,[l_*-1],\Sigma_t^{\epsilon_0}} &\leqslant C_l\delta_0(1+t)^{-1}(1+\log(1+t)) \\ \|{}^{(T)}\not{\pi}'\|_{\infty,[m,l_*],\Sigma_t^{\epsilon_0}} &\leqslant C_l\delta_0(1+t)^{-1}(1+\log(1+t))\end{aligned}$$

and (if $l_* \geqslant 2$)

$$\max_i \|{}^{(R_i)}\not{\pi}_1\|_{\infty,[l_*-2],\Sigma_t^{\epsilon_0}} \leqslant C_l\delta_0(1+t)^{-2}(1+\log(1+t))$$

Then if $\delta_0$ is suitably small (depending on $l$) we have

$$\begin{aligned}\|{}^{(T)}\not{\pi}_1\|_{2,[m,l-1],\Sigma_t^{\epsilon_0}} \leqslant& C_l(1+t)^{-1}(\mathcal{T}_{[m,l]} + \delta_0(1+t)^{-1}(1+\log(1+t)) \\ &\cdot (\max_i \|{}^{(R_i)}\not{\pi}\|_{2,[l-2],\Sigma_t^{\epsilon_0}} + (1+t)\max_i \|{}^{(R_i)}\not{\pi}_1\|_{2,[l-2],\Sigma_t^{\epsilon_0}}))\end{aligned}$$

*Proof*. The proof is quite similar to that of Lemma 11.11. The only difference is that here we must use Lemma 11.12 to connect the $L^2$ estimates of ${}^{(T)}\not{\pi}_1$ and ${}^{(T)}(\not{\pi}_1)_b$. □

We now apply Lemma 11.11 and 11.13 to obtain estimates for the commutators ${}^{(i_1\cdots i_n)}c_{m,n}[\xi]$, defined by (11.83).

**Lemma 11.14** Let $k, l$ be positive integers, $k \leqslant l$. Let also hypothesis **H2′** hold. Suppose that

$$\|\lambda'\|_{\infty,[k-1,l],\Sigma_t^{\epsilon_0}} \leqslant C_l\delta_0(1+\log(1+t))$$
$$\max_i \|{}^{(R_i)}\not\pi\|_{\infty,[l-1],\Sigma_t^{\epsilon_0}} \leqslant C_l\delta_0(1+t)^{-1}(1+\log(1+t))$$
$$\|{}^{(T)}\not\pi'\|_{\infty,[k-1,l],\Sigma_t^{\epsilon_0}} \leqslant C_l\delta_0(1+t)^{-1}(1+\log(1+t))$$

and

$$\max_i \|{}^{(R_i)}\not\pi_1\|_{\infty,[l-1],\Sigma_t^{\epsilon_0}} \leqslant C_l\delta_0(1+t)^{-2}(1+\log(1+t))$$

Then for any $S_{t,u}$ 1-form $\xi$ we have

$$\max_{m\leqslant k, n\leqslant l-m}\max_{i_1\cdots i_n} \|{}^{(i_1\cdots i_n)}c_{m,n}[\xi]\|_{L^\infty(\Sigma_t^{\epsilon_0})}$$
$$\leqslant C_l\delta_0(1+t)^{-2}(1+\log(1+t))\|\xi\|_{\infty,[k,l-1],\Sigma_t^{\epsilon_0}}$$

provided that $\delta_0$ is suitably small (depending on $l$).

*Proof.* The assumptions of the present lemma include those of Lemma 11.11 with $k-1$ in the role of $m$. Therefore the conclusion of Lemma 11.11 holds with $m$ replaced by $k-1$, that is, we have

$$\|{}^{(T)}\not\pi_1\|_{\infty,[k-1,l-1],\Sigma_t^{\epsilon_0}} \leqslant C_l\delta_0(1+t)^{-2}(1+\log(1+t)) \tag{11.95}$$

Then we just use the expressions given by Lemma 11.9. After a direct calculation, the lemma is seen to follow. □

**Lemma 11.15** Let $k, l$ be positive integers, $k \leqslant l$. Let also hypothesis **H2′** hold. Suppose that

$$\|\lambda'\|_{\infty,[k-1,l_*],\Sigma_t^{\epsilon_0}} \leqslant C_l\delta_0(1+\log(1+t))$$
$$\max_i \|{}^{(R_i)}\not\pi\|_{\infty,[l_*-1],\Sigma_t^{\epsilon_0}} \leqslant C_l\delta_0(1+t)^{-1}(1+\log(1+t))$$
$$\|{}^{(T)}\not\pi'\|_{\infty,[k-1,l_*],\Sigma_t^{\epsilon_0}} \leqslant C_l\delta_0(1+t)^{-1}(1+\log(1+t))$$

and

$$\max_i \|{}^{(R_i)}\not\pi_1\|_{\infty,[l_*-1],\Sigma_t^{\epsilon_0}} \leqslant C_l\delta_0(1+t)^{-2}(1+\log(1+t))$$

Then for any $S_{t,u}$ 1-form $\xi$ we have

$$\max_{m\leqslant k, n\leqslant l-m}\max_{i_1\cdots i_n} \|{}^{(i_1\cdots i_n)}c_{m,n}[\xi]\|_{L^2(\Sigma_t^{\epsilon_0})}$$
$$\leqslant C_l\delta_0(1+t)^{-2}(1+\log(1+t))\|\xi\|_{2,[k,l-1],\Sigma_t^{\epsilon_0}}$$
$$+ C_l(1+t)^{-1}\|\xi\|_{\infty,[k,l_*],\Sigma_t^{\epsilon_0}}(\mathcal{T}_{[k-1,l]} + (1+t)\max_i \|{}^{(R_i)}\not\pi_1\|_{2,[l-1],\Sigma_t^{\epsilon_0}}$$
$$+ \delta_0(1+t)^{-1}(1+\log(1+t))\max_i \|{}^{(R_i)}\not\pi\|_{2,[l-2],\Sigma_t^{\epsilon_0}})$$

provided that $\delta_0$ is suitably small (depending on $l$).

*Proof.* The assumptions of the present lemma include those of Lemma 11.11 with $(k-1, l_*)$ in the role of $(m, l)$. Therefore we have

$$\|{}^{(T)}\not{\pi}_1\|_{\infty,[k-1,l_*-1],\Sigma_t^{\epsilon_0}} \leqslant C_l\delta_0(1+t)^{-2}(1+\log(1+t)) \tag{11.96}$$

Also, the assumptions of the present lemma include those of Lemma 11.13 with $k-1$ in role of $m$. Therefore we have

$$\begin{aligned}&\|{}^{(T)}\not{\pi}_1\|_{2,[k-1,l-1],\Sigma_t^{\epsilon_0}}\\&\leqslant C_l\delta_0(1+t)^{-1}(\mathcal{T}_{[k-1,l]}+\delta_0(1+t)^{-1}(1+\log(1+t))\\&\quad\cdot(\max_i\|{}^{(R_i)}\not{\pi}\|_{2,[l-2],\Sigma_t^{\epsilon_0}}+(1+t)\max_i\|{}^{(R_i)}\not{\pi}_1\|_{2,[l-2],\Sigma_t^{\epsilon_0}}))\end{aligned} \tag{11.97}$$

Then, as in establishing Lemma 11.14, we just use the expressions given by Lemma 11.9 and proceed to a direct calculation. The lemma then follows. □

### 11.1.2 $L^\infty$ Estimates for $T\hat{T}^i$

Given non-negative integers $m, n$, we denote by $\mathbf{E}^Q_{m,n}$ the bootstrap assumption that there is a constant $C$ independent of $s$ such that for all $t \in [0, s]$,

$$\mathbf{E}^Q_{m,n}: \max_{\alpha;i_1\cdots i_n}\|R_{i_n}\cdots R_{i_1}(T)^m Q\psi_\alpha\|_{L^\infty(\Sigma_t^{\epsilon_0})} \leqslant C\delta_0(1+t)^{-1}$$

We then denote by $\mathbf{E}^Q_{\{l\}}$ the conjunction of the assumptions $\mathbf{E}^Q_{m,n}$ corresponding to the triangle (11.8). The constant $C$ then depends on $l$ only and shall be denoted by $C_l$. Note that the assumptions $\mathbf{E}^Q_{0,n}$ coincide with the assumptions $\not{\mathbf{E}}^Q_n$, and $\mathbf{E}^Q_{\{l\}}$ contain $\not{\mathbf{E}}^Q_{[l]}$.

Given non-negative integers $m, l$, $m \leqslant l$, we denote by $\mathbf{M}_{[m,l]}$ the bootstrap assumption that there is a constant $C_l$ independent of $s$ such that for all $t \in [0, s]$,

$$\mathbf{M}_{[m,l]}: \|\mu-1\|_{\infty,[m,l],\Sigma_t^{\epsilon_0}} \leqslant C_l\delta_0(1+\log(1+t))$$

This is equivalent, modulo $\mathbf{E}_{\{l\}}$, to the assumption that there is a constant $C_l$ independent of $s$ such that for all $t \in [0, s]$,

$$\|\kappa-1\|_{\infty,[m,l],\Sigma_t^{\epsilon_0}} \leqslant C_l\delta_0(1+\log(1+t)) \tag{11.98}$$

**Proposition 11.1** Let hypotheses **H0**, **H1**, **H2$'$** and the estimate (6.177) hold. Let also the bootstrap assumptions $\mathbf{E}_{\{l+1\}}$, $\mathbf{E}^Q_{\{l\}}$, and $\not{\mathbf{X}}_{[l]}$, hold for some non-negative integer $l$. Moreover, let the bootstrap assumption $\mathbf{M}_{[m,l+1]}$ hold for some non-negative integer $m \leqslant l$. Then if $\delta_0$ is suitably small (depending on $l$) we have

$$\max_i\|(T)^{k+1}\hat{T}^i\|_{\infty,[l-k],\Sigma_t^{\epsilon_0}} \leqslant C_l\delta_0(1+t)^{-1}(1+\log(1+t))$$

for all $k = 0, \cdots, m$.

*Proof*. The assumptions of the present proposition include those of Proposition 10.1 with $l+1$ in the role of $l$. Therefore by the corollaries of Proposition 10.1, we have, with $l+1$ in the role of $l$,

$$\|\psi_L\|_{\infty,[l+1],\Sigma_t^{\epsilon_0}} \leqslant C_l\delta_0(1+t)^{-1} \tag{11.99}$$
$$\|\psi_{\hat{T}}\|_{\infty,[l+1],\Sigma_t^{\epsilon_0}} \leqslant C_l\delta_0(1+t)^{-1} \tag{11.100}$$
$$\|\psi\!\!\!/\|_{\infty,[l+1],\Sigma_t^{\epsilon_0}} \leqslant C_l\delta_0(1+t)^{-1} \tag{11.101}$$

and

$$\|\omega_{L\hat{T}}\|_{\infty,[l],\Sigma_t^{\epsilon_0}} \leqslant C_l\delta_0(1+t)^{-2} \tag{11.102}$$
$$\|\psi\!\!\!/_L\|_{\infty,[l],\Sigma_t^{\epsilon_0}} \leqslant C_l\delta_0(1+t)^{-2} \tag{11.103}$$
$$\|\psi\!\!\!/\|_{\infty,[l],\Sigma_t^{\epsilon_0}} \leqslant C_l\delta_0(1+t)^{-2} \tag{11.104}$$

Moreover, from the definition

$$\psi\!\!\!/_{\hat{T}} = \hat{T}^i d\!\!\!/\psi_i$$

we readily obtain

$$\|\psi\!\!\!/_{\hat{T}}\|_{\infty,[l],\Sigma_t^{\epsilon_0}} \leqslant C_l\delta_0(1+t)^{-2} \tag{11.105}$$
$$\|\omega_{T\hat{T}}\|_{\infty,[l],\Sigma_t^{\epsilon_0}} \leqslant C_l\delta_0(1+t)^{-1} \tag{11.106}$$

By the assumption $\mathbf{M}_{[0,l+1]}$, we have

$$\|(q_T)_b\|_{\infty,[l],\Sigma_t^{\epsilon_0}} \leqslant C_l\delta_0(1+t)^{-1}(1+\log(1+t)) \tag{11.107}$$

(provided that $\delta_0$ is suitably small depending on $l$). This together with Corollary 10.1.d implies

$$\|q_T\|_{\infty,[l],\Sigma_t^{\epsilon_0}} \leqslant C_l\delta_0(1+t)^{-1}(1+\log(1+t)) \tag{11.108}$$

In view of (11.12), the statement of the proposition for $k=0$ follows.

We proceed by induction on $k$ for fixed $m$ and $l$. Let $k \in \{1, \cdots, m\}$ and let

$$\max_i \|(T)^{k'+1}\hat{T}^i\|_{\infty,[l-k'],\Sigma_t^{\epsilon_0}} \leqslant C_l\delta_0(1+t)^{-1}(1+\log(1+t)) \tag{11.109}$$

hold for all $k' = 0, \cdots, k-1$.

By this induction hypothesis and $\mathbf{E}_{\{l+1\}}$ we can easily obtain

$$\|(T)^{k'}\psi_{\hat{T}}\|_{\infty,[l+1-k'],\Sigma_t^{\epsilon_0}} \leqslant C_l\delta_0(1+t)^{-1} \tag{11.110}$$

for all $k' \in \{1, \cdots, k\}$.

This together with (11.100) implies

$$\|\psi_{\hat{T}}\|_{\infty,[k,l+1],\Sigma_t^{\epsilon_0}} \leqslant C_l\delta_0(1+t)^{-1} \tag{11.111}$$

Since

$$\omega_{T\hat{T}} = \hat{T}^i(T\psi_i), \quad \omega_{L\hat{T}} = \hat{T}^i(L\psi_i), \quad \not{\psi}_{\hat{T}} = \hat{T}^i(\not{d}\psi_i)$$

these are analogous to $\psi_{\hat{T}}$ with $T\psi_i$, $L\psi_i$, $\not{d}\psi_i$, in the role of $\psi_i$, respectively. We thus obtain

$$\|\omega_{T\hat{T}}\|_{\infty,[k,l],\Sigma_t^{\epsilon_0}} \leqslant C_l\delta_0(1+t)^{-1} \tag{11.112}$$

$$\|\omega_{L\hat{T}}\|_{\infty,[k,l],\Sigma_t^{\epsilon_0}} \leqslant C_l\delta_0(1+t)^{-2} \tag{11.113}$$

$$\|\not{\psi}_{\hat{T}}\|_{\infty,[k,l],\Sigma_t^{\epsilon_0}} \leqslant C_l\delta_0(1+t)^{-2} \tag{11.114}$$

Similarly, by the relation

$$L^i = -\alpha\hat{T}^i - \psi_i \tag{11.115}$$

and $\mathbf{E}_{\{l+1\}}$, we have

$$\max_i \|(T)^{k'+1}L^i\|_{\infty,[l-k'],\Sigma_t^{\epsilon_0}} \leqslant C_l\delta_0(1+t)^{-1}(1+\log(1+t)) \tag{11.116}$$

for all $k' = 0, \cdots, k-1$.
Then we have

$$\|(T)^{k'}\psi_L\|_{\infty,[l+1-k'],\Sigma_t^{\epsilon_0}} \leqslant C_l\delta_0(1+t)^{-1} \tag{11.117}$$

for all $k' = 0, \cdots, k$ which, together with (11.99), yields

$$\|\psi_L\|_{\infty,[k,l+1],\Sigma_t^{\epsilon_0}} \leqslant C_l\delta_0(1+t)^{-1} \tag{11.118}$$

and then

$$\|\not{\psi}_L\|_{\infty,[k,l],\Sigma_t^{\epsilon_0}} \leqslant C_l\delta_0(1+t)^{-2} \tag{11.119}$$

Also we have

$$\|\not{\psi}\|_{\infty,[k,l+1],\Sigma_t^{\epsilon_0}} \leqslant C_l\delta_0(1+t)^{-1} \tag{11.120}$$

$$\|\not{\psi}\|_{\infty,[k,l],\Sigma_t^{\epsilon_0}} \leqslant C_l\delta_0(1+t)^{-2} \tag{11.121}$$

Then we have

$$\|\not{k}\|_{\infty,[k,l],\Sigma_t^{\epsilon_0}} \leqslant C_l\delta_0(1+t)^{-2} \tag{11.122}$$

$$\|\kappa^{-1}\zeta\|_{\infty,[k,l],\Sigma_t^{\epsilon_0}} \leqslant C_l\delta_0(1+t)^{-2} \tag{11.123}$$

Therefore

$$\|\kappa\not{k}\|_{\infty,[k,l],\Sigma_t^{\epsilon_0}} \leqslant C_l\delta_0(1+t)^{-2}(1+\log(1+t)) \tag{11.124}$$

$$\|\zeta\|_{\infty,[k,l],\Sigma_t^{\epsilon_0}} \leqslant C_l\delta_0(1+t)^{-2}(1+\log(1+t)) \tag{11.125}$$

By $\mathbf{E}_{\{l+1\}}$ and $\mathbf{M}_{[k,l+1]}$ we obtain

$$\|(q_T)_b\|_{\infty,[k,l],\Sigma_t^{\epsilon_0}} \leqslant C_l\delta_0(1+t)^{-1}(1+\log(1+t)) \tag{11.126}$$

To complete the inductive step we need a similar estimate for $q_T$. This estimate would follow from Lemma 11.7 if we can show that

$$\|{}^{(T)}\not{\pi}'\|_{\infty,[k-1,l-1],\Sigma_t^{\epsilon_0}} \leqslant C_l\delta_0(1+t)^{-1}(1+\log(1+t)) \tag{11.127}$$

because the assumptions of Lemma 11.7, which are those of Lemma 11.5, hold by virtue of Corollary 10.1.d and $\mathbf{M}_{[k,l+1]}$. From the definition (11.51) and the estimate (11.124) we see that (11.127) would in turn follow if we can show that

$$\|\chi'\|_{\infty,[k-1,l-1],\Sigma_t^{\epsilon_0}} \leqslant C_l\delta_0(1+t)^{-1} \tag{11.128}$$

A stronger estimate is in fact provided by the following lemma:

**Lemma 11.16** Let the assumptions of Proposition 11.1 hold. Let also the inductive hypothesis (11.109) hold, for some $k \in \{1,\cdots,m\}$. Then, provided that $\delta_0$ is suitably small (depending on $l$), the following estimate holds:

$$\|\chi'\|_{\infty,[k,l],\Sigma_t^{\epsilon_0}} \leqslant C_l\delta_0(1+t)^{-2}(1+\log(1+t))$$

*Proof*. As remarked above, under the present assumptions Proposition 10.1 and all its corollaries hold with $l+1$ in the role of $l$. From the definition (11.50) and Corollary 10.1.d we obtain

$$\|\chi'\|_{\infty,[l],\Sigma_t^{\epsilon_0}} \leqslant C_l\delta_0(1+t)^{-2}(1+\log(1+t)) \tag{11.129}$$

We proceed, by induction on $k'$, to show that

$$\|(\not{\mathcal{L}}_T)^{k'}\chi'\|_{\infty,[l-k'],\Sigma_t^{\epsilon_0}} \leqslant C_l\delta_0(1+t)^{-2}(1+\log(1+t)) \tag{11.130}$$

for all $k' \in \{1,\cdots,k\}$ which, together with (11.129), yields the conclusion of the lemma. In fact we shall show that (11.109) together with the estimate

$$\|\chi'\|_{\infty,[k-1,l],\Sigma_t^{\epsilon_0}} \leqslant C_l\delta_0(1+t)^{-2}(1+\log(1+t)) \tag{11.131}$$

implies

$$\|(\not{\mathcal{L}}_T)^{k}\chi'\|_{\infty,[l-k],\Sigma_t^{\epsilon_0}} \leqslant C_l\delta_0(1+t)^{-2}(1+\log(1+t)) \tag{11.132}$$

From the definition (11.50) we have

$$\not{\mathcal{L}}_T\chi' = \not{\mathcal{L}}_T\chi - \frac{{}^{(T)}\not{\pi}}{1-u+t} - \frac{\not{g}}{(1-u+t)^2} \tag{11.133}$$

Now, $\not{\mathcal{L}}_T\chi$ is given by the formula (3.124)—(3.125). Substituting in (11.133) this formula, and also substituting for ${}^{(T)}\not{\pi}$ in terms of $\chi'$ from (11.51)—(11.52), we obtain a formula for $\not{\mathcal{L}}_T\chi'$. We apply $(\not{\mathcal{L}}_T)^{k-1}$ to this formula and then take the

$\| \quad \|_{\infty,[l-k],\Sigma_t^{\epsilon_0}}$ norm. The contribution of any given term on the right-hand side, will then be bounded by the $\| \quad \|_{\infty,[k-1,l-1],\Sigma_t^{\epsilon_0}}$ norm of that term. Now from (11.51), (11.131) together with (11.124) and $\mathbf{M}_{[k,l+1]}$ imply

$$\|{}^{(T)}\not{\pi}'\|_{\infty,[k-1,l],\Sigma_t^{\epsilon_0}} \leqslant C_l\delta_0(1+t)^{-2}(1+\log(1+t))^2 \tag{11.134}$$

By Lemma 11.7 with $k-1$ in the role of $m$ we then obtain

$$\begin{aligned}
\|{}^{(T)}\not{\pi} + 2(1-u+t)^{-1}\not{g}\|_{\infty,[l],\Sigma_t^{\epsilon_0}} &\leqslant C_l\delta_0(1+t)^{-1}(1+\log(1+t)) \\
\|\not{\mathcal{L}}_T{}^{(T)}\not{\pi} - 2(1-u+t)^{-1}\not{g}\|_{\infty,[l-1],\Sigma_t^{\epsilon_0}} &\leqslant C_l\delta_0(1+t)^{-1}(1+\log(1+t)) \\
\max_{2\leqslant k'\leqslant k-1}\|(\not{\mathcal{L}}_T)^{k'}{}^{(T)}\not{\pi}\|_{\infty,[l-k'],\Sigma_t^{\epsilon_0}} &\leqslant C_l\delta_0(1+t)^{-1}(1+\log(1+t))
\end{aligned} \tag{11.135}$$

These estimates, together with Corollary 10.1.d, imply that for any $S_{t,u}$ tensorfield $\xi$ we have

$$\|\xi\cdot\not{g}\|_{\infty,[k-1,l-1],\Sigma_t^{\epsilon_0}}, \|\xi\cdot\not{g}^{-1}\|_{\infty,[k-1,l-1],\Sigma_t^{\epsilon_0}} \leqslant C_l\|\xi\|_{\infty,[k-1,l-1],\Sigma_t^{\epsilon_0}} \tag{11.136}$$

provided that $\delta_0$ is suitably small (depending on $l$). This together with the estimates from (11.99) to (11.125) as well as Lemma 11.14 implies (11.132). So the lemma follows. □

We now resume the proof of Proposition 11.1. Using Lemma 11.16 we obtain the estimate

$$\|q_T\|_{\infty,[k,l],\Sigma_t^{\epsilon_0}} \leqslant C_l\delta_0(1+t)^{-1}(1+\log(1+t)) \tag{11.137}$$

Using this and $\mathbf{M}_{[k,l+1]}$ we deduce, from the first statement of Lemma 11.4, the estimates

$$\|p_{T,j}\|_{\infty,[k-j,l-j],\Sigma_t^{\epsilon_0}} \leqslant C_l\delta_0^2(1+t)^{-2}(1+\log(1+t))^2 \tag{11.138}$$

for all $j \in \{0,\cdots,k\}$, and

$$\|q_{T,j}\|_{\infty,[k-j,l-j],\Sigma_t^{\epsilon_0}} \leqslant C_l\delta_0(1+t)^{-1}(1+\log(1+t)) \tag{11.139}$$

for all $j \in \{0,\cdots,k\}$.

We write the second statement of Lemma 11.4 in the form

$$r_{T,k}^n = \sum_{i=0}^{k-1-n} N_i(n) \sum_{j'=0}^{k-1-n-i} (\not{\mathcal{L}}_T)^{k-1-n-j'}(\kappa q_{T,j'}) \tag{11.140}$$

Here $n \in \{0,\cdots,k-1\}$. Taking the $\| \quad \|_{\infty,[l-k+n],\Sigma_t^{\epsilon_0}}$ norm of this we obtain, using (11.139) and $\mathbf{M}_{[k,l+1]}$,

$$\begin{aligned}
\|r_{T,k}^n\|_{\infty,[l-k],\Sigma_t^{\epsilon_0}} &\leqslant \sum_{i=0}^{k-1-n} N_i(n) \sum_{j'=0}^{k-1-n-i} \|\kappa q_{T,j'}\|_{\infty,[k-1-n-j',l-1-j'],\Sigma_t^{\epsilon_0}} \\
&\leqslant C_l \max_{j'\in\{0,\cdots,k-1-n\}} \|\kappa q_{T,j'}\|_{\infty,[k-1-n-j',l-1-j'],\Sigma_t^{\epsilon_0}} \\
&\leqslant C_l\delta_0(1+t)^{-1}(1+\log(1+t))^2
\end{aligned} \tag{11.141}$$

for all $n \in \{0, \cdots, k-1\}$.

Now according to Lemma 11.3,

$$(T)^{k+1}\hat{T}^i = p_{T,k}\hat{T}^i + q_{T,k} \cdot \not{d}x^i + \sum_{n=0}^{k-1} r^n_{T,k} \cdot \not{d}(T)^n\hat{T}^i \tag{11.142}$$

We take the $\| \quad \|_{\infty,[l-k],\Sigma_t^{\epsilon_0}}$ norm of this. From the above estimates, we have

$$\max_i \|p_{T,k}\hat{T}\|_{\infty,[l-k],\Sigma_t^{\epsilon_0}} \leqslant C_l\delta_0^2(1+t)^{-2}(1+\log(1+t))^2 \tag{11.143}$$

$$\max_i \|q_{T,k} \cdot \not{d}x^i\|_{\infty,[l-k],\Sigma_t^{\epsilon_0}} \leqslant C_l\delta_0(1+t)^{-1}(1+\log(1+t)) \tag{11.144}$$

$$\max_i \|\sum_{n=0}^{k-1} r^n_{T,k} \cdot \not{d}(T)^n\hat{T}^i\|_{\infty,[l-k],\Sigma_t^{\epsilon_0}} \leqslant C_l\delta_0(1+t)^{-2}(1+\log(1+t))^3 \tag{11.145}$$

These imply

$$\max_i \|(T)^{k+1}\hat{T}^i\|_{\infty,[l-k],\Sigma_t^{\epsilon_0}} \leqslant C_l\delta_0(1+t)^{-1}(1+\log(1+t)) \tag{11.146}$$

This completes the proof of Proposition 11.1. □

The foregoing proposition has the following corollaries.

**Corollary 11.1.a** Under the assumptions of Proposition 11.1 the coefficients of the expression for $(T)^{k+1}\hat{T}^i$, $k = 0, \cdots, m$, of Lemma 11.3 satisfy

$$\|p_{T,k}\|_{\infty,[m-k,l-k],\Sigma_t^{\epsilon_0}} \leqslant C_l\delta_0^2(1+t)^{-2}(1+\log(1+t))^2$$

for all $k \in \{0, \cdots, m\}$,

$$\|q_{T,k}\|_{\infty,[m-k,l-k],\Sigma_t^{\epsilon_0}} \leqslant C_l\delta_0(1+t)^{-1}(1+\log(1+t))$$

for all $k \in \{0, \cdots, m\}$, and

$$\|r^n_{T,k}\|_{\infty,[l-k+n],\Sigma_t^{\epsilon_0}} \leqslant C_l\delta_0(1+t)^{-1}(1+\log(1+t))^2$$

for all $k \in \{0, \cdots, m\}$, $n \in \{0, \cdots, k-1\}$.

Also,

$$\max_i \|(T)^{k+1}L^i\|_{\infty,[l-k],\Sigma_t^{\epsilon_0}} \leqslant C_l\delta_0(1+t)^{-1}(1+\log(1+t))$$

for all $k = 0, \cdots, m$.

**Corollary 11.1.b** Under the assumptions of Proposition 11.1 the following estimates hold:

$$\|\psi_{\hat{T}}\|_{\infty,[m+1,l+1],\Sigma_t^{\epsilon_0}} \leqslant C_l\delta_0(1+t)^{-1}$$

$$\|\psi_L\|_{\infty,[m+1,l+1],\Sigma_t^{\epsilon_0}} \leqslant C_l\delta_0(1+t)^{-1}$$

$$\|\psi\|_{\infty,[m+1,l+1],\Sigma_t^{\epsilon_0}} \leqslant C_l\delta_0(1+t)^{-1}$$

and

$$
\begin{aligned}
\|\omega_{T\hat{T}}\|_{\infty,[m,l],\Sigma_t^{\epsilon_0}} &\leqslant C_l\delta_0(1+t)^{-1}\\
\|\omega_{L\hat{T}}\|_{\infty,[m+1,l+1],\Sigma_t^{\epsilon_0}} &\leqslant C_l\delta_0(1+t)^{-2}\\
\|\not\psi_{\hat{T}}\|_{\infty,[m,l],\Sigma_t^{\epsilon_0}} &\leqslant C_l\delta_0(1+t)^{-2}\\
\|\not\psi_{L}\|_{\infty,[m,l],\Sigma_t^{\epsilon_0}} &\leqslant C_l\delta_0(1+t)^{-2}\\
\|\not\psi\|_{\infty,[m,l],\Sigma_t^{\epsilon_0}} &\leqslant C_l\delta_0(1+t)^{-2}
\end{aligned}
$$

and

$$
\begin{aligned}
\|\not{k}\|_{\infty,[m,l],\Sigma_t^{\epsilon_0}} &\leqslant C_l\delta_0(1+t)^{-2}\\
\|\kappa^{-1}\zeta\|_{\infty,[m,l],\Sigma_t^{\epsilon_0}} &\leqslant C_l\delta_0(1+t)^{-2}\\
\|\kappa\not{k}\|_{\infty,[m,l],\Sigma_t^{\epsilon_0}} &\leqslant C_l\delta_0(1+t)^{-2}(1+\log(1+t))\\
\|\zeta\|_{\infty,[m,l],\Sigma_t^{\epsilon_0}} &\leqslant C_l\delta_0(1+t)^{-2}(1+\log(1+t))
\end{aligned}
$$

The next corollary is obtained from Lemma 11.16 by substituting Proposition 11.1 itself in place of the inductive hypothesis (11.109), and then substituting the result in Lemma 11.7 and Lemma 11.11.

**Corollary 11.1.c** Under the assumptions of Proposition 11.1 we have

$$
\|\chi'\|_{\infty,[m,l],\Sigma_t^{\epsilon_0}} \leqslant C_l\delta_0(1+t)^{-2}(1+\log(1+t))
$$

Moreover, we have

$$
\begin{aligned}
\|{}^{(T)}\not\pi + 2(1-u+t)^{-1}\not{g}\|_{\infty,[l],\Sigma_t^{\epsilon_0}} &\leqslant C_l\delta_0(1+t)^{-1}(1+\log(1+t))\\
\|\not{\mathcal{L}}_T{}^{(T)}\not\pi - 2(1-u+t)^{-2}\not{g}\|_{\infty,[l-1],\Sigma_t^{\epsilon_0}} &\leqslant C_l\delta_0(1+t)^{-1}(1+\log(1+t))\\
\max_{2\leqslant k\leqslant m}\|(\not{\mathcal{L}}_T)^{k}{}^{(T)}\not\pi\|_{\infty,[l-k],\Sigma_t^{\epsilon_0}} &\leqslant C_l\delta_0(1+t)^{-1}(1+\log(1+t))
\end{aligned}
$$

Also

$$
\|\Lambda\|_{\infty,[m,l],\Sigma_t^{\epsilon_0}} \leqslant C_l\delta_0(1+t)^{-1}(1+\log(1+t))
$$

and

$$
\|{}^{(T)}\not\pi_1\|_{\infty,[m,l-1],\Sigma_t^{\epsilon_0}} \leqslant C_l\delta_0(1+t)^{-2}(1+\log(1+t))
$$

### 11.1.3 $L^2$ Estimates for $T\hat{T}^i$

We proceed with the $L^2$ estimates. Given non-negative integers $m$, $n$, we denote by $\mathcal{W}^Q_{m,n}$ the quantity

$$
\mathcal{W}^Q_{m,n} = \max_{\alpha;i_1\cdots i_n}\|R_{i_n}\cdots R_{i_1}(T)^m Q\psi_\alpha\|_{L^2(\Sigma_t^{\epsilon_0})} \tag{11.147}
$$

We then denote by $\mathcal{W}^Q_{\{l\}}$ the sum of the quantities $\mathcal{W}^Q_{m,n}$ corresponding to the triangle (11.8). Note that the quantities $\mathcal{W}^Q_{0,n}$ coincide with the quantities $\mathcal{W}^Q_n$, and $\mathcal{W}^Q_{\{l\}}$ dominate $\mathcal{W}^Q_{[l]}$. Moreover, we have

$$\max_\alpha \|L\psi_\alpha\|_{2,\{l\},\Sigma_t^{\epsilon_0}} \leqslant (1+t)^{-1}\mathcal{W}^Q_{\{l\}} \tag{11.148}$$

Given non-negative integers $m$, $l$, $m \leqslant l$ we denote

$$\mathcal{B}_{[m,l]} = \|\mu - 1\|_{2,[m,l],\Sigma_t^{\epsilon_0}} \tag{11.149}$$

Note that $\mathcal{B}_{[0,l]}$ coincide with $\mathcal{B}_{[l]}$. Since $\mu = \alpha\kappa$, it follows from Lemma 11.1 that under the bootstrap assumptions $\mathbf{E}_{\{l_*\}}$ and $\mathbf{M}_{[m,l_*]}$ we have

$$\|\kappa - 1\|_{2,[m,l],\Sigma_t^{\epsilon_0}} \leqslant C_l(\mathcal{B}_{[m,l]} + (1+\log(1+t))\mathcal{W}_{\{l\}}) \tag{11.150}$$

**Proposition 11.2** Let the hypotheses **H0**, **H1**, **H2**$'$ and the estimate (6.177) hold. Let also the bootstrap assumptions $\mathbf{E}_{\{l_*+1\}}$, $\mathbf{E}^Q_{\{l_*\}}$, and $\mathbf{X}_{[l_*]}$, hold for some non-negative integer $l$. Moreover, let the bootstrap assumption $\mathbf{M}_{[m,l_*+1]}$ hold for some non-negative integer $m \leqslant l$ (where by definition $\mathbf{M}_{[m,l_*+1]}$ coincides with $\mathbf{M}_{\{l_*+1\}} = \mathbf{M}_{[l_*+1,l_*+1]}$ for $m \geqslant l_* + 1$). Then if $\delta_0$ is suitably small (depending on $l$) we have

$$\begin{aligned}
&\sum_{k=0}^{m} \max_i \|(T)^{k+1}\hat{T}^i\|_{2,[l-k],\Sigma_t^{\epsilon_0}} \\
&\leqslant C_l(1+t)^{-1}(\mathcal{B}_{[m,l+1]} + \delta_0(1+\log(1+t))(\mathcal{Y}_0 + (1+t)\mathcal{A}_{[l-1]}) \\
&\quad + (1+\log(1+t))(\mathcal{W}_{\{l+1\}} + \delta_0(1+t)^{-2}(1+\log(1+t))^2\mathcal{W}^Q_{\{l-1\}}))
\end{aligned}$$

*Proof*. The assumptions of the present proposition coincide with those of Proposition 11.1 with $(m, l_*)$ in the role of $(m, l)$. Therefore under the present assumptions Proposition 11.1 and its corollaries hold with $(m, l_*)$ in the role of $(m, l)$. Moreover, the present assumptions include those of Proposition 10.1 with $l_* + 1$ in the role of $l$, as well as those of Proposition 10.2 with $l + 1$ in the role of $l$. Therefore under the present assumptions Proposition 10.1 and its all corollaries follow with $l_* + 1$ in the role of $l$, and Proposition 10.2 and its all corollaries follow with $l + 1$ in the role of $l$. In particular, we have (Corollaries 10.2.b and 10.2.g with $l$ replaced by $l + 1$):

$$\|\psi_L\|_{2,[l+1],\Sigma_t^{\epsilon_0}} \leqslant C_l(\mathcal{W}_{[l+1]} + \delta_0(1+t)^{-1}(\mathcal{Y}_0 + (1+t)\mathcal{A}_{[l]})) \tag{11.151}$$

$$\|\psi_{\hat{T}}\|_{2,[l+1],\Sigma_t^{\epsilon_0}} \leqslant C_l(\mathcal{W}_{[l+1]} + \delta_0(1+t)^{-1}(\mathcal{Y}_0 + (1+t)\mathcal{A}_{[l]})) \tag{11.152}$$

$$\|\not{\psi}\|_{2,[l+1],\Sigma_t^{\epsilon_0}} \leqslant C_l(\mathcal{W}_{[l+1]} + \delta_0(1+t)^{-1}(\mathcal{Y}_0 + (1+t)\mathcal{A}_{[l]})) \tag{11.153}$$

and

$$\|\omega_{L\hat{T}}\|_{2,[l],\Sigma_t^{\epsilon_0}} \leqslant C_l(1+t)^{-1}(\mathcal{W}^Q_{[l]} + \delta_0(1+t)^{-1}(\mathcal{Y}_0 + (1+t)\mathcal{A}_{[l-1]} + \mathcal{W}_{[l]})) \tag{11.154}$$

$$\|\not{\varphi}_L\|_{2,[l],\Sigma_t^{\epsilon_0}} \leqslant C_l(1+t)^{-1}(\mathcal{W}_{[l+1]} + \delta_0(1+t)^{-1}(\mathcal{Y}_0 + (1+t)\mathcal{A}_{[l-1]})) \tag{11.155}$$

$$\|\not{\varphi}\|_{2,[l],\Sigma_t^{\epsilon_0}} \leqslant C_l(1+t)^{-1}(\mathcal{W}_{[l+1]} + \delta_0(1+t)^{-1}(\mathcal{Y}_0 + (1+t)\mathcal{A}_{[l-1]})) \tag{11.156}$$

Moreover, we readily obtain

$$\|\not{\psi}_{\hat{T}}\|_{2,[l],\Sigma_t^{\epsilon_0}} \leqslant C_l(1+t)^{-1}(\mathcal{W}_{[l+1]} + \delta_0(1+t)^{-1}(\mathcal{Y}_0 + (1+t)\mathcal{A}_{[l-1]})) \quad (11.157)$$

$$\|\omega_{T\hat{T}}\|_{2,[l],\Sigma_t^{\epsilon_0}} \leqslant C_l(\mathcal{W}_{[l+1]} + \delta_0(1+t)^{-1}(\mathcal{Y}_0 + (1+t)\mathcal{A}_{[l-1]})) \quad (11.158)$$

Let us define

$$\mathcal{U}_{k,l} = \sum_{k'=0}^{k} \max_i \|(T)^{k'+1}\hat{T}^i\|_{2,[l-k'],\Sigma_t^{\epsilon_0}} \quad (11.159)$$

By the definition (11.14),

$$\|(\not{q}_T)_b\|_{2,[l],\Sigma_t^{\epsilon_0}} \leqslant C(1+t)^{-1}\|\kappa - 1\|_{2,[l+1],\Sigma_t^{\epsilon_0}} \quad (11.160)$$

This together with (10.145) as well as the $L^\infty$ estimates (11.126) and Corollary 10.1.d, in turn yields

$$\begin{aligned}\|\not{q}_T\|_{2,[l],\Sigma_t^{\epsilon_0}} \leqslant & C_l(1+t)^{-1}(\|\kappa - 1\|_{2,[l+1],\Sigma_t^{\epsilon_0}} \\ & + \delta_0(1+\log(1+t))(\mathcal{Y}_0 + (1+t)\mathcal{A}_{[l-1]} + \mathcal{W}_{[l]}))\end{aligned} \quad (11.161)$$

From this and Corollary 10.2.a,

$$\begin{aligned}\mathcal{U}_{0,l} \leqslant & C_l(1+t)^{-1}(\|\kappa - 1\|_{2,[l+1],\Sigma_t^{\epsilon_0}} \\ & + \delta_0(1+\log(1+t))(\mathcal{Y}_0 + (1+t)\mathcal{A}_{[l-1]} + \mathcal{W}_{[l]}))\end{aligned} \quad (11.162)$$

We shall prove the proposition by deriving a recursive, in $k$, inequality, for the quantities $\mathcal{U}_{k,l}$, for fixed $l$. We first establish $L^2$ estimates for the $T$-derivatives of up to the $k$th order of the $\psi$ and $\omega$ components, in terms of $\mathcal{U}_{k-1,l}$. As we did in proving Proposition 11.1, we make use of the following expressions:

$$\begin{aligned}(T)^{k'}\psi_{\hat{T}} = & \hat{T}^i(T)^{k'}\psi_i \\ & + \sum_{j=0}^{k'-1} \frac{k'!}{(j+1)!(k'-1-j)!}((T)^{j+1}\hat{T}^i)((T)^{k'-1-j}\psi_i)\end{aligned} \quad (11.163)$$

$$\begin{aligned}(T)^{k'}\psi_L = & (T)^{k'}\psi_0 + L^i(T)^{k'}\psi_i \\ & + \sum_{j=0}^{k'-1} \frac{k'!}{(j+1)!(k'-1-j)!}((T)^{j+1}L^i)((T)^{k'-1-j}\psi_i)\end{aligned} \quad (11.164)$$

$$\begin{aligned}(\not{\mathcal{L}}_T)^{k'}\not{\psi} = & (\not{d}x^i)(T)^{k'}\psi_i \\ & + \sum_{j=0}^{k'-1} \frac{k'!}{(j+1)!(k'-1-j)!}(\not{d}(T)^{j+1}x^i)((T)^{k'-1-j}\psi_i)\end{aligned} \quad (11.165)$$

We now take the $\| \quad \|_{2,[l+1-k'],\Sigma_t^{\epsilon_0}}$ norm of the above. This together with (11.151)–(11.158), Proposition 10.2 as well as its corollaries and Proposition 11.1 implies

$$\|\psi_{\hat{T}}\|_{2,[k,l+1],\Sigma_t^{\epsilon_0}} \leqslant C_l(\mathcal{W}_{\{l+1\}} + \delta_0(1+t)^{-1}(\mathcal{U}_{k-1,l} + \mathcal{Y}_0 + (1+t)\mathcal{A}_{[l]})) \quad (11.166)$$

and

$$\|\omega_{T\hat{T}}\|_{2,[k,l],\Sigma_t^{\epsilon_0}} \leqslant C_l(\mathcal{W}_{\{l+1\}} + \delta_0(1+t)^{-1}(\mathcal{U}_{k-1,l-1} + \mathcal{Y}_0 + (1+t)\mathcal{A}_{[l-1]})) \quad (11.167)$$

$$\|\omega_{L\hat{T}}\|_{2,[k,l],\Sigma_t^{\epsilon_0}} \leqslant C_l(1+t)^{-1}(\mathcal{W}^Q_{\{l\}} + \delta_0(1+t)^{-1}(\mathcal{U}_{[k-1,l-1]} + \mathcal{Y}_0 + (1+t)\mathcal{A}_{[l-1]}) + \mathcal{W}_{\{l\}}) \quad (11.168)$$

$$\|\phi_{\hat{T}}\|_{2,[k,l],\Sigma_t^{\epsilon_0}} \leqslant C_l(1+t)^{-1}(\mathcal{W}_{\{l+1\}} + \delta_0(1+t)^{-1}(\mathcal{U}_{[k-1,l-1]} + \mathcal{Y}_0 + (1+t)\mathcal{A}_{[l-1]})) \quad (11.169)$$

Also

$$\|\psi_L\|_{2,[k,l+1],\Sigma_t^{\epsilon_0}} \leqslant C_l(\mathcal{W}_{\{l+1\}} + \delta_0(1+t)^{-1}(\mathcal{U}_{k-1,l} + \mathcal{Y}_0 + (1+t)\mathcal{A}_{[l]})) \quad (11.170)$$

$$\|\phi_L\|_{2,[k,l],\Sigma_t^{\epsilon_0}} \leqslant C_l(1+t)^{-1}(\mathcal{W}_{\{l+1\}} + \delta_0(1+t)^{-1}(\mathcal{U}_{k-1,l-1} + \mathcal{Y}_0 + (1+t)\mathcal{A}_{[l-1]})) \quad (11.171)$$

and finally

$$\begin{aligned}\|\psi\|_{2,[k,l+1],\Sigma_t^{\epsilon_0}} \leqslant C_l(&\mathcal{W}_{\{l+1\}} + \delta_0(1+t)^{-2}\|\kappa-1\|_{2,[k-1,l+1],\Sigma_t^{\epsilon_0}} \\ &+ \delta_0(1+t)^{-2}(1+\log(1+t))\mathcal{U}_{k-2,l} \\ &+ \delta_0(1+t)^{-1}(\mathcal{Y}_0 + (1+t)\mathcal{A}_{[l]}))\end{aligned} \quad (11.172)$$

as well as

$$\begin{aligned}\|\phi\|_{2,[k,l],\Sigma_t^{\epsilon_0}} \leqslant C_l(1+t)^{-1}(&\mathcal{W}_{\{l+1\}} + \delta_0(1+t)^{-2}\|\kappa-1\|_{2,[k-1,l],\Sigma_t^{\epsilon_0}} \\ &+ \delta_0(1+t)^{-2}(1+\log(1+t))\mathcal{U}_{k-2,l-1} \\ &+ \delta_0(1+t)^{-1}(\mathcal{Y}_0 + (1+t)\mathcal{A}_{[l-1]}))\end{aligned} \quad (11.173)$$

(11.173) together with Lemma 11.2.a implies that

$$\begin{aligned}\|\not{k}\|_{2,[k,l],\Sigma_t^{\epsilon_0}} \leqslant C_l(1+t)^{-1}(&\mathcal{W}_{\{l+1\}} + \delta_0(1+t)^{-2}\|\kappa-1\|_{2,[k-1,l],\Sigma_t^{\epsilon_0}} \\ &+ \delta_0(1+t)^{-2}(1+\log(1+t))\mathcal{U}_{k-2,l-1} \\ &+ \delta_0(1+t)^{-1}(\mathcal{Y}_0 + (1+t)\mathcal{A}_{[l-1]}))\end{aligned} \quad (11.174)$$

Similarly, (11.169) implies

$$\begin{aligned}\|\kappa^{-1}\zeta\|_{2,[k,l],\Sigma_t^{\epsilon_0}} \leqslant C_l(1+t)^{-1}(&\delta_0(1+t)^{-1}\mathcal{U}_{k-1,l-1} \\ &+ \delta_0(1+t)^{-1}(\mathcal{Y}_0 + (1+t)\mathcal{A}_{[l-1]}) + \mathcal{W}_{\{l+1\}})\end{aligned} \quad (11.175)$$

So we have

$$\begin{aligned}\|\kappa\not{k}\|_{2,[k,l],\Sigma_t^{\epsilon_0}} \leqslant & C_l(1+t)^{-1}(\delta_0(1+t)^{-2}(1+\log(1+t))^2\mathcal{U}_{[k-2,l-1]} \\ & + \delta_0(1+t)^{-1}\|\kappa-1\|_{2,[k,l],\Sigma_t^{\epsilon_0}} \\ & + \delta_0(1+t)^{-1}(1+\log(1+t))(\mathcal{Y}_0+(1+t)\mathcal{A}_{[l-1]}) \\ & + (1+\log(1+t))\mathcal{W}_{\{l+1\}})\end{aligned} \tag{11.176}$$

$$\begin{aligned}\|\zeta\|_{2,[k,l],\Sigma_t^{\epsilon_0}} \leqslant & C_l(1+t)^{-1}(\delta_0(1+t)^{-1}(1+\log(1+t))\mathcal{U}_{k-1,l-1} \\ & + \delta_0(1+t)^{-1}\|\kappa-1\|_{2,[k,l],\Sigma_t^{\epsilon_0}} \\ & + \delta_0(1+t)^{-1}(1+\log(1+t))(\mathcal{Y}_0+(1+t)\mathcal{A}_{[l-1]}) \\ & + (1+\log(1+t))\mathcal{W}_{\{l+1\}})\end{aligned} \tag{11.177}$$

From the definition, we have

$$\|(q_T)_b\|_{2,[k,l],\Sigma_t^{\epsilon_0}} \leqslant C_l(1+t)^{-1}\|\kappa-1\|_{2,[k,l+1],\Sigma_t^{\epsilon_0}} \tag{11.178}$$

To derive a recursive inequality for the quantities $\mathcal{U}_{[k,l]}$, we need to derive a similar estimate for $q_T$ in terms of $\mathcal{U}_{[k-1,l-1]}$. This requires an appropriate estimate for the quantity $\mathcal{T}_{[k-1,l-1]}$ defined in the statement of Lemma 11.8. Now, the assumptions of Lemma 11.8, which are those of Lemma 11.6, all follow under the assumptions of the present proposition, by Corollary 10.1.d, with $l_*$ in the role of $l$, assumption $\mathbf{M}_{[m,l_*]}$, and Corollary 11.1.c with $l_*$ in the role of $l$. Lemma 11.8 gives a bound for the quantity $\mathcal{T}_{[k,l]}$ in terms of $\|{}^{(T)}\not{\pi}'\|_{2,[k,l],\Sigma_t^{\epsilon_0}}$, $\|\lambda'\|_{2,[k,l],\Sigma_t^{\epsilon_0}}$ and $\max_i \|{}^{(R_i)}\not{\pi}\|_{2,[l-1],\Sigma_t^{\epsilon_0}}$. From (11.51) we obtain, using the estimate (11.176) and Corollary 11.1.c with $l_*$ in the role of $l$,

$$\begin{aligned}&\|{}^{(T)}\not{\pi}'\|_{2,[k,l],\Sigma_t^{\epsilon_0}} \\ \leqslant & C(1+\log(1+t))\|\chi'\|_{2,[k,l],\Sigma_t^{\epsilon_0}} \\ & + C_l(1+t)^{-1}(1+\log(1+t))(\delta_0(1+t)^{-2}(1+\log(1+t))\mathcal{U}_{k-2,l-1} \\ & + \delta_0(1+t)^{-1}(\|\kappa-1\|_{2,[k,l],\Sigma_t^{\epsilon_0}}+\mathcal{Y}_0+(1+t)\mathcal{A}_{[l-1]})+\mathcal{W}_{\{l+1\}})\end{aligned} \tag{11.179}$$

Also, by Lemma 11.1 and Lemma 11.2.a,

$$\|\lambda'\|_{2,[k,l],\Sigma_t^{\epsilon_0}} \leqslant C_l(\|\kappa-1\|_{2,[k,l],\Sigma_t^{\epsilon_0}}+(1+\log(1+t))\mathcal{W}_{\{l\}}) \tag{11.180}$$

Also, by (10.145) we have

$$\max_i \|{}^{(R_i)}\not{\pi}\|_{2,[l-1],\Sigma_t^{\epsilon_0}} \leqslant C_l(\mathcal{Y}_0+(1+t)\mathcal{A}_{[l-1]}+\mathcal{W}_{[l]}) \tag{11.181}$$

Substituting in the conclusion of Lemma 11.8 the estimates (11.179)—(11.181) yields the following bound:

$$\begin{aligned}\mathcal{T}_{[k,l]} \leqslant & C_l((1+\log(1+t))\|\chi'\|_{2,[k,l],\Sigma_t^{\epsilon_0}} \\ & + \delta_0(1+t)^{-3}(1+\log(1+t))^2\mathcal{U}_{k-2,l-1} \\ & + (1+t)^{-1}\|\kappa-1\|_{2,[k,l],\Sigma_t^{\epsilon_0}} \\ & + \delta_0(1+t)^{-1}(1+\log(1+t))(\mathcal{Y}_0+(1+t)\mathcal{A}_{[l-1]}) \\ & + (1+t)^{-1}(1+\log(1+t))\mathcal{W}_{\{l+1\}})\end{aligned} \tag{11.182}$$

So now we need an estimate for $\|\chi'\|_{2,[k,l],\Sigma_t^{\epsilon_0}}$, which can be proved similarly as Lemma 11.16:

**Lemma 11.17** Let the assumptions of Proposition 11.2 hold. Then, provided that $\delta_0$ is suitably small (depending on $l$), the following estimate holds, for each $k \in \{0, \cdots, m\}$:

$$\begin{aligned}\|\chi'\|_{2,[k,l],\Sigma_t^{\epsilon_0}} \leqslant\, & C_l(1+t)^{-1}(\delta_0(1+t)^{-2}(1+\log(1+t))\mathcal{U}_{[k-2,l-1]} \\ & + (1+t)^{-1}\|\kappa - 1\|_{2,[k-1,l+1],\Sigma_t^{\epsilon_0}} + (\mathcal{Y}_0 + (1+t)\mathcal{A}_{[l]}) \\ & + \mathcal{W}_{\{l+1\}} + (1+t)^{-1}(1+\log(1+t))\mathcal{W}^Q_{\{l\}})\end{aligned} \tag{11.183}$$

We now resume the proof of Proposition 11.2. Substituting the result of Lemma 11.17, yields the following estimate for $\mathcal{T}_{[k,l]}$:

$$\begin{aligned}\mathcal{T}_{[k,l]} \leqslant\, & C_l(1+t)^{-1}(\delta_0(1+t)^{-2}(1+\log(1+t))^2\mathcal{U}_{k-2,l-1} \\ & + \|\kappa - 1\|_{2,[k,l+1],\Sigma_t^{\epsilon_0}} + (1+\log(1+t))(\mathcal{Y}_0 + (1+t)\mathcal{A}_{[l]}) \\ & + (1+\log(1+t))(\mathcal{W}_{\{l+1\}} + (1+t)^{-1}(1+\log(1+t))\mathcal{W}^Q_{\{l\}}))\end{aligned} \tag{11.184}$$

Consider now Lemma 11.12 with $(k+1, l+1)$ in the role of $(k, l)$. Since $(l+1)_* \leqslant l_* + 1$, the assumptions hold if

$$\begin{aligned}\|\lambda'\|_{\infty,[k,l_*],\Sigma_t^{\epsilon_0}} &\leqslant C_l\delta_0(1+\log(1+t)) \\ \max_i \|{}^{(R_i)}\not{\pi}\|_{\infty,[l_*],\Sigma_t^{\epsilon_0}} &\leqslant C_l\delta_0(1+t)^{-1}(1+\log(1+t)) \\ \|{}^{(T)}\not{\pi}'\|_{\infty,[k,l_*],\Sigma_t^{\epsilon_0}} &\leqslant C_l\delta_0(1+t)^{-1}(1+\log(1+t))\end{aligned} \tag{11.185}$$

These will follow from the assumptions of Proposition 11.2. So we have

$$\begin{aligned}& \|\xi \cdot \not{g}\|_{2,[k,l],\Sigma_t^{\epsilon_0}}, \|\xi \cdot \not{g}^{-1}\|_{2,[k,l],\Sigma_t^{\epsilon_0}} \\ & \leqslant C_l(\|\xi\|_{2,[k,l],\Sigma_t^{\epsilon_0}} + \|\xi\|_{\infty,[k,l_*],\Sigma_t^{\epsilon_0}}(\max_i \|{}^{(R_i)}\not{\pi}\|_{2,[l-1],\Sigma_t^{\epsilon_0}} + \mathcal{T}_{[k-1,l-1]}))\end{aligned} \tag{11.186}$$

Substituting from (10.145) and the bound (11.184) this becomes

$$\begin{aligned}& \|\xi \cdot \not{g}\|_{2,[k,l],\Sigma_t^{\epsilon_0}}, \|\xi \cdot \not{g}^{-1}\|_{2,[k,l],\Sigma_t^{\epsilon_0}} \\ & \leqslant C_l\|\xi\|_{2,[k,l],\Sigma_t^{\epsilon_0}} + C_l\|\xi\|_{\infty,[k,l_*],\Sigma_t^{\epsilon_0}} \\ & \cdot (\delta_0(1+t)^{-3}(1+\log(1+t))^2\mathcal{U}_{k-3,l-2} \\ & + (1+t)^{-1}\|\kappa - 1\|_{2,[k-1,l],\Sigma_t^{\epsilon_0}} + (\mathcal{Y}_0 + (1+t)\mathcal{A}_{[l-1]}) \\ & + \mathcal{W}_{\{l\}} + (1+t)^{-2}(1+\log(1+t))^2\mathcal{W}^Q_{\{l-1\}})\end{aligned} \tag{11.187}$$

Applying this to $\xi = (q_T)_b$, we obtain, by (11.178) and (11.107),

$$\begin{aligned}& \|q_T\|_{2,[k,l],\Sigma_t^{\epsilon_0}} \\ & \leqslant C_l(1+t)^{-1}(\delta_0^2(1+t)^{-1}\mathcal{U}_{k-1,l-1} + \|\kappa - 1\|_{2,[k,l+1],\Sigma_t^{\epsilon_0}} \\ & + \delta_0(1+\log(1+t))(\mathcal{Y}_0 + (1+t)\mathcal{A}_{[l-1]}) \\ & + \delta_0(1+\log(1+t))(\mathcal{W}_{\{l\}} + (1+t)^{-2}(1+\log(1+t))^2\mathcal{W}^Q_{\{l-1\}}))\end{aligned} \tag{11.188}$$

Consider now the first two recursion relations of Lemma 11.3, replacing $m$ by $j \in \{0, \cdots, k\}$. Taking the $\| \quad \|_{2,[k-j,l-j],\Sigma_t^{\epsilon_0}}$ norm we obtain

$$\begin{aligned}\|p_{T,j}\|_{2,[k-j,l-j],\Sigma_t^{\epsilon_0}} \leqslant \|p_{T,j-1}\|_{2,[k+1-j,l+1-j],\Sigma_t^{\epsilon_0}} \\ + C_l(\|\not{d}\kappa\|_{\infty,[k-j,l_*],\Sigma_t^{\epsilon_0}}\|q_{T,j-1}\|_{2,[k-j,l-j],\Sigma_t^{\epsilon_0}} \\ + \|\not{d}\kappa\|_{2,[k-j,l-j],\Sigma_t^{\epsilon_0}}\|q_{T,j-1}\|_{\infty,[k-j,l_*-j],\Sigma_t^{\epsilon_0}})\end{aligned} \tag{11.189}$$

$$\begin{aligned}\|q_{T,j}\|_{2,[k-j,l-j],\Sigma_t^{\epsilon_0}} \leqslant \|q_{T,j-1}\|_{2,[k+1-j,l+1-j],\Sigma_t^{\epsilon_0}} \\ + C_l(\|q_T\|_{\infty,[k-j,l_*],\Sigma_t^{\epsilon_0}}\|p_{T,j-1}\|_{2,[k-j,l-j],\Sigma_t^{\epsilon_0}} \\ + \|q_T\|_{2,[k-j,l-j],\Sigma_t^{\epsilon_0}}\|p_{T,j-1}\|_{\infty,[k-j,l_*-j],\Sigma_t^{\epsilon_0}})\end{aligned} \tag{11.190}$$

Setting, for fixed $(k, l)$ and with $j \in \{0, \cdots, k\}$,

$$\begin{aligned}x_j &= \|p_{T,j}\|_{2,[k-j,l-j],\Sigma_t^{\epsilon_0}} \\ y_j &= \|q_{T,j}\|_{2,[k-j,l-j],\Sigma_t^{\epsilon_0}}\end{aligned} \tag{11.191}$$

and

$$\begin{aligned}a &= x_0 = 0 \\ b &= \|q_T\|_{2,[k,l],\Sigma_t^{\epsilon_0}} = y_0 \\ c &= \|\not{d}\kappa\|_{2,[k,l],\Sigma_t^{\epsilon_0}}\end{aligned} \tag{11.192}$$

(11.189)—(11.190) imply by assumptions $\mathbf{M}_{[k,l_*+1]}$ and $\mathbf{E}_{\{l_*+1\}}$ and Corollary 11.1.a with $l_*$ in the role of $l$,

$$\begin{aligned}x_j &\leqslant x_{j-1} + \epsilon(y_{j-1} + c) \\ y_j &\leqslant y_{j-1} + \epsilon(x_{j-1} + \epsilon b)\end{aligned} \tag{11.193}$$

where

$$\epsilon = C_l\delta_0(1+t)^{-1}(1 + \log(1+t)) \tag{11.194}$$

The inequalities (11.193) can be written in the form

$$\begin{bmatrix} x_j \\ y_j \end{bmatrix} \leqslant (\mathbf{I} + \mathbf{E}) \begin{bmatrix} x_{j-1} \\ y_{j-1} \end{bmatrix} + \mathbf{a} \tag{11.195}$$

where

$$\mathbf{E} = \begin{bmatrix} 0 & \epsilon \\ \epsilon & 0 \end{bmatrix} \quad \mathbf{a} = \begin{bmatrix} \epsilon c \\ \epsilon^2 b \end{bmatrix} \tag{11.196}$$

It follows that

$$\begin{bmatrix} x_j \\ y_j \end{bmatrix} \leqslant (\mathbf{I} + \mathbf{E})^j \begin{bmatrix} x_0 \\ y_0 \end{bmatrix} + (\sum_{i=0}^{j-1}(\mathbf{I} + \mathbf{E})^i)\mathbf{a} \tag{11.197}$$

Now, in matrix norm,

$$|\mathbf{E}| \leqslant \epsilon$$

and (11.197) implies, for $\epsilon \leqslant 1$,

$$\begin{aligned} x_j &\leqslant x_0 + C_j\epsilon(x_0 + y_0) + C_j(\epsilon c + \epsilon^2 b) \\ y_j &\leqslant y_0 + C_j\epsilon(x_0 + y_0) + C_j(\epsilon c + \epsilon^2 b) \end{aligned} \tag{11.198}$$

In view of (11.191) and (11.192), substituting for $b$ the bounds (11.188), and noting that

$$c \leqslant C(1+t)^{-1}\|\kappa - 1\|_{2,[k,l+1],\Sigma_t^{\epsilon_0}}$$

yields the estimates

$$\begin{aligned} &\|p_{T,j}\|_{2,[k-j,l-j],\Sigma_t^{\epsilon_0}} \\ &\leqslant C_l\delta_0(1+t)^{-2}(1+\log(1+t))(\delta_0(1+t)^{-1}\mathcal{U}_{[k-1,l-1]} \\ &\quad + \|\kappa - 1\|_{2,[k,l+1],\Sigma_t^{\epsilon_0}} + \delta_0(1+\log(1+t))(\mathcal{Y}_0 + (1+t)\mathcal{A}_{[l-1]}) \\ &\quad + \delta_0(1+\log(1+t))(\mathcal{W}_{\{l\}} + (1+t)^{-2}(1+\log(1+t))^2\mathcal{W}^Q_{\{l-1\}})) \end{aligned} \tag{11.199}$$

for all $j \in \{0, \cdots, k\}$.

$$\begin{aligned} &\|q_{T,j}\|_{2,[k-j,l-j],\Sigma_t^{\epsilon_0}} \\ &\leqslant C_l(1+t)^{-1}(\delta_0(1+t)^{-1}\mathcal{U}_{[k-1,l-1]} \\ &\quad + \|\kappa - 1\|_{2,[k,l+1],\Sigma_t^{\epsilon_0}} + \delta_0(1+\log(1+t))(\mathcal{Y}_0 + (1+t)\mathcal{A}_{[l-1]}) \\ &\quad + \delta_0(1+\log(1+t))(\mathcal{W}_{\{l\}} + (1+t)^{-2}(1+\log(1+t))^2\mathcal{W}^Q_{\{l-1\}})) \end{aligned} \tag{11.200}$$

for all $j \in \{0, \cdots, k\}$.

Next, we consider the formula (11.140). Taking the $\| \quad \|_{2,[l-k+n],\Sigma_t}$ norm we obtain, using (11.200),

$$\begin{aligned} &\|r^n_{T,k}\|_{2,[l-k+n],\Sigma_t^{\epsilon_0}} \\ &\leqslant \sum_{i=0}^{k-1-n} N_i(n) \sum_{j'=0}^{k-n-1-i} \|\kappa q_{T,j'}\|_{2,[k-1-n-j',l-1-n],\Sigma_t^{\epsilon_0}} \\ &\leqslant C_l \max_{j' \in \{0,\cdots,k-1-n\}} \|\kappa q_{T,j'}\|_{2,[k-1-j',l-1-j'],\Sigma_t^{\epsilon_0}} \\ &\leqslant C_l(1+t)^{-1}(1+\log(1+t))(\delta_0(1+t)^{-1}\mathcal{U}_{[k-2,l-2]} \\ &\quad + \|\kappa - 1\|_{2,[k-1,l],\Sigma_t^{\epsilon_0}} + \delta_0(1+\log(1+t))(\mathcal{Y}_0 + (1+t)\mathcal{A}_{[l-2]}) \\ &\quad + \delta_0(1+\log(1+t))(\mathcal{W}_{\{l-1\}} + (1+t)^{-2}(1+\log(1+t))^2\mathcal{W}^Q_{\{l-2\}})) \end{aligned} \tag{11.201}$$

for all $n \in \{0, \cdots, k-1\}$.

Finally, we take the $\| \quad \|_{2,[l-k],\Sigma_t^{\epsilon_0}}$ norm of (11.142). We then obtain on the left

$$\|(T)^{k+1}\hat{T}^i\|_{2,[l-k],\Sigma_t^{\epsilon_0}} = \mathcal{U}_{k,l} - \mathcal{U}_{k-1,l}$$

By (11.199) and (11.200) with $j = k$ and Proposition 10.2 together with Proposition 10.1 and Corollary 11.1.a with $l_*$ in the role of $l$,

$$\begin{aligned} &\max_i \|p_{T,k}\hat{T}^i\|_{2,[l-k],\Sigma_t^{\epsilon_0}} \\ &\leqslant C_l\delta_0(1+t)^{-2}(1+\log(1+t))(\delta_0(1+t)^{-1}\mathcal{U}_{[k-1,l-1]} \\ &\quad + \|\kappa - 1\|_{2,[k,l+1],\Sigma_t^{\epsilon_0}} + \delta_0(1+\log(1+t))(\mathcal{Y}_0 + (1+t)\mathcal{A}_{[l-1]}) \\ &\quad + \delta_0(1+\log(1+t))(\mathcal{W}_{\{l\}} + (1+t)^{-2}(1+\log(1+t))^2\mathcal{W}^Q_{\{l-1\}})) \end{aligned} \tag{11.202}$$

$$\begin{aligned} &\max_i \|q_{T,k}\cdot \not{d}x^i\|_{2,[l-k],\Sigma_t^{\epsilon_0}} \\ &\leqslant C_l(1+t)^{-1}(\delta_0(1+t)^{-1}\mathcal{U}_{[k-1,l-1]} \\ &\quad + \|\kappa - 1\|_{2,[k,l+1],\Sigma_t^{\epsilon_0}} + \delta_0(1+\log(1+t))(\mathcal{Y}_0 + (1+t)\mathcal{A}_{[l-1]}) \\ &\quad + \delta_0(1+\log(1+t))(\mathcal{W}_{\{l\}} + (1+t)^{-2}(1+\log(1+t))^2\mathcal{W}^Q_{\{l-1\}})) \end{aligned} \tag{11.203}$$

By (11.201) with $n = 0$ and Proposition 10.2 ($k \geqslant 1$) together with Proposition 10.1 with $l_* + 1$ in the role of $l$ and Corollary 11.1.a with $l_*$ in the role of $l$,

$$\begin{aligned} &\|r^0_{T,k}\cdot \not{d}\hat{T}^i\|_{2,[l-k],\Sigma_t^{\epsilon_0}} \\ &\leqslant C_l(1+t)^2(1+\log(1+t))\cdot(\delta_0^2(1+t)^{-1}\mathcal{U}_{k-2,l-2} + \|\kappa - 1\|_{2,[k-1,l],\Sigma_t^{\epsilon_0}} \\ &\quad + \delta_0(1+\log(1+t))(\mathcal{Y}_0 + (1+t)\mathcal{A}_{[l-1]}) \\ &\quad + \delta_0(1+\log(1+t))(\mathcal{W}_{\{l\}} + (1+t)^{-2}(1+\log(1+t))^2\mathcal{W}^Q_{\{l-1\}})) \end{aligned} \tag{11.204}$$

Also by (11.201) together with Proposition 11.1 and Corollary 11.1.a with $l_*$ in the role of $l$,

$$\begin{aligned} &\max_i \|\sum_{n=1}^{k-1} r^n_{T,k}\cdot \not{d}(T)^n\hat{T}^i\|_{2,[l-k],\Sigma_t^{\epsilon_0}} \\ &\leqslant C_l(1+t)^{-1}(\max_{n\in\{1,\cdots,k-1\}}\max_i \|r^n_{T,k}\|_{\infty,[l_*-k+n],\Sigma_t^{\epsilon_0}}\mathcal{U}_{k-2,l-1} \\ &\quad + (\sum_{n=1}^{k-1}\|r^n_{T,k}\|_{2,[l-k],\Sigma_t^{\epsilon_0}})\max_{n\in\{1,\cdots,k-1\}}\|(T)^n\hat{T}^i\|_{\infty,[l_*+1-n],\Sigma_t^{\epsilon_0}}) \\ &\leqslant C_l\delta_0(1+t)^{-2}(1+\log(1+t))^2\mathcal{U}_{k-2,l-1} \\ &\quad + C_l\delta_0(1+t)^{-3}(1+\log(1+t))^2(\|\kappa - 1\|_{2,[k-1,l-1],\Sigma_t^{\epsilon_0}} \\ &\quad + \delta_0(1+\log(1+t))(\mathcal{Y}_0 + (1+t)\mathcal{A}_{[l-2]}) \\ &\quad + \delta_0(1+\log(1+t))(\mathcal{W}_{\{l-1\}} + (1+t)^{-2}(1+\log(1+t))^2\mathcal{W}^Q_{\{l-2\}})) \end{aligned} \tag{11.205}$$

Combining (11.204) and (11.205) we obtain

$$
\begin{aligned}
&\max_i \|\sum_{n=0}^{k-1} r^n_{T,k} \cdot \not{d}(T)^n \hat{T}^i\|_{2,[l-k],\Sigma_t^{\epsilon_0}} \\
&\leqslant C_l(1+t)^{-2}(1+\log(1+t)) \qquad (11.206)\\
&\quad \cdot (\delta_0(1+\log(1+t))\mathcal{U}_{k-2,l-1} + \|\kappa - 1\|_{2,[k-1,l],\Sigma_t^{\epsilon_0}} \\
&\quad + \delta_0(1+\log(1+t))(\mathcal{Y}_0 + (1+t)\mathcal{A}_{[l-1]}) \\
&\quad + \delta_0(1+\log(1+t))(\mathcal{W}_{\{l\}} + (1+t)^{-2}(1+\log(1+t))^2\mathcal{W}^Q_{\{l-1\}}))
\end{aligned}
$$

In view of (11.202), (11.203) and (11.206), we arrive at the following recursive inequality:

$$
\mathcal{U}_{k,l} - \mathcal{U}_{k-1,l} \leqslant C_l\delta_0(1+t)^{-2}(1+\log(1+t))^2\mathcal{U}_{k-1,l-1} + b_{k,l} \qquad (11.207)
$$

where

$$
\begin{aligned}
b_{k,l} = &C_l(1+t)^{-1}(\|\kappa - 1\|_{2,[k,l+1],\Sigma_t^{\epsilon_0}} \qquad (11.208)\\
&+ \delta_0(1+\log(1+t))(\mathcal{Y}_0 + (1+t)\mathcal{A}_{[l-1]} \\
&+ \mathcal{W}_{\{l\}} + (1+t)^{-2}(1+\log(1+t))^2\mathcal{W}^Q_{\{l-1\}}))
\end{aligned}
$$

(11.207) implies

$$
\mathcal{U}_{k,l} \leqslant (1 + C_l\delta_0(1+t)^{-2}(1+\log(1+t))^2)\mathcal{U}_{k-1,l} + b_{k,l} \qquad (11.209)
$$

which yields

$$
\begin{aligned}
\mathcal{U}_{k,l} \leqslant &(1 + C_l\delta_0(1+t)^{-2}(1+\log(1+t))^2)^k\mathcal{U}_{0,l} \qquad (11.210)\\
&+ \sum_{j=1}^{k}(1 + C_l\delta_0(1+t)^{-2}(1+\log(1+t))^2)^{k-j}b_{j,l}
\end{aligned}
$$

Since $b_{k,l}$ are non-decreasing in $k$, this in turn implies

$$
\mathcal{U}_{k,l} \leqslant C_l(\mathcal{U}_{0,l} + b_{k,l}) \qquad (11.211)
$$

Substituting finally in (11.211) the estimate (11.162) and the definition (11.208) we obtain

$$
\begin{aligned}
\mathcal{U}_{k,l} \leqslant &C_l(1+t)^{-1}(\|\kappa - 1\|_{2,[k,l+1],\Sigma_t^{\epsilon_0}} \qquad (11.212)\\
&+ \delta_0(1+\log(1+t))(\mathcal{Y}_0 + (1+t)\mathcal{A}_{[l-1]} \\
&+ \mathcal{W}_{\{l\}} + (1+t)^{-2}(1+\log(1+t))^2\mathcal{W}^Q_{\{l-1\}}))
\end{aligned}
$$

Since we have

$$
\|\kappa - 1\|_{2,[k,l+1],\Sigma_t^{\epsilon_0}} \leqslant C_l(\mathcal{B}_{[k,l+1]} + (1+\log(1+t))\mathcal{W}_{\{l+1\}}) \qquad (11.213)
$$

The proposition follows. □

As before, we have many corollaries.

**Corollary 11.2.a** Under the assumptions of Proposition 11.2 the coefficients of the expression for $(T)^{k+1}\hat{T}^i$, $k = 0, \cdots, m$, of Lemma 11.3 satisfy

$$\begin{aligned}
&\|p_{T,k}\|_{2,[m-k,l-k],\Sigma_t^{\epsilon_0}} \\
&\leqslant C_l\delta_0(1+t)^{-2}(1+\log(1+t))(\mathcal{B}_{[m,l+1]} + \delta_0(1+\log(1+t))(\mathcal{Y}_0 + (1+t)\mathcal{A}_{[l-1]}) \\
&\quad + (1+\log(1+t))(\mathcal{W}_{\{l+1\}} + \delta_0(1+t)^{-2}(1+\log(1+t))^2\mathcal{W}^Q_{\{l-1\}}))
\end{aligned}$$

for all $k \in \{0, \cdots, m\}$,

$$\begin{aligned}
&\|q_{T,k}\|_{2,[m-k,l-k],\Sigma_t^{\epsilon_0}} \\
&\leqslant C_l(1+t)^{-1}(\mathcal{B}_{[m,l+1]} + \delta_0(1+\log(1+t))(\mathcal{Y}_0 + (1+t)\mathcal{A}_{[l-1]}) \\
&\quad + (1+\log(1+t))(\mathcal{W}_{\{l+1\}} + \delta_0(1+t)^{-2}(1+\log(1+t))^2\mathcal{W}^Q_{\{l-1\}}))
\end{aligned}$$

for all $k \in \{0, \cdots, m\}$, and

$$\begin{aligned}
&\|r^n_{T,k}\|_{2,[l-k+n],\Sigma_t^{\epsilon_0}} \\
&\leqslant C_l(1+t)^{-1}(1+\log(1+t))(\mathcal{B}_{[m,l]} + \delta_0(1+\log(1+t))(\mathcal{Y}_0 + (1+t)\mathcal{A}_{[l-2]}) \\
&\quad + (1+\log(1+t))(\mathcal{W}_{\{l\}} + \delta_0(1+t)^{-2}(1+\log(1+t))^2\mathcal{W}^Q_{\{l-2\}}))
\end{aligned}$$

for all $k \in \{0, \cdots, m\}$, $n \in \{0, \cdots, k-1\}$.

Also,

$$\begin{aligned}
&\sum_{k=0}^{m} \max_i \|(T)^{k+1}L^i\|_{2,[l-k],\Sigma_t^{\epsilon_0}} \\
&\leqslant C_l(1+t)^{-1}(\mathcal{B}_{[m,l+1]} + \delta_0(1+\log(1+t))(\mathcal{Y}_0 + (1+t)\mathcal{A}_{[l-1]}) \\
&\quad + (1+\log(1+t))(\mathcal{W}_{\{l+1\}} + \delta_0(1+t)^{-2}(1+\log(1+t))^2\mathcal{W}^Q_{\{l-1\}}))
\end{aligned}$$

**Corollary 11.2.b** Under the assumptions of Proposition 11.2 the following estimates hold:

$$\begin{aligned}
&\|\psi_{\hat{T}}\|_{2,[m+1,l+1],\Sigma_t^{\epsilon_0}} \\
&\leqslant C_l(\mathcal{W}_{\{l+1\}} + \delta_0(1+t)^{-1}((1+t)^{-1}\mathcal{B}_{[m,l+1]} + \mathcal{Y}_0 \\
&\quad + (1+t)\mathcal{A}_{[l]} + \delta_0(1+t)^{-2}\mathcal{W}^Q_{\{l\}})) \\
&\|\psi_L\|_{2,[m+1,l+1],\Sigma_t^{\epsilon_0}} \\
&\leqslant C_l(\mathcal{W}_{\{l+1\}} + \delta_0(1+t)^{-1}((1+t)^{-1}\mathcal{B}_{[m,l+1]} + \mathcal{Y}_0 \\
&\quad + (1+t)\mathcal{A}_{[l]} + \delta_0(1+t)^{-2}\mathcal{W}^Q_{\{l\}})) \\
&\|\psi\|_{2,[m+1,l+1],\Sigma_t^{\epsilon_0}} \\
&\leqslant C_l(\mathcal{W}_{\{l+1\}} + \delta_0(1+t)^{-1}((1+t)^{-1}\mathcal{B}_{[m,l+1]} + \mathcal{Y}_0 \\
&\quad + (1+t)\mathcal{A}_{[l]} + \delta_0(1+t)^{-2}\mathcal{W}^Q_{\{l\}}))
\end{aligned}$$

and

$$\begin{aligned}
&\|\omega_{T\hat{T}}\|_{2,[m,l],\Sigma_T^{\epsilon_0}} \\
&\leqslant C_l(\mathcal{W}_{\{l+1\}} + \delta_0(1+t)^{-1}((1+t)^{-1}\mathcal{B}_{[m-1,l]} + \mathcal{Y}_0 \\
&\quad + (1+t)\mathcal{A}_{[l-1]} + \delta_0(1+t)^{-2}\mathcal{W}^Q_{\{l-1\}})) \\
&\|\omega_{L\hat{T}}\|_{2,[m,l],\Sigma_t^{\epsilon_0}} \\
&\leqslant C_l(1+t)^{-1}(\mathcal{W}^Q_{\{l\}} + \delta_0(1+t)^{-1}((1+t)^{-1}\mathcal{B}_{[m-1,l]} + \mathcal{Y}_0 \\
&\quad + (1+t)\mathcal{A}_{[l-1]} + \mathcal{W}_{\{l\}})) \\
&\|\not{\psi}_{\hat{T}}\|_{2,[m,l],\Sigma_t^{\epsilon_0}} \\
&\leqslant C_l(1+t)^{-1}(\mathcal{W}_{\{l+1\}} + \delta_0(1+t)^{-1}((1+t)^{-1}\mathcal{B}_{[m-1,l]} + \mathcal{Y}_0 \\
&\quad + (1+t)\mathcal{A}_{[l-1]} + \delta_0(1+t)^{-2}\mathcal{W}^Q_{\{l-1\}})) \\
&\|\not{\psi}_L\|_{2,[m,l],\Sigma_t^{\epsilon_0}} \\
&\leqslant C_l(1+t)^{-1}(\mathcal{W}_{\{l+1\}} + \delta_0(1+t)^{-1}((1+t)^{-1}\mathcal{B}_{[m-1,l]} + \mathcal{Y}_0 \\
&\quad + (1+t)\mathcal{A}_{[l-1]} + \delta_0(1+t)^{-2}\mathcal{W}^Q_{\{l-1\}})) \\
&\|\not{\psi}\|_{2,[m,l],\Sigma_t^{\epsilon_0}} \\
&\leqslant C_l(1+t)^{-1}(\mathcal{W}_{\{l+1\}} + \delta_0(1+t)^{-1}((1+t)^{-1}\mathcal{B}_{[m-1,l]} + \mathcal{Y}_0 \\
&\quad + (1+t)\mathcal{A}_{[l-1]} + \delta_0(1+t)^{-2}\mathcal{W}^Q_{\{l-1\}}))
\end{aligned}$$

Also, we have

$$\begin{aligned}
\|\not{k}\|_{2,[m,l],\Sigma_t^{\epsilon_0}} &\leqslant C_l(1+t)^{-1}(\mathcal{W}_{\{l+1\}} + \delta_0(1+t)^{-2}\mathcal{W}^Q_{\{l\}} \\
&\quad + \delta_0(1+t)^{-1}((1+t)^{-1}\mathcal{B}_{[m-1,l]} + \mathcal{Y}_0 + (1+t)\mathcal{A}_{[l-1]})) \\
\|\kappa^{-1}\zeta\|_{2,[m,l],\Sigma_t^{\epsilon_0}} &\leqslant C_l(1+t)^{-1}(\mathcal{W}_{\{l+1\}} + \delta_0(1+t)^{-2}\mathcal{W}^Q_{\{l\}} \\
&\quad + \delta_0(1+t)^{-1}((1+t)^{-1}\mathcal{B}_{[m-1,l]} + \mathcal{Y}_0 + (1+t)\mathcal{A}_{[l-1]})) \\
\|\kappa\not{k}\|_{2,[m,l],\Sigma_t^{\epsilon_0}} &\leqslant C_l(1+t)^{-1}((1+\log(1+t))(\mathcal{W}_{\{l+1\}} + \delta_0(1+t)^{-2}\mathcal{W}^Q_{\{l\}}) \\
&\quad + \delta_0(1+t)^{-1}(\mathcal{B}_{[m,l]} + (1+\log(1+t))(\mathcal{Y}_0 + (1+t)\mathcal{A}_{[l-1]}))) \\
\|\zeta\|_{2,[m,l],\Sigma_t^{\epsilon_0}} &\leqslant C_l(1+t)^{-1}((1+\log(1+t))(\mathcal{W}_{\{l+1\}} + \delta_0(1+t)^{-2}\mathcal{W}^Q_{\{l\}}) \\
&\quad + \delta_0(1+t)^{-1}(\mathcal{B}_{[m,l]} + (1+\log(1+t))(\mathcal{Y}_0 + (1+t)\mathcal{A}_{[l-1]})))
\end{aligned}$$

The next Corollary is obtained from Lemma 11.17 by substituting Proposition 11.2.

**Corollary 11.2.c** Under the assumptions of Proposition 11.2 we have

$$\begin{aligned}
\|\chi'\|_{2,[m,l],\Sigma_t^{\epsilon_0}} &\leqslant C_l(1+t)^{-1}((1+t)^{-1}\mathcal{B}_{[m-1,l+1]} + \mathcal{Y}_0 + (1+t)\mathcal{A}_{[l]} \\
&\quad + \mathcal{W}_{\{l+1\}} + (1+t)^{-1}(1+\log(1+t))\mathcal{W}^Q_{\{l\}})
\end{aligned}$$

Moreover by (11.184) we have

$$\mathcal{T}_{[m,l]} \leqslant C_l(1+t)^{-1}(\mathcal{B}_{[m,l+1]} + (1+\log(1+t))(\mathcal{Y}_0 + (1+t)\mathcal{A}_{[l]} + \mathcal{W}_{\{l+1\}} + (1+t)^{-1}(1+\log(1+t))\mathcal{W}^Q_{\{l\}}))$$

and obviously,

$$\|{}^{(T)}\not{\pi} + 2(1-u+t)^{-1}\not{g}\|_{2,[m,l],\Sigma_t^{\epsilon_0}} \leqslant C_l \mathcal{T}_{[m,l]}$$

Also from the estimate for $\zeta$ and (11.187),

$$\|\Lambda\|_{2,[m,l],\Sigma_t^{\epsilon_0}} \leqslant C_l(1+t)^{-1}(\mathcal{B}_{[m,l+1]} + \delta_0(1+\log(1+t))(\mathcal{Y}_0 + (1+t)\mathcal{A}_{[l-1]}) + (1+\log(1+t))(\mathcal{W}_{\{l+1\}} + \delta_0(1+t)^{-2}(1+\log(1+t))\mathcal{W}^Q_{\{l\}}))$$

and by Lemma 11.13 and Lemma 10.11,

$$\|{}^{(T)}\not{\pi}_1\|_{2,[m,l-1],\Sigma_t^{\epsilon_0}} \leqslant C_l(1+t)^{-2}(\mathcal{B}_{[m,l+1]} + (1+\log(1+t))(\mathcal{Y}_0 + (1+t)\mathcal{A}_{[l]} + \mathcal{W}_{\{l+1\}} + (1+t)^{-1}(1+\log(1+t))\mathcal{W}^Q_{\{l\}}))$$

We now revisit Lemma 11.12. Under the assumptions of Proposition 11.2 the following bounds hold:

$$\begin{aligned} \|\lambda'\|_{\infty,[m,l_*],\Sigma_t^{\epsilon_0}} &\leqslant C_l\delta_0(1+\log(1+t)) \\ \max_i \|{}^{(R_i)}\not{\pi}\|_{\infty,[l_*],\Sigma_t^{\epsilon_0}} &\leqslant C_l\delta_0(1+t)^{-1}(1+\log(1+t)) \\ \|{}^{(T)}\not{\pi}'\|_{\infty,[m,l_*],\Sigma_t^{\epsilon_0}} &\leqslant C_l\delta_0(1+t)^{-1}(1+\log(1+t)) \end{aligned} \tag{11.214}$$

These bounds imply the assumptions of Lemma 11.12 with $(m,l)$ in the role of $(k-1,l-1)$. Then we obtain

$$\|\xi\cdot\not{g}\|_{2,[m,l],\Sigma_t^{\epsilon_0}}, \|\xi\cdot\not{g}^{-1}\|_{2,[m,l],\Sigma_t^{\epsilon_0}} \leqslant C_l(\|\xi\|_{2,[m,l],\Sigma_t^{\epsilon_0}} + \|\xi\|_{\infty,[m,l_*-1],\Sigma_t^{\epsilon_0}}(\max_i \|{}^{(R_i)}\not{\pi}\|_{2,[l-1],\Sigma_t^{\epsilon_0}} + \mathcal{T}_{[m-1,l-1]})) \tag{11.215}$$

So by Corollary 10.1.d and Corollary 11.2.c, we obtain

**Corollary 11.2.d** Under the assumptions of Proposition 11.2, we have

$$\|\xi\cdot\not{g}\|_{2,[m,l],\Sigma_t^{\epsilon_0}}, \|\xi\cdot\not{g}^{-1}\|_{2,[m,l],\Sigma_t^{\epsilon_0}} \leqslant C_l(\|\xi\|_{2,[m,l],\Sigma_t^{\epsilon_0}} + \|\xi\|_{\infty,[m,l_*-1],\Sigma_t^{\epsilon_0}}(\mathcal{W}_{\{l\}} + (1+t)^{-2}(1+\log(1+t))\mathcal{W}^Q_{\{l-1\}} + (1+t)^{-1}\mathcal{B}_{[m-1,l]} + \mathcal{Y}_0 + (1+t)\mathcal{A}_{[l-1]}))$$

## 11.2 Bounds for Quantities $Q'_{m,l}$ and $P'_{m,l}$

The object of this section is to obtain appropriate bounds for the quantities ${}^{(i_1\cdots i_l)}Q'_{m,l}$ and ${}^{(i_1\cdots i_l)}P'_{m,l}$, which through ${}^{(i_1\cdots i_l)}B'_{m,l}$, enter the final estimates for the functions ${}^{(i_1\cdots i_l)}x'_{m,l}$.

### 11.2.1 Bounds for $Q'_{m,l}$

The quantities ${}^{(i_1\cdots i_l)}Q'_{m,l}(t)$ are defined by (9.256). We first estimate the last term:

$$\|{}^{(i_1\cdots i_l)}\dot{g}'_{m,l}(t)\|_{L^2([0,\epsilon_0]\times S^2)} \tag{11.216}$$

${}^{(i_1\cdots i_l)}\dot{g}'_{m,l}$ are defined in terms of ${}^{(i_1\cdots i_l)}g'_{m,l}$ by (9.242) and (9.243). ${}^{(i_1\cdots i_l)}g'_{m,l}$ are in turn defined in Proposition 9.1 and Proposition 9.2. The first term in the expression for ${}^{(i_1\cdots i_l)}g'_{m,l}$ is

$$R_{i_l}\cdots R_{i_1}(T)^m\breve{g}'$$

where $\breve{g}'$ is defined by (9.44). Note that for $m = l = 0$ we have

$$g'_{0,0} = \breve{g}' \tag{11.217}$$

while according to (9.242),

$$\dot{g}'_{0,0} = g'_{0,0} - \xi \cdot x_0 \tag{11.218}$$

where $\xi$ is defined by (9.59). Defining

$$\dot{\breve{g}}' := \dot{g}'_{0,0} \tag{11.219}$$

Our objective in the following is to estimate

$$\|\dot{\breve{g}}'\|_{2,[m,l],\Sigma_t^{\epsilon_0}}$$

Recalling from Chapter 8 that

$$x_0 = \mu \not{d}\mathrm{tr}\chi + \not{d}\check{f}$$

we can write

$$\dot{\breve{g}}' = \breve{g}' - \xi \cdot (\mu\not{d}\mathrm{tr}\chi + \not{d}\check{f}) \tag{11.220}$$

Substituting the expression (9.44) for $\breve{g}'$ in (11.220):

$$\begin{aligned}\dot{\breve{g}}' = &-\xi \cdot \not{d}\check{f} - 2\mu(\not{d}\mu)\cdot(i - \not{d}e) + \frac{1}{2}\frac{d\log\Omega}{dh}(Lh)f'_0 \\ &-\frac{1}{2}(\mu\mathrm{tr}\chi + 2m + 2\mu e)\mu f'_1 + \dot{n}''_0 + \mu(\dot{n}'_1 + \dot{n}'_2)\end{aligned} \tag{11.221}$$

where

$$\dot{n}''_0 = -\frac{1}{2}L(\frac{dH}{dh})(\underline{L}Th) + \frac{1}{2}\frac{dH}{dh}(\ddot{\tau} + \underline{\nu}(LTh) + 2\zeta\cdot\not{d}Th) \tag{11.222}$$

$\ddot{\tau}$ is defined in terms of

$$\dot{\upsilon}' = \upsilon' + \Omega\kappa^2(\not{d}h)\cdot\not{d}\mathrm{tr}\chi \tag{11.223}$$

that is

$$\ddot{\tau} = \Omega^{-1}(\dot{\upsilon}' + T\tilde{\tau} + {}^{(T)}\delta\tilde{\tau}) \tag{11.224}$$

We shall state the definition of $\upsilon'$ later. By the definition of $\upsilon'$, which will be introduced later, we see there is no acoustical principal part in $n''_0$. Also $\dot{n}'_1$ is just $n'_1$, which is defined by

$$n'_1 = \frac{1}{2}\frac{dH}{dh}n_1 + \not{d}(\frac{dH}{dh})\cdot\not{d}T(h) + \frac{1}{2}T(h)\not{\Delta}(\frac{dH}{dh})$$

while $\dot{n}'_2$ is defined by

$$\dot{n}'_2 = \mu\dot{n}_2 - L(\frac{\mu}{2\eta^2}(\frac{\rho}{\rho'})')\not{\Delta}h - L(\mu\eta^{-1}\hat{T}^i)\not{\Delta}\psi_i \tag{11.225}$$

with $\dot{n}_2$ defined by

$$\dot{n}_2 = n_2 - (\frac{1}{2\eta^2}(\frac{\rho}{\rho'})'\not{d}h + \eta^{-1}\not{d}\psi_{\hat{T}} - \eta^{-2}(L\psi_i)\not{d}x^i)\cdot\not{d}\mathrm{tr}\chi \tag{11.226}$$

and

$$\begin{aligned} n_2 = & -\frac{1}{2\alpha^2}(\frac{\rho}{\rho'})'[L,\not{\Delta}]h - \alpha^{-1}\hat{T}^i[L,\not{\Delta}]\psi_i \\ & + \not{d}(\frac{1}{\alpha^2}(\frac{\rho}{\rho'})')\cdot\not{d}Lh + 2\not{d}(\alpha^{-1}\hat{T}^i)\cdot\not{d}(L\psi_i) \\ & + \not{\Delta}(\frac{1}{2\alpha^2}(\frac{\rho}{\rho'})')Lh + \not{\Delta}(\alpha^{-1}\hat{T}^i)(L\psi_i) \end{aligned}$$

$\dot{n}_2$ does not contain principal acoustical parts.

Now $\upsilon'$ is defined by

$$\upsilon' = \upsilon - \Omega(Th)\not{\Delta}\mu \tag{11.227}$$

We thus have

$$\dot{\upsilon}' = \upsilon - \Omega(Th)\not{\Delta}\mu + \Omega\kappa^2(\not{d}h)\cdot\not{d}\mathrm{tr}\chi \tag{11.228}$$

$\upsilon$ is given by

$$\upsilon = \upsilon_1 + \upsilon_2 + \upsilon_3 \tag{11.229}$$

where

$$\begin{aligned} \upsilon_1 = & (1/2)\mathrm{tr}^{(T)}\tilde{\not{\pi}}(L\underline{L}h + \nu\underline{L}h) \\ & + (1/4)(\mu^{-1(T)}\tilde{\pi}_{L\underline{L}})L^2h \\ & - {}^{(T)}\tilde{Z}\cdot\not{d}\underline{L}h - {}^{(T)}\tilde{\underline{Z}}\cdot\not{d}Lh \\ & + (1/2)^{(T)}\tilde{\pi}_{L\underline{L}}\not{\Delta}h + \mu^{(T)}\hat{\tilde{\not{\pi}}}\cdot\not{D}^2h \\ \upsilon_2 = & (1/4)L(\mathrm{tr}^{(T)}\tilde{\not{\pi}})\underline{L}h + (1/4)\underline{L}(\mathrm{tr}^{(T)}\tilde{\not{\pi}})Lh \\ & + (1/4)L(\mu^{-1(T)}\tilde{\pi}_{L\underline{L}})Lh \end{aligned}$$

$$
\begin{aligned}
&-(1/2)\not{\mathcal{L}}_L{}^{(T)}\underline{\tilde{Z}}\cdot\not{d}h-(1/2)\not{\mathcal{L}}_{\underline{L}}{}^{(T)}\tilde{Z}\cdot\not{d}h\\
&-(1/2)\mathrm{d}\not{\mathrm{i}}\mathrm{v}^{(T)}\tilde{Z}(\underline{L}h)-(1/2)\mathrm{d}\not{\mathrm{i}}\mathrm{v}^{(T)}\underline{\tilde{Z}}(Lh)\\
&+(1/2)\not{d}^{(T)}\tilde{\pi}_{L\underline{L}}\cdot\not{d}h+\mathrm{d}\not{\mathrm{i}}\mathrm{v}(\mu^{(T)}\hat{\tilde{\not{\pi}}})\cdot\not{d}h
\end{aligned}
$$

and

$$v_3=v_3^L Lh+v_3^{\underline{L}}\underline{L}h+\not{v}_3\cdot\not{d}h$$

where

$$
\begin{aligned}
v_3^L&=(1/4)\mathrm{tr}\underline{\chi}\mathrm{tr}^{(T)}\tilde{\not{\pi}}+(1/4)\mathrm{tr}\chi(\mu^{-1(T)}\tilde{\pi}_{\underline{L}\underline{L}})+(1/2)^{(T)}\tilde{Z}\cdot\not{d}(\alpha^{-1}\kappa)\\
v_3^{\underline{L}}&=-(1/4)(L\log\Omega)\mathrm{tr}^{(T)}\tilde{\not{\pi}}\\
\not{v}_3&=-(1/2)(\mathrm{tr}^{(T)}\tilde{\not{\pi}})\Lambda-(1/2)(\mathrm{tr}\underline{\chi}+L(\alpha^{-1}\kappa))^{(T)}\tilde{Z}-(1/2)\mathrm{tr}\chi^{(T)}\underline{\tilde{Z}}
\end{aligned}
$$

Defining

$$
\begin{aligned}
\dot{v}_1&=\Omega^{-1}v_1\\
\dot{v}_2&=\Omega^{-1}v_2-(Th)\not{\Delta}\mu+\kappa^2(\not{d}h)\cdot\not{d}\mathrm{tr}\chi\\
\dot{v}_3&=\Omega^{-1}v_3
\end{aligned}
\tag{11.230}
$$

and

$$\dot{v}=\dot{v}_1+\dot{v}_2+\dot{v}_3 \tag{11.231}$$

Then according to the expression of $v_1, v_2, v_3, \dot{v}_1$, $\dot{v}_2$ and $\dot{v}_3$ do not contain principal acoustical parts, and we have

$$\dot{v}'=\Omega\dot{v} \tag{11.232}$$

Our first objective is to estimate

$$\|\dot{v}\|_{2,[m,l],\Sigma_t^{\epsilon_0}} \tag{11.233}$$

Here we just state the results:

**Lemma 11.18** Under the assumptions of Proposition 11.2 augmented by the assumptions $\mathbf{E}_{\{l_*+2\}}$, $\mathbf{E}^{Q}_{\{l_*+1\}}$, $\mathbf{E}^{QQ}_{\{l_*\}}$ and $\mathbf{M}_{[m+1,l_*+1]}$ we have the $L^2$ estimate

$$
\begin{aligned}
\|\dot{v}\|_{2,[m,l],\Sigma_t^{\epsilon_0}}\leqslant\;&C_l(1+t)^{-2}((1+\log(1+t))(\mathcal{W}_{\{l+2\}}+\mathcal{W}^{Q}_{\{l+1\}}+\mathcal{W}^{QQ}_{\{l\}})\\
&+\delta_0(1+t)^{-1}(\mathcal{B}_{[m+1,l+1]}+(1+\log(1+t))(\mathcal{Y}_0+(1+t)\mathcal{A}_{[l]})))
\end{aligned}
$$

as well as $L^\infty$ estimate

$$\|\dot{v}\|_{\infty,[m,l_*],\Sigma_t^{\epsilon_0}}\leqslant C_l\delta_0(1+t)^{-3}(1+\log(1+t))$$

where we denote by $\mathbf{E}^{QQ}_{m,n}$ the bootstrap assumption that there is a constant $C$ independent of $s$ such that for all $t\in[0,s]$,

$$\mathbf{E}^{QQ}_{m,n} : \max_{\alpha; i_1 \cdots i_n} \| R_{i_n} \cdots R_{i_1} (T)^m (Q)^2 \psi_\alpha \|_{L^\infty(\Sigma_t^{\epsilon_0})} \leqslant C\delta_0 (1+t)^{-1}$$

We then denote by $\mathbf{E}^{QQ}_{\{l\}}$ the conjunction of $\mathbf{E}^{QQ}_{m,n}$ corresponding to (11.8). The constant $C$ then depends on $l$ only. Here we also have introduced the quantities $\mathcal{W}^{QQ}_{m,n}$:

$$\mathcal{W}^{QQ}_{m,n} = \max_{\alpha; i_1 \cdots i_n} \| R_{i_n} \cdots R_{i_1} (T)^m (Q)^2 \psi_\alpha \|_{L^2(\Sigma_t^{\epsilon_0})} \tag{11.234}$$

We then denote by $\mathcal{W}^{QQ}_{\{l\}}$ the sum of $\mathcal{W}^{QQ}_{m,n}$ corresponding to (11.8). Moreover, we have

$$\max_\alpha \| (L)^2 \psi_\alpha \|_{2,\{l\},\Sigma_t^{\epsilon_0}} \leqslant (1+t)^{-2} (\mathcal{W}^{QQ}_{\{l\}} + \mathcal{W}^{Q}_{\{l\}}) \tag{11.235}$$

We now consider $\ddot{\tau}$, defined by (11.224). Setting

$$\dot{\tau} = \Omega^{-1} \tilde{\tau} \tag{11.236}$$

(11.224) becomes, in view of (11.232),

$$\ddot{\tau} = \dot{v} + T\dot{\tau} + (T \log \Omega + {}^{(T)}\delta) \dot{\tau} \tag{11.237}$$

By (9.37) we have

$$\dot{\tau} = -\sum_i (-(L\psi_i)(\underline{L}\psi_i) + \mu \not{d}\psi_i \cdot \not{d}\psi_i) \tag{11.238}$$

Here we must use Lemma 11.12 with $(k,l)$ replaced by $(m+2, l+2)$ to estimate the second term on the right of (11.237):

$$\begin{aligned} \| \dot{v} \|_{2,[m+1,l+1],\Sigma_t^{\epsilon_0}} \leqslant & C_l \delta_0 (1+t)^{-2} (\mathcal{W}_{\{l+2\}} + \mathcal{W}^Q_{\{l+1\}} \\ & + \delta_0 (1+t)^{-2} (\mathcal{B}_{[m+1,l+1]} \\ & + (1+\log(1+t))(\mathcal{Y}_0 + (1+t)\mathcal{A}_{[l]}))) \end{aligned} \tag{11.239}$$

Also, we have

$$\| \dot{\tau} \|_{\infty,[m+1,l_*+1],\Sigma_t^{\epsilon_0}} \leqslant C_l \delta_0^2 (1+t)^{-3} \tag{11.240}$$

Then combining with Lemma 11.18, we obtain

$$\begin{aligned} \| \ddot{\tau} \|_{2,[m,l],\Sigma_t^{\epsilon_0}} \leqslant & C_l (1+t)^{-2} ((1+\log(1+t))(\mathcal{W}_{\{l+2\}} + \mathcal{W}^Q_{\{l+1\}} + \mathcal{W}^{QQ}_{\{l\}}) \\ & + \delta_0 (1+t)^{-1} (\mathcal{B}_{[m+1,l+1]} \\ & + (1+\log(1+t))(\mathcal{Y}_0 + (1+t)\mathcal{A}_{[l]}))) \end{aligned} \tag{11.241}$$

as well as

$$\| \ddot{\tau} \|_{2,[m,l_*],\Sigma_t^{\epsilon_0}} \leqslant C_l \delta_0 (1+t)^{-3} (1+\log(1+t)) \tag{11.242}$$

By the above results and a direct calculation, we get an estimate:

$$\begin{aligned}\|\dot{n}_0''\|_{2,[m,l],\Sigma_t^{\epsilon_0}} \leqslant & C_l(1+t)^{-2}((1+\log(1+t))(\mathcal{W}_{\{l+2\}}+\mathcal{W}^Q_{\{l+1\}}+\mathcal{W}^{QQ}_{\{l\}}) \quad (11.243)\\ &+\delta_0(1+t)^{-1}(\mathcal{B}_{[m+1,l+1]}\\ &+(1+\log(1+t))(\mathcal{Y}_0+(1+t)\mathcal{A}_{[l]})))\end{aligned}$$

We proceed to estimate $\dot{n}_1'$ and $\dot{n}_2'$, given by (9.52) and (11.225). The estimate for $\dot{n}_1'$ is straightforward. We use Lemma 11.12 to obtain

$$\begin{aligned}\|\dot{n}_1'\|_{2,[m,l],\Sigma_t^{\epsilon_0}} \leqslant & C_l\delta_0(1+t)^{-3}(\mathcal{W}_{\{l+2\}}+\delta_0(1+t)^{-1} \quad (11.244)\\ &\cdot((1+t)^{-1}\mathcal{B}_{[m-1,l+1]}+\mathcal{Y}_0+(1+t)\mathcal{A}_{[l]}\\ &+(1+t)^{-2}(1+\log(1+t))\mathcal{W}^Q_{\{l+1\}}))\end{aligned}$$

and also, obviously,

$$\|\dot{n}_1'\|_{2,[m,l_*],\Sigma_t^{\epsilon_0}} \leqslant C_l\delta_0^2(1+t)^{-4} \quad (11.245)$$

To estimate $\dot{n}_2'$ is a bit more complicated, because we have the term $\not{\Delta}\hat{T}^i$ in the expression for $n_2$. We must first derive an estimate for this. Let us consider

$$U^i = \not{\Delta}\hat{T}^i + \eta^{-1}\not{d}\mathrm{tr}\chi\cdot\not{d}x^i + (1-u+t)^{-1}\eta^{-1}\not{\Delta}x^i \quad (11.246)$$

The last term in the above comes from the term $\not{g}\cdot\not{D}^2x^i$. We can write

$$U^i = b_1\cdot\not{d}x^i + b_2\cdot\not{D}^2x^i \quad (11.247)$$

where

$$b_1 = \not{\mathrm{div}}\not{k} - \eta^{-1}i + \eta^{-2}\chi\cdot(\not{d}\eta) \quad (11.248)$$

and

$$b_2 = \not{k} - \eta^{-1}\chi' \quad (11.249)$$

To estimate

$$\|U^i\|_{2,[m,l],\Sigma_t^{\epsilon_0}}$$

we need the following two lemmas:

**Lemma 11.19** Under the assumptions of Proposition 11.2, for an arbitrary $S_{t,u}$ tensorfield $\xi$ we have

$$\begin{aligned}\|\xi\cdot\not{d}x^i\|_{2,[m,l],\Sigma_t^{\epsilon_0}} \leqslant & C_l(\|\xi\|_{2,[m,l],\Sigma_t^{\epsilon_0}}+\|\xi\|_{\infty,[m,l_*-1],\Sigma_t^{\epsilon_0}}((1+t)^{-1}\mathcal{B}_{[m-1,l]}+\mathcal{Y}_0\\ &+(1+t)\mathcal{A}_{[l-1]}+\mathcal{W}_{\{l\}}))\end{aligned}$$

**Lemma 11.20** Under the assumptions of Proposition 11.2, for an arbitrary $S_{t,u}$ tensorfield $\xi$ we have

$$\begin{aligned}\|\xi\cdot \not{D}^2 x^i\|_{2,[m,l],\Sigma_t^{\epsilon_0}} \leqslant & C_l(1+t)^{-1}(\|\xi\|_{2,[m,l],\Sigma_t^{\epsilon_0}} \\ & + \|\xi\|_{\infty,[k,l_*-1],\Sigma_t^{\epsilon_0}}((1+t)^{-1}\mathcal{B}_{[m-1,l+1]} + \mathcal{Y}_0 + (1+t)\mathcal{A}_{[l]} \\ & + \mathcal{W}_{\{l+1\}} + (1+t)^{-2}(1+\log(1+t))^2\mathcal{W}^Q_{\{l\}}))\end{aligned}$$

By Propositions 10.1/11.1, 10.2/11.2 as well as their corollaries, the proof of these two lemmas is almost a direct calculation.

Using these two lemmas and (11.247)—(11.249), we deduce

**Lemma 11.21** Under the assumptions of Proposition 11.2 augmented by the assumptions $\mathbf{E}_{\{l_*+2\}}$, $\mathbf{E}^Q_{\{l_*+1\}}$, we have

$$\begin{aligned}\|U^i\|_{2,[m,l],\Sigma_t^{\epsilon_0}} \leqslant & C_l(1+t)^{-2}\cdot(\mathcal{W}_{\{l+2\}} + (1+t)^{-1}(1+\log(1+t))\mathcal{W}^Q_{\{l+1\}} \\ & + (1+t)^{-1}\mathcal{B}_{[m-1,l+1]} + \mathcal{Y}_0 + (1+t)\mathcal{A}_{[l]})\end{aligned}$$

and

$$\|U^i\|_{\infty,[m,l_*],\Sigma_t^{\epsilon_0}} \leqslant C_l\delta_0(1+t)^{-3}(1+\log(1+t))$$

Next we consider the function $n'_2$ given by (11.226). Substituting in (11.226) we obtain the following expression for $\dot{n}_2$:

$$\begin{aligned}\dot{n}_2 = & \frac{1}{2\eta^2}(\frac{\rho}{\rho'})'(\mathrm{tr}\chi \not{\Delta} h + 2\hat{\chi}\cdot \hat{\not{D}}^2 h + 2i\cdot \not{d}h) \\ & + \eta^{-1}(\mathrm{tr}\chi \hat{T}^i \not{\Delta}\psi_i + 2\hat{\chi}\cdot \hat{T}^i \hat{\not{D}}^2\psi_i + 2i\cdot \hat{T}^i \not{d}\psi_i) \\ & + \not{d}(\frac{1}{\eta^2}(\frac{\rho}{\rho'})')\cdot \not{d}Lh + 2\not{d}(\eta^{-1}\hat{T}^i)\cdot \not{d}(L\psi_i) \\ & + \not{\Delta}(\frac{1}{2\eta^2}(\frac{\rho}{\rho'})')Lh + \not{\Delta}(\eta^{-1}\hat{T}^i)(L\psi_i) \\ & + \eta^{-2}(L\psi_i)\not{d}x^i\cdot \not{d}\mathrm{tr}\chi\end{aligned} \tag{11.250}$$

By Lemma 11.21 and (11.250), we obtain the following estimates for $\dot{n}_2$:

$$\begin{aligned}\|\dot{n}_2\|_{2,[m,l],\Sigma_t^{\epsilon_0}} \leqslant & C_l(1+t)^{-3}(\mathcal{W}_{\{l+2\}} + \mathcal{W}^Q_{\{l+1\}} \\ & + \delta_0(1+t)^{-1}((1+t)^{-1}\mathcal{B}_{[m-1,l+1]} + \mathcal{Y}_0 \\ & + (1+t)\mathcal{A}_{[l]}))\end{aligned} \tag{11.251}$$

and

$$\|\dot{n}_2\|_{\infty,[m,l_*],\Sigma_t^{\epsilon_0}} \leqslant C_l\delta_0(1+t)^{-4} \tag{11.252}$$

We turn to the function $\dot{n}'_2$ given in terms of $\dot{n}_2$ by (11.225). We just use Corollary 11.1.b and Corollary 11.2.b as well as a direct calculation to obtain

$$\begin{aligned}\|\dot{n}_2'\|_{2,[m,l],\Sigma_t^{\epsilon_0}} \leqslant & C_l(1+t)^{-3}((1+\log(1+t))(\mathcal{W}_{\{l+2\}}+\mathcal{W}^Q_{\{l+1\}}) \\ & +\delta_0(1+t)^{-1}(\mathcal{B}_{[m,l]}+(1+\log(1+t))(\mathcal{Y}_0+(1+t)\mathcal{A}_{[l]})))\end{aligned} \tag{11.253}$$

and

$$\|\dot{n}_2'\|_{\infty,[m,l_*],\Sigma_t^{\epsilon_0}} \leqslant C_l(1+t)^{-4}(1+\log(1+t)) \tag{11.254}$$

We now turn to the remaining terms in the expression (11.221) for $\check{g}'$. We first consider the functions $f_0'$ and $f_1'$, given by (9.26) and (9.36) respectively:

$$f_0' = \frac{1}{2}\frac{dH}{dh}(\underline{L}Th) \tag{11.255}$$

$$f_1' = \frac{1}{2\eta^2}(\frac{\rho}{\rho'})'(\Delta h) + \eta^{-1}\hat{T}^i(\Delta\psi_i) \tag{11.256}$$

Obviously, by the expression

$$\underline{L}Th = 2(T)^2 h + \eta^{-1}\kappa TLh + \eta^{-1}\kappa\Lambda\cdot \not{d}h$$

we have

$$\|f_0'\|_{\infty,[m,l_*],\Sigma_t} \leqslant C_l\delta_0(1+t)^{-1} \tag{11.257}$$

and

$$\begin{aligned}\|f_0'\|_{2,[m,l],\Sigma_t^{\epsilon_0}} \leqslant & C_l(\mathcal{W}_{\{l+2\}}+(1+t)^{-1}(1+\log(1+t))\mathcal{W}^Q_{\{l+1\}} \\ & +\delta_0(1+t)^{-2}(\mathcal{B}_{[m,l+1]}+\delta_0(\mathcal{Y}_0+(1+t)\mathcal{A}_{[l-1]})))\end{aligned} \tag{11.258}$$

Using Proposition 11.2 we deduce

$$\begin{aligned}\|f_1'\|_{2,[m,l],\Sigma_t^{\epsilon_0}} \leqslant & C_l(1+t)^{-2}(\mathcal{W}_{\{l+2\}}+\delta_0(1+t)^{-3}(1+\log(1+t))^2\mathcal{W}^Q_{\{l\}} \\ & +\delta_0(1+t)^{-1}((1+t)^{-1}\mathcal{B}_{[m-1,l+1]}+\mathcal{Y}_0+(1+t)\mathcal{A}_{[l]}))\end{aligned} \tag{11.259}$$

and also

$$\|f_1'\|_{\infty,[m,l_*],\Sigma_t^{\epsilon_0}} \leqslant C_l\delta_0(1+t)^{-3} \tag{11.260}$$

With the help of (11.257)—(11.260) we can estimate, in $\| \quad \|_{2,[m,l],\Sigma_t^{\epsilon_0}}$ norm, the third and fourth terms on the right in (11.221) in a straightforward manner. Concerning the second term, we just use the estimates for $e$ and $i$ which we can obtain by using Propositions 11.1 and 11.2:

$$\begin{aligned}\|e\|_{2,[m+1,l+1],\Sigma_t^{\epsilon_0}} \leqslant & C_l(1+t)^{-1}(\mathcal{W}^Q_{\{l+1\}} \\ & +\delta_0(1+t)^{-1}((1+t)^{-1}\mathcal{B}_{[m,l+1]} \\ & +\mathcal{Y}_0+(1+t)\mathcal{A}_{[l]}+\mathcal{W}_{\{l+1\}}))\end{aligned} \tag{11.261}$$

$$\|e\|_{\infty,[m,l_*],\Sigma_t^{\epsilon_0}} \leqslant C_l\delta_0(1+t)^{-2} \tag{11.262}$$

and

$$\|i\|_{2,[m,l],\Sigma_t^{\epsilon_0}} \leqslant C_l(1+t)^{-2}(\mathcal{W}_{\{l+2\}} + \delta_0(1+t)^{-2}(1+\log(1+t))\mathcal{W}^Q_{\{l+1\}} + \delta_0(1+t)^{-1}((1+t)^{-1}\mathcal{B}_{[m-1,l+1]} + \mathcal{Y}_0 + (1+t)\mathcal{A}_{[l]})) \tag{11.263}$$

$$\|i\|_{\infty,[m,l_*],\Sigma_t^{\epsilon_0}} \leqslant C_l\delta_0^2(1+t)^{-4} \tag{11.264}$$

Finally, we have

$$-\xi \cdot \not{d}\check{f}$$

the first term on the right in (11.221). Here $\xi$ is given by (9.59):

$$\xi = -\not{d}\mu + (\frac{\mu}{2\eta^2}(\frac{\rho}{\rho'})' - \frac{1}{2}\frac{dH}{dh}\eta^{-1}\kappa)\not{d}h + T^i\not{d}\psi_i - \eta^{-1}\kappa(L\psi_i)\not{d}x^i \tag{11.265}$$

and $\check{f}$ is the function defined by (8.27):

$$\check{f} = -\frac{1}{2}\frac{dH}{dh}\tau_{\underline{L}} \tag{11.266}$$

So we obtain

$$\|\not{d}\check{f}\|_{2,[m,l],\Sigma_t^{\epsilon_0}} \leqslant C_l(1+t)^{-1}(\mathcal{W}_{\{l+2\}} + (1+t)^{-1}(1+\log(1+t))\mathcal{W}^Q_{\{l+1\}} + \delta_0(1+t)^{-2}\mathcal{B}_{[m,l+1]}) \tag{11.267}$$

and also

$$\|\not{d}\check{f}\|_{\infty,[m,l_*],\Sigma_t^{\epsilon_0}} \leqslant C_l\delta_0(1+t)^{-2} \tag{11.268}$$

Next, using Corollary 11.1.b with $l_*$ in the role of $l$ and Corollary 11.2.b we deduce

$$\|\xi\|_{\infty,[m,l_*],\Sigma_t^{\epsilon_0}} \leqslant C_l(1+t)^{-1}(1+\log(1+t)) \tag{11.269}$$

and

$$\|\xi\|_{2,[m,l],\Sigma_t^{\epsilon_0}} \leqslant C_l(1+t)^{-1}((1+\log(1+t))(\mathcal{W}_{\{l+1\}} + \delta_0^2(1+t)^{-2}\mathcal{W}^Q_{\{l\}}) + \mathcal{B}_{[m,l+1]} + \delta_0(1+t)^{-1}(\mathcal{Y}_0 + (1+t)\mathcal{A}_{[l-1]})) \tag{11.270}$$

To get the final estimate, we need Corollary 11.2.d to see that

$$\|\xi \cdot \not{g}^{-1}\|_{2,[m,l],\Sigma_t^{\epsilon_0}} \leqslant C_l(1+t)^{-1}((1+\log(1+t))(\mathcal{W}_{\{l+1\}} + \delta_0(1+t)^{-2}(1+\log(1+t))^2\mathcal{W}^Q_{\{l\}}) + \mathcal{B}_{[m,l+1]} + \delta_0(1+\log(1+t))(\mathcal{Y}_0 + (1+t)\mathcal{A}_{[l-1]})) \tag{11.271}$$

and also

$$\|\xi \cdot \not{g}^{-1}\|_{\infty,[m,l_*],\Sigma_t^{\epsilon_0}} \leqslant C_l \delta_0 (1+t)^{-1}(1+\log(1+t)) \tag{11.272}$$

Then the first term on the right of (11.221) can be estimated by applying (11.267), (11.268), (11.271) and (11.272).

Combining finally the above results with the earlier results (11.243), (11.244), (11.245), (11.251), (11.252) and also (11.257)—(11.260), yields, through (11.221), the following proposition.

**Proposition 11.3** Let the hypotheses **H0**, **H1**, **H2$'$** and the estimate (6.177) hold. Let also the bootstrap assumptions $\mathbf{E}_{\{l_*+2\}}$, $\mathbf{E}^Q_{\{l_*+1\}}$, $\mathbf{E}^{QQ}_{\{l_*\}}$ and $\not{\mathbf{X}}_{[l_*]}$ hold for some non-negative integer $l$. Moreover, let the bootstrap assumption $\mathbf{M}_{[m+1,l_*+1]}$ hold for some non-negative integer $m \leqslant l$. Then if $\delta_0$ is suitably small (depending on $l$) we have

$$\begin{aligned}
&\|\breve{g}'\|_{2,[m,l],\Sigma_t^{\epsilon_0}} \\
&\leqslant C_l(1+t)^{-2}((1+\log(1+t))(\mathcal{W}_{\{l+2\}} + \mathcal{W}^Q_{\{l+1\}} + \mathcal{W}^{QQ}_{\{l\}}) \\
&\quad + \delta_0(1+t)^{-1}(\mathcal{B}_{[m+1,l+1]} + (1+\log(1+t))(\mathcal{Y}_0 + (1+t)\mathcal{A}_{[l]})))
\end{aligned} \tag{11.273}$$

We proceed to estimate in $L^2(\Sigma_t^{\epsilon_0})$ the functions ${}^{(i_1\cdots i_n)}\dot{g}'_{m,n}$, defined by (9.242)—(9.243), for $n = l - m$. Note that $l$ have different meanings here and in Chapter 9. We shall now re-express the functions ${}^{(i_1\cdots i_n)}\dot{g}'_{m,n}$ in terms of the function $\breve{g}'$, which has been estimated by Proposition 11.3. Let us define the functions

$${}^{(i_1\cdots i_n)}w'_{m,n} = \sum_{k=0}^{m-1} R_{i_n}\cdots R_{i_1}(T)^k \Lambda x'_{m-k-1,0} - m\Lambda^{(i_1\cdots i_n)}x'_{m-1,n} \tag{11.274}$$

$$\begin{aligned}
{}^{(i_1\cdots i_n)}w''_{m,n} =& \sum_{k=0}^{n-1} R_{i_n}\cdots R_{i_{n-k+1}} {}^{(R_{i_{n-k}})}Z^{(i_1\cdots i_{n-k-1})}x'_{m,n-k-1} \\
&- \sum_{k=0}^{n-1} {}^{(R_{i_{n-k}})}Z^{(i_1 \overset{>i_{n-k}<}{\cdots} i_n)}x'_{m,n-1}
\end{aligned} \tag{11.275}$$

According to (9.123), (9.131) and (9.134), (9.137), the functions ${}^{(i_1\cdots i_n)}w'_{m,n}$, ${}^{(i_1\cdots i_n)}w''_{m,n}$ do not contain terms of the top order $l+2$. From Proposition 9.1 and 9.2 we then have

$$\begin{aligned}
&{}^{(i_1\cdots i_n)}g'_{m,n} - m\Lambda^{(i_1\cdots i_n)}x'_{m-1,n} \\
&- \sum_{k=0}^{n-1} {}^{(R_{i_{n-k}})}Z^{(i_1 \overset{>i_{n-k}<}{\cdots} i_n)}x'_{m,n-1} \\
&= R_{i_n}\cdots R_{i_1}(T)^m \breve{g}' + {}^{(i_1\cdots i_n)}w'_{m,n} + {}^{(i_1\cdots i_n)}w''_{m,n} \\
&+ \sum_{k=0}^{m-1} R_{i_n}\cdots R_{i_1}(T)^k y'_{m-k,0} \\
&+ \sum_{k=0}^{n-1} R_{i_n}\cdots R_{i_{n-k+1}} {}^{(i_1\cdots i_{n-k})}y'_{m,n-k}
\end{aligned} \tag{11.276}$$

We now substitute for $\breve{g}'$ in terms of $\check{\breve{g}}'$ from (11.220). Consider first the case $m = 0$. We have

$$R_{i_n} \cdots R_{i_1} \breve{g}' = R_{i_n} \cdots R_{i_1} \check{\breve{g}}' + \xi \cdot {}^{(i_1 \cdots i_n)} x_n + {}^{(i_1 \cdots i_n)} u_n \tag{11.277}$$

where

$$^{(i_1 \cdots i_n)} u_n = \xi \cdot \sum_{s_1 \neq \emptyset} ((R)^{s_1} \mu) \not{d} (R)^{s_2} \mathrm{tr}\chi + \sum_{s_1 \neq \emptyset} (\not{\mathcal{L}}_R)^{s_1} \xi \cdot (\not{\mathcal{L}}_R)^{s_2} x_0 \tag{11.278}$$

Then comparing with (9.242) we obtain

$$\begin{aligned} {}^{(i_1 \cdots i_n)} \dot{g}'_{0,n} = & R_{i_n} \cdots R_{i_1} \check{\breve{g}}' + {}^{(i_1 \cdots i_n)} u_n + {}^{(i_1 \cdots i_n)} w''_{0,n} \\ & + \sum_{k=0}^{n-1} R_{i_n} \cdots R_{i_{n-k+1}} {}^{(i_1 \cdots i_{n-k})} y'_{0,n-k} \end{aligned} \tag{11.279}$$

A similar calculation yields

$$\begin{aligned} {}^{(i_1 \cdots i_n)} \dot{g}'_{m,n} = & R_{i_n} \cdots R_{i_1} (T)^m \check{\breve{g}}' + {}^{(i_1 \cdots i_n)} u'_{m-1,n} \\ & + {}^{(i_1 \cdots i_n)} w'_{m,n} + {}^{(i_1 \cdots i_n)} w''_{m,n} \\ & + \sum_{k=0}^{m-1} R_{i_n} \cdots R_{i_1} (T)^k y'_{m-k,0} \\ & + \sum_{k=0}^{n-1} R_{i_n} \cdots R_{i_{n-k+1}} {}^{(i_1 \cdots i_{n-k})} y'_{m,n-k} \end{aligned} \tag{11.280}$$

where

$$^{(i_1 \cdots i_n)} u'_{m-1,n} = R_{i_n} \cdots R_{i_1} (T)^{m-1} u'_{0,0} + {}^{(i_1 \cdots i_n)} u''_{m-1,n} \tag{11.281}$$

$$\begin{aligned} {}^{(i_1 \cdots i_n)} u''_{m-1,n} = & R_{i_n} \cdots R_{i_1} u''_{m-1,0} \\ & + \xi \cdot \sum_{s_1 \neq \emptyset} \not{d}(((R)^{s_1} \mu)(R)^{s_2} (T)^{m-1} \Delta\mu) \\ & + \sum_{s_1 \neq \emptyset} (\not{\mathcal{L}}_R)^{s_1} \xi \cdot \not{d}(R)^{s_2} x'_{m-1,0} \end{aligned} \tag{11.282}$$

$$\begin{aligned} u''_{m-1,0} = & \sum_{k=1}^{m-1} \frac{(m-1)!}{k!(m-1-k)!} (\xi \cdot \not{d}((T)^k \mu)(T)^{m-1-k} \Delta\mu \\ & + ((\not{\mathcal{L}}_T)^k \xi) \cdot \not{d}(T)^{m-1-k} x'_{0,0}) \end{aligned} \tag{11.283}$$

$$u'_{0,0} = \xi \cdot \dot{x}_0 + (\not{\mathcal{L}}_T \xi) \cdot x_0 \tag{11.284}$$

where

$$\dot{x}_0 = \mu \not{d}(T \mathrm{tr}\chi' - \Delta\mu) + (T\mu) \not{d} \mathrm{tr}\chi' - (\not{d}\mu) \Delta\mu + \not{d}(T \check{f} + \check{f}') \tag{11.285}$$

Then by the earlier results in this chapter, we finally obtain

**Proposition 11.4** Under the assumptions of Proposition 11.3, augmented by the assumptions $\not{X}_{[(l+1)_*]}$ and $\mathbf{M}_{[m,(l+1)_*+1]}$ we have

$$\max_{i_1\cdots i_{l-m}} \|^{(i_1\cdots i_{l-m})}\dot{g}'_{m,l-m}\|_{L^2(\Sigma_t^{\epsilon_0})}$$
$$\leqslant C_l(1+t)^{-2}((1+\log(1+t))(\mathcal{W}_{\{l+2\}}+\mathcal{W}^Q_{\{l+1\}}+\mathcal{W}^{QQ}_{\{l\}}) \tag{11.286}$$
$$+\delta_0((1+t)^{-1}\mathcal{B}_{[m+1,l+1]}+\mathcal{Y}_0+(1+t)\mathcal{A}_{[l]}))$$

In deriving this proposition, we made use of the following lemma:

**Lemma 11.22** Let $A_1,\cdots,A_n$ and $B$ be linear operators acting on some space $X$. We then have

$$[A_n\cdots A_1,B]=\sum_{m=0}^{n-1} A_n\cdots A_{n-m+1}[A_{n-m},B]A_{n-m-1}\cdots A_1$$

*Proof.* We just applying induction on $m$, and then use Proposition 8.2. □

Now we can give an estimate for $^{(i_1\cdots i_{l-m})}\mathcal{Q}'_{m,l-m}$ defined by (9.256). Here we just need to estimate the third and the fourth term on the right of (9.256). This is straightforward. We obtain

**Proposition 11.5** Under the assumptions of Proposition 11.3, augmented by the assumptions $\not{X}_{[(l+1)_*]}$ and $\mathbf{M}_{[m,(l+1)_*+1]}$, we have

$$\max_{(i_1\cdots i_{l-m})} \mathcal{Q}'_{m,l-m}(t)\leqslant C_l(1+t)^{-3}((1+\log(1+t))(\mathcal{W}_{\{l+2\}}+\mathcal{W}^Q_{\{l+1\}}+\mathcal{W}^{QQ}_{\{l\}})$$
$$+\delta_0((1+t)^{-1}\mathcal{B}_{[m+1,l+1]}+\mathcal{Y}_0+(1+t)\mathcal{A}_{[l]}))$$

### 11.2.2 Bounds for $P'_{m,l}$

We now consider the principal part in the defining expression (9.259) for the quantity $^{(i_1\cdots i_n)}B'_{m,n}$, namely, the term

$$C(1+t)^{-1}(^{(i_1\cdots i_n)}\bar{P}'^{(0)}_{m,n,a}(t)+(1+t)^{-1/2(i_1\cdots i_n)}\bar{P}'^{(1)}_{m,n,a}(t))\bar{\mu}_m^{-a}(t)$$

Here, the quantities $^{(i_1\cdots i_n)}\bar{P}'^{(0)}_{m,n,a}$, $^{(i_1\cdots i_n)}\bar{P}'^{(1)}_{m,n,a}$ are defined by (9.215) and (9.216), in terms of $^{(i_1\cdots i_n)}P'^{(0)}_{m,n}$ and $^{(i_1\cdots i_n)}P'^{(1)}_{m,n}$, whose sum bounds the quantity $^{(i_1\cdots i_n)}P'_{m,n}$ itself defined by (9.213):

$$^{(i_1\cdots i_n)}P'_{m,n}(t)=(1+t)\|^{(i_1\cdots i_n)}\check{f}'_{m,n}(t)\|_{L^2([0,\epsilon_0]\times S^2)} \tag{11.287}$$

We shall derive a bound for $^{(i_1\cdots i_n)}P'_{m,n}(t)$, with $n=l-m$. Since

$$^{(i_1\cdots i_n)}\check{f}'_{m,n}=R_{i_n}\cdots R_{i_1}(T)^m\check{f}'$$

by (8.333) we have

$$\begin{aligned} {}^{(i_1\cdots i_{l-m})}P'_{m,l-m} &\leqslant C\|{}^{(i_1\cdots i_{l-m})}\check{f}'_{m,l-m}\|_{L^2(\Sigma_t^{\epsilon_0})} \\ &= C\|R_{i_{l-m}}\cdots R_{i_1}(T)^m\check{f}'\|_{L^2(\Sigma_t^{\epsilon_0})} \end{aligned} \tag{11.288}$$

Let us recall (9.40):

$$\check{f}' = f'_0 + \mu^2 f'_1$$

where

$$\begin{aligned} f'_0 &= \frac{1}{2}\frac{dH}{dh}(\underline{L}Th) \\ f'_1 &= \frac{1}{2\eta^2}(\frac{\rho}{\rho'})'(\not{\Delta}h) + \eta^{-1}\hat{T}^i(\not{\Delta}\psi_i) \end{aligned}$$

First, we consider the contribution from $f'_0$. Since

$$\begin{aligned} Th =& T\psi_0 - \sum_i \psi_i T\psi_i \\ \underline{L}Th =& \underline{L}T\psi_0 - \sum_i \underline{L}\psi_i T\psi_i - \sum_i \psi_i \underline{L}T\psi_i \end{aligned}$$

defining as in Chapter 10,

$$m_T^0 = \frac{1}{2}\frac{dH}{dh}, \quad m_T^i = -\frac{1}{2}\frac{dH}{dh}\psi_i \tag{11.289}$$

we have

$$m = m_T^\alpha(T\psi_\alpha) \tag{11.290}$$

We decompose

$$f'_0 = f'_{0,P} + f'_{0,N} \tag{11.291}$$

where $f'_{0,P}$ is the principal part of $f'_0$:

$$f'_{0,P} = m_T^\alpha(\eta^{-1}\kappa LT\psi_\alpha + 2T^2\psi_\alpha) \tag{11.292}$$

and the lower order part

$$f'_{0,N} = -\frac{1}{2}\frac{dH}{dh}\sum_i \underline{L}\psi_i T\psi_i \tag{11.293}$$

We have

$$\begin{aligned} R_{i_{l-m}}\cdots R_{i_1}(T)^m f'_0 =& 2m_T^\alpha R_{i_{l-m}}\cdots R_{i_1}(T)^{m+2}\psi_\alpha \\ &+ \alpha^{-1}\kappa m_T^\alpha R_{i_{l-m}}\cdots R_{i_1}(T)^m LT\psi_\alpha \\ &+ {}^{(i_1\cdots i_{l-m})}n'_{m,l-m} \end{aligned} \tag{11.294}$$

where ${}^{(i_1\cdots i_{l-m})}n'_{m,l-m}$, of order $l+1$, is given by

$$
\begin{aligned}
&{}^{(i_1\cdots i_{l-m})}n'_{m,l-m}\\
&=2R_{i_{l-m}}\cdots R_{i_1}\Big(\sum_{k=1}^{m}\frac{m!}{k!(m-k)!}((T)^k m_T^\alpha)((T)^{m-k+2}\psi_\alpha)\Big) \qquad (11.295)\\
&+2\sum_{s_1\neq\emptyset}((R)^{s_1}m_T^\alpha)((R)^{s_2}(T)^{m+2}\psi_\alpha)\\
&+R_{i_{l-m}}\cdots R_{i_1}\Big(\sum_{k=1}^{m}\frac{m!}{k!(m-k)!}((T)^k(\alpha^{-1}\kappa m_T^\alpha))((T)^{m-k}LT\psi_\alpha)\Big)\\
&+\sum_{s_1\neq\emptyset}((R)^{s_1}(\alpha^{-1}\kappa m_T^\alpha))((R)^{s_2}(T)^m LT\psi_\alpha)\\
&+R_{i_{l-m}}\cdots R_{i_1}(T)^m f'_{0,N}
\end{aligned}
$$

From (11.294), taking into account (10.184), we have

$$
\begin{aligned}
&\|R_{i_{l-m}}\cdots R_{i_1}(T)^m f'_0\|_{L^2(\Sigma_t^{\epsilon_0})}\\
&\leqslant(|\ell|+C\delta_0(1+t)^{-1})\sum_\alpha\|R_{i_{l-m}}\cdots R_{i_1}(T)^{m+2}\psi_\alpha\|_{L^2(\Sigma_t^{\epsilon_0})} \qquad (11.296)\\
&+C\sum_\alpha\|\mu R_{i_{l-m}}\cdots R_{i_1}(T)^m LT\psi_\alpha\|_{L^2(\Sigma_t^{\epsilon_0})}\\
&+\|{}^{(i_1\cdots i_{l-m})}n'_{m,l-m}\|_{L^2(\Sigma_t^{\epsilon_0})}
\end{aligned}
$$

Here, the constant $C$ does not depend on $\ell$. Moreover, using the estimate for $\Lambda$ in Corollary 11.2.c, we readily deduce

$$
\begin{aligned}
&\max_{i_1\cdots i_{l-m}}\|{}^{(i_1\cdots i_{l-m})}n'_{m,l-m}\|_{L^2(\Sigma_t^{\epsilon_0})}\\
&\leqslant C_l\delta_0(1+t)^{-1}(\mathcal{W}_{\{l+1\}}+(1+\log(1+t))\mathcal{W}^Q \qquad (11.297)\\
&\quad+(1+t)^{-1}(\mathcal{B}_{[m,l]}+\delta_0(\mathcal{Y}_0+(1+t)\mathcal{A}_{[l-1]})))
\end{aligned}
$$

Consider next the contribution of the term $\mu^2 f'_1$ in $\check{f}'$ to $R_{i_{l-m}}\cdots R_{i_1}(T)^m\check{f}'$. Also, we only consider the principal part. Since

$$
\begin{aligned}
\not{d}h&=\not{d}\psi_0-\sum_i\psi_i\not{d}\psi_i\\
\not{\Delta}h&=\not{\Delta}\psi_0-\sum_i\psi_i\not{\Delta}\psi_i-\sum_i\not{d}\psi_i\cdot\not{d}\psi_i
\end{aligned}
$$

we have

$$\mu^2 f'_1=\frac{\mu^2}{2\eta^2}\Big(\frac{\rho}{\rho'}\Big)'\Big(\not{\Delta}\psi_0-\sum_i\psi_i\not{\Delta}\psi_i-\sum_i\not{d}\psi_i\cdot\not{d}\psi_i\Big)+\eta^{-1}\mu^2\hat{T}^i(\not{\Delta}\psi_i) \qquad (11.298)$$

if we define

$$\tilde{m}^0 = \frac{1}{2\eta^2}(\frac{\rho}{\rho'})', \quad \tilde{m}^i = -\frac{1}{2\eta^2}(\frac{\rho}{\rho'})'\psi_i + \eta^{-1}\hat{T}^i \tag{11.299}$$

We have the decomposition

$$f'_1 = f'_{1,P} + f'_{1,N} \tag{11.300}$$

The principal part of $f'_1$ as follows:

$$f'_{1,P} = \tilde{m}^\alpha \Delta\psi_\alpha \tag{11.301}$$

and $f'_{1,N}$ is the lower order term

$$f'_{1,N} = -\frac{1}{2\eta^2}(\frac{\rho}{\rho'})' \sum_i \not{d}\psi_i \cdot \not{d}\psi_i \tag{11.302}$$

We then obtain

$$R_{i_{l-m}} \cdots R_{i_1}(T)^m(\mu^2 f'_1) = \mu^2 \tilde{m}^\alpha R_{i_{l-m}} \cdots R_{i_1}(T)^m \Delta\psi_\alpha + {}^{(i_1\cdots i_{l-m})}n''_{m,l-m} \tag{11.303}$$

where ${}^{(i_1\cdots i_{l-m})}n''_{m,l-m}$ is of order $l+1$:

$$\begin{aligned} &{}^{(i_1\cdots i_{l-m})}n''_{m,l-m} \\ &= R_{i_{l-m}} \cdots R_{i_1}(\sum_{k=1}^{m} \frac{m!}{k!(m-k)!}((T)^k(\mu^2\tilde{m}^\alpha))((T)^{m-k}\Delta\psi_\alpha)) \\ &+ \sum_{s_1\neq\emptyset}((R)^{s_1}(\mu^2\tilde{m}^\alpha))((R)^{s_2}(T)^m\Delta\psi_\alpha) \\ &+ R_{i_{l-m}} \cdots R_{i_1}(T)^m(\mu^2 f'_{1,N}) \end{aligned} \tag{11.304}$$

In view of (10.303) we have

$$\begin{aligned} &\|R_{i_{l-m}} \cdots R_{i_1}(T)^m(\mu^2 f'_1)\|_{L^2(\Sigma_t^{\epsilon_0})} \\ &\leqslant C(1+\log(1+t))\sum_\alpha \|\mu R_{i_{l-m}} \cdots R_{i_1}(T)^m \Delta\psi_\alpha\|_{L^2(\Sigma_t^{\epsilon_0})} \\ &+ \|{}^{(i_1\cdots i_{l-m})}n''_{m,l-m}\|_{L^2(\Sigma_t^{\epsilon_0})} \end{aligned} \tag{11.305}$$

where $C$ is independent of $\ell$.

The lower order term ${}^{(i_1\cdots i_{l-m})}n''_{m,l-m}$ can be estimated straightforwardly, by using Proposition 11.1 and Proposition 11.2, as well as their corollaries:

$$\begin{aligned} &\max_{i_1\cdots i_{l-m}} \|{}^{(i_1\cdots i_{l-m})}n''_{m,l-m}\|_{L^2(\Sigma_t^{\epsilon_0})} \\ &\leqslant C_l(1+t)^{-2}(1+\log(1+t)) \\ &\cdot((1+\log(1+t))(\mathcal{W}_{\{l+1\}} + \delta_0(1+t)^{-3}(1+\log(1+t))^2\mathcal{W}^Q_{\{l-1\}}) \\ &+ \delta_0(1+t)^{-1}(\mathcal{B}_{[m,l]} + (1+\log(1+t))(\mathcal{Y}_0 + (1+t)\mathcal{A}_{[l-1]}))) \end{aligned} \tag{11.306}$$

We now return to (11.296). Our object is to express

$$R_{i_{l-m}} \cdots R_{i_1}(T)^{m+2}\psi_\alpha \quad \text{and} \quad R_{i_{l-m}} \cdots R_{i_1}(T)^m LT\psi_\alpha$$

as

$$TR_{i_{l-m}} \cdots R_{i_1}(T)^{m+1}\psi_\alpha \quad \text{and} \quad LR_{i_{l-m}} \cdots R_{i_1}(T)^{m+1}\psi_\alpha$$

respectively, plus lower order terms. Then by (10.217), we can bound the first two terms on the right in (11.296) by

$$\begin{aligned}&(|\ell| + C\delta_0(1+t)^{-1})\sqrt{\sum_\alpha \mathcal{E}_0[R_{i_{l-m}} \cdots R_{i_1}(T)^{m+1}\psi_\alpha]}\\&+ C(1+t)^{-1}(1+\log(1+t))^{1/2}\sqrt{\sum_\alpha \mathcal{E}_1'[R_{i_{l-m}} \cdots R_{i_1}(T)^{m+1}\psi_\alpha]}\\&+ C(1+t)^{-1}(1+\log(1+t))\mathcal{W}_{\{l+1\}} \qquad (11.307)\end{aligned}$$

up to a bound for the lower order terms of a form similar to (11.297).

In view of (10.205) and (10.207),

$$\begin{aligned}&TR_{i_n} \cdots R_{i_1} - R_{i_n} \cdots R_{i_1}T\\&= \sum_{k=0}^{n-1} R_{i_n} \cdots R_{i_{n-k+1}} {}^{(R_{i_{n-k}})}\Theta R_{i_{n-k-1}} \cdots R_{i_1} \qquad (11.308)\end{aligned}$$

we have

$$\begin{aligned}&R_{i_{l-m}} \cdots R_{i_1}(T)^{m+2}\psi_\alpha\\&= TR_{i_{l-m}} \cdots R_{i_1}(T)^{m+1}\psi_\alpha \qquad (11.309)\\&- \sum_{k=0}^{l-m-1} R_{i_{l-m}} \cdots R_{i_{l-m-k+1}} {}^{(R_{i_{l-m-k}})}\Theta R_{i_{l-m-k-1}} \cdots R_{i_1}(T)^{m+1}\psi_\alpha\end{aligned}$$

To estimate in $L^2(\Sigma_t^{\epsilon_0})$ the sum on the right, we observe that it is given by

$$\sum_{k=0}^{l-m-1} \sum_{|s_1|+|s_2|=k} (\not{\mathcal{L}}_R)^{s_1} {}^{(R_{i_{l-m-k}})}\Theta \cdot \not{d}(R)^{s_2'}(T)^{m+1}\psi_\alpha \qquad (11.310)$$

Here, in the inner sum, we are considering all ordered partitions $\{s_1, s_2\}$ of the set $\{l-m-k+1, \cdots, l-m\}$ into two ordered subsets $s_1$, $s_2$. Also,

$$s_2' = s_2 \bigcup \{1, \cdots, l-m-k-1\}$$

To estimate in $L^2(\Sigma_t^{\epsilon_0})$ a term in the inner sum in (11.310), we use Lemma 10.20. So we obtain that (11.310) is bounded by

$$C_l\delta_0(1+t)^{-2}((1+\log(1+t))^2\mathcal{W}_{\{l+1\}}+\max_i\|^{(R_i)}\Theta\|_{2,[l-1],\Sigma_t^{\epsilon_0}})$$
$$\leqslant C_l\delta_0(1+t)^{-2}(1+\log(1+t))\cdot((1+\log(1+t))\mathcal{W}_{\{l+1\}}+\delta_0(1+t)^{-1}\mathcal{B}_{[0,l]}+\mathcal{Y}_0+(1+t)\mathcal{A}_{[l-1]}) \tag{11.311}$$

Then it follows through (11.309) that

$$\sum_\alpha\|R_{i_{l-m}}\cdots R_{i_1}(T)^{m+2}\psi_\alpha\|_{L^2(\Sigma_t^{\epsilon_0})}$$
$$\leqslant\sum_\alpha\|TR_{i_{l-m}}\cdots R_{i_1}(T)^{m+1}\psi_\alpha\|_{L^2(\Sigma_t^{\epsilon_0})}+C_l\delta_0(1+t)^{-2}(1+\log(1+t))\cdot((1+\log(1+t))\mathcal{W}_{\{l+1\}}+\delta_0(1+t)^{-1}\mathcal{B}_{[0,l]}+\mathcal{Y}_0+(1+t)\mathcal{A}_{[l-1]}) \tag{11.312}$$

Similarly, we have

$$LR_{i_n}\cdots R_{i_1}-R_{i_n}\cdots R_{i_1}L=\sum_{k=0}^{n-1}R_{i_n}\cdots R_{i_{n-k+1}}{}^{(R_{i_{n-k}})}ZR_{i_{n-k-1}}\cdots R_{i_1} \tag{11.313}$$

It follows that, taking $n=l-m$,

$$R_{i_{l-m}}\cdots R_{i_1}L(T)^{m+1}\psi_\alpha=LR_{i_{l-m}}\cdots R_{i_1}(T)^{m+1}\psi_\alpha-\sum_{k=0}^{l-m-1}R_{i_{l-m}}\cdots R_{i_{l-m-k+1}}{}^{(R_{i_{l-m-k}})}ZR_{i_{l-m-k-1}}\cdots R_{i_1}(T)^{m+1}\psi_\alpha \tag{11.314}$$

By Lemma 11.22 we have

$$[(T)^m,L]=-\sum_{k=0}^{m-1}(T)^k\Lambda(T)^{m-k-1} \tag{11.315}$$

hence

$$R_{i_{l-m}}\cdots R_{i_1}(T)^mLT\psi_\alpha=R_{i_{l-m}}\cdots R_{i_1}L(T)^{m+1}\psi_\alpha-\sum_{k=0}^{m-1}R_{i_{l-m}}\cdots R_{i_1}(T)^k\Lambda(T)^{m-k}\psi_\alpha \tag{11.316}$$

Now, $\mu$ times the sum on the right in (11.314) is bounded in $L^2(\Sigma_t^{\epsilon_0})$ by

$$C(1+\log(1+t))\sum_{k=0}^{l-m-1}\|^{(R_{i_{l-m-k}})}Z\cdot \not{d}R_{i_{l-m-k-1}}\cdots R_{i_1}(T)^{m+1}\psi_\alpha\|_{2,[k],\Sigma_t^{\epsilon_0}} \tag{11.317}$$

Using Corollary 10.1.i and Corollary 10.2.i we obtain that this is bounded by

$$C_l\delta_0(1+t)^{-2}(1+\log(1+t))((1+\log(1+t))\mathcal{W}_{\{l+1\}}+\mathcal{Y}_0+(1+t)\mathcal{A}_{[l-1]}) \tag{11.318}$$

Also, $\mu$ times the sum on the right in (11.316) is bounded in $L^2(\Sigma_t^{\epsilon_0})$ by

$$C(1+\log(1+t))\sum_{k=0}^{m-1}\|\Lambda\cdot\not{d}(T)^{m-k}\psi_\alpha\|_{2,[k,l-m+k],\Sigma_t^{\epsilon_0}} \tag{11.319}$$

Using Corollary 11.1.c and Corollary 11.2.c we obtain that the above is bounded by

$$\begin{aligned}&C_l\delta_0(1+t)^{-2}(1+\log(1+t))\\&\cdot((1+\log(1+t))(\mathcal{W}_{\{l+1\}}+\delta_0(1+t)^{-3}(1+\log(1+t))\mathcal{W}^Q_{\{l-1\}})\\&+(1+t)^{-1}(\mathcal{B}_{[m-1,l]}+\delta_0(1+\log(1+t))(\mathcal{Y}_0+(1+t)\mathcal{A}_{[l-2]})))\end{aligned} \tag{11.320}$$

In view of the estimates (11.318) and (11.320), it follows through (11.314), (11.319) that

$$\begin{aligned}&\sum_\alpha\|\mu R_{i_{l-m}}\cdots R_{i_1}(T)^m LT\psi_\alpha\|_{L^2(\Sigma_t^{\epsilon_0})}\\&\leqslant\sum_\alpha\|\mu LR_{i_{l-m}}\cdots R_{i_1}(T)^{m+1}\psi_\alpha\|_{L^2(\Sigma_t^{\epsilon_0})}\\&\quad+C_l\delta_0(1+t)^{-2}(1+\log(1+t))\\&\quad\cdot((1+\log(1+t))\mathcal{W}_{\{l+1\}}+\delta_0(1+t)^{-3}(1+\log(1+t))^2\mathcal{W}^Q_{\{l-1\}}\\&\quad+(1+t)^{-1}\mathcal{B}_{[m-1,l]}+\mathcal{Y}_0+(1+t)\mathcal{A}_{[l-1]})\end{aligned} \tag{11.321}$$

In view of (11.307), inequalities (11.312), (11.321), together with the estimate (11.297) imply through (11.296) that

$$\begin{aligned}&\|R_{i_{l-m}}\cdots R_{i_1}(T)^m f_0'\|_{L^2(\Sigma_t^{\epsilon_0})}\\&\leqslant(|\ell|+\delta_0(1+t)^{-1})\sqrt{\sum_\alpha\mathcal{E}_0[R_{i_{l-m}}\cdots R_{i_1}(T)^{m+1}\psi_\alpha]}\\&\quad+C(1+t)^{-1}(1+\log(1+t))^{1/2}\sqrt{\sum_\alpha\mathcal{E}_1'[R_{i_{l-m}}\cdots R_{i_1}(T)^{m+1}\psi_\alpha]}\\&\quad+C(1+t)^{-1}(1+\log(1+t))\mathcal{W}_{\{l+1\}}+C_l\delta_0(1+t)^{-1}((1+\log(1+t))\mathcal{W}^Q_{\{l\}}\\&\quad+(1+t)^{-1}(\mathcal{B}_{[m,l]}+(1+\log(1+t))(\mathcal{Y}_0+(1+t)\mathcal{A}_{[l-1]})))\end{aligned} \tag{11.322}$$

We finally turn to (11.305). Our object here is to express $R_{i_{l-m}}\cdots R_{i_1}(T)^m\not{\Delta}\psi_\alpha$ as $\not{\Delta}(R_{i_{l-m}}\cdots R_{i_1}(T)^m\psi_\alpha)$ plus lower order terms. Since

$$\not{\Delta}\psi_\alpha=\mathrm{tr}(\not{g}^{-1})\cdot(\not{D}^2\psi_\alpha)$$

we have

$$\begin{aligned}&R_{i_{l-m}}\cdots R_{i_1}(T)^m\not{\Delta}\psi_\alpha\\&=R_{i_{l-m}}\cdots R_{i_1}(T)^m(\not{g}^{-1})^{AB}(\not{D}^2\psi_\alpha)_{AB}\\&=(\not{g}^{-1})^{AB}(\not{\mathcal{L}}_{R_{i_{l-m}}}\cdots\not{\mathcal{L}}_{R_{i_1}}(\not{\mathcal{L}}_T)^m(\not{D}^2\psi_\alpha)_{AB})\end{aligned} \tag{11.323}$$

$$+ \not{\mathcal{L}}_{R_{i_{l-m}}} \cdots \not{\mathcal{L}}_{R_{i_1}} \sum_{k=1}^{m} \frac{m!}{k!(m-k)!} (((\not{\mathcal{L}}_T)^k (\not{g}^{-1}))^{AB} (\not{\mathcal{L}}_T)^{m-k} (\not{D}^2 \psi_\alpha)_{AB})$$
$$\sum_{|s_1|+|s_2|=l-m, s_1 \neq \emptyset} (((\not{\mathcal{L}}_R)^{s_1} (\not{g}^{-1}))^{AB} (\not{\mathcal{L}}_R)^{s_2} (\not{\mathcal{L}}_T)^m (\not{D}^2 \psi_\alpha)_{AB})$$

Applying formula (11.83) to the $S_{t,u}$ 1-form $\not{d}\psi_\alpha$ we obtain

$$\begin{aligned} &\not{\mathcal{L}}_{R_{i_{l-m}}} \cdots \not{\mathcal{L}}_{R_{i_1}} (\not{\mathcal{L}}_T)^m \not{D}^2 \psi_\alpha \\ &= \not{D}^2 (R_{i_{l-m}} \cdots R_{i_1} (T)^m \psi_\alpha) + {}^{(i_1 \cdots i_{l-m})} c_{m,l-m} [\not{d}\psi_\alpha] \end{aligned} \tag{11.324}$$

Hence the first term on the right in (11.323) is given by

$$\begin{aligned} &(\not{g}^{-1})^{AB} (\not{\mathcal{L}}_{R_{i_{l-m}}} \cdots \not{\mathcal{L}}_{R_{i_1}} (\not{\mathcal{L}}_T)^m (\not{D}^2 \psi_\alpha)_{AB}) \\ &= \not{\Delta}(R_{i_{l-m}} \cdots R_{i_1} (T)^m \psi_\alpha) + (\not{g}^{-1})^{AB} ({}^{(i_1 \cdots i_{l-m})} c_{m,l-m} [\not{d}\psi_\alpha])_{AB} \end{aligned} \tag{11.325}$$

Taking into account of the fact that

$$\not{\mathcal{L}}_T (\not{g}^{-1}) = -\not{g}^{-1} \cdot {}^{(T)}\not{\pi} \cdot \not{g}^{-1}$$

and using Corollary 10.2.d, Corollary 11.1.c, Corollary 11.2.c and Corollary 11.2.d, we deduce that $\mu$ times the first sum on the right in (11.323) is bounded in $L^2(\Sigma_t^{\epsilon_0})$ by

$$\begin{aligned} &C_l (1+\log(1+t))((1+t)^{-1}(1+\log(1+t)) \|\not{D}^2 \psi_\alpha\|_{2,[m-1,l-1],\Sigma_t^{\epsilon_0}} \\ &+ \delta_0 (1+t)^{-3} (\mathcal{T}_{[m,l]} + (1+t)^{-2} \|{}^{(R_j)}\not{\pi}\|_{2,[l-1],\Sigma_t^{\epsilon_0}})) \\ &\leqslant C_l (1+t)^{-3} (1+\log(1+t)) \\ &\cdot ((1+\log(1+t)) \mathcal{W}_{\{l+1\}} + \delta_0 (1+t)^{-2} (1+\log(1+t))^2 \mathcal{W}^Q_{\{l-1\}} \\ &+ \delta_0 (1+t)^{-1} (\mathcal{B}_{[m-1,l]} + (1+\log(1+t))(\mathcal{Y}_0 + (1+t)\mathcal{A}_{[l-1]}))) \end{aligned} \tag{11.326}$$

Similarly, taking into account of the fact that

$$\not{\mathcal{L}}_{R_i} (\not{g}^{-1}) = -\not{g}^{-1} \cdot {}^{(R_i)}\not{\pi} \cdot \not{g}^{-1}$$

and using Corollary 10.1.d, Corollary 10.2.d and Corollary 11.2.d, we deduce that $\mu$ times the second sum on the right in (11.323) is bounded in $L^2(\Sigma_t^{\epsilon_0})$ by

$$\begin{aligned} &C_l (1+\log(1+t))(\delta_0 (1+t)^{-1}(1+\log(1+t)) \|\not{D}^2 \psi_\alpha\|_{2,[m,l-1],\Sigma_t^{\epsilon_0}} \\ &+ \delta_0 (1+t)^{-3} \max_i \|{}^{(R_i)}\not{\pi}\|_{2,[l-1],\Sigma_t^{\epsilon_0}}) \\ &\leqslant C_l \delta_0 (1+t)^{-3} (1+\log(1+t))((1+\log(1+t)) \mathcal{W}_{\{l+1\}} \\ &+ \delta_0 (1+t)^{-2} (1+\log(1+t))^2 \mathcal{W}^Q_{\{l-1\}} \\ &+ \delta_0 (1+t)^{-2} (1+\log(1+t)) \mathcal{B}_{[m-1,l]} + \mathcal{Y}_0 + (1+t)\mathcal{A}_{[l-1]}) \end{aligned} \tag{11.327}$$

Finally, to estimate in $L^2(\Sigma_t^{\epsilon_0})$ $\mu$ times the second term on the right in (11.325), we use the expression in Lemma 11.9 as well as Lemma 10.10, Lemma 10.11, Corollary 11.1.c and Corollary 11.2.c, to obtain

$$\begin{aligned}
&\max_{k\leqslant m, n\leqslant l-k}\max_{i_1\cdots i_n}\|{}^{(i_1\cdots i_n)}c_{k,n}[\not{d}\psi_\alpha]\|_{L^2(\Sigma_t^{\epsilon_0})} \\
&\leqslant C_l(1+t)^{-1}(\delta_0(1+t)^{-1}(1+\log(1+t))\|\not{d}\psi_\alpha\|_{2,[m,l-1],\Sigma_t^{\epsilon_0}} \\
&\quad+\|\not{d}\psi_\alpha\|_{\infty,[m,l_*],\Sigma_t^{\epsilon_0}}(\mathcal{W}_{\{l+1\}}+(1+t)^{-2}(1+\log(1+t))^2\mathcal{W}^Q_{\{l\}} \\
&\quad+(1+t)^{-1}\mathcal{B}_{[m-1,l+1]}+\mathcal{Y}_0+(1+t)\mathcal{A}_{[l]}))
\end{aligned}$$

which gives a bound by

$$\begin{aligned}
&C_l\delta_0(1+t)^{-3}(1+\log(1+t)) \\
&\cdot((1+\log(1+t))(\mathcal{W}_{\{l+1\}}+(1+t)^{-2}(1+\log(1+t))\mathcal{W}^Q_{\{l\}}) \\
&+(1+t)^{-1}\mathcal{B}_{[m-1,l+1]}+\mathcal{Y}_0+(1+t)\mathcal{A}_{[l]})
\end{aligned}\tag{11.328}$$

In view of (11.326)—(11.328), it follows through (11.323), (11.325) that

$$\begin{aligned}
&\sum_\alpha\|\mu R_{i_{l-m}}\cdots R_{i_1}(T)^m\not{\Delta}\psi_\alpha\|_{L^2(\Sigma_t^{\epsilon_0})} \\
&\leqslant\sum_\alpha\|\mu\not{\Delta}(R_{i_{l-m}}\cdots R_{i_1}(T)^m\psi_\alpha)\|_{L^2(\Sigma_t^{\epsilon_0})} \qquad (11.329)\\
&+C_l(1+t)^{-3}(1+\log(1+t)) \\
&\cdot((1+\log(1+t))(\mathcal{W}_{\{l+1\}}+\delta_0(1+t)^{-2}(1+\log(1+t))\mathcal{W}^Q_{\{l\}}) \\
&+\delta_0((1+t)^{-1}\mathcal{B}_{[m,l+1]}+\mathcal{Y}_0+(1+t)\mathcal{A}_{[l]}))
\end{aligned}$$

Moreover, by **H1** the first term on the right in (11.329) is bounded by

$$\begin{aligned}
&\sum_\alpha\|\mu\not{\Delta}(R_{i_{l-m}}\cdots R_{i_1}(T)^m\psi_\alpha)\|_{L^2(\Sigma_t^{\epsilon_0})} \\
&\leqslant C(1+t)^{-1}\sum_{\alpha,j}\|\mu\not{d}(R_jR_{i_{l-m}}\cdots R_{i_1}(T)^m\psi_\alpha)\|_{L^2(\Sigma_t^{\epsilon_0})} \\
&\leqslant C(1+t)^{-2}(1+\log(1+t))^{1/2}\sqrt{\sum_{\alpha,j}\mathcal{E}'_1[R_jR_{i_{l-m}}\cdots R_{i_1}(T)^m\psi_\alpha]}
\end{aligned}\tag{11.330}$$

The bounds (11.329)—(11.330) together with (11.306) imply through (11.303) that

$$\begin{aligned}
&\|R_{i_{l-m}}\cdots R_{i_1}(T)^m(\mu^2f'_1)\|_{L^2(\Sigma_t^{\epsilon_0})} \\
&\leqslant C(1+t)^{-2}(1+\log(1+t))^{3/2}\sqrt{\sum_{\alpha,j}\mathcal{E}'_1[R_jR_{i_{l-m}}\cdots R_{i_1}(T)^m\psi_\alpha]} \qquad (11.331)\\
&+C_l(1+t)^{-2}(1+\log(1+t)) \\
&\cdot((1+\log(1+t))(\mathcal{W}_{\{l+1\}}+\delta_0(1+t)^{-3}(1+\log(1+t))^2\mathcal{W}^Q_{\{l\}}) \\
&+\delta_0(1+t)^{-1}(\mathcal{B}_{[m,l+1]}+(1+\log(1+t))(\mathcal{Y}_0+(1+t)\mathcal{A}_{[l]})))
\end{aligned}$$

The bounds (11.322) and (11.331) yield finally the following proposition:

**Proposition 11.6** Under the assumptions of Proposition 11.5 we have

$$
{}^{(i_1\cdots i_{l-m})}P'_{m,l-m} \leqslant {}^{(i_1\cdots i_{l-m})}P'^{(0)}_{m,l-m} + {}^{(i_1\cdots i_{l-m})}P'^{(1)}_{m,l-m}
$$

where

$$
{}^{(i_1\cdots i_{l-m})}P'^{(0)}_{m,l-m} = |\ell|\sqrt{\sum_\alpha \mathcal{E}_0[R_{i_{l-m}}\cdots R_{i_1}(T)^{m+1}\psi_\alpha]}
$$

and

$$
\begin{aligned}
&{}^{(i_1\cdots i_{l-m})}P'^{(1)}_{m,l-m}\\
&= C\delta_0(1+t)^{-1}\sqrt{\sum_\alpha \mathcal{E}_0[R_{i_{l-m}}\cdots R_{i_1}(T)^{m+1}\psi_\alpha]}\\
&\quad + C(1+t)^{-1}(1+\log(1+t))^{1/2}\sqrt{\sum_\alpha \mathcal{E}'_1[R_{i_{l-m}}\cdots R_{i_1}(T)^{m+1}\psi_\alpha]}\\
&\quad + C(1+t)^{-2}(1+\log(1+t))^{3/2}\sqrt{\sum_{\alpha,j} \mathcal{E}'_1[R_j R_{i_{l-m}}\cdots R_{i_1}(T)^{m}\psi_\alpha]}\\
&\quad + (C(1+t)^{-1}(1+\log(1+t)) + C_l(1+t)^{-2}(1+\log(1+t))^2)\mathcal{W}_{\{l+1\}}\\
&\quad + C_l\delta_0(1+t)^{-1}((1+\log(1+t))\mathcal{W}^Q_{\{l\}} + (1+t)^{-1}(\mathcal{B}_{[m,l+1]}\\
&\quad + (1+\log(1+t))(\mathcal{Y}_0 + (1+t)\mathcal{A}_{[l]})))
\end{aligned}
$$

# 12

# Recovery of the Acoustical Assumptions. Estimates for Up to the Next to the Top Order Angular Derivatives of $\chi$ and Spatial Derivatives of $\mu$

In the first part of the present chapter we shall show that assumptions $\mathbf{E}_{\{1\}}$ and $\mathbf{E}^{Q}_{\{0\}}$ on $W^s_{\epsilon_0}$, together with certain primitive assumptions on the initial conditions on $\Sigma_0^{\epsilon_0}$, imply pointwise estimates for $\kappa$, the functions $\lambda_i, y'_i, y_i$, as well as sharp upper and lower bounds for $r$. Moreover, we shall show that the same assumptions imply hypothesis **H0** on $W^s_{\epsilon_0}$.

In the second part of the present chapter we shall show that the assumptions $\mathbf{E}_{\{l+2\}}, \mathbf{E}^{Q}_{\{l+1\}}$ and $\mathbf{E}^{QQ}_{\{l\}}$, on $W^s_{\epsilon_0}$, together with appropriate assumptions on the initial conditions on $\Sigma_0^{\epsilon_0}$, imply the acoustical assumptions $\mathbf{X}_{[l]}$, as well as $\mathbf{H1}, \mathbf{H2}$ and $\mathbf{H2}'$. Moreover, we shall show that under some appropriate assumptions on the initial conditions, the acoustical assumptions $\mathbf{M}_{[m,l+1]}$, for $m = 0, \cdots, l+1$, also follow.

Finally, in the third part of this chapter we shall derive estimates for the quantities $\mathcal{A}_{[l]}$ and $\mathcal{B}_{[m+1,l+1]}$, for $m = 0, \cdots, l$, under the assumptions $\mathbf{E}_{\{l_*+2\}}, \mathbf{E}^{Q}_{\{l_*+1\}}$, $\mathbf{E}^{QQ}_{\{l_*\}}$, on $W^s_{\epsilon_0}$, together with appropriate assumptions on the initial conditions on $\Sigma_0^{\epsilon_0}$.

## 12.1 Estimates for $\lambda_i$, $y'_i$, $y_i$ and $r$. Establishing the Hypothesis H0

**Proposition 12.1** Let assumptions $\mathbf{E}_{\{1\}}$ and $\mathbf{E}^{Q}_{\{0\}}$ hold on $W^s_{\epsilon_0}$. Then provided that $\delta_0$ is suitably small, for all $t \in [0, s]$ we have

$$\|\kappa - 1\|_{L^\infty(\Sigma_t^{\epsilon_0})} \leqslant C\delta_0(1 + \log(1+t))$$

where the constant $C$ is independent of $s$.

*Proof.* Recall from Chapter 3, we have

$$L\kappa = \eta^{-1}m + \kappa e' \tag{12.1}$$

where

$$m = \frac{1}{2}\frac{dH}{dh}T(h) \tag{12.2}$$

and

$$e' = \frac{1}{\eta}\frac{d\eta}{dh}Lh - \alpha^{-1}\hat{T}^i(L\psi_i) - \alpha^{-1}L\alpha \tag{12.3}$$

Assumption $\mathbf{E}_{\{0\}}$ implies

$$\|h\|_{L^\infty(\Sigma_t^{\epsilon_0})} \leqslant C\delta_0(1+t)^{-1} \tag{12.4}$$

and

$$\|\eta^{-1} - 1\|_{L^\infty(\Sigma_t^{\epsilon_0})} \leqslant C\delta_0(1+t)^{-1} \tag{12.5}$$

From Chapter 6, we have

$$|\hat{T}^i| \leqslant 1, \quad |L^i| \leqslant C \tag{12.6}$$

So $\mathbf{E}^Q_{\{0\}}$ implies

$$\|\hat{T}^i(L\psi_i)\|_{L^\infty(\Sigma_t^{\epsilon_0})} \leqslant C\delta_0(1+t)^{-2} \tag{12.7}$$

Also from $\mathbf{E}_{\{1\}}$ and $\mathbf{E}^Q_{\{0\}}$ we have

$$\|Th\|_{L^\infty(\Sigma_t^{\epsilon_0})} \leqslant C\delta_0(1+t)^{-1}, \quad \|Lh\|_{L^\infty(\Sigma_t^{\epsilon_0})} \leqslant C\delta_0(1+t)^{-2} \tag{12.8}$$

It then follows that

$$\alpha^{-1}|m| \leqslant C\delta_0(1+t)^{-1}, \quad |e'| \leqslant C\delta_0(1+t)^{-2} \tag{12.9}$$

on $W^s_{\epsilon_0}$.

Writing

$$L(\kappa - 1) = e'(\kappa - 1) + \alpha^{-1}m + e' \tag{12.10}$$

and noting that for any function $f$,

$$L|f| \leqslant |Lf|$$

we obtain from (12.9)

$$L|\kappa - 1| \leqslant C\delta_0(1+t)^{-2}|\kappa - 1| + C\delta_0(1+t)^{-1} \tag{12.11}$$

Integrating this along the integral curves of $L$ and recalling that on $\Sigma_0$,

$$\kappa = 1$$

the proposition follows. □

Recall from Chapter 6:

$$\lambda_i = \bar{g}(\mathring{R}_i, \hat{T}) \tag{12.12}$$

**Lemma 12.1** Let assumptions $\mathbf{E}_{\{1\}}, \mathbf{E}_{\{0\}}$ hold on $W^s_{\epsilon_0}$. Moreover, let hypothesis **H0** hold on $W^{s_0}_{\epsilon_0}$ for some $s_0 \in (0, s]$. Then, provided that $\delta_0$ is suitably small, we have, for all $t \in [0, s_0]$,

$$\max_i \|\lambda_i\|_{L^\infty(\Sigma_t^{\epsilon_0})} \leqslant C\delta_0(1 + \log(1 + t))$$

*Proof.* The result follows in the exactly the same way which we used to derive (6.143). Note that to derive an estimate for $\kappa^{-1}\zeta$, we must appeal to the hypothesis **H0**. □

Since the result of Proposition 12.1 coincides with (6.97), so we conclude that under the assumptions of Proposition 12.1, the lower bounds (6.128) and (6.129) holds:

$$r \geqslant 1 - u + t - C\delta_0 u(1 + \log(1 + t)) \tag{12.13}$$

and

$$r \geqslant C^{-1}(1 + t) \tag{12.14}$$

if $\delta_0$ is suitably small.

Recall from Chapter 6 the projection operator $\Sigma$:

$$\Sigma^i_j = \delta^i_j - r^{-2} x^i x^j \tag{12.15}$$

and the fact that

$$|\Sigma\hat{T}|^2 = r^{-2} \sum_i (\lambda_i)^2 \tag{12.16}$$

Then Lemma 12.1 together with (12.14) implies

$$|\Sigma\hat{T}| \leqslant C\delta_0(1 + t)^{-1}(1 + \log(1 + t)) \tag{12.17}$$

in $W^{s_0}_{\epsilon_0}$.

Recall also from Chapter 6:

$$N = \frac{x^i}{r}\frac{\partial}{\partial x^i} \tag{12.18}$$

and the fact that

$$0 \leqslant 1 + \bar{g}(N, \hat{T}) \leqslant C\delta_0^2(1 + t)^{-2}(1 + \log(1 + t))^2 \tag{12.19}$$

Moreover, recall from Chapter 6 that

$$y' = \Sigma\hat{T} + (1 + \bar{g}(N, \hat{T}))N$$

Following the argument in Chapter 6 ((6.169)—(6.177)) we deduce

$$|r - 1 - t + u| \leqslant C\delta_0 u(1 + \log(1 + t)) \tag{12.20}$$

$$|\frac{1}{r} - \frac{1}{1-u+t}| \leqslant C\delta_0(1+t)^{-2}(1+\log(1+t)) \tag{12.21}$$

and

$$|y'| \leqslant C\delta_0(1+t)^{-1}(1+\log(1+t)) \tag{12.22}$$

$$\max_i \|y^i\|_{L^\infty(\Sigma_t^{\epsilon_0})} \leqslant C\delta_0(1+t)^{-1}(1+\log(1+t)) \tag{12.23}$$

Note that we have only used the assumptions in Proposition 12.1 and Lemma 12.1 (valid for $t \in [0, s_0]$) to get these results.

Now we begin the argument which leads to our establishing hypothesis **H0**.

**Proposition 12.2** Consider a surface $S_{t,u}$ for which the following hold:

$$\inf_{S_{t,u}} r > 0, \quad \sup_{S_{t,u}} |y'| < 1$$

Then for every $S_{t,u}$ 1-form $\xi$ we have

$$|\xi|^2 \leqslant (1 - \sup_{S_{t,u}} |y'|^2)^{-1} (\inf_{S_{t,u}} r)^{-2} \sum_{i=1}^{3} (\xi(R_i))^2$$

*Proof*. Since $R_i = \Pi \mathring{R}_i$, we have

$$(R_i)^a = \Pi_b^a (\mathring{R}_i)^b = \Pi_b^a \epsilon_{ikb} x^k \tag{12.24}$$

hence

$$\begin{aligned} \sum_{i=1}^{3} (R_i)^a (R_i)^c &= \sum_{i=1}^{3} \Pi_b^a \epsilon_{ikb} x^k \Pi_d^c \epsilon_{ild} x^l \\ &= (\delta_{kl}\delta_{bd} - \delta_{kd}\delta_{bl}) x^k x^l \Pi_b^a \Pi_d^c \\ &= (r^2 \delta_{bd} - x^b x^d) \Pi_b^a \Pi_d^c \end{aligned} \tag{12.25}$$

where we have used the fact that

$$\sum_{i=1}^{3} \epsilon_{ikb}\epsilon_{ild} = \delta_{kl}\delta_{bd} - \delta_{kd}\delta_{bl} \tag{12.26}$$

It follows that

$$\sum_{i=1}^{3} (\xi(R_i))^2 = (\sum_{i=1}^{3} (R_i)^a (R_i)^c) \xi_a \xi_c = (r^2\delta_{bd} - x^b x^d)(\Pi_b^a \xi_a)(\Pi_d^c \xi_c) \tag{12.27}$$

Since $\xi$ is an $S_{t,u}$ 1-form, we have

$$\Pi_b^a \xi_a = \xi_b$$

We thus obtain

$$\sum_{i=1}^{3}(\xi(R_i))^2 = (r^2\delta_{bd} - x^b x^d)\xi_b\xi_d = r^2(|\xi|^2 - (\xi(N))^2) \tag{12.28}$$

Recall (6.174),

$$y' = \hat{T} + N \tag{12.29}$$

In view of this and the fact that $\xi(\hat{T}) = 0$, we have

$$\xi(N) = \xi(y') = y'^a\xi_a \tag{12.30}$$

hence

$$|\xi(N)| = |\xi(y')| \leqslant |y'||\xi| \tag{12.31}$$

and

$$|\xi|^2 - (\xi(N))^2 \geqslant (1 - \sup_{S_{t,u}}|y'|^2)|\xi|^2 \tag{12.32}$$

Substituting in (12.28) yields

$$\sum_{i=1}^{3}(\xi(R_i))^2 \geqslant (\inf_{S_{t,u}} r)^2(1 - \sup_{S_{t,u}}|y'|^2)|\xi|^2 \tag{12.33}$$

The proposition follows. □

**Corollary 12.2.a** Suppose that there are positive constants $\epsilon_1$ and $\epsilon_2 < 1$ such that

$$\inf_{\Sigma_t^{\epsilon_0}} r \geqslant \epsilon_1(1+t), \quad \sup_{\Sigma_t^{\epsilon_0}}|y'|^2 \leqslant 1 - \epsilon_2$$

Then **H0** holds on $\Sigma_t^{\epsilon_0}$. In fact, for any differentiable function $f$ on $S_{t,u}$, we have, pointwise on $\Sigma_t^{\epsilon_0}$,

$$|\not{d}f|^2 \leqslant C(1+t)^{-2}\sum_i(R_i f)^2$$

where

$$C = \frac{1}{\epsilon_2\epsilon_1^2}$$

*Proof*. This follows by applying Proposition 12.2 to the $S_{t,u}$ 1-form $\xi = \not{d}f$. □

Since

$$\sup_{\Sigma_t^{\epsilon_0}} u = \epsilon_0 \leqslant \frac{1}{2}$$

fixing any

$$\epsilon_1 < \frac{1}{2}$$

then (12.13) implies

$$\inf_{\Sigma_t^{\epsilon_0}} r > \epsilon_1(1+t) \tag{12.34}$$

for all $t \in [0, s_0]$, provided that $\delta_0$ is suitably small. Also, fixing any $\epsilon_2 < 1$, (12.22) implies

$$\sup_{\Sigma_t^{\epsilon_0}} |y'|^2 < 1 - \epsilon_2 \tag{12.35}$$

for all $t \in [0, s_0]$, provided that $\delta_0$ is suitably small.

Here we have used Lemma 12.1, which is based on **H0** to derive (12.34)—(12.35). We shall show that **H0** actually holds on $[0, s]$. Let $\bar{s}_0$ be the maximal value of $s_0 \in [0, s]$ such that **H0** holds on $W_{\epsilon_0}^{s_0}$. We show that in fact $\bar{s}_0 = s$. For, suppose that $\bar{s}_0 < s$. Since (12.17), (12.19) and (12.22) hold on $W_{\epsilon_0}^{\bar{s}_0}$, hence also (12.34) and (12.35) hold for all $t \in [0, \bar{s}_0]$, in particular at $t = \bar{s}_0$. However, (12.34) and (12.35) are strict inequalities, by continuity they will hold on some small interval $[\bar{s}_0, s_*) \subset [\bar{s}_0, s]$. This contradicts the maximality of $\bar{s}_0$. So **H0** holds on $W_{\epsilon_0}^s$. From Lemma 12.1 with $s_0 = s$, we get the following proposition:

**Proposition 12.3** Let $\mathbf{E}_{\{1\}}, \mathbf{E}_{\{0\}}^Q$ hold on $W_{\epsilon_0}^s$. Then if $\delta_0$ is suitably small, **H0** holds on $W_{\epsilon_0}^s$. Moreover, we have, on $W_{\epsilon_0}^s$,

$$1 + t - u - C\delta_0 u(1 + \log(1+t)) \leqslant r \leqslant 1 + t + \min(0, -u + C\delta_0 u(1 + \log(1+t)))$$
$$|y'| \leqslant C\delta_0(1+t)^{-1}(1 + \log(1+t))$$

and, for all $t \in [0, s]$,

$$\|\kappa - 1\|_{L^\infty(\Sigma_t^{\epsilon_0})} \leqslant C\delta_0(1 + \log(1+t))$$
$$\max_i \|\lambda_i\|_{L^\infty(\Sigma_t^{\epsilon_0})} \leqslant C\delta_0(1 + \log(1+t))$$
$$\max_i \|y^i\|_{L^\infty(\Sigma_t^{\epsilon_0})} \leqslant C\delta_0(1+t)^{-1}(1 + \log(1+t))$$

## 12.2 The Coercivity Hypothesis H1, H2 and H2′. Estimates for $\chi'$

We start this section with a proposition analogous to Proposition 12.2.

**Proposition 12.4** Consider a surface $S_{t,u}$ for which the following hold:

$$\inf_{S_{t,u}} r > 0, \quad \sup_{S_{t,u}} |y'| < 1$$

Then for every symmetric 2-covariant $S_{t,u}$ tensorfield $\vartheta$ we have

$$|\vartheta|^2 \leqslant (1 - \sup_{S_{t,u}} |y'|^2)^{-2} (\inf_{S_{t,u}} r)^{-4} \sum_{i,j=1}^{3} (\vartheta(R_i, R_j))^2$$

pointwise on $S_{t,u}$.

*Proof*. We have

$$\sum_{i,j}(\vartheta(R_i,R_j))^2 = (\sum_{i=1}^{3}(R_i)^a(R_i)^c)(\sum_{j=1}^{3}(R_j)^b(R_j)^d)\vartheta_{ab}\vartheta_{cd} \tag{12.36}$$

We substitute (12.25) in the first two factors on the right. Since $x^b = rN^b = r(y'^b - \hat{T}^b)$, and $\hat{T}^b \Pi_b^a = 0$, we have

$$x^b \Pi_b^a = r y'^b \Pi_b^a \tag{12.37}$$

(12.25) becomes

$$\sum_{i=1}^{3}(R_i)^a(R_i)^c = r^2(\delta_{bd} - y'^b y'^d)\Pi_b^a \Pi_d^c \tag{12.38}$$

Substituting (12.38) in (12.36) we get

$$\begin{aligned}\sum_{i,j=1}^{3}(\vartheta(R_i,R_j))^2 &= r^4(\delta_{ac} - y'^a y'^c)(\delta_{bd} - y'^b y'^d)\vartheta_{ab}\vartheta_{cd} \\ &= r^4(|\vartheta|^2 - 2|i_{y'}\vartheta|^2 + (\vartheta(y',y'))^2)\end{aligned} \tag{12.39}$$

where

$$(i_{y'}\vartheta)_b = y'^a \vartheta_{ab} \tag{12.40}$$

Consider the expression

$$|\vartheta|^2 - 2|i_{y'}\vartheta|^2 + (\vartheta(y',y'))^2$$

At a given point, by a suitable rotation of the rectangular coordinates we can arrange that $y'^2 = y'^3 = 0$. Then this expression becomes

$$\begin{aligned}&(1-(y'^1)^2)^2(\vartheta_{11})^2 + (\vartheta_{22})^2 + (\vartheta_{33})^2 + 2(1-(y'^1)^2)((\vartheta_{12})^2 + (\vartheta_{13})^2) + 2(\vartheta_{23})^2 \\ &\geqslant (1-(y'^1)^2)^2((\vartheta_{11})^2 + (\vartheta_{22})^2 + (\vartheta_{33})^2 + 2(\vartheta_{12})^2 + 2(\vartheta_{13})^2 + 2(\vartheta_{23})^2) \\ &= (1-|y'|^2)^2|\vartheta|^2\end{aligned}$$

We conclude that

$$|\vartheta|^2 - 2|i_{y'}\vartheta|^2 + (\vartheta(y',y'))^2 \geqslant (1-|y'|^2)^2|\vartheta|^2 \tag{12.41}$$

So (12.39) implies

$$\sum_{i,j=1}^{3}(\vartheta(R_i,R_j))^2 \geqslant r^4(1-|y'|^2)^2|\vartheta|^2 \tag{12.42}$$

The proposition follows. □

We now return to the propagation equation for $\chi$ in Chapter 3:

$$L\chi_{AB} = e\chi_{AB} + \chi_A^C \chi_{BC} - \alpha'_{AB} \tag{12.43}$$

by direct calculation (see Chapter 4),

$$\alpha'_{AB} = \alpha'^{[P]}_{AB} + \alpha^{[N]}_{AB} \tag{12.44}$$

where

$$\alpha'^{[P]}_{AB} = -\frac{1}{2}\frac{dH}{dh}\not{D}^2 h(X_A, X_B) \tag{12.45}$$

and $\alpha^{[N]}_{AB}$ is given by (4.25) and (4.26).

Taking into account of the fact that

$$\begin{aligned} \not{\mathcal{L}}_L \not{g} &= 2\chi \\ \chi' &= \chi - \frac{\not{g}}{1-u+t} \end{aligned} \tag{12.46}$$

we obtain the following propagation equation for $\chi'$:

$$L\chi'_{AB} = e\chi'_{AB} + (\not{g}^{-1})^{CD}\chi'_{AD}\chi'_{BC} + \frac{e\not{g}_{AB}}{1-u+t} - \alpha'_{AB} \tag{12.47}$$

**Lemma 12.2** Let assumptions $\mathbf{E}_{\{2\}}, \mathbf{E}^{Q}_{\{1\}}, \mathbf{E}^{QQ}_{\{0\}}$ hold on $W^s_{\epsilon_0}$, and let the initial data satisfy

$$\|\chi'\|_{L^\infty(\Sigma^{\epsilon_0}_0)} \leqslant C\delta_0$$

Moreover, let **H1** hold on $W^{s_0}_{\epsilon_0}$ for some $s_0 \in (0, s]$. Then, provided $\delta_0$ is suitably small, we have, for all $t \in [0, s_0]$,

$$\|\chi'\|_{L^\infty(\Sigma^{\epsilon_0}_t)} \leqslant C\delta_0(1+t)^{-2}(1+\log(1+t))$$

*Proof*. From the assumptions and the expression for $e$, we deduce

$$\|e\|_{L^\infty(\Sigma^{\epsilon_0}_t)} \leqslant C\delta_0(1+t)^{-2} \tag{12.48}$$

and from the hypotheses **H0** and **H1** and the expression for $\alpha'_{AB}$, we deduce

$$\|\alpha'\|_{L^\infty(\Sigma^{\epsilon_0}_t)} \leqslant C\delta_0(1+t)^{-3} \tag{12.49}$$

Here we have used the fact that the magnitude of the symmetric 2-covariant $S_{t,u}$ tensorfield $\not{\psi}$ is bounded by $C\delta_0(1+t)^{-2}$. In fact, since

$$\not{\psi}_{AB} = X^i_A(X_B\psi_i)$$

we have

$$\begin{aligned} |\not{\psi}|^2 &= (\not{g}^{-1})^{AC}(\not{g}^{-1})^{BD}X^i_A(X_B\psi_i)X^i_C(X_D\psi_i) \\ &= (g^{-1})^{ij}(X_B\psi_i)(X_D\psi_j)(\not{g}^{-1})^{BD} - \hat{T}^i\hat{T}^j(X_B\psi_i)(X_D\psi_j)(\not{g}^{-1})^{BD} \\ &= \sum_i |\not{d}\psi_i|^2 - |\not{\psi}_{\hat{T}}|^2 \leqslant \sum_i |\not{d}\psi_i|^2 \leqslant C\delta_0^2(1+t)^{-4} \end{aligned}$$

Also, we have used the fact that by **H1**,

$$|\not{D}^2 h| \leqslant C(1+t)^{-1} \max_i |\not{\mathcal{L}}_{R_i} \not{d} h| \quad \text{on} \quad W_{\epsilon_0}^{s_0}$$

hence since $\not{\mathcal{L}}_{R_i} \not{d} h = \not{d} R_i h$, we have, by **H0**,

$$|\not{\mathcal{L}}_{R_i} \not{d} h| \leqslant C(1+t)^{-1} \max_j |R_j R_i h| \quad \text{on} \quad W_{\epsilon_0}^{s}$$

Thus,

$$|\not{D}^2 h| \leqslant C(1+t)^{-2} \max_{i,j} |R_i R_j h| \quad \text{on} \quad W_{\epsilon_0}^{s_0}$$

Therefore, by $\mathbf{E}_{\{2\}}$,

$$\|\not{D}^2 h\|_{L^\infty(\Sigma_t^{\epsilon_0})} \leqslant C\delta_0 (1+t)^{-3} \quad \text{for all} \quad t \in [0, s_0]$$

We can write Equation (12.47) in the form

$$\not{\mathcal{L}}_L \chi' = \chi' \cdot \chi' + e\chi' + \frac{\not{g} e}{1-u+t} - \alpha' \tag{12.50}$$

where

$$(\chi' \cdot \chi')_{AB} = \chi'_{AC} \chi'_{BD} (\not{g}^{-1})^{CD}$$

Let now $\vartheta, \vartheta'$ be a pair of symmetric 2-covariant $S_{t,u}$ tensorfields. Their pointwise inner product, with respect to the induced acoustical metric $\not{g}$, is given by

$$(\vartheta, \vartheta') = (\not{g}^{-1})^{AC} (\not{g}^{-1})^{BD} \vartheta_{AB} \vartheta'_{CD} \tag{12.51}$$

Since

$$\not{\mathcal{L}}_L (\not{g}^{-1})^{AB} = -2\chi^{AB}$$

we have

$$L(\vartheta, \vartheta') = (\not{\mathcal{L}}_L \vartheta, \vartheta') + (\vartheta, \not{\mathcal{L}}_L \vartheta') - 4\mathrm{tr}(\vartheta \cdot \chi \cdot \vartheta') \tag{12.52}$$

Taking in particular $\vartheta' = \vartheta$ we obtain

$$|\vartheta| L |\vartheta| = \frac{1}{2} L(\vartheta, \vartheta) = (\vartheta, \not{\mathcal{L}}_L \vartheta) - 2\mathrm{tr}(\vartheta \cdot \chi \cdot \vartheta) \tag{12.53}$$

Writing

$$\chi = \frac{\not{g}}{1-u+t} + \chi'$$

we have

$$\mathrm{tr}(\vartheta \cdot \chi \cdot \vartheta) = \frac{|\vartheta|^2}{1-u+t} + \mathrm{tr}(\vartheta \cdot \chi' \cdot \vartheta)$$

and (12.53) can be written as

$$|\vartheta| L|\vartheta| = (\vartheta, \not{\mathcal{L}}_L \vartheta) - \frac{2|\vartheta|^2}{1-u+t} - 2\mathrm{tr}(\vartheta \cdot \chi' \cdot \vartheta) \tag{12.54}$$

Now, the following inequality holds:

$$|\mathrm{tr}(\vartheta \cdot \chi' \cdot \vartheta)| \leqslant |\chi'||\vartheta|^2 \tag{12.55}$$

To see this, we can work in an orthonormal frame on $S_{t,u}$ relative to $\not{g}$ which is a frame of eigenvectors of $\chi'$. Let $\lambda_1, \lambda_2$ be the eigenvalues of $\chi'$, then at a given point,

$$\mathrm{tr}(\vartheta \cdot \chi' \cdot \vartheta) = \lambda_1(\vartheta_{11})^2 + (\lambda_1 + \lambda_2)(\vartheta_{12})^2 + \lambda_2(\vartheta_{22})^2$$

Since

$$|\vartheta|^2 = (\vartheta_{11})^2 + 2(\vartheta_{12})^2 + (\vartheta_{22})^2$$

It follows that

$$|\mathrm{tr}(\vartheta \cdot \chi' \cdot \vartheta)| \leqslant \max\{|\lambda_1|, |\lambda_2|\}|\vartheta|^2$$

In view of the fact that

$$\max\{|\lambda_1|, |\lambda_2|\} \leqslant \sqrt{(\lambda_1)^2 + (\lambda_2)^2} = |\chi'|$$

then (12.55) follows.

Taking into account (12.55) as well as the fact that

$$|(\vartheta, \not{\mathcal{L}}_L \vartheta)| \leqslant |\vartheta||\not{\mathcal{L}}_L \vartheta|$$

we deduce from (12.54) the inequality

$$L((1-u+t)^2|\vartheta|) \leqslant (1-u+t)^2(2|\chi'||\vartheta| + |\not{\mathcal{L}}_L \vartheta|) \tag{12.56}$$

We apply this to the case $\vartheta = \chi'$. In view of the fact that

$$|\chi' \cdot \chi'| = \sqrt{(\lambda_1)^4 + (\lambda_2)^4} \leqslant (\lambda_1)^2 + (\lambda_2)^2 = |\chi'|^2$$

we obtain, substituting for $\not{\mathcal{L}}_L \chi'$ from (12.50),

$$L((1-u+t)^2|\chi'|) \leqslant (1-u+t)^2((3|\chi'| + |e|)|\chi'| + |b|) \tag{12.57}$$

where

$$b = \frac{\not{g} e}{1-u+t} - \alpha'$$

Let $\mathcal{P}(t)$ be the property

$$\mathcal{P}(t) : \|\chi'\|_{L^\infty(\Sigma_{t'}^{\epsilon_0})} \leqslant C_0\delta_0(1+t')^{-2}(1+\log(1+t'))$$

for all $t' \in [0, t]$.

Choosing $C_0$ suitably large, we have, by the assumption on the initial data,

$$\|\chi'\|_{L^\infty(\Sigma_0^{\epsilon_0})} < C_0\delta_0 \tag{12.58}$$

It follows by continuity that $\mathcal{P}(t)$ is true for sufficiently small positive $t$. Let $t_0$ be the least upper bound of the set of values of $t \in [0, s_0]$ for which $\mathcal{P}(t)$ holds. Then by continuity $\mathcal{P}(t_0)$ is true. Hence, in $W_{\epsilon_0}^{t_0}$, we have, in view of (12.48),

$$3|\chi'| + |e| \leqslant (3C_0 + C)\delta_0(1+t)^{-2}(1+\log(1+t)) \tag{12.59}$$

in $W_{\epsilon_0}^{t_0}$.

Substituting in (12.57) and taking into account the estimates (12.48) and (12.49) we obtain

$$L((1-u+t)^2|\chi'|) \leqslant (3C_0+C)\delta_0(1+t)^{-2}(1+\log(1+t))((1-u+t)^2|\chi'|)+C\delta_0(1+t)^{-1} \tag{12.60}$$

Setting along an integral curve of $L$

$$x(t) = (1-u+t)^2|\chi'| \tag{12.61}$$

then (12.60) takes the form

$$\frac{dx}{dt} \leqslant fx + g \tag{12.62}$$

where

$$f(t) = (3C_0 + C)\delta_0(1+t)^{-2}(1+\log(1+t)), \quad g(t) = C\delta_0(1+t)^{-1} \tag{12.63}$$

Integrating from $t = 0$ yields

$$x(t) \leqslant e^{\int_0^t f(t')dt'}\left(x(0) + \int_0^t e^{-\int_0^{t'} f(t'')dt''} g(t')dt'\right) \tag{12.64}$$

Since

$$x(0) \leqslant \|\chi'\|_{L^\infty(\Sigma_0^{\epsilon_0})}, \tag{12.65}$$

while

$$x(t) \geqslant \frac{1}{4}(1+t)^2|\chi'|, \tag{12.66}$$

taking into account the facts that

$$\int_0^t f(t')dt' \leqslant \int_0^\infty f(t')dt' \leqslant 2(3C_0+C)\delta_0, \quad \int_0^t g(t')dt' = C\delta_0 \log(1+t) \tag{12.67}$$

(12.64) yields the bound

$$\frac{1}{4}(1+t)^2|\chi'| \leqslant e^{2(3C_0+C)\delta_0}(\|\chi'\|_{L^\infty(\Sigma_0^{\epsilon_0})} + C\delta_0\log(1+t)) \tag{12.68}$$

So we get on $\Sigma_t^{\epsilon_0}$

$$\|\chi'\|_{L^\infty(\Sigma_t^{\epsilon_0})} \leqslant 4e^{2(3C_0+C)\delta_0}(\|\chi'\|_{L^\infty(\Sigma_0^{\epsilon_0})} + C\delta_0 \log(1+t))(1+t)^{-2} \tag{12.69}$$

This holds for all $t \in [0, t_0]$. Let us now fix $C_0$ large enough so that

$$C_0\delta_0 \geqslant 8\|\chi'\|_{L^\infty(\Sigma_0^{\epsilon_0})}, \quad C_0 > 8C \tag{12.70}$$

Then provided $\delta_0$ satisfies

$$\delta_0 \leqslant \frac{\log 2}{2(3C_0+C)} \tag{12.71}$$

(12.69) implies

$$\|\chi'\|_{L^\infty(\Sigma_t^{\epsilon_0})} < C_0\delta_0(1+t)^{-2}(1+\log(1+t)) \tag{12.72}$$

for all $t \in [0, t_0]$.

It follows by continuity that $\mathcal{P}(t)$ is true for some $t > t_0$. So we must have $t_0 = s_0$, and the lemma follows. □

Next, we investigate **H1**. Consider $\not{\mathcal{L}}_{R_i}\xi$ for an arbitrary $S_{t,u}$ 1-form $\xi$. We have, in rectangular coordinates,

$$(\not{\mathcal{L}}_{R_i}\xi)_a = (\not{D}_{R_i}\xi)_a + \xi_m(\not{D}R_i)_a^m, \quad (\not{D}_{R_i}\xi)_a = (R_i)^m(\not{D}\xi)_{ma} \tag{12.73}$$

where we have used that for any $S_{t,u}$ vectorfield $X$,

$$(\not{\mathcal{L}}_{R_i}\xi)(X) = (\not{D}_{R_i}\xi)(X) + \xi(\not{D}_X R_i)$$

Hence, we have

$$\begin{aligned}\sum_i |\not{\mathcal{L}}_{R_i}\xi|^2 = \sum_i |\not{D}_{R_i}\xi|^2 + 2\sum_i (\bar{g}^{-1})^{ab}(\not{D}_{R_i}\xi)_a\xi_n(\not{D}R_i)_b^n \\ + \sum_i (\bar{g}^{-1})^{ab}\xi_m\xi_n(\not{D}R_i)_a^m(\not{D}R_i)_b^n\end{aligned} \tag{12.74}$$

Consider the first term on the right of (12.74). We have

$$\begin{aligned}\sum_i |\not{D}_{R_i}\xi|^2 &= \sum_i (\bar{g}^{-1})^{ab}(\not{D}_{R_i}\xi)_a(\not{D}_{R_i}\xi)_b \\ &= (\bar{g}^{-1})^{ab}(\sum_i (R_i)^m(R_i)^n)(\not{D}\xi)_{ma}(\not{D}\xi)_{nb}\end{aligned} \tag{12.75}$$

Substituting from (12.38) and noting that

$$\Pi_c^m \Pi_d^n(\not{D}\xi)_{ma}(\not{D}\xi)_{nb} = (\not{D}\xi)_{ca}(\not{D}\xi)_{db}$$

we obtain

$$\sum_i |\not{D}_{R_i}\xi|^2 = r^2(\bar{g}^{-1})^{ab}(\delta_{cd} - y'^c y'^d)(\not{D}\xi)_{ca}(\not{D}\xi)_{db} \tag{12.76}$$

Consider the symmetric 2-covariant $S_{t,u}$ tensorfield $\vartheta$:

$$\vartheta_{cd} = (\bar{g}^{-1})^{ab}(\not{D}\xi)_{ca}(\not{D}\xi)_{db} \tag{12.77}$$

Let $\mu_1, \mu_2, \mu_3$ be the eigenvalues of $\vartheta$ relative to $\bar{g}$. Since $\vartheta$ is positive semi-definite, we have

$$\mu_i \geqslant 0, \quad i = 1, 2, 3 \tag{12.78}$$

It follows that

$$\max_i \mu_i \leqslant \sum_i \mu_i = \mathrm{tr}\vartheta \tag{12.79}$$

where

$$\mathrm{tr}\vartheta = (\bar{g}^{-1})^{cd}\vartheta_{cd} = \delta_{cd}\vartheta_{cd} \tag{12.80}$$

Then (12.79) implies

$$y'^c y'^d \vartheta_{cd} \leqslant |y'|^2 \mathrm{tr}\vartheta \tag{12.81}$$

We then have

$$(\bar{g}^{-1})^{ab}(\delta_{cd} - y'^c y'^d)(\not{D}\xi)_{ca}(\not{D}\xi)_{db} = (\delta_{cd} - y'^c y'^d)\vartheta_{cd} \geqslant (1 - |y'|^2)\mathrm{tr}\vartheta \tag{12.82}$$

Thus, since

$$\mathrm{tr}\vartheta = (\bar{g}^{-1})^{cd}(\bar{g}^{-1})^{ab}(\not{D}\xi)_{ca}(\not{D}\xi)_{db} = |\not{D}\xi|^2 \tag{12.83}$$

(12.76) implies

$$\sum_i |\not{D}_{R_i}\xi|^2 \geqslant r^2(1 - |y'|^2)|\not{D}\xi|^2 \tag{12.84}$$

Consider next the third term on the right of (12.74). We should first obtain a suitable expression for $\not{D}R_i$. Recall from (6.45) and (6.56), we have

$$g(\not{D}_{X_A}R_i, X_B) = X^l_A \epsilon_{ilm} X^m_B + \lambda_i(\eta^{-1}\chi_{AB} - \not{k}_{AB}) \tag{12.85}$$

Let us denote

$$(\nu_i)_{AB} = X^l_A \epsilon_{ilm} X^m_B \tag{12.86}$$

The rectangular components of $\nu_i$ are given by

$$(\nu_i)_{lm} = \Pi^{l'}_l \Pi^{m'}_m \epsilon_{il'm'} \tag{12.87}$$

Let us also denote

$$\tau_i = \lambda_i(\eta^{-1}\chi - \not{k}) \tag{12.88}$$

Then by (12.85) the rectangular components of $\not{D}R_i$ are given by

$$(\not{D}R_i)^k_l = \Pi^k_m \Pi^{l'}_l \epsilon_{il'm} + (\tau_i)^k_l \tag{12.89}$$

Using (12.89) we obtain, in view of (12.26),

$$\begin{aligned}\sum_i (\not{D}R_i)^m_a (\not{D}R_i)^n_b &= \Pi^m_{m'}\Pi^n_{n'}\Pi^{a'}_a \Pi^{b'}_b(\delta_{a'b'}\delta_{m'n'} - \delta_{a'n'}\delta_{b'm'}) \\ &\quad + \sum_i (\tau_i)^m_a \Pi^n_{n'}\Pi^{b'}_b \epsilon_{ib'n'} + \sum_i (\tau_i)^n_b \Pi^m_{m'}\Pi^{a'}_a \epsilon_{ia'm'}\end{aligned} \tag{12.90}$$

$$+\sum_i (\tau_i)_a^m (\tau_i)_b^n$$

Substituting this in the third term on the right in (12.74) yields

$$\sum_i (\bar{g}^{-1})^{ab}\xi_m\xi_n(\not{D}R_i)_a^m(\not{D}R_i)_b^n$$
$$= \delta_{a'b'}(\bar{g}^{-1})^{ab}\Pi_a^{a'}\Pi_b^{b'}|\xi|^2 - |\xi|^2 + 2\sum_i (\bar{g}^{-1})^{ab}(\tau_i\cdot\xi)_a(v_i)_b + \sum_i |\tau_i\cdot\xi|^2 \quad (12.91)$$

where $v_i$ is the $S_{t,u}$ 1-form:

$$(v_i)_b = \Pi_b^{b'}\epsilon_{ib'n'}\xi_{n'} \quad (12.92)$$

Using (12.26) and (12.92) we obtain

$$\begin{aligned}\sum_i |v_i|^2 &= \sum_i (\bar{g}^{-1})^{ab}\Pi_a^{a'}\Pi_b^{b'}\epsilon_{ia'm'}\epsilon_{ib'n'}\xi_{m'}\xi_{n'}\\ &= (\bar{g}^{-1})^{ab}\Pi_a^{a'}\Pi_b^{b'}(\delta_{a'b'}\delta_{m'n'} - \delta_{a'n'}\delta_{b'm'})\xi_{m'}\xi_{n'}\\ &= \delta_{a'b'}(\bar{g}^{-1})^{ab}\Pi_a^{a'}\Pi_b^{b'}|\xi|^2 - |\xi|^2\end{aligned} \quad (12.93)$$

Since

$$(\bar{g}^{-1})^{ab}\Pi_a^{a'}\Pi_b^{b'} = (\bar{g}^{-1})^{a'b'} - \hat{T}^{a'}\hat{T}^{b'} \quad (12.94)$$

we have

$$\delta_{a'b'}(\bar{g}^{-1})^{ab}\Pi_a^{a'}\Pi_b^{b'} = 2 \quad (12.95)$$

Hence (12.93) reduces to

$$\sum_i |v_i|^2 = |\xi|^2$$

while

$$2\sum_i (\bar{g}^{-1})^{ab}(\tau_i\cdot\xi)_a(v_i)_b \geqslant -2\sum_i |\tau_i\cdot\xi||v_i| \geqslant -2|\xi|\sqrt{\sum_i |\tau_i|^2}\sqrt{\sum_i |v_i|^2}$$

So from (12.91) we have

$$\begin{aligned}\sum_i (\bar{g}^{-1})^{ab}\xi_m\xi_n(\not{D}R_i)_a^m(\not{D}R_i)_b^n &= \sum_i |v_i|^2 + 2\sum_i (\tau_i\cdot\xi)\cdot v_i + \sum_i |\tau_i\cdot\xi|^2\\ &\geqslant \sum_i |v_i|^2 - 2\sum_i |\tau_i\cdot\xi||v_i|\\ &\geqslant \sum_i |v_i|^2 - 2\sqrt{\sum_i |\tau_i\cdot\xi|^2}\sqrt{\sum_i |v_i|^2}\end{aligned} \quad (12.96)$$

i.e.

$$\sum_i (\bar{g}^{-1})^{ab}\xi_m\xi_n(\not{D}R_i)^m_a(\not{D}R_i)^n_b \geqslant |\xi|^2(1-2\sqrt{\sum_i(\tau_i)^2}) \tag{12.97}$$

Finally, we consider the second term on the right of (12.74)

$$2\sum_i(\bar{g}^{-1})^{ab}(\not{D}_{R_i}\xi)_a\xi_n(\not{D}R_i)^n_b = 2\sum_i(\bar{g}^{-1})^{ab}(R_i)^m(\not{D}\xi)_{ma}\xi_n(\not{D}R_i)^n_b \tag{12.98}$$

From (12.24), (12.26) and (12.89) we have

$$\begin{aligned}
&\sum_i (R_i)^m(\not{D}R_i)^n_b \\
&=\sum_i \Pi^m_{m'}\epsilon_{ikm'}x^k(\Pi^n_{n'}\Pi^{b'}_b\epsilon_{ib'n'}+(\tau_i)^n_b) \\
&=(\delta_{kb'}\delta_{m'n'}-\delta_{kn'}\delta_{m'b'})\Pi^m_{m'}\Pi^n_{n'}\Pi^{b'}_b x^k + \Pi^m_{m'}x^k\sum_i\epsilon_{ikm'}(\tau_i)^n_b \\
&=\Pi^m_{n'}\Pi^n_{n'}(\Pi^{b'}_b x^{b'}) - \Pi^m_b(\Pi^n_{n'}x^{n'}) + \Pi^m_{m'}x^k\sum_i\epsilon_{ikm'}(\tau_i)^n_b
\end{aligned} \tag{12.99}$$

Consider the first term on the right of (12.99). We have

$$\Pi^{b'}_b x^{b'} = r\Pi^{b'}_b N^{b'} = r\bar{g}_{b'c}\Pi^{b'}_b N^c$$

By (12.29) and the fact that $\bar{g}_{b'c}\Pi^{b'}_b\hat{T}^c = 0$, we have

$$\bar{g}_{b'c}\Pi^{b'}_b N^c = \bar{g}_{b'c}\Pi^{b'}_b y'^c$$

Hence, we obtain

$$\Pi^{b'}_b x^{b'} = r\Pi^{b'}_b(\bar{g}_{b'c}y'^c) \tag{12.100}$$

Introducing the $S_{t,u}$-tangential vectorfield

$$\not{y}' = \Pi\cdot y' \tag{12.101}$$

then (12.100) takes the form

$$\Pi^{b'}_b x^{b'} = r\not{y}'^b \tag{12.102}$$

Similarly, we have

$$\Pi^n_{n'}x^{n'} = r\not{y}'^n \tag{12.103}$$

In the above, we have used the fact that $\bar{g}_{ab} = \delta_{ab}$.

In view of (12.102), (12.103), (12.99) reduces to

$$\sum_i(R_i)^m(\not{D}R_i)^n_b = r(\Pi^m_{n'}\Pi^n_{n'}\not{y}'^b - \Pi^m_b\not{y}'^n + \Pi^m_{m'}N^k\sum_i\epsilon_{ikm'}(\tau_i)^n_b) \tag{12.104}$$

Hence, from (12.98), the second term on the right of (12.74) is given by

$$2r(\zeta_{n'}\xi_{n'} - \mathrm{tr}(\not{D}\xi)(\xi\cdot\not{y}') + \sum_i\eta_i\cdot(\tau_i\cdot\xi)) \tag{12.105}$$

where

$$\zeta_{n'} = (\not{D}\xi)_{n'a}\not{y}'^a \tag{12.106}$$

and

$$(\eta_i)_a = N^k \epsilon_{ikm'}(\not{D}\xi)_{m'a} \tag{12.107}$$

Now we have

$$|\zeta_{n'}\xi_{n'}| \leqslant |\zeta||\xi| \leqslant |\not{D}\xi||\xi||\not{y}'| \tag{12.108}$$

Since for any 2-covariant $S_{t,u}$ tensorfield $M$ we have

$$\frac{1}{2}(\mathrm{tr}M)^2 \leqslant |M|^2$$

hence

$$|\mathrm{tr}(\not{D}\xi)(\xi \cdot \not{y}')| \leqslant \sqrt{2}|\not{D}\xi||\xi||\not{y}'| \tag{12.109}$$

Finally,

$$|\sum_i \eta_i \cdot (\tau_i \cdot \xi)| \leqslant \sum_i |\eta_i||\tau_i||\xi| \leqslant |\xi|\sqrt{\sum_i |\tau_i|^2}\sqrt{\sum_i |\eta_i|^2} \tag{12.110}$$

Using (12.26) we obtain

$$\begin{aligned}\sum_i |\eta_i|^2 &= \sum_i (\bar{g}^{-1})^{ab} N^k N^l \epsilon_{ikm'}\epsilon_{iln'}(\not{D}\xi)_{m'a}(\not{D}\xi)_{n'b} \\ &= (\bar{g}^{-1})^{ab} N^k N^l (\delta_{kl}\delta_{m'n'} - \delta_{kn'}\delta_{lm'})(\not{D}\xi)_{m'a}(\not{D}\xi)_{n'b} \\ &= (\bar{g}^{-1})^{ab}((\not{D}\xi)_{n'a}(\not{D}\xi)_{n'b} - N^{m'}N^{n'}(\not{D}\xi)_{m'a}(\not{D}\xi)_{n'b}) \\ &\leqslant (\bar{g}^{-1})^{ab}(\not{D}\xi)_{n'a}(\not{D}\xi)_{n'b} = |\not{D}\xi|^2\end{aligned} \tag{12.111}$$

Combining (12.108), (12.109) and (12.111) we conclude that the second term on the right of (12.74) is bounded by

$$2r((1+\sqrt{2})|\not{y}'| + \sqrt{\sum_i |\tau_i|^2})|\xi||\not{D}\xi| \tag{12.112}$$

Combining (12.84), (12.97) and (12.112) results in the following proposition:

**Proposition 12.5** Consider a surface $S_{t,u}$ for which the following hold:

$$\sup_{S_{t,u}} |y'| \leqslant 1$$

Then for every $S_{t,u}$ 1-form $\xi$ we have

$$\begin{aligned}\sum_i |\not{\mathcal{L}}_{R_i}\xi|^2 \geqslant\, & r^2(1-|y'|^2)|\not{D}\xi|^2 + (1 - 2\sqrt{\sum_i |\tau_i|^2})|\xi|^2 \\ & - 2r((1+\sqrt{2})|\not{y}'| + \sqrt{\sum_i |\tau_i|^2})|\xi||\not{D}\xi|\end{aligned} \tag{12.113}$$

pointwise on $S_{t,u}$.

**Corollary 12.5.a** Suppose that

$$\sup_{\Sigma_t^{\epsilon_0}} |y'| \leqslant C\delta_0, \quad \sup_{\Sigma_t^{\epsilon_0}} \sqrt{\sum_i |\tau_i|^2} \leqslant C\delta_0$$

for some fixed positive constant $C$. Then if $\delta_0$ is suitably small, for any $S_{t,u}$ 1-form $\xi$, differentiable on $S_{t,u}$, we have, pointwise on $\Sigma_t^{\epsilon_0}$

$$\sum_i |\not{\mathcal{L}}_{R_i}\xi|^2 \geqslant \frac{1}{2}(r^2|\not{D}\xi|^2 + |\xi|^2)$$

If also

$$\inf_{\Sigma_t^{\epsilon_0}} r \geqslant \epsilon_1(1+t)$$

then **H1** holds on $\Sigma_t^{\epsilon_0}$. In fact, we have

$$(1+t)^{-2}|\xi|^2 + |\not{D}\xi|^2 \leqslant C(1+t)^{-2}\sum_i |\not{\mathcal{L}}_{R_i}\xi|^2$$

where

$$C = \frac{2}{\epsilon_1^2}$$

*Proof*. From the assumptions, we can choose $\delta_0$ small enough such that

$$(1+\sqrt{2})|\not{y}'| + \sqrt{\sum_i |\tau_i|^2} \leqslant \frac{1}{4}$$

and

$$1 - |y'|^2 \geqslant \frac{3}{4}, \quad (1 - 2\sqrt{\sum_i |\tau_i|})^2 \geqslant \frac{3}{4}$$

then the corollary follows. □

Let us now return to Lemma 12.2 and let $\bar{s}_0$ be the maximal value of $s_0 \in [0, s]$ such that **H1** holds on $W_{\epsilon_0}^{s_0}$. We shall show that in fact $\bar{s}_0 = s$. Suppose that $\bar{s}_0 < s$. By Proposition 12.3, all the assumptions of Corollary 12.5.a except the one about $\tau_i$ hold on $W_{\epsilon_0}^{s}$. So if we can prove that the assumption about $\tau_i$ holds on $W_{\epsilon_0}^{s}$, then **H1** holds on $W_{\epsilon_0}^{s}$. To do this, we use a continuity argument. From (12.88), **H0**, $\mathbf{E}_{\{1\}}$ and Proposition 12.3, we have

$$\max_i \sup_{\Sigma_t^{\epsilon_0}} |\tau_i - \lambda_i \eta^{-1}\chi| \leqslant C\delta_0^2(1+t)^{-2}(1+\log(1+t)) \tag{12.114}$$

for all $t \in [0, s]$.

Thus, if for some $t \in [0, s]$, $\chi'$ satisfies on $\Sigma_t^{\epsilon_0}$ the estimate

$$\|\chi'\|_{L^\infty(\Sigma_t^{\epsilon_0})} \leqslant C_0(1+t)^{-1} \tag{12.115}$$

for some fixed positive constant $C_0$, then we have

$$\max_i \|\tau_i\|_{L^\infty(\Sigma_t^{\epsilon_0})} \leqslant C\delta_0(1+t)^{-1}(1+\log(1+t)) \tag{12.116}$$

so the assumptions about $\tau_i$ holds as well. On the other hand, taking $s_0 = \bar{s}_0$ in Lemma 12.2, yields the estimate

$$\|\chi'\|_{L^\infty(\Sigma_t^{\epsilon_0})} \leqslant C\delta_0(1+t)^{-2}(1+\log(1+t)) \tag{12.117}$$

for all $t \in [0, \bar{s}_0]$. If $\delta_0$ is suitably small, this implies that

$$\|\chi'\|_{L^\infty(\Sigma_t^{\epsilon_0})} \leqslant \frac{1}{2}C_0(1+t)^{-1}$$

at $t = \bar{s}_0$.

Therefore by continuity (12.115) holds in a suitably small interval $[\bar{s}_0, s_*] \subset [\bar{s}_0, s]$. So all the assumptions of Corollary 12.5.a hold for $t \in [0, s_*]$, hence **H1** holds on $W_{\epsilon_0}^{s_*}$, contradicting the maximality of $\bar{s}_0$. We conclude that **H1** holds on $W_{\epsilon_0}^s$. So from Lemma 12.2 with $s_0 = s$, we have

**Proposition 12.6** Let assumptions $\mathbf{E}_{\{2\}}$, $\mathbf{E}_{\{1\}}^Q$, $\mathbf{E}_{\{0\}}^{QQ}$ hold on $W_{\epsilon_0}^s$, and let the initial data satisfy

$$\|\chi'\|_{L^\infty(\Sigma_0^{\epsilon_0})} \leqslant C\delta_0$$

Then, if $\delta_0$ is suitably small, **H1** holds on $W_{\epsilon_0}^s$. Moreover, for all $t \in [0, s]$,

$$\|\chi'\|_{L^\infty(\Sigma_t^{\epsilon_0})} \leqslant C\delta_0(1+t)^{-2}(1+\log(1+t))$$

Next, we shall investigate **H2** and **H2**$'$. We consider $\not{\mathcal{L}}_{R_i}\vartheta$ for any symmetric 2-covariant $S_{t,u}$ tensorfield $\vartheta$. We have

$$\begin{aligned}(\not{\mathcal{L}}_{R_i}\vartheta)_{ab} &= (\not{D}_{R_i}\vartheta)_{ab} + \vartheta_{mb}(\not{D}R_i)_a^m + \vartheta_{am}(\not{D}R_i)_b^m \\ (\not{D}_{R_i}\vartheta)_{ab} &= (R_i)^m(\not{D}\vartheta)_{mab}\end{aligned} \tag{12.118}$$

Hence we have

$$\begin{aligned}&\sum_i |\not{\mathcal{L}}_{R_i}\vartheta|^2 \\ &= \sum_i |\not{D}_{R_i}\vartheta|^2 + 2\sum_i (\bar{g}^{-1})^{ac}(\bar{g}^{-1})^{bd}(\vartheta_{mb}(\not{D}R_i)_a^m + \vartheta_{am}(\not{D}R_i)_b^m)(\not{D}_{R_i}\vartheta)_{cd} \\ &\quad + \sum_i (\bar{g}^{-1})^{ac}(\bar{g}^{-1})^{bd}(\vartheta_{mb}(\not{D}R_i)_a^m + \vartheta_{am}(\not{D}R_i)_b^m)(\vartheta_{nd}(\not{D}R_i)_c^n + \vartheta_{cn}(\not{D}R_i)_d^n)\end{aligned} \tag{12.119}$$

Consider the first term on the right of (12.119). We have

$$\sum_i |\not{D}_{R_i}\vartheta|^2 = \sum_i (\bar{g}^{-1})^{ac}(\bar{g}^{-1})^{bd}(\not{D}_{R_i}\vartheta)_{ab}(\not{D}_{R_i}\vartheta)_{cd}$$
$$= (\bar{g}^{-1})^{ac}(\bar{g}^{-1})^{bd}(\sum_i (R_i)^m (R_i)^n)(\not{D}\vartheta)_{mab}(\not{D}\vartheta)_{ncd} \tag{12.120}$$

Substituting from (12.38) and noting that

$$\Pi_k^m \Pi_l^n (\not{D}\vartheta)_{mab}(\not{D}\vartheta)_{ncd} = (\not{D}\vartheta)_{kab}(\not{D}\vartheta)_{lcd}$$

we obtain

$$\sum_i |\not{D}_{R_i}\vartheta|^2 = r^2(\bar{g}^{-1})^{ac}(\bar{g}^{-1})^{bd}(\delta_{kl} - y'^k y'^l)(\not{D}\vartheta)_{kab}(\not{D}\vartheta)_{lcd} \tag{12.121}$$

Consider the symmetric 2-covariant $S_{t,u}$ tensorfield $\varphi$:

$$\varphi_{kl} = (\bar{g}^{-1})^{ac}(\bar{g}^{-1})^{bd}(\not{D}\vartheta)_{kab}(\not{D}\vartheta)_{lcd} \tag{12.122}$$

Since $\varphi$ is positive semi-definite, we have as in (12.81)

$$y'^k y'^l \varphi_{kl} \leqslant |y'|^2 \mathrm{tr}\varphi \tag{12.123}$$

Assuming that $|y'| \leqslant 1$ and noting that

$$\mathrm{tr}\varphi = (\bar{g}^{-1})^{kl}(\bar{g}^{-1})^{ac}(\bar{g}^{-1})^{bd}(\not{D}\vartheta)_{kab}(\not{D}\vartheta)_{lcd} = |\not{D}\vartheta|^2 \tag{12.124}$$

(12.121) implies

$$\sum_i |\not{D}_{R_i}\vartheta|^2 \geqslant r^2(1 - |y'|^2)|\not{D}\vartheta|^2 \tag{12.125}$$

Consider next the third term on the right of (12.119). This is

$$2A_1 + 2A_2$$

where

$$A_1 = (\bar{g}^{-1})^{ac}(\bar{g}^{-1})^{bd}\vartheta_{mb}\vartheta_{nd}(\sum_i (\not{D}R_i)_a^m (\not{D}R_i)_c^n) \tag{12.126}$$

$$A_2 = (\bar{g}^{-1})^{ac}(\bar{g}^{-1})^{bd}\vartheta_{am}\vartheta_{nd}(\sum_i (\not{D}R_i)_b^m (\not{D}R_i)_c^n) \tag{12.127}$$

Substituting (12.90) in (12.126) we find

$$A_1 = (\bar{g}^{-1})^{ac}(\bar{g}^{-1})^{bd}\vartheta_{m'b}\vartheta_{n'd}(\delta_{a'c'}\Pi_a^{a'}\Pi_c^{c'}\delta_{m'n'} - \Pi_a^{n'}\Pi_c^{m'})$$
$$+ 2\sum_i (\bar{g}^{-1})^{ac}(\bar{g}^{-1})^{bd}(w_i)_{ab}(\tau_i \cdot \vartheta)_{cd} + \sum_i |\tau_i \cdot \vartheta|^2 \tag{12.128}$$

where $w_i$ are the 2-covariant $S_{t,u}$ tensorfields:

$$(w_i)_{ab} = \Pi_a^{a'} \epsilon_{ia'm'} \vartheta_{m'b} \tag{12.129}$$

and

$$(\tau_i \cdot \vartheta)_{ab} = (\tau_i)_a^m \vartheta_{mb} \tag{12.130}$$

Substituting (12.96) in the first term on the right of (12.128) this term becomes

$$2\delta_{m'n'}(\bar{g}^{-1})^{bd}\vartheta_{m'b}\vartheta_{n'd} - |\vartheta|^2 = 2|\vartheta|^2 - |\vartheta|^2 = |\vartheta|^2 \tag{12.131}$$

The second term on the right of (12.128) is bounded in absolute value by

$$2\sum_i |w_i||\tau_i \cdot \vartheta| \leqslant 2|\vartheta|\sqrt{\sum_i |w_i|^2}\sqrt{\sum_i |\tau_i|^2} \tag{12.132}$$

Using (12.26) we obtain

$$\begin{aligned}\sum_i |w_i|^2 &= \sum_i (\bar{g}^{-1})^{ac}(\bar{g}^{-1})^{bd} \Pi_a^{a'} \Pi_c^{c'} \epsilon_{ia'm'}\epsilon_{ic'n'}\vartheta_{m'b}\vartheta_{n'd} \\ &= (\bar{g}^{-1})^{ac}(\bar{g}^{-1})^{bd} \Pi_a^{a'} \Pi_c^{c'} (\delta_{a'c'}\delta_{m'n'} - \delta_{a'n'}\delta_{c'm'})\vartheta_{m'b}\vartheta_{n'd} \\ &= (\bar{g}^{-1})^{bd}\delta_{a'c'}(\bar{g}^{-1})^{ac} \Pi_a^{a'} \Pi_c^{c'} \delta_{m'n'}\vartheta_{m'b}\vartheta_{n'd} - |\vartheta|^2 \end{aligned} \tag{12.133}$$

hence, by (12.96),

$$\sum_i |w_i|^2 = 2\delta_{m'n'}(\bar{g}^{-1})^{bd}\vartheta_{m'b}\vartheta_{n'd} - |\vartheta|^2 = |\vartheta|^2 \tag{12.134}$$

So we conclude that

$$A_1 \geqslant (1 - 2\sqrt{\sum_i |\tau_i|^2})|\vartheta|^2 + \sum_i |\tau_i \cdot \vartheta|^2 \tag{12.135}$$

Substituting (12.90) in (12.127) we find

$$\begin{aligned} A_2 = &(\bar{g}^{-1})^{ac}(\bar{g}^{-1})^{bd}\vartheta_{am'}\vartheta_{n'd}(\delta_{b'c'}\delta_{m'n'}\Pi_b^{b'}\Pi_c^{c'} - \Pi_b^{n'}\Pi_c^{m'}) \\ &+ 2\sum_i (\bar{g}^{-1})^{ac}(\bar{g}^{-1})^{bd}(\tau_i \tilde{\cdot} \vartheta)_{ab}(w_i)_{cd} \\ &+ \sum_i (\bar{g}^{-1})^{ac}(\bar{g}^{-1})^{bd}(\tau_i \tilde{\cdot} \vartheta)_{ab}(\tau_i \cdot \vartheta)_{cd} \end{aligned} \tag{12.136}$$

where

$$(\tau_i \tilde{\cdot} \vartheta)_{ab} = (\tau_i \cdot \vartheta)_{ba}$$

The first term on the right of (12.136) is given by

$$(\bar{g}^{-1})^{ac}(\bar{g}^{-1})^{bd}\vartheta_{am'}\vartheta_{n'd}(\Pi_c^b\delta_{m'n'} - \Pi_b^{n'}\Pi_c^{m'}) = |\vartheta|^2 - (\mathrm{tr}\vartheta)^2$$

Here, we have used the fact that

$$\delta_{b'c'}\Pi_b^{b'}\Pi_c^{c'} = \Pi_c^b$$

Since

$$|\tau_i \tilde{\cdot} \vartheta| = |\tau_i \cdot \vartheta|$$

the second term on the right of (12.136) is bounded in absolute value in the same way as (12.132). The third term is bounded in absolute value by

$$\sum_i |\tau_i \cdot \vartheta|^2 \tag{12.137}$$

We conclude that

$$A_2 \geqslant (1 - 2\sqrt{\sum_i |\tau_i|^2})|\vartheta|^2 - (\mathrm{tr}\vartheta)^2 - \sum_i |\tau_i \cdot \vartheta|^2 \tag{12.138}$$

Combining (12.135) and (12.138) we get

$$A_1 + A_2 \geqslant 2(1 - 2\sqrt{\sum_i |\tau_i|^2})|\vartheta|^2 - (\mathrm{tr}\vartheta)^2 \tag{12.139}$$

Consider finally the second term on the right of (12.119). This is

$$4B$$

where

$$B = (\bar{g}^{-1})^{ac}(\bar{g}^{-1})^{bd}\vartheta_{mb}(\sum_i (\not{D}R_i)_a^m (R_i)^k)(\not{D}\vartheta)_{kcd} \tag{12.140}$$

Substituting (12.104), by direct calculation we get

$$\begin{aligned} B = & r((\bar{g}^{-1})^{bd}\vartheta_{m'b}\iota_{m'd} - (\bar{g}^{-1})^{bd}(\vartheta \cdot \not{y}')_b(\mathrm{tr}\not{D}\vartheta)_d \\ & + \sum_i (\bar{g}^{-1})^{ac}(\bar{g}^{-1})^{bd}(\tau_i \cdot \vartheta)_{ab}(\varpi_i)_{cd}) \end{aligned} \tag{12.141}$$

where

$$\iota_{m'd} = (\not{D}\vartheta)_{m'cd}\not{y}'^c \tag{12.142}$$

$$(\varpi_i)_{cd} = N'^l \epsilon_{ilk'}(\not{D}\vartheta)_{k'cd} \tag{12.143}$$

Also

$$(\mathrm{tr}\not{D}\vartheta)_d = (\bar{g}^{-1})^{ac}(\not{D}\vartheta)_{acd}, \quad (\vartheta \cdot \not{y}')_b = \vartheta_{bm}\not{y}'^m$$

Since

$$(\bar{g}^{-1})^{m'n'} = \delta_{m'n'}$$

we have

$$|(\bar{g}^{-1})^{bd}\vartheta_{m'b}\iota_{m'd}| \leqslant |\vartheta||\iota| \tag{12.144}$$

Since

$$|\iota| \leqslant |\not{D}\vartheta||\not{y}'| \tag{12.145}$$

the first term in parenthesis in (12.141) is bounded in absolute value by

$$|\vartheta||\not{D}\vartheta||\not{y}'| \tag{12.146}$$

Since

$$\frac{1}{2}|\mathrm{tr}\not{D}\vartheta|^2 \leqslant |\not{D}\vartheta|^2$$

the second term in parenthesis in (12.141) is bounded in absolute value by

$$\sqrt{2}|\not{y}'||\vartheta||\not{D}\vartheta| \tag{12.147}$$

Finally the third term in parenthesis in (12.141) is bounded in absolute value by

$$\sum_i |\vartheta||\tau_i||\varpi_i| \leqslant |\vartheta|\sqrt{\sum_i |\tau_i|^2}\sqrt{\sum_i |\varpi_i|^2} \tag{12.148}$$

Moreover, using (12.26) we obtain

$$\begin{aligned}
\sum_i |\varpi_i|^2 &= \sum_i (\bar{g}^{-1})^{ac}(\bar{g}^{-1})^{bd}N^k N^l \epsilon_{ikm'}\epsilon_{iln'}(\not{D}\vartheta)_{m'ab}(\not{D}\vartheta)_{n'cd} \\
&= (\bar{g}^{-1})^{ac}(\bar{g}^{-1})^{bd}N^k N^l(\delta_{kl}\delta_{m'n'} - \delta_{kn'}\delta_{lm'})(\not{D}\vartheta)_{m'ab}(\not{D}\vartheta)_{n'cd} \\
&= (\bar{g}^{-1})^{ac}(\bar{g}^{-1})^{bd}((\not{D}\vartheta)_{m'ab}(\not{D}\vartheta)_{m'cd} - N^{m'}N^{n'}(\not{D}\vartheta)_{m'ab}(\not{D}\vartheta)_{n'cd}) \\
&\leqslant (\bar{g}^{-1})^{ac}(\bar{g}^{-1})^{bd}(\not{D}\vartheta)_{m'ab}(\not{D}\vartheta)_{m'cd} = |\not{D}\vartheta|^2
\end{aligned} \tag{12.149}$$

So we conclude that

$$|B| \leqslant r((1+\sqrt{2})|\not{y}'| + \sqrt{\sum_i |\tau_i|^2})|\not{D}\vartheta||\vartheta| \tag{12.150}$$

Combining (12.125), (12.139) and (12.150) we get the following proposition:

**Proposition 12.7** Consider a surface $S_{t,u}$ for which the following hold:

$$\sup_{S_{t,u}} |y'| \leqslant 1$$

Then for every symmetric 2-covariant $S_{t,u}$ tensorfield $\vartheta$ we have

$$\begin{aligned}
\sum_i |\not{\mathcal{L}}_{R_i}\vartheta|^2 \geqslant & \, r^2(1-|y'|^2)|\not{D}\vartheta|^2 \\
& + 4(1 - 2\sqrt{\sum_i |\tau_i|^2})|\vartheta|^2 - 2(\mathrm{tr}\vartheta)^2 \\
& - 4r((1+\sqrt{2})|\not{y}'| + \sqrt{\sum_i |\tau_i|^2})|\not{D}\vartheta||\vartheta|
\end{aligned}$$

pointwise on $S_{t,u}$.

**Corollary 12.7.a** Suppose that

$$\sup_{\Sigma_t^{\epsilon_0}} |y'| \leqslant C\delta_0 \quad \text{and} \quad \sup_{\Sigma_t^{\epsilon_0}} \sqrt{\sum_i |\tau_i|^2} \leqslant C\delta_0$$

for some fixed constant $C$. Then if $\delta_0$ is suitably small, for any symmetric 2-covariant $S_{t,u}$ tensorfield $\vartheta$, differentiable on $S_{t,u}$, we have

$$\sum_i |\not{\mathcal{L}}_{R_i}\vartheta|^2 \geqslant \frac{1}{2} r^2 |\not{D}\vartheta|^2 + 2|\vartheta|^2 - 2(\mathrm{tr}\vartheta)^2$$

If also

$$\inf_{\Sigma_t^{\epsilon_0}} r \geqslant \epsilon_1(1+t)$$

then **H2** and **H2$'$** hold on $\Sigma_t^{\epsilon_0}$. In fact, we have, pointwise on $\Sigma_t^{\epsilon_0}$,

$$|\not{D}\vartheta|^2 \leqslant C(1+t)^{-2}\Big(\sum_i |\not{\mathcal{L}}_{R_i}\vartheta|^2 + 2|\vartheta|^2\Big)$$

where

$$C = \frac{2}{\epsilon_1^2}$$

Moreover, if $\vartheta$ is trace-free, we have

$$|\not{D}\vartheta|^2 \leqslant C(1+t)^{-2} \sum_i |\not{\mathcal{L}}_{R_i}\vartheta|^2$$

*Proof.* Choosing $\delta_0$ sufficiently small, the second result is straightforward. Concerning the first result, we just use the fact that

$$(\mathrm{tr}\vartheta)^2 \leqslant 2|\vartheta|^2$$

□

We end this part with coercivity inequalities for Lie derivatives.

**Proposition 12.8** Let the assumptions of Proposition 12.6 hold and let $X$ be an arbitrary $S_{t,u}$ -tangential vectorfield defined on $W^s_{\epsilon_0}$ and differentiable on $S_{t,u}$. Then, for any 1-form $\xi$ we have, pointwise,

$$|\not{\mathcal{L}}_X\xi| \leqslant C(1+t)^{-1}\big(|X|(\max_i |\not{\mathcal{L}}_{R_i}\xi| + |\xi|) + |\xi| \max_i |\not{\mathcal{L}}_{R_i}X|\big)$$

Also, for any symmetric 2-covariant $S_{t,u}$ tensorfield $\vartheta$ we have, pointwise,

$$|\not{\mathcal{L}}_X\vartheta| \leqslant C(1+t)^{-1}\big(|X|(\max_i |\not{\mathcal{L}}_{R_i}\vartheta| + |\vartheta|) + |\vartheta| \max_i |\not{\mathcal{L}}_{R_i}X|\big)$$

*Proof.* Consider first the case of $\xi$. By Proposition 12.2, we have

$$|\not{\mathcal{L}}_X\xi| \leqslant C(1+t)^{-1}\sum_i |(\not{\mathcal{L}}_X\xi)(R_i)| \tag{12.151}$$

Since

$$(\not{\mathcal{L}}_X\xi)(R_i) = X(\xi(R_i)) - \xi([X,R_i]) = X\cdot \not{d}(\xi(R_i)) + \xi(\not{\mathcal{L}}_{R_i}X) \tag{12.152}$$

we have

$$|(\not{\mathcal{L}}_X\xi)(R_i)| \leqslant |X||\not{d}(\xi(R_i))| + |\xi||\not{\mathcal{L}}_{R_i}X| \tag{12.153}$$

By **H0**,

$$|\not{d}(\xi(R_i))| \leqslant C(1+t)^{-1}\sum_j |R_j(\xi(R_i))| \tag{12.154}$$

and we have

$$R_j(\xi(R_i)) = (\not{\mathcal{L}}_{R_j}\xi)(R_i) + \xi([R_i,R_j]) \tag{12.155}$$

hence

$$|R_j(\xi(R_i))| \leqslant |\not{\mathcal{L}}_{R_j}\xi||R_i| + |\xi||[R_i,R_j]| \tag{12.156}$$

By Corollary 10.1.e,

$$|R_i|, \quad |[R_i,R_j]| \leqslant C(1+t) \tag{12.157}$$

It follows that

$$|R_j(\xi(R_i))| \leqslant C(1+t)(|\not{\mathcal{L}}_{R_j}\xi| + |\xi|) \tag{12.158}$$

Substituting this in (12.154) we obtain

$$|\not{d}(\xi(R_i))| \leqslant C(\max_j |\not{\mathcal{L}}_{R_j}\xi| + |\xi|) \tag{12.159}$$

Substituting this in (12.153), the result for $\xi$ follows through (12.151).

Consider next the case of a symmetric 2-covariant $S_{t,u}$ tensorfield $\vartheta$. By Proposition 12.4,

$$|\not{\mathcal{L}}_X\vartheta| \leqslant C(1+t)^{-2}\sum_{i,j} |(\not{\mathcal{L}}_X\vartheta)(R_i,R_j)| \tag{12.160}$$

Now, we have

$$\begin{aligned}(\not{\mathcal{L}}_X\vartheta)(R_i,R_j) &= X(\vartheta(R_i,R_j)) - \vartheta([X,R_i],R_j) - \vartheta(R_i,[X,R_j]) \\ &= X\cdot \not{d}(\vartheta(R_i,R_j)) + \vartheta(\not{\mathcal{L}}_{R_i}X,R_j) + \vartheta(R_i,\not{\mathcal{L}}_{R_j}X)\end{aligned} \tag{12.161}$$

hence

$$\begin{aligned}|(\not{\mathcal{L}}_X\vartheta)(R_i,R_j)| &\leqslant |X||\not{d}(\vartheta(R_i,R_j))| + |\vartheta|(|R_i||\not{\mathcal{L}}_{R_j}X| + |R_j||\not{\mathcal{L}}_{R_i}X|) \\ &\leqslant |X||\not{d}(\vartheta(R_i,R_j))| + C(1+t)|\vartheta|\max_k |\not{\mathcal{L}}_{R_k}X|\end{aligned} \tag{12.162}$$

in view of the first of (12.157). By **H0**,

$$|\not{d}(\vartheta(R_i,R_j))| \leqslant C(1+t)^{-1}\sum_k |R_k(\vartheta(R_i,R_j))| \tag{12.163}$$

and from

$$R_k(\vartheta(R_i, R_j)) = (\not{\mathcal{L}}_{R_k}\vartheta)(R_i, R_j) + \vartheta([R_k, R_i], R_j) + \vartheta(R_i, [R_k, R_j]) \qquad (12.164)$$

as well as (12.157), we have

$$|R_k(\vartheta(R_i, R_j))| \leqslant C(1+t)^2(|\not{\mathcal{L}}_{R_k}\vartheta| + |\vartheta|) \qquad (12.165)$$

Substituting this in (12.163) we have

$$|\not{d}(\vartheta(R_i, R_j))| \leqslant C(1+t)(\max_k |\not{\mathcal{L}}_{R_k}\vartheta| + |\vartheta|) \qquad (12.166)$$

Substituting this in (12.162), the result for $\vartheta$ follows through (12.160). □

## 12.3 Estimates for Higher Order Derivatives of $\chi'$ and $\mu$

**Lemma 12.3** Let $Y$ be an arbitrary $S_{t,u}$-tangential vectorfield and $\vartheta$ an arbitrary 2-covariant $S_{t,u}$ tensorfield, defined on $W^s_{\epsilon_0}$. Then we have

$$\not{\mathcal{L}}_L \not{\mathcal{L}}_Y \vartheta - \not{\mathcal{L}}_Y \not{\mathcal{L}}_L \vartheta = \not{\mathcal{L}}_{^{(Y)}Z}\vartheta$$

*Proof*. We can restrict ourselves on $C_u$. We extend $\vartheta$ to $TC_u$ by the condition

$$\vartheta(L, W) = \vartheta(W, L) = 0$$

for any $W \in TC_u$. Thus, for arbitrary $W \in TC_u$ we have

$$(\mathcal{L}_L\vartheta)(L, W) = L(\vartheta(L, W)) - \vartheta([L, L], W) - \vartheta(L, [L, W]) = 0$$

(Note that $[L, W]$ is tangential to $C_u$). Similarly, $(\mathcal{L}_L\vartheta)(W, L) = 0$. It follows that

$$\not{\mathcal{L}}_L\vartheta = \mathcal{L}_L\vartheta \qquad (12.167)$$

On the other hand, for any $S_{t,u}$-tangential vectorfield $X$, by Lemma 8.2 we have

$$\begin{aligned}(\mathcal{L}_Y\vartheta)(L, X) &= Y(\vartheta(L, X)) - \vartheta([Y, L], X) - \vartheta(L, [Y, X]) \\ &= \vartheta([L, Y], X) = \vartheta(^{(Y)}Z, X)\end{aligned}$$

Similarly, $(\mathcal{L}_Y\vartheta)(X, L) = \vartheta(X, {}^{(Y)}Z)$. Also,

$$(\mathcal{L}_Y\vartheta)(L, L) = Y(\vartheta(L, L)) - \vartheta([Y, L], L) - \vartheta(L, [Y, L]) = 0$$

It follows that, on the manifold $C_u$,

$$\not{\mathcal{L}}_Y\vartheta = \mathcal{L}_Y\vartheta - (\vartheta \cdot {}^{(Y)}Z) \otimes dt - dt \otimes ({}^{(Y)}Z \cdot \vartheta) \qquad (12.168)$$

where, by definition,

$$\begin{aligned}(\vartheta \cdot {}^{(Y)}Z)(X) &= \vartheta(X, {}^{(Y)}Z) \\ ({}^{(Y)}Z \cdot \vartheta)(X) &= \vartheta({}^{(Y)}Z, X) \quad \text{for all} \quad X \in TS_{t,u}\end{aligned} \tag{12.169}$$

Consider now the evaluation of

$$\not{\mathcal{L}}_L \not{\mathcal{L}}_Y \vartheta - \not{\mathcal{L}}_Y \not{\mathcal{L}}_L \vartheta$$

on the $S_{t,u}$ frame vectorfields $X_A$. This evaluation is

$$(\mathcal{L}_L(\not{\mathcal{L}}_Y \vartheta) - \mathcal{L}_Y(\not{\mathcal{L}}_L \vartheta))(X_A, X_B) \tag{12.170}$$

Substituting (12.167) and (12.168) this becomes

$$\begin{aligned}&(\mathcal{L}_L\mathcal{L}_Y\vartheta - \mathcal{L}_L(\vartheta \cdot {}^{(Y)}Z) \otimes dt - dt \otimes \mathcal{L}_L({}^{(Y)}Z \cdot \vartheta) - \mathcal{L}_Y\mathcal{L}_L\vartheta)(X_A, X_B) \\ &= (\mathcal{L}_L\mathcal{L}_Y\vartheta - \mathcal{L}_Y\mathcal{L}_L\vartheta)(X_A, X_B)\end{aligned} \tag{12.171}$$

where we have used the facts that $\mathcal{L}_L dt = d(Lt) = 0$ and $dt(X_A) = dt(X_B) = 0$. Therefore (12.170) reduces to

$$(\mathcal{L}_L\mathcal{L}_Y\vartheta - \mathcal{L}_Y\mathcal{L}_L\vartheta)(X_A, X_B) = (\mathcal{L}_{[L,Y]}\vartheta)(X_A, X_B) = (\mathcal{L}_{{}^{(Y)}Z}\vartheta)(X_A, X_B) \tag{12.172}$$

Thus, the lemma follows. □

Given a positive integer $l$ and a multi-index $(i_1 \cdots i_l)$, we define the symmetric 2-covariant $S_{t,u}$ tensorfield:

$${}^{(i_1\cdots i_l)}\chi'_l = \not{\mathcal{L}}_{R_{i_l}} \cdots \not{\mathcal{L}}_{R_{i_1}} \chi' \tag{12.173}$$

We shall derive from (12.47), a propagation equation for ${}^{(i_1\cdots i_l)}\chi'$. Using Lemma 11.22, we obtain

$$[\not{\mathcal{L}}_{R_{i_l}} \cdots \not{\mathcal{L}}_{R_{i_1}}, \not{\mathcal{L}}_L]\chi' = \sum_{k=0}^{l-1} \not{\mathcal{L}}_{R_{i_l}} \cdots \not{\mathcal{L}}_{R_{i_{l-k+1}}} [\not{\mathcal{L}}_{R_{i_{l-k}}}, \not{\mathcal{L}}_L] \not{\mathcal{L}}_{R_{i_{l-k-1}}} \cdots \not{\mathcal{L}}_{R_{i_1}} \chi' \tag{12.174}$$

By Lemma 12.3 applied to the case $Y = R_{i_{l-k}}$, $\vartheta = \not{\mathcal{L}}_{R_{i_{l-k-1}}} \cdots \not{\mathcal{L}}_{R_{i_1}} \chi'$, we obtain

$$[\not{\mathcal{L}}_{R_{i_{l-k}}}, \not{\mathcal{L}}_L] \not{\mathcal{L}}_{R_{i_{l-k-1}}} \cdots \not{\mathcal{L}}_{R_{i_1}} \chi' = -\not{\mathcal{L}}_{{}^{(R_{i_{l-k}})}Z} \not{\mathcal{L}}_{R_{i_{l-k-1}}} \cdots \not{\mathcal{L}}_{R_{i_1}} \chi' \tag{12.175}$$

Substituting in (12.174) then yields

$$\begin{aligned}\not{\mathcal{L}}_L {}^{(i_1\cdots i_l)}\chi'_l = &\sum_{k=0}^{l-1} \not{\mathcal{L}}_{R_{i_l}} \cdots \not{\mathcal{L}}_{R_{i_{l-k+1}}} \not{\mathcal{L}}_{{}^{(R_{i_{l-k}})}Z} \not{\mathcal{L}}_{R_{i_{l-k-1}}} \cdots \not{\mathcal{L}}_{R_{i_1}} \chi' \\ &+ \not{\mathcal{L}}_{R_{i_l}} \cdots \not{\mathcal{L}}_{R_{i_1}} (\not{\mathcal{L}}_L \chi')\end{aligned} \tag{12.176}$$

We shall express the last term using (12.47). First, we express

$$\not{\mathcal{L}}_{R_{i_l}} \cdots \not{\mathcal{L}}_{R_{i_1}} (\chi' \cdot \chi') = {}^{(i_1\cdots i_l)}\chi'_l \cdot \chi' + \chi' \cdot {}^{(i_1\cdots i_l)}\chi'_l + {}^{(i_1\cdots i_l)}r_l \tag{12.177}$$

where

$$^{(i_1\cdots i_l)}r_l = \sum_{|s_1|+|s_2|+|s_3|=l;|s_1|,|s_3|<l} (\not{\mathcal{L}}_R)^{s_1}\chi' \cdot (\not{\mathcal{L}}_R)^{s_2}\not{g}^{-1} \cdot (\not{\mathcal{L}}_R)^{s_3}\chi' \tag{12.178}$$

Next, we have

$$\not{\mathcal{L}}_{R_{i_l}} \cdots \not{\mathcal{L}}_{R_{i_1}}(e\chi') = e^{(i_1\cdots i_l)}\chi'_l + {}^{(i_1\cdots i_l)}s_l \tag{12.179}$$

where

$$^{(i_1\cdots i_l)}s_l = \sum_{|s_1|+|s_2|=l,|s_2|<l} (\not{\mathcal{L}}_R)^{s_1} e (\not{\mathcal{L}}_R)^{s_2}\chi' \tag{12.180}$$

We then obtain from (12.47), we get

$$\not{\mathcal{L}}_{R_{i_l}} \cdots \not{\mathcal{L}}_{R_{i_1}}(\not{\mathcal{L}}_L\chi') = e^{(i_1\cdots i_l)}\chi'_l + {}^{(i_1\cdots i_l)}\chi'_l \cdot \chi' + \chi' \cdot {}^{(i_1\cdots i_l)}\chi'_l + {}^{(i_1\cdots i_l)}b_l \tag{12.181}$$

where

$$^{(i_1\cdots i_l)}b_l = {}^{(i_1\cdots i_l)}s_l + {}^{(i_1\cdots i_l)}r_l + \not{\mathcal{L}}_{R_{i_l}} \cdots \not{\mathcal{L}}_{R_{i_1}} b \tag{12.182}$$

with

$$b = \frac{e\not{\phi}}{1-u+t} - \alpha' \tag{12.183}$$

Given a non-negative integer $l$ let us denote by $\mathbf{X}'_l$ the statement that there is a constant $C$ independent of $s$ such that for all $t \in [0, s]$

$$\mathbf{X}'_l : \max_{i_1\cdots i_l} \|\not{\mathcal{L}}_{R_{i_l}} \cdots \not{\mathcal{L}}_{R_{i_1}}\chi'\|_{L^\infty(\Sigma_0^{\epsilon_0})} \leqslant C\delta_0(1+t)^{-2}(1+\log(1+t))$$

We then denote by $\mathbf{X}'_{[l]}$ the statement:

$$\mathbf{X}'_{[l]} : \mathbf{X}'_0 \quad \text{and} \cdots \text{and} \quad \mathbf{X}'_l$$

**Lemma 12.4** Let **H0** and (12.23) hold. Let also the bootstrap assumptions $\not{\mathbf{E}}_{[l]}, \not{\mathbf{E}}^Q_{[l-1]}$ and $\mathbf{X}'_{[l-1]}$ hold, for some positive integer $l$. Then the assumption $\not{\mathbf{X}}_{[l-1]}$ (assumption of Proposition 10.1) also holds.

*Proof.* The lemma is trivial for $l = 1$. Let the lemma hold with $l$ replaced by $l-1$, for some $l \geqslant 2$. That is, let $\not{\mathbf{X}}'_{[l-2]}$ together with $\not{\mathbf{E}}_{[l-1]}, \not{\mathbf{E}}^Q_{[l-2]}$, **H0** and (12.23), imply $\not{\mathbf{X}}_{[l-2]}$. Then by Corollary 10.1.d with $l$ replaced by $l-1$, we have

$$\max_i \|^{(R_i)}\not{\pi}\|_{\infty,[l-2],\Sigma_t^{\epsilon_0}} \leqslant C_l\delta_0(1+t)^{-1}(1+\log(1+t)) \quad \text{for all} \quad t \in [0, s] \tag{12.184}$$

Then for any positive integer $n$ we have

$$\not{\mathcal{L}}_{R_{i_n}} \cdots \not{\mathcal{L}}_{R_{i_1}}\chi = \not{\mathcal{L}}_{R_{i_n}} \cdots \not{\mathcal{L}}_{R_{i_1}}\chi' + \frac{1}{1-u+t}\not{\mathcal{L}}_{R_{i_n}} \cdots \not{\mathcal{L}}_{R_{i_2}}{}^{(R_{i_1})}\not{\pi} \tag{12.185}$$

hence

$$\max_{i_1\cdots i_n} \|\not{\mathcal{L}}_{R_{i_n}} \cdots \not{\mathcal{L}}_{R_{i_1}} \chi\|_{L^\infty(\Sigma_t^{\epsilon_0})}$$
$$\leqslant \|\not{\mathcal{L}}_{R_{i_n}} \cdots \not{\mathcal{L}}_{R_{i_1}} \chi'\|_{L^\infty(\Sigma_t^{\epsilon_0})} + C(1+t)^{-1} \max_j \|{}^{(R_j)}\not{\pi}\|_{\infty,[n-1],\Sigma_t^{\epsilon_0}} \tag{12.186}$$

Taking $n = l-1$, the lemma follows. □

**Proposition 12.9** Let assumptions of Proposition 12.6 hold. Let also $\mathbf{E}_{\{l+2\}}$, $\mathbf{E}^Q_{\{l+1\}}$, $\mathbf{E}^{QQ}_{\{l\}}$ hold in $W^s_{\epsilon_0}$ for some non-negative integer $l$, and let the initial data satisfy

$$\|\chi'\|_{\infty,[l],\Sigma_0^{\epsilon_0}} \leqslant C_l\delta_0$$

Then there is a constant $C_l$ independent of $s$ such that for all $t \in [0,s]$ we have

$$\|\chi'\|_{\infty,[l],\Sigma_t^{\epsilon_0}} \leqslant C_l\delta_0(1+t)^{-2}(1+\log(1+t))$$

that is, $\mathbf{X}'_{[l]}$ holds in $W^s_{\epsilon_0}$.

*Proof*. For $l = 0$ the proposition reduces to Proposition 12.6. We proceed by induction. Let the proposition hold with $l$ replaced by $l-1$, for some $l \geqslant 1$. Then $\mathbf{X}'_{[l-1]}$ holds on $W^s_{\epsilon_0}$, hence by Lemma 12.4, $\mathbf{X}_{[l-1]}$ holds on $W^s_{\epsilon_0}$ as well. Then the Proposition 10.1 and all its corollaries hold.

By Proposition 12.6 and the estimate (12.48) we have

$$\|e + 2\chi'\|_{L^\infty(\Sigma_t^{\epsilon_0})} \leqslant C\delta_0(1+t)^{-2}(1+\log(1+t)) \quad \text{for all} \quad t\in[0,s] \tag{12.187}$$

We shall estimate ${}^{(i_1\cdots i_l)}b_l$ using the corollaries of Proposition 10.1. From the expression of $e, \kappa^{-1}\zeta$ and the assumptions of the present proposition, we have

$$\|e\|_{\infty,[l],\Sigma_t^{\epsilon_0}}, \|\kappa^{-1}\zeta\|_{\infty,[l],\Sigma_t^{\epsilon_0}}, \|\not{k}\|_{\infty,[l],\Sigma_t^{\epsilon_0}} \leqslant C_l\delta_0(1+t)^{-2} \quad \text{for all} \quad t\in[0,s] \tag{12.188}$$

Here, we have used Corollary 10.1.a and Corollary 10.1.b.

Similarly, from the expression of $\alpha'$, we have

$$\|\alpha'\|_{\infty,[l],\Sigma_t^{\epsilon_0}} \leqslant C\delta_0(1+t)^{-3}(1+\log(1+t)) \max_j |\not{\mathcal{L}}_{R_j}\not{\mathcal{L}}_{R_{i_l}} \cdots \not{\mathcal{L}}_{R_{i_2}}\chi'| \tag{12.189}$$
$$+ C_l\delta_0(1+t)^{-3}$$

for all $t\in[0,s]$.

To see how the principal part in $\alpha'$, namely $\not{D}^2 h$, is estimated in the $\| \quad \|_{\infty,[l],\Sigma_t^{\epsilon_0}}$, we appeal to Lemma 10.9, **H0**, **H1**, **H2** and the fact that

$${}^{(R_i)}\pi_{AB} = -2\lambda_i(-\eta^{-1}\chi_{AB} + \not{k}_{AB})$$

(see (3.27) and (6.59)). So we get

$$|\not{\mathcal{L}}_{R_{i_l}} \cdots \not{\mathcal{L}}_{R_{i_1}} b|$$
$$\leqslant C\delta_0(1+t)^{-3}(1+\log(1+t)) \max_j |\not{\mathcal{L}}_{R_j}\not{\mathcal{L}}_{R_{i_l}} \cdots \not{\mathcal{L}}_{R_{i_2}}\chi'| C_l\delta_0(1+t)^{-3} \tag{12.190}$$

pointwise on $W^s_{\epsilon_0}$

From the inductive hypothesis and Corollary 10.1.d,

$$\max_{i;i_1\cdots i_k} \|\not{\mathcal{L}}_{R_{i_k}}\cdots\not{\mathcal{L}}_{R_{i_1}}{}^{(R_i)}\not{\pi}\|_{L^\infty(\Sigma_t^{\epsilon_0})} \leqslant C_l\delta_0(1+t)^{-1}(1+\log(1+t))$$
$$\text{for all}\quad k=0,1,\cdots,l-1$$

which is needed to estimate ${}^{(i_1\cdots i_l)}r_l$ given by (12.178). We then deduce

$$\|{}^{(i_1\cdots i_l)}s_l\|_{L^\infty(\Sigma_t^{\epsilon_0})} \leqslant C_l\delta_0^2(1+t)^{-4}(1+\log(1+t)) \quad \text{for all}\quad t\in[0,s] \tag{12.191}$$

and

$$\|{}^{(i_1\cdots i_l)}r_l\|_{L^\infty(\Sigma_t^{\epsilon_0})} \leqslant C_l\delta_0^2(1+t)^{-4}(1+\log(1+t))^2 \quad \text{for all}\quad t\in[0,s] \tag{12.192}$$

It follows that

$$\begin{aligned}|{}^{(i_1\cdots i_l)}b_l| \leqslant\, & C\delta_0(1+t)^{-3}(1+\log(1+t))\max_j|\not{\mathcal{L}}_{R_j}\not{\mathcal{L}}_{R_{i_l}}\cdots\not{\mathcal{L}}_{R_{i_2}}\chi'| \\ &+C_l\delta_0(1+t)^{-3}\quad\text{for all}\quad t\in[0,s]\end{aligned} \tag{12.193}$$

From (12.181), (12.187) and (12.193) we then conclude that

$$\begin{aligned}|\not{\mathcal{L}}_{R_{i_l}}\cdots\not{\mathcal{L}}_{R_{i_1}}(\not{\mathcal{L}}_L\chi')| \leqslant\, & C\delta_0(1+t)^{-2}(1+\log(1+t))\max_{j_1\cdots j_l}|{}^{(j_1\cdots j_l)}\chi_l'| \\ &+C_l\delta_0(1+t)^{-3}\quad\text{pointwise on}\quad W_{\epsilon_0}^s\end{aligned} \tag{12.194}$$

What remains to be estimated is the sum on the right of (12.176). Consider a given term in this sum, corresponding to $k\in\{0,\cdots,l-1\}$. We use Lemma 8.5, by taking $\not{\mathcal{L}}_{R_{i_{l-k-1}}}\cdots\not{\mathcal{L}}_{R_{i_1}}\chi'$ in the role of $\xi$, and the index set $\{l-k+1,\cdots,l\}$ in the role of the index set $\{1,\cdots,l\}$; to express the term in question in the form

$$\begin{aligned}&\not{\mathcal{L}}_{R_{i_l}}\cdots\not{\mathcal{L}}_{R_{i_{l-k+1}}}\not{\mathcal{L}}_{{}^{(R_{i_{l-k}})}Z}\not{\mathcal{L}}_{R_{i_{l-k-1}}}\cdots\not{\mathcal{L}}_{R_{i_1}}\chi' \\ &=\not{\mathcal{L}}_{{}^{(R_{i_{l-k}})}Z}{}^{(i_1\overset{>i_{l-k}<}{\cdots}i_l)}\chi_{l-1}' \\ &+\sum_{p=1}^{k}\sum_{m_1<\cdots<m_p=l-k+1}^{l}\not{\mathcal{L}}_{{}^{(i_{m_1}\cdots i_{m_p};i_{l-k})}Z}{}^{(i_1\overset{>i_{m_1}\cdots i_{m_p}i_{l-k}<}{\cdots}i_l)}\chi_{l-1-p}'\end{aligned} \tag{12.195}$$

where

$${}^{(i_{m_1}\cdots i_{m_p};i_{l-k})}Z=\not{\mathcal{L}}_{R_{i_{m_p}}}\cdots\not{\mathcal{L}}_{R_{i_{m_1}}}{}^{(R_{i_{l-k}})}Z \tag{12.196}$$

(noting that Lemma 8.5 holds for an arbitrary $S_{t,u}$ tensorfields $\xi$).

To estimate the terms on the right of (12.195), we apply the second statement of Proposition 12.8. We then obtain, for the first term,

$$\begin{aligned}&|\not{\mathcal{L}}_{{}^{(R_{i_{l-k}})}Z}{}^{(i_1\overset{>i_{l-k}<}{\cdots}i_l)}\chi_{l-1}'| \\ &\leqslant C(1+t)^{-1}(|{}^{(R_{i_{l-k}})}Z|(\max_j|{}^{(i_1\overset{>i_{l-k}<}{\cdots}i_lj)}\chi_l'|+|{}^{(i_1\overset{>i_{l-k}<}{\cdots}i_l)}\chi_{l-1}'|) \\ &+|{}^{(i_1\overset{>i_{l-k}<}{\cdots}i_l)}\chi_{l-1}'|\max_j|{}^{(j;i_{l-k})}Z|)\end{aligned} \tag{12.197}$$

In the case of $l \geqslant 2$, Corollary 10.1.i together with the inductive hypothesis $\mathbf{X}'_{[l-1]}$ implies that the right hand side of (12.197) is bounded pointwise by

$$C_l\delta_0(1+t)^{-2}(1+\log(1+t))\max_{j_1\cdots j_l}|^{(j_1\cdots j_l)}\chi'_l| + C_l\delta_0^2(1+t)^{-4}(1+\log(1+t))^2 \tag{12.198}$$

In the case of $l=1$, Corollary 10.1.i can only give the estimate for ${}^{(R_i)}Z$, so the term ${}^{(j;i_{l-k})}Z$ can not be estimated in this way. Recalling that ${}^{(R_i)}Z = {}^{(R_i)}\not{\pi}_L \cdot \not{g}^{-1}$, we use (6.70) to express

$${}^{(R_i)}\not{\pi}_L + \chi' \cdot R_i = \epsilon_{ijm} z^j \not{d}x^m + \lambda_i(\kappa^{-1}\zeta) \tag{12.199}$$

By Corollary 10.1.g, Corollary 10.1.a for $(R)^s x^j$ and $\lambda_i$, we have the following estimates:

$$\max_i \|\lambda_i\|_{\infty,[l],\Sigma_t^{\epsilon_0}} \leqslant C_l\delta_0(1+\log(1+t)), \quad \max_i \|\not{d}x^i\|_{\infty,[l],\Sigma_t^{\epsilon_0}} \leqslant C,$$
$$\max_j \|z^j\|_{\infty,[l],\Sigma_t^{\epsilon_0}} \leqslant C\delta_0(1+t)^{-1}(1+\log(1+t)) \quad \text{for all} \quad t\in[0,s]$$

In view also of the second of (12.188), we then obtain

$$\|{}^{(R_i)}\not{\pi}_L + \chi' \cdot R_i\|_{\infty,[1],\Sigma_t^{\epsilon_0}} \leqslant C\delta_0(1+t)^{-1}(1+\log(1+t)) \quad \text{for all} \quad t\in[0,s] \tag{12.200}$$

Moreover, we have, by Corollary 10.1.e with $l=1$ and $\mathbf{X}'_0$,

$$\begin{aligned}|\not{\mathcal{L}}_{R_j}(\chi' \cdot R_i)| &\leqslant |\not{\mathcal{L}}_{R_j}\chi'||R_i| + |\chi'||\not{\mathcal{L}}_{R_j}R_i| \\ &\leqslant C(1+t)|\not{\mathcal{L}}_{R_j}\chi'| + C\delta_0(1+t)^{-1}(1+\log(1+t))\end{aligned} \tag{12.201}$$

Hence

$$|\not{\mathcal{L}}_{R_j}{}^{(R_i)}\not{\pi}_L| \leqslant C(1+t)|\not{\mathcal{L}}_{R_j}\chi'| + C\delta_0(1+t)^{-1}(1+\log(1+t)) \tag{12.202}$$

therefore by Corollary 10.1.d, we have

$$|\not{\mathcal{L}}_{R_j}{}^{(R_i)}Z| \leqslant C(1+t)|\not{\mathcal{L}}_{R_j}\chi'| + C\delta_0(1+t)^{-1}(1+\log(1+t)) \tag{12.203}$$

pointwise on $W^s_{\epsilon_0}$. Using this, we can estimate the last term on the right of (12.197). Thus, also in the case $l=1$, an estimate of the form (12.198) holds.

Consider next a term in the double sum in (12.195), corresponding to an ordered subset $\{m_1,\cdots,m_p\}\subset\{l-k+1,\cdots,l\}, p\in\{1,\cdots,k\}$. This double sum is present only for $l\geqslant 2$. Applying the second statement of Proposition 12.8 we obtain

$$\begin{aligned}&|\not{\mathcal{L}}_{{}^{(i_{m_1}\cdots i_{m_p};i_{l-k})}Z}\,{}^{(i_1 \overset{>i_{m_1}\cdots i_{m_p} i_{l-k}<}{\cdots} i_l)}\chi'_{l-1-p}| \\ &\leqslant C(1+t)^{-1}\cdot(|{}^{(i_{m_1}\cdots i_{m_p};i_{l-k})}Z|(\max_j|{}^{(i_1 \overset{>i_{m_1}\cdots i_{m_p} i_{l-k}<}{\cdots} i_l j)}\chi'_{l-p}| \\ &\quad + |{}^{(i_1 \overset{>i_{m_1}\cdots i_{m_p} i_{l-k}<}{\cdots} i_l)}\chi'_{l-1-p}|) \\ &\quad + |{}^{(i_1 \overset{>i_{m_1}\cdots i_{m_p} i_{l-k}<}{\cdots} i_l)}\chi'_{l-1-p}|\max_j|{}^{(i_{m_1}\cdots i_{m_p} j;i_{l-k})}Z|)\end{aligned} \tag{12.204}$$

Since $p \geqslant 1$, by the inductive hypothesis $\mathbf{X}'_{[l-1]}$,

$$\|{}^{(i_1 \overset{>i_{m_1}\cdots i_{m_p} i_{l-k}<}{\cdots} i_l)}\chi'_{l-1-p}\|_{L^\infty(\Sigma_t^{\epsilon_0})} \leqslant C_l\delta_0(1+t)^{-2}(1+\log(1+t))$$

$$\max_j \|{}^{(i_1 \overset{>i_{m_1}\cdots i_{m_p} i_{l-k}<}{\cdots} i_l j)}\chi'_{l-p}\|_{L^\infty(\Sigma_t^{\epsilon_0})} \leqslant C_l\delta_0(1+t)^{-2}(1+\log(1+t)) \quad (12.205)$$
$$\text{for all} \quad t \in [0,s]$$

Also, since $p \leqslant k \leqslant l-1$, we have, by Corollary 10.1.i,

$$\|{}^{(i_{m_1}\cdots i_{m_p};i_{l-k})}Z\|_{L^\infty(\Sigma_t^{\epsilon_0})} \leqslant C_l\delta_0(1+t)^{-1}(1+\log(1+t)) \quad \text{for all} \quad t \in [0,s] \quad (12.206)$$

But we can not estimate ${}^{(i_{m_1}\cdots i_{m_p} j;i_{l-k})}Z$ when $p = k = l-1$ in this way for the same reason as above. However, using (12.199) and the estimates below (12.199), as well as the second of (12.188) and results in Chapter 10 we obtain

$$\|{}^{(R_i)}\not\pi_L + \chi' \cdot R_i\|_{\infty,[l],\Sigma_t^{\epsilon_0}} \leqslant C_l\delta_0(1+t)^{-1}(1+\log(1+t)) \quad \text{for all} \quad t \in [0,s] \quad (12.207)$$

Moreover, by Corollary 10.1.e and $\mathbf{X}'_{[l-1]}$,

$$\begin{aligned} &|\not{\mathcal{L}}_{R_{j_l}} \cdots \not{\mathcal{L}}_{R_{j_1}}(\chi' \cdot R_i)| \\ &\leqslant |\not{\mathcal{L}}_{R_{j_l}} \cdots \not{\mathcal{L}}_{R_{j_1}}\chi'||R_i| + \sum_{|s_1|+|s_2|=l, s_2 \geqslant 1} |(\not{\mathcal{L}}_R)^{s_1}\chi'||(\not{\mathcal{L}}_R)^{s_2}R_i| \\ &\leqslant C(1+t)|{}^{(j_1\cdots j_l)}\chi'_l| + C_l\delta_0(1+t)^{-1}(1+\log(1+t)) \end{aligned} \quad (12.208)$$

Combining (12.207) and (12.208) yields

$$|\not{\mathcal{L}}_{R_{j_1}} \cdots \not{\mathcal{L}}_{R_{j_1}} {}^{(R_i)}\not\pi_L| \leqslant C(1+t)|{}^{(j_1\cdots j_l)}\chi'_l| + C_l\delta_0(1+t)^{-1}(1+\log(1+t)) \quad (12.209)$$

hence, by Corollary 10.1.d, also

$$|{}^{(j_1\cdots j_l;i)}Z| \leqslant C(1+t)|{}^{(j_1\cdots j_l)}\chi'_l| + C_l\delta_0(1+t)^{-1}(1+\log(1+t)) \quad (12.210)$$

Combining the above results, we obtain:
For $p = k = l-1$,

$$\begin{aligned} &|\not{\mathcal{L}}_{{}^{(i_{m_1}\cdots i_{m_p};i_{l-k})}Z} {}^{(i_1 \overset{>i_{m_1}\cdots i_{m_p} i_{l-k}<}{\cdots} i_l)}\chi'_{l-1-p}| \\ &\leqslant C_l\delta_0(1+t)^{-2}(1+\log(1+t)) \max_{j_1\cdots j_l} |{}^{(j_1\cdots j_l)}\chi'_l| \\ &\quad + C_l\delta_0^2(1+t)^{-4}(1+\log(1+t))^2 \end{aligned} \quad (12.211)$$

Otherwise,

$$|\not{\mathcal{L}}_{{}^{(i_{m_1}\cdots i_{m_p};i_{l-k})}Z} {}^{(i_1 \overset{>i_{m_1}\cdots i_{m_p} i_{l-k}<}{\cdots} i_l)}\chi'_{l-k-p}| \leqslant C_l\delta_0^2(1+t)^{-4}(1+\log(1+t))^2 \quad (12.212)$$

Combining these with the estimate (12.198) for the first term on the right in (12.195) we conclude that

$$\begin{aligned}&|\sum_{k=0}^{l-1}\not{\mathcal{L}}_{R_{i_l}}\cdots\not{\mathcal{L}}_{R_{i_{l-k+1}}}\not{\mathcal{L}}_{({R_{i_{l-k}}})_Z}\not{\mathcal{L}}_{R_{i_{l-k-1}}}\cdots\not{\mathcal{L}}_{R_{i_1}}\chi'|\\&\leqslant C_l\delta_0(1+t)^{-2}(1+\log(1+t))\max_{j_1\cdots j_l}|^{(j_1\cdots j_l)}\chi_l'|\\&+C_l\delta_0^2(1+t)^{-4}(1+\log(1+t))^2\end{aligned}\tag{12.213}$$

This together with (12.194) yields

$$\begin{aligned}|\not{\mathcal{L}}_L{}^{(i_1\cdots i_l)}\chi_l'|\leqslant& C_l\delta_0(1+t)^{-2}(1+\log(1+t))\max_{j_1\cdots j_l}|^{(j_1\cdots j_l)}\chi_l'|\\&+C_l\delta_0(1+t)^{-3}\end{aligned}\tag{12.214}$$

Applying then (12.56), taking $\vartheta={}^{(i_1\cdots i_l)}\chi_l'$, and using $\mathbf{X}_0'$ we deduce

$$\begin{aligned}&L((1-u+t)^2|^{(i_1\cdots i_l)}\chi_l'|)\\&\leqslant C_l\delta_0(1+t)^{-2}(1+\log(1+t))((1-u+t)^2\max_{j_1\cdots j_l}|^{(j_1\cdots j_l)}\chi_l'|)\\&+C_l\delta_0(1+t)^{-1}\end{aligned}\tag{12.215}$$

Setting along a given integral curve of $L$,

$${}^{(i_1\cdots i_l)}x_l(t)=(1-u+t)^2|^{(i_1\cdots i_l)}\chi_l'|\tag{12.216}$$

(12.215) becomes

$$\frac{d^{(i_1\cdots i_l)}x_l}{dt}\leqslant f_l\max_{j_1\cdots j_l}{}^{(j_1\cdots j_l)}x_l+g_l\tag{12.217}$$

where

$$f_l(t)=C_l\delta_0(1+t)^{-2}(1+\log(1+t)),\quad g_l(t)=C_l\delta_0(1+t)^{-1}\tag{12.218}$$

Integrating from $t=0$ yields

$${}^{(i_1\cdots i_l)}x_l(t)\leqslant{}^{(i_1\cdots i_l)}x_l(0)+\int_0^t(f_l(t')\max_{j_1\cdots j_l}{}^{(j_1\cdots j_l)}x_l(t')+g_l(t'))dt'\tag{12.219}$$

Taking the maximum over $i_1\cdots i_l$, on both sides, and setting

$$\bar{x}_l(t)=\max_{i_1\cdots i_l}{}^{(i_1\cdots i_l)}x_l(t)\tag{12.220}$$

then we obtain

$$\bar{x}_l(t)\leqslant\bar{x}_l(0)+\int_0^t(f_l(t')\bar{x}_l(t')+g_l(t'))dt'\tag{12.221}$$

which implies

$$\bar{x}_l(t)\leqslant e^{\int_0^t f_l(t')dt'}(\bar{x}_l(0)+\int_0^t g_l(t')dt')\tag{12.222}$$

Since

$$\bar{x}_l(0) \leqslant \max_{i_1\cdots i_l} \|^{(i_1\cdots i_l)}\chi_l'\|_{L^\infty(\Sigma_0^{\epsilon_0})} \leqslant C_l\delta_0 \tag{12.223}$$

while

$$\bar{x}_l(t) \geqslant \frac{1}{4}(1+t)^2 \max_{i_1\cdots i_l} |^{(i_1\cdots i_l)}\chi_l'| \quad (u \leqslant \epsilon_0 \leqslant \frac{1}{2}) \tag{12.224}$$

taking into account the facts that

$$\int_0^t f_l(t')dt' \leqslant \int_0^\infty f_l(t')dt' = 2C_l, \qquad \int_0^t g_l(t')dt' = C_l\delta_0 \log(1+t)$$

then taking the supremum on $\Sigma_t^{\epsilon_0}$, we conclude that

$$\max_{i_1\cdots i_l} \|^{(i_1\cdots i_l)}\chi_l'\|_{L^\infty(\Sigma_t^{\epsilon_0})} \leqslant C_l\delta_0(1+t)^{-2}(1+\log(1+t)) \quad \text{for all} \quad t \in [0,s] \tag{12.225}$$

This, together with the inductive hypothesis $\mathbf{X}'_{[l-1]}$ implies $\mathbf{X}'_{[l]}$, completing the inductive step and therefore the proof of the proposition. □

From Lemma 12.4 with $l+1$ in the role of $l$ we conclude that under the assumptions of Proposition 12.9, $\mathbf{X}_{[l]}$ holds on $W^s_{\epsilon_0}$. So Proposition 10.1 and all its corollaries hold with $l+1$ in the role of $l$.

**Proposition 12.10** Let the assumptions of Proposition 12.9 hold for some non-negative integer $l$. Let, in addition the initial data satisfy

$$\|\mu - 1\|_{\infty,[m,l+1],\Sigma_0^{\epsilon_0}} \leqslant C_l\delta_0$$

for some $m \in \{0,\cdots,l+1\}$. Then there is a constant $C_l$ independent of $s$ such that for all $t \in [0,s]$ we have

$$\|\mu - 1\|_{\infty,[m,l+1],\Sigma_t^{\epsilon_0}} \leqslant C_l\delta_0(1+\log(1+t))$$

that is, $\mathbf{M}_{[m,l+1]}$ holds on $W^s_{\epsilon_0}$.

*Proof.* We consider first the case $m=0$. Since $M_{[0,0]}$ follows directly from Proposition 12.1, we apply induction on $l$. Assuming then $\mathbf{M}_{[0,l]}$ we are to establish $\mathbf{M}_{[0,l+1]}$ under the assumptions of the proposition for $m=0$.

We consider the equation

$$L\mu = m + \mu e \tag{12.226}$$

We apply $R_{i_{l+1}}\cdots R_{i_1}$ to this equation. By Lemma 11.22, we obtain

$$[R_{i_{l+1}}\cdots R_{i_1}, L] = -\sum_{k=0}^{l} R_{i_{l+1}}\cdots R_{i_{l+2-k}}{}^{(R_{i_{l+1-k}})}Z R_{i_{l-k}}\cdots R_{i_1} \tag{12.227}$$

Defining

$$^{(i_1\cdots i_{l+1})}\mu_{0,l+1} = R_{i_{l+1}}\cdots R_{i_1}\mu \tag{12.228}$$

We get

$$
\begin{aligned}
L^{(i_1\cdots i_{l+1})}\mu_{0,l+1} = & \sum_{k=0}^{l} R_{i_{l+1}}\cdots R_{i_{l+2-k}}({}^{(R_{i_{l+1-k}})}Z\cdot \not{d}^{(i_1\cdots i_{l-k})}\mu_{0,l-k}) \\
& + e^{(i_1\cdots i_{l+1})}\mu_{0,l+1} + R_{i_{l+1}}\cdots R_{i_1}m + {}^{(i_1\cdots i_{l+1})}r'_{0,l+1}
\end{aligned}
\tag{12.229}
$$

where

$$
{}^{(i_1\cdots i_{l+1})}r'_{0,l+1} = \sum_{|s_1|+|s_2|=l+1, s_1\geqslant 1} ((R)^{s_1}e)((R)^{s_2}\mu) \tag{12.230}
$$

Now by $\mathbf{E}_{\{l+2\}}, \mathbf{E}^Q_{\{l+1\}}$ and Corollary 10.1.a (remember now that Proposition 10.1 is valid for $l+1$),

$$
\begin{aligned}
\|Th\|_{\infty,[l+1],\Sigma_t^{\epsilon_0}} &\leqslant C_l\delta_0(1+t)^{-1} \\
\|Lh\|_{\infty,[l+1],\Sigma_t^{\epsilon_0}} &\leqslant C_l\delta_0(1+t)^{-2} \\
\|\omega_{L\hat{T}}\|_{\infty,[l+1],\Sigma_t^{\epsilon_0}} &\leqslant C_l\delta_0(1+t)^{-2} \quad \text{for all} \quad t\in[0,s]
\end{aligned}
\tag{12.231}
$$

Thus using the expression for $e$ we obtain

$$
\|e\|_{\infty,[l+1],\Sigma_t^{\epsilon_0}} \leqslant C_l\delta_0(1+t)^{-2} \quad \text{for all} \quad t\in[0,s] \tag{12.232}
$$

Using the first of (12.231) we obtain

$$
\|m\|_{\infty,[l+1],\Sigma_t^{\epsilon_0}} \leqslant C_l\delta_0(1+t)^{-1} \quad \text{for all} \quad t\in[0,s] \tag{12.233}
$$

The estimate (12.232) together with the induction hypothesis $\mathbf{M}_{[0,l]}$ imply

$$
\|{}^{(i_1\cdots i_{l+1})}r'_{0,l+1}\|_{L^\infty(\Sigma_t^{\epsilon_0})} \leqslant C_l\delta_0(1+t)^{-2}(1+\log(1+t)) \quad \text{for all} \quad t\in[0,s] \tag{12.234}
$$

Consider finally a term in the sum on the right in (12.229), corresponding to $k\in\{0,\cdots,l\}$. We express the term as

$$
\begin{aligned}
& R_{i_{l+1}}\cdots R_{i_{l+2-k}}({}^{(R_{i_{l+1-k}})}Z\cdot \not{d}^{(i_1\cdots i_{l-k})}\mu_{0,l-k}) \\
&= {}^{(R_{i_{l+1-k}})}Z\cdot \not{d}^{(i_1 \overset{>i_{l+1-k}<}{\cdots} i_{l+1})}\mu_{0,l} \\
&+ \sum_{|s_1|+|s_2|=k, |s_1|\geqslant 1} ((\not{\mathcal{L}}_R)^{s_1}\,{}^{(R_{i_{l+1-k}})}Z)\cdot \not{d}(R)^{s_2}\,{}^{(i_1\cdots i_{l-k})}\mu_{0,l-k}
\end{aligned}
\tag{12.235}
$$

Consider first the sum on the right of (12.235). By Corollary 10.1.i with $l+1$ in the role of $l$,

$$
\max_i \|{}^{(R_i)}Z\|_{\infty,[l],\Sigma_t^{\epsilon_0}} \leqslant C_l\delta_0(1+t)^{-1}(1+\log(1+t)) \quad \text{for all} \quad t\in[0,s] \tag{12.236}
$$

Since $|s_1|\leqslant k\leqslant l$, this allows us to bound the first factor in the sum. Moreover, by **H0** and the induction hypothesis $\mathbf{M}_{[0,l]}$, the second factor in the sum can be bounded in $L^\infty(\Sigma_t^{\epsilon_0})$ by

$$
C_l\delta_0(1+t)^{-1}(1+\log(1+t))
$$

since $1 + |s_2| + l - k \leqslant l$. So the sum in (12.235) is bounded in $L^\infty(\Sigma_t^{\epsilon_0})$ by

$$C_l \delta_0 (1+t)^{-2}(1+\log(1+t))^2 \tag{12.237}$$

On the other hand, by **H0** and (12.236), the first term on the right of (12.235) is bounded pointwise by

$$C\delta_0(1+t)^{-2}(1+\log(1+t)) \max_j |{}^{(i_1 \overset{>i_{l+1-k}<}{\cdots} i_{l+1} j)}\mu_{0,l+1}| \tag{12.238}$$

Combining (12.237) and (12.238), the sum on the right of (12.229) is bounded by

$$C_l\delta_0(1+t)^{-2}(1+\log(1+t)) \max_{j_1 \cdots j_l} |{}^{(j_1\cdots j_{l+1})}\mu_{0,l+1}| + C_l\delta_0(1+t)^{-2}(1+\log(1+t))^2 \tag{12.239}$$

The estimates (12.232), (12.233), (12.234) and (12.239) imply through (12.229)

$$\begin{aligned} &L(|{}^{(i_1\cdots i_{l+1})}\mu_{0,l+1}|) \\ &\leqslant C_l\delta_0(1+t)^{-2}(1+\log(1+t)) \max_{j_1\cdots j_{l+1}} |{}^{(j_1\cdots j_{l+1})}\mu_{0,l+1}| + C_l\delta_0(1+t)^{-1} \end{aligned} \tag{12.240}$$

Defining, along a given integral curve of $L$,

$${}^{(i_1\cdots i_{l+1})}x'_{0,l+1}(t) = |{}^{(i_1\cdots i_{l+1})}\mu_{0,l+1}| \tag{12.241}$$

(12.240) takes the form

$$\frac{d\,{}^{(i_1\cdots i_{l+1})}x'_{0,l+1}}{dt} \leqslant f_l \max_{j_1\cdots j_{l+1}} {}^{(j_1\cdots j_{l+1})}x'_{0,l+1} + g_l \tag{12.242}$$

where

$$f_l(t) = C_l\delta_0(1+t)^{-2}(1+\log(1+t)), \quad g_l(t) = C_l\delta_0(1+t)^{-1} \tag{12.243}$$

This is similar to (12.217)—(12.218). Setting then

$$\bar{x}'_{0,l+1}(t) = \max_{i_1\cdots i_{l+1}} {}^{(i_1\cdots i_{l+1})}x'_{0,l+1}(t) \tag{12.244}$$

and using the initial condition

$$\bar{x}'_{0,l+1}(0) \leqslant C_l\delta_0$$

we deduce

$$\max_{i_1\cdots i_{l+1}} \|{}^{(i_1\cdots i_{l+1})}\mu_{0,l+1}\|_{L^\infty(\Sigma_t^{\epsilon_0})} \leqslant C_l\delta_0(1+\log(1+t)) \quad \text{for all} \quad t\in[0,s] \tag{12.245}$$

This proves the proposition in the case of $m = 0$.

To prove the proposition for $m \in \{1, \cdots, l+1\}$ we apply induction on $m$. We assume $\mathbf{M}_{[m,l+1]}$ for some $m \in \{0, \cdots, l\}$, and we will establish $\mathbf{M}_{[m+1,l+1]}$ under the assumption

$$\|\mu - 1\|_{\infty,[m+1,l+1],\Sigma_0^{\epsilon_0}} \leqslant C_l\delta_0 \tag{12.246}$$

We first apply $T^{m+1}$ to $L\mu$ to obtain, using Lemma 11.22,

$$L(T)^{m+1}\mu = T^{m+1}L\mu + \sum_{k=0}^{m}(T)^k \Lambda (T)^{m-k}\mu \tag{12.247}$$

We then apply $R_{i_n}\cdots R_{i_1}$ to this equation, to obtain, using again Lemma 11.22,

$$\begin{aligned} &LR_{i_n}\cdots R_{i_1}(T)^{m+1}\mu \\ &= R_{i_n}\cdots R_{i_1}(T)^{m+1}L\mu + \sum_{k=0}^{m} R_{i_n}\cdots R_{i_1}(T)^k\Lambda(T)^{m-k}\mu \\ &+ \sum_{k=0}^{n-1} R_{i_n}\cdots R_{i_{n-k+1}}{}^{(R_{i_{n-k}})}Z R_{i_{n-k-1}}\cdots R_{i_1}(T)^{m+1}\mu \end{aligned} \tag{12.248}$$

Let us define as in Chapter 9

$$^{(i_1\cdots i_n)}\mu_{m+1,n} = R_{i_n}\cdots R_{i_1}(T)^{m+1}\mu \tag{12.249}$$

Applying $R_{i_n}\cdots R_{i_1}(T)^{m+1}$ to Equation (12.226), we obtain

$$\begin{aligned} R_{i_n}\cdots R_{i_1}(T)^{m+1}L\mu = &e^{(i_1\cdots i_n)}\mu_{m+1,n} + R_{i_n}\cdots R_{i_1}(T)^{m+1}m \\ &+ {}^{(i_1\cdots i_n)}r'_{m+1,n} \end{aligned} \tag{12.250}$$

where

$$\begin{aligned} {}^{(i_1\cdots i_n)}r'_{m+1,n} = &\sum_{|s_1|+|s_2|=n, s_1\geqslant 1} ((R)^{s_1}e)((R)^{s_2}(T)^{m+1}\mu) \\ &+ R_{i_n}\cdots R_{i_1}\Big(\sum_{k=1}^{m+1}\frac{(m+1)!}{k!(m+1-k)!}((T)^k e)((T)^{m+1-k}\mu)\Big) \end{aligned} \tag{12.251}$$

Substituting (12.250) in (12.248) we get

$$\begin{aligned} &L^{(i_1\cdots i_n)}\mu_{m+1,n} \\ &= e^{(i_1\cdots i_n)}\mu_{m+1,n} + R_{i_n}\cdots R_{i_1}(T)^{m+1}m + {}^{(i_1\cdots i_n)}r'_{m+1,n} \\ &+ \sum_{k=0}^{m} R_{i_n}\cdots R_{i_1}(T)^k\Lambda(T)^{m-k}\mu \\ &+ \sum_{k=0}^{n-1} R_{i_n}\cdots R_{i_{n-k+1}}{}^{(R_{i_{n-k}})}Z R_{i_{n-k-1}}\cdots R_{i_1}(T)^{m+1}\mu \end{aligned} \tag{12.252}$$

By assumptions $\mathbf{E}_{\{l+2\}}, \mathbf{E}^Q_{\{l+1\}}$

$$\begin{aligned} &\|Th\|_{\infty,[m+1,l+1],\Sigma_t^{\epsilon_0}} \leqslant C_l\delta_0(1+t)^{-1} \\ &\|Lh\|_{\infty,[m+1,l+1],\Sigma_t^{\epsilon_0}} \leqslant C_l\delta_0(1+t)^{-2} \quad \text{for all} \quad t\in[0,s] \end{aligned} \tag{12.253}$$

By Corollary 11.1.b, we have

$$\|\omega_{L\hat{T}}\|_{\infty,[m+1,l+1],\Sigma_t^{\epsilon_0}} \leqslant C_l\delta_0(1+t)^{-2} \quad \text{for all} \quad t \in [0,s] \tag{12.254}$$

The bounds (12.253) and (12.254) imply

$$\|e\|_{\infty,[m+1,l+1],\Sigma_t^{\epsilon_0}} \leqslant C_l\delta_0(1+t)^{-2} \quad \text{for all} \quad t \in [0,s] \tag{12.255}$$

and (12.253) implies

$$\|m\|_{\infty,[m+1,l+1],\Sigma_t^{\epsilon_0}} \leqslant C_l\delta_0(1+t)^{-1} \quad \text{for all} \quad t \in [0,s] \tag{12.256}$$

Also, (12.255) together with $\mathbf{M}_{[m,l+1]}$ imply

$$\|^{(i_1\cdots i_n)}r'_{m+1,n}\|_{L^\infty(\Sigma_t^{\epsilon_0})} \leqslant C_l\delta_0(1+t)^{-2}(1+\log(1+t)) \quad \text{for} \quad n \leqslant l-m \tag{12.257}$$

provided that either $n=0$, so that the first sum on the right in (12.251) vanishes, or $\mathbf{M}_{m+1,m+n}$ holds. Moreover, by Corollary 11.1.c,

$$\|\Lambda\|_{\infty,[m,l],\Sigma_t^{\epsilon_0}} \leqslant C_l\delta_0(1+t)^{-1}(1+\log(1+t)) \quad \text{for all} \quad t \in [0,s] \tag{12.258}$$

To establish $\mathbf{M}_{[m+1,l+1]}$, we establish $\mathbf{M}_{[m+1,m+1+n]}$ for $n=0,\cdots,l-m$ by induction on $n$. We begin with the case $n=0$. In this case, Equation (12.252) reduces to

$$L\mu_{m+1,0} = e\mu_{m+1,0} + (T)^{m+1}m + r'_{m+1,0} + \sum_{k=0}^{m}(T)^k\Lambda(T)^{m-k}\mu \tag{12.259}$$

Using (12.258) with $l=m$ and $\mathbf{M}_{[m,m+1]}$, the sum in (12.259) is bounded in $L^\infty(\Sigma_t^{\epsilon_0})$ by

$$\begin{aligned}\sum_{k=0}^{m}\|\Lambda\cdot \not{d}(T)^{m-k}\mu\|_{\infty,[k,k],\Sigma_t^{\epsilon_0}} &\leqslant C_m(1+t)^{-1}\|\Lambda\|_{\infty,[m,m],\Sigma_t^{\epsilon_0}}\|\mu-1\|_{\infty,[m,m+1],\Sigma_t^{\epsilon_0}} \\ &\leqslant C_m\delta_0^2(1+t)^{-2}(1+\log(1+t))^2\end{aligned} \tag{12.260}$$

Appealing also to (12.48) and (12.256) with $l=m$, (12.257) with $n=0$, we deduce

$$L(|\mu_{m+1,0}|) \leqslant C\delta_0(1+t)^{-2}|\mu_{m+1,0}| + C_m\delta_0(1+t)^{-1} \tag{12.261}$$

which together with

$$\|\mu-1\|_{\infty,[m+1,m+1],\Sigma_0^{\epsilon_0}} \leqslant C_m\delta_0 \tag{12.262}$$

yields

$$\|\mu_{m+1,0}\|_{L^\infty(\Sigma_t^{\epsilon_0})} \leqslant C_m\delta_0(1+\log(1+t)) \quad \text{for all} \quad t \in [0,s] \tag{12.263}$$

This estimate together with $\mathbf{M}_{[m,m+1]}$ yields $\mathbf{M}_{[m+1,m+1]}$.

Next, let $\mathbf{M}_{[m+1,m+n]}$ hold for some $n \in \{1,\cdots,l-m\}$. We shall show that $\mathbf{M}_{[m+1,m+1+n]}$ holds. Consider the first sum on the right in (12.252). Since $n \leqslant l-m$ this is bounded in $L^\infty(\Sigma_t^{\epsilon_0})$ by

$$\begin{aligned}
&\sum_{k=0}^{m} \|\Lambda \cdot \not{d}(T)^{m-k}\mu\|_{\infty,[k,l+k-m],\Sigma_t^{\epsilon_0}} \\
&\leqslant C_l(1+t)^{-1}\|\Lambda\|_{\infty,[m,l],\Sigma_t^{\epsilon_0}}\|\mu-1\|_{\infty,[m,l+1],\Sigma_t^{\epsilon_0}} \\
&\leqslant C_l\delta_0^2(1+t)^{-2}(1+\log(1+t))^2
\end{aligned} \tag{12.264}$$

by (12.258) and $\mathbf{M}_{[m,l+1]}$.

Consider next the second sum on the right of (12.252). Consider a term in this sum corresponding to $k \in \{0, \cdots, n-1\}$. We express the term as

$$\begin{aligned}
&R_{i_n} \cdots R_{i_{n-k+1}}({}^{(R_{i_{n-k}})}Z \cdot \not{d}R_{i_{n-k-1}} \cdots R_{i_1}(T)^{m+1}\mu) \\
&= {}^{(R_{i_{n-k}})}Z \cdot \not{d}^{(i_1 \overset{>i_{n-k}<}{\cdots} i_n)}\mu_{m+1,n-1} \\
&+ \sum_{|s_1|+|s_2|=k, |s_1|\geqslant 1} ((\not{\mathcal{L}}_R)^{s_1}\,{}^{(R_{i_{n-k}})}Z) \cdot \not{d}(R)^{s_2'}(T)^{m+1}\mu
\end{aligned} \tag{12.265}$$

where the sum on the right is over all ordered partitions $\{s_1, s_2\}$ of the set $\{n-k+1, \cdots, n\}$ into two ordered subsets $s_1, s_2$, with $s_1$ non-empty. Also

$$s_2' = s_2 \bigcup \{1, \cdots, n-k-1\}$$

Since $|s_1| \leqslant k \leqslant n-1 \leqslant l-m-1$, the first factor in the sum is bounded by (12.236). Also, since $|s_2| \leqslant k-1$, we have

$$|s_2'| = |s_2| + n - k - 1 \leqslant n-2 \quad \text{and} \quad 1 + |s_2'| + m + 1 \leqslant n + m$$

Therefore, using **H0** and $\mathbf{M}_{m+1,m+n}$ the sum is seen to be bounded by

$$C_l\delta_0^2(1+t)^{-2}(1+\log(1+t))^2 \tag{12.266}$$

On the other hand, by (12.236) and **H0** the first term on the right of (12.265) is bounded by

$$C_l\delta_0(1+t)^{-2}(1+\log(1+t)) \max_j |{}^{(i_1 \overset{>i_{n-k}<}{\cdots} i_n j)}\mu_{m+1,n}| \tag{12.267}$$

Combining (12.266) and (12.267) we conclude that the second sum on the right of (12.252) is bounded by

$$\begin{aligned}
&C_l\delta_0(1+t)^{-2}(1+\log(1+t)) \max_j |{}^{(i_1 \overset{>i_{n-k}<}{\cdots} i_n j)}\mu_{m+1,n}| \\
&+ C_l\delta_0^2(1+t)^{-2}(1+\log(1+t))^2
\end{aligned} \tag{12.268}$$

The estimates (12.255)—(12.257), (12.264) and (12.268) imply through (12.252)

$$\begin{aligned}
&L(|{}^{(i_1 \cdots i_l)}\mu_{m+1,n}|) \\
&\leqslant C_l\delta_0(1+t)^{-2}(1+\log(1+t)) \max_{j_1 \cdots j_n} |{}^{(j_1 \cdots j_n)}\mu_{m+1,n}| + C_l\delta_0(1+t)^{-1}
\end{aligned} \tag{12.269}$$

Setting along an integral curve of $L$,

$$ {}^{(i_1\cdots i_n)}x'_{m+1,n}(t) = |{}^{(i_1\cdots i_n)}\mu_{m+1,n}| \tag{12.270} $$

(12.269) takes the form

$$ \frac{d^{(i_1\cdots i_n)}x'_{m+1,n}}{dt} \leqslant f_l \max_{j_1\cdots j_n} {}^{(j_1\cdots j_n)}x'_{m+1,n} + g_l \tag{12.271} $$

where

$$ f_l(t) = C_l\delta_0(1+t)^{-2}(1+\log(1+t)), \quad g_l(t) = C_l\delta_0(1+t)^{-1} \tag{12.272} $$

This is identical in form to (12.217)—(12.218). Setting then

$$ \bar{x}'_{m+1,n}(t) = \max_{i_1\cdots i_n} {}^{(i_1\cdots i_n)}x'_{m+1,n}(t) \tag{12.273} $$

in view of the fact that

$$ \bar{x}'_{m+1,n}(0) \leqslant C_l\delta_0 $$

we conclude that

$$ \max_{i_1\cdots i_n} \|{}^{(i_1\cdots i_n)}\mu_{m+1,n}\|_{L^\infty(\Sigma_t^{\epsilon_0})} \leqslant C_l\delta_0(1+\log(1+t)) \quad \text{for all} \quad t\in[0,s] \tag{12.274} $$

This together with $\mathbf{M}_{[m+1,m+n]}$ yields $\mathbf{M}_{[m+1,m+1+n]}$, so the proposition follows. □

Let us now define, for any non-negative integer $k$, the quantities

$$ \mathcal{A}'_k = \max_{i_1\cdots i_k} \|\not{\mathcal{L}}_{R_{i_k}}\cdots\not{\mathcal{L}}_{R_{i_1}}\chi'\|_{L^2(\Sigma_t^{\epsilon_0})} \tag{12.275} $$

and, for any non-negative integer $l$, the quantities

$$ \mathcal{A}'_{[l]} = \sum_{k=0}^{l} \mathcal{A}'_k \tag{12.276} $$

Let us recall the definitions

$$ \mathcal{A}_0 = \|\chi - \frac{\not{g}}{1-u+t}\|_{L^2(\Sigma_t^{\epsilon_0})} \tag{12.277} $$

and for $k>0$,

$$ \mathcal{A}_k = \max_{i_1\cdots i_k} \|\not{\mathcal{L}}_{R_{i_k}}\cdots\not{\mathcal{L}}_{R_{i_1}}\chi\|_{L^2(\Sigma_t^{\epsilon_0})} \tag{12.278} $$

and

$$ \mathcal{A}_{[l]} = \sum_{k=0}^{l} \mathcal{A}_k \tag{12.279} $$

Let us also recall

$$\mathcal{W}_0 = \max_{\alpha}(\|\psi_\alpha\|_{L^2(\Sigma_t^{\epsilon_0})}) \tag{12.280}$$

and for $k > 0$,

$$\mathcal{W}_k = \max_{\alpha;i_1\cdots i_k} \|R_{i_k} \cdots R_{i_1}\psi_\alpha\|_{L^2(\Sigma_t^{\epsilon_0})} \tag{12.281}$$

and

$$\mathcal{W}_{[l]} = \sum_{k=0}^{l} \mathcal{W}_k \tag{12.282}$$

$$\mathcal{W}_k^Q = \max_{\alpha;i_1\cdots i_k} \|R_{i_k} \cdots R_{i_1} Q\psi_\alpha\|_{L^2(\Sigma_t^{\epsilon_0})}, \quad \mathcal{W}_{[l]}^Q = \sum_{k=0}^{l} \mathcal{W}_k^Q \tag{12.283}$$

For all $k = 0, \cdots, l$,

$$\mathcal{Y}_k = \max_{j;i_1\cdots i_k} \|R_{i_k} \cdots R_{i_1} y^i\|_{L^2(\Sigma_t^{\epsilon_0})}, \quad \mathcal{Y}_{[l]} = \sum_{k=0}^{l} \mathcal{Y}_k \tag{12.284}$$

Recall also that given a positive integer $n$ we denote

$$n_* = \begin{cases} \dfrac{n}{2}, & \text{if } n \text{ is even} \\[2ex] \dfrac{n-1}{2}, & \text{if } n \text{ is odd} \end{cases} \tag{12.285}$$

**Lemma 12.5** Let **H0** and (12.23) hold. Let $l$ be a positive integer and let the assumptions $\not{\mathbf{E}}_{[l_*]}, \not{\mathbf{E}}^Q_{[l_*-1]}$ as well as $\not{\mathbf{X}}'_{[l_*-1]}$ hold. Then we have

$$\mathcal{A}_{[l]} \leqslant \mathcal{A}'_{[l]} + C_l \mathcal{A}'_{[l-1]} + C_l(1+t)^{-1}(\mathcal{Y}_0 + \mathcal{W}_{[l]})$$

*Proof*. From (12.185) we obtain, for $k \geqslant 1$,

$$\mathcal{A}_k \leqslant \mathcal{A}'_k + \frac{1}{1-u+t} \max_i \max_{j_1\cdots j_{k-1}} \|\not{\mathcal{L}}_{R_{j_{k-1}}} \cdots \not{\mathcal{L}}_{R_{j_1}} {}^{(R_i)}\not{\pi}\|_{L^2(\Sigma_t^{\epsilon_0})} \tag{12.286}$$

Summing over $k \in \{1, \cdots, l\}$ and adding

$$\mathcal{A}_0 = \mathcal{A}'_0 \tag{12.287}$$

we deduce, for $l \geqslant 1$,

$$\mathcal{A}_{[l]} \leqslant \mathcal{A}'_{[l]} + \frac{3}{1-u+t} \max_i \|{}^{(R_i)}\not{\pi}\|_{2,[l-1],\Sigma_t^{\epsilon_0}} \tag{12.288}$$

By Lemma 12.4 with $l_*$ in the role of $l$, the assumptions of the lemma imply $\not{\mathbf{X}}_{[l_*-1]}$, therefore Proposition 10.2 holds. Then Corollary 10.2.d implies

$$\max_i \|{}^{(R_i)}\not{\pi}\|_{2,[l-1],\Sigma_t^{\epsilon_0}} \leqslant C_l(\mathcal{Y}_0 + (1+t)\mathcal{A}_{[l-1]} + \mathcal{W}_{[l]}) \tag{12.289}$$

Substituting this in (12.288) we get

$$\begin{aligned}\mathcal{A}_{[l]} &\leqslant \mathcal{A}'_{[l]} + C_l\mathcal{A}_{[l-1]} + C_l(1+t)^{-1}(\mathcal{Y}_0 + \mathcal{W}_{[l]}) \quad \text{for all} \quad l \geqslant 1 \\ \mathcal{A}_{[0]} &= \mathcal{A}'_{[0]}\end{aligned} \tag{12.290}$$

The lemma follows. □

**Proposition 12.11** Let the assumptions of Lemma 12.4 hold. Let $l$ be a non-negative integer and let the assumptions $\mathbf{E}_{\{l_*+2\}}, \mathbf{E}^Q_{\{l_*+1\}}, \mathbf{E}^{QQ}_{\{l_*\}}$, hold on $W^s_{\epsilon_0}$. Moreover let the initial data satisfy

$$\|\chi'\|_{\infty,[l_*],\Sigma_0^{\epsilon_0}} \leqslant C_l\delta_0$$

Then $\mathbf{X}'_{[l_*]}$ holds on $W^s_{\epsilon_0}$ and there is a constant $C_l$ independent of $s$ such that for all $t \in [0,s]$ we have

$$\mathcal{A}'_{[l]}(t) \leqslant C_l(1+t)^{-1}(\mathcal{A}'_{[l]}(0) + \delta_0\mathcal{Y}_0(0) + \int_0^t (1+t')^{-1}(\mathcal{W}_{[l+2]}(t') + \mathcal{W}^Q_{[l+1]}(t'))dt')$$

*Proof*. The assumptions of this proposition include those of Proposition 12.9 with $l_*$ in the role of $l$, therefore $\mathbf{X}'_{[l_*]}$ holds on $W^s_{\epsilon_0}$. Then from Lemma 12.4, $\mathbf{X}_{[l_*]}$ holds on $W^s_{\epsilon_0}$. Thus, Proposition 10.1 holds with $l_*+1$ in the role of $l$, then Proposition 10.2 holds with $l+1$ in the role of $l$.

We shall estimate ${}^{(i_1\cdots i_l)}b_l$ (see (12.182)) in $L^2(\Sigma_t^{\epsilon_0})$. By assumptions $\mathbf{E}^Q_{\{l_*\}}$, $\mathbf{E}_{\{l_*+1\}}$, we have

$$\|L\psi_\mu\|_{\infty,[l_*],\Sigma_t^{\epsilon_0}}, \quad \|\not{d}\psi_\mu\|_{\infty,[l_*],\Sigma_t^{\epsilon_0}} \leqslant C_l\delta_0(1+t)^{-2} \quad \text{for all} \quad t \in [0,s] \tag{12.291}$$

as well as

$$\|Lh\|_{\infty,[l_*],\Sigma_t^{\epsilon_0}}, \quad \|\not{d}h\|_{\infty,[l_*],\Sigma_t^{\epsilon_0}} \leqslant C_l\delta_0(1+t)^{-2} \quad \text{for all} \quad t \in [0,s] \tag{12.292}$$

So by Corollary 10.1.b we have

$$\|e\|_{\infty,[l_*],\Sigma_t^{\epsilon_0}}, \quad \|\kappa^{-1}\zeta\|_{\infty,[l_*],\Sigma_t^{\epsilon_0}} \leqslant C_l\delta_0(1+t)^{-2} \quad \text{for all} \quad t \in [0,s] \tag{12.293}$$

Obviously, by the assumptions of the present proposition we have

$$\begin{aligned}\|Lh\|_{2,[l],\Sigma_t^{\epsilon_0}} &\leqslant C_l(1+t)^{-1}(\mathcal{W}^Q_{[l]} + \delta_0(1+t)^{-1}\mathcal{W}_{[l]}) \\ \|\not{d}h\|_{2,[l],\Sigma_t^{\epsilon_0}} &\leqslant C_l(1+t)^{-1}\mathcal{W}_{[l+1]} \quad \text{for all} \quad t \in [0,s]\end{aligned} \tag{12.294}$$

so by Corollary 10.1.g and Corollary 10.2.g with $l$ replaced by $l+1$, we have

$$\begin{aligned}&\|e\|_{2,[l],\Sigma_t^{\epsilon_0}} \\ &\leqslant C_l(1+t)^{-1}(\mathcal{W}^Q_{[l]} + \delta_0(1+t)^{-1}(\mathcal{Y}_0 + (1+t)\mathcal{A}_{[l-1]} + \mathcal{W}_{[l]})) \quad \text{for all} \quad t \in [0,s]\end{aligned} \tag{12.295}$$

Also from the expression

$$\kappa^{-1}\zeta = \alpha\epsilon - \not{d}\alpha$$

and Corollary 10.2.h, we have

$$\|\kappa^{-1}\zeta\|_{2,[l],\Sigma_t^{\epsilon_0}} \leqslant C_l(1+t)^{-1}(\mathcal{W}_{[l+1]}+\delta_0(1+t)^{-1}(\mathcal{Y}_0+(1+t)\mathcal{A}_{[l-1]})) \quad \text{for all} \quad t \in [0,s] \tag{12.296}$$

From (12.45), using the expression for $L^i$:

$$L^i = -\eta\hat{T}^i + \psi_i$$

and Corollary 10.2.g, we have

$$\|\alpha'\|_{2,[l],\Sigma_t^{\epsilon_0}} \leqslant C_l\delta_0(1+t)^{-2}(\mathcal{W}_{[l+2]} + \delta_0(1+t)^{-1}\mathcal{W}^Q_{[l+1]} + (1+t)^{-1}\mathcal{Y}_0 + \mathcal{A}_{[l]}) \tag{12.297}$$

where we have used Lemma 10.8 and Lemma 10.11 to estimate the commutator as well as **H0, H1, H2** to estimate the $S_{t,u}$ -tangential derivatives. This and (12.295), together with the estimates for ${}^{(R_i)}\not{\pi}$ in Chapter 10, (10.145):

$$\begin{aligned}\|{}^{(R_i)}\not{\pi}\|_{\infty,[l],\Sigma_t^{\epsilon_0}} &\leqslant C_l\delta_0(1+t)^{-1}(1+\log(1+t))\\ \|{}^{(R_i)}\not{\pi}\|_{2,[l],\Sigma_t^{\epsilon_0}} &\leqslant C_l(\mathcal{Y}_0+(1+t)\mathcal{A}_{[l]}+\mathcal{W}_{[l+1]})\end{aligned} \tag{12.298}$$

yield the following bound for $b$ defined by (12.183):

$$\|b\|_{2,[l],\Sigma_t^{\epsilon_0}} \leqslant C_l\delta_0(1+t)^{-2}(\mathcal{W}_{[l+2]} + (1+t)^{-1}\mathcal{Y}_0 + \mathcal{A}_{[l]} + \mathcal{W}^Q_{[l+1]}) \tag{12.299}$$

Next, we shall estimate ${}^{(i_1\cdots i_l)}s_l$ and ${}^{(i_1\cdots i_l)}r_l$ in $L^2(\Sigma_t^{\epsilon_0})$. From the expression (12.178), bounds for ${}^{(R_i)}\not{\pi}$ (12.298) as well as the fact that $\mathbf{X}'_{[l_*]}$ holds, we get

$$\begin{aligned}\|{}^{(i_1\cdots i_l)}r_l\|_{L^2(\Sigma_t^{\epsilon_0})} \leqslant\ & C_l\delta_0(1+t)^{-2}(1+\log(1+t))\mathcal{A}'_{[l-1]}\\ &+ C_l\delta_0^2(1+t)^{-4}(1+\log(1+t))^2(\mathcal{Y}_0+\mathcal{W}_{[l]})\end{aligned} \tag{12.300}$$

Also from (12.180) and the estimates for $e$ we get

$$\begin{aligned}\|{}^{(i_1\cdots i_l)}s_l\|_{L^2(\Sigma_t^{\epsilon_0})} \leqslant\ & C_l(1+t)^{-2}\mathcal{A}'_{[l-1]}+C_l\delta_0(1+t)^{-3}(1+\log(1+t))(\mathcal{W}^Q_{[l]}\\ &+ \delta_0(1+t)^{-1}(\mathcal{Y}_0+\mathcal{W}_{[l]}))\end{aligned} \tag{12.301}$$

Combining (12.299)—(12.301) and Lemma 12.5, we get

$$\begin{aligned}\|{}^{(i_1\cdots i_l)}b_l\|_{L^2(\Sigma_t^{\epsilon_0})} \leqslant\ & C_l\delta_0((1+t)^{-2}(1+\log(1+t))\mathcal{A}'_{[l]}\\ &+ (1+t)^{-2}(\mathcal{W}_{[l+2]}+(1+t)^{-1}\mathcal{Y}_0+\mathcal{W}^Q_{[l+1]}))\end{aligned} \tag{12.302}$$

Next, we shall estimate the $L^2(\Sigma_t^{\epsilon_0})$ norm of the sum on the right of (12.176). A term $k \in \{0,\cdots,l-1\}$ in this sum is expressed by (12.195). For the first term on the right of (12.195), we have the pointwise estimate (12.197). This implies

$$
\begin{aligned}
&\|\not{\mathcal{L}}_{{}^{(R_{i_{l-k}})}Z} {}^{(i_1 \overset{>i_{l-k}<}{\cdots} i_l)}\chi'_{l-1}\|_{L^2(\Sigma_t^{\epsilon_0})} \\
&\leqslant C(1+t)^{-1}\cdot\big(\|{}^{(R_{i_{l-k}})}Z\|_{L^\infty(\Sigma_t^{\epsilon_0})}(3\max_j \|{}^{(i_1 \overset{i_{l-k}}{\cdots} i_l j)}\chi'_l\|_{L^2(\Sigma_t^{\epsilon_0})} \qquad (12.303)\\
&+\|{}^{(i_1 \overset{i_{l-k}}{\cdots} i_l)}\chi'_{l-1}\|_{L^2(\Sigma_t^{\epsilon_0})}) + \|{}^{(i_1 \overset{i_{l-k}}{\cdots} i_l)}\chi'_{l-1}\|_{L^2(\Sigma_t^{\epsilon_0})}\max_j \|{}^{(j;i_{l-k})}Z\|_{L^\infty(\Sigma_t^{\epsilon_0})}\big)
\end{aligned}
$$

If $l \geqslant 2$, so that $l_* \geqslant 1$, we apply Corollary 10.1.i with $l_*+1$ in the role of $l$ to get

$$
\begin{aligned}
&\|{}^{(R_{i_{l-k}})}Z\|_{L^\infty(\Sigma_t^{\epsilon_0})}, \quad \max_j \|{}^{(j;i_{l-k})}Z\|_{L^\infty(\Sigma_t^{\epsilon_0})} \\
&\leqslant C\delta_0(1+t)^{-1}(1+\log(1+t)) \qquad (12.304)
\end{aligned}
$$

We then obtain

$$
\begin{aligned}
&\|\not{\mathcal{L}}_{{}^{(R_{i_{l-k}})}Z} {}^{(i_1 \overset{i_{l-k}}{\cdots} i_l)}\chi'_{l-1}\|_{L^2(\Sigma_t^{\epsilon_0})} \\
&\leqslant C\delta_0(1+t)^{-2}(1+\log(1+t))\mathcal{A}'_{[l]}, \quad l \geqslant 2 \qquad (12.305)
\end{aligned}
$$

On the other hand, if $l = 1$, then $l_* = 0$, we estimate the last term in parenthesis in (12.197) in $L^2$ by

$$
3\|\chi'\|_{L^\infty(\Sigma_t^{\epsilon_0})}\max_j \|{}^{(j;i_1)}Z\|_{L^2(\Sigma_t^{\epsilon_0})} \qquad (12.306)
$$

By Corollary 10.2.i with $l+1$ in the role of $l$, we have

$$
\max_i \|{}^{(R_i)}Z\|_{2,[l],\Sigma_t^{\epsilon_0}} \leqslant C_l(\mathcal{Y}_0 + (1+t)\mathcal{A}'_{[l]} + \mathcal{W}_{[l+1]}) \qquad (12.307)
$$

where we have used Lemma 12.5 to express $\mathcal{A}_{[l]}$ in terms of $\mathcal{A}'_{[l]}$. Then in view of Proposition 12.6 and (12.307) with $l = 1$, (12.306) is bounded by

$$
C\delta_0(1+t)^{-2}(1+\log(1+t))(\mathcal{Y}_0 + (1+t)\mathcal{A}'_{[1]} + \mathcal{W}_{[2]}) \qquad (12.308)
$$

Thus, in the case of $l = 1$, we have

$$
\begin{aligned}
\|\not{\mathcal{L}}_{{}^{(R_i)}Z}\chi'\|_{L^2(\Sigma_t^{\epsilon_0})} &\leqslant C\delta_0(1+t)^{-2}(1+\log(1+t))\mathcal{A}'_{[1]} \qquad (12.309)\\
&+ C\delta_0(1+t)^{-3}(1+\log(1+t))(\mathcal{Y}_0 + \mathcal{W}_{[2]})
\end{aligned}
$$

Next, we should consider a term in the double sum in (12.195), corresponding to an ordered subset $\{m_1, \cdots, m_p\} \subset \{l-k+1, \cdots, l\}$, $p \in \{1, \cdots, k\}$. We have the pointwise estimate (12.204).

We distinguish three cases:

$$
\text{Case 1}: p \leqslant l_* - 1, \quad \text{Case 2}: p = l_*, \quad \text{Case 3}: p \geqslant l_* + 1
$$

In Case 1, Corollary 10.1.i with $l_*+1$ in the role of $l$ implies

$$
\begin{aligned}
&\|{}^{(i_{m_1}\cdots i_{m_p};i_{l-k})}Z\|_{L^\infty(\Sigma_t^{\epsilon_0})}, \quad \max_j \|{}^{(i_{m_1}\cdots i_{m_p}j;i_{l-k})}Z\|_{L^\infty(\Sigma_t^{\epsilon_0})} \\
&\leqslant C_l\delta_0(1+t)^{-1}(1+\log(1+t)) \qquad (12.310)
\end{aligned}
$$

then (12.204) implies

$$
\begin{aligned}
&\|\not{\mathcal{L}}_{{}^{(i_{m_1}\cdots i_{m_p};i_{l-k})}Z}{}^{(i_1 \overset{>i_{m_1}\cdots i_{m_p} i_{l-k}<}{\cdots} i_l)}\chi'_{l-1-p}\|_{L^2(\Sigma_t^{\epsilon_0})} \\
&\leqslant C(1+t)^{-1}\cdot\Big(\|{}^{(i_{m_1}\cdots i_{m_p};i_{l-k})}Z\|_{L^\infty(\Sigma_t^{\epsilon_0})}\big(3\max_j\|{}^{(i_1 \overset{>i_{m_1}\cdots i_{m_p} i_{l-k}<}{\cdots} i_l j)}\chi'_{l-p}\|_{L^2(\Sigma_t^{\epsilon_0})} \\
&\quad\cdot\|{}^{(i_1 \overset{>i_{m_1}\cdots i_{m_p} i_{l-k}<}{\cdots} i_l)}\chi'_{l-1-p}\|_{L^2(\Sigma_t^{\epsilon_0})}\big) \\
&\quad+\max_j\|{}^{(i_{m_1}\cdots i_{m_p} j;i_{l-k})}Z\|_{L^\infty(\Sigma_t^{\epsilon_0})}\|{}^{(i_1 \overset{>i_{m_1}\cdots i_{m_p} i_{l-k}<}{\cdots} i_l)}\chi'_{l-1-p}\|_{L^2(\Sigma_t^{\epsilon_0})}\Big) \\
&\leqslant C_l\delta_0(1+t)^{-2}(1+\log(1+t))\mathcal{A}'_{[l-1]} \qquad (12.311)
\end{aligned}
$$

where we have used the fact that $p \geqslant 1$.

In Case 2, Corollary 10.1.i with $l_*+1$ in the role of $l$ implies the first of (12.310), but not the second. We use instead Corollary 10.2.i with $l+1$ in the role of $l$, to get

$$\max_j\|{}^{(i_{m_1}\cdots i_{m_p} j;i_{l-k})}Z\|_{L^2(\Sigma_t^{\epsilon_0})} \leqslant C_l(\mathcal{Y}_0+(1+t)\mathcal{A}'_{[l]}+\mathcal{W}_{[l+1]}) \qquad (12.312)$$

where we have used the fact that $p \leqslant k \leqslant l-1$.

Since in Case 2 we have $l-1-p = l-1-l_* \leqslant l_*$, we may use $\mathbf{X}'_{[l_*]}$ to estimate

$$\|{}^{(i_1 \overset{>i_{m_1}\cdots i_{m_p} i_{l-k}<}{\cdots} i_l)}\chi'_{l-1-p}\|_{L^\infty(\Sigma_t^{\epsilon_0})} \leqslant C_l\delta_0(1+t)^{-2}(1+\log(1+t)) \qquad (12.313)$$

In view of (12.312) and (12.313) we get in Case 2

$$
\begin{aligned}
&\|\not{\mathcal{L}}_{{}^{(i_{m_1}\cdots i_{m_p};i_{l-k})}Z}{}^{(i_1 \overset{>i_{m_1}\cdots i_{m_p} i_{l-k}<}{\cdots} i_l)}\chi'_{l-1-p}\|_{L^2(\Sigma_t^{\epsilon_0})} \\
&\leqslant C(1+t)^{-1}\cdot\Big(\|{}^{(i_{m_1}\cdots i_{m_p};i_{l-k})}Z\|_{L^\infty(\Sigma_t^{\epsilon_0})}\big(3\max_j\|{}^{(i_1 \overset{>i_{m_1}\cdots i_{m_p} i_{l-k}<}{\cdots} i_l j)}\chi'_{l-p}\|_{L^2(\Sigma_t^{\epsilon_0})} \\
&\quad\cdot\|{}^{(i_1 \overset{>i_{m_1}\cdots i_{m_p} i_{l-k}<}{\cdots} i_l)}\chi'_{l-1-p}\|_{L^2(\Sigma_t^{\epsilon_0})}\big) \\
&\quad+3\|{}^{(i_1 \overset{>i_{m_1}\cdots i_{m_p} i_{l-k}<}{\cdots} i_l)}\chi'_{l-1-p}\|_{L^\infty(\Sigma_t^{\epsilon_0})}\max_j\|{}^{(i_{m_1}\cdots i_{m_p} j;i_{l-k})}Z\|_{L^2(\Sigma_t^{\epsilon_0})}\Big) \\
&\leqslant C_l\delta_0(1+t)^{-3}(1+\log(1+t))(\mathcal{Y}_0+(1+t)\mathcal{A}'_{[l]}+\mathcal{W}_{[l+1]}) \qquad (12.314)
\end{aligned}
$$

In Case 3, we apply (12.312) and also (12.307) with $l$ replaced by $l-1$ to estimate

$$\|{}^{(i_{m_1}\cdots i_{m_p};i_{l-k})}Z\|_{L^2(\Sigma_t^{\epsilon_0})} \leqslant C_l(\mathcal{Y}_0+(1+t)\mathcal{A}'_{[l-1]}+\mathcal{W}_{[l]}) \qquad (12.315)$$

Since in Case 3 we have $l-p \leqslant l-l_*-1 \leqslant l_*$, we may use $\mathbf{X}'_{[l_*]}$ to estimate

$$\max_j\|{}^{(i_1 \overset{i_{m_1}\cdots i_{m_p} i_{l-k}}{\cdots} i_l j)}\chi'_{l-p}\|_{L^\infty(\Sigma_t^{\epsilon_0})} \leqslant C_l\delta_0(1+t)^{-2}(1+\log(1+t)) \qquad (12.316)$$

as well as (12.313).

So in Case 3 we have

$$
\begin{aligned}
&\|\not{\mathcal{L}}_{{}^{(i_{m_1}\cdots i_{m_p};i_{l-k})}Z}\, {}^{(i_1 \overset{>i_{m_1}\cdots i_{m_p} i_{l-k}<}{\cdots} i_l)}\chi'_{l-1-p}\|_{L^2(\Sigma_t^{\epsilon_0})} \\
&\leqslant C(1+t)^{-1}\cdot\Big(\|{}^{(i_{m_1}\cdots i_{m_p};i_{l-k})}Z\|_{L^2(\Sigma_t^{\epsilon_0})}\Big(3\max_j \|{}^{(i_1 \overset{>i_{m_1}\cdots i_{m_p} i_{l-k}<}{\cdots} i_l j)}\chi'_{l-p}\|_{L^\infty(\Sigma_t^{\epsilon_0})} \\
&\quad\cdot \|{}^{(i_1 \overset{>i_{m_1}\cdots i_{m_p} i_{l-k}<}{\cdots} i_l)}\chi'_{l-1-p}\|_{L^\infty(\Sigma_t^{\epsilon_0})}\Big) \\
&\quad + 3\|{}^{(i_1 \overset{>i_{m_1}\cdots i_{m_p} i_{l-k}<}{\cdots} i_l)}\chi'_{l-1-p}\|_{L^\infty(\Sigma_t^{\epsilon_0})}\max_j \|{}^{(i_{m_1}\cdots i_{m_p}j;i_{l-k})}Z\|_{L^2(\Sigma_t^{\epsilon_0})}\Big) \\
&\leqslant C_l\delta_0(1+t)^{-3}(1+\log(1+t))(\mathcal{Y}_0+(1+t)\mathcal{A}'_{[l]}+\mathcal{W}_{[l+1]}) \qquad (12.317)
\end{aligned}
$$

Combining (12.311), (12.314) and (12.317) we conclude that the double sum in (12.195) is bounded in $L^2(\Sigma_t^{\epsilon_0})$ by

$$C_l\delta_0(1+t)^{-3}(1+\log(1+t))(\mathcal{Y}_0+(1+t)\mathcal{A}'_{[l]}+\mathcal{W}_{[l+1]}) \qquad (12.318)$$

Combining this result with (12.305) and (12.309) we then conclude that the sum on the right of (12.176) is bounded in $L^2(\Sigma_t^{\epsilon_0})$ by

$$
\begin{aligned}
&\|\sum_{k=0}^{l-1}\not{\mathcal{L}}_{R_{i_l}}\cdots\not{\mathcal{L}}_{R_{i_{l-k+1}}}\not{\mathcal{L}}_{{}^{(R_{i_{l-k}})}Z}\not{\mathcal{L}}_{R_{i_{l-k-1}}}\cdots\not{\mathcal{L}}_{R_{i_1}}\chi'\|_{L^2(\Sigma_t^{\epsilon_0})} \\
&\leqslant C_l\delta_0(1+t)^{-3}(1+\log(1+t))(\mathcal{Y}_0+(1+t)\mathcal{A}'_{[l]}+\mathcal{W}_{[l+1]}) \qquad (12.319)
\end{aligned}
$$

In view of (12.302) and (12.319), we obtain through (12.176) and (12.181), recalling also (12.187)

$$
\begin{aligned}
\|\not{\mathcal{L}}_L{}^{(i_1\cdots i_l)}\chi'_l\|_{L^2(\Sigma_t^{\epsilon_0})} &\leqslant C_l\delta_0(1+t)^{-3}(1+\log(1+t))(\mathcal{Y}_0+(1+t)\mathcal{A}'_{[l]}) \qquad (12.320)\\
&\quad +(1+t)^{-2}(\mathcal{W}_{[l+2]}+\mathcal{W}^Q_{[l+1]})
\end{aligned}
$$

Let us define the non-negative functions

$${}^{(i_1\cdots i_l)}\phi_l=(1-u+t)^2|{}^{(i_1\cdots i_l)}\chi'_l| \qquad (12.321)$$

$${}^{(i_1\cdots i_l)}\rho_l=(1-u+t)^2(2|\chi'||{}^{(i_1\cdots i_l)}\chi'_l|+|\not{\mathcal{L}}_L{}^{(i_1\cdots i_l)}\chi'_l|) \qquad (12.322)$$

From (12.56) we have

$$L\,{}^{(i_1\cdots i_l)}\phi_l\leqslant {}^{(i_1\cdots i_l)}\rho_l \qquad (12.323)$$

We pull back ${}^{(i_1\cdots i_l)}\phi_l$ and ${}^{(i_1\cdots i_l)}\rho_l$ to $S^2$:

$${}^{(i_1\cdots i_l)}\phi_l(t,u)={}^{(i_1\cdots i_l)}\phi_l\circ\Phi_{t,u},\qquad {}^{(i_1\cdots i_l)}\rho_l(t,u)={}^{(i_1\cdots i_l)}\rho_l\circ\Phi_{t,u} \qquad (12.324)$$

where $\Phi_{t,u}$ is the diffeomorphism of $S^2$ onto $S_{t,u}$ defined in Chapter 8. Then (12.324) reads

$$\frac{\partial}{\partial t}{}^{(i_1\cdots i_l)}\phi_l(t,u)\leqslant {}^{(i_1\cdots i_l)}\rho_l(t,u)$$

Integrating we obtain

$$ {}^{(i_1\cdots i_l)}\phi_l(t,u) \leqslant {}^{(i_1\cdots i_l)}\phi_l(0,u) + \int_0^t {}^{(i_1\cdots i_l)}\rho_l(t',u)dt' \tag{12.325} $$

Taking $L^2$ norm on $[0,\epsilon_0]\times S^2$ yields

$$ \begin{aligned} &\|{}^{(i_1\cdots i_l)}\phi_l(t)\|_{L^2([0,\epsilon_0]\times S^2)} \\ \leqslant &\|{}^{(i_1\cdots i_l)}\phi_l(0)\|_{L^2([0,\epsilon_0]\times S^2)} + \int_0^t \|{}^{(i_1\cdots i_l)}\rho_l(t')\|_{L^2([0,\epsilon_0]\times S^2)}dt' \end{aligned} \tag{12.326} $$

In view of the comparison inequalities (8.333) we then obtain

$$ \begin{aligned} &(1+t)^{-1}\|{}^{(i_1\cdots i_l)}\phi_l\|_{L^2(\Sigma_t^{\epsilon_0})} \\ \leqslant &C(\|{}^{(i_1\cdots i_l)}\phi_l\|_{L^2(\Sigma_0^{\epsilon_0})} + \int_0^t (1+t')^{-1}\|{}^{(i_1\cdots i_l)}\rho_l\|_{L^2(\Sigma_{t'}^{\epsilon_0})}dt') \end{aligned} \tag{12.327} $$

Replacing $l$ by $k\in\{0,\cdots,l\}$, taking on both sides the maximum over $i_1,\cdots,i_k$ and then summing over $k\in\{0,\cdots,l\}$ yields

$$ \begin{aligned} &(1+t)^{-1}\sum_{k=0}^{l}\max_{i_1\cdots i_k}\|{}^{(i_1\cdots i_k)}\phi_k\|_{L^2(\Sigma_t^{\epsilon_0})} \\ \leqslant &C(\sum_{k=0}^{l}\max_{i_1\cdots i_k}\|{}^{(i_1\cdots i_k)}\phi_k\|_{L^2(\Sigma_0^{\epsilon_0})} \\ &+\int_0^t (1+t')^{-1}\sum_{k=0}^{l}\max_{i_1\cdots i_k}\|{}^{(i_1\cdots i_k)}\rho_k\|_{L^2(\Sigma_{t'}^{\epsilon_0})}dt') \end{aligned} \tag{12.328} $$

From (12.321) we have

$$ \begin{aligned} &\sum_{k=0}^{l}\max_{i_1\cdots i_l}\|{}^{(i_1\cdots i_k)}\phi_k\|_{L^2(\Sigma_t^{\epsilon_0})} \geqslant C^{-1}(1+t)^2\mathcal{A}'_{[l]}(t) \\ &\sum_{k=0}^{l}\max_{i_1\cdots i_k}\|{}^{(i_1\cdots i_k)}\phi_k\|_{L^2(\Sigma_0^{\epsilon_0})} \leqslant \mathcal{A}'_{[l]}(0) \end{aligned} \tag{12.329} $$

and from (12.322), (12.320) and $\mathbf{X}'_{[0]}$ we have

$$ \begin{aligned} &\sum_{k=0}^{l}\max_{i_1\cdots i_k}\|{}^{(i_1\cdots i_k)}\rho_k\|_{L^2(\Sigma_t^{\epsilon_0})} \\ \leqslant &C\delta_0(1+\log(1+t))\mathcal{A}'_{[l]}(t) + C(1+t)^2\|\not{\mathcal{L}}_L{}^{(i_1\cdots i_l)}\chi'_l\|_{L^2(\Sigma_t^{\epsilon_0})} \\ \leqslant &C_l\delta_0(1+t)^{-1}(1+\log(1+t))(\mathcal{Y}_0+(1+t)\mathcal{A}'_{[l]}(t)) \\ &+C_l(\mathcal{W}_{[l+2]}(t)+\mathcal{W}^Q_{[l+1]}(t)) \end{aligned} \tag{12.330} $$

Substituting (12.329) and (12.330) in (12.328) yields

$$(1+t)\mathcal{A}'_{[l]}(t) \leqslant C\mathcal{A}'_{[l]}(0) + \int_0^t C_l\delta_0(1+t')^{-2}(1+\log(1+t')) \\ \cdot(1+t')\mathcal{A}'_{[l]}(t')dt' + \int_0^t C_l\delta_0(1+t')^{-2}(1+\log(1+t'))\mathcal{Y}_0(t')dt' \\ + C_l\int_0^t (1+t)^{-1}(\mathcal{W}_{[l+2]}(t') + \mathcal{W}^Q_{[l+1]}(t'))dt' \tag{12.331}$$

This implies

$$(1+t)\mathcal{A}'_{[l]}(t) \leqslant C\mathcal{A}'_{[l]}(0) + \int_0^t C_l\delta_0(1+t')^{-2}(1+\log(1+t'))\mathcal{Y}_0(t')dt' \\ + C_l\int_0^t (1+t')^{-1}(\mathcal{W}_{[l+2]}(t') + \mathcal{W}^Q_{[l+1]}(t'))dt' \tag{12.332}$$

Finally, we must estimate $\mathcal{Y}_0(t)$ in terms of $\mathcal{Y}_0(0)$. To do this, we must derive a propagation equation for $y^i$. According to (6.60),

$$\hat{T}^i = -\frac{x^i}{1-u+t} + y^i \tag{12.333}$$

hence

$$L\hat{T}^i = -\frac{L^i}{1-u+t} + \frac{x^i}{(1-u+t)^2} + Ly^i \tag{12.334}$$

Also from (6.64) and (6.65),

$$L^i = \frac{x^i}{1-u+t} + \omega^i - \eta y^i \tag{12.335}$$

where

$$\omega^i = \frac{(\eta-1)x^i}{1-u+t} - \psi_i \tag{12.336}$$

Substituting (12.335) in (12.334) we get

$$L\hat{T}^i = \frac{\eta y^i - \omega^i}{1-u+t} + Ly^i \tag{12.337}$$

On the other hand, recalling from Chapter 3 that

$$L(\hat{T}^i) = q_L \cdot \not{d}x^i \tag{12.338}$$

we conclude $y^i$ satisfies the propagation equation

$$Ly^i + \frac{y^i}{1-u+t} = \tilde{\omega}^i \tag{12.339}$$

where

$$\tilde{\omega}^i = \frac{\omega^i - (\eta - 1)y^i}{1 - u + t} + q_L \cdot \not{d}x^i \tag{12.340}$$

From (12.23) and (12.336), the first term on the right of (12.340) is bounded in $L^2(\Sigma_t^{\epsilon_0})$ by

$$C(1+t)^{-1}\mathcal{W}_{[0]}(t) \tag{12.341}$$

From (3.182) and (12.333), we deduce

$$\|q_L\|_{L^2(\Sigma_t^{\epsilon_0})} \leqslant C(1+t)^{-1}\mathcal{W}_{[1]}(t) \tag{12.342}$$

hence we obtain

$$\max_i \|\tilde{\omega}^i\|_{L^2(\Sigma_t^{\epsilon_0})} \leqslant C(1+t)^{-1}\mathcal{W}_{[1]}(t) \tag{12.343}$$

Integrating (12.339) we get, in acoustical coordinates,

$$(1 - u + t)y^i(t, u, \vartheta) = (1 - u)y^i(0, u, \vartheta) + \int_0^t (1 - u + t')\tilde{\omega}^i(t', u, \vartheta)dt' \tag{12.344}$$

Taking $L^2$ norm on $[0, \epsilon_0] \times S^2$ we then obtain

$$\begin{aligned}&(1+t)\|y^i(t)\|_{L^2([0,\epsilon_0]\times S^2)}\\ &\leqslant C(\|y^i(0)\|_{L^2([0,\epsilon_0]\times S^2)} + \int_0^t (1+t')\|\tilde{\omega}^i\|_{L^2([0,\epsilon_0]\times S^2)}dt')\end{aligned} \tag{12.345}$$

This is equivalent to

$$\|y^i\|_{L^2(\Sigma_t^{\epsilon_0})} \leqslant C(\|y^i\|_{L^2(\Sigma_0^{\epsilon_0})} + \int_0^t \|\tilde{\omega}^i\|_{L^2(\Sigma_{t'}^{\epsilon_0})}dt') \tag{12.346}$$

Taking the maximum over $i = 1, 2, 3$, yields

$$\mathcal{Y}_0(t) \leqslant C(\mathcal{Y}_0(0) + \int_0^t (1+t')^{-1}\mathcal{W}_{[1]}(t')dt') \tag{12.347}$$

This implies

$$\begin{aligned}&\int_0^t (1+t')^{-2}(1+\log(1+t'))\mathcal{Y}_0(t')dt'\\ &\leqslant C(\mathcal{Y}_0(0) + \int_0^t (1+t')^{-2}(1+\log(1+t'))\mathcal{W}_{[1]}(t')dt')\end{aligned} \tag{12.348}$$

In fact, let

$$I_t(t') = -\int_{t'}^t (1+s)^{-2}(1+\log(1+s))ds$$

Then

$$\frac{d}{dt'}I_t(t') = (1+t')^{-2}(1+\log(1+t'))$$

Let also

$$J(t') = \int_0^{t'} (1+s)^{-1} \mathcal{W}_{[1]}(s) ds$$

Then

$$\begin{aligned} &\int_0^t (1+t')^{-2}(1+\log(1+t')) \int_0^{t'} (1+s)^{-1}\mathcal{W}_{[1]}(s)dsdt' \\ =& I_t(t)J(t) - I_t(0)J(0) - \int_0^t I_t(t')\frac{dJ}{dt'}(t')dt' \end{aligned}$$

But

$$I_t(t) = J(0) = 0$$

So this reduces to

$$-\int_0^t I_t(t')\frac{dJ}{dt'}(t')dt'$$

Now,

$$\begin{aligned} -I_t(t') =& \int_{t'}^t (1+s)^{-2}(1+\log(1+s))ds \leqslant \int_{t'}^{\infty} (1+s)^{-2}(1+\log(1+s))ds \\ =& [-\frac{2}{1+s} - \frac{\log(1+s)}{1+s}]_{t'}^{\infty} = \frac{2+\log(1+t')}{1+t'} \end{aligned}$$

Therefore

$$-\int_0^t I_t(t')\frac{dJ}{dt'}(t')dt' \leqslant 2\int_0^t (1+t')^{-2}(1+\log(1+t'))\mathcal{W}_{[1]}(t')dt'$$

Substituting (12.348) in the right in (12.332), the integral on the right in (12.348) is absorbed into the last integral on the right in (12.332). So the proposition follows. □

**Proposition 12.12** Let the assumptions of Proposition 12.11 hold for some non-negative integer $l$. Let $m \in \{0, \cdots, l+1\}$, and let the initial data satisfy

$$\|\mu - 1\|_{\infty,[m,l_*+1],\Sigma_0^{\epsilon_0}} \leqslant C_l\delta_0$$

Then $\mathbf{M}_{[m,l_*+1]}$ holds on $W^s_{\epsilon_0}$ and there is a constant $C_l$ independent of $s$ such that for all $t \in [0, s]$ we have

$$\begin{aligned} (1+t)^{-1}\mathcal{B}_{[m,l+1]}(t) \leqslant& C\mathcal{B}_{[m,l+1]}(0) + C_l(\mathcal{A}'_{[l]}(0) + \delta_0\mathcal{Y}_0(0) \\ &+ \int_0^t (1+t')^{-1}(\mathcal{W}_{\{l+2\}}(t') + \mathcal{W}^Q_{\{l+1\}}(t'))dt') \end{aligned}$$

*Proof*. The assumptions of this proposition include those of Proposition 12.10 with $l_*$ in the role of $l$, therefore $\mathbf{M}_{[m,l_*+1]}$ holds on $W^s_{\epsilon_0}$. Thus, the assumptions of Proposition 11.1 hold with $l_*$ in the role of $l$, hence also the assumptions of Proposition 11.2 hold. Therefore, the conclusions of Proposition 11.1 and of all its corollaries hold with $l_*$ in the role of $l$, and the conclusions of Proposition 11.2 and of all its corollaries hold as well.

We consider the propagation equation (12.252), replacing $m+1$ by $m$, so that $m \in \{0, \cdots, l+1\}$. We must estimate in $L^2(\Sigma_t^{\epsilon_0})$ the terms on the right side. From the definition of $m$ and $e$, we have

$$\|m\|_{2,[m,l+1],\Sigma_t^{\epsilon_0}} \leqslant C_l \mathcal{W}_{\{l+2\}} \tag{12.349}$$

$$\begin{aligned} \|e\|_{2,[m,l+1],\Sigma_t^{\epsilon_0}} \leqslant & C_l(1+t)^{-1}(\mathcal{W}^Q_{\{l+1\}} + \delta_0(1+t)^{-1}((1+t)^{-1}\mathcal{B}_{[m-1,l+1]} \\ & + \mathcal{Y}_0 + (1+t)\mathcal{A}_{[l]} + \mathcal{W}_{\{l+1\}})) \end{aligned} \tag{12.350}$$

Here, we have also used Corollary 11.2.b (the estimate for $\omega_{L\hat{T}}$; note that actually, only $\mathbf{M}_{[m,l_*]}$ is used in proving this estimate).

Also from Corollary 11.1.b with $l_*$ in the role of $l$ we obtain

$$\|e\|_{\infty,[m,l_*],\Sigma_t^{\epsilon_0}} \leqslant C_l \delta_0 (1+t)^{-2} \tag{12.351}$$

Consider next ${}^{(i_1 \cdots i_l)}r'_{m,n}$, given by (12.251) with $m+1$ replaced by $m$. Here $n \leqslant l+1-m$. Consider the summand

$$((R)^{s_1} e)((R)^{s_2}(T)^m \mu), \quad |s_1| + |s_2| = n, |s_1| \geqslant 1$$

If $|s_2| \leqslant l_* + 1 - m$ we appeal to $\mathbf{M}_{[m,l_*+1]}$ to place the second factor in $L^\infty(\Sigma_t^{\epsilon_0})$. Otherwise, we have $|s_1| \leqslant n - (l_* + 2 - m) \leqslant l - l_* - 1 \leqslant l_*$ and we appeal to (12.351) to place the first factor in $L^\infty(\Sigma_t^{\epsilon_0})$. Noting that $|s_2| \leqslant n - 1 \leqslant l - m$, we then obtain an $L^2(\Sigma_t^{\epsilon_0})$ for the first sum by

$$C_l \delta_0 (1+t)^{-2} \mathcal{B}_{[m,l]} + C_l(1 + \log(1+t))\|e\|_{2,[m,l+1],\Sigma_t^{\epsilon_0}} \tag{12.352}$$

Consider next the summand

$$R_{i_n} \cdots R_{i_1}(((T)^k e)((T)^{m-k}\mu)), \quad k \in \{1, \cdots, m\}$$

of the second sum. If at most $l_* + 1 + k - m$ of the $n$ angular derivatives fall on $(T)^{m-k}\mu$ we appeal to $\mathbf{M}_{[m,l_*+1]}$ to place the corresponding factor in $L^\infty(\Sigma_t^{\epsilon_0})$. Otherwise, at most

$$n - (l_* + 2 + k - m) \leqslant l - l_* - 1 - k \leqslant l_* - k$$

angular derivatives fall on $(T)^k e$ and we appeal to (12.351) to place the corresponding factor in $L^\infty(\Sigma_t^{\epsilon_0})$. Noting that $k \geqslant 1$, the second term on the right of (12.251) is bounded by

$$C_l \delta_0 (1+t)^{-2} \mathcal{B}_{[m-1,l]} + C_l(1 + \log(1+t))\|e\|_{2,[m,l+1],\Sigma_t^{\epsilon_0}} \tag{12.353}$$

Combining (12.352) and (12.353), and substituting the bound (12.350), we obtain

$$
\begin{aligned}
&\max_{n\leqslant l+1-m}\max_{i_1\cdots i_n}\|^{(i_1\cdots i_n)}r'_{m,n}\|_{L^2(\Sigma_t^{\epsilon_0})}\\
&\leqslant C_l(1+t)^{-1}(\delta_0(1+t)^{-1}\mathcal{B}_{[m,l+1]}+(1+\log(1+t))(\mathcal{W}^Q_{\{l+1\}} \qquad (12.354)\\
&+\delta_0(1+t)^{-1}(\mathcal{Y}_0+(1+t)\mathcal{A}_{[l]}+\mathcal{W}_{\{l+1\}})))
\end{aligned}
$$

Next, we must estimate the two sums on the right of (12.252) with $m+1$ replaced by $m$. The first sum is bounded in $L^2(\Sigma_t^{\epsilon_0})$ by

$$
\sum_{k=0}^{m-1}\|\Lambda\cdot d\!\!/(T)^{m-1-k}\mu\|_{2,[k,l+k+1-m],\Sigma_t^{\epsilon_0}} \qquad (12.355)
$$

Here, $\Lambda, d\!\!/(T)^{m-1-k}\mu$, receives at most $k$ $T$-derivatives, thus $\mu$ receives at most $m-1$ $T$-derivatives, and there are $l+k+1-m$ spatial derivatives. By Corollary 11.1.c with $l_*$ in the role of $l$,

$$
\|\Lambda\|_{\infty,[m,l_*],\Sigma_t^{\epsilon_0}}\leqslant C_l\delta_0(1+t)^{-1}(1+\log(1+t)) \qquad (12.356)
$$

If at most $l_*$ spatial derivatives fall on $\Lambda$ we appeal to this estimate to place the corresponding factor in $L^\infty(\Sigma_t^{\epsilon_0})$. Otherwise, at most

$$
(l+k+1-m)-(l_*+1)\leqslant l_*+k+1-m
$$

spatial derivatives fall on $d\!\!/(T)^{m-1-k}\mu$, for a total of at most $l_*+1$ spatial derivatives falling on $\mu$, and we appeal to $\mathbf{M}_{[m-1,l_*+1]}$ to place the corresponding factor in $L^\infty(\Sigma_t^{\epsilon_0})$. In this way, we deduce that (12.355) is bounded by

$$
\begin{aligned}
&C_l\delta_0(1+t)^{-2}(1+\log(1+t))\mathcal{B}_{[m-1,l+1]}\\
&+C_l\delta_0(1+t)^{-1}(1+\log(1+t))\|\Lambda\|_{2,[m-1,l],\Sigma_t^{\epsilon_0}}\\
&\leqslant C_l\delta_0(1+t)^{-2}(1+\log(1+t)) \qquad (12.357)\\
&\cdot(\mathcal{B}_{[m-1,l+1]}+\delta_0(1+\log(1+t))(\mathcal{Y}_0+(1+t)\mathcal{A}_{[l-1]})\\
&+(1+\log(1+t))(\mathcal{W}_{\{l+1\}}+\delta_0(1+t)^{-2}(1+\log(1+t))^2\mathcal{W}^Q_{\{l\}}))
\end{aligned}
$$

by Corollary 11.2.c with $m$ replaced by $m-1$.

Finally, we must consider the second sum on the right of (12.252) with $m+1$ replaced by $m$. Consider a term in this sum, corresponding to $k\in\{0,\cdots,n-1\}$,

$$
\begin{aligned}
&R_{i_n}\cdots R_{i_{n-k+1}}(^{(R_{i_{n-k}})}Z\cdot d\!\!/R_{i_{n-k-1}}\cdots R_{i_1}(T)^m\mu)\\
&=\sum_{|s_1|+|s_2|=k}((\not{\mathcal{L}}_R)^{s_1}{}^{(R_{i_{n-k}})}Z)\cdot d\!\!/(R)^{s'_2}(T)^m\mu \qquad (12.358)\\
&s'_2=s_2\bigcup\{1,\cdots,n-k-1\}
\end{aligned}
$$

If $|s_1|\leqslant l_*$ we appeal to Corollary 10.1.i with $l_*+1$ in the role of $l$ to place the first factor in $L^\infty(\Sigma_t^{\epsilon_0})$. Otherwise, $|s_2|\leqslant k-l_*-1$ and

$$
|s'_2|=|s_2|+n-k-1\leqslant n-l_*-2\leqslant l-l_*-1-m\leqslant l_*-m
$$

hence $1 + |s_2'| + m \leqslant l_* + 1$ and we appeal to $\mathbf{M}_{[m,l_*+1]}$ to place the second factor in $L^\infty(\Sigma_t^{\epsilon_0})$. Since also $|s_1| \leqslant k \leqslant n-1 \leqslant l-m$ and

$$1 + |s_2'| + m = |s_2| + n - k + m \leqslant n + m \leqslant l + 1$$

we obtain in this way that the second sum in (12.252) is bounded by

$$\begin{aligned} & C_l\delta_0(1+t)^{-2}(1+\log(1+t))\mathcal{B}_{[m,l+1]} \\ & + C_l\delta_0(1+t)^{-1}(1+\log(1+t))\max_i \|{}^{(R_i)}Z\|_{2,[l-m],\Sigma_t^{\epsilon_0}} \\ \leqslant\ & C_l\delta_0(1+t)^{-1}(1+\log(1+t)) \\ & \cdot((1+t)^{-1}\mathcal{B}_{[m,l+1]} + \mathcal{Y}_0 + (1+t)\mathcal{A}_{[l]} + \mathcal{W}_{\{l+1\}}) \end{aligned} \tag{12.359}$$

by Corollary 10.2.i with $l+1$ in the role of $l$.

In view of (12.347), (12.352), (12.355) and (12.357) and also substituting for $\mathcal{A}_{[l]}$ in terms of $\mathcal{A}'_{[l]}$ from Lemma 12.5 we deduce

$$\begin{aligned} & \|L^{(i_1\cdots i_n)}\mu_{m,n}\|_{L^2(\Sigma_t^{\epsilon_0})} \\ \leqslant\ & C_l\delta_0(1+t)^{-2}(1+\log(1+t))\mathcal{B}_{[m,l+1]} \\ & + C_l(\mathcal{W}_{\{l+2\}} + (1+t)^{-1}(1+\log(1+t))(\mathcal{W}^Q_{\{l+1\}} + \delta_0(\mathcal{Y}_0 + (1+t)\mathcal{A}'_{[l]}))) \end{aligned} \tag{12.360}$$

for $n \leqslant l + 1 - m$.

Defining

$${}^{(i_1\cdots i_n)}\phi'_{m,n} = |{}^{(i_1\cdots i_n)}\mu_{m,n}| \tag{12.361}$$

$${}^{(i_1\cdots i_n)}\rho'_{m,n} = |L^{(i_1\cdots i_n)}\mu_{m,n}| \tag{12.362}$$

obviously, we have

$$L^{(i_1\cdots i_n)}\phi'_{m,n} \leqslant {}^{(i_1\cdots i_n)}\rho'_{m,n} \tag{12.363}$$

Following an argument similar to that leading to (12.327), we deduce

$$\begin{aligned} & (1+t)^{-1}\|{}^{(i_1\cdots i_n)}\phi'_{m,n}\|_{L^2(\Sigma_t^{\epsilon_0})} \\ \leqslant\ & C(\|{}^{(i_1\cdots i_n)}\phi'_{m,n}\|_{L^2(\Sigma_0^{\epsilon_0})} + \int_0^t (1+t')^{-1}\|{}^{(i_1\cdots i_n)}\rho'_{m,n}\|_{L^2(\Sigma_{t'}^{\epsilon_0})}dt') \end{aligned} \tag{12.364}$$

Taking on both sides the maximum over $i_1\cdots i_n$, replacing $m$ by $k \in \{0,\cdots,m\}$ and then summing over $n \in \{0,\cdots,l+1-k\}$ and over $k \in \{0,\cdots,m\}$ yields

$$\begin{aligned} & (1+t)^{-1}\sum_{k=0}^{m}\sum_{n=0}^{l+1-k}\max_{i_1\cdots i_n}\|{}^{(i_1\cdots i_n)}\phi'_{k,n}\|_{L^2(\Sigma_t^{\epsilon_0})} \\ \leqslant\ & C(\sum_{k=0}^{m}\sum_{n=0}^{l+1-k}\|{}^{(i_1\cdots i_n)}\phi'_{k,n}\|_{L^2(\Sigma_0^{\epsilon_0})} \\ & + \int_0^t (1+t')^{-1}\sum_{k=0}^{m}\sum_{n=0}^{l+1-k}\|{}^{(i_1\cdots i_n)}\rho'_{k,n}\|_{L^2(\Sigma_{t'}^{\epsilon_0})}dt') \end{aligned} \tag{12.365}$$

From (12.361) we have

$$\sum_{k=0}^{m}\sum_{n=0}^{l+1-k}\max_{i_1\cdots i_n}\|^{(i_1\cdots i_n)}\phi'_{k,n}\|_{L^2(\Sigma_t^{\epsilon_0})}=\mathcal{B}_{[m,l+1]}(t) \tag{12.366}$$

From (12.360) and (12.362) we have

$$\begin{aligned}&\sum_{k=0}^{m}\sum_{n=0}^{l+1-k}\|^{(i_1\cdots i_n)}\rho'_{k,n}\|_{L^2(\Sigma_t^{\epsilon_0})}\\ &\leqslant C_l\delta_0(1+t)^{-2}(1+\log(1+t))\mathcal{B}_{[m,l+1]}\\ &\quad+C_l(\mathcal{W}_{\{l+2\}}+(1+t)^{-1}(1+\log(1+t))(\mathcal{W}^Q_{\{l+1\}}+\delta_0(\mathcal{Y}_0+(1+t)\mathcal{A}'_{[l]})))\end{aligned} \tag{12.367}$$

Substitute (12.366) and (12.367) in (12.365) we obtain

$$\begin{aligned}&(1+t)^{-1}\mathcal{B}_{[m,l+1]}(t)\\ &\leqslant C\mathcal{B}_{[m,l+1]}(0)+\int_0^t C_l\delta_0(1+t')^{-2}(1+\log(1+t'))\\ &\quad\cdot(1+t')^{-1}\mathcal{B}_{[m,l+1]}(t')dt'\\ &\quad+\int_0^t C_l\delta_0(1+t')^{-2}(1+\log(1+t'))\mathcal{Y}_0(t')dt'\\ &\quad+\int_0^t C_l(1+t')^{-1}(\mathcal{W}_{\{l+2\}}(t')+(1+t')^{-1}(1+\log(1+t'))\mathcal{W}^Q_{\{l+1\}}(t'))dt'\\ &\quad+\int_0^t C_l\delta_0(1+t')^{-1}(1+\log(1+t'))\mathcal{A}'_{[l]}(t')dt'\end{aligned} \tag{12.368}$$

This implies

$$\begin{aligned}&(1+t)^{-1}\mathcal{B}_{[m,l+1]}(t)\\ &\leqslant C\mathcal{B}_{[m,l+1]}(0)+\int_0^t C_l\delta_0(1+t')^{-2}(1+\log(1+t'))\mathcal{Y}_0(t')dt'\\ &\quad+\int_0^t C_l(1+t')^{-1}(\mathcal{W}_{\{l+2\}}(t')+(1+t')^{-1}(1+\log(1+t'))\mathcal{W}^Q_{\{l+1\}}(t'))dt'\\ &\quad+\int_0^t C_l\delta_0(1+t')^{-1}(1+\log(1+t'))\mathcal{A}'_{[l]}(t')dt'\end{aligned} \tag{12.369}$$

From Proposition 12.11, we have

$$\begin{aligned}&\int_0^t(1+t')^{-1}(1+\log(1+t'))\mathcal{A}'_{[l]}(t')dt'\\ &\leqslant C_l(\mathcal{A}'_{[l]}(0)+\delta_0\mathcal{Y}_0(0)+\int_0^t(1+t')^{-1}(\mathcal{W}_{[l+1]}(t')+\mathcal{W}^Q_{[l]}(t'))dt')\end{aligned} \tag{12.370}$$

Using then also (12.348), the proposition follows. □

# 13

# Derivation of the Basic Properties of $\mu$

In this chapter we shall establish the assumptions **C1**, **C2**, **C3** introduced in Chapter 5, on which Theorem 5.1 relies. To do this, we shall use Proposition 8.6. The assumption of this proposition are those of Proposition 8.5 together with assumption **I0** in Chapter 6, which coincide with the assumption of Proposition 12.1 on the initial data. While the assumptions of Proposition 8.5 are the assumptions of Lemma 8.10 together with the bootstrap assumptions $\mathbf{E_{LT}3}$, $\mathbf{E_{LL}3}$. Finally, the assumptions of Lemma 8.10 are basic bootstrap assumptions **A1**, **A2**, **A3**, introduced in Chapter 5, the bootstrap assumptions **E1**, **E2**, **F2** of Chapter 6, together with the additional bootstrap assumption $\not{\mathbf{E}}\mathbf{3}_0$. Now, **A1** and **A2** follow from **E1**, while **A3** follows from **A2** and Proposition 12.1. **F2** coincides with $\not{\mathbf{X}}'_0$ which has been established by Proposition 12.6. **E1** is $\mathbf{E}_{\{0\}}$ while **E2** follows from $\mathbf{E}_{\{1\}}, \mathbf{E}^Q_{\{0\}}$, and **H0**, which has been established by Corollary 12.2.a. $\mathbf{E_{LL}3}$ follows from $\mathbf{E}^{QQ}_{\{0\}}$ and $\mathbf{E}^Q_{\{0\}}$, while $\mathbf{E_{LT}3}$ follows from $\mathbf{E}^Q_{\{1\}}, \mathbf{E}_{\{1\}}$, and the estimates for $\|\Lambda\|_{L^\infty(\Sigma_t^{\epsilon_0})}$ of Corollary 11.1.c with $m = l = 0$ as well as **H0**. The assumptions of Corollary 11.1.c are those of Proposition 11.1 and follow from Proposition 12.10. Finally, $\not{\mathbf{E}}\mathbf{3}_0$ follows from $\mathbf{E}_{\{2\}}$ and **H0** and **H1**, **H1** having been established by Corollary 12.5. We conclude from above that the assumptions of Proposition 8.6 all follow from the assumptions of Proposition 12.10 with $m = l = 0$, namely the assumptions of Proposition 12.9 with $l = 0$, which coincide with those of Proposition 12.6, together with

$$\|\mu - 1\|_{\infty,[0,1],\Sigma_0^{\epsilon_0}} \leqslant C\delta_0 \tag{13.1}$$

**Proposition 13.1** Let the assumptions of Proposition 12.6 hold and let the initial data satisfy (13.1). Then on $W^s_{\epsilon_0}$ we have

$$\mu^{-1}(L\mu)_+ \leqslant (1+t)^{-1}(1+\log(1+t))^{-1} + A(t)$$

where

$$A(t) = C\delta_0(1+t)^{-2}(1+\log(1+t))$$

Since

$$\int_0^s A(t)dt \leqslant C\delta_0$$

$C$ being a constant independent of $s$, we conclude that **C1** holds on $W^s_{\epsilon_0}$.

*Proof.* We appeal to Proposition 8.6 to express, in acoustical coordinates,

$$\mu^{-1}(L\mu)_+ = \frac{1}{\hat{\mu}_s}(\frac{\partial\hat{\mu}_s}{\partial t})_+ = \frac{(\hat{E}_s(u,\vartheta)(1+t)^{-1} + \hat{Q}_{1,s}(t,u,\vartheta))_+}{1+\hat{E}_s(u,\vartheta)\log(1+t) + \hat{Q}_{0,s}(t,u,\vartheta)} \tag{13.2}$$

Consider a given $(u,\vartheta) \in [0,\epsilon_0] \times S^2$. There are three cases to consider:
Case 1: $\hat{E}_s(u,\vartheta) = 0$, Case 2: $\hat{E}_s(u,\vartheta) > 0$, Case 3: $\hat{E}_s(u,\vartheta) < 0$.
In Case 1, (13.2) reduces to

$$\frac{1}{\hat{\mu}_s}(\frac{\partial\hat{\mu}_s}{\partial t})_+ = \frac{(\hat{Q}_{1,s}(t,u,\vartheta))_+}{1+\hat{Q}_{0,s}(t,u,\vartheta)} \tag{13.3}$$

From Proposition 8.6,

$$|\hat{Q}_{0,s}(t,u,\vartheta)| \leqslant C\delta_0\frac{1+\log(1+t)}{1+t} \tag{13.4}$$

$$|\hat{Q}_{1,s}(t,u,\vartheta)| \leqslant C\delta_0\frac{1+\log(1+t)}{(1+t)^2} \tag{13.5}$$

these imply

$$\frac{1}{\hat{\mu}_s}(\frac{\partial\hat{\mu}_s}{\partial t})_+ \leqslant C\delta_0\frac{1+\log(1+t)}{(1+t)^2} \tag{13.6}$$

In Case 2, we have

$$(\hat{E}_s(u,\vartheta)(1+t)^{-1} + \hat{Q}_{1,s}(t,u,\vartheta))_+ \leqslant \hat{E}_s(u,\vartheta)(1+t)^{-1} + |\hat{Q}_{1,s}(t,u,\vartheta)|$$

hence, substituting in (13.2),

$$\begin{aligned}\frac{1}{\hat{\mu}_s}(\frac{\partial\hat{\mu}_s}{\partial t})_+ \leqslant &\frac{\hat{E}_s(u,\vartheta)(1+t)^{-1} + |\hat{Q}_{1,s}(t,u,\vartheta)|}{1+\hat{E}_s(u,\vartheta)\log(1+t) + \hat{Q}_{0,s}(t,u,\vartheta)}\\ = &\frac{\hat{E}_s(u,\vartheta)(1+t)^{-1}}{1+\hat{E}_s(u,\vartheta)\log(1+t)}\\ &+ \frac{\hat{E}_s(u,\vartheta)}{1+t}(\frac{1}{1+\hat{E}_s(u,\vartheta)\log(1+t) + \hat{Q}_{0,s}(t,u,\vartheta)}\\ &- \frac{1}{1+\hat{E}_s(u,\vartheta)\log(1+t)})\\ &+ \frac{|\hat{Q}_{1,s}(t,u,\vartheta)|}{1+\hat{E}_s(u,\vartheta)\log(1+t) + \hat{Q}_{0,s}(t,u,\vartheta)}\end{aligned} \tag{13.7}$$

Since, for suitably small $\delta_0$,

$$|\hat{E}_s(u,\vartheta)| \leqslant C\delta_0 \leqslant 1 \tag{13.8}$$

the first term on the right of (13.7) is bounded by

$$\frac{1}{(1+t)(1+\log(1+t))} \tag{13.9}$$

The factor in parenthesis in the second term on the right in (13.7) is bounded in absolute value by

$$\frac{|\hat{Q}_{0,s}(t,u,\vartheta)|}{(1+\hat{E}_s(u,\vartheta)\log(1+t)+\hat{Q}_{0,s}(t,u,\vartheta))(1+\hat{E}_s(u,\vartheta)\log(1+t))}$$
$$\leqslant \frac{1}{(1+\hat{E}_s(u,\vartheta)\log(1+t))}\frac{|\hat{Q}_{0,s}(t,u,\vartheta)|}{(1-|\hat{Q}_{0,s}(t,u,\vartheta)|)}$$

therefore, from (13.4), taking $\delta_0$ suitably small, the second term on the right of (13.7) is bounded in absolute value by

$$\frac{1}{(1+t)(1+\log(1+t))}\frac{|\hat{Q}_{0,s}(t,u,\vartheta)|}{1-|\hat{Q}_{0,s}(t,u,\vartheta)|} \leqslant \frac{C\delta_0}{(1+t)^2} \tag{13.10}$$

Also, by (13.5) and (13.4), for small $\delta_0$, the third term on the right in (13.7) is bounded in absolute value by

$$C\delta_0\frac{1+\log(1+t)}{(1+t)^2} \tag{13.11}$$

Combining (13.9)—(13.11), we conclude in Case 2

$$\frac{1}{\hat{\mu}_s}(\frac{\partial\hat{\mu}_s}{\partial t})_+ \leqslant \frac{1}{(1+t)(1+\log(1+t))} + C\delta_0\frac{1+\log(1+t)}{(1+t)^2} \tag{13.12}$$

Finally, we consider Case 3. In this case we define

$$t_1(u,\vartheta) = e^{-\frac{1}{2\hat{E}_s(u,\vartheta)}} - 1 \tag{13.13}$$

There are two subcases to consider:
Subcase 3a: $t \leqslant t_1(u,\vartheta)$, Subcase 3b: $t > t_1(u,\vartheta)$.
In Subcase 3a we have

$$\log(1+t) \leqslant -\frac{1}{2\hat{E}_s(u,\vartheta)}$$

hence

$$1+\hat{E}_s(u,\vartheta)\log(1+t) \geqslant \frac{1}{2} \tag{13.14}$$

From (13.4), for suitably small $\delta_0$,

$$|\hat{Q}_{0,s}(t,u,\vartheta)| \leqslant \frac{1}{4} \quad \text{for all} \quad t\in[0,s] \tag{13.15}$$

it follows that

$$\hat{\mu}_s(t,u,\vartheta) = 1 + \hat{E}_s(u,\vartheta)\log(1+t) + \hat{Q}_{0,s}(t,u,\vartheta) \geqslant \frac{1}{4} \tag{13.16}$$

hence, from (13.2),

$$\begin{aligned} \frac{1}{\hat{\mu}_s}\left(\frac{\partial \hat{\mu}_s}{\partial t}\right)_+ &\leqslant 4\left(\frac{\hat{E}_s(u,\vartheta)}{1+t} + \hat{Q}_{1,s}(t,u,\vartheta)\right)_+ \\ &\leqslant 4|\hat{Q}_{1,s}(t,u,\vartheta)| \\ &\leqslant C\delta_0 \frac{1+\log(1+t)}{(1+t)^2} \end{aligned} \tag{13.17}$$

by (13.5).

In Subcase 3b we have, by (13.5),

$$\begin{aligned} \frac{\partial \hat{\mu}_s}{\partial t}(t,u,\vartheta) &= \frac{\hat{E}_s(u,\vartheta)}{1+t} + \hat{Q}_{1,s}(t,u,\vartheta) \\ &\leqslant \frac{1}{1+t}\left(\hat{E}_s(u,\vartheta) + C\delta_0 \frac{1+\log(1+t)}{1+t}\right) \end{aligned} \tag{13.18}$$

Let us set

$$\tau = \log(1+t) \quad \text{so that} \quad t = e^\tau - 1 \tag{13.19}$$

The mapping $t \mapsto \tau$ is an orientation preserving diffeomorphism of $[0,\infty)$ onto itself. Also,

$$\frac{1+\log(1+t)}{1+t} = e^{-\tau}(1+\tau) := f(\tau) \tag{13.20}$$

is a decreasing function of $\tau$. Thus $t > t_1(u,\vartheta)$ corresponds to $\tau > \tau_1(u,\vartheta)$, where

$$\tau_1(u,\vartheta) = -\frac{1}{2\hat{E}_s(u,\vartheta)} \tag{13.21}$$

and to

$$f(\tau) < f(\tau_1(u,\vartheta)) = \left(1 - \frac{1}{2\hat{E}_s(u,\vartheta)}\right)e^{\frac{1}{2\hat{E}_s(u,\vartheta)}} \tag{13.22}$$

It then follows from (13.18) that in Subcase 3b

$$\begin{aligned} \frac{\partial \hat{\mu}_s}{\partial t}(t,u,\vartheta) &< \frac{1}{1+t}\left(\hat{E}_s(u,\vartheta) + C\delta_0\left(1 - \frac{1}{2\hat{E}_s(u,\vartheta)}\right)e^{\frac{1}{2\hat{E}_s(u,\vartheta)}}\right) \\ &= \frac{\hat{E}_s(u,\vartheta)}{1+t}(1 - 2C\delta_0 g(x)) \end{aligned} \tag{13.23}$$

where we set

$$x = -2\hat{E}_s(u,\vartheta) > 0 \quad \text{and} \quad g(x) = \frac{1}{x}\left(1+\frac{1}{x}\right)e^{-\frac{1}{x}} \tag{13.24}$$

Now, $0 < x \leqslant C\delta_0$, the function $g(x)$ is bounded on $(0,1]$, and we have

$$\lim_{x \to 0_+} g(x) = 0 \tag{13.25}$$

It follows that, for small $\delta_0$

$$1 - 2C\delta_0 g(x) > 0$$

hence from (13.23),

$$\frac{\partial \hat{\mu}_s}{\partial t}(t, u, \vartheta) < 0 \tag{13.26}$$

Therefore in Subcase 3b,

$$\frac{1}{\hat{\mu}_s}(\frac{\partial \hat{\mu}_s}{\partial t})_+ = 0 \tag{13.27}$$

In view of (13.17) and (13.27), we conclude that in Case 3,

$$\frac{1}{\hat{\mu}_s}(\frac{\partial \hat{\mu}_s}{\partial t})_+ \leqslant C\delta_0 \frac{1 + \log(1+t)}{(1+t)^2} \tag{13.28}$$

Combining (13.6), (13.12), (13.28), we conclude that in $W^s_{\epsilon_0}$,

$$\frac{1}{\hat{\mu}_s}(\frac{\partial \hat{\mu}_s}{\partial t})_+ \leqslant \frac{1}{(1+t)(1+\log(1+t))} + A(t) \tag{13.29}$$

where

$$A(t) = C\delta_0 \frac{1 + \log(1+t)}{(1+t)^2} \tag{13.30}$$

So the proposition follows. □

**Proposition 13.2** Let the assumptions of Proposition 12.10 hold for $m = 2, l = 1$. Then on $W^s_{\epsilon_0}$ we have

$$\mu^{-1}(T\mu)_+ \leqslant B_s(t)$$

Here

$$B_s(t) = C\sqrt{\delta_0}\frac{1+\tau}{\sqrt{\sigma - \tau}} + C\delta_0(1+\tau)$$

where $\tau = \log(1+t), \sigma = \log(1+s)$. We have

$$\int_0^s (1+t)^{-2}(1+\log(1+t))^4 B_s(t)dt \leqslant C\sqrt{\delta_0}$$

where $C$ is a constant independent of $s$. Moreover, by **C1**, which we have just proved, **C2** holds on $W^s_{\epsilon_0}$.

*Proof*. We shall again use Proposition 8.6. Denote

$$\hat{E}_{s,m} = \min_{(u,\vartheta)\in[0,\epsilon_0]\times S^2} \hat{E}_s(u, \vartheta) \tag{13.31}$$

There are two cases, according as to whether $\hat{E}_{s,m} \geqslant 0$, or $\hat{E}_{s,m} < 0$. The first case is easier to treat. In the second case, we set

$$\hat{E}_{s,m} = -\delta_1, \quad \delta_1 > 0 \tag{13.32}$$

We define

$$\mathcal{V}'_{s-} = \{(u,\vartheta) \in [0,\epsilon_0] \times S^2 : \hat{E}_s(u,\vartheta) < -\frac{\delta_1}{2}\} \tag{13.33}$$

We denote by $\mathcal{U}'_{s,t-}$ the corresponding open subset of $\Sigma_t^{\epsilon_0}$, namely the set of points on $\Sigma_t^{\epsilon_0}$ with acoustical coordinates $(t,u,\vartheta)$, such that $(u,\vartheta) \in \mathcal{V}'_{s-}$. Now, from Proposition 8.6,

$$\hat{\mu}_s(s,u,\vartheta) = 1 + \hat{E}_s(u,\vartheta)\log(1+s) \tag{13.34}$$

hence, with $\sigma = \log(1+s)$,

$$0 \leqslant \min_{(u,\vartheta)\in[0,\epsilon_0]\times S^2} \hat{\mu}_s(s,u,\vartheta) = 1 - \delta_1\sigma \tag{13.35}$$

Consider a point $p \in \Sigma_t^{\epsilon_0} \setminus \mathcal{U}'_{s,t-}$. The acoustical coordinates of $p$ are $(t,u,\vartheta)$ with $(u,\vartheta) \in ([0,\epsilon_0] \times S^2) \setminus \mathcal{V}'_{s-}$. With $\tau = \log(1+t)$ we then have, by Proposition 8.6 and (13.35), since $\tau \leqslant \sigma$,

$$\begin{aligned}\hat{\mu}_s(t,u,\vartheta) =& 1 + \hat{E}_s(u,\vartheta)\log(1+t) + \hat{Q}_{0,s}(t,u,\vartheta)\\ \geqslant& 1 - \frac{1}{2}\delta_1\tau - |\hat{Q}_{0,s}(t,u,\vartheta)| \geqslant \frac{1}{2} - |\hat{Q}_{0,s}(t,u,\vartheta)| \geqslant \frac{1}{4}\end{aligned} \tag{13.36}$$

by (13.15). Since

$$\mu(p) = \mu_{[1],s}(u,\vartheta)\hat{\mu}_s(t,u,\vartheta) \tag{13.37}$$

and by Proposition 8.6,

$$\inf_{(u,\vartheta)\in[0,\epsilon_0]\times S^2} \mu_{[1],s}(u,\vartheta) \geqslant \frac{1}{2} \tag{13.38}$$

it follows that

$$\mu(p) \geqslant \frac{1}{8} \tag{13.39}$$

In the case that $\hat{E}_{s,m} \geqslant 0$, by Proposition 8.6 and (13.4) we have in this case

$$\hat{\mu}_s(t,u,\vartheta) \geqslant 1 + \hat{Q}_{0,s}(t,u,\vartheta) \geqslant 1 - |\hat{Q}_{0,s}(t,u,\vartheta)| \geqslant \frac{3}{4} \tag{13.40}$$

hence by (13.37) and (13.38),

$$\mu(p) \geqslant \frac{3}{8} \tag{13.41}$$

According to Proposition 12.10 with $m = 2, l = 1, \mathbf{M}_{[2,2]}$ holds on $W_{\epsilon_0}^s$. In particular, we have

$$\sup_{\Sigma_t^{\epsilon_0}} |T\mu| \leqslant C\delta_0(1+\log(1+t)) \quad \text{for all} \quad t \in [0,s] \tag{13.42}$$

In view of (13.39), (13.41) and (13.42), it follows that

$$(\mu^{-1}(T\mu)_+)(p) \leqslant C\delta_0(1+\log(1+t)) \tag{13.43}$$

for any $p \in \Sigma_t^{\epsilon_0}, t \in [0,s]$, in the case that $\hat{E}_{s,m} \geqslant 0$, for any $p \in \Sigma_t^{\epsilon_0} \setminus \mathcal{U}'_{s,t-}, t \in [0,s]$, in the case that $\hat{E}_{s,m} < 0$.

It remains for us to consider the case where $\hat{E}_{s,m} < 0$ and $p \in \mathcal{U}'_{s,t-}$. Let us set

$$\tau_1 = \frac{1}{2\delta_1}, \quad t_1 = e^{\tau_1} - 1 \tag{13.44}$$

There are two subcases to consider, according as to whether $t \leqslant t_1$ or $t > t_1$. In the first subcase, which corresponds to $\tau \leqslant \tau_1$, we have, by Proposition 8.6 and (13.15),

$$\begin{aligned}\hat{\mu}_s(t,u,\vartheta) &= 1 + \hat{E}_s(u,\vartheta)\log(1+t) + \hat{Q}_{0,s}(t,u,\vartheta) \\ &\geqslant 1 - \delta_1\tau - |\hat{Q}_{0,s}(t,u,\vartheta)| \geqslant \frac{1}{2} - |\hat{Q}_{0,s}(t,u,\vartheta)| \geqslant \frac{1}{4}\end{aligned} \tag{13.45}$$

By (13.37), (13.38), the lower bound (13.39) follows, which, together with (13.42) yields

$$(\mu^{-1}(T\mu)_+)(p) \leqslant C\delta_0(1+\log(1+t)) \tag{13.46}$$

for $p \in \mathcal{U}'_{s,t-}$, in the subcase $t \leqslant t_1$.

Finally, we must consider the subcase $p \in \mathcal{U}'_{s,t-}, t > t_1$. If $(T\mu)(p) \leqslant 0$, then $((T\mu))_+(p) = 0$ and the estimate is trivial. Otherwise $(T\mu)(p) > 0$, and we consider the integral curve of $T$, on $\Sigma_t^{\epsilon_0}$, through $p$. The intersection of $\mathcal{U}'_{s,t-}$ with this integral curve is an open subset of the integral curve, corresponding to an open subset $\mathcal{J}$ of the parameter interval $(0,\epsilon_0]$ (the integral curves of $T$ are parametrized by $u$). Thus $\mathcal{J}$ is a countable union of open intervals. The point 0 is not a boundary point of $\mathcal{J}$. Let the point $p$ correspond to the parameter value $u_*$. Then $u_* \in \mathcal{J}$. Consider the interval $\mathcal{I}$, the component of $\mathcal{J}$ to which $u_*$ belongs. Now $\mathcal{I}$ is of the form $\mathcal{I} = (a,b), a > 0, b < \epsilon_0$, if $\mathcal{I}$ is not the rightmost interval, while either $\mathcal{I} = (a,\epsilon_0]$ or $\mathcal{I} = (a,\epsilon_0), a > 0$, if $\mathcal{I}$ is the rightmost interval. Consider the function $\mu$ along the given integral curve as a function of $u$. Then $T\mu$ is the function $\frac{d\mu}{du}$, and $T^2\mu$ is the function $\frac{d^2\mu}{du^2}$ along the same integral curve. Consider the subinterval $(a,u_*]$ of $\mathcal{I}$. Then either (Case 1) $\frac{d\mu}{du} > 0$ throughout $(a,u_*]$, or (Case 2) there is a rightmost value of $u$ in $(a,u_*]$ where $\frac{d\mu}{du} = 0$, and, denoting this rightmost value by $\bar{u}$, we have $\frac{d\mu}{du} > 0$ on $(\bar{u},u_*]$.

In Case 1 we have

$$\mu(u_*) > \mu(a) \tag{13.47}$$

Now $a \notin \mathcal{J}$, therefore the point $q$ at parameter value $a$ along the integral curve in question belongs to $\Sigma_t^{\epsilon_0} \setminus \mathcal{U}'_{s,t-}$. Consequently the lower bound (13.39) holds at $q$:

$$\mu(q) \geqslant \frac{1}{8} \tag{13.48}$$

Since according to (13.47) $\mu(u_*) > \mu(a)$, this implies

$$\mu(p) > \frac{1}{8} \tag{13.49}$$

which together with (13.42) yields

$$(\mu^{-1}(T\mu)_+)(p) \leqslant C\delta_0(1 + \log(1 + t)) \tag{13.50}$$

In Case 2 the mean value of $\frac{d^2\mu}{du^2}$ on $[\bar{u}, u_*]$ is positive. Denoting then by

$$\gamma(t) = \sup_{\Sigma_t^{\epsilon_0}} (T)^2\mu \tag{13.51}$$

we have $\gamma(t) > 0$. Consider then some $u \in [\bar{u}, u_*]$. We have

$$\frac{d\mu}{du}(u_*) - \frac{d\mu}{du}(u) = \int_u^{u_*} \frac{d^2\mu}{du^2}(u')du'$$

that is

$$(T\mu)(u_*) - (T\mu)(u) = \int_u^{u_*} ((T)^2\mu)(u')du' \leqslant \gamma(t)(u_* - u) \tag{13.52}$$

Therefore

$$(T\mu)(u) \geqslant (T\mu)(u_*) - \gamma(t)(u_* - u) \geqslant \frac{1}{2}(T\mu)(u_*) \quad \text{for all} \quad u \in [u_1, u_*] \tag{13.53}$$

where

$$u_1 = u_* - \frac{(T\mu)(u_*)}{2\gamma(t)} \tag{13.54}$$

This shows in particular that $\bar{u} < u_1$, for otherwise (13.53) would apply in the case $u = \bar{u}$, yielding $(T\mu)(\bar{u}) \geqslant \frac{1}{2}(T\mu)(u_*) > 0$, contradicting the fact that $(T\mu)(\bar{u}) = 0$. Since $T\mu > 0$ throughout $(\bar{u}, u_*]$, in particular on $(\bar{u}, u_1]$, we have

$$\mu(u_1) > \mu(\bar{u}) \tag{13.55}$$

Moreover, by (13.53), (13.54), we have

$$\mu(u_*) - \mu(u_1) = \int_{u_1}^{u_*} (T\mu)(u)du \geqslant \frac{1}{2}(T\mu)(u_*)(u_* - u_1) = \frac{((T\mu)(u_*))^2}{4\gamma(t)} \tag{13.56}$$

Denoting

$$\mu_m(t) = \min_{\Sigma_t^{\epsilon_0}} \mu \tag{13.57}$$

we have

$$\mu(\bar{u}) \geqslant \mu_m(t) \tag{13.58}$$

Combining (13.55), (13.56) and (13.58), we conclude that

$$\mu(p) - \mu_m(t) \geqslant \frac{((T\mu)(p))^2}{4\gamma(t)} \tag{13.59}$$

We thus obtain

$$(\mu^{-1}(T\mu)_+)(p) = \frac{(T\mu)(p)}{\mu(p)} \leqslant \frac{(T\mu)(p)}{\frac{((T\mu)(p))^2}{4\gamma(t)} + \mu_m(t)} = 2\sqrt{\gamma(t)} f_\epsilon(x) \tag{13.60}$$

where

$$x = \frac{(T\mu)(p)}{2\sqrt{\gamma(t)}} > 0, \quad \epsilon = \sqrt{\mu_m(t)} > 0 \tag{13.61}$$

$f_\epsilon(x)$ is the function

$$f_\epsilon(x) = \frac{x}{x^2 + \epsilon^2} \tag{13.62}$$

Now, the function $f_\epsilon(x)$ achieves its maximum at $x = \epsilon$ and the maximum value is $\frac{1}{2\epsilon}$. Thus

$$f_\epsilon(x) \leqslant \frac{1}{2\epsilon} \tag{13.63}$$

hence from (13.60),

$$(\mu^{-1}(T\mu)_+)(p) \leqslant \frac{\sqrt{\gamma(t)}}{\sqrt{\mu_m(t)}} \tag{13.64}$$

Consider now $\mu_m(t)$. From Proposition 8.6 we have, by (13.37), (13.38),

$$\begin{aligned} \mu(t, u, \vartheta) &\geqslant \frac{1}{2}\hat{\mu}_s(t, u, \vartheta) = \frac{1}{2}(1 + \hat{E}_s(u, \vartheta)\log(1 + t) + \hat{Q}_{0,s}(t, u, \vartheta)) \\ &\geqslant \frac{1}{2}(1 - \delta_1 \tau - |\hat{Q}_{0,s}|) \end{aligned} \tag{13.65}$$

Since $\delta_1 \sigma \leqslant 1$, this implies

$$\mu_m(t) \geqslant \frac{1}{2}(\delta_1(\sigma - \tau) - |\hat{Q}_{0,s}|) \tag{13.66}$$

From Proposition 8.6 we have

$$|\hat{Q}_{0,s}(t, u, \vartheta)| \leqslant C\delta_0 b(t, s) \tag{13.67}$$

where

$$b(t, s) = \frac{1 + \log(1 + t)}{1 + t} \frac{s - t}{1 + s} \tag{13.68}$$

According to (8.288),

$$0 \leqslant b(t, s) \leqslant \frac{1 + \tau}{e^\tau}(\sigma - \tau) \tag{13.69}$$

Substituting (13.67), (13.69), in (13.66) we obtain

$$\mu_m(t) \geqslant \frac{1}{2}\delta_1(1 - C\frac{\delta_0}{\delta_1}\frac{1+\tau}{e^\tau})(\sigma - \tau) \tag{13.70}$$

Since we are considering the subcase $\tau > \tau_1$, we have

$$\frac{1+\tau}{e^\tau} < \frac{1+\tau_1}{e^{\tau_1}} = \frac{1+\frac{1}{2\delta_1}}{e^{\frac{1}{2\delta_1}}}$$

The factor

$$\frac{1}{\delta_1}(1 + \frac{1}{2\delta_1})e^{-\frac{1}{2\delta_1}}$$

is bounded and tends to zero as $\delta_1 \to 0$. Therefore, if $\delta_0$ is suitably small we have

$$1 - C\frac{\delta_0}{\delta_1}\frac{1+\tau}{e^\tau} \geqslant \frac{1}{2}$$

Since also $2\delta_1 \geqslant \frac{1}{\tau}$, it follows from (13.70) that

$$\mu_m(t) \geqslant \frac{1}{8}\frac{\sigma - \tau}{1+\tau} \tag{13.71}$$

On the other hand, by $\mathbf{M}_{[2,2]}$ we have

$$\gamma(t) \leqslant C\delta_0(1+\tau) \tag{13.72}$$

In view of (13.71), (13.72), we conclude from (13.64)

$$(\mu^{-1}(T\mu)_+)(p) \leqslant C\sqrt{\delta_0}\frac{1+\tau}{\sqrt{\sigma - \tau}} \tag{13.73}$$

This is the bound in Case 2. Combining this with (13.50), we conclude that for any $p \in \mathcal{U}'_{s,t-}, t \in (t_1, s]$, we have

$$(\mu^{-1}(T\mu)_+)(p) \leqslant C\sqrt{\delta_0}\frac{1+\tau}{\sqrt{\sigma - \tau}} + C\delta_0(1+\tau) \tag{13.74}$$

Combining this with (13.46) and (13.43), we conclude that, in general,

$$\mu^{-1}(T\mu)_+ \leqslant B_s(t) \tag{13.75}$$

where

$$B_s(t) = C\sqrt{\delta_0}\frac{1+\tau}{\sqrt{\sigma - \tau}} + C\delta_0(1+\tau) \tag{13.76}$$

We have

$$\int_0^s (1+t)^{-2}(1+\log(1+t))^4 B_s(t)dt = \int_0^\sigma (1+\tau)^4 B_s(t)e^{-\tau}d\tau$$

and

$$\int_0^{\frac{\sigma}{2}} \frac{(1+\tau)^5}{\sqrt{\sigma-\tau}} e^{-\tau} d\tau \leqslant C \int_0^{\frac{\sigma}{2}} \frac{(1+\tau)^5}{\sqrt{\tau}} e^{-\tau} d\tau$$
$$\leqslant C \int_0^{\infty} \frac{(1+\tau)^5}{\sqrt{\tau}} e^{-\tau} d\tau = C \quad \text{independent of} \quad s$$

while

$$\int_{\frac{\sigma}{2}}^{\sigma} \frac{(1+\tau)^5}{\sqrt{\sigma-\tau}} e^{-\tau} d\tau \leqslant (1+\sigma)^5 e^{-\frac{\sigma}{2}} \int_{\frac{\sigma}{2}}^{\sigma} \frac{d\tau}{\sqrt{\sigma-\tau}}$$
$$= \sqrt{2\sigma}(1+\sigma)^5 e^{-\frac{\sigma}{2}} \leqslant C \quad \text{independent of} \quad s$$

It follows that

$$\int_0^s (1+t)^{-2}(1+\log(1+t))^4 B_s(t) dt \leqslant C\sqrt{\delta_0}$$

where $C$ is a constant independent of $s$.

Consider now $\mu^{-1}(L\mu + \underline{L}\mu)_+$. Since $\underline{L} = \eta^{-2}\mu L + 2T$, we have

$$\mu^{-1}(L\mu + \underline{L}\mu)_+ \leqslant \eta^{-2}|L\mu| + \mu^{-1}(L\mu)_+ + 2\mu^{-1}(T\mu)_+ \tag{13.77}$$

By the propagation equation for $\mu$, the first term on the right in (13.77) is bounded by $C\delta_0(1+t)^{-1}$. The second term on the right is bounded by Proposition 13.1, but we treat Case 2 in a different manner, namely, we use (13.8) to estimate the first term on the right of (13.7) by

$$C\delta_0(1+t)^{-1}$$

We then obtain in Case 2 ($\hat{E}_s(u, \vartheta) > 0$) the bound

$$\frac{1}{\hat{\mu}_s}\left(\frac{\partial \hat{\mu}_s}{\partial t}\right)_+ \leqslant \frac{C\delta_0}{1+t} \tag{13.78}$$

so we have

$$\mu^{-1}(L\mu)_+ \leqslant C\delta_0(1+t)^{-1} \tag{13.79}$$

Together with (13.75) this yields, through (13.77),

$$\mu^{-1}(L\mu + \underline{L}\mu)_+ \leqslant B'_s(t) \quad \text{on} \quad W^s_{\epsilon_0} \tag{13.80}$$

where

$$B'_s(t) = 2B_s(t) + C\delta_0(1+t)^{-1} \tag{13.81}$$

Noting that

$$\int_0^s (1+t)^{-2}(1+\log(1+t))^4 B'_s(t) dt \leqslant C\sqrt{\delta_0} \tag{13.82}$$

where $C$ is a constant independent of $s$, we conclude that **C2** holds on $W^s_{\epsilon_0}$. So Proposition 13.2 holds. □

**Proposition 13.3** Let the assumptions of Proposition 13.1 hold and let $\mathcal{U}$ be the region

$$\mathcal{U} = \{x \in W^*_{\epsilon_0} : \mu < \frac{1}{4}\}$$

Then there is a positive constant $C$ independent of $s$ such that in $\mathcal{U} \bigcap W^s_{\epsilon_0}$ we have

$$L\mu \leqslant -C^{-1}(1+t)^{-1}(1+\log(1+t))^{-1}$$

that is, property **C3** holds on $W^s_{\epsilon_0}$.

*Proof*. Consider a point $p \in \mathcal{U} \bigcap \Sigma_t^{\epsilon_0}, t \in [0, s]$, and let its acoustical coordinates be $(t, u, \vartheta)$. First we show that $\hat{E}_s(u, \vartheta) < 0$. For, otherwise we have, from Proposition 8.6, for suitably small $\delta_0$,

$$\hat{\mu}_s(t, u, \vartheta) \geqslant 1 + \hat{Q}_{0,s}(t, u, \vartheta) \geqslant 1 - C\delta_0 \geqslant \frac{1}{2}$$

hence

$$\mu(t, u, \vartheta) = \mu_{[1],s}(u, \vartheta)\hat{\mu}_s(t, u, \vartheta) \geqslant \frac{1}{4}$$

contradicts with the fact that $p \in \mathcal{U}$. Next we show that $\log(1+t) = \tau > \tau_2$, where

$$\tau_2 = \frac{1}{4\delta}, \quad \delta = -\hat{E}_s(u, \vartheta) > 0 \tag{13.83}$$

For, otherwise we have, from Proposition 8.6, for suitably small $\delta_0$,

$$\hat{\mu}_s(t, u, \vartheta) = 1 - \delta\tau + \hat{Q}_{0,s}(t, u, \vartheta) \geqslant \frac{3}{4} - C\delta_0 \geqslant \frac{1}{2}$$

hence again $\mu(t, u, \vartheta) \geqslant \frac{1}{4}$ in contradiction with the fact that $p \in \mathcal{U}$,

Now, from Proposition 8.6,

$$L\mu = \mu_{[1],s}\frac{\partial \hat{\mu}_s}{\partial t} \tag{13.84}$$

and

$$\frac{\partial \hat{\mu}_s}{\partial t}(t, u, \vartheta) = -\frac{\delta}{1+t} + \hat{Q}_{1,s}(t, u, \vartheta) \tag{13.85}$$

From (13.5),

$$\frac{\partial \hat{\mu}_s}{\partial t} \leqslant -\frac{\delta}{1+t}(1 - \frac{C\delta_0}{\delta}\frac{1+\tau}{e^\tau}) \tag{13.86}$$

Now $\tau > \tau_2$ implies

$$\frac{1+\tau}{e^\tau} < \frac{1+\tau_2}{e^{\tau_2}} = (1 + \frac{1}{4\delta})e^{-\frac{1}{4\delta}}$$

Since the function

$$\frac{1}{\delta}(1 + \frac{1}{4\delta})e^{-\frac{1}{4\delta}}$$

is bounded and tends to zero as $\delta \to 0$, it follows that

$$\frac{C\delta_0}{\delta}\frac{1+\tau}{e^\tau} \leqslant \frac{1}{2}$$

provided that $\delta_0$ is suitably small. Hence, from (13.86),

$$\frac{\partial \hat{\mu}_s}{\partial t}(t,u,\vartheta) \leqslant -\frac{1}{2}\frac{\delta}{1+t} \tag{13.87}$$

Since $\log(1+t) = \tau > \tau_2 = \frac{1}{4\delta}$, this implies

$$\frac{\partial \hat{\mu}_s}{\partial t}(t,u,\vartheta) \leqslant -\frac{1}{8}\frac{1}{(1+t)(1+\log(1+t))} \tag{13.88}$$

Then by (13.84) and the fact that, by Proposition 8.6, $\mu_{[1],s} \geqslant \frac{1}{2}$, it follows that

$$L\mu \leqslant -C^{-1}(1+t)^{-1}(1+\log(1+t))^{-1} \quad \text{with} \quad C = 16 \tag{13.89}$$

So the proposition follows. □

Next, we shall define a function $\omega$, as needed in Chapter 5.

**Proposition 13.4** Let the assumptions of Proposition 12.6 hold and let the initial data satisfy (13.1). Let us define

$$\omega = 2(1+t)$$

Then the function $\omega$ has the properties **D1**, **D2**, **D3**, **D4** and **D5**, required in Chapter 5.

*Proof*. **D1** is obvious. We have

$$L\omega = 2, \quad T\omega = 0, \quad \not{d}\omega = 0 \tag{13.90}$$

By the first of (13.90),

$$L\omega - \frac{\omega}{1-u+t} = 2(1 - \frac{1+t}{1-u+t}) = \frac{-2u}{1-u+t} \tag{13.91}$$

hence

$$|L\omega - \frac{\omega}{1-u+t}| \leqslant \frac{C}{1+t} \tag{13.92}$$

On the other hand, by (5.9),

$$\nu = \frac{1}{2}(\mathrm{tr}\chi + L\log\Omega) = \frac{1}{1-u+t} + \frac{1}{2}(\mathrm{tr}\chi' + L\log\Omega) \tag{13.93}$$

hence, by Proposition 12.6,

$$|\nu - \frac{1}{1-u+t}| \leqslant C\delta_0\frac{1+\log(1+t)}{(1+t)^2} \tag{13.94}$$

Then by (13.92) and (13.94),

$$|L\omega - \nu\omega| \leqslant C(1+t)^{-1}(1+\log(1+t)) \tag{13.95}$$

This implies **D2**. By the first two of (13.90),

$$\underline{L}\omega = 2\eta^{-1}\kappa \tag{13.96}$$

hence, by Proposition 12.1,

$$|\underline{L}\omega| \leqslant C(1+\log(1+t)) \tag{13.97}$$

This implies **D3**. **D4** is obvious from the third of (13.90).

It remains for us to verify **D5**. To compute $\mu\Box_{\tilde{g}}\omega$ we appeal to (3.154) with $f$ replaced by $\omega$. By the third of (13.90), this reduces to

$$-\mu\Omega\Box_{\tilde{g}}\omega = L(\underline{L}\omega) + \nu\underline{L}\omega + \underline{\nu}L\omega \tag{13.98}$$

From (13.96),

$$L(\underline{L}\omega) = 2L(\eta^{-1}\kappa) \tag{13.99}$$

Also, from (13.96) and the first of (13.90),

$$\begin{aligned} \nu\underline{L}\omega + \underline{\nu}L\omega &= 2(\nu\eta^{-1}\kappa + \underline{\nu}) \\ &= (\eta^{-1}\kappa\mathrm{tr}\chi + \mathrm{tr}\underline{\chi} + \eta^{-1}\kappa L\log\Omega + \underline{L}\log\Omega) \\ &= 2(\kappa\mathrm{tr}\not{k} + \eta^{-1}\kappa L\log\Omega + T\log\Omega) \end{aligned} \tag{13.100}$$

where we have used the definition of $\nu, \underline{\nu}$ and the fact

$$\eta^{-1}\kappa\chi + \underline{\chi} = 2\kappa\not{k}$$

We thus obtain

$$-\mu\Omega\Box_{\tilde{g}}\omega = 2(L(\eta^{-1}\kappa) + \kappa\mathrm{tr}\not{k} + \eta^{-1}\kappa L\log\Omega + T\log\Omega) \tag{13.101}$$

Proposition 12.6 then implies

$$\sup_{\Sigma_t^{\epsilon_0}}(\mu|\Box_{\tilde{g}}\omega|) \leqslant C\delta_0(1+t)^{-1} \tag{13.102}$$

This implies **D5**. So the proposition follows. □

The remaining assumptions on which Theorem 5.1 relies, assumptions **B**, readily follow from assumptions $\mathbf{X}_{[1]}$ and $\mathbf{M}_{[1,2]}$, established by Proposition 12.9 with $l = 1$ and 12.10 with $m = l = 1$, respectively. Propositions 10.5, 10.6, 11.5, 11.6, require the assumptions $\mathbf{X}_{[(l+1)_*]}$ and $\mathbf{M}_{\{(l+1)_*+1\}}$. These assumptions are established by Propositions 12.9 and 12.10 on the basis of the bootstrap assumptions $\mathbf{E}_{\{(l+1)_*+2\}}, \mathbf{E}^{Q}_{\{(l+1)_*+1\}}, \mathbf{E}^{QQ}_{\{(l+1)_*\}}$. We now make these assumptions more precise by setting the constant appearing on the right equal to 1. We define, more generally, the bootstrap assumptions $\mathbf{E}^{Q\cdots Q}_{\{q\}}$, where the length of the string of $Q$'s is $p$, for $p = 0, \cdots, q$ as follows. First, the bootstrap assumption $\mathbf{E}_{0,0}$ holds on $W^s_{\epsilon_0}$ if for all $t \in [0, s]$,

$$\mathbf{E}_{0,0} : \max_{\alpha} \|\psi_\alpha\|_{L^\infty(\Sigma_t^{\epsilon_0})} \leqslant \delta_0(1+t)^{-1}$$

For non-negative integers $p, m, n$ not all zero we say that the bootstrap assumption

$$\mathbf{E}_{m,n}^{Q\cdots Q}$$

holds on $W_{\epsilon_0}^s$ if for all $t \in [0, s]$,

$$\mathbf{E}_{m,n}^{Q\cdots Q} : \max_{\alpha;i_1\cdots i_n} \|R_{i_n}\cdots R_{i_1}(T)^m(Q)^p\psi_\alpha\|_{L^\infty(\Sigma_t^{\epsilon_0})} \leqslant \delta_0(1+t)^{-1}$$

We then define $\mathbf{E}_{\{q\}}^{Q\cdots Q}$ to be the conjunction of the assumptions $\mathbf{E}_{m,n}^{Q\cdots Q}$ corresponding to the triangle

$$\{(m,n) : m, n \geqslant 0, \quad m+n \leqslant q\}$$

Finally we define the bootstrap assumption $\mathbf{E}_{\{\{k\}\}}$ to be the conjunction of the assumptions

$$\mathbf{E}_{\{k-p\}}^{Q\cdots Q} : p = 0, \cdots, k$$

From now on by **bootstrap assumption** we mean the assumption $\mathbf{E}_{\{\{(l+1)_*+2\}\}}$. We have seen that all the assumptions which we have used so far follow from the bootstrap assumption together with appropriate assumptions on the initial data.

# 14

# The Error Estimates Involving the Top Order Spatial Derivatives of the Acoustical Entities

## 14.1 The Error Terms Involving the Top Order Spatial Derivatives of the Acoustical Entities

In this chapter we deal with the error estimates involving the top order spatial derivatives of the acoustical entities $\chi$ and $\mu$. In view of the fact that $\not{\mathcal{L}}_L\chi$ and $L\mu$ are expressed by the basic propagation equation in terms of $\chi$ and $\mu$ respectively, the top order derivatives of $\chi$ and $\mu$ are in fact the top order spatial derivatives of these entities.

In the following, given a positive integer $n$, we denote by ${}^{(\alpha;I_1\cdots I_{n-1})}\psi_n$ the $n$th order variation:

$$ {}^{(\alpha;I_1\cdots I_{n-1})}\psi_n = Y_{I_{n-1}}\cdots Y_{I_1}{}^{(\alpha)}\psi_1 \tag{14.1} $$

where

$$ {}^{(\alpha)}\psi_1 = \psi_\alpha \tag{14.2} $$

Here the indices $I_1, \cdots, I_{n-1}$ range over the set $\{1,2,3,4,5\}$ and we have

$$ Y_1 = T, \quad Y_2 = Q, \quad Y_{i+2} = R_i, \quad i = 1,2,3 \tag{14.3} $$

For $n \geqslant 2$ we may express (14.1) as

$$ {}^{(\alpha;I_1\cdots I_{n-1})}\psi_n = Y_{I_{n-1}}\cdots Y_{I_1}\psi_\alpha \quad \text{for} \quad n \geqslant 2 \tag{14.4} $$

We shall require the multi-indices $(I_1\cdots I_{n-1})$ to be one of the following form. First, there should be a string of 2's (which may be empty). Then, there should be a string of 1's (which may also be empty). Finally, there should be a multi-index from the set $\{3,4,5\}$. If the string of 2's has length $p$, the string of 1's has length $m$, and the multi-index from the set $\{3,4,5\}$ has length $n$, then we have

$$ (I_1\cdots I_{p+m+n}) = (2\cdots 21\cdots 1 i_1+2\cdots i_n+2) \tag{14.5} $$

where $(i_1\cdots i_n)$ is a multi-index from the set $\{1,2,3\}$. Thus, if $p+m+n>0$ we have

$$^{(\alpha;I_1\cdots I_{p+m+n})}\psi_{p+m+n+1} = R_{i_n}\cdots R_{i_1}(T)^m(Q)^p\psi_\alpha \tag{14.6}$$

We shall also use in the following, in place of (14.1), the simpler notation:

$$^{(I_1\cdots I_{n-1})}\psi_n = Y_{I_{n-1}}\cdots Y_{I_1}\psi_1 \tag{14.7}$$

when we do not care about which of the 1st order variations we are considering.

We have the inhomogeneous wave equations

$$\Box_{\tilde{g}}{}^{(I_1\cdots I_{n-1})}\psi_n = {}^{(I_1\cdots I_{n-1})}\rho_n \tag{14.8}$$

Also as in Chapter 7, we define

$$^{(I_1\cdots I_{n-1})}\tilde{\rho}_n = \Omega^2\mu^{(I_1\cdots I_{n-1})}\rho_n \tag{14.9}$$

According to (7.30), for $n \geqslant 2$ we have the recursion formula

$$^{(I_1\cdots I_{n-1})}\tilde{\rho}_n = (Y_{I_{n-1}} + {}^{(Y_{I_{n-1}})}\delta)^{(I_1\cdots I_{n-2})}\tilde{\rho}_{n-1} + {}^{(Y_{I_{n-1}};I_1\cdots I_{n-2})}\sigma_{n-1} \tag{14.10}$$

and we have

$$\tilde{\rho}_1 = 0 \tag{14.11}$$

In (14.10) the function $^{(Y_{I_{n-1}})}\delta$ is defined by (7.33) and depends only on the commutation field $Y_{I_{n-1}}$, while the function $^{(Y_{I_{n-1}};I_1\cdots I_{n-2})}\sigma_{n-1}$ is defined by (7.31) and (7.32) and depends on $Y_{I_{n-1}}$ as well as $^{(I_1\cdots I_{n-2})}\psi_{n-1}$. We shall apply Proposition 8.2 to (14.10). Replacing $n$ by $n+1$ we have

$$^{(I_1\cdots I_n)}\tilde{\rho}_{n+1} = (Y_{I_n} + {}^{(Y_{I_n})}\delta)^{(I_1\cdots I_{n-1})}\tilde{\rho}_n + {}^{(Y_{I_n};I_1\cdots I_{n-1})}\sigma_n \tag{14.12}$$

for $n = 1, 2, \cdots$. Applying then Proposition 8.2 yields

$$^{(I_1\cdots I_n)}\tilde{\rho}_{n+1} = \sum_{m=0}^{n-1}(Y_{I_n} + {}^{(Y_{I_n})}\delta)\cdots(Y_{I_{n-m+1}} + {}^{(Y_{I_{n-m+1}})}\delta)^{(Y_{I_{n-m}};I_1\cdots I_{n-m-1})}\sigma_{n-m} \tag{14.13}$$

or, setting $n = l+1, m = k$,

$$^{(I_1\cdots I_{l+1})}\tilde{\rho}_{l+2} = \sum_{k=0}^{l}(Y_{I_{l+1}} + {}^{(Y_{I_{l+1}})}\delta)\cdots(Y_{I_{l-k+2}} + {}^{(Y_{I_{l-k+2}})}\delta)^{(Y_{I_{l-k+1}};I_1\cdots I_{l-k})}\sigma_{l+1-k} \tag{14.14}$$

We shall focus our attention on the contribution of top order spatial derivatives of the acoustical entities, i.e. the $(l+1)$st order spatial derivatives of $\chi$ and the $(l+2)$nd order spatial derivatives of $\mu$. These appear only in the $(l+1)$st spatial derivatives of $^{(Y)}\tilde{\pi}$. From (7.59) we have

$$^{(Y)}\sigma_{n-1} = {}^{(Y)}\sigma_{1,n-1} + {}^{(Y)}\sigma_{2,n-1} + {}^{(Y)}\sigma_{3,n-1} \tag{14.15}$$

where $^{(Y)}\sigma_{1,n-1}, {}^{(Y)}\sigma_{2,n-1}, {}^{(Y)}\sigma_{3,n-1}$ are given by (7.74)—(7.79). Of these only $^{(Y)}\sigma_{2,n-1}$ contains the spatial derivatives of $^{(Y)}\tilde{\pi}$.

We conclude that the derivatives of top order, $l+1$, of the ${}^{(Y)}\tilde{\pi}$, occur only in the term $k = l$ in the sum in (14.14):

$$(Y_{I_{l+1}} + {}^{(Y_{I_{l+1}})}\delta) \cdots (Y_{I_2} + {}^{(Y_{I_2})}\delta)^{(Y_{I_1})}\sigma_1 \tag{14.16}$$

more precisely, in

$$Y_{I_{l+1}} \cdots Y_{I_2}{}^{(Y_{I_1})}\sigma_{2,1} \tag{14.17}$$

Furthermore, we can replace ${}^{(Y_{I_1})}\sigma_{2,1}$ by its principal spatial derivative part:

$$\begin{aligned}&(1/2)T(\mathrm{tr}^{(Y_{I_1})}\tilde{\not{\pi}})L\psi_1 - \not{\mathcal{L}}_T{}^{(Y_{I_1})}\tilde{Z}\cdot \not{d}\psi_1 - (1/2)\not{\mathrm{div}}^{(Y_{I_1})}\tilde{Z}\underline{L}\psi_1\\ &- (1/2)\not{\mathrm{div}}^{(Y_{I_1})}\tilde{\underline{Z}}L\psi_1 + (1/2)\not{d}^{(Y_{I_1})}\tilde{\pi}_{L\underline{L}}\cdot(\not{d}\psi_1) + \not{\mathrm{div}}(\mu^{(Y_{I_1})}\hat{\tilde{\not{\pi}}})\cdot(\not{d}\psi_1)\end{aligned} \tag{14.18}$$

Here we have expressed the derivatives with respect to $\underline{L}$ in terms of the derivatives with respect to $T$. We remark that the only expressions of the form (14.17), which may contain top order spatial derivatives of acoustical entities are those for which $(I_2 \cdots I_{l+1})$ does not contain any 2's, therefore, either $p = 0$, in which case $Y_{I_1}$ is one of $T, R_j, j = 1,2,3$, or $p = 1$, in which case $Y_{I_1} = Q$. In the following, we denote by $[\quad]_{P.A.}$, the principal acoustical part, expressed in terms of $\chi'$ and $\mu$, the difference $\chi - \chi' = (1-u+t)^{-1}\not{g}$ being of lower order.

Consider first the case $p = 1, Y_{I_1} = Q$. The components of ${}^{(Q)}\tilde{\pi}$ are given in Chapter 6. Noting that

$$[\nu]_{P.A.} = \frac{1}{2}\mathrm{tr}\chi', \quad [\Lambda]_{P.A.} = -(\not{d}\mu)^\sharp \tag{14.19}$$

we have

$$\begin{aligned}[{}^{(Q)}\tilde{\pi}_{L\underline{L}}]_{P.A.} &= 0\\ [{}^{(Q)}\tilde{Z}]_{P.A.} &= 0\\ [{}^{(Q)}\tilde{\underline{Z}}]_{P.A.} &= 2\Omega(1+t)(\not{d}\mu)^\sharp\\ [\mathrm{tr}^{(Q)}\tilde{\not{\pi}}]_{P.A.} &= 2\Omega(1+t)\mathrm{tr}\chi'\\ [{}^{(Q)}\hat{\tilde{\not{\pi}}}]_{P.A.} &= 2\Omega(1+t)\hat{\chi}'\end{aligned} \tag{14.20}$$

From (14.18) we then obtain

$$[{}^{(Q)}\sigma_{2,1}]_{P.A.} = [\Omega(1+t)((T\mathrm{tr}\chi' - \not{\Delta}\mu)L\psi_1 + 2\mu(\not{\mathrm{div}}\hat{\chi}')\cdot(\not{d}\psi_1))]_{P.A.} \tag{14.21}$$

By (8.99),

$$[2\not{\mathrm{div}}\hat{\chi}' - \not{d}\mathrm{tr}\chi']_{P.A.} = 0 \tag{14.22}$$

Also by (3.124)—(3.125),

$$[T\mathrm{tr}\chi' - \not{\Delta}\mu]_{P.A.} = 0 \tag{14.23}$$

In view of (14.22) and (14.23), (14.21) reduces to

$$[{}^{(Q)}\sigma_{2,1}]_{P.A.} = \Omega(1+t)\mu(\not{d}\mathrm{tr}\chi')\cdot(\not{d}\psi_1) \tag{14.24}$$

Consider next the case $p=0, Y_{I_1}=R_j$. By (6.8) and (6.180),

$$^{(R_j)}\tilde{\pi}_{L\underline{L}} = -2\Omega R_j\mu - 2\mu R_j\Omega \tag{14.25}$$

hence

$$[{}^{(R_j)}\tilde{\pi}_{L\underline{L}}]_{P.A.} = -2\Omega R_j\mu \tag{14.26}$$

From (6.67),

$$[{}^{(R_j)}\not{\pi}_L]_{P.A.} = -R_j\cdot\chi' \tag{14.27}$$

Since

$$^{(R_j)}\tilde{Z} = \Omega^{(R_j)}Z$$

it follows that

$$[{}^{(R_j)}\tilde{Z}]_{P.A.} = -\Omega R_j\cdot\chi'^{\sharp} \tag{14.28}$$

We have

$$^{(R_j)}\not{\pi}_{\underline{L}} = \eta^{-1}\kappa^{(R_j)}\not{\pi}_L + 2^{(R_j)}\not{\pi}_T$$

From (6.63) and the fact that

$$\chi = \eta(\not{k}-\theta)$$

together with (14.28), we deduce

$$[{}^{(R_j)}\underline{\tilde{Z}}]_{P.A.} = \Omega\eta^{-1}(\kappa R_j\cdot\chi'^{\sharp} + 2\lambda_j(\not{d}\mu)^{\sharp}) \tag{14.29}$$

Next, from (6.8) we have

$$[{}^{(R_j)}\tilde{\not{\pi}}]_{P.A.} = [\Omega^{(R_j)}\not{\pi}]_{P.A.} \tag{14.30}$$

From (6.59) and (3.27),

$$[{}^{(R_j)}\not{\pi}]_{P.A.} = 2\lambda_j\eta^{-1}\chi' \tag{14.31}$$

Hence

$$[{}^{(R_j)}\tilde{\not{\pi}}]_{P.A.} = 2\Omega\lambda_j\eta^{-1}\chi' \tag{14.32}$$

Then

$$[\mathrm{tr}^{(R_j)}\tilde{\not{\pi}}]_{P.A.} = 2\Omega\lambda_j\eta^{-1}\mathrm{tr}\chi' \tag{14.33}$$

and

$$[{}^{(R_j)}\hat{\tilde{\not{\pi}}}]_{P.A.} = 2\Omega\lambda_j\eta^{-1}\hat{\chi}' \tag{14.34}$$

Substituting (14.26), (14.28), (14.29), (14.33), (14.34) in (14.18) we obtain

$$\begin{aligned}[{}^{(R_j)}\sigma_{2,1}]_{P.A.} = &[\Omega\lambda_j\alpha^{-1}(T\mathrm{tr}\chi')L\psi_1 + \Omega(R_j\cdot\not{\mathcal{L}}_T\chi')\cdot(\not{d}\psi_1) \\ &+ (1/2)\Omega\not{\mathrm{div}}(R_j\cdot\chi')\underline{L}\psi_1 \\ &- (1/2)\Omega\alpha^{-1}(\kappa\not{\mathrm{div}}(R_j\cdot\chi') + 2\lambda_j(\not{\Delta}\mu))L\psi_1\end{aligned} \tag{14.35}$$

$$- \Omega(\not{d}R_j\mu)\cdot(\not{d}\psi_1) + 2\Omega\lambda_j\kappa\not{\mathrm{d}}\mathrm{iv}\hat{\chi}'\cdot(\not{d}\psi_1)]_{P.A.}$$

Using (14.22), (14.23) and the fact by (3.124)—(3.125),

$$[\not{\mathcal{L}}_T\chi' - \not{D}^2\mu]_{P.A.} = 0 \tag{14.36}$$

(14.35) reduces to

$$\begin{aligned}[{}^{(R_j)}\sigma_{2,1}]_{P.A.} = &[\Omega\not{\mathrm{d}}\mathrm{iv}(R_j\cdot\chi')T\psi_1 + \Omega\lambda_j\kappa\not{d}\mathrm{tr}\chi'\cdot(\not{d}\psi_1)\\ &+ \Omega(R_j\cdot\not{D}^2\mu - \not{d}R_j\mu)\cdot(\not{d}\psi_1)]_{P.A.}\end{aligned} \tag{14.37}$$

Consider $\not{\mathrm{d}}\mathrm{iv}(R_j\cdot\chi')$ in components in an arbitrary local frame field for $S_{t,u}$:

$$\not{\mathrm{d}}\mathrm{iv}(R_j\cdot\chi') = \not{D}_A(R_j^B\chi_B'^A) = R_j^B(\not{D}_A\chi_B'^A) + (\not{D}_A R_j^B)\chi_B'^A$$

and with $R_{jB} = \not{g}_{BC}R_j^C$:

$$(\not{D}_A R_j^B)\chi_B'^A = (\not{D}_A R_{jB})\chi'^{AB} = \frac{1}{2}(\not{D}_A R_{jB} + \not{D}_B R_{jA})\chi'^{AB} = \frac{1}{2}{}^{(R_j)}\not{\pi}_{AB}\chi'^{AB}$$

We conclude that

$$\not{\mathrm{d}}\mathrm{iv}(R_j\cdot\chi') = R_j\cdot\not{\mathrm{d}}\mathrm{iv}\chi' + \frac{1}{2}\mathrm{tr}({}^{(R_j)}\not{\pi}\cdot\chi') \tag{14.38}$$

The second term is of lower order. In view of (14.22) it follows that

$$[\not{\mathrm{d}}\mathrm{iv}(R_j\cdot\chi')]_{P.A.} = [R_j\cdot\not{\mathrm{d}}\mathrm{iv}\chi']_{P.A.} = R_j\mathrm{tr}\chi' \tag{14.39}$$

Consider finally $(R_j\cdot\not{D}^2\mu - \not{d}R_j\mu)\cdot(\not{d}\psi_1)$. In terms of the $S_{t,u}$-tangential vectorfield $X = (\not{d}\psi_1)^\sharp$ we have

$$R_j\cdot\not{D}^2\mu\cdot(\not{d}\psi_1) = \not{D}^2\mu\cdot(X, R_j) \tag{14.40}$$

and

$$\begin{aligned}\not{D}^2\mu(X, R_j) &= XR_j\mu - \not{d}\mu\cdot(\not{D}_X R_j)\\ &= XR_j\mu - \not{d}\mu\cdot(\not{D}_{R_j}X) + \not{d}\mu\cdot[R_j, X]\end{aligned} \tag{14.41}$$

Therefore

$$\begin{aligned}(R_j\cdot\not{D}^2\mu - \not{d}R_j\mu)\cdot(\not{d}\psi_1) &= \not{D}^2\mu\cdot(X, R_j) - XR_j\mu\\ &= \not{d}\mu\cdot\not{\mathcal{L}}_{R_j}X - \not{d}\mu\cdot(\not{D}_{R_j}X)\end{aligned} \tag{14.42}$$

Obviously, (14.42) does not contain acoustical terms of order 2. It follows that

$$[\Omega(R_j\cdot\not{D}^2\mu - \not{d}R_j\mu)\cdot(\not{d}\psi_1)]_{P.A.} = 0 \tag{14.43}$$

In view of (14.39) and (14.43), (14.37) reduces finally to

$$[{}^{(R_j)}\sigma_{2,1}]_{P.A.} = \Omega(R_j\mathrm{tr}\chi')T\psi_1 + \Omega\kappa\lambda_j \not{d}\mathrm{tr}\chi' \cdot (\not{d}\psi_1) \tag{14.44}$$

Finally, we consider the case $p = 0, Y_{I_1} = T$. The components of ${}^{(T)}\tilde{\pi}$ are given in Chapter 6. We have

$$ {}^{(T)}\tilde{Z} = {}^{(T)}\tilde{\not{\pi}}_L \cdot \not{g}^{-1} = \Omega\Lambda, \quad {}^{(T)}\tilde{\underline{Z}} = {}^{(T)}\tilde{\not{\pi}}_{\underline{L}} \cdot \not{g}^{-1} = \Omega\eta^{-1}\kappa\Lambda \tag{14.45}$$

From the fact that $\underline{\chi} = -\eta^{-1}\kappa\chi + 2\kappa\not{k}$ we have

$$[\underline{\nu}]_{P.A.} = [\frac{1}{2}\mathrm{tr}\underline{\chi}]_{P.A.} = -\frac{1}{2}\eta^{-1}\kappa\mathrm{tr}\chi', \quad [\hat{\underline{\chi}}]_{P.A.} = -\eta^{-1}\kappa\hat{\chi}' \tag{14.46}$$

Using (14.19), we obtain

$$\begin{aligned} [{}^{(T)}\tilde{\pi}_{L\underline{L}}]_{P.A.} &= -2\Omega T\mu \\ [{}^{(T)}\tilde{Z}]_{P.A.} &= -\Omega(\not{d}\mu) \\ [{}^{(T)}\tilde{\underline{Z}}]_{P.A.} &= -\Omega\eta^{-1}\kappa(\not{d}\mu) \\ [\mathrm{tr}{}^{(T)}\tilde{\not{\pi}}]_{P.A.} &= -2\Omega\eta^{-1}\kappa\mathrm{tr}\chi' \\ [{}^{(T)}\hat{\tilde{\not{\pi}}}]_{P.A.} &= -2\Omega\eta^{-1}\kappa\hat{\chi}' \end{aligned} \tag{14.47}$$

Substituting above in (14.18) we obtain

$$\begin{aligned} [{}^{(T)}\sigma_{2,1}]_{P.A.} =& [-\Omega\alpha^{-1}\kappa(T\mathrm{tr}\chi')L\psi_1 + \Omega(\not{d}T\mu)\cdot(\not{d}\psi_1) \\ &+ (1/2)\Omega(\not{\Delta}\mu)\underline{L}\psi_1 + (1/2)\Omega\eta^{-1}\kappa(\not{\Delta}\mu)L\psi_1 \\ &- \Omega(\not{d}T\mu)\cdot(\not{d}\psi_1) - 2\Omega\kappa^2\not{\mathrm{d}}\mathrm{iv}\hat{\chi}'\cdot(\not{d}\psi_1)]_{P.A.} \end{aligned} \tag{14.48}$$

Using (14.22) and (14.23), (14.48) reduces to

$$[{}^{(T)}\sigma_{2,1}]_{P.A.} = \Omega(\not{\Delta}\mu)T\psi_1 - \Omega\kappa^2\not{d}\mathrm{tr}\chi'\cdot(\not{d}\psi_1) \tag{14.49}$$

We conclude from the above that the top order variations ${}^{(\alpha;I_1\cdots I_{l+1})}\psi_{l+2}$ to which the re-scaled sources ${}^{(\alpha;I_1\cdots I_{l+1})}\tilde{\rho}_{l+2}$ containing the top order spatial derivatives of the acoustical entities correspond are as follows.

In the case $Y_{I_1} = Q$, the variations

$${}^{(\alpha;21\cdots1i_1+2\cdots i_n+2)}\psi_{l+2} = R_{i_n}\cdots R_{i_1}(T)^m Q\psi_\alpha, \quad m+n=l \tag{14.50}$$

where there are $m$ 1's in the superscript. By (14.17) and (14.24) the principal acoustical part of the corresponding re-scaled sources is

$$\begin{aligned} &[{}^{(\alpha;21\cdots1i_1+2\cdots i_n+2)}\tilde{\rho}_{l+2}]_{P.A.} \\ &= [\Omega(1+t)\mu(\not{d}R_{i_n}\cdots R_{i_1}(T)^m\mathrm{tr}\chi')\cdot(\not{d}\psi_\alpha)]_{P.A.} \\ &= \begin{cases} \Omega(1+t)\mu(\not{d}R_{i_n}\cdots R_{i_1}\mathrm{tr}\chi')\cdot(\not{d}\psi_\alpha), & m=0 \\ \Omega(1+t)\mu(\not{d}R_{i_n}\cdots R_{i_1}(T)^{m-1}\not{\Delta}\mu)\cdot(\not{d}\psi_\alpha), & m\geqslant 1 \end{cases} \end{aligned} \tag{14.51}$$

where again there are $m$ 1's in the superscript, and in the case $m \geqslant 1$ we have used (14.23).

In the case $Y_{I_1} = R_j$, setting $i_1 = j$ we have

$$I_1 \cdots I_{l+1} = i_1 + 2 \cdots i_{l+1} + 2$$

the variations

$$^{(\alpha;i_1+2\cdots i_l+2)}\psi_{l+2} = R_{i_{l+1}} \cdots R_{i_1}\psi_\alpha \tag{14.52}$$

By (14.17) and (14.44) the principal acoustical part of the corresponding re-scaled sources is

$$\begin{aligned}&[^{(\alpha;i_1+2\cdots i_{l+1}+2)}\tilde{\rho}_{l+2}]_{P.A.}\\ &= \Omega(R_{i_{l+1}} \cdots R_{i_1}\mathrm{tr}\chi')T\psi_\alpha + \Omega\kappa\lambda_{i_1}(\not{d}R_{i_{l+1}} \cdots R_{i_2}\mathrm{tr}\chi') \cdot (\not{d}\psi_\alpha)\end{aligned} \tag{14.53}$$

In the case $Y_{I_1} = T$, the variations

$$^{(\alpha;1\cdots 1 i_1+2\cdots i_n+2)}\psi_{l+2} = R_{i_n} \cdots R_{i_1}(T)^{m+1}\psi_\alpha, \quad m + n = l \tag{14.54}$$

where there are $m+1$ 1's in the superscript. By (14.17) and (14.49) the principal acoustical part of the corresponding re-scaled sources is

$$\begin{aligned}&[^{(\alpha;i_1+2\cdots i_{l+1}+2)}\tilde{\rho}_{l+2}]_{P.A.}\\ &= \begin{cases} \Omega(R_{i_n} \cdots R_{i_1}(T)^m \Delta\mu)T\psi_\alpha - \Omega\kappa^2(\not{d}R_{i_n} \cdots R_{i_1}\mathrm{tr}\chi') \cdot (\not{d}\psi_\alpha), & m = 0 \\ \Omega(R_{i_n} \cdots R_{i_1}(T)^m \Delta\mu)T\psi_\alpha - \Omega\kappa^2(\not{d}R_{i_n} \cdots R_{i_1}(T)^{m-1}\Delta\mu) \cdot (\not{d}\psi_\alpha), & m \geqslant 1 \end{cases}\end{aligned} \tag{14.55}$$

where again there are $m+1$ 1's in the superscript, and in the case $m \geqslant 1$ we have used (14.23).

We see from (14.51), (14.53), (14.55), that the most difficult error integrals are which result in terms

$$\Omega(R_{i_{l+1}} \cdots R_{i_1}\mathrm{tr}\chi')T\psi_\alpha \tag{14.56}$$

in (14.53), and

$$\Omega(R_{i_n} \cdots R_{i_1}(T)^m \Delta\mu)T\psi_\alpha, \quad m + n = l \tag{14.57}$$

in (14.55). These are the terms proportional to $T\psi_\alpha$ in (14.53) and (14.55). The other terms in these two expressions are proportional to $\not{d}\psi_\alpha$. Besides containing at least one factor of $\kappa$, which makes the estimates much easier in the region of small $\mu$, these terms have a decay factor of $(1+t)^{-2}(1+\log(1+t))^2$ relative to the principal parts (14.56) and (14.57). The terms in (14.51) are also proportional to $\not{d}\psi_\alpha$, and, besides containing the factor $\mu$, these terms have a decay factor of $(1+t)^{-1}(1+\log(1+t))$ relative to (14.56) ($m = 0$) and (14.57) ($m \geqslant 1$). So we may confine ourselves to (14.56) and (14.57).

## 14.2 The Borderline Error Integrals

We recall from Chapter 7 that the error integrals corresponding to an $n$th order variation $\psi_n$ are the integrals

$$-\int_{W_u^t} \tilde{\rho}_n(K_0\psi_n)dt'du'd\mu_{\not g} \tag{14.58}$$

associated to $K_0$, and

$$-\int_{W_u^t} \tilde{\rho}_n(K_1\psi_n + \omega\psi_n)dt'du'd\mu_{\not g} \tag{14.59}$$

associated to $K_1$.

Since

$$K_0\psi_n = \underline{L}\psi_n + (1+\eta^{-1}\kappa)L\psi_n \tag{14.60}$$

while

$$K_1\psi_n + \omega\psi_n = (\omega/\nu)(L\psi_n + \nu\psi_n) \tag{14.61}$$

the coefficient of $L\psi_n$ in (14.60) has a decay factor of $(1+t)^{-2}(1+\log(1+t))$ relative to the coefficient of the same term in (14.61). Therefore it suffices to bound the error integral (14.59) and the contribution of $\underline{L}\psi_n$ in (14.60) to the error integral (14.58). This contribution is bounded in absolute value by

$$\int_{W_u^t} |\tilde{\rho}_n||\underline{L}\psi_n|dt'dud\mu_{\not g} \tag{14.62}$$

We are thus to estimate the contributions of (14.56) and (14.57) to (14.62) and (14.59).

We begin with the contribution of (14.56) to (14.62). The contribution is

$$\begin{aligned}&\int_{W_{\epsilon_0}^t} \Omega|R_{i_{l+1}}\cdots R_{i_1}\mathrm{tr}\chi'||T\psi_\alpha||\underline{L}R_{i_{l+1}}\cdots R_{i_1}\psi_\alpha|dt'dud\mu_{\not g}\\ &\leqslant C\int_0^t \sup_{\Sigma_{t'}^{\epsilon_0}}(\mu^{-1}|T\psi_\alpha|)\|\mu R_{i_{l+1}}\cdots R_{i_1}\mathrm{tr}\chi'\|_{L^2(\Sigma_{t'}^{\epsilon_0})}\|\underline{L}R_{i_{l+1}}\cdots R_{i_1}\psi_\alpha\|_{L^2(\Sigma_{t'}^{\epsilon_0})}dt'\end{aligned} \tag{14.63}$$

Now, in view of (8.40), (8.63), (8.335) and (8.346) we have

$$\begin{aligned}&\|\mu R_{i_{l+1}}\cdots R_{i_1}\mathrm{tr}\chi'\|_{L^2(\Sigma_t^{\epsilon_0})}\\ &\leqslant C(1+t)\|\mu \not d R_{i_l}\cdots R_{i_1}\mathrm{tr}\chi'\|_{L^2(\Sigma_t^{\epsilon_0})}\\ &\leqslant C(1+t)^2(\|\|^{(i_1\cdots i_l)}x_l(t)\|\|_{L^2([0,\epsilon_0]\times S^2)} + \|\|\not d^{(i_1\cdots i_l)}\check{f}_l(t)\|\|_{L^2([0,\epsilon_0]\times S^2)})\\ &= C((1+t)^2\|\|^{(i_1\cdots i_l)}x_l(t)\|\|_{L^2([0,\epsilon_0]\times S^2)} + {}^{(i_1\cdots i_l)}P_l(t))\end{aligned} \tag{14.64}$$

Substituting (8.413) we obtain

$$\|\mu R_{i_{l+1}} \cdots R_{i_1} \mathrm{tr}\chi'\|_{L^2(\Sigma_t^{\epsilon_0})}$$
$$\leqslant C(1+t)^2((1+t)^{-2} {}^{(i_1\cdots i_l)}P_l(t) + {}^{(i_1\cdots i_l)}B_l(t)) \tag{14.65}$$
$$+ C_l\delta_0(1+t)^{-1}(1+\log(1+t))^2 \int_0^t (1+t')(1+\log(1+t'))B_l(t')dt'$$

Here, according to (8.403),

$$ {}^{(i_1\cdots i_l)}B_l(t) = C(1+t)^{-2}({}^{(i_1\cdots i_l)}\bar{P}^{(0)}_{l,a}(t) + (1+t)^{-1/2} {}^{(i_1\cdots i_l)}\bar{P}^{(1)}_{l,a}(t))\bar{\mu}_m^{-a}(t) \tag{14.66}$$
$$+ C(1+t)^{-3}(1+\log(1+t))^2 \int_0^t (1+t')^3 {}^{(i_1\cdots i_l)}Q_l(t')dt'$$
$$+ C(1+t)^{-3}(1+\log(1+t))^2 \| {}^{(i_1\cdots i_l)}x_l(0)\|_{L^2(\Sigma_0^{\epsilon_0})}$$

Also (see (8.405))

$$B_l(t) = \max_{i_1\cdots i_l} {}^{(i_1\cdots i_l)}B_l(t) \tag{14.67}$$

According to (8.348) and (8.349),

$$ {}^{(i_1\cdots i_l)}\bar{P}^{(0)}_{l,a}(t) = \sup_{t'\in[0,t]} (\bar{\mu}_m^a(t') {}^{(i_1\cdots i_l)}P^{(0)}_l(t')) \tag{14.68}$$
$$ {}^{(i_1\cdots i_l)}\bar{P}^{(1)}_{l,a}(t) = \sup_{t'\in[0,t]} ((1+t')^{1/2}\bar{\mu}_m^a(t') {}^{(i_1\cdots i_l)}P^{(1)}_l(t')) \tag{14.69}$$

where ${}^{(i_1\cdots i_l)}P^{(0)}_l$, ${}^{(i_1\cdots i_l)}P^{(1)}_l$ are defined in the statement of Proposition 10.6 and we have

$$ {}^{(i_1\cdots i_l)}P_l(t) \leqslant {}^{(i_1\cdots i_l)}P^{(0)}_l(t) + {}^{(i_1\cdots i_l)}P^{(1)}_l(t) \tag{14.70}$$

Note that ${}^{(i_1\cdots i_l)}\bar{P}^{(0)}_{l,a}(t)$ and ${}^{(i_1\cdots i_l)}\bar{P}^{(1)}_{l,a}(t)$ are non-decreasing.

The leading contribution to the right hand side of (14.63) comes from the first term on the right in (14.65). That is, from $(1+t)^{-2} {}^{(i_1\cdots i_l)}P_l(t)$ and the first term in the expression (14.66) for ${}^{(i_1\cdots i_l)}B_l(t)$. So what we consider here is

$$C({}^{(i_1\cdots i_l)}\bar{P}^{(0)}_{l,a}(t) + (1+t)^{-1/2} {}^{(i_1\cdots i_l)}\bar{P}^{(1)}_{l,a}(t))\bar{\mu}_m^{-a}(t) \tag{14.71}$$

The actual *borderline contribution* is the contribution from

$$ {}^{(i_1\cdots i_l)}P^{(0)}_l(t) = |\ell| \sqrt{\sum_{j,\alpha} \mathcal{E}_0[R_j R_{i_l} \cdots R_{i_1}\psi_\alpha](t)} \tag{14.72}$$

## 14.3 Assumption J

To estimate the borderline contributions, we shall use the assumption **J** below. Recall the operators

$$S = x^\alpha \frac{\partial}{\partial x^\alpha} - 1, \quad \mathring{R}_i = \frac{1}{2}\epsilon_{ijk}\frac{\partial}{\partial x^k}, \quad i = 1,2,3 \tag{14.73}$$

associated to the background Galilean spacetime. The assumption is that there is a positive constant $C$ independent of $s$ such that in $W^s_{\epsilon_0}$

$$\mathbf{J}: |S\phi|, |TS\phi| \leqslant C\delta_0(1+t)^{-1}, \quad |\mathring{R}_i\phi|, |T\mathring{R}_i\phi| \leqslant C\delta_0(1+t)^{-1}, \quad i=1,2,3 \tag{14.74}$$

where $\phi$ is the wave function. We shall establish this assumption in the sequel, on the basis of the bootstrap assumption. We shall presently use assumption **J** to derive a pointwise estimate for $T\psi_\alpha$ in terms of $L\mu$.

Let us consider the $\Sigma_t$-tangential vectorfield

$$V = \sum_{i=1}^{3} (T\psi_i)\frac{\partial}{\partial x^i} \tag{14.75}$$

Recalling from Chapter 6 the Euclidean outward unit normal $N$ to the Euclidean coordinate spheres:

$$N = \sum_{i=1}^{3} \frac{x^i}{r}\frac{\partial}{\partial x^i} \tag{14.76}$$

We decompose $V$ into its components which are tangential and orthogonal, relative to $\bar{g}$, to the Euclidean spheres:

$$V = V^{\parallel} + V^{\perp}, \quad V^{\perp} = \bar{g}(V,N)N, \quad |V|^2 = |V^{\parallel}|^2 + |V^{\perp}|^2 \tag{14.77}$$

Consider the vector $N \times V$. We have

$$|N \times V|^2 = |V^{\parallel}|^2 \tag{14.78}$$

The $i$th rectangular component of $N \times V$ is given by

$$r(N \times V)^i = \sum_{j,k=1}^{3} \epsilon_{ijk} x^j T(\partial_k\phi) = \sum_{j,k=1}^{3} \epsilon_{ijk}(T(x^j\partial_k\phi) - (Tx^j)\partial_k\phi) \tag{14.79}$$

Thus we have

$$r(N \times V)^i = T\mathring{R}_i\phi - \sum_{j,k=1}^{3} \epsilon_{ijk} T^j \psi_k \tag{14.80}$$

By **J**, Proposition 12.1 and (12.6),

$$r|N \times V| \leqslant C\delta_0(1+t)^{-1}(1+\log(1+t)) \quad \text{in} \quad W^s_{\epsilon_0} \tag{14.81}$$

From (12.14) and (14.78) we then obtain

$$\sup_{\Sigma^{\epsilon_0}_t} |V^{\parallel}| \leqslant C\delta_0(1+t)^{-2}(1+\log(1+t)) \tag{14.82}$$

By (14.75)—(14.77) and (14.82) it follows that

$$\sup_{\Sigma^{\epsilon_0}_t} |T\psi_i - \frac{x^i}{r}\bar{g}(V,N)| \leqslant C\delta_0(1+t)^{-2}(1+\log(1+t)) \tag{14.83}$$

Consider next

$$S\phi = x^\alpha \partial_\alpha \phi - \phi = t\psi_0 + \sum_{i=1}^{3} x^i \psi_i - \phi \tag{14.84}$$

We have

$$TS\phi = tT\psi_0 + \sum_{i=1}^{3} x^i T\psi_i + \sum_{i=1}^{3} \psi_i Tx^i - T\phi$$

The last two terms cancel and we obtain

$$TS\phi = tT\psi_0 + \sum_{i=1}^{3} x^i T\psi_i = tT\psi_0 + r\bar{g}(V, N) \tag{14.85}$$

By virtue of **J** we then have

$$\sup_{\Sigma_t^{\epsilon_0}} |\frac{t}{r} T\psi_0 + \bar{g}(V, N)| \leqslant C\delta_0(1+t)^{-2} \tag{14.86}$$

Combining (14.83) and (14.86) we conclude that

$$\sup_{\Sigma_t^{\epsilon_0}} |T\psi_i + \frac{tx^i}{r^2} T\psi_0| \leqslant C\delta_0(1+t)^{-2}(1+\log(1+t)) \tag{14.87}$$

Recalling the proof of Proposition 8.5, we have

$$L\mu = m + e\mu$$

and according to (8.209),

$$m = m_0 + m_1, \quad m_0 = \frac{1}{2}\ell T\psi_0$$

while according to (8.214), (8.218) and (8.220),

$$|m_1| \leqslant C\delta_0(1+t)^{-2}, \quad |e| \leqslant C\delta_0(1+t)^{-2}$$

It follows that

$$\sup_{\Sigma_t^{\epsilon_0}} |L\mu - \frac{1}{2}\ell T\psi_0| \leqslant C\delta_0(1+t)^{-2}(1+\log(1+t)) \tag{14.88}$$

In view of (14.87), (12.14) and (14.88), we obtain

$$\begin{aligned}\max_\alpha(\mu^{-1}|T\psi_\alpha|) &\leqslant C\mu^{-1}|T\psi_0| + C\mu^{-1}\delta_0(1+t)^{-2}(1+\log(1+t)) \\ &\leqslant \frac{C}{|\ell|}(\mu^{-1}|L\mu| + C\mu^{-1}\delta_0(1+t)^{-2}(1+\log(1+t)))\end{aligned}$$

Taking supremum on $\Sigma_t^{\epsilon_0}$ yields the desired estimate:

$$\begin{aligned}&\max_{\alpha}\sup_{\Sigma_t^{\epsilon_0}}(\mu^{-1}|T\psi_\alpha|)\\&\leqslant\frac{C}{|\ell|}(\sup_{\Sigma_t^{\epsilon_0}}(\mu^{-1}|L\mu|)+C\bar{\mu}_m^{-1}\delta_0(1+t)^{-2}(1+\log(1+t)))\end{aligned}\tag{14.89}$$

We shall use this to estimate the borderline contribution (14.72), through (14.68), (14.71) and (14.65), to the integral on the right of (14.63). On the other hand, in estimating all other contributions, through (14.65), to (14.63), we simply use

$$\max_{\alpha}\sup_{\Sigma_t^{\epsilon_0}}(\mu^{-1}|T\psi_\alpha|)\leqslant C\bar{\mu}_m^{-1}\delta_0(1+t)^{-1}\tag{14.90}$$

implied by $\mathbf{E}_{\{1\}}$.

## 14.4 The Borderline Estimates Associated to $K_0$

### 14.4.1 Estimates for the Contribution of (14.56)

Let us define, for non-negative real numbers $a$ and $p$, the quantities

$$\begin{aligned}&{}^{(i_1\cdots i_l)}\mathcal{G}_{0,l+2;a,p}(t)\\&=\sup_{t'\in[0,t]}((1+\log(1+t'))^{-2p}\bar{\mu}_m^{2a}(t')\sum_{j,\alpha}\mathcal{E}_0[R_jR_{i_l}\cdots R_{i_1}\psi_\alpha](t'))\end{aligned}\tag{14.91}$$

These quantities are non-decreasing functions of $t$ and we have

$$\begin{aligned}&\sqrt{\sum_{j,\alpha}\mathcal{E}_0[R_jR_{i_l}\cdots R_{i_1}\psi_\alpha](t')}\\&\leqslant\bar{\mu}_m^{-a}(t')(1+\log(1+t'))^p\sqrt{{}^{(i_1\cdots i_l)}\mathcal{G}_{0,l+2;a,p}(t)}\quad\text{for all}\quad t'\in[0,t]\end{aligned}\tag{14.92}$$

hence, in view of (14.68), the borderline contribution (14.72) to (14.71) is bounded by

$$C|\ell|\bar{\mu}_m^{-a}(t)(1+\log(1+t))^p\sqrt{{}^{(i_1\cdots i_l)}\mathcal{G}_{0,l+2;a,p}(t)}\tag{14.93}$$

Also, in regard to the last factor on the right in (14.63) we have

$$\begin{aligned}\|\underline{L}R_{i_{l+1}}\cdots R_{i_1}\psi_\alpha\|_{L^2(\Sigma_t^{\epsilon_0})}&\leqslant\sqrt{\mathcal{E}_0[R_{i_{l+1}}\cdots R_{i_1}\psi_\alpha](t)}\\&\leqslant\bar{\mu}_m^{-a}(t)(1+\log(1+t))^p\sqrt{{}^{(i_1\cdots i_l)}\mathcal{G}_{0,l+2;a,p}(t)}\end{aligned}\tag{14.94}$$

Substituting (14.89), (14.93) and (14.94) in the integral on the right of (14.63), the factors $|\ell|$ cancel and we obtain that the borderline contribution to the integral in question is bounded by

$$C\int_0^t(\sup_{\Sigma_{t'}^{\epsilon_0}}(\mu^{-1}|L\mu|)+C\bar{\mu}_m^{-1}\delta_0(1+t')^{-2}(1+\log(1+t')))$$

$$\cdot \bar{\mu}_m^{-2a}(t')(1+\log(1+t'))^{2p(i_1\cdots i_l)}\mathcal{G}_{0,l+2;a,p}(t')dt' \tag{14.95}$$

Now, the partial contribution of the second term in the first factor is actually not borderline. We shall show how to estimate contribution of this type afterwards, in connection with the estimate for the contribution of the term

$$C(1+t)^{-1}(1+\log(1+t))^{1/2}\sqrt{\sum_{j,\alpha}\mathcal{E}'_1[R_j R_{i_l}\cdots R_{i_1}\psi_\alpha]} \tag{14.96}$$

in the expression for ${}^{(i_1\cdots i_l)}P_l^{(1)}$ of Proposition 10.6, through (14.69), (14.71) and (14.65), to the integral on the right of (14.63). For the present we focus attention on the *borderline integral*:

$$C\int_0^t \sup_{\Sigma_{t'}^{\epsilon_0}}(\mu^{-1}|L\mu|)\bar{\mu}_m^{-2a}(t')(1+\log(1+t'))^{2p(i_1\cdots i_l)}\mathcal{G}_{0,l+2;a,p}(t')dt' \tag{14.97}$$

Since

$$|L\mu| = \max\{-(L\mu)_-, (L\mu)_+\} \leqslant -(L\mu)_- + (L\mu)_+$$

we have

$$\sup_{\Sigma_{t'}^{\epsilon_0}}(\mu^{-1}|L\mu|) \leqslant \sup_{\Sigma_{t'}^{\epsilon_0}}(-\mu^{-1}(L\mu)_-) + \sup_{\Sigma_{t'}^{\epsilon_0}}(\mu^{-1}(L\mu)_+) \tag{14.98}$$

Substituting (14.98) in (14.97), we first consider the first term on the right of (14.98). According to (8.249),

$$\sup_{\Sigma_t^{\epsilon_0}}(-\mu^{-1}(L\mu)_-) = M(t) \tag{14.99}$$

therefore the contribution in question is

$$\begin{aligned} &C\int_0^t M(t')\bar{\mu}_m^{-2a}(t')(1+\log(1+t'))^{2p(i_1\cdots i_l)}\mathcal{G}_{0,l+2;a,p}(t')dt' \\ &\leqslant C(1+\log(1+t))^{2p(i_1\cdots i_l)}\mathcal{G}_{0,l+2;a,p}(t)I_{2a}(t) \end{aligned} \tag{14.100}$$

where $I_a(t)$ is the integral of Lemma 8.11. According to Lemma 8.11 we have

$$I_{2a}(t) \leqslant C(2a)^{-1}\bar{\mu}_m^{-2a}(t) \tag{14.101}$$

where $C$ is a constant independent of $a$, provided that $a \geqslant 2$ and $\delta_0$ is suitably small depending on $a$. We conclude that (14.100) is bounded by

$$\frac{C}{2a}\bar{\mu}_m^{-2a}(t)(1+\log(1+t))^{2p(i_1\cdots i_l)}\mathcal{G}_{0,l+2;a,p}(t) \tag{14.102}$$

Consider next the contribution of the second term on the right in (14.98). Proposition 13.1 implies

$$\sup_{\Sigma_t^{\epsilon_0}}(\mu^{-1}(L\mu)_+) \leqslant C(1+t)^{-1}(1+\log(1+t))^{-1} \tag{14.103}$$

therefore the contribution in question is bounded by

$$C\int_0^t \bar{\mu}_m^{-2a}(t')(1+t')^{-1}(1+\log(1+t'))^{2p-1}{}^{(i_1\cdots i_l)}\mathcal{G}_{0,l+2;a,p}(t')dt' \tag{14.104}$$

By Corollary 2 of Lemma 8.11,

$$\bar{\mu}_m^{-2a}(t') \leqslant C\bar{\mu}_m^{-2a}(t) \tag{14.105}$$

where $C$ is independent of $a$, also provided that $a \geqslant 2$ and $\delta_0$ is suitably small depending on $a$. Since ${}^{(i_1\cdots i_l)}\mathcal{G}_{0,l+2;a,p}(t)$ is a non-decreasing function of $t$, it follows that (14.104) is bounded by

$$C\bar{\mu}_m^{-2a}(t){}^{(i_1\cdots i_l)}\mathcal{G}_{0,l+2;a,p}(t)\int_0^t (1+t')^{-1}(1+\log(1+t'))^{2p-1}dt' \tag{14.106}$$

Since

$$\int_0^t (1+t')^{-1}(1+\log(1+t'))^{2p-1}dt' = \frac{1}{2p}((1+\log(1+t))^{2p}-1)$$

this is bounded by

$$\frac{C}{2p}\bar{\mu}_m^{-2a}(t)(1+\log(1+t))^{2p}{}^{(i_1\cdots i_l)}\mathcal{G}_{0,l+2;a,p}(t) \tag{14.107}$$

We have thus estimated the borderline contributions to the integral in the right of (14.63). We proceed to estimate the remaining contributions.

Let us define

$$\mathcal{E}_{0,[n]}(t) = \sum_{m=1}^{n} \mathcal{E}_{0,m}(t) \tag{14.108}$$

where $\mathcal{E}_{0,n}$ represents the sum of the energies associated to the vectorfield $K_0$ of all the $n$th order variations. We then define

$$\mathcal{G}_{0,[n];a,p}(t) = \sup_{t'\in[0,t]} ((1+\log(1+t'))^{-2p}\bar{\mu}_m^{2a}(t')\mathcal{E}_{0,[n]}(t')) \tag{14.109}$$

Let us also define

$$\mathcal{E}'_{1,[n]}(t) = \sum_{m=1}^{n} \mathcal{E}'_{1,m}(t) \tag{14.110}$$

where $\mathcal{E}'_{1,n}$ represents the sum of the energies associated to the vectorfield $K_1$ of all $n$th order variations. We then define, for $q \geqslant p$,

$$\mathcal{G}'_{1,[n];a,q}(t) = \sup_{t'\in[0,t]} ((1+\log(1+t'))^{-2q}\bar{\mu}_m^{2a}(t')\mathcal{E}'_{1,[n]}(t')) \tag{14.111}$$

We now consider the contribution of the term (14.96), which comes from the second term on the right of definition for ${}^{(i_1\cdots i_l)}P_l^{(1)}$, through (14.69) and (14.65), to the integral on the right of (14.63). From (14.110) and (14.111), we have

$$\sqrt{\sum_{j,\alpha}\mathcal{E}'_1[R_jR_{i_l}\cdots R_{i_1}\psi_\alpha](t')} \leqslant \sqrt{\mathcal{E}'_{1,[l+2]}(t')}$$

$$\leqslant \bar{\mu}_m^{-a}(t')(1+\log(1+t'))^q\sqrt{\mathcal{G}'_{1,[l+2];a,q}(t)} \quad \text{for all} \quad t'\in[0,t] \tag{14.112}$$

hence, in view of (14.69), the contribution of (14.96) to (14.71) is bounded by

$$C_q(1+t)^{-1/2}\bar{\mu}_m^{-a}(t)\sqrt{\mathcal{G}'_{1,[l+2];a,q}(t)} \tag{14.113}$$

where $C_q$ is a constant depending on $q$:

$$C_q = C\sup_{t\in[0,\infty)}((1+t)^{-1/2}(1+\log(1+t))^{q+1/2})$$

Also, by (14.94) and the definitions (14.108), (14.109) we have

$$\|\underline{L}R_{i_{l+1}}\cdots R_{i_1}\psi_\alpha\|_{L^2(\Sigma_t^{\epsilon_0})} \leqslant \bar{\mu}_m^{-a}(t)(1+\log(1+t))^p\sqrt{\mathcal{G}_{0,[l+2];a,p}(t)} \tag{14.114}$$

Hence, by (14.90), the contribution of (14.96) through (14.65) to the integral on the right in (14.63) is bounded by

$$C_q\delta_0\int_0^t(1+t')^{-3/2}(1+\log(1+t'))^p\bar{\mu}_m^{-2a-1}(t')\sqrt{\mathcal{G}_{0,[l+2];a,p}(t')\mathcal{G}'_{1,[l+2];a,q}(t')}dt' \tag{14.115}$$

To estimate this we consider the two cases occurring in the proof of Lemma 8.11:

Case 1: $\hat{E}_{s,m}\geqslant 0$.

In this case we have the lower bound (8.258):

$$\bar{\mu}_m(t)\geqslant 1-C\delta_0 \tag{14.116}$$

which implies

$$\bar{\mu}_m^{-2a-1}(t)\leqslant C \tag{14.117}$$

provided that $\delta_0 a$ is suitably small (see (8.260)). It follows that in Case 1, (14.115) is bounded by

$$C_q\delta_0\int_0^t(1+t')^{-3/2}(1+\log(1+t'))^p\sqrt{\mathcal{G}_{0,[l+2];a,p}(t')\mathcal{G}'_{1,[l+2];a,q}(t')}dt' \tag{14.118}$$

In Case 2, we set

$$\hat{E}_{s,m} = -\delta_1, \quad \delta_1>0 \tag{14.119}$$

Following the proof of Lemma 8.11 with $a$ replaced by $2a$, we then set

$$t_1 = e^{\frac{1}{4a\delta_1}}-1 \tag{14.120}$$

and we consider the two subcases:

Subcase 2a: $t'\leqslant t_1$, Subcase 2b: $t'>t_1$.

In Subcase 2a we have the lower bound (8.273) with $a$ replaced by $2a$:

$$\bar{\mu}_m(t') \geqslant 1 - \frac{1}{a} \tag{14.121}$$

Since $(1-\frac{1}{a})^{-2a-1}$ is bounded for $a \in [2, \infty)$, if $t \leqslant t_1$ (14.115) is bounded by

$$C_q\delta_0 \int_0^t (1+t')^{-3/2}(1+\log(1+t'))^p \sqrt{\mathcal{G}_{0,[l+2];a,p}(t')\mathcal{G}'_{1,[l+2];a,q}(t')}dt' \tag{14.122}$$

and if $t > t_1$ we have

$$\begin{aligned}&\int_0^{t_1} (1+t')^{-3/2}(1+\log(1+t'))^p \bar{\mu}_m^{-2a-1}(t') \sqrt{\mathcal{G}_{0,[l+2];a,p}(t')\mathcal{G}'_{1,[l+2];a,q}(t')}dt' \\ &\leqslant C \int_0^{t_1} (1+t')^{-3/2}(1+\log(1+t'))^p \sqrt{\mathcal{G}_{0,[l+2];a,p}(t')\mathcal{G}'_{1,[l+2];a,q}(t')}dt'\end{aligned} \tag{14.123}$$

In Subcase 2b we have the lower bound (8.303) with $a$ replaced by $2a$:

$$\bar{\mu}_m(t') \geqslant (1 - \frac{1}{a})(1-\delta_1\tau'), \quad \tau' = \log(1+t') \tag{14.124}$$

In view of the fact that $\mathcal{G}_{0,[n];a,p}(t)$, $\mathcal{G}'_{1,[n];a,q}(t)$ are non-decreasing functions of $t$ it follows that

$$\begin{aligned}&\int_{t_1}^{t} (1+t')^{-3/2}(1+\log(1+t'))^p \bar{\mu}_m^{-2a-1}(t') \sqrt{\mathcal{G}_{0,[l+2];a,p}(t')\mathcal{G}'_{1,[l+2];a,q}(t')}dt' \\ &\leqslant C \sqrt{\mathcal{G}_{0,[l+2];a,p}(t)\mathcal{G}'_{1,[l+2];a,q}(t)} \cdot \int_{t_1}^{t} (1+t')^{-3/2}(1+\log(1+t'))^p (1-\delta_1\tau')^{-2a-1}dt'\end{aligned} \tag{14.125}$$

The integral in (14.125) is

$$\begin{aligned}&\int_{t_1}^{t} (1+t')^{-3/2}(1+\log(1+t'))^p (1-\delta_1\tau')^{-2a-1}dt' \\ &\leqslant (1+t_1)^{-1/2}(1+\log(1+t_1))^p \int_{\tau_1}^{\tau} (1-\delta_1\tau')^{-2a-1}d\tau' \\ &\leqslant (1+t_1)^{-1/2}(1+\log(1+t_1))^p \frac{1}{2a\delta_1}(1-\delta_1\tau)^{-2a} \\ &\leqslant 2\varphi_{1+p}(4a\delta_1)(1-\delta_1\tau)^{-2a}\end{aligned} \tag{14.126}$$

Here for any real number $r$ we denote by $\varphi_r(x)$ the following function on the positive real line:

$$\varphi_r(x) = e^{-\frac{1}{2x}}(1+\frac{1}{x})^r \tag{14.127}$$

The function $\varphi_r(x)$ decreases to 0 exponentially as $x \to 0$. Since $\delta_1 \leqslant C\delta_0$, we have

$$\varphi_r(4a\delta_1) \leqslant \varphi_r(Ca\delta_0) \tag{14.128}$$

provided that $Ca\delta_0$ is suitably small. In view also of the lower bound (8.314) with $a$ replaced by $2a$:

$$\bar{\mu}_m^{-2a}(t) \geqslant \frac{1}{C}(1-\delta_1\tau)^{-2a} \tag{14.129}$$

we conclude that the right hand side of (14.125) is bounded by

$$C\varphi_{1+p}(Ca\delta_0)\bar{\mu}_m^{-2a}(t)\sqrt{\mathcal{G}_{0,[l+2];a,p}(t)\mathcal{G}'_{1,[l+2];a,q}(t)} \tag{14.130}$$

Combining with (14.123) we obtain that if $t > t_1$, (14.115) is bounded by

$$\begin{aligned} &C_q\delta_0\bar{\mu}_m^{-2a}(t)(\varphi_{1+p}(Ca\delta_0)\sqrt{\mathcal{G}_{0,[l+2];a,p}(t)\mathcal{G}'_{1,[l+2];a,q}(t)} \\ &+\int_0^t (1+t')^{-3/2}(1+\log(1+t'))^p\sqrt{\mathcal{G}_{0,[l+2];a,p}(t')\mathcal{G}'_{1,[l+2];a,q}(t')}dt') \end{aligned} \tag{14.131}$$

Combining finally with the earlier results (14.122), for the subcase $t \leqslant t_1$, and (14.118), for Case 1, we conclude that the contribution of (14.96) through (14.65) to the integral on the right in (14.63) is bounded by

$$\begin{aligned} &C_q\delta_0\bar{\mu}_m^{-2a}(t)(\varphi_{1+p}(Ca\delta_0)\sqrt{\mathcal{G}_{0,[l+2];a,p}(t)\mathcal{G}'_{1,[l+2];a,q}(t)} \\ &+\int_0^t (1+t')^{-3/2}(1+\log(1+t'))^p\sqrt{\mathcal{G}_{0,[l+2];a,p}(t')\mathcal{G}'_{1,[l+2];a,q}(t')}dt') \end{aligned} \tag{14.132}$$

Now, we consider the contribution of the second term in the expression (14.66) for ${}^{(i_1\cdots i_l)}B_l(t)$ to the first term on the right of (14.65), namely the contribution of

$$C(1+t)^{-1}(1+\log(1+t))^2\int_0^t (1+t')^{3\,(i_1\cdots i_l)}Q_l(t')dt' \tag{14.133}$$

to the estimate for $\|\mu R_{i_{l+1}}\cdots R_{i_1}\mathrm{tr}\chi'\|_{L^2(\Sigma_t^{\epsilon_0})}$.

Here, we will use the estimate for $\max_{i_1\cdots i_l}{}^{(i_1\cdots i_l)}Q_l$ of Proposition 10.5. The principal part of the right-hand-side is

$$C_l(1+t)^{-4}(1+\log(1+t))(\mathcal{W}^Q_{[l+1]}+\delta_0(\mathcal{W}^T_{[l+1]}+\mathcal{W}_{[l+2]})) \tag{14.134}$$

We shall estimate the contribution of this principal part, through (14.133) to the integral on the right in (14.63).

We have, for each non-negative integer $n$,

$$\mathcal{W}_{n+1} \leqslant C(1+t)\sqrt{\sum_{i_1\cdots i_n,\alpha}\|\not{d}R_{i_n}\cdots R_{i_1}\psi_\alpha\|^2_{L^2(\Sigma_t^{\epsilon_0})}} \leqslant C\bar{\mu}_m^{-1/2}\sqrt{\mathcal{E}'_{1,n+1}} \tag{14.135}$$

It follows that

$$\mathcal{W}_{[l+2]} \leqslant C_l\bar{\mu}_m^{-1/2}\sqrt{\mathcal{E}'_{1,[l+2]}}+\mathcal{W}_0 \tag{14.136}$$

By Lemma 5.1, we have

$$\mathcal{W}_0 \leqslant \sqrt{\epsilon_0}\max_\alpha\sup_{u\in[0,\epsilon_0]}\|\psi_\alpha\|_{L^2(S_{t,u})} \leqslant C\epsilon_0\sqrt{\sum_\alpha\mathcal{E}_0[\psi_\alpha]} = C\epsilon_0\sqrt{\mathcal{E}_{0,[1]}} \tag{14.137}$$

Similarly, for each non-negative integer $n$,

$$\mathcal{W}^T_{n+1} \leqslant C(1+t)\sqrt{\sum_{i_1\cdots i_n,\alpha} \|\not{d}R_{i_n}\cdots R_{i_1}T\psi_\alpha\|^2_{L^2(\Sigma_t^{\epsilon_0})}} \leqslant C\bar{\mu}_m^{-1/2}\sqrt{\mathcal{E}'_{1,n+2}} \quad (14.138)$$

It follows that

$$\mathcal{W}^T_{[l+1]} \leqslant C_l\bar{\mu}_m^{-1/2}\sqrt{\mathcal{E}'_{1,[l+2]}} + \mathcal{W}^T_0 \quad (14.139)$$

by Lemma 5.1,

$$\mathcal{W}^T_0 \leqslant \sqrt{\epsilon_0}\max_\alpha \sup_{u\in[0,\epsilon_0]} \|T\psi_\alpha\|_{L^2(S_{t,u})} \leqslant C\epsilon_0\sqrt{\sum_\alpha \mathcal{E}_0[T\psi_\alpha]} \leqslant C\epsilon_0\sqrt{\mathcal{E}_{0,[2]}} \quad (14.140)$$

Similarly, we have

$$\mathcal{W}^Q_{[l+1]} \leqslant C_l\bar{\mu}_m^{-1/2}\sqrt{\mathcal{E}'_{1,[l+2]}} + \mathcal{W}^Q_0 \quad (14.141)$$

and by Lemma 5.1,

$$\mathcal{W}^Q_0 \leqslant \sqrt{\epsilon_0}\max_\alpha \sup_{u\in[0,\epsilon_0]} \|Q\psi_\alpha\|_{L^2(S_{t,u})} \leqslant C\epsilon_0\sqrt{\sum_\alpha \mathcal{E}_0[Q\psi_\alpha]} \leqslant C\epsilon_0\sqrt{\mathcal{E}_{0,[2]}} \quad (14.142)$$

The above inequalities imply that (14.134) is bounded by

$$C_l(1+t)^{-4}(1+\log(1+t))(\bar{\mu}_m^{-1/2}\sqrt{\mathcal{E}'_{1,[l+2]}} + C\epsilon_0\sqrt{\mathcal{E}_{0,[2]}}) \quad (14.143)$$

therefore, the contribution of (14.134) to (14.133) is bounded by

$$C_l(1+t)^{-1}(1+\log(1+t))^2 \cdot \int_0^t (1+t')^{-1}(1+\log(1+t'))(\bar{\mu}_m^{-1/2}(t')\sqrt{\mathcal{E}'_{1,[l+2]}(t')}+C\epsilon_0\sqrt{\mathcal{E}_{0,[2]}(t')})dt' \quad (14.144)$$

We shall first estimate the contribution of the principal part in the last factor of the above, namely the term $\bar{\mu}_m^{-1/2}(t')\sqrt{\mathcal{E}'_{1,[l+2]}(t')}$. From the definition (14.111) we have

$$\bar{\mu}_m^{-1/2}(t')\sqrt{\mathcal{E}'_{1,[l+2]}(t')} \leqslant \bar{\mu}_m^{-a-1/2}(t')(1+\log(1+t'))^q\sqrt{\mathcal{G}'_{1,[l+2];a,q}(t)} \quad (14.145)$$

Therefore, the contribution in question is bounded by

$$C_l(1+t)^{-1}(1+\log(1+t))^2 J_{a,q}(t)\sqrt{\mathcal{G}'_{1,[l+2];a,q}(t)} \quad (14.146)$$

where

$$J_{a,q}(t) = \int_0^t (1+t')^{-1}\bar{\mu}_m^{-a-1/2}(t')(1+\log(1+t'))^{q+1}dt' \quad (14.147)$$

To estimate $J_{a,q}(t)$, we consider again the two cases occurring in the proof of Lemma 8.11, as in the estimate of (14.115).

In Case 1 we have the lower bound (14.116) hence also (14.117), which implies

$$J_{a,q}(t) \leqslant C\int_0^t (1+t')^{-1}(1+\log(1+t'))^{q+1}dt' = C\int_0^\tau (1+\tau')^{q+1}d\tau'$$
$$\leqslant \frac{C}{q+2}(1+\tau)^{q+2} = \frac{C}{q+2}(1+\log(1+t))^{q+2} \tag{14.148}$$

Similarly, in Subcase 2a, we obtain if $t \leqslant t_1$

$$J_{a,q}(t) \leqslant \frac{C}{q+2}(1+\log(1+t))^{q+2} \tag{14.149}$$

and if $t > t_1$,

$$J_{a,q}(t_1) \leqslant \frac{C}{q+2}(1+\log(1+t_1))^{q+2} < \frac{C}{q+2}(1+\log(1+t))^{q+2} \tag{14.150}$$

In Subcase 2b we have the lower bound (14.124) which implies

$$J_{a,q}(t) - J_{a,q}(t_1) \leqslant C(1+\log(1+t))^{q+1}\int_{t_1}^t (1-\delta_1\tau')^{-a-1/2}(1+t')^{-1}dt'$$
$$= C(1+\tau)^{q+1}\int_{\tau_1}^\tau (1-\delta_1\tau')^{-a-1/2}d\tau'$$
$$\leqslant \frac{C}{\delta_1}\frac{(1+\tau)^{q+1}}{a-\frac{1}{2}}(1-\delta_1\tau)^{-a+1/2}$$
$$\leqslant \frac{C}{a\delta_1}(1+\tau)^{q+1}\overline{\mu}_m^{-a+1/2}(t) \tag{14.151}$$

where in the last step we have used (8.312) which implies

$$\overline{\mu}_m^{-a+1/2}(t) \geqslant \frac{1}{C}(1-\delta_1\tau)^{-a+1/2}$$

Since $\tau \geqslant \tau_1 = 1/4a\delta_1$, (14.151) implies

$$J_{a,q}(t) - J_{a,q}(t_1) \leqslant C(1+\log(1+t))^{q+2}\overline{\mu}_m^{-a+1/2}(t) \tag{14.152}$$

In view of (14.149), (14.150), (14.152), we conclude that, in general,

$$J_{a,q}(t) \leqslant C(1+\log(1+t))^{q+2}\overline{\mu}_m^{-a+1/2}(t) \tag{14.153}$$

hence (14.146) is bounded by

$$C_l(1+t)^{-1}(1+\log(1+t))^{q+4}\overline{\mu}_m^{-a+1/2}(t)\sqrt{\mathcal{G}'_{1,[l+2];a,q}(t)} \tag{14.154}$$

This bounds the contribution in question through (14.133), to the estimate for

$$\|\mu R_{i_{l+1}}\cdots R_{i_1}\mathrm{tr}\chi'\|_{L^2(\Sigma_t^{\epsilon_0})}$$

In view of (14.114) and (14.90), the corresponding contribution to the integral on the right of (14.63) is then bounded by

$$C_l\delta_0\int_0^t(1+t')^{-2}(1+\log(1+t'))^{p+q+4}\bar{\mu}_m^{-2a-1/2}(t')\sqrt{\mathcal{G}_{0,[l+2];a,p}(t')\mathcal{G}'_{1,[l+2];a,q}(t')}dt' \tag{14.155}$$

This is estimated by following an argument similar to that used to estimate (14.115). We obtain in this way a bound by

$$C_l\delta_0\bar{\mu}_m^{-2a+1/2}(t)(1+\log(1+t))^{2p}(\varphi'_{5+q-p}(Ca\delta_0)\sqrt{\mathcal{G}_{0,[l+2];a,p}(t)\mathcal{G}'_{1,[l+2];a,q}(t)}$$
$$+\int_0^t(1+t')^{-2}(1+\log(1+t'))^{4+q-p}\sqrt{\mathcal{G}_{0,[l+2];a,p}(t')\mathcal{G}'_{1,[l+2];a,q}(t')}dt') \tag{14.156}$$

Here, for any real number $r$ we denote by $\varphi'_r(x)$ the following function on the positive real line:

$$\varphi'_r(x)=e^{-\frac{1}{x}}(1+\frac{1}{x})^r \tag{14.157}$$

Note that $\varphi'_r(x)$ decreases to 0 exponentially as $x\to 0$, and that, since $\delta_1\leqslant C\delta_0$,

$$\varphi'_r(4a\delta_1)\leqslant\varphi'_r(Ca\delta_0)$$

provided that $Ca\delta_0$ is suitably small.

Finally, we consider the contribution of the last term in the expression (14.66) for ${}^{(i_1\cdots i_l)}B_l(t)$ to the first term on the right in (14.65), namely, the contribution of

$$C(1+t)^{-1}(1+\log(1+t))^2\|{}^{(i_1\cdots i_l)}x_l(0)\|_{L^2(\Sigma_0^{\epsilon_0})} \tag{14.158}$$

to the estimate for $\|\mu R_{i_{l+1}}\cdots R_{i_1}\text{tr}\chi'\|_{L^2(\Sigma_t^{\epsilon_0})}$. In view of (14.90) and (14.114), the corresponding contribution to the integral on the right in (14.63) is bounded by

$$C\delta_0\|{}^{(i_1\cdots i_l)}x_l(0)\|_{L^2(\Sigma_0^{\epsilon_0})}\int_0^t(1+t')^{-2}(1+\log(1+t'))^{p+2}\bar{\mu}_m^{-a-1}(t')\sqrt{\mathcal{G}_{0,[l+2];a,p}(t')}dt' \tag{14.159}$$

This is again estimated by following an argument similar to that used to estimate (14.115). We obtain a bound by

$$C_p\delta_0\|{}^{(i_1\cdots i_l)}x_l(0)\|_{L^2(\Sigma_0^{\epsilon_0})}\bar{\mu}_m^{-a}(t)(\varphi'_{3+p}(Ca\delta_0)\sqrt{\mathcal{G}_{0,[l+2];a,p}(t)}$$
$$+\int_0^t(1+t')^{-2}(1+\log(1+t'))^{2+p}\sqrt{\mathcal{G}_{0,[l+2];a,p}(t')}dt') \tag{14.160}$$

In regard to the contribution of the last term on the right in (14.65), we remark that it is bounded by

$$C_l\delta_0(1+t)^{-1}(1+\log(1+t))^4\sup_{t'\in[0,t]}((1+t')^2\bar{\mu}_m^a(t')B_l(t'))\bar{\mu}_m^{-a}(t) \tag{14.161}$$

Thus, relative to the term $(1+t)^{2(i_1\cdots i_l)}B_l$ there is an extra factor of $C_l\delta_0(1+t)^{-1}(1+\log(1+t))^4$, consequently the contribution in question is absorbed in the estimates already made.

### 14.4.2 Estimates for the Contribution of (14.57)

We now consider the contribution of (14.57) to the corresponding integral (14.62). Recalling that this is associated to the variation (14.54), the contribution in question is

$$\begin{aligned}&\int_{W^t_{\epsilon_0}} \Omega|R_{i_{l-m}}\cdots R_{i_1}(T)^m\Delta\mu||T\psi_\alpha||\underline{L}R_{i_{l-m}}\cdots R_{i_1}(T)^{m+1}\psi_\alpha|dt'dud\mu_{\not{g}}\\&\leqslant C\int_0^t \sup_{\Sigma^{\epsilon_0}_{t'}}(\mu^{-1}|T\psi_\alpha|)\|\mu R_{i_{l-m}}\cdots R_{i_1}(T)^m\Delta\mu\|_{L^2(\Sigma^{\epsilon_0}_{t'})}\\&\cdot\|\underline{L}R_{i_{l-m}}\cdots R_{i_1}(T)^{m+1}\psi_\alpha\|_{L^2(\Sigma^{\epsilon_0}_{t'})}dt'\end{aligned}\tag{14.162}$$

$m=0,\cdots,l$.

Now in view of (9.60), (9.70) and (9.213), we have

$$\begin{aligned}&\|\mu R_{i_{l-m}}\cdots R_{i_1}(T)^m\Delta\mu\|_{L^2(\Sigma^{\epsilon_0}_t)}\\&\leqslant C(1+t)(\||{}^{(i_1\cdots i_{l-m})}x'_{m,l-m}(t)\||_{L^2([0,\epsilon_0]\times S^2)}\\&+\||{}^{(i_1\cdots i_{l-m})}\check{f}'_{m,l-m}(t)\||_{L^2([0,\epsilon_0]\times S^2)})\\&=C((1+t)\||{}^{(i_1\cdots i_{l-m})}x'_{m,l-m}(t)\||_{L^2([0,\epsilon_0]\times S^2)}\\&+{}^{(i_1\cdots i_{l-m})}P'_{m,l-m}(t))\end{aligned}\tag{14.163}$$

Substituting (9.268) and (9.270) we then obtain

$$\begin{aligned}&\|\mu R_{i_{l-m}}\cdots R_{i_1}(T)^m\Delta\mu\|_{L^2(\Sigma^{\epsilon_0}_t)}\\&\leqslant C(1+t)((1+t)^{-1}{}^{(i_1\cdots i_{l-m})}P'_{m,l-m}(t)+{}^{(i_1\cdots i_{l-m})}B'_{m,l-m}(t))\\&+C_l\delta_0(1+t)^{-1}(1+\log(1+t))^2\Big(\int_0^t(1+t')(1+\log(1+t'))B_l(t')dt'\\&+\sum_{k=0}^m\int_0^t(1+\log(1+t'))B'_{k,l-k}(t')dt'\Big)\end{aligned}\tag{14.164}$$

for $m=0,\cdots,l$.

Here, the quantities ${}^{(i_1\cdots i_{l-m})}B'_{m,l-m}(t)$ are defined by (9.259):

$$\begin{aligned}&{}^{(i_1\cdots i_{l-m})}B'_{m,l-m}(t)\\&=C(1+t)^{-1}({}^{(i_1\cdots i_{l-m})}\bar{P}'^{(0)}_{m,l-m,a}(t)\\&+(1+t)^{-1/2}{}^{(i_1\cdots i_{l-m})}\bar{P}'^{(1)}_{m,l-m,a}(t))\bar{\mu}^{-a}_m(t)\\&+C(1+t)^{-2}(1+\log(1+t))^2\int_0^t(1+t')^2\,{}^{(i_1\cdots i_{l-m})}Q'_{m,l-m}(t')dt'\\&+C(1+t)^{-2}(1+\log(1+t))^2\|{}^{(i_1\cdots i_{l-m})}x'_{m,l-m}(0)\|_{L^2(\Sigma^{\epsilon_0}_0)}\end{aligned}\tag{14.165}$$

Also (see (9.261))

$$B'_{m,l-m}(t) = \max_{i_1\cdots i_{l-m}} {}^{(i_1\cdots i_{l-m})}B'_{m,l-m}(t) \tag{14.166}$$

The quantities ${}^{(i_1\cdots i_{l-m})}\bar{P}'^{(0)}_{m,l-m,a}(t)$, ${}^{(i_1\cdots i_{l-m})}\bar{P}'^{(1)}_{m,l-m,a}(t)$ are defined by (9.215) and (9.216):

$${}^{(i_1\cdots i_{l-m})}\bar{P}'^{(0)}_{m,l-m,a}(t) = \sup_{t'\in[0,t]} (\bar{\mu}^a_m(t')^{(i_1\cdots i_{l-m})}P'^{(0)}_{m,l-m}(t')) \tag{14.167}$$

$${}^{(i_1\cdots i_{l-m})}\bar{P}'^{(1)}_{m,l-m,a}(t) = \sup_{t'\in[0,t]} ((1+t')^{1/2}\bar{\mu}^a_m(t')^{(i_1\cdots i_{l-m})}P'^{(1)}_{m,l-m}(t')) \tag{14.168}$$

The quantities ${}^{(i_1\cdots i_{l-m})}P'^{(0)}_{m,l-m}(t)$, ${}^{(i_1\cdots i_{l-m})}P'^{(1)}_{m,l-m}(t)$ are defined in the statement of Proposition 11.6, and we have

$${}^{(i_1\cdots i_{l-m})}P'_{m,l-m}(t) \leqslant {}^{(i_1\cdots i_{l-m})}P'^{(0)}_{m,l-m}(t) + {}^{(i_1\cdots i_{l-m})}P'^{(1)}_{m,l-m}(t) \tag{14.169}$$

Note that the quantities ${}^{(i_1\cdots i_{l-m})}\bar{P}'^{(0)}_{m,l-m,a}(t)$, ${}^{(i_1\cdots i_{l-m})}\bar{P}'^{(1)}_{m,l-m,a}(t)$ are non-decreasing.

The leading contribution to the right hand side of (14.164) comes from the first term on the right in (14.164). That is, from $(1+t)^{-1(i_1\cdots i_{l-m})}P'_{m,l-m}(t)$ and the first term in the expression (14.165) for ${}^{(i_1\cdots i_{l-m})}B'_{m,l-m}(t)$. So what we consider here is

$$C({}^{(i_1\cdots i_{l-m})}\bar{P}'^{(0)}_{m,l-m,a}(t) + (1+t)^{-1/2(i_1\cdots i_{l-m})}\bar{P}'^{(1)}_{m,l-m,a}(t))\bar{\mu}^{-a}_m(t) \tag{14.170}$$

The actual *borderline contribution* is the contribution from

$${}^{(i_1\cdots i_{l-m})}P'^{(0)}_{m,l-m} = |\ell|\sqrt{\sum_\alpha \mathcal{E}_0[R_{i_{l-m}}\cdots R_{i_1}(T)^{m+1}\psi_\alpha]} \tag{14.171}$$

We shall appeal to (14.89) in estimating the borderline contribution (14.171), through (14.167), (14.170) and (14.165), to the integral on the right in (14.162). On the other hand, in estimating all other contributions, through (14.165), to the same integral, we simply use (14.90).

Let us define, for non-negative real numbers $a$ and $p$, the quantities

$$\begin{aligned}&{}^{(i_1\cdots i_{l-m})}\mathcal{G}_{0,m,l+2;a,p}(t)\\ &= \sup_{t'\in[0,t]} ((1+\log(1+t'))^{-2p}\bar{\mu}^{2a}_m(t')\sum_\alpha \mathcal{E}_0[R_{i_{l-m}}\cdots R_{i_1}(T)^{m+1}\psi_\alpha](t'))\end{aligned} \tag{14.172}$$

These quantities are non-decreasing functions of $t$ and we have

$$\begin{aligned}&\sqrt{\sum_\alpha \mathcal{E}_0[R_{i_{l-m}}\cdots R_{i_1}(T)^{m+1}\psi_\alpha](t')}\\ &\leqslant \bar{\mu}^{-a}_m(t')(1+\log(1+t'))^p\sqrt{{}^{(i_1\cdots i_{l-m})}\mathcal{G}_{0,m,l+2;a,p}(t)}\end{aligned} \tag{14.173}$$

for all $t' \in [0, t]$. Hence, in view of (14.167), the borderline contribution (14.171) to (14.170) is bounded by

$$|\ell|\overline{\mu}_m^{-a}(t)(1+\log(1+t))^p \sqrt{{}^{(i_1\cdots i_{l-m})}\mathcal{G}_{0,m,l+2;a,p}(t)} \tag{14.174}$$

Also, in regard to the last factor in the integral on the right in (14.162) we have

$$\begin{aligned}
&\|\underline{L}R_{i_{l-m}}\cdots R_{i_1}(T)^{m+1}\psi_\alpha\|_{L^2(\Sigma_t^{\epsilon_0})} \\
&\leqslant \sqrt{\mathcal{E}_0[R_{i_{l-m}}\cdots R_{i_1}(T)^{m+1}\psi_\alpha](t)} \\
&\leqslant \overline{\mu}_m^{-a}(t)(1+\log(1+t))^p \sqrt{{}^{(i_1\cdots i_{l-m})}\mathcal{G}_{0,m,l+2;a,p}(t)}
\end{aligned} \tag{14.175}$$

Substituting (14.89), (14.174) and (14.175) in the integral on the right in (14.162), the factors $|\ell|$ cancel and we obtain that the borderline contribution to the integral in question is bounded by

$$\begin{aligned}
&C\int_0^t (\sup_{\Sigma_{t'}^{\epsilon_0}}(\mu^{-1}|L\mu|) + C\overline{\mu}_m^{-1}\delta_0(1+t')^{-2}(1+\log(1+t'))) \\
&\cdot \overline{\mu}_m^{-2a}(t')(1+\log(1+t'))^{2p}\, {}^{(i_1\cdots i_{l-m})}\mathcal{G}_{0,m,l+2;a,p}(t')dt'
\end{aligned} \tag{14.176}$$

Now, the partial contribution of the second term in the first factor is actually not borderline. We shall show how to estimate this afterwards, in connection with the estimate for the contribution of the terms

$$\begin{aligned}
&C(1+t)^{-1}(1+\log(1+t))^{1/2}\sqrt{\sum_\alpha \mathcal{E}_1'[R_{i_{l-m}}\cdots R_{i_1}(T)^{m+1}\psi_\alpha]} \\
&+C(1+t)^{-2}(1+\log(1+t))^{3/2}\sqrt{\sum_{\alpha,j}\mathcal{E}_1'[R_jR_{i_{l-m}}\cdots R_{i_1}(T)^m\psi_\alpha]}
\end{aligned} \tag{14.177}$$

in the expression for ${}^{(i_1\cdots i_{l-m})}P'^{(1)}_{m,l-m}$ of Proposition 11.6. For the present we focus on the *borderline* integral

$$C\int_0^t \sup_{\Sigma_{t'}^{\epsilon_0}}(\mu^{-1}|L\mu|)\overline{\mu}_m^{-2a}(t')(1+\log(1+t'))^{2p}\, {}^{(i_1\cdots i_{l-m})}\mathcal{G}_{0,m,l+2;a,p}(t')dt' \tag{14.178}$$

This is estimated in exactly the same way as the integral (14.97). We obtain that it is bounded by (see (14.102) and (14.107))

$$C(\frac{1}{2a}+\frac{1}{2p})\overline{\mu}_m^{-2a}(t)(1+\log(1+t))^{2p}\, {}^{(i_1\cdots i_{l-m})}\mathcal{G}_{0,m,l+2;a,p}(t) \tag{14.179}$$

We proceed to estimate the remaining contributions to the right of (14.162). We first consider the contribution of (14.179). From the definitions (14.110) and (14.111) we have

$$\sqrt{\sum_\alpha \mathcal{E}_1'[R_{i_{l-m}} \cdots R_{i_1}(T)^{m+1}\psi_\alpha](t')}$$
$$\leqslant \sqrt{\mathcal{E}'_{1,[l+2]}(t')} \leqslant \bar{\mu}_m^{-a}(t')(1+\log(1+t'))^q \sqrt{\mathcal{G}'_{1,[l+2];a,q}(t)}$$
$$\sqrt{\sum_\alpha \mathcal{E}_1'[R_j R_{i_{l-m}} \cdots R_{i_1}(T)^{m}\psi_\alpha](t')}$$
$$\leqslant \sqrt{\mathcal{E}'_{1,[l+2]}(t')} \leqslant \bar{\mu}_m^{-a}(t')(1+\log(1+t'))^q \sqrt{\mathcal{G}'_{1,[l+2];a,q}(t)} \tag{14.180}$$

for all $t' \in [0,t]$. Hence, in view of the definition (14.168), the contribution of (14.177) to (14.170) is bounded by

$$C_q(1+t)^{-1/2}\bar{\mu}_m^{-a}(t)\sqrt{\mathcal{G}'_{1,[l+2];a,q}(t)} \tag{14.181}$$

Also, we have

$$\|\underline{L}R_{i_{l-m}} \cdots R_{i_1}(T)^{m+1}\psi_\alpha\|_{L^2(\Sigma_t^{\epsilon_0})} \leqslant \bar{\mu}_m^{-a}(t)(1+\log(1+t))^p\sqrt{\mathcal{G}_{0,[l+2];a,p}(t)} \tag{14.182}$$

Using (14.90) we conclude that the contribution of (14.177) through (14.164) to the integral on the right in (14.162) is bounded by

$$C_q\delta_0 \int_0^t (1+t')^{-3/2}(1+\log(1+t'))^p \bar{\mu}_m^{-2a-1}(t')\sqrt{\mathcal{G}_{0,[l+2];a,p}(t')\mathcal{G}'_{1,[l+2];a,q}(t')}dt' \tag{14.183}$$

This is identical in the form to (14.115) and it therefore bounded by (14.122).

Next, we consider the contribution of the second term in the expression (14.165) to the first term on the right of (14.164), namely the contribution of

$$C(1+t)^{-1}(1+\log(1+t))^2 \int_0^t (1+t')^{2}\,{}^{(i_1\cdots i_{l-m})}Q'_{m,l-m}(t')dt' \tag{14.184}$$

to the estimate for $\|\mu R_{i_{l-m}} \cdots R_{i_1}(T)^m \Delta\mu\|_{L^2(\Sigma_t^{\epsilon_0})}$. Here, we shall use Proposition 11.5, namely the estimate for $\max_{i_1\cdots i_{l-m}} {}^{(i_1\cdots i_{l-m})}Q'_{m,l-m}$. The principal part is

$$C_l(1+t)^{-3}(1+\log(1+t))(\mathcal{W}_{\{l+2\}} + \mathcal{W}^Q_{\{l+1\}} + \mathcal{W}^{QQ}_{\{l\}}) \tag{14.185}$$

We shall estimate the contribution of this principal part, through (14.184) and (14.164), to the integral on the right in (14.162).

For each pair of non-negative integers $m, n$ we have

$$\mathcal{W}_{m,n+1} \leqslant C(1+t)\sqrt{\sum_{i_1\cdots i_n,\alpha} \|\not{d}R_{i_n} \cdots R_{i_1}(T)^m\psi_\alpha\|^2_{L^2(\Sigma_t^{\epsilon_0})}}$$
$$\leqslant C\bar{\mu}_m^{-1/2}\sqrt{\mathcal{E}'_{1,m+n+1}} \tag{14.186}$$

and, for each non-negative integer $m$ we have

$$\mathcal{W}_{m+1,0} \leqslant \sqrt{\sum_\alpha \|(T)^{m+1}\psi_\alpha\|^2_{L^2(\Sigma_t^{\epsilon_0})}} \leqslant C\sqrt{\mathcal{E}_{0,m+1}} \tag{14.187}$$

It follows that

$$\begin{aligned}\mathcal{W}_{\{l+2\}} &= \sum_{n+m\leqslant l+2} \mathcal{W}_{m,n} = \sum_{n+m\leqslant l+1} \mathcal{W}_{m,n+1} + \sum_{m\leqslant l+1} \mathcal{W}_{m+1,0} + \mathcal{W}_0 \\ &\leqslant C_l(\bar{\mu}_m^{-1/2}\sqrt{\mathcal{E}'_{1,[l+2]}} + \sqrt{\mathcal{E}_{0,[l+2]}}) + \mathcal{W}_0 \end{aligned} \tag{14.188}$$

Similarly, we have

$$\mathcal{W}^Q_{\{l+1\}} \leqslant C_l(\bar{\mu}_m^{-1/2}\sqrt{\mathcal{E}'_{1,[l+2]}} + \sqrt{\mathcal{E}_{0,[l+2]}}) + \mathcal{W}^Q_0 \tag{14.189}$$

$$\mathcal{W}^{QQ}_{\{l\}} \leqslant C_l(\bar{\mu}_m^{-1/2}\sqrt{\mathcal{E}'_{1,[l+2]}} + \sqrt{\mathcal{E}_{0,[l+2]}}) + \mathcal{W}^{QQ}_0 \tag{14.190}$$

We then conclude that (14.185) is bounded by

$$C_l(1+t)^{-3}(1+\log(1+t))(\bar{\mu}_m^{-1/2}\sqrt{\mathcal{E}'_{1,[l+2]}} + \sqrt{\mathcal{E}_{0,[l+2]}}) \tag{14.191}$$

(as we may assume that $l \geqslant 1$). Consequently, the contribution of (14.185) to (14.184) is bounded by

$$\begin{aligned}&C_l(1+t)^{-1}(1+\log(1+t))^2 \\ &\cdot \int_0^t (1+t')^{-1}(1+\log(1+t'))(\bar{\mu}_m^{-1/2}(t')\sqrt{\mathcal{E}'_{1,[l+2]}(t')} + \sqrt{\mathcal{E}_{0,[l+2]}(t')})dt' \end{aligned} \tag{14.192}$$

From the definition (14.109),

$$\sqrt{\mathcal{E}_{0,[l+2]}(t')} \leqslant \bar{\mu}_m^{-a}(t')(1+\log(1+t'))^p\sqrt{\mathcal{G}_{0,[l+2];a,p}(t)} \tag{14.193}$$

This together with (14.145) implies that (14.192) is bounded by, recalling the definition (14.147),

$$C_l(1+t)^{-1}(1+\log(1+t))^2(J_{a,q}(t)\sqrt{\mathcal{G}'_{1,[l+2];a,q}(t)} + J_{a-1/2,p}(t)\sqrt{\mathcal{G}_{0,[l+2];a,p}(t)}) \tag{14.194}$$

Substituting (14.153) and the same with $(a,q)$ replaced by $(a-\frac{1}{2},p)$ we conclude that the contribution of (14.185), through (14.184) to the estimate for

$$\|\mu R_{i_{l-m}} \cdots R_{i_1}(T)^m \Delta\mu\|_{L^2(\Sigma_t^{\epsilon_0})}$$

is bounded by

$$C_l(1+t)^{-1}(1+\log(1+t))^{q+4}\bar{\mu}_m^{-a+1/2}(t)\sqrt{\mathcal{G}'_{1,[l+2];a,q}(t)}$$
$$+C_l(1+t)^{-1}(1+\log(1+t))^{p+4}\bar{\mu}_m^{-a+1}(t)\sqrt{\mathcal{G}_{0,[l+2];a,p}(t)} \quad (14.195)$$

In view of (14.182) and (14.90), the corresponding contribution to the integral on the right in (14.162) is then bounded by

$$C_l\delta_0\int_0^t(1+t')^{-2}(1+\log(1+t'))^{p+q+4}\bar{\mu}_m^{-2a-1/2}(t')\sqrt{\mathcal{G}_{0,[l+2];a,p}(t')\mathcal{G}'_{1,[l+2];a,q}(t')}dt'$$
$$+C_l\delta_0\int_0^t(1+t')^{-2}(1+\log(1+t'))^{2p+4}\bar{\mu}_m^{-2a}(t')\mathcal{G}_{0,[l+2];a,p}(t')dt' \quad (14.196)$$

The first terms in (14.196) coincides with (14.155) and is estimated by (14.156) while the second is estimated in a similar manner by

$$C_l\delta_0\bar{\mu}_m^{-2a+1}(1+\log(1+t))^{2p}(\varphi'_5(Ca\delta_0)\mathcal{G}_{0,[l+2];a,p}(t)$$
$$+\int_0^t(1+t')^{-2}(1+\log(1+t'))^4\mathcal{G}_{0,[l+2];a,p}(t')dt') \quad (14.197)$$

Let us finally consider the contribution of the last term in the expression (14.167) to the first term on the right in (14.164), namely the contribution of

$$C(1+t)^{-1}(1+\log(1+t))^2\|^{(i_1\cdots i_{l-m})}x'_{m,l-m}(0)\|_{L^2(\Sigma_0^{\epsilon_0})} \quad (14.198)$$

In view of (14.182) and the estimate (14.90), the contribution in question is bounded by

$$C\delta_0\|^{(i_1\cdots i_{l-m})}x'_{m,l-m}(0)\|_{L^2(\Sigma_0^{\epsilon_0})}$$
$$\cdot\int_0^t(1+t')^{-2}(1+\log(1+t'))^{p+2}\bar{\mu}_m^{-a-1}(t')\sqrt{\mathcal{G}_{0,[l+2];a,p}(t')}dt' \quad (14.199)$$

This is similar to (14.159), and is bounded by

$$C_p\delta_0\|^{(i_1\cdots i_{l-m})}x'_{m,l-m}(0)\|_{L^2(\Sigma_0^{\epsilon_0})}\bar{\mu}_m^{-a}(t)(\varphi'_{3+p}(Ca\delta_0)\sqrt{\mathcal{G}_{0,[l+2];a,p}(t)}$$
$$+\int_0^t(1+t')^{-2}(1+\log(1+t'))^{2+p}\sqrt{\mathcal{G}_{0,[l+2];a,p}(t')}dt') \quad (14.200)$$

In regard to the contribution of the last term on the right in (14.164), we remark that it is bounded by

$$C_l\delta_0(1+t)^{-1}(1+\log(1+t))^4$$
$$\cdot(\sup_{t'\in[0,t]}((1+t')^2\bar{\mu}_m^a(t')B_l(t'))+\sum_{k=0}^m\sup_{t'\in[0,t]}((1+t')\bar{\mu}_m^a(t')B'_{k,l-k}))\bar{\mu}_m^{-a}(t) \quad (14.201)$$

Thus, relative to the terms $(1+t)^{2(i_1\cdots i_l)}B_l$ and $(1+t)^{(i_1\cdots i_{l-m})}B'_{m,l-m}$ there is an extra factor of $C_l\delta_0(1+t)^{-1}(1+\log(1+t))^4$, consequently the contribution in question is absorbed in the estimates already made.

## 14.5 The Borderline Estimates Associated to $K_1$

### 14.5.1 Estimates for the Contribution of (14.56)

We now consider the contribution of (14.56) to the corresponding integral (14.59). Recalling that this is associated to the variation (14.52), the contribution in question is

$$-\int_{W_u^t}(\omega/\nu)\Omega(R_{i_{l+1}}\cdots R_{i_1}\mathrm{tr}\chi')(T\psi_\alpha)((L+\nu)R_{i_{l+1}}\cdots R_{i_1}\psi_\alpha)dt'du'd\mu_{\not g} \quad (14.202)$$

Here, using the flux $\mathcal{F}_1'[R_{i_{l+1}}\cdots R_{i_1}\psi_\alpha]$ does not lead to an appropriate estimate. Instead, we proceed as follows. First, since

$$\tilde{\not g}=\Omega\not g,\quad d\mu_{\tilde{\not g}}=\Omega d\mu_{\not g}$$

the integral (14.202) is

$$-\int_{W_u^t}(\omega/\nu)(T\psi_\alpha)(R_{i_{l+1}}\cdots R_{i_1}\mathrm{tr}\chi')((L+\nu)R_{i_{l+1}}\cdots R_{i_1}\psi_\alpha)dt'du'd\mu_{\tilde{\not g}} \quad (14.203)$$

Let us consider, for an arbitrary function $f$, the integral

$$\int_{W_u^t}(Lf+2\nu f)dt'du'd\mu_{\tilde{\not g}}$$

Let

$$F(t,u)=\int_{S_{t,u}}fd\mu_{\tilde{\not g}} \quad (14.204)$$

Then we have

$$\frac{\partial F}{\partial t}=\int_{S_{t,u}}(Lf+2\nu f)d\mu_{\tilde{\not g}} \quad (14.205)$$

We have used this formula in Chapter 5. Therefore

$$\int_{W_u^t}(Lf+2\nu f)dt'du'd\mu_{\tilde{\not g}}=\int_0^u\int_0^t\frac{\partial F}{\partial t'}(t',u')dt'du'$$
$$=\int_0^u(F(t,u')-F(0,u'))du'=\int_{\Sigma_t^u}fdu'd\mu_{\tilde{\not g}}-\int_{\Sigma_0^u}fdu'd\mu_{\tilde{\not g}} \quad (14.206)$$

Going back to (14.203), we write the integrand in the form

$$\begin{aligned}&-(\omega/\nu)(T\psi_\alpha)(R_{i_{l+1}}\cdots R_{i_1}\mathrm{tr}\chi')((L+\nu)R_{i_{l+1}}\cdots R_{i_1}\psi_\alpha)\\&=-(L+2\nu)((\omega/\nu)(T\psi_\alpha)(R_{i_{l+1}}\cdots R_{i_1}\mathrm{tr}\chi')(R_{i_{l+1}}\cdots R_{i_1}\psi_\alpha))\\&\quad+(R_{i_{l+1}}\cdots R_{i_1}\psi_\alpha)(L+\nu)((\omega/\nu)(T\psi_\alpha)(R_{i_{l+1}}\cdots R_{i_1}\mathrm{tr}\chi'))\end{aligned}$$

By (14.206) with the function $(\omega/\nu)(T\psi_\alpha)(R_{i_{l+1}}\cdots R_{i_1}\mathrm{tr}\chi')(R_{i_{l+1}}\cdots R_{i_1}\psi_\alpha)$ in the role of the function $f$, we then conclude that (14.203) equals

$$-\int_{\Sigma_t^u}(\omega/\nu)(T\psi_\alpha)(R_{i_{l+1}}\cdots R_{i_1}\mathrm{tr}\chi')(R_{i_{l+1}}\cdots R_{i_1}\psi_\alpha)du'd\mu_{\tilde{\not g}}$$
$$+\int_{\Sigma_0^u}(\omega/\nu)(T\psi_\alpha)(R_{i_{l+1}}\cdots R_{i_1}\mathrm{tr}\chi')(R_{i_{l+1}}\cdots R_{i_1}\psi_\alpha)du'd\mu_{\tilde{\not g}}$$
$$+\int_{W_u^t}(R_{i_{l+1}}\cdots R_{i_1}\psi_\alpha)(L+\nu)((\omega/\nu)(T\psi_\alpha)(R_{i_{l+1}}\cdots R_{i_1}\mathrm{tr}\chi'))dt'du'd\mu_{\tilde{\not g}} \quad (14.207)$$

We first consider the hypersurface integral

$$-\int_{\Sigma_t^u}(\omega/\nu)(T\psi_\alpha)(R_{i_{l+1}}\cdots R_{i_1}\mathrm{tr}\chi')(R_{i_{l+1}}\cdots R_{i_1}\psi_\alpha)du'd\mu_{\tilde{\not g}} \quad (14.208)$$

(The other hypersurface integral can be expressed in terms of initial data.) Let $f$, $g$ be arbitrary functions defined on $S_{t,u}$ and $X$ an arbitrary vectorfield tangent to $S_{t,u}$. We have

$$\int_{S_{t,u}} f(Xg)d\mu_{\tilde{\not g}} = \int_{S_{t,u}} X(fg)d\mu_{\tilde{\not g}} - \int_{S_{t,u}} g(Xf)d\mu_{\tilde{\not g}}$$

and

$$\int_{S_{t,u}} X(fg)d\mu_{\tilde{\not g}} = \int_{S_{t,u}} \tilde{\not{\mathrm{d}}\mathrm{iv}}(fgX)d\mu_{\tilde{\not g}} - \int_{S_{t,u}} (\tilde{\not{\mathrm{d}}\mathrm{iv}}X)fgd\mu_{\tilde{\not g}}$$

Since

$$\int_{S_{t,u}} \tilde{\not{\mathrm{d}}\mathrm{iv}}(fgX)d\mu_{\tilde{\not g}} = 0, \quad \tilde{\not{\mathrm{d}}\mathrm{iv}}X = \frac{1}{2}\mathrm{tr}\,{}^{(X)}\tilde{\not\pi}$$

we obtain

$$\int_{S_{t,u}} f(Xg)d\mu_{\tilde{\not g}} = -\int_{S_{t,u}} (g(Xf) + \frac{1}{2}\mathrm{tr}\,{}^{(X)}\tilde{\not\pi} fg)d\mu_{\tilde{\not g}} \quad (14.209)$$

Applying (14.209), taking

$$X = R_{i_{l+1}}, \quad g = R_{i_l}\cdots R_{i_1}\mathrm{tr}\chi', \quad f = (\omega/\nu)(T\psi_\alpha)(R_{i_{l+1}}\cdots R_{i_1}\psi_\alpha)$$

we conclude that (14.208) equals the sum

$$H_0 + H_1 + H_2 \quad (14.210)$$

with

$$H_0 = \int_{\Sigma_t^u}(\omega/\nu)(T\psi_\alpha)(R_{i_l}\cdots R_{i_1}\mathrm{tr}\chi')(R_{i_{l+1}}R_{i_{l+1}}\cdots R_{i_1}\psi_\alpha)du'd\mu_{\tilde{\not g}} \quad (14.211)$$

$$H_1 = \int_{\Sigma_t^u} (\omega/\nu)(R_{i_{l+1}} T\psi_\alpha)(R_{i_l} \cdots R_{i_1} \mathrm{tr}\chi')(R_{i_{l+1}} \cdots R_{i_1} \psi_\alpha) du' d\mu_{\tilde{g}} \tag{14.212}$$

$$H_2 = \int_{\Sigma_t^u} (T\psi_\alpha)(R_{i_l} \cdots R_{i_1} \mathrm{tr}\chi')(R_{i_{l+1}} \cdots R_{i_1} \psi_\alpha)(R_{i_{l+1}}(\omega/\nu) + \frac{1}{2}\mathrm{tr}^{(R_{i_{l+1}})}\tilde{\not{\pi}}(\omega/\nu)) du' d\mu_{\tilde{g}} \tag{14.213}$$

Now, by Proposition 12.9 and Corollary 10.1.d with $l = 1$, recalling that $\omega$ is constant on each $S_{t,u}$,

$$(\nu/\omega)|R_{i_{l+1}}(\omega/\nu)| + \frac{1}{2}|\mathrm{tr}^{(R_{i_{l+1}})}\tilde{\not{\pi}}| \leqslant C\delta_0(1+t)^{-1}(1+\log(1+t)) \tag{14.214}$$

It follows that $H_2$ has an extra decay factor of $\delta_0(1+t)^{-1}(1+\log(1+t))$ relative to $H_1$. So we confine ourselves to $H_0$ and $H_1$.

We first consider $H_0$. We have

$$|H_0| \leqslant C \int_{\Sigma_t^{\epsilon_0}} (1+t)^3 |T\psi_\alpha||R_{i_l} \cdots R_{i_1} \mathrm{tr}\chi'||\not{d} R_{i_{l+1}} \cdots R_{i_1} \psi_\alpha| du d\mu_{\not{g}} \tag{14.215}$$

Although we have an $L^2$ bound for $R_{i_l} \cdots R_{i_1} \mathrm{tr}\chi'$ from Proposition 12.11, here we need a more precise estimate. Since

$$L\mathrm{tr}\chi' = \mathrm{tr}(\not{\mathcal{L}}_L \chi') - 2\mathrm{tr}(\chi \cdot \chi') = \mathrm{tr}(\not{\mathcal{L}}_L \chi') - \frac{2}{1-u+t}\mathrm{tr}\chi' - 2|\chi'|^2 \tag{14.216}$$

taking the trace of (12.47), we obtain

$$L\mathrm{tr}\chi' + \frac{2}{1-u+t}\mathrm{tr}\chi' = \rho_0 \tag{14.217}$$

where

$$\rho_0 = e\mathrm{tr}\chi - |\chi'|^2 + \mathrm{tr}b \tag{14.218}$$

Applying $R_{i_l} \cdots R_{i_1}$ to this equation and using the Lemma 11.22 then yields the propagation equation

$$LR_{i_l} \cdots R_{i_1} \mathrm{tr}\chi' + \frac{2}{1-u+t} R_{i_l} \cdots R_{i_1} \mathrm{tr}\chi' = {}^{(i_1 \cdots i_l)}\rho_l \tag{14.219}$$

where

$${}^{(i_1 \cdots i_l)}\rho_l = R_{i_l} \cdots R_{i_1} \rho_0 + \sum_{k=0}^{l-1} R_{i_l} \cdots R_{i_{l-k+1}} {}^{(R_{i_{l-k}})}Z R_{i_{l-k-1}} \cdots R_{i_1} \mathrm{tr}\chi' \tag{14.220}$$

Obviously, the principal part of $\rho_0$ is contained in $-\mathrm{tr}\alpha'$, and is $\frac{1}{2}\frac{dH}{dh}\not{\Delta} h$. We can write this in a more precise way:

$$\frac{1}{2}\frac{dH}{dh}\not{\Delta} h = \frac{1}{2}\frac{dH}{dh}(\not{\Delta}\psi_0 - \sum_i \psi_i \not{\Delta}\psi_i - \sum_i \not{d}\psi_i \cdot \not{d}\psi_i)$$

So we can express

$$-R_{i_l}\cdots R_{i_1}(\mathrm{tr}\alpha'^{[P]}) = \frac{1}{2}R_{i_l}\cdots R_{i_1}[\frac{dH}{dh}(\Delta\psi_0 - \sum_i \psi_i\Delta\psi_i - \sum_i \not{d}\psi_i\cdot\not{d}\psi_i)]$$
$$= m_T^\alpha R_{i_l}\cdots R_{i_1}\Delta\psi_\alpha + {}^{(i_1\cdots i_l)}\tilde{n}_l \tag{14.221}$$

Here, $m_T^\alpha$ are defined in Chapter 10, and ${}^{(i_1\cdots i_l)}\tilde{n}_l$ is a lower order term (of order $l+1$):

$${}^{(i_1\cdots i_l)}\tilde{n}_l = \sum_{|s_1|>0}((R)^{s_1}m_T^\alpha)((R)^{s_2}\Delta\psi_\alpha) + R_{i_l}\cdots R_{i_1}\tilde{n}_0 \tag{14.222}$$

where

$$\tilde{n}_0 = -\frac{1}{2}\frac{dH}{dh}\sum_i \not{d}\psi_i\cdot\not{d}\psi_i \tag{14.223}$$

From the bootstrap assumption, we have

$$\|{}^{(i_1\cdots i_l)}\tilde{n}_l\|_{L^2(\Sigma_t^{\epsilon_0})} \leqslant C_l\delta_0(1+t)^{-3}\mathcal{W}_{\{l+1\}} \tag{14.224}$$

According to (10.185),

$$|m_T^0 - \frac{1}{2}\ell| \leqslant C\delta_0(1+t)^{-1}, \quad |m_T^i| \leqslant C\delta_0(1+t)^{-1} \tag{14.225}$$

Let us define the functions

$${}^{(i_1\cdots i_l)}\rho_l^{(0)} = \frac{1}{2}\ell\Delta R_{i_l}\cdots R_{i_1}\psi_0 \tag{14.226}$$
$${}^{(i_1\cdots i_l)}\rho_l^{(1)} = (m_T^0 - \frac{1}{2}\ell)\Delta R_{i_l}\cdots R_{i_1}\psi_0 + m_T^i\Delta R_{i_l}\cdots R_{i_1}\psi_i \tag{14.227}$$

We can then express (14.221) as

$$-R_{i_l}\cdots R_{i_1}(\mathrm{tr}\alpha'^{[P]}) = {}^{(i_1\cdots i_l)}\rho_l^{(0)} + {}^{(i_1\cdots i_l)}\rho_l^{(1)} + {}^{(i_1\cdots i_l)}\tilde{n}_l \tag{14.228}$$

Let us define ${}^{(i_1\cdots i_l)}\rho_l^{(2)}$ by

$${}^{(i_1\cdots i_l)}\rho_l = {}^{(i_1\cdots i_l)}\rho_l^{(0)} + {}^{(i_1\cdots i_l)}\rho_l^{(1)} + {}^{(i_1\cdots i_l)}\rho_l^{(2)} \tag{14.229}$$

We shall estimate the $L^2(\Sigma_t^{\epsilon_0})$ norm of ${}^{(i_1\cdots i_l)}\rho_l^{(2)}$. First, we estimate the contribution from

$$\sum_{k=0}^{l-1} R_{i_l}\cdots R_{i_{l-k+1}}{}^{(R_{i_{l-k}})}ZR_{i_{l-k-1}}\cdots R_{i_1}\mathrm{tr}\chi'$$

Obviously, this contribution is bounded by

$$(1+t)^{-2}(1+\log(1+t))\mathcal{A}'_{[l]}(t) \tag{14.230}$$

where we have used the bounds for ${}^{(R_i)}Z$ in Corollary 10.1.i.

Next, we should estimate the contribution from $R_{i_l}\cdots R_{i_1}\rho_0$. Recall the definition

$$\rho_0 = e\mathrm{tr}\chi' + \frac{2e}{1-u+t} - |\chi'|^2 - \mathrm{tr}\alpha'$$

By Proposition 12.6, (12.293) and (12.295), the contribution from the first term of above is bounded by

$$C_l(1+t)^{-3}(1+\log(1+t))\mathcal{W}^Q_{\{l\}} + \delta_0(1+t)^{-3}(\mathcal{Y}_0 + (1+t)\mathcal{A}'_{[l]} + \mathcal{W}_{[l]}) \quad (14.231)$$

Also by (12.295), the contribution from the second term is bounded by

$$C_l(1+t)^{-2}(\mathcal{W}^Q_{\{l\}} + \delta_0(1+t)^{-1}(\mathcal{Y}_0 + (1+t)\mathcal{A}'_{[l]} + \mathcal{W}_{\{l\}})) \quad (14.232)$$

By Proposition 12.6, the contribution from the third term is bounded by

$$(1+t)^{-2}(1+\log(1+t))(\mathcal{A}'_{[l]} + \delta_0(1+t)^{-2}(1+\log(1+t))\cdot(\mathcal{W}_{[l]} + \mathcal{Y}_0)) \quad (14.233)$$

where we have used Corollary 10.2.d.

Finally, by the expression of $\alpha^{[N]}_{AB}$ and Corollary 10.1.g and Corollary 10.2.g, the contribution from the fourth term is bounded by

$$(1+t)^{-3}(\mathcal{W}_{\{l+1\}} + \mathcal{W}^Q_{\{l\}} + (1+t)^{-1}(\mathcal{Y}_0 + (1+t)\mathcal{A}'_{[l]})) \quad (14.234)$$

So we obtain the following estimate for ${}^{(i_1\cdots i_l)}\rho^{(2)}_l$:

$$\begin{aligned}&\|{}^{(i_1\cdots i_l)}\rho^{(2)}_l\|_{L^2(\Sigma^{\epsilon_0}_t)}\\ &\leqslant C_l(1+t)^{-2}\mathcal{W}^Q_{\{l\}} + \delta_0(1+t)^{-3}(1+\log(1+t))(\mathcal{W}_{\{l+1\}} + \mathcal{Y}_0 + (1+t)\mathcal{A}'_{[l]})\end{aligned} \quad (14.235)$$

On the other hand, by **H1** we have, pointwise

$$|{}^{(i_1\cdots i_l)}\rho^{(0)}_l| \leqslant C(1+t)^{-1}|\ell|\sqrt{\sum_{j,\alpha}|\not{d}R_jR_{i_l}\cdots R_{i_1}\psi_\alpha|^2} \quad (14.236)$$

and

$$|{}^{(i_1\cdots i_l)}\rho^{(1)}_l| \leqslant C\delta_0(1+t)^{-2}\sqrt{\sum_{j,\alpha}|\not{d}R_jR_{i_l}\cdots R_{i_1}\psi_\alpha|^2} \quad (14.237)$$

These inequalities imply

$$\|{}^{(i_1\cdots i_l)}\rho^{(0)}_l\|_{L^2(\Sigma^{\epsilon_0}_t)} \leqslant C|\ell|(1+t)^{-2}\bar{\mu}_m^{-1/2}(t)\sqrt{\sum_{j,\alpha}\mathcal{E}'_1[R_jR_{i_l}\cdots R_{i_1}\psi_\alpha](t)} \quad (14.238)$$

and

$$\|{}^{(i_1\cdots i_l)}\rho^{(1)}_l\|_{L^2(\Sigma^{\epsilon_0}_t)} \leqslant C\delta_0(1+t)^{-3}\bar{\mu}_m^{-1/2}(t)\sqrt{\sum_{j,\alpha}\mathcal{E}'_1[R_jR_{i_l}\cdots R_{i_1}\psi_\alpha](t)} \quad (14.239)$$

while

$$\|\mu^{1/2(i_1\cdots i_l)}\rho_l^{(0)}\|_{L^2(\Sigma_t^{\epsilon_0})} \leqslant C|\ell|(1+t)^{-2}\sqrt{\sum_{j,\alpha}\mathcal{E}_1'[R_jR_{i_l}\cdots R_{i_1}\psi_\alpha](t)} \tag{14.240}$$

and

$$\|\mu^{1/2(i_1\cdots i_l)}\rho_l^{(1)}\|_{L^2(\Sigma_t^{\epsilon_0})} \leqslant C\delta_0(1+t)^{-3}\sqrt{\sum_{j,\alpha}\mathcal{E}_1'[R_jR_{i_l}\cdots R_{i_1}\psi_\alpha](t)} \tag{14.241}$$

We express the propagation equation (14.219) in acoustical coordinates $(t,u,\vartheta)$, so that $L$ becomes $\frac{\partial}{\partial t}$, and integrate it along each integral curve of $L$ from $\Sigma_0^{\epsilon_0}$ to obtain

$$\begin{aligned}&(1-u+t)^2(R_{i_l}\cdots R_{i_1}\mathrm{tr}\chi')(t,u,\vartheta)\\&=(1-u)^2(R_{i_l}\cdots R_{i_1}\mathrm{tr}\chi')(0,u,\vartheta)+\int_0^t(1-u+t')^{2(i_1\cdots i_l)}\rho_l(t',u,\vartheta)dt'\end{aligned} \tag{14.242}$$

It follows that

$$\begin{aligned}&|R_{i_l}\cdots R_{i_1}\mathrm{tr}\chi'|(t,u,\vartheta)\\&\leqslant C(1+t)^{-2}\Big(|R_{i_l}\cdots R_{i_1}\mathrm{tr}\chi'|(0,u,\vartheta)+\sum_{k=0}^{2}{}^{(i_1\cdots i_l)}A_l^{(k)}(t,u,\vartheta)\Big)\end{aligned} \tag{14.243}$$

where

$${}^{(i_1\cdots i_l)}A^{(k)}(t,u,\vartheta)=\int_0^t(1+t')^2|{}^{(i_1\cdots i_l)}\rho_l^{(k)}|(t',u,\vartheta)dt', \quad k=0,1,2 \tag{14.244}$$

By (8.333) we have

$$\begin{aligned}\|{}^{(i_1\cdots i_l)}A_l^{(k)}(t)\|_{L^2([0,\epsilon_0]\times S^2)}&\leqslant\int_0^t(1+t')^2\|{}^{(i_1\cdots i_l)}\rho_l^{(k)}(t')\|_{L^2([0,\epsilon_0]\times S^2)}dt'\\&\leqslant C\int_0^t(1+t')\|{}^{(i_1\cdots i_l)}\rho_l^{(k)}\|_{L^2(\Sigma_{t'}^{\epsilon_0})}dt'\end{aligned} \tag{14.245}$$

We now substitute the pointwise estimate (14.243) in (14.215). The *borderline* contribution is from ${}^{(i_1\cdots i_l)}A_l^{(0)}$. This contribution is the *borderline* integral

$$C\int_{\Sigma_t^{\epsilon_0}}(1+t)|T\psi_\alpha||{}^{(i_1\cdots i_l)}A_l^{(0)}||\not{d}R_{i_{l+1}}\cdots R_{i_1}\psi_\alpha|dud\mu_{\not{g}} \tag{14.246}$$

Recalling from Chapter 8 the partition of $[0,\epsilon_0]\times S^2$ into the subsets $\mathcal{V}_{s-}$, $\mathcal{V}_{s+}$, defined by (8.337), (8.338), respectively, we shall consider separately the integrals over the corresponding regions $\mathcal{U}_{s,t}^-$, $\mathcal{U}_{s,t}^+$ of $\Sigma_t^{\epsilon_0}$:

$$\mathcal{U}_{s,t}^-=\{(t,u,\vartheta)\in\Sigma_t^{\epsilon_0},(u,\vartheta)\in\mathcal{V}_{s-}\} \tag{14.247}$$

$$\mathcal{U}_{s,t}^+=\{(t,u,\vartheta)\in\Sigma_t^{\epsilon_0},(u,\vartheta)\in\mathcal{V}_{s+}\} \tag{14.248}$$

We consider first the integral over $\mathcal{U}_{s,t}^-$. We have

$$\begin{aligned}
&\int_{\mathcal{U}_{s,t}^-}(1+t)|T\psi_\alpha||^{(i_1\cdots i_l)}A_l^{(0)}||\not{d}R_{i_{l+1}}\cdots R_{i_1}\psi_\alpha|dud\mu_{\not{g}} \\
&\leqslant (1+t)\sup_{\mathcal{U}_{s,t}^-}|T\psi_\alpha|\|^{(i_1\cdots i_l)}A_l^{(0)}\|_{L^2(\mathcal{U}_{s,t}^-)}\|\not{d}R_{i_{l+1}}\cdots R_{i_1}\psi_\alpha\|_{L^2(\Sigma_t^{\epsilon_0})} \\
&\leqslant C(1+t)\overline{\mu}_m^{-1/2}(t)\sup_{\mathcal{U}_{s,t}^-}|T\psi_\alpha|\|^{(i_1\cdots i_l)}A_l^{(0)}(t)\|_{L^2(\mathcal{V}_{s-})}\sqrt{\mathcal{E}_1'[R_{i_{l+1}}\cdots R_{i_1}\psi_\alpha](t)} \qquad (14.249)
\end{aligned}$$

by (8.333). Now by (14.87) and (14.88), we have, pointwise on $\Sigma_t^{\epsilon_0}$,

$$\max_\alpha |T\psi_\alpha| \leqslant \frac{C}{|\ell|}(|L\mu| + C\delta_0(1+t)^{-2}(1+\log(1+t))) \qquad (14.250)$$

In estimating the integral over $\mathcal{U}_{s,t}^-$ we can assume that $\mathcal{V}_{s-}$ is non-empty, thus, recalling the definitions (8.251) and (8.261),

$$\min_{(u,\vartheta)\in\mathcal{V}_{s-}} \hat{E}_s(u,\vartheta) = \hat{E}_{s,m} = -\delta_1, \quad \delta_1 > 0 \qquad (14.251)$$

Now, from Proposition 8.6 we have

$$(L\mu)(t,u,\vartheta) = \mu_{[1],s}(u,\vartheta)(\frac{\hat{E}_s(u,\vartheta)}{1+t} + \hat{Q}_{1,s}(t,u,\vartheta)) \qquad (14.252)$$

It follows that

$$\sup_{\mathcal{U}_{s,t}^-}|L\mu| \leqslant C(\frac{\delta_1}{1+t} + \frac{C\delta_0(1+\log(1+t))}{(1+t)^2}) \qquad (14.253)$$

hence, substituting in (14.250),

$$\max_\alpha \sup_{\mathcal{U}_{s,t}^-}|T\psi_\alpha| \leqslant \frac{C}{|\ell|}(\frac{\delta_1}{1+t} + \frac{C\delta_0(1+\log(1+t))}{(1+t)^2}) \qquad (14.254)$$

On the other hand, from (14.245) for $k=0$ and (14.238),

$$\begin{aligned}
&\|^{(i_1\cdots i_l)}A_l^{(0)}(t)\|_{L^2([0,\epsilon_0]\times S^2)} \\
&\leqslant C|\ell|\int_0^t (1+t')^{-1}\overline{\mu}_m^{-1/2}(t')\sqrt{\sum_{j,\alpha}\mathcal{E}_1'[R_jR_{i_l}\cdots R_{i_1}\psi_\alpha](t')dt'} \qquad (14.255)
\end{aligned}$$

Substituting (14.254) and (14.255) in (14.249), the factors $|\ell|$ cancel and we obtain that (14.249) is bounded by

$$\begin{aligned}
&C\overline{\mu}_m^{-1/2}(t)(\delta_1 + \frac{C\delta_0(1+\log(1+t))}{1+t})\sqrt{\mathcal{E}_1'[R_{i_{l+1}}\cdots R_{i_1}\psi_\alpha](t)} \\
&\cdot\int_0^t (1+t')^{-1}\overline{\mu}_m^{-1/2}(t')\sqrt{\sum_{j,\alpha}\mathcal{E}_1'[R_jR_{i_l}\cdots R_{i_1}\psi_\alpha](t')dt'} \qquad (14.256)
\end{aligned}$$

We now define

$$
\begin{aligned}
&{}^{(i_1\cdots i_l)}\mathcal{G}'_{1,l+2;a,q}(t)\\
&= \sup_{t'\in[0,t]}((1+\log(1+t))^{-2q}\bar{\mu}_m^{2a}(t')\sum_{j,\alpha}\mathcal{E}'_1[R_jR_{i_l}\cdots R_{i_1}\psi_\alpha](t'))
\end{aligned} \tag{14.257}
$$

Then (14.256) is bounded by

$$
\begin{aligned}
&C(1+\log(1+t))^{2q}\bar{\mu}_m^{-a-1/2}(t)(\delta_1+\frac{C\delta_0(1+\log(1+t))}{1+t})\\
&\cdot\int_0^t\bar{\mu}_m^{-a-1/2}(t')\frac{dt'}{1+t'}\cdot{}^{(i_1\cdots i_l)}\mathcal{G}'_{1,l+2;a,q}(t)
\end{aligned} \tag{14.258}
$$

To estimate the integral in (14.258), we follow the argument of Lemma 8.11. Since we are assuming that $\mathcal{V}_{s-}$ is non-empty, we only have Case 2 to consider. Setting again, as in (8.267),

$$
t_1=e^{\frac{1}{2a\delta_1}}-1
$$

we have two subcases of Case 2 as in the proof of Lemma 8.11. In Subcase 2a the lower bound (8.273) holds, hence $\bar{\mu}_m^{-a-1/2}(t')\leqslant C$ and

$$
\int_0^t\bar{\mu}_m^{-a-1/2}(t')\frac{dt'}{1+t'}\leqslant C\int_0^t\frac{dt'}{1+t'}=C\log(1+t) \tag{14.259}
$$

therefore

$$
\begin{aligned}
&(\delta_1+\frac{C\delta_0(1+\log(1+t))}{1+t})\int_0^t\bar{\mu}_m^{-a-1/2}(t')\frac{dt'}{1+t'}\\
&\leqslant C(\delta_1\log(1+t)+C\delta_0\frac{(1+\log(1+t))^2}{1+t})\leqslant\frac{C}{a}
\end{aligned} \tag{14.260}
$$

for, in Subcase 2a

$$
\delta_1\log(1+t)\leqslant\delta_1\log(1+t_1)=\frac{1}{2a}
$$

while

$$
\frac{(1+\log(1+t))^2}{1+t}\leqslant C\quad\text{and}\quad C\delta_0\leqslant\frac{1}{a}
$$

In subcase 2b we first estimate the contribution of

$$
\int_0^{t_1}\bar{\mu}_m^{-a-1/2}(t')\frac{dt'}{1+t'}\leqslant C\log(1+t_1)=\frac{C}{2a\delta_1}
$$

to the left-hand side of (14.260). This contribution is then bounded by (note that $t\geqslant t_1$)

$$
\begin{aligned}
(\delta_1 + C\delta_0 \frac{(1+\log(1+t))}{1+t})\frac{C}{2a\delta_1} &\leqslant (\delta_1 + C\delta_0 \frac{(1+\log(1+t_1))}{1+t_1})\frac{C}{2a\delta_1} \\
&\leqslant \frac{C}{2a}(1 + \frac{C\delta_0}{\delta_1}(1+\frac{1}{2a\delta_1})e^{-\frac{1}{2a\delta_1}}) \\
&\leqslant \frac{C}{2a}(1 + C\varphi_2'(2a\delta_1)) \leqslant \frac{C}{a} \qquad (14.261)
\end{aligned}
$$

(see (14.157)). Next, we estimate the contribution of

$$
\int_{t_1}^{t} \bar{\mu}_m^{-a-1/2}(t')\frac{dt'}{1+t'} \qquad (14.262)
$$

By the lower bound (8.303),

$$
\begin{aligned}
\int_{t_1}^{t} \bar{\mu}_m^{-a-1/2}(t')\frac{dt'}{1+t'} &\leqslant C\int_{\tau_1}^{\tau}(1-\delta_1\tau')^{-a-1/2}d\tau' \\
&\leqslant \frac{C}{\delta_1}\frac{(1-\delta_1\tau)^{-a+1/2}}{a-1/2} \leqslant \frac{C}{\delta_1}\frac{\bar{\mu}_m^{-a+1/2}(t)}{a-1/2} \qquad (14.263)
\end{aligned}
$$

where in the last step, we have used (8.312). It follows that the contribution of (14.262) to the left-hand side of (14.260) is bounded by

$$
\begin{aligned}
&(\delta_1 + C\delta_0\frac{(1+\log(1+t))}{1+t})\frac{C}{\delta_1}\frac{\bar{\mu}_m^{-a+1/2}(t)}{a-1/2} \\
\leqslant\ &\frac{C\bar{\mu}_m^{-a+1/2}(t)}{a-1/2}(1 + \frac{C\delta_0}{\delta_1}\frac{(1+\log(1+t_1))}{1+t_1}) \\
\leqslant\ &\frac{C\bar{\mu}_m^{-a+1/2}(t)}{a-1/2}(1+C\varphi_2'(2a\delta_1)) \leqslant \frac{C\bar{\mu}_m^{-a+1/2}(t)}{a-1/2} \qquad (14.264)
\end{aligned}
$$

Combining the results (14.260), (14.264) of the two subcases we conclude that, in general:

$$
(\delta_1 + C\delta_0\frac{(1+\log(1+t))}{1+t})\int_0^t \bar{\mu}_m^{-a-1/2}(t')\frac{dt'}{1+t'} \leqslant C\frac{\bar{\mu}_m^{-a+1/2}(t)}{a-1/2} \qquad (14.265)
$$

hence (14.258), therefore also (14.256) and the integral on the left in (14.249), is bounded by

$$
\frac{C}{a-1/2}\bar{\mu}_m^{-2a}(t)(1+\log(1+t))^{2q(i_1\cdots i_l)}\mathcal{G}'_{1,l+2;a,q}(t) \qquad (14.266)
$$

We now consider the integral over $\mathcal{U}_{s,t}^{+}$:

$$
\int_{\mathcal{U}_{s,t}^{+}}(1+t)|T\psi_\alpha|^{(i_1\cdots i_l)}A_l^{(0)}|\not{d}R_{i_{l+1}}\cdots R_{i_1}\psi_\alpha|dud\mu_{\not{g}} \qquad (14.267)
$$

We estimate this by

$$
\begin{aligned}
&(1+t)\sup_{\mathcal{U}^+_{s,t}}(\mu^{-1}|T\psi_\alpha|)\|\mu^{1/2(i_1\cdots i_l)}A_l^{(0)}\|_{L^2(\mathcal{U}^+_{s,t})}\|\mu^{1/2}\not{d}R_{i_{l+1}}\cdots R_{i_1}\psi_\alpha\|_{L^2(\Sigma_t^{\epsilon_0})}\\
\leqslant&C(1+t)\sup_{\mathcal{U}^+_{s,t}}(\mu^{-1}|T\psi_\alpha|)\|(\mu^{1/2(i_1\cdots i_l)}A_l^{(0)})(t)\|_{L^2(\mathcal{V}_{s+})}\sqrt{\mathcal{E}'_1[R_{i_{l+1}}\cdots R_{i_1}\psi_\alpha](t)}
\end{aligned}
\tag{14.268}
$$

Since by Proposition 8.6, in $\mathcal{U}^+_{s,t}$ we have $\mu \geqslant C^{-1}$, then the pointwise estimate (14.250) implies

$$
\max_\alpha \sup_{\mathcal{U}^+_{s,t}}(\mu^{-1}|T\psi_\alpha|) \leqslant \frac{C}{|\ell|}(\sup_{\mathcal{U}^+_{s,t}}(\mu^{-1}|L\mu|) + C\delta_0(1+t)^{-2}(1+\log(1+t))) \tag{14.269}
$$

Let us define

$$
\hat{E}_{s,M} = \max_{(u,\vartheta)\in[0,\epsilon_0]\times S^2} \hat{E}_s(u,\vartheta) \tag{14.270}
$$

Setting

$$
\delta_2 = \hat{E}_{s,M} \tag{14.271}
$$

we can assume in estimating (14.269) that $\delta_2 > 0$. From Proposition 8.6 we have

$$
\mu^{-1}L\mu = \frac{\hat{E}_s(u,\vartheta)(1+t)^{-1} + \hat{Q}_{1,s}(t,u,\vartheta)}{1+\hat{E}_s(u,\vartheta)\log(1+t) + \hat{Q}_{0,s}(t,u,\vartheta)} \tag{14.272}
$$

In view of the fact that $\hat{E}_s(u,\vartheta) \geqslant 0$ in $\mathcal{V}_{s+}$, (14.272) implies that in $\mathcal{U}^+_{s,t}$

$$
\begin{aligned}
\mu^{-1}|L\mu| \leqslant& \frac{\hat{E}_s(u,\vartheta)(1+t)^{-1} + |\hat{Q}_{1,s}(t,u,\vartheta)|}{1+\hat{E}_s(u,\vartheta)\log(1+t) - |\hat{Q}_{0,s}(t,u,\vartheta)|}\\
\leqslant& \frac{\hat{E}_s(u,\vartheta)(1+t)^{-1} + C\delta_0(1+t)^{-2}(1+\log(1+t))}{1+\hat{E}_s(u,\vartheta)\log(1+t) - C\delta_0(1+t)^{-1}(1+\log(1+t))}\\
\leqslant& \frac{\hat{E}_s(u,\vartheta)(1+t)^{-1}}{1+\hat{E}_s(u,\vartheta)\log(1+t) - C\delta_0(1+t)^{-1}(1+\log(1+t))}\\
&+C'\delta_0\frac{1+\log(1+t)}{(1+t)^2}
\end{aligned}
\tag{14.273}
$$

With

$$
\eta = \hat{E}_s(u,\vartheta)\log(1+t) \geqslant 0, \quad \epsilon = C\delta_0(1+t)^{-1}(1+\log(1+t)) > 0
$$

we write

$$
\frac{1}{1+\eta-\epsilon} = \frac{1}{1+\eta} + \frac{\epsilon}{(1+\eta-\epsilon)(1+\eta)} \leqslant \frac{1}{1+\eta} + C\epsilon
$$

then since

$$
\hat{E}_s(u,\vartheta)(1+t)^{-1}\epsilon \leqslant C''\delta_0^2(1+t)^{-2}(1+\log(1+t))
$$

(14.273) implies that in $\mathcal{U}^+_{s,t}$

$$\mu^{-1}|L\mu| \leqslant \frac{\hat{E}_s(u,\vartheta)(1+t)^{-1}}{1+\hat{E}_s(u,\vartheta)\log(1+t)} + C\delta_0(1+t)^{-2}(1+\log(1+t)) \tag{14.274}$$

The first term on the right in (14.274) is

$$\frac{1}{(1+t)\log(1+t)}\frac{\eta}{1+\eta} \tag{14.275}$$

This is an increasing function of $\eta$, and achieves its maximum in $\mathcal{U}^+_{s,t}$ where $\eta$ achieves its maximum value

$$\eta_M = \delta_2 \log(1+t) \tag{14.276}$$

in $\mathcal{V}_{s+}$. The supremum of (14.277) in $\mathcal{U}^+_{s,t}$ is therefore

$$\frac{1}{(1+t)\log(1+t)}\frac{\eta_M}{1+\eta_M} = \frac{1}{1+t}\frac{\delta_2}{1+\delta_2\log(1+t)} \tag{14.277}$$

It follows that

$$\sup_{\mathcal{U}^+_{s,t}} \mu^{-1}|L\mu| \leqslant \frac{1}{1+t}\frac{\delta_2}{1+\delta_2\log(1+t)} + C\delta_0\frac{1+\log(1+t)}{(1+t)^2} \tag{14.278}$$

Substituting this in (14.269) and the result in (14.268), we conclude that the integral (14.267) is bounded by

$$\begin{aligned}&\frac{C}{|\ell|}\left(\frac{\delta_2}{1+\delta_2\log(1+t)} + C\delta_0\frac{1+\log(1+t)}{1+t}\right)\cdot\\ &\cdot\|\mu^{1/2(i_1\cdots i_l)}A_l^{(0)}(t)\|_{L^2(\mathcal{V}_{s+})}\sqrt{\mathcal{E}_1'[R_{i_{l+1}}\cdots R_{i_1}\psi_\alpha](t)}\end{aligned} \tag{14.279}$$

Now from (14.244) for $k=0$ we have

$$\begin{aligned}&(\mu^{1/2(i_1\cdots i_l)}A_l^{(0)})(t,u,\vartheta)\\ &=\int_0^t (1+t')^2\left(\frac{\mu(t,u,\vartheta)}{\mu(t',u,\vartheta)}\right)^{1/2}|(\mu^{1/2(i_1\cdots i_l)}\rho_l^{(0)})(t',u,\vartheta)|dt'\end{aligned} \tag{14.280}$$

It follows that

$$\begin{aligned}&\|(\mu^{1/2(i_1\cdots i_l)}A_l^{(0)})(t)\|_{L^2(\mathcal{V}_{s+})}\\ &\leqslant \int_0^t (1+t')^2\left(\sup_{(u,\vartheta)\in\mathcal{V}_{s+}}\left(\frac{\mu(t,u,\vartheta)}{\mu(t',u,\vartheta)}\right)\right)^{1/2}\|(\mu^{1/2(i_1\cdots i_l)}\rho_l^{(0)})(t')\|_{L^2(\mathcal{V}_{s+})}dt'\\ &\leqslant C\int_0^t (1+t')\left(\sup_{(u,\vartheta)\in\mathcal{V}_{s+}}\left(\frac{\mu(t,u,\vartheta)}{\mu(t',u,\vartheta)}\right)\right)^{1/2}\|\mu^{1/2(i_1\cdots i_l)}\rho_l^{(0)}\|_{L^2(\Sigma_{t'}^{\epsilon_0})}dt'\end{aligned} \tag{14.281}$$

where in the last step, we have used (8.333). From Proposition 8.6 we have, for $(u,\vartheta)\in\mathcal{V}_{s+}$,

$$\begin{aligned}\frac{\mu(t,u,\vartheta)}{\mu(t',u,\vartheta)}&=\frac{1+\hat{E}_s(u,\vartheta)\log(1+t)+\hat{Q}_{0,s}(t,u,\vartheta)}{1+\hat{E}_s(u,\vartheta)\log(1+t')+\hat{Q}_{0,s}(t',u,\vartheta)}\\&\leqslant\frac{1+\hat{E}_s(u,\vartheta)\log(1+t)+C\delta_0(1+t)^{-1}(1+\log(1+t))}{1+\hat{E}_s(u,\vartheta)\log(1+t')-C\delta_0(1+t')^{-1}(1+\log(1+t'))}\end{aligned}\tag{14.282}$$

Since

$$C\delta_0\frac{1+\log(1+t')}{1+t'}\leqslant C\delta_0\leqslant\frac{1}{2}$$

the denominator in the fraction on the right in (14.282) is

$$\geqslant\frac{1}{2}+\hat{E}_s(u,\vartheta)\log(1+t')\geqslant\frac{1}{2}(1+\hat{E}_s(u,\vartheta)\log(1+t'))$$

Similarly, since

$$C\delta_0\frac{1+\log(1+t)}{1+t}\leqslant C\delta_0\leqslant\frac{1}{2}$$

the numerator in the fraction on the right in (14.282) is

$$\leqslant\frac{3}{2}+\hat{E}_s(u,\vartheta)\log(1+t)\leqslant\frac{3}{2}(1+\hat{E}_s(u,\vartheta)\log(1+t))$$

Therefore, for $(u,\vartheta)\in\mathcal{V}_{s+}$ it holds that

$$\frac{\mu(t,u,\vartheta)}{\mu(t',u,\vartheta)}\leqslant 3\frac{1+\hat{E}_s(u,\vartheta)\log(1+t)}{1+\hat{E}_s(u,\vartheta)\log(1+t')}\tag{14.283}$$

Now for $a>b>0, x\geqslant 0$, the function

$$f(x)=\frac{1+ax}{1+bx}$$

is increasing in $x$. Hence, taking $a=\log(1+t)$, $b=\log(1+t')$, $x=\hat{E}_s(u,\vartheta)$, the ratio on the right in (14.283) achieves its maximum in $\mathcal{V}_{s+}$ where $x$ achieves its maximum $\delta_2$ in $\mathcal{V}_{s+}$. Therefore

$$\sup_{(u,\vartheta)\in\mathcal{V}_{s+}}\frac{\mu(t,u,\vartheta)}{\mu(t',u,\vartheta)}\leqslant 3\frac{1+\delta_2\log(1+t)}{1+\delta_2\log(1+t')}\tag{14.284}$$

Substituting this and (14.240) in (14.241) we obtain

$$\begin{aligned}&\|(\mu^{1/2(i_1\cdots i_l)}A_l^{(0)})(t)\|_{L^2(\mathcal{V}_{s+})}\\&\leqslant C|\ell|\int_0^t\sqrt{\frac{1+\delta_2\log(1+t)}{1+\delta_2\log(1+t')}}\sqrt{\sum_{j,\alpha}\mathcal{E}_1'[R_jR_{i_l}\cdots R_{i_1}\psi_\alpha](t')}\frac{dt'}{1+t'}\end{aligned}$$

$$
\begin{aligned}
&\leqslant C|\ell|\sqrt{{}^{(i_1\cdots i_l)}\mathcal{G}'_{1,l+2;a,q}}\int_0^t\sqrt{\frac{1+\delta_2\log(1+t)}{1+\delta_2\log(1+t')}}(1+\log(1+t'))^q\bar{\mu}_m^{-a}(t')\frac{dt'}{1+t'}\\
&\leqslant C|\ell|\bar{\mu}_m^{-a}(t)\sqrt{{}^{(i_1\cdots i_l)}\mathcal{G}'_{1,l+2;a,q}}\int_0^t\sqrt{\frac{1+\delta_2\log(1+t)}{1+\delta_2\log(1+t')}}(1+\log(1+t'))^q\frac{dt'}{1+t'}
\end{aligned}
\tag{14.285}
$$

where in the last step we have used the fact that by Corollary 2 of Lemma 8.11, $\bar{\mu}_m^{-a}(t')\leqslant C\bar{\mu}_m^{-a}(t)$. Substituting (14.285) in (14.279) and noting that

$$
\sqrt{\mathcal{E}'_1[R_{i_{l+1}}\cdots R_{i_1}\psi_\alpha](t)}\leqslant\bar{\mu}_m^{-a}(t)(1+\log(1+t))^q\sqrt{{}^{(i_1\cdots i_l)}\mathcal{G}'_{1,l+2;a,q}(t)}
$$

the factors $|\ell|$ cancel and we obtain that (14.279) hence also (14.267) is bounded by

$$
\begin{aligned}
&C\bar{\mu}_m^{-2a}(t)(1+\log(1+t))^{q\,(i_1\cdots i_l)}\mathcal{G}'_{1,l+2;a,q}(t)\\
&\cdot\left(\frac{\delta_2}{\sqrt{1+\delta_2\log(1+t)}}+C\delta_0\frac{(1+\log(1+t))^{3/2}}{1+t}\right)I_{q;\delta_2}(t)
\end{aligned}
\tag{14.286}
$$

where

$$
I_{q;\delta_2}(t)=\int_0^t\frac{(1+\log(1+t'))^q}{\sqrt{1+\delta_2\log(1+t')}}\frac{dt'}{1+t'}
\tag{14.287}
$$

Setting $x=1+\log(1+t')$, the integral $I_{q;\delta_2}(t)$ takes the form

$$
I_{q;\delta_2}(t)=\int_1^{1+\log(1+t)}\frac{x^q}{\sqrt{1+\delta_2(x-1)}}dx
\tag{14.288}
$$

Since $\delta_2\leqslant 1$ we have

$$
1+\delta_2(x-1)\geqslant\delta_2+\delta_2(x-1)=\delta_2 x
$$

hence

$$
I_{q;\delta_2}(t)\leqslant\frac{1}{\sqrt{\delta_2}}\int_0^{1+\log(1+t)}x^{q-1/2}dx=\frac{1}{\sqrt{\delta_2}}\frac{(1+\log(1+t))^{q+1/2}}{q+1/2}
\tag{14.289}
$$

Also, since the denominator of the integrand in (14.288) is $\geqslant 1$,

$$
I_{q;\delta_2}(t)\leqslant\int_0^{1+\log(1+t)}x^q dx=\frac{(1+\log(1+t))^{q+1}}{q+1}
\tag{14.290}
$$

We use (14.289) to estimate the product:

$$
\begin{aligned}
&\frac{\delta_2}{\sqrt{1+\delta_2\log(1+t)}}I_{q;\delta_2}(t)\\
&\leqslant\frac{\delta_2}{\sqrt{1+\delta_2\log(1+t)}}\frac{(1+\log(1+t))^{q+1/2}}{\sqrt{\delta_2}(q+1/2)}\\
&=\frac{\sqrt{\delta_2}\sqrt{1+\log(1+t)}}{\sqrt{1+\delta_2\log(1+t)}}\frac{(1+\log(1+t))^q}{(q+1/2)}\leqslant\frac{(1+\log(1+t))^q}{(q+1/2)}
\end{aligned}
\tag{14.291}
$$

where in the last step we have used the fact that $\delta_2 \leqslant 1$. We use (14.290) to estimate the product:

$$\begin{aligned}
&C\delta_0 \frac{(1+\log(1+t))^{3/2}}{1+t} I_{q;\delta_2}(t) \\
\leqslant &C\delta_0 \frac{(1+\log(1+t))^{3/2}}{1+t} \frac{(1+\log(1+t))^{q+1}}{q+1} \\
= &C\delta_0 \frac{(1+\log(1+t))^{5/2}}{1+t} \frac{(1+\log(1+t))^{q}}{q+1} \leqslant \frac{C'\delta_0}{q+1}(1+\log(1+t))^q
\end{aligned} \tag{14.292}$$

We conclude that (14.286) hence also (14.267) is bounded by

$$\frac{C}{q+1/2}\bar{\mu}_m^{-2a}(t)(1+\log(1+t))^{2q}\,{}^{(i_1\cdots i_l)}\mathcal{G}'_{1,l+2;a,q}(t) \tag{14.293}$$

Thus, we have estimated the borderline integral (14.246). We proceed to estimate the remaining contributions to the hypersurface integral $H_0$. These are the contributions from (14.243). To estimate these contributions, we simply use the bound

$$\max_{\alpha} \sup_{\Sigma_t^{\epsilon_0}} |T\psi_\alpha| \leqslant C\delta_0(1+t)^{-1} \tag{14.294}$$

These contributions are then bounded by, for $k=1,2$,

$$\begin{aligned}
&C\int_{\Sigma_t^{\epsilon_0}} (1+t)|T\psi_\alpha|\,|{}^{(i_1\cdots i_l)}A_l^{(k)}|\,|\not{d}R_{i_{l+1}}\cdots R_{i_1}\psi_\alpha| du d\mu_{\not{g}} \\
\leqslant &C\delta_0 \int_{\Sigma_t^{\epsilon_0}} |{}^{(i_1\cdots i_l)}A_l^{(k)}|\,|\not{d}R_{i_{l+1}}\cdots R_{i_1}\psi_\alpha| du d\mu_{\not{g}} \\
\leqslant &C\delta_0 \|{}^{(i_1\cdots i_l)}A_l^{(k)}\|_{L^2(\Sigma_t^{\epsilon_0})} \|\not{d}R_{i_{l+1}}\cdots R_{i_1}\psi_\alpha\|_{L^2(\Sigma_t^{\epsilon_0})} \\
\leqslant &C\delta_0 \bar{\mu}_m^{-1/2}(t) \|{}^{(i_1\cdots i_l)}A_l^{(k)}(t)\|_{L^2([0,\epsilon_0]\times S^2)} \sqrt{\mathcal{E}'_{1,[l+2]}(t)}
\end{aligned} \tag{14.295}$$

by (8.333).

Consider first the contribution of ${}^{(i_1\cdots i_l)}A_l^{(1)}$. From (14.245) and the fact that by the estimate (14.239),

$$\|{}^{(i_1\cdots i_l)}\rho_l^{(1)}\|_{L^2(\Sigma_t^{\epsilon_0})} \leqslant C\bar{\mu}_m^{-1/2}(t)\delta_0(1+t)^{-3}\sqrt{\mathcal{E}'_{1,[l+2]}(t)} \tag{14.296}$$

we obtain

$$\begin{aligned}
\|{}^{(i_1\cdots i_l)}A_l^{(1)}(t)\|_{L^2([0,\epsilon_0]\times S^2)} &\leqslant C\delta_0 \int_0^t \bar{\mu}_m^{-1/2}(t')(1+t')^{-2}\sqrt{\mathcal{E}'_{1,[l+2]}(t')}dt' \\
&\leqslant C\delta_0 \sqrt{\mathcal{G}'_{1,[l+2];a,q}(t)}\,J'_{a,q}(t)
\end{aligned} \tag{14.297}$$

where

$$J'_{a,q}(t) = \int_0^t \bar{\mu}_m^{-a-1/2}(t')(1+t')^{-2}(1+\log(1+t'))^q dt' \tag{14.298}$$

To estimate this integral we consider again the two cases occurring in the proof of Lemma 8.11. In Case 1 we have the upper bound (14.117), which implies

$$J'_{a,q}(t) \leqslant C \int_0^t (1+t')^{-2}(1+\log(1+t'))^q dt' \leqslant C_q \tag{14.299}$$

Similarly, in Subcase 2a,

$$J'_{a,q}(t_1) \leqslant C_q \tag{14.300}$$

In Subcase 2b we have the lower bound (14.124), which, with

$$C_q(t_1) = \sup_{t' \geqslant t_1} ((1+t')^{-1}(1+\log(1+t'))^q) \tag{14.301}$$

implies

$$\begin{aligned} J'_{a,q}(t) - J'_{a,q}(t_1) &\leqslant C \cdot C_q(t_1) \int_{t_1}^t (1+t')^{-1}(1-\delta_1\tau')^{-a-1/2} dt' \\ &\leqslant C \cdot \frac{C_q(t_1)}{a\delta_1} \bar{\mu}_m^{-a+1/2}(t) \end{aligned} \tag{14.302}$$

In the last step we have used the bound (8.312).

We have

$$C_q(t_1) = \sup_{\tau' \geqslant \tau_1} (e^{-\tau'}(1+\tau')^q), \quad \tau_1 = \log(1+t_1)$$

Now, the function

$$f(x) = e^{-x}x^q \quad \text{on} \quad [1,\infty)$$

is decreasing in $x$ for $x \geqslant q$. Hence if $\tau_1 = 1/2a\delta_1 \geqslant q$, which is the case if $aq\delta_0$ is suitably small, then

$$C_q(t_1) = e^{-\tau_1}(1+\tau_1)^q = e^{-\frac{1}{2a\delta_1}}(1+\frac{1}{2a\delta_1})^q \tag{14.303}$$

Therefore, from (14.302),

$$\begin{aligned} J'_{a,q}(t) - J'_{a,q}(t_1) &\leqslant \frac{C}{2a\delta_1}(1+\frac{1}{2a\delta_1})^q e^{-\frac{1}{2a\delta_1}} \bar{\mu}_m^{-a+1/2}(t) \\ &\leqslant C\varphi'_{q+1}(2a\delta_1)\bar{\mu}_m^{-a+1/2}(t) \leqslant C\varphi'_{q+1}(Ca\delta_0)\bar{\mu}_m^{-a+1/2}(t) \end{aligned} \tag{14.304}$$

We conclude that in general,

$$J'_{a,q}(t) \leqslant C_q + C\varphi'_{q+1}(Ca\delta_0)\bar{\mu}_m^{-a+1/2}(t) \tag{14.305}$$

Substituting in (14.297) and the result in (14.295) for $k = 1$, then yields

$$C \int_{\Sigma_t^{\epsilon_0}} (1+t)|T\psi_\alpha|^{(i_1 \cdots i_l)} A_l^{(1)} |\not{d} R_{i_{l+1}} \cdots R_{i_1} \psi_\alpha| du d\mu_{\not{g}}$$

$$\leqslant C\delta_0^2 \mathcal{G}'_{1,[l+2];a,q}(t)(C_q \bar{\mu}_m^{-a-1/2}(t) + C\varphi'_{q+1}(Ca\delta_0)\bar{\mu}_m^{-2a}(t))(1+\log(1+t))^q \quad (14.306)$$

Next, we consider the contribution of ${}^{(i_1 \cdots i_l)}A_l^{(2)}$. We have (14.245) for $k = 2$ and for $\|{}^{(i_1 \cdots i_l)}\rho_l^{(2)}\|_{L^2(\Sigma_t^{\epsilon_0})}$ we have the estimate (14.235). Here we shall only consider the contribution of the principal part on the right of (14.235) namely the term $C_l(1+t)^{-2}\mathcal{W}^Q_{\{l\}}$. The contribution of this term to (14.245) is

$$C_l \int_0^t (1+t')^{-1} \mathcal{W}^Q_{\{l\}}(t') dt' \quad (14.307)$$

By Lemma 5.1 we have

$$\mathcal{W}^Q_{\{l\}}(t') \leqslant C\epsilon_0 \sqrt{\mathcal{E}_{0,[l+2]}(t')}$$
$$\leqslant C\epsilon_0 \bar{\mu}_m^{-a}(t')(1+\log(1+t'))^p \sqrt{\mathcal{G}_{0,[l+2];a,p}} \quad (14.308)$$

hence (14.307) is bounded by

$$C_l \epsilon_0 (1+\log(1+t))^p \sqrt{\mathcal{G}_{0,[l+2];a,p}(t)} \int_0^t (1+t')^{-1} \bar{\mu}_m^{-a}(t') dt' \quad (14.309)$$

The last integral coincides with the integral $J_{a-1/2,-1}(t)$ (see (14.147)), therefore by (14.153) it is bounded by

$$C(1+\log(1+t))\bar{\mu}_m^{-a+1}(t)$$

Consequently (14.309), hence also (14.307), are bounded by

$$C_l \epsilon_0 (1+\log(1+t))^{p+1} \bar{\mu}_m^{-a+1}(t) \sqrt{\mathcal{G}_{0,[l+2];a,p}(t)} \quad (14.310)$$

We conclude that the contribution to (14.295) for $k = 2$ is bounded by

$$C_l \epsilon_0 \delta_0 \bar{\mu}_m^{-2a+1/2}(t)(1+\log(1+t))^{p+q+1} \sqrt{\mathcal{G}_{0,[l+2];a,p}(t)\mathcal{G}'_{1,[l+2];a,q}(t)} \quad (14.311)$$

Finally, we consider the contribution of the hypersurface integral $H_1$, given by (14.212). We have

$$|H_1| \leqslant C \int_{\Sigma_t^{\epsilon_0}} (1+t)^2 |R_{i_{l+1}} T\psi_\alpha||R_{i_l} \cdots R_{i_1} \mathrm{tr}\chi'||R_{i_{l+1}} \cdots R_{i_1} \psi_\alpha| du d\mu_{\not{g}} \quad (14.312)$$

By virtue of the bound

$$\max_\alpha \sup_{\Sigma_t^{\epsilon_0}} |R_{i_{l+1}} T\psi_\alpha| \leqslant C\delta_0(1+t)^{-1} \quad (14.313)$$

we have

$$\begin{aligned}|H_1| &\leqslant C\delta_0 \int_{\Sigma_t^{\epsilon_0}} (1+t)|R_{i_l} \cdots R_{i_1} \mathrm{tr}\chi'||R_{i_{l+1}} \cdots R_{i_1}\psi_\alpha| du d\mu_{\not{g}} \\ &\leqslant C\delta_0(1+t)\|R_{i_l} \cdots R_{i_1} \mathrm{tr}\chi'\|_{L^2(\Sigma_t^{\epsilon_0})}\|R_{i_{l+1}} \cdots R_{i_1}\psi_\alpha\|_{L^2(\Sigma_t^{\epsilon_0})} \\ &\leqslant C\delta_0(1+t)^2\|(R_{i_l} \cdots R_{i_1} \mathrm{tr}\chi')(t)\|_{L^2([0,\epsilon_0]\times S^2)}\mathcal{W}_{\{l+1\}}(t) \qquad (14.314)\end{aligned}$$

Now by (14.243), (14.245),

$$\begin{aligned}&(1+t)^2\|(R_{i_l} \cdots R_{i_1} \mathrm{tr}\chi')(t)\|_{L^2([0,\epsilon_0]\times S^2)} \\ &\leqslant C(\|R_{i_l} \cdots R_{i_1} \mathrm{tr}\chi'\|_{L^2(\Sigma_0^{\epsilon_0})} + \sum_{k=0}^{2} \|{}^{(i_1 \cdots i_l)}A_l^{(k)}(t)\|_{L^2([0,\epsilon_0]\times S^2)}) \\ &\leqslant C(\|R_{i_l} \cdots R_{i_1} \mathrm{tr}\chi'\|_{L^2(\Sigma_0^{\epsilon_0})} + \sum_{k=0}^{2} \int_0^t (1+t')\|{}^{(i_1 \cdots i_l)}\rho_l^{(k)}\|_{L^2(\Sigma_{t'}^{\epsilon_0})} dt') \qquad (14.315)\end{aligned}$$

Here we need only consider the leading principal contribution, namely that of $\|{}^{(i_1 \cdots i_l)}\rho_l^{(0)}\|_{L^2(\Sigma_{t'}^{\epsilon_0})}$. This contribution to (14.315) is bounded by

$$\begin{aligned}&C|\ell| \int_0^t (1+t')^{-1}\bar{\mu}_m^{-1/2}(t')\sqrt{\sum_{j,\alpha} \mathcal{E}_1'[R_j R_{i_l} \cdots R_{i_1}\psi_\alpha](t')} dt' \\ &\leqslant C|\ell|\sqrt{\mathcal{G}'_{1,[l+2];a,q}(t)} \int_0^t (1+t')^{-1}\bar{\mu}_m^{-a-1/2}(t')(1+\log(1+t'))^q dt' \qquad (14.316)\end{aligned}$$

The last integral coincides with $J_{a,q-1}$ (see (14.147)), therefore by (14.153) it is bounded by

$$C(1+\log(1+t))^{q+1}\bar{\mu}_m^{-a+1/2}(t)$$

Consequently the leading principal contribution to (14.315) is bounded by

$$C|\ell|\bar{\mu}_m^{-a+1/2}(t)(1+\log(1+t))^{q+1}\sqrt{\mathcal{G}'_{1,[l+2];a,q}(t)} \qquad (14.317)$$

On the other hand, as in (14.308) we have

$$\mathcal{W}_{\{l+1\}} \leqslant C\epsilon_0\sqrt{\mathcal{E}_{0,[l+2]}(t)} \leqslant C\epsilon_0\bar{\mu}_m^{-a}(t)(1+\log(1+t))^p\sqrt{\mathcal{G}_{0,[l+2];a,p}(t)} \qquad (14.318)$$

Substituting (14.317) and (14.318) in (14.314), we conclude that the leading principal contribution to $H_1$ is bounded by

$$C|\ell|\epsilon_0\delta_0\bar{\mu}_m^{-2a+1/2}(t)(1+\log(1+t))^{p+q+1}\sqrt{\mathcal{G}_{0,[l+2];a,p}(t)\mathcal{G}'_{1,[l+2];a,q}(t)} \qquad (14.319)$$

the same in the form as (14.311).

We turn to the spacetime integral in (14.207) namely

$$\int_{W^t_u} (R_{i_{l+1}} \cdots R_{i_1}\psi_\alpha)(L+\nu)((\omega/\nu)(T\psi_\alpha)(R_{i_{l+1}} \cdots R_{i_1}\mathrm{tr}\chi'))dt'du'd\mu_{\tilde{\not g}} \qquad (14.320)$$

Here we shall use the fact that $(L+\nu)T\psi_\alpha$ decays faster than $(1+t)^{-2}$. To establish this fact we consider the wave equation satisfied by $\psi_\alpha$. In analogy with (8.180) and (8.181) we have for $\alpha = 0, 1, 2, 3$

$$(L+\nu)\underline{L}\psi_\alpha = \rho_\alpha \qquad (14.321)$$

where

$$\rho_\alpha = \mu\not\Delta\psi_\alpha - \underline{\nu}L\psi_\alpha - 2\zeta \cdot \not d\psi_\alpha + \mu\frac{d\log\Omega}{dh}\not dh \cdot \not d\psi_\alpha \qquad (14.322)$$

Under the same assumptions as those of Lemma 8.10 an estimate similar to (8.182) holds for each $\alpha = 0, 1, 2, 3$, that is, we have

$$\max_\alpha |\rho_\alpha| \leqslant C\delta_0(1+t)^{-3}(1+\log(1+t)) \qquad (14.323)$$

Since $2T\psi_\alpha = \underline{L}\psi_\alpha - \eta^{-1}\kappa L\psi_\alpha$, (14.321) implies

$$(L+\nu)T\psi_\alpha = \tau_\alpha \qquad (14.324)$$

where

$$2\tau_\alpha = \rho_\alpha - (L+\nu)(\eta^{-1}\kappa L\psi_\alpha) \qquad (14.325)$$

Now we have

$$(L+\nu)(\eta^{-1}\kappa L\psi_\alpha) = \eta^{-1}\kappa(L)^2\psi_\alpha + ((L+\nu)(\eta^{-1}\kappa))L\psi_\alpha$$

Using the bound for $L\mu$ and the assumptions $\mathbf{E}^{QQ}_{\{0\}}$, $\mathbf{E}^{Q}_{\{0\}}$ we deduce

$$\max_\alpha |(L+\nu)(\eta^{-1}\kappa L\psi_\alpha)| \leqslant C\delta_0(1+t)^{-3}(1+\log(1+t)) \qquad (14.326)$$

Combining (14.323) and (14.326) we obtain

$$\max_\alpha |\tau_\alpha| \leqslant C\delta_0(1+t)^{-3}(1+\log(1+t)) \qquad (14.327)$$

This estimate actually relies only on the assumptions of Proposition 12.6. By direct calculation, under the assumptions of Proposition 12.9 with $l = 1$ together with those of Proposition 12.10 with $m = l = 0$, we deduce

$$\max_\alpha \|\tau_\alpha\|_{\infty,[1],\Sigma^{\epsilon_0}_t} \leqslant C\delta_0(1+t)^{-3}(1+\log(1+t)) \qquad (14.328)$$

Next, we consider the factor $\omega/\nu$. Setting

$$\nu' = \frac{1}{2}(\mathrm{tr}\chi' + L\log\Omega) \qquad (14.329)$$

we have

$$\nu = \frac{1}{1-u+t} + \nu' \tag{14.330}$$

Recalling that $\omega = 2(1+t)$ we then obtain

$$(L - 2\nu)(\omega/\nu) = \gamma\nu^{-2}\omega \tag{14.331}$$

where

$$\gamma = \frac{-u}{(1+t)(1-u+t)^2} + (\frac{1}{1+t} - \frac{4}{1-u+t})\nu' - 2\nu'^2 - L\nu' \tag{14.332}$$

Proposition 12.9 with $l = 1$ implies

$$\|\gamma\|_{\infty,[1],\Sigma_t^{\epsilon_0}} \leqslant C\delta_0(1+t)^{-3}(1+\log(1+t)) \tag{14.333}$$

Here we also have used (14.217) and (14.219).

By (14.324) and (14.331) we have,

$$\begin{aligned}&(L+\nu)((\omega/\nu)(T\psi_\alpha)(R_{i_{l+1}}\cdots R_{i_1}\mathrm{tr}\chi'))\\ &= (\omega/\nu)((T\psi_\alpha)(L+2\nu)(R_{i_{l+1}}\cdots R_{i_1}\mathrm{tr}\chi') + \tilde{\tau}_\alpha(R_{i_{l+1}}\cdots R_{i_1}\mathrm{tr}\chi'))\end{aligned} \tag{14.334}$$

where

$$\tilde{\tau}_\alpha = \tau_\alpha + \gamma\nu^{-1}T\psi_\alpha \tag{14.335}$$

By (14.328) and (14.333) and the estimate for $\nu'$ resulting from Proposition 12.9 with $l = 1$ we have

$$\max_\alpha \|\tilde{\tau}_\alpha\|_{\infty,[1],\Sigma_t^{\epsilon_0}} \leqslant C\delta_0(1+t)^{-3}(1+\log(1+t)) \tag{14.336}$$

Substituting (14.334) in (14.320), we are thus to estimate the spacetime integral:

$$\begin{aligned}&\int_{W_u^t}(R_{i_{l+1}}\cdots R_{i_1}\psi_\alpha)(\omega/\nu)\\ &\cdot((T\psi_\alpha)(L+2\nu)(R_{i_{l+1}}\cdots R_{i_1}\mathrm{tr}\chi') + \tilde{\tau}_\alpha(R_{i_{l+1}}\cdots R_{i_1}\mathrm{tr}\chi'))dt'du'd\mu_{\tilde{g}}\end{aligned} \tag{14.337}$$

Writing

$$\begin{aligned}&(L+2\nu)R_{i_{l+1}}\cdots R_{i_1}\mathrm{tr}\chi'\\ &= R_{i_{l+1}}(L+2\nu)R_{i_l}\cdots R_{i_1}\mathrm{tr}\chi' + {}^{(R_{i_{l+1}})}ZR_{i_l}\cdots R_{i_1}\mathrm{tr}\chi'\\ &\quad - 2(R_{i_{l+1}}\nu)(R_{i_l}\cdots R_{i_1}\mathrm{tr}\chi')\end{aligned} \tag{14.338}$$

we integrate by parts on each $S_{t,u}$ using (14.209), first taking $X = R_{i_{l+1}}$ to obtain

$$\begin{aligned}&\int_{S_{t,u}}(\omega/\nu)(R_{i_{l+1}}\cdots R_{i_1}\psi_\alpha)(T\psi_\alpha)R_{i_{l+1}}((L+2\nu)R_{i_l}\cdots R_{i_1}\mathrm{tr}\chi')d\mu_{\tilde{g}}\\ &= -\int_{S_{t,u}}(\omega/\nu)(T\psi_\alpha)(R_{i_{l+1}}R_{i_{l+1}}\cdots R_{i_1}\psi_\alpha)(L+2\nu)R_{i_l}\cdots R_{i_1}\mathrm{tr}\chi' d\mu_{\tilde{g}}\end{aligned} \tag{14.339}$$

$$
\begin{aligned}
&-\int_{S_{t,u}}(\omega/\nu)(R_{i_{l+1}}T\psi_\alpha)(R_{i_{l+1}}\cdots R_{i_1}\psi_\alpha)(L+2\nu)R_{i_l}\cdots R_{i_1}\mathrm{tr}\chi' d\mu_{\tilde{\not g}}\\
&-\int_{S_{t,u}}(T\psi_\alpha)(R_{i_{l+1}}(\omega/\nu)+\frac{1}{2}(\omega/\nu)\mathrm{tr}\,{}^{(R_{i_{l+1}})}\tilde{\not\pi})\\
&\cdot(R_{i_{l+1}}\cdots R_{i_1}\psi_\alpha)(L+2\nu)R_{i_l}\cdots R_{i_1}\mathrm{tr}\chi' d\mu_{\tilde{\not g}}
\end{aligned}
$$

and then taking $X = {}^{(R_{i_{l+1}})}Z$ to obtain

$$
\begin{aligned}
&\int_{S_{t,u}}(\omega/\nu)(T\psi_\alpha)(R_{i_{l+1}}\cdots R_{i_1}\psi_\alpha)^{(R_{i_{l+1}})}Z(R_{i_l}\cdots R_{i_1}\mathrm{tr}\chi')d\mu_{\tilde{\not g}}\\
&=-\int_{S_{t,u}}({}^{(R_{i_{l+1}})}Z\cdot \not d R_{i_{l+1}}\cdots R_{i_1}\psi_\alpha)(\omega/\nu)(T\psi_\alpha)(R_{i_l}\cdots R_{i_1}\mathrm{tr}\chi')d\mu_{\tilde{\not g}} \qquad (14.340)\\
&-\int_{S_{t,u}}(\omega/\nu)({}^{(R_{i_{l+1}})}Z\cdot \not d T\psi_\alpha)(R_{i_{l+1}}\cdots R_{i_1}\psi_\alpha)(R_{i_l}\cdots R_{i_1}\mathrm{tr}\chi')d\mu_{\tilde{\not g}}\\
&-\int_{S_{t,u}}(T\psi_\alpha)({}^{(R_{i_{l+1}})}Z\cdot \not d(\omega/\nu)+(\omega/\nu)\tilde{\not{\mathrm{div}}}\,{}^{(R_{i_{l+1}})}Z)\\
&\cdot(R_{i_{l+1}}\cdots R_{i_1}\psi_\alpha)(R_{i_l}\cdots R_{i_1}\mathrm{tr}\chi')d\mu_{\tilde{\not g}}
\end{aligned}
$$

Also, taking again $X = R_{i_{l+1}}$,

$$
\begin{aligned}
&\int_{S_{t,u}}(R_{i_{l+1}}\cdots R_{i_1}\psi_\alpha)(\omega/\nu)\tilde{\tau}_\alpha R_{i_{l+1}}\cdots R_{i_1}\mathrm{tr}\chi' d\mu_{\tilde{\not g}}\\
&=-\int_{S_{t,u}}(\omega/\nu)\tilde{\tau}_\alpha(R_{i_{l+1}}R_{i_{l+1}}\cdots R_{i_1}\psi_\alpha)(R_{i_l}\cdots R_{i_1}\mathrm{tr}\chi')d\mu_{\tilde{\not g}} \qquad (14.341)\\
&-\int_{S_{t,u}}(\omega/\nu)(R_{i_{l+1}}\tilde{\tau}_\alpha)(R_{i_{l+1}}\cdots R_{i_1}\psi_\alpha)(R_{i_l}\cdots R_{i_1}\mathrm{tr}\chi')d\mu_{\tilde{\not g}}\\
&-\int_{S_{t,u}}\tilde{\tau}_\alpha(R_{i_{l+1}}\cdots R_{i_1}\psi_\alpha)(R_{i_{l+1}}(\omega/\nu)+\frac{1}{2}(\omega/\nu)\mathrm{tr}\,{}^{(R_{i_{l+1}})}\tilde{\not\pi})(R_{i_l}\cdots R_{i_1}\mathrm{tr}\chi')d\mu_{\tilde{\not g}}
\end{aligned}
$$

In view of (14.339)—(14.341) the spacetime integral (14.337) becomes

$$
-V_0 - V_1 - V_2 - V_3 \qquad (14.342)
$$

where

$$
V_0 = V_{0,0} + V_{0,1} \qquad (14.343)
$$

$$
V_{0,0} = \int_{W^t_u}(\omega/\nu)(T\psi_\alpha)(R_{i_{l+1}}R_{i_{l+1}}\cdots R_{i_1}\psi_\alpha)(L+2\nu)R_{i_l}\cdots R_{i_1}\mathrm{tr}\chi' dt'du'd\mu_{\tilde{\not g}} \qquad (14.344)
$$

$$
\begin{aligned}
V_{0,1} = &\int_{W^t_u}(\omega/\nu)(\tilde{\tau}_\alpha(R_{i_{l+1}}R_{i_{l+1}}\cdots R_{i_1}\psi_\alpha) \qquad (14.345)\\
&+(T\psi_\alpha)^{(R_{i_{l+1}})}Z\cdot \not d R_{i_{l+1}}\cdots R_{i_1}\psi_\alpha)R_{i_l}\cdots R_{i_1}\mathrm{tr}\chi' dt'du'd\mu_{\tilde{\not g}}
\end{aligned}
$$

$$
V_1 = V_{1,0} + V_{1,1} \qquad (14.346)
$$

$$V_{1,0} = \int_{W_u^t} (\omega/\nu)(R_{i_{l+1}} T\psi_\alpha)(R_{i_{l+1}} \cdots R_{i_1}\psi_\alpha)(L+2\nu)R_{i_l}\cdots R_{i_1}\mathrm{tr}\chi' dt' du' d\mu_{\tilde{\not g}} \tag{14.347}$$

$$V_{1,1} = \int_{W_u^t} (\omega/\nu)(R_{i_{l+1}}\tilde{\tau}_\alpha + {}^{(R_{i_{l+1}})}Z \cdot \not d T\psi_\alpha) \tag{14.348}$$
$$\cdot (R_{i_{l+1}}\cdots R_{i_1}\psi_\alpha)R_{i_l}\cdots R_{i_1}\mathrm{tr}\chi' dt' du' d\mu_{\tilde{\not g}}$$

$$V_2 = V_{2,0} + V_{2,1} \tag{14.349}$$

$$V_{2,0} = \int_{W_u^t} (T\psi_\alpha)(R_{i_{l+1}}(\omega/\nu) + \frac{1}{2}(\omega/\nu)\mathrm{tr}{}^{(R_{i_{l+1}})}\tilde{\not\pi}) \tag{14.350}$$
$$\cdot (R_{i_{l+1}}\cdots R_{i_1}\psi_\alpha)(L+2\nu)R_{i_l}\cdots R_{i_1}\mathrm{tr}\chi' dt' du' d\mu_{\tilde{\not g}}$$

$$V_{2,1} = \int_{W_u^t} (R_{i_{l+1}}\cdots R_{i_1}\psi_\alpha)(R_{i_l}\cdots R_{i_1}\mathrm{tr}\chi') \tag{14.351}$$
$$\cdot (\tilde{\tau}_\alpha(R_{i_{l+1}}(\omega/\nu) + \frac{1}{2}(\omega/\nu)\mathrm{tr}{}^{(R_{i_{l+1}})}\tilde{\not\pi})$$
$$+ (T\psi_\alpha)({}^{(R_{i_{l+1}})}Z \cdot \not d(\omega/\nu) + (\omega/\nu)\tilde{\not{\mathrm{d}}}\mathrm{iv}{}^{(R_{i_{l+1}})}Z))dt' du' d\mu_{\tilde{\not g}}$$

and

$$V_3 = \int_{W_u^t} 2(\omega/\nu)(R_{i_{l+1}}\cdots R_{i_1}\psi_\alpha)(T\psi_\alpha)(R_{i_{l+1}}\nu)(R_{i_l}\cdots R_{i_1}\mathrm{tr}\chi')dt' du' d\mu_{\tilde{\not g}} \tag{14.352}$$

The last term is a lower order integral which can be easily estimated.

In the following we shall focus attention on the two leading integrals $V_{0,0}$ and $V_{1,0}$, the other integrals containing decay factors compared to these two leading integrals. In fact, comparing $V_{0,0}$ and $V_{0,1}$ as well as $V_{1,0}$ and $V_{1,1}$ by using (14.336) and the estimate

$$\max_j \|{}^{(R_j)}Z\|_{\infty,[1],\Sigma_t^{\epsilon_0}} \leqslant C\delta_0(1+t)^{-1}(1+\log(1+t)) \tag{14.353}$$

of Corollary 10.1.i, we see that $V_{0,1}$ and $V_{1,1}$ contain a decay factor of $\delta_0(1+t)^{-1}(1+\log(1+t))$ as compared to $V_{0,0}$ and $V_{1,0}$ respectively.

We first consider $V_{0,0}$. We have

$$|V_{0,0}| \leqslant C\int_{W_{\epsilon_0}^t} (1+t')^3|T\psi_\alpha||\not d R_{i_{l+1}}\cdots R_{i_1}\psi_\alpha| \tag{14.354}$$
$$\cdot |(L+2\nu)R_{i_l}\cdots R_{i_1}\mathrm{tr}\chi'|dt' du d\mu_{\not g}$$

From the propagation equation (14.219) we have

$$(L+2\nu)R_{i_l}\cdots R_{i_1}\mathrm{tr}\chi' = {}^{(i_1\cdots i_l)}\tilde{\rho}_l \tag{14.355}$$

where

$${}^{(i_1\cdots i_l)}\tilde{\rho}_l = {}^{(i_1\cdots i_l)}\rho_l + 2\nu' R_{i_l}\cdots R_{i_1}\mathrm{tr}\chi' \tag{14.356}$$

Now we can see that $V_{0,1}$ and $V_{1,1}$ enjoy the same bounds as the contribution of the second term on the right of above to $V_{0,1}$ and $V_{1,0}$.

We write, as in (14.229),

$$ {}^{(i_1\cdots i_l)}\tilde{\rho}_l = {}^{(i_1\cdots i_l)}\rho_l^{(0)} + {}^{(i_1\cdots i_l)}\rho_l^{(1)} + {}^{(i_1\cdots i_l)}\tilde{\rho}_l^{(2)} \tag{14.357} $$

where

$$ {}^{(i_1\cdots i_l)}\tilde{\rho}_l^{(2)} = {}^{(i_1\cdots i_l)}\rho_l^{(2)} + 2\nu' R_{i_l}\cdots R_{i_1}\mathrm{tr}\chi' \tag{14.358} $$

Since

$$ |\nu'| \leqslant C\delta_0(1+t)^{-2}(1+\log(1+t)) $$

${}^{(i_1\cdots i_l)}\tilde{\rho}_l^{(2)}$ enjoys the same bound as ${}^{(i_1\cdots i_l)}\rho_l^{(2)}$. The *borderline* contribution to (14.354) is the contribution from ${}^{(i_1\cdots i_l)}\rho_l^{(0)}$. This is the *borderline* integral

$$ \begin{aligned} &C\int_{W_u^t}(1+t')^3|T\psi_\alpha||\not{d}R_{i_{l+1}}\cdots R_{i_1}\psi_\alpha||{}^{(i_1\cdots i_l)}\rho_l^{(0)}|dt'dud\mu_{\not{g}} \\ &\leqslant C\int_0^t(1+t')^3\sup_{\Sigma_{t'}^{\epsilon_0}}(\mu^{-1}|T\psi_\alpha|) \\ &\quad\cdot\|\mu^{1/2}\not{d}R_{i_{l+1}}\cdots R_{i_1}\psi_\alpha\|_{L^2(\Sigma_{t'}^{\epsilon_0})}\|\mu^{1/2}{}^{(i_1\cdots i_l)}\rho_l^{(0)}\|_{L^2(\Sigma_{t'}^{\epsilon_0})}dt' \end{aligned} \tag{14.359} $$

Here we substitute (14.89) for $\sup_{\Sigma_{t'}^{\epsilon_0}}(\mu^{-1}|T\psi_\alpha|)$ and the estimate (14.240) for $\|\mu^{1/2(i_1\cdots i_l)}\rho_l^{(0)}\|_{L^2(\Sigma_{t'}^{\epsilon_0})}$. The factors $|\ell|$ then cancel. Now the partial contribution of the second term on the right of (14.89) is actually not borderline. We shall show how to estimate contribution of this type afterwards, in connection with the estimate for the contribution of ${}^{(i_1\cdots i_l)}\rho_l^{(1)}$. For the present we focus on the *borderline* integral

$$ \begin{aligned} &C\int_0^t\sup_{\Sigma_{t'}^{\epsilon_0}}(\mu^{-1}|L\mu|)\sqrt{\mathcal{E}_1'[R_{i_{l+1}}\cdots R_{i_1}\psi_\alpha](t')}\sqrt{\sum_{j,\alpha}\mathcal{E}_1'[R_jR_{i_l}\cdots R_{i_1}\psi_\alpha](t')dt'} \\ &\leqslant C\int_0^t\sup_{\Sigma_{t'}^{\epsilon_0}}(\mu^{-1}|L\mu|)\sum_{j,\alpha}\mathcal{E}_1'[R_jR_{i_l}\cdots R_{i_1}\psi_\alpha](t')dt' \\ &\leqslant C\int_0^t\sup_{\Sigma_{t'}^{\epsilon_0}}(\mu^{-1}|L\mu|)\bar{\mu}_m^{-2a}(t')(1+\log(1+t'))^{2q}{}^{(i_1\cdots i_l)}\mathcal{G}'_{1,l+2;a,q}(t')dt' \end{aligned} \tag{14.360} $$

This has the same form as (14.97), with $q$ replacing $p$ and ${}^{(i_1\cdots i_l)}\mathcal{G}'_{1,l+2;a,q}$ replacing ${}^{(i_1\cdots i_l)}\mathcal{G}_{0,l+2;a,p}$. Thus, from (14.102) and (14.107) we conclude that (14.360) is bounded by

$$ C(\frac{1}{2a}+\frac{1}{2q})\bar{\mu}_m^{-2a}(t)(1+\log(1+t))^{2q}{}^{(i_1\cdots i_l)}\mathcal{G}'_{1,l+2;a,q}(t) \tag{14.361} $$

We proceed to consider the contribution of ${}^{(i_1\cdots i_l)}\rho_l^{(1)}$ to (14.354). To estimate this and all remaining contributions we simply use (14.90). Using (14.241) we obtain that the contribution in question is bounded by

$$\begin{aligned}
&C\int_{W^t_{\epsilon_0}}(1+t')^3|T\psi_\alpha||\not{d}R_{i_{l+1}}\cdots R_{i_1}\psi_\alpha||{}^{(i_1\cdots i_l)}\rho_l^{(1)}|dt'dud\mu_{\not{g}}\\
&\leqslant C\delta_0\int_0^t(1+t')^2\bar{\mu}_m^{-1}(t')\|\mu^{1/2}\not{d}R_{i_{l+1}}\cdots R_{i_1}\psi_\alpha\|_{L^2(\Sigma^{\epsilon_0}_{t'})}\|\mu^{1/2}{}^{(i_1\cdots i_l)}\rho_l^{(1)}\|_{L^2(\Sigma^{\epsilon_0}_{t'})}dt'\\
&\leqslant C\delta_0^2\int_0^t(1+t')^{-2}\bar{\mu}_m^{-1}(t')\mathcal{E}'_{1,[l+2]}(t')dt'\\
&\leqslant C\delta_0^2\int_0^t(1+t')^{-2}(1+\log(1+t'))^{2q}\bar{\mu}_m^{-1-2a}(t')\mathcal{G}'_{1,[l+2];a,q}(t')dt'
\end{aligned}\tag{14.362}$$

The last integral is similar as (14.115). We find (14.362) is bounded by

$$\begin{aligned}
&C\delta_0^2\bar{\mu}_m^{-2a}(t)(1+\log(1+t))^{2q}\cdot(\varphi_1(Ca\delta_0)\mathcal{G}'_{1,[l+2];a,q}(t)\\
&+\int_0^t(1+t')^{-2}\mathcal{G}'_{1,[l+2];a,q}(t')dt')
\end{aligned}\tag{14.363}$$

We consider next the contribution to (14.354) of the principal part $C_l(1+t)^{-2}\mathcal{W}^Q_{\{l\}}$ in the estimate for $\|{}^{(i_1\cdots i_l)}\tilde{\rho}_l^{(2)}\|_{L^2(\Sigma^{\epsilon_0}_t)}$. This contribution is bounded by

$$\begin{aligned}
&C_l\delta_0\int_0^t\bar{\mu}_m^{-1/2}(t')\|\mu^{1/2}\not{d}R_{i_{l+1}}\cdots R_{i_1}\psi_\alpha\|_{L^2(\Sigma^{\epsilon_0}_{t'})}\mathcal{W}^Q_{\{l\}}(t')dt'\\
&\leqslant C_l\epsilon_0\delta_0\int_0^t(1+t')^{-1}\bar{\mu}_m^{-2a-1/2}(t')(1+\log(1+t'))^{p+q}\\
&\quad\cdot\sqrt{\mathcal{G}'_{1,[l+2];a,q}(t')\mathcal{G}_{0,[l+2];a,p}(t')}dt'\\
&\leqslant C_l\epsilon_0\delta_0J_{2a,p+q-1}(t)\sqrt{\mathcal{G}'_{1,[l+2];a,q}(t)\mathcal{G}_{0,[l+2];a,p}(t)}
\end{aligned}\tag{14.364}$$

where $J_{a,q}(t)$ is the integral (14.147). By (14.153) we have

$$J_{2a,p+q-1}(t)\leqslant C(1+\log(1+t))^{p+q+1}\bar{\mu}_m^{-2a+1/2}(t)\tag{14.365}$$

hence (14.364) is bounded by

$$C_l\epsilon_0\delta_0\bar{\mu}_m^{-2a+1/2}(t)(1+\log(1+t))^{p+q+1}\sqrt{\mathcal{G}'_{1,[l+2];a,q}(t)\mathcal{G}_{0,[l+2];a,p}(t)}\tag{14.366}$$

Finally, we consider the integral $V_{1,0}$, given by (14.347). By the bound (14.213) we have

$$\begin{aligned}
|V_{1,0}|&\leqslant C\delta_0\int_{W^t_u}(1+t')|R_{i_{l+1}}\cdots R_{i_1}\psi_\alpha||{}^{(i_1\cdots i_l)}\tilde{\rho}_l|dt'dud\mu_{\not{g}}\\
&\leqslant C\delta_0\int_0^t(1+t')\mathcal{W}_{\{l+1\}}(t')\|{}^{(i_1\cdots i_l)}\tilde{\rho}_l\|_{L^2(\Sigma^{\epsilon_0}_{t'})}dt'
\end{aligned}\tag{14.367}$$

Here we need only consider the leading principal contribution, namely that of $\|^{(i_1\cdots i_l)}\tilde{\rho}_l^{(0)}\|_{L^2(\Sigma_{t'}^{\epsilon_0})}$. By (14.240) and (14.318) this contribution is bounded by

$$C\epsilon_0\delta_0\int_0^t (1+t')^{-1}\bar{\mu}_m^{-2a-1/2}(t')(1+\log(1+t'))^{p+q}\sqrt{\mathcal{G}_{0,[l+2];a,p}(t')\mathcal{G}'_{1,[l+2];a,q}(t')}dt' \tag{14.368}$$

This is identical to (14.364), which has been bounded by (14.366). This completes the estimates for the spacetime integrals $V_0, V_1, V_2$. Thus, the estimate of the contribution of (14.56) to (14.59) is completed.

### 14.5.2 Estimates for the Contribution of (14.57)

We now consider the contribution of (14.57) to (14.59). Recalling that this is associated to the variation (14.54), the contribution in question is

$$-\int_{W_u^t}(\omega/\nu)(T\psi_\alpha)(R_{i_{l-m}}\cdots R_{i_1}(T)^m\Delta\mu)(L+\nu)(R_{i_{l-m}}\cdots R_{i_1}(T)^{m+1}\psi_\alpha)dt'du'd\mu_{\tilde{g}} \tag{14.369}$$

We deal with integral in the same way as we dealt with (14.202). That is, we write the integrand in the form

$$\begin{aligned}
&-(\omega/\nu)(T\psi_\alpha)(R_{i_{l-m}}\cdots R_{i_1}(T)^m\Delta\mu)((L+\nu)R_{i_{l-m}}\cdots R_{i_1}(T)^{m+1}\psi_\alpha)\\
=&-(L+2\nu)((\omega/\nu)(T\psi_\alpha)(R_{i_{l-m}}\cdots R_{i_1}(T)^m\Delta\mu)(R_{i_{l-m}}\cdots R_{i_1}(T)^{m+1}\psi_\alpha))\\
&+(R_{i_{l-m}}\cdots R_{i_1}(T)^{m+1}\psi_\alpha)(L+\nu)((\omega/\nu)(T\psi_\alpha)(R_{i_{l-m}}\cdots R_{i_1}(T)^m\Delta\mu))
\end{aligned}$$

By (14.206) with the function

$$(\omega/\nu)(T\psi_\alpha)(R_{i_{l-m}}\cdots R_{i_1}(T)^m\Delta\mu)(R_{i_{l-m}}\cdots R_{i_1}(T)^{m+1}\psi_\alpha)$$

in the role of the function $f$, we conclude that (14.369) equals

$$\begin{aligned}
&-\int_{\Sigma_t^u}(\omega/\nu)(T\psi_\alpha)(R_{i_{l-m}}\cdots R_{i_1}(T)^m\Delta\mu)(R_{i_{l-m}}\cdots R_{i_1}(T)^{m+1}\psi_\alpha)du'd\mu_{\tilde{g}}\\
&+\int_{\Sigma_0^u}(\omega/\nu)(T\psi_\alpha)(R_{i_{l-m}}\cdots R_{i_1}(T)^m\Delta\mu)(R_{i_{l-m}}\cdots R_{i_1}(T)^{m+1}\psi_\alpha)du'd\mu_{\tilde{g}}\\
&+\int_{W_u^t}(R_{i_{l-m}}\cdots R_{i_1}(T)^{m+1}\psi_\alpha)(L+\nu)\\
&\cdot((\omega/\nu)(T\psi_\alpha)(R_{i_{l-m}}\cdots R_{i_1}(T)^m\Delta\mu))dt'du'd\mu_{\tilde{g}}
\end{aligned} \tag{14.370}$$

We first consider the hypersurface integral

$$-\int_{\Sigma_t^u}(\omega/\nu)(T\psi_\alpha)(R_{i_{l-m}}\cdots R_{i_1}(T)^m\Delta\mu)(R_{i_{l-m}}\cdots R_{i_1}(T)^{m+1}\psi_\alpha)du'd\mu_{\tilde{g}} \tag{14.371}$$

We define the $S_{t,u}$-tangential vectorfields

$$^{(i_1\cdots i_{l-m})}Y_{m,l-m} = (\not{d}R_{i_{l-m}}\cdots R_{i_1}(T)^m\mu)\cdot(\not{g}^{-1}) \tag{14.372}$$

Then with

$$^{(i_1\cdots i_{l-m})}r_{m,l-m} = R_{i_{l-m}}\cdots R_{i_1}(T)^m\not{\Delta}\mu - \not{\Delta}R_{i_{l-m}}\cdots R_{i_1}(T)^m\mu \tag{14.373}$$

in view of the fact that, since $\tilde{\not{g}} = \Omega\not{g}$, for any $S_{t,u}$-tangential vectorfield $X$:

$$\tilde{\not{\mathrm{div}}}X = \not{\mathrm{div}}X + X\cdot\not{d}\log\Omega, \tag{14.374}$$

we have

$$R_{i_{l-m}}\cdots R_{i_1}(T)^m\not{\Delta}\mu = \tilde{\not{\mathrm{div}}}\,{}^{(i_1\cdots i_{l-m})}Y_{m,l-m} + {}^{(i_1\cdots i_{l-m})}\tilde{r}_{m,l-m} \tag{14.375}$$

where

$$^{(i_1\cdots i_{l-m})}\tilde{r}_{m,l-m} = {}^{(i_1\cdots i_{l-m})}r_{m,l-m} - {}^{(i_1\cdots i_{l-m})}Y_{m,l-m}\cdot\not{d}\log\Omega \tag{14.376}$$

In analogy with (11.323)—(11.325) we deduce

$$\begin{aligned}
&^{(i_1\cdots i_{l-m})}r_{m,l-m}\\
&= \mathrm{tr}({}^{(i_1\cdots i_{l-m})}c_{m,l-m}[\not{d}\mu])\cdot(\not{g}^{-1})\\
&+ \mathrm{tr}(\not{\mathcal{L}}_{R_{i_{l-m}}}\cdots\not{\mathcal{L}}_{R_{i_1}}\sum_{k=1}^{m}\frac{m!}{k!(m-k)!}((\not{\mathcal{L}}_T)^k(\not{g}^{-1}))\cdot(\not{\mathcal{L}}_T)^{m-k}\not{D}^2\mu)\\
&+ \mathrm{tr}(\sum_{|s_1|+|s_2|=l-m,|s_1|>0}((\not{\mathcal{L}}_R)^{s_1}(\not{g}^{-1}))\cdot(\not{\mathcal{L}}_R)^{s_2}(\not{\mathcal{L}}_T)^m\not{D}^2\mu)
\end{aligned} \tag{14.377}$$

Using Corollaries 10.1.d, 10.2.d, 11.1.c, 11.2.c, 11.2.g, we obtain

$$\begin{aligned}
&\|{}^{(i_1\cdots i_{l-m})}r_{m,l-m}\|_{L^2(\Sigma_t^{\epsilon_0})}\\
&\leqslant C_l(1+t)^{-2}(1+\log(1+t))((1+t)^{-1}\mathcal{B}_{[m,l+1]} + \delta_0(\mathcal{Y}_0 + (1+t)\mathcal{A}'_{[l]}\\
&+ \mathcal{W}_{\{l+1\}} + (1+t)^{-2}(1+\log(1+t))^2\mathcal{W}^Q_{\{l\}}))
\end{aligned} \tag{14.378}$$

Moreover, since

$$\|{}^{(i_1\cdots i_{l-m})}Y_{m,l-m}\|_{L^2(\Sigma_t^{\epsilon_0})} \leqslant C(1+t)^{-1}\mathcal{B}_{[m,l+1]} \tag{14.379}$$

while

$$\|\not{d}\log\Omega\|_{L^\infty(\Sigma_t^{\epsilon_0})} \leqslant C\delta_0(1+t)^{-2} \tag{14.380}$$

$^{(i_1\cdots i_{l-m})}\tilde{r}_{m,l-m}$ enjoys the same bounds as $^{(i_1\cdots i_{l-m})}r_{m,l-m}$.

Now substituting (14.375) in (14.371) the hypersurface integral becomes

$$- \int_{\Sigma_t^u} (\omega/\nu)(T\psi_\alpha) \tilde{\text{d}\!\!/\text{iv}}^{(i_1 \cdots i_{l-m})} Y_{m,l-m}(R_{i_{l-m}} \cdots R_{i_1}(T)^{m+1}\psi_\alpha) du' d\mu_{\tilde{g\!\!/}}$$
$$- \int_{\Sigma_t^u} (\omega/\nu)(T\psi_\alpha)^{(i_1 \cdots i_{l-m})} \tilde{r}_{m,l-m}(R_{i_{l-m}} \cdots R_{i_1}(T)^{m+1}\psi_\alpha) du' d\mu_{\tilde{g\!\!/}} \tag{14.381}$$

If $f$ is an arbitrary function defined on $S_{t,u}$ and $X$ an arbitrary vectorfield tangential to $S_{t,u}$, we have

$$\int_{S_{t,u}} f \tilde{\text{d}\!\!/\text{iv}} X d\mu_{\tilde{g\!\!/}} = - \int_{S_{t,u}} X \cdot d\!\!/ f d\mu_{\tilde{g\!\!/}} \tag{14.382}$$

Applying (14.382) to the first of (14.381), taking $X = {}^{(i_1 \cdots i_{l-m})}Y_{m,l-m}$ and

$$f = (\omega/\nu)(T\psi_\alpha)(R_{i_{l-m}} \cdots R_{i_1}(T)^{m+1}\psi_\alpha)$$

we obtain that (14.381) equals

$$H_0' + H_1' + H_2' \tag{14.383}$$

where

$$H_0' = \int_{\Sigma_t^u} (\omega/\nu)(T\psi_\alpha)^{(i_1 \cdots i_{l-m})} Y_{m,l-m} \cdot d\!\!/(R_{i_{l-m}} \cdots R_{i_1}(T)^{m+1}\psi_\alpha) du' d\mu_{\tilde{g\!\!/}} \tag{14.384}$$

$$H_1' = \int_{\Sigma_t^u} (\omega/\nu)(d\!\!/ T\psi_\alpha) \cdot {}^{(i_1 \cdots i_{l-m})} Y_{m,l-m}(R_{i_{l-m}} \cdots R_{i_1}(T)^{m+1}\psi_\alpha) du' d\mu_{\tilde{g\!\!/}} \tag{14.385}$$

$$H_2' = \int_{\Sigma_t^u} (T\psi_\alpha)(d\!\!/(\omega/\nu) \cdot {}^{(i_1 \cdots i_{l-m})} Y_{m,l-m} - (\omega/\nu)^{(i_1 \cdots i_{l-m})} \tilde{r}_{m,l-m})(R_{i_{l-m}} \cdots R_{i_1}(T)^{m+1}\psi_\alpha) du' d\mu_{\tilde{g\!\!/}} \tag{14.386}$$

Now by (14.378) and (14.379) (also (14.214)),

$$\begin{aligned} &\|(\nu/\omega)^{(i_1 \cdots i_{l-m})} Y_{m,l-m} \cdot d\!\!/(\omega/\nu) - {}^{(i_1 \cdots i_{l-m})} \tilde{r}_{m,l-m}\|_{L^2(\Sigma_t^{\epsilon_0})} \\ &\leqslant C_l (1+t)^{-2}(1+\log(1+t))((1+t)^{-1}\mathcal{B}_{[m,l+1]} \\ &\quad + \delta_0(\mathcal{Y}_0 + (1+t)\mathcal{A}'_{[l]} + \mathcal{W}_{\{l+1\}} + (1+t)^{-2}(1+\log(1+t))^2 \mathcal{W}^Q_{\{l\}})) \end{aligned} \tag{14.387}$$

Comparing on one hand (14.386) with (14.385) and on the other hand (14.387) with (14.379), we see that $H_2'$ has an extra decay factor of $(1+t)^{-1}(1+\log(1+t))$ relative to $H_1'$. So we confine attention to $H_0'$ and $H_1'$.

We first consider the integral $H_0'$. We have, recalling the definition (14.372),

$$\begin{aligned} &|H_0'| \\ &\leqslant C \int_{\Sigma_t^{\epsilon_0}} (1+t)^2 |T\psi_\alpha| |d\!\!/ R_{i_{l-m}} \cdots R_{i_1}(T)^m \mu| |d\!\!/ R_{i_{l-m}} \cdots R_{i_1}(T)^{m+1}\psi_\alpha| du d\mu_{g\!\!/} \\ &\leqslant C \int_{\Sigma_t^{\epsilon_0}} (1+t)|T\psi_\alpha| (\sum_j |R_j R_{i_{l-m}} \cdots R_{i_1}(T)^m \mu|) |d\!\!/ R_{i_{l-m}} \cdots R_{i_1}(T)^{m+1}\psi_\alpha| du d\mu_{g\!\!/} \end{aligned} \tag{14.388}$$

Let us set $j = i_{l-m+1}$.

Here we must obtain a sharper estimate for $R_{i_{l-m+1}} \cdots R_{i_1}(T)^m \mu$ than that of Proposition 12.12. Writing

$$ {}^{(i_1 \cdots i_{l-m+1})}\mu_{m,l-m+1} = R_{i_{l-m+1}} \cdots R_{i_1}(T)^m \mu \quad (m = 0, \cdots, l) \tag{14.389} $$

we have the propagation equations (12.229) for $m = 0$, and (12.252) for $m \geqslant 1$, which we can write in combined form as

$$ L^{(i_1 \cdots i_{l-m+1})}\mu_{m,l-m+1} = {}^{(i_1 \cdots i_{l-m+1})}\rho'_{m,l-m+1} \tag{14.390} $$

where

$$ \begin{aligned} &{}^{(i_1 \cdots i_{l-m+1})}\rho'_{m,l-m+1} \\ &= e^{(i_1 \cdots i_{l-m+1})}\mu_{m,l-m+1} + R_{i_{l-m+1}} \cdots R_{i_1}(T)^m m \\ &+ \mu R_{i_{l-m+1}} \cdots R_{i_1}(T)^m e + {}^{(i_1 \cdots i_{l-m+1})}r''_{m,l-m+1} \\ &+ \sum_{k=0}^{m-1} R_{i_{l-m+1}} \cdots R_{i_1}(T)^k \Lambda(T)^{m-1-k}\mu \\ &+ \sum_{k=0}^{l-m} R_{i_{l-m+1}} \cdots R_{i_{l-m-k+2}} {}^{(R_{i_{l-m-k+1}})}Z R_{i_{l-m-k}} \cdots R_{i_1}(T)^m \mu \end{aligned} \tag{14.391} $$

Here we have defined the functions

$$ {}^{(i_1 \cdots i_{l-m+1})}r''_{m,l-m+1} = {}^{(i_1 \cdots i_{l-m+1})}r'_{m,l-m+1} - (R_{i_{l-m+1}} \cdots R_{i_1}(T)^m e)\mu \tag{14.392} $$

where ${}^{(i_1 \cdots i_{l-m+1})}r'_{m,l-m+1}$ are given by (12.251). The functions ${}^{(i_1 \cdots i_{l-m+1})}r''_{m,l-m+1}$ are of order $l+1$ and contain spatial derivatives of $\mu$ of order at most $l$ of which at most $m$ are $T$-derivatives. Explicitly, for $m = 0$,

$$ {}^{(i_1 \cdots i_{l+1})}r''_{0,l+1} = \sum_{|s_1|+|s_2|=l+1, |s_1|, |s_2|>0} ((R)^{s_1} e)((R)^{s_2}\mu) \tag{14.393} $$

and for $m \geqslant 1$,

$$ \begin{aligned} {}^{(i_1 \cdots i_{l-m+1})}r''_{m,l-m+1} = &\sum_{|s_1|+|s_2|=l-m+1, |s_1|>0} ((R)^{s_1} e)((R)^{s_2}(T)^m \mu) \\ &+ \sum_{|s_1|+|s_2|=l-m+1, |s_2|>0} ((R)^{s_1}(T)^m e)((R)^{s_2}\mu) \\ &+ R_{i_{l-m+1}} \cdots R_{i_1}\Big(\sum_{k=1}^{m-1} \frac{m!}{k!(m-k)!}((T)^k e)((T)^{m-k}\mu)\Big) \end{aligned} \tag{14.394} $$

As in Chapter 10 we write

$$ m = m_T^{\alpha}(T\psi_\alpha) \tag{14.395} $$

and we express

$$R_{i_{l-m+1}}\cdots R_{i_1}(T)^m m = m_T^\alpha R_{i_{l-m+1}}\cdots R_{i_1}(T)^{m+1}\psi_\alpha + {}^{(i_1\cdots i_{l-m+1})}\tilde{n}'^{(0)}_{m,l-m+1} \tag{14.396}$$

where

$$\begin{aligned}&{}^{(i_1\cdots i_{l-m+1})}\tilde{n}'^{(0)}_{m,l-m+1}\\ &= \sum_{|s_1|+|s_2|=l-m+1,|s_1|>0}((R)^{s_1}m_T^\alpha)((R)^{s_2}(T)^{m+1}\psi_\alpha)\\ &+ R_{i_{l-m+1}}\cdots R_{i_1}(\sum_{k=1}^m\frac{m!}{k!(m-k)!}((T)^k m_T^\alpha)((T)^{m-k+1}\psi_\alpha))\end{aligned} \tag{14.397}$$

Also, we write

$$e = e_L^\alpha(L\psi_\alpha) \tag{14.398}$$

where

$$\begin{aligned}e_L^0 &= \frac{1}{2\eta^2}(\eta^2)'\\ e_L^i &= \eta^{-1}\hat{T}^i - \frac{1}{2\eta^2}(\eta^2)'\psi_i\end{aligned} \tag{14.399}$$

Obviously,

$$|e_L^\alpha| \leqslant C \tag{14.400}$$

We then express

$$R_{i_{l-m+1}}\cdots R_{i_1}(T)^m e = e_L^\alpha R_{i_{l-m+1}}\cdots R_{i_1}(T)^m L\psi_\alpha + {}^{(i_1\cdots i_{l-m+1})}\tilde{n}'^{(1)}_{m,l-m+1} \tag{14.401}$$

where

$$\begin{aligned}&{}^{(i_1\cdots i_{l-m+1})}\tilde{n}'^{(1)}_{m,l-m+1}\\ &= \sum_{|s_1|+|s_2|=l-m+1,|s_1|>0}((R)^{s_1}e_L^\alpha)((R)^{s_2}(T)^m L\psi_\alpha)\\ &+ R_{i_{l-m+1}}\cdots R_{i_1}(\sum_{k=1}^m\frac{m!}{k!(m-k)!}((T)^k e_L^\alpha)((T)^{m-k}L\psi_\alpha))\end{aligned} \tag{14.402}$$

We define

$${}^{(i_1\cdots i_{l-m+1})}\rho'^{(0)}_{m,l-m+1} = \frac{1}{2}\ell R_{i_{l-m+1}}\cdots R_{i_1}(T)^{m+1}\psi_0 \tag{14.403}$$

and

$$\begin{aligned}{}^{(i_1\cdots i_{l-m+1})}\rho'^{(1)}_{m,l-m+1} &= (m_T^0 - \frac{1}{2}\ell)R_{i_{l-m+1}}\cdots R_{i_1}(T)^{m+1}\psi_0\\ &+ m_T^i R_{i_{l-m+1}}\cdots R_{i_1}(T)^{m+1}\psi_i\\ &+ (1+t)^{-1}\mu e_L^\alpha R_{i_{l-m+1}}\cdots R_{i_1}(T)^m Q\psi_\alpha\end{aligned} \tag{14.404}$$

We then have

$$\begin{aligned} &R_{i_{l-m+1}} \cdots R_{i_1}(T)^m m + \mu R_{i_{l-m+1}} \cdots R_{i_1}(T)^m e \\ &= {}^{(i_1 \cdots i_{l-m})}\rho'^{(0)}_{m,l-m+1} + {}^{(i_1 \cdots i_{l-m+1})}\rho'^{(1)}_{m,l-m+1} + {}^{(i_1 \cdots i_{l-m+1})}\tilde{n}'_{m,l-m+1} \end{aligned} \tag{14.405}$$

where

$${}^{(i_1 \cdots i_{l-m+1})}\tilde{n}'_{m,l-m+1} = {}^{(i_1 \cdots i_{l-m+1})}\tilde{n}'^{(0)}_{m,l-m+1} + \mu^{(i_1 \cdots i_{l-m+1})}\tilde{n}'^{(1)}_{m,l-m+1} \tag{14.406}$$

and (14.391) takes the form

$$\begin{aligned} &{}^{(i_1 \cdots i_{l-m+1})}\rho'_{m,l-m+1} \\ &= {}^{(i_1 \cdots i_{l-m+1})}\rho'^{(0)}_{m,l-m+1} + {}^{(i_1 \cdots i_{l-m+1})}\rho'^{(1)}_{m,l-m+1} + {}^{(i_1 \cdots i_{l-m+1})}\rho'^{(2)}_{m,l-m+1} \end{aligned} \tag{14.407}$$

where

$$\begin{aligned} &{}^{(i_1 \cdots i_{l-m+1})}\rho'^{(2)}_{m,l-m+1} \\ &= e^{(i_1 \cdots i_{l-m+1})}\mu_{m,l-m+1} + {}^{(i_1 \cdots i_{l-m+1})}\tilde{n}'_{m,l-m+1} + {}^{(i_1 \cdots i_{l-m+1})}r''_{m,l-m+1} \\ &+ \sum_{k=0}^{m-1} R_{i_{l-m+1}} \cdots R_{i_1}(T)^k \Lambda(T)^{m-1-k}\mu \\ &+ \sum_{k=0}^{l-m} R_{i_{l-m+1}} \cdots R_{i_{l-m-k+2}}{}^{(R_{i_{l-m-k+1}})}Z R_{i_{l-m-k}} \cdots R_{i_1}(T)^m \mu \end{aligned} \tag{14.408}$$

We shall now obtain an $L^2(\Sigma_t^{\epsilon_0})$ estimate for ${}^{(i_1 \cdots i_{l-m+1})}\rho'^{(2)}_{m,l-m+1}$. The first term on the right in (14.408) is bounded in $L^2(\Sigma_t^{\epsilon_0})$ by

$$C\delta_0(1+t)^{-2}\mathcal{B}_{[m,l+1]} \tag{14.409}$$

From (14.397) we deduce

$$\|{}^{(i_1 \cdots i_{l-m+1})}\tilde{n}'^{(0)}_{m,l-m+1}\|_{L^2(\Sigma_t^{\epsilon_0})} \leqslant C\delta_0(1+t)^{-1}\mathcal{W}_{\{l+1\}} \tag{14.410}$$

Using Corollary 11.2.b, we deduce that $\|{}^{(i_1 \cdots i_{l-m+1})}\tilde{n}'^{(1)}_{m,l-m+1}\|_{L^2(\Sigma_t^{\epsilon_0})}$ is bounded by

$$C_l\delta_0(1+t)^{-1}(\mathcal{W}_{\{l+1\}} + \mathcal{W}^Q_{\{l\}} + \delta_0(1+t)^{-1}((1+t)^{-1}\mathcal{B}_{[m-1,l+1]} + \mathcal{Y}_0 + (1+t)\mathcal{A}'_{[l]})) \tag{14.411}$$

We turn to the third term on the right in (14.408). Using Corollary 11.2.b we deduce through (14.393), (14.394),

$$\begin{aligned} &\|{}^{(i_1 \cdots i_{l-m})}r''_{m,l-m+1}\|_{L^2(\Sigma_t^{\epsilon_0})} \\ &\leqslant C\delta_0((1+\log(1+t))(1+t)^{-1}(\mathcal{W}^Q_{\{l\}} + \delta_0(1+t)^{-1}(\mathcal{Y}_0 + (1+t)\mathcal{A}'_{[l-1]} \\ &+ \mathcal{W}_{\{l\}})) + (1+t)^{-2}\mathcal{B}_{[m,l]}) \end{aligned} \tag{14.412}$$

Finally, the two sums on the right in (14.408) have already been estimated in $L^2(\Sigma_t^{\epsilon_0})$ in (12.357) and (12.359). So we conclude that

$$\sum_j \|^{(i_1\cdots i_{l-m}j)}\rho'^{(2)}_{m,l-m+1}\|_{L^2(\Sigma_t^{\epsilon_0})} \leqslant C_l\delta_0(1+t)^{-1}(1+\log(1+t)) \cdot (\mathcal{W}_{\{l+1\}} + \mathcal{W}^Q_{\{l\}} + (1+t)^{-1}\mathcal{B}_{[m,l+1]} + \mathcal{Y}_0 + (1+t)\mathcal{A}'_{[l]}) \quad (14.413)$$

Moreover, from (14.403) and (14.404), similarly as before, we have

$$\sum_j \|^{(i_1\cdots i_{l-m}j)}\rho'^{(0)}_{m,l-m+1}\|_{L^2(\Sigma_t^{\epsilon_0})} \leqslant C|\ell|\bar{\mu}_m^{-1/2}\sqrt{\sum_\alpha \mathcal{E}'_1[R_{i_{l-m}}\cdots R_{i_1}(T)^{m+1}\psi_0](t)} \quad (14.414)$$

$$\sum_j \|^{(i_1\cdots i_{l-m}j)}\rho'^{(1)}_{m,l-m+1}\|_{L^2(\Sigma_t^{\epsilon_0})} \leqslant C\delta_0(1+t)^{-1}\bar{\mu}_m^{-1/2}\sqrt{\sum_\alpha \mathcal{E}'_1[R_{i_{l-m}}\cdots R_{i_1}(T)^{m+1}\psi_\alpha](t)} + C\delta_0(1+t)^{-1}(1+\log(1+t))^{1/2}\sqrt{\sum_\alpha \mathcal{E}'_1[R_{i_{l-m}}\cdots R_{i_1}(T)^{m}Q\psi_\alpha](t)} \quad (14.415)$$

and

$$\sum_j \|\mu^{1/2(i_1\cdots i_{l-m}j)}\rho'^{(0)}_{m,l-m+1}\|_{L^2(\Sigma_t^{\epsilon_0})} \leqslant C|\ell|\sqrt{\sum_\alpha \mathcal{E}'_1[R_{i_{l-m}}\cdots R_{i_1}(T)^{m+1}\psi_\alpha](t)} \quad (14.416)$$

$$\sum_j \|\mu^{1/2(i_1\cdots i_{l-m}j)}\rho'^{(1)}_{m,l-m+1}\|_{L^2(\Sigma_t^{\epsilon_0})} \leqslant C\delta_0(1+t)^{-1}\sqrt{\sum_\alpha \mathcal{E}'_1[R_{i_{l-m}}\cdots R_{i_1}(T)^{m+1}\psi_\alpha](t)} + C\delta_0(1+t)^{-1}(1+\log(1+t))\sqrt{\sum_\alpha \mathcal{E}'_1[R_{i_{l-m}}\cdots R_{i_1}(T)^{m}Q\psi_\alpha](t)} \quad (14.417)$$

We now integrate the propagation equation (14.390) along each integral curve of $L$ from $\Sigma_0^{\epsilon_0}$ to obtain

$$\sum_j |^{(i_1\cdots i_{l-m}j)}\mu_{m,l-m+1}(t,u,\vartheta)| \leqslant C(\sum_j |^{(i_1\cdots i_{l-m}j)}\mu_{m,l-m+1}(0,u,\vartheta)| + \sum_{k=0}^{2} {}^{(i_1\cdots i_{l-m}j)}A'^{(k)}_{m,l-m}(t,u,\vartheta)) \quad (14.418)$$

where

$$ {}^{(i_1\cdots i_{l-m})}A'^{(k)}_{m,l-m}(t,u,\vartheta) = \int_0^t \sum_j |{}^{(i_1\cdots i_{l-m}j)}\rho'^{(k)}_{m,l-m+1}|(t',u,\vartheta)dt' \tag{14.419} $$

for $k = 0,1,2$.

By (8.333) we have

$$ \begin{aligned} &\|{}^{(i_1\cdots i_{l-m})}A'^{(k)}_{m,l-m}(t)\|_{L^2([0,\epsilon_0]\times S^2)} \\ &\leqslant \int_0^t \sum_j \|{}^{(i_1\cdots i_{l-m})}\rho'^{(k)}_{m,l-m+1}(t')\|_{L^2([0,\epsilon_0]\times S^2)}dt' \\ &\leqslant \int_0^t (1+t')^{-1}\sum_j \|{}^{(i_1\cdots i_{l-m}j)}\rho'^{(k)}_{m,l-m+1}\|_{L^2(\Sigma^{\epsilon_0}_{t'})}dt' \end{aligned} \tag{14.420} $$

We now substitute (14.418) in (14.388), noting that

$$ \sum_j |R_j R_{i_{l-m}}\cdots R_{i_1}(T)^m\mu| = \sum_j |{}^{(i_1\cdots i_{l-m}j)}\mu_{m,l-m+1}| $$

The *borderline* contribution is from ${}^{(i_1\cdots i_{l-m})}A'^{(0)}_{m,l-m}$. It is the *borderline* integral

$$ C\int_{\Sigma^{\epsilon_0}_t}(1+t)|T\psi_\alpha|{}^{(i_1\cdots i_{l-m})}A'^{(0)}_{m,l-m}|\not{d}R_{i_{l-m}}\cdots R_{i_1}(T)^{m+1}\psi_\alpha|dud\mu_{\not{g}} \tag{14.421} $$

This is similar to (14.246) and it is estimated in exactly the same manner. Defining (see (14.257))

$$ \begin{aligned} &{}^{(i_1\cdots i_{l-m})}\mathcal{G}'_{1,m,l+2;a,q}(t) \\ &= \sup_{t'\in[0,t]}((1+\log(1+t'))^{-2q}\bar{\mu}^{2a}_m(t')\sum_\alpha \mathcal{E}'_1[R_{i_{l-m}}\cdots R_{i_1}(T)^{m+1}\psi_\alpha](t')) \end{aligned} \tag{14.422} $$

then similarly, we obtain (see (14.266), (14.293)) that the borderline integral (14.421) is bounded by

$$ C(\frac{1}{a-1/2}+\frac{1}{q+1/2})\bar{\mu}^{-2a}_m(t)(1+\log(1+t))^{2q}\,{}^{(i_1\cdots i_{l-m})}\mathcal{G}'_{1,m,l+2;a,q}(t) \tag{14.423} $$

The remaining contributions to (14.388) are estimated in exactly the same manner as the remaining contributions to (14.215). In view of (14.294) these contributions are bounded by

$$ \begin{aligned} &C\int_{\Sigma^{\epsilon_0}_t}(1+t)|T\psi_\alpha|{}^{(i_1\cdots i_{l-m})}A'^{(k)}_{m,l-m}|\not{d}R_{i_{l-m}}\cdots R_{i_1}(T)^{m+1}\psi_\alpha|dud\mu_{\not{g}} \\ &\leqslant C\delta_0\int_{\Sigma^{\epsilon_0}_t}{}^{(i_1\cdots i_{l-m})}A'^{(k)}_{m,l-m}|\not{d}R_{i_{l-m}}\cdots R_{i_1}(T)^{m+1}\psi_\alpha|dud\mu_{\not{g}} \\ &\leqslant C\delta_0\|{}^{(i_1\cdots i_{l-m})}A'^{(k)}_{m,l-m}\|_{L^2(\Sigma^{\epsilon_0}_t)}\|\not{d}R_{i_{l-m}}\cdots R_{i_1}(T)^{m+1}\psi_\alpha\|_{L^2(\Sigma^{\epsilon_0}_t)} \\ &\leqslant C\delta_0\bar{\mu}^{-1/2}_m(t)\|{}^{(i_1\cdots i_{l-m})}A'^{(k)}_{m,l-m}(t)\|_{L^2([0,\epsilon_0]\times S^2)}\sqrt{\mathcal{E}'_{1,[l+2]}(t)} \end{aligned} \tag{14.424} $$

for $k=1,2$, which is similar to (14.297). Here we need only consider the contribution of ${}^{(i_1\cdots i_{l-m})}A'^{(1)}_{m,l-m}$, because ${}^{(i_1\cdots i_{l-m})}A'^{(2)}_{m,l-m}$ besides being of lower order does not contain any principal parts, as is evident by comparing (14.413) to (14.237). By (14.415),

$$\sum_j \|{}^{(i_1\cdots i_{l-m}j)}\rho'^{(1)}_{m,l-m+1}\|_{L^2(\Sigma_t^{\epsilon_0})}$$
$$\leqslant C\delta_0(1+t)^{-1}(1+\log(1+t))^{1/2}\bar{\mu}_m^{-1/2}(t)\sqrt{\mathcal{E}'_{1,[l+2]}(t)} \qquad (14.425)$$

we obtain

$$\|{}^{(i_1\cdots i_{l-m})}A'^{(1)}_{m,l-m}(t)\|_{L^2([0,\epsilon_0]\times S^2)}$$
$$\leqslant C\delta_0\int_0^t \bar{\mu}_m^{-1/2}(t')(1+t')^{-2}(1+\log(1+t'))^{1/2}\sqrt{\mathcal{E}'_{1,[l+2]}(t')}dt'$$
$$\leqslant C\delta_0 J'_{a,q+1/2}(t)\sqrt{\mathcal{G}'_{1,[l+2];a,q}(t)} \qquad (14.426)$$

where $J'_{a,q}(t)$ is the integral (14.298). Substituting (14.305) with $q$ replaced by $q+1/2$ yields

$$C\int_{\Sigma_t^{\epsilon_0}}(1+t)|T\psi_\alpha|{}^{(i_1\cdots i_{l-m})}A'^{(1)}_{m,l-m}|\not{d}R_{i_{l-m}}\cdots R_{i_1}(T)^{m+1}\psi_\alpha|dud\mu_{\not{g}}$$
$$\leqslant C\delta_0^2\mathcal{G}'_{1,[l+2];a,q}(t)(C_q\bar{\mu}_m^{-a-1/2}(t)+C\varphi'_{q+3/2}(Ca\delta_0)\bar{\mu}_m^{-2a}(t))(1+\log(1+t))^q \qquad (14.427)$$

Finally, we consider the hypersurface integral $H'_1$, given by (14.387). We have

$$|H'_1|\leqslant C\int_{\Sigma_t^{\epsilon_0}}(1+t)|\not{d}T\psi_\alpha|\sum_j|R_jR_{i_{l-m}}\cdots R_{i_1}(T)^m\mu||R_{i_{l-m}}\cdots R_{i_1}(T)^{m+1}\psi_\alpha|dud\mu_{\not{g}} \qquad (14.428)$$

Using

$$\max_\alpha\sup_{\Sigma_t^{\epsilon_0}}|\not{d}T\psi_\alpha|\leqslant C\delta_0(1+t)^{-2} \qquad (14.429)$$

we have

$$|H'_1|$$
$$\leqslant C\delta_0\int_{\Sigma_t^{\epsilon_0}}(1+t)^{-1}\sum_j|R_jR_{i_{l-m}}\cdots R_{i_1}(T)^m\mu||R_{i_{l-m}}\cdots R_{i_1}(T)^{m+1}\psi_\alpha|dud\mu_{\not{g}}$$
$$\leqslant C\delta_0(1+t)^{-1}\sum_j\|R_jR_{i_{l-m}}\cdots R_{i_1}(T)^m\mu\|_{L^2(\Sigma_t^{\epsilon_0})}\|R_{i_{l-m}}\cdots R_{i_1}(T)^{m+1}\psi_\alpha\|_{L^2(\Sigma_t^{\epsilon_0})}$$
$$\leqslant C\delta_0\sum_j\|(R_jR_{i_{l-m}}\cdots R_{i_1}(T)^m\mu)(t)\|_{L^2([0,\epsilon_0]\times S^2)}\mathcal{W}_{\{l+1\}}(t) \qquad (14.430)$$

From (14.418) and (14.420),

$$\begin{aligned}
&\sum_j \|(R_j R_{i_{l-m}} \cdots R_{i_1}(T)^m \mu)(t)\|_{L^2([0,\epsilon_0]\times S^2)} \\
\leqslant\, & C(\sum_j \|R_j R_{i_{l-m}} \cdots R_{i_1}(T)^m \mu\|_{L^2(\Sigma_0^{\epsilon_0})} \\
&+ \sum_{k=0}^{2} \|^{(i_1\cdots i_{l-m})} A'^{(k)}_{m,l-m}(t)\|_{L^2([0,\epsilon_0]\times S^2)}) \\
\leqslant\, & C(\sum_j \|R_j R_{i_{l-m}} \cdots R_{i_1}(T)^m \mu\|_{L^2(\Sigma_0^{\epsilon_0})} \\
&+ \sum_{k=0}^{2} \int_0^t (1+t')^{-1} \sum_j \|^{(i_1\cdots i_{l-m} j)} \rho'^{(k)}_{m,l-m+1}\|_{L^2(\Sigma_{t'}^{\epsilon_0})} dt')
\end{aligned} \tag{14.431}$$

Here we need only consider the leading principal contribution, namely that of

$$\sum_j \|^{(i_1\cdots i_{l-m} j)} \rho'^{(0)}_{m,l-m+1}\|_{L^2(\Sigma_{t'}^{\epsilon_0})}$$

This contribution to (14.431) is bounded by (see (14.414))

$$\begin{aligned}
& C|\ell| \int_0^t (1+t')^{-1} \bar{\mu}_m^{-1/2}(t') \sqrt{\sum_\alpha \mathcal{E}_1'[R_{i_{l-m}} \cdots R_{i_1}(T)^{m+1}\psi_\alpha](t')} dt' \\
\leqslant\, & C|\ell| \sqrt{\mathcal{G}'_{1,[l+2];a,q}(t)} \int_0^t (1+t')^{-1} \bar{\mu}_m^{-a-1/2}(t')(1+\log(1+t'))^q dt' \\
=\, & C|\ell| J_{a,q-1}(t) \sqrt{\mathcal{G}'_{1,[l+2];a,q}(t)}
\end{aligned} \tag{14.432}$$

Substituting the bound (14.153) with $q$ replaced by $q-1$ and also the bound (14.318) for $\mathcal{W}_{\{l+1\}}$, we obtain that the leading contribution to $H_1'$ is bounded by

$$C\epsilon_0 \delta_0 \bar{\mu}_m^{-2a+1/2}(t)(1+\log(1+t))^{p+q+1} \sqrt{\mathcal{G}_{0,[l+2];a,p}(t)\mathcal{G}'_{1,[l+2];a,q}(t)} \tag{14.433}$$

the same in form as (14.319).

We turn to the spacetime integral in (14.370), namely

$$\int_{W_u^t} (R_{i_{l-m}} \cdots R_{i_1}(T)^{m+1}\psi_\alpha)(L+\nu)((\omega/\nu)(T\psi_\alpha)(R_{i_{l-m}} \cdots R_{i_1}(T)^m \Delta\mu)) dt' du' d\mu_{\tilde{g}} \tag{14.434}$$

As in (14.334) we have, in regard to the integrand in (14.434),

$$\begin{aligned}
&(L+\nu)((\omega/\nu)(T\psi_\alpha)(R_{i_{l-m}} \cdots R_{i_1}(T)^m \Delta\mu)) \\
=\, & (\omega/\nu)((T\psi_\alpha)(L+2\nu)(R_{i_{l-m}} \cdots R_{i_1}(T)^m \Delta\mu) + \tilde{\tau}_\alpha(R_{i_{l-m}} \cdots R_{i_1}(T)^m \Delta\mu))
\end{aligned} \tag{14.435}$$

We substitute on the right-hand side the expression (14.375) for $R_{i_{l-m}} \cdots R_{i_1}(T)^m \not{\Delta}\mu$. Here we apply the following: Let $X$ be an arbitrary $S_{t,u}$-tangential vectorfield. We then have

$$(L + 2\nu)\tilde{\not{\text{d}}\text{iv}}X = \tilde{\not{\text{d}}\text{iv}}(\not{\mathcal{L}}_L X + 2\nu X) \tag{14.436}$$

To establish this we work in acoustical coordinates $(t, u, \vartheta^A, A = 1, 2)$. We then have

$$\begin{aligned} X &= X^A \frac{\partial}{\partial \vartheta^A} \\ \not{\text{d}}\text{iv}X &= \frac{1}{\sqrt{\det \not{g}}} \frac{\partial}{\partial \vartheta^A}(\sqrt{\det \not{g}} X^A) = \frac{\partial X^A}{\partial \vartheta^A} + X^A \frac{\partial}{\partial \vartheta^A} \log \sqrt{\det \not{g}} \end{aligned} \tag{14.437}$$

and

$$\begin{aligned} L\not{\text{d}}\text{iv}X &= \frac{\partial}{\partial t}\not{\text{d}}\text{iv}X \\ &= \frac{\partial^2 X^A}{\partial t \partial \vartheta^A} + \frac{\partial X^A}{\partial t} \frac{\partial}{\partial \vartheta^A} \log \sqrt{\det \not{g}} + X^A \frac{\partial^2}{\partial t \partial \vartheta^A} \log \sqrt{\det \not{g}} \end{aligned} \tag{14.438}$$

Now

$$\not{\mathcal{L}}_L X = [L, X] = \frac{\partial X^A}{\partial t} \frac{\partial}{\partial \vartheta^A} \tag{14.439}$$

and

$$\frac{\partial}{\partial t} \log \sqrt{\det \not{g}} = \text{tr}\chi \tag{14.440}$$

We thus obtain

$$L\not{\text{d}}\text{iv}X = \not{\text{d}}\text{iv}(\not{\mathcal{L}}_L X) + X \cdot \not{d}\text{tr}\chi \tag{14.441}$$

In view of (14.374) and the definition of $\nu$, this is equivalent to

$$L\tilde{\not{\text{d}}\text{iv}}X = \tilde{\not{\text{d}}\text{iv}}(\not{\mathcal{L}}_L X) + 2X \cdot \not{d}\nu \tag{14.442}$$

which is equivalent to (14.436).

We appeal to (14.436) taking $X = {}^{(i_1 \cdots i_{l-m})}Y_{m,l-m}$ after substituting the expression (14.375) in (14.437). Substituting then the result in (14.434) and integrating by parts on each $S_{t,u}$ we obtain

$$\begin{aligned} &\int_{W^t_u} (\omega/\nu)(T\psi_\alpha)(R_{i_{l-m}} \cdots R_{i_1}(T)^{m+1}\psi_\alpha)(L + 2\nu)\tilde{\not{\text{d}}\text{iv}}{}^{(i_1 \cdots i_{l-m})}Y_{m,l-m} dt' du' d\mu_{\tilde{\not{g}}} \\ &= -\int_{W^t_u} (\omega/\nu)(T\psi_\alpha)((\not{\mathcal{L}}_L + 2\nu)^{(i_1 \cdots i_{l-m})}Y_{m,l-m}) \\ &\quad \cdot \not{d}R_{i_{l-m}} \cdots R_{i_1}(T)^{m+1}\psi_\alpha dt' du' d\mu_{\tilde{\not{g}}} \\ &\quad - \int_{W^t_u} (\omega/\nu)(R_{i_{l-m}} \cdots R_{i_1}(T)^{m+1}\psi_\alpha)((\not{\mathcal{L}}_L + 2\nu)^{(i_1 \cdots i_{l-m})}Y_{m,l-m}) \\ &\quad \cdot \not{d}T\psi_\alpha dt' du' d\mu_{\tilde{\not{g}}} \end{aligned} \tag{14.443}$$

$$- \int_{W_u^t} (T\psi_\alpha)(R_{i_{l-m}} \cdots R_{i_1}(T)^{m+1}\psi_\alpha) \cdot ((\not{\mathcal{L}}_L + 2\nu)^{(i_1\cdots i_{l-m})} Y_{m,l-m}) \not{d}(\omega/\nu) dt' du' d\mu_{\tilde{\not{g}}}$$

and we are left with

$$\int_{W_u^t} (\omega/\nu)(T\psi_\alpha)(R_{i_{l-m}} \cdots R_{i_1}(T)^{m+1}\psi_\alpha)(L+2\nu)^{(i_1\cdots i_{l-m})} \tilde{r}_{m,l-m} dt' du' d\mu_{\tilde{\not{g}}} \tag{14.444}$$

We also integrate by parts on each $S_{t,u}$:

$$\begin{aligned} &\int_{W_u^t} (\omega/\nu)\tilde{\tau}_\alpha (R_{i_{l-m}} \cdots R_{i_1}(T)^{m+1}\psi_\alpha) \tilde{\not{d}\text{iv}}^{(i_1\cdots i_{l-m})} Y_{m,l-m} dt' du' d\mu_{\tilde{\not{g}}} \\ &= - \int_{W_u^t} (\omega/\nu) \tilde{\tau}_\alpha{}^{(i_1\cdots i_{l-m})} Y_{m,l-m} \cdot \not{d}(R_{i_{l-m}} \cdots R_{i_1}(T)^{m+1}\psi_\alpha) dt' du' d\mu_{\tilde{\not{g}}} \quad (14.445) \\ &\quad - \int_{W_u^t} (\omega/\nu)(R_{i_{l-m}} \cdots R_{i_1}(T)^{m+1}\psi_\alpha)^{(i_1\cdots i_{l-m})} Y_{m,l-m} \not{d}\tilde{\tau}_\alpha dt' du' d\mu_{\tilde{\not{g}}} \\ &\quad - \int_{W_u^t} \tilde{\tau}_\alpha (R_{i_{l-m}} \cdots R_{i_1}(T)^{m+1}\psi_\alpha)^{(i_1\cdots i_{l-m})} Y_{m,l-m} \not{d}(\omega/\nu) dt' du' d\mu_{\tilde{\not{g}}} \end{aligned}$$

and we are left with

$$\int_{W_u^t} (\omega/\nu)\tilde{\tau}_\alpha (R_{i_{l-m}} \cdots R_{i_1}(T)^{m+1}\psi_\alpha)^{(i_1\cdots i_{l-m})} \tilde{r}_{m,l-m} dt' du' d\mu_{\tilde{\not{g}}} \tag{14.446}$$

In view of (14.443)—(14.446) the spacetime integral (14.434) becomes

$$-V_0' - V_1' - V_2' \tag{14.447}$$

where

$$V_0' = V_{0,0}' + V_{0,1}' \tag{14.448}$$

$$V_{0,0}' = \int_{W_u^t} (\omega/\nu)(T\psi_\alpha)((\not{\mathcal{L}}_L + 2\nu)^{(i_1\cdots i_{l-m})} Y_{m,l-m}) \cdot \not{d}R_{i_{l-m}} \cdots R_{i_1}(T)^{m+1}\psi_\alpha dt' du' d\mu_{\tilde{\not{g}}} \tag{14.449}$$

$$V_{0,1}' = \int_{W_u^t} (\omega/\nu) \tilde{\tau}_\alpha{}^{(i_1\cdots i_{l-m})} Y_{m,l-m} \cdot \not{d}R_{i_{l-m}} \cdots R_{i_1}(T)^{m+1}\psi_\alpha dt' du' d\mu_{\tilde{\not{g}}} \tag{14.450}$$

$$V_1' = V_{1,0}' + V_{1,1}' \tag{14.451}$$

$$V_{1,0}' = \int_{W_u^t} (\omega/\nu)(R_{i_{l-m}} \cdots R_{i_1}(T)^{m+1}\psi_\alpha) \cdot ((\not{\mathcal{L}}_L + 2\nu)^{(i_1\cdots i_{l-m})} Y_{m,l-m}) \not{d}T\psi_\alpha dt' du' d\mu_{\tilde{\not{g}}} \tag{14.452}$$

$$V_{1,1}' = \int_{W_u^t} (\omega/\nu)(R_{i_{l-m}} \cdots R_{i_1}(T)^{m+1}\psi_\alpha)^{(i_1\cdots i_{l-m})} Y_{m,l-m} \not{d}\tilde{\tau}_\alpha dt' du' d\mu_{\tilde{\not{g}}} \tag{14.453}$$

and

$$V_2' = V_{2,0}' + V_{2,1}' \tag{14.454}$$

$$V_{2,0}' = \int_{W_u^t} (T\psi_\alpha)(R_{i_{l-m}} \cdots R_{i_1}(T)^{m+1}\psi_\alpha) \cdot (((\not{\mathcal{L}}_L + 2\nu)^{(i_1\cdots i_{l-m})}Y_{m,l-m}) \cdot \not{d}(\omega/\nu) - (\omega/\nu)(L+2\nu)^{(i_1\cdots i_{l-m})}\tilde{r}_{m,l-m})dt'du'd\mu_{\tilde{\not{g}}} \tag{14.455}$$

$$V_{2,1}' = \int_{W_u^t} (R_{i_{l-m}} \cdots R_{i_1}(T)^{m+1}\psi_\alpha)\tilde{\tau}_\alpha \cdot ({}^{(i_1\cdots i_{l-m})}Y_{m,l-m} \cdot (\not{d}(\omega/\nu)) - (\omega/\nu)^{(i_1\cdots i_{l-m})}\tilde{r}_{m,l-m})dt'du'd\mu_{\tilde{\not{g}}} \tag{14.456}$$

In the following, we estimate $V_{0,0}'$ and $V_{1,0}'$, seeing that the other terms are of lower order.

We have

$$|V_{0,0}'| \leqslant C\int_{W_{\epsilon_0}^t} (1+t')^2|T\psi_\alpha||\not{d}R_{i_{l-m}} \cdots R_{i_1}(T)^{m+1}\psi_\alpha| \cdot |(\not{\mathcal{L}}_L + 2\nu)^{(i_1\cdots i_{l-m})}Y_{m,l-m}|dt'dud\mu_{\not{g}} \tag{14.457}$$

Since according to (14.372),

$$\not{d}R_{i_{l-m}} \cdots R_{i_1}(T)^m\mu = \not{g} \cdot {}^{(i_1\cdots i_{l-m})}Y_{m,l-m} \tag{14.458}$$

and $\not{\mathcal{L}}_L\not{g} = 2\chi$, we have

$$\begin{aligned}\not{d}LR_{i_{l-m}} \cdots R_{i_1}(T)^m\mu &= \not{\mathcal{L}}_L\not{d}R_{i_{l-m}} \cdots R_{i_1}(T)^m\mu \\ &= 2\chi \cdot {}^{(i_1\cdots i_{l-m})}Y_{m,l-m} + \not{g} \cdot \not{\mathcal{L}}_L{}^{(i_1\cdots i_l)}Y_{m,l-m}\end{aligned} \tag{14.459}$$

Since

$$\chi = \frac{1}{2}\mathrm{tr}\chi\not{g} + \hat{\chi} = (\nu - \frac{1}{2}L\log\Omega)\not{g} + \hat{\chi}'$$

defining

$$\tilde{\chi}' = \hat{\chi}' - \frac{1}{2}(L\log\Omega)\not{g} \tag{14.460}$$

we then obtain

$$\not{g} \cdot (\not{\mathcal{L}}_L + 2\nu)^{(i_1\cdots i_{l-m})}Y_{m,l-m} = \not{d}LR_{i_{l-m}} \cdots R_{i_1}(T)^m\mu - 2\tilde{\chi}' \cdot {}^{(i_1\cdots i_{l-m})}Y_{m,l-m} \tag{14.461}$$

Substituting for $LR_{i_{l-m}} \cdots R_{i_1}(T)^m\mu$ from (14.390) with $l+1$ replaced by $l$ then yields

$$\not{g} \cdot (\not{\mathcal{L}}_L + 2\nu)^{(i_1\cdots i_{l-m})}Y_{m,l-m} = \not{d}^{(i_1\cdots i_{l-m})}\rho_{m,l-m}' - 2\tilde{\chi}' \cdot \not{d}R_{i_{l-m}} \cdots R_{i_1}(T)^m\mu \tag{14.462}$$

which implies

$$|(\not{\mathcal{L}}_L + 2\nu)^{(i_1\cdots i_{l-m})}Y_{m,l-m}| \leqslant \sum_{k=0}^{2} |\not{d}^{(i_1\cdots i_{l-m})}\rho'^{(k)}_{m,l-m}| + 2|\tilde{\chi}'||\not{d}R_{i_{l-m}}\cdots R_{i_1}(T)^m\mu| \tag{14.463}$$

Now from (14.403) with $l+1$ replaced by $l$

$$\not{d}^{(i_1\cdots i_{l-m})}\rho'^{(0)}_{m,l-m} = \frac{1}{2}\ell\not{d}R_{i_{l-m}}\cdots R_{i_1}(T)^{m+1}\psi_0 \tag{14.464}$$

hence

$$\|\mu^{1/2}\not{d}^{(i_1\cdots i_l)}\rho'^{(0)}_{m,l-m}\|_{L^2(\Sigma_t^{\epsilon_0})} \leqslant C|\ell|(1+t)^{-1}\sqrt{\sum_\alpha \mathcal{E}'_1[R_{i_{l-m}}\cdots R_{i_1}(T)^{m+1}\psi_\alpha](t)} \tag{14.465}$$

Also, from (14.404) with $l+1$ replaced by $l$

$$\begin{aligned}\not{d}^{(i_1\cdots i_{l-m})}\rho'^{(1)}_{m,l-m} = &(m^0_T - \frac{1}{2}\ell)\not{d}R_{i_{l-m}}\cdots R_{i_1}(T)^{m+1}\psi_0\\ &+ m^i_T\not{d}R_{i_{l-m}}\cdots R_{i_1}(T)^{m+1}\psi_i\\ &+(1+t)^{-1}\mu e^\alpha_L\not{d}R_{i_{l-m}}\cdots R_{i_1}(T)^m Q\psi_\alpha\\ &+(\not{d}m^\alpha_T)(R_{i_{l-m}}\cdots R_{i_1}(T)^{m+1}\psi_\alpha)\\ &+(1+t)^{-1}(\not{d}(\mu e^\alpha_L))(R_{i_{l-m}}\cdots R_{i_1}(T)^m Q\psi_\alpha)\end{aligned} \tag{14.466}$$

hence, in view of the facts that

$$\sup_{\Sigma_t^{\epsilon_0}}|\not{d}m^\alpha_T| \leqslant C\delta_0(1+t)^{-2}, \quad \sup_{\Sigma_t^{\epsilon_0}}|\not{d}e^\alpha_L| \leqslant C\delta_0(1+t)^{-1} \tag{14.467}$$

we obtain

$$\begin{aligned}&\|\mu^{1/2}\not{d}^{(i_1\cdots i_{l-m})}\rho'^{(1)}_{m,l-m}\|_{L^2(\Sigma_t^{\epsilon_0})}\\ &\leqslant C\delta_0(1+t)^{-2}\sqrt{\sum_\alpha \mathcal{E}'_1[R_{i_{l-m}}\cdots R_{i_1}(T)^{m+1}\psi_\alpha](t)}\\ &\quad + C\delta_0(1+t)^{-2}(1+\log(1+t))\sqrt{\sum_\alpha \mathcal{E}'_1[R_{i_{l-m}}\cdots R_{i_1}(T)^m Q\psi_\alpha](t)}\end{aligned} \tag{14.468}$$

Note that (14.465) and (14.468) are compatible with (14.416) and (14.417) respectively, and from (14.413) we have

$$\begin{aligned}&\||\not{d}^{(i_1\cdots i_{l-m})}\rho'^{(2)}_{m,l-m}| + 2|\tilde{\chi}'||\not{d}R_{i_{l-m}}\cdots R_{i_1}(T)^m\mu|\|_{L^2(\Sigma_t^{\epsilon_0})}\\ &\leqslant C_l\delta_0(1+t)^{-2}(1+\log(1+t))(\mathcal{W}_{\{l+1\}} + \mathcal{W}^Q_{\{l\}}\\ &\quad + (1+t)^{-1}\mathcal{B}_{[m,l+1]} + \mathcal{Y}_0 + (1+t)\mathcal{A}'_{[l]})\end{aligned} \tag{14.469}$$

The *borderline* contribution to (14.457) is the contribution from $\not{d}^{(i_1\cdots i_{l-m})}\rho'^{(0)}_{m,l-m}$. This is the *borderline* integral

$$
\begin{aligned}
&C\int_{W_u^t}(1+t')^2|T\psi_\alpha||\not{d}R_{i_{l-m}}\cdots R_{i_1}(T)^{m+1}\psi_\alpha||\not{d}^{(i_1\cdots i_{l-m})}\rho'^{(0)}_{m,l-m}|dt'dud\mu_{\not{g}}\\
&\leqslant C\int_0^t(1+t')^2\sup_{\Sigma_{t'}^{\epsilon_0}}(\mu^{-1}|T\psi_\alpha|)\|\mu^{1/2}\not{d}R_{i_{l-m}}\cdots R_{i_1}(T)^{m+1}\psi_\alpha\|_{L^2(\Sigma_{t'}^{\epsilon_0})}\\
&\cdot\|\mu^{1/2}\not{d}^{(i_1\cdots i_{l-m})}\rho'^{(0)}_{m,l-m}\|_{L^2(\Sigma_{t'}^{\epsilon_0})}dt'
\end{aligned}
\tag{14.470}
$$

Here we substitute (14.89) for $\sup_{\Sigma_{t'}^{\epsilon_0}}(\mu^{-1}|T\psi_\alpha|)$ and (14.465) for

$$\|\mu^{1/2}\not{d}^{(i_1\cdots i_{l-m})}\rho'^{(0)}_{m,l-m}\|_{L^2(\Sigma_{t'}^{\epsilon_0})}$$

The factors $|\ell|$ then cancel. Now the contribution of the second term on the right in (14.89) is in fact not borderline. The actual *borderline* contribution is

$$
\begin{aligned}
&C\int_0^t\sup_{\Sigma_{t'}^{\epsilon_0}}(\mu^{-1}|L\mu|)\sqrt{\mathcal{E}_1'[R_{i_{l-m}}\cdots R_{i_1}(T)^{m+1}\psi_\alpha](t')}\\
&\cdot\sqrt{\sum_\alpha\mathcal{E}_1'[R_{i_{l-m}}\cdots R_{i_1}(T)^{m+1}\psi_\alpha](t')}dt'\\
&\leqslant C(\frac{1}{2a}+\frac{1}{2q})\bar{\mu}_m^{-2a}(t)(1+\log(1+t))^{2q\,(i_1\cdots i_{l-m})}\mathcal{G}'_{1,m,l+2;a,q}(t)
\end{aligned}
\tag{14.471}
$$

(similar to (14.360) and (14.361)).

We proceed to consider the contribution of $\not{d}^{(i_1\cdots i_{l-m})}\rho'^{(1)}_{m,l-m}$ to (14.457). Here we simply appeal to (14.90) for $\sup_{\Sigma_{t'}^{\epsilon_0}}(\mu^{-1}|T\psi_\alpha|)$. Substituting also (14.468) for $\|\mu^{1/2}\not{d}^{(i_1\cdots i_{l-m})}\rho'^{(1)}_{m,l-m}\|_{L^2(\Sigma_{t'}^{\epsilon_0})}$, we obtain that the contribution in question is bounded by

$$
\begin{aligned}
&C\int_{W_{\epsilon_0}^t}(1+t')^2|T\psi_\alpha||\not{d}R_{i_{l-m}}\cdots R_{i_1}(T)^{m+1}\psi_\alpha||\not{d}^{(i_1\cdots i_{l-m})}\rho'^{(1)}_{m,l-m}|dt'dud\mu_{\not{g}}\\
&\leqslant C\delta_0\int_0^t(1+t')\bar{\mu}_m^{-1}(t')\|\mu^{1/2}\not{d}R_{i_{l-m}}\cdots R_{i_1}(T)^{m+1}\psi_\alpha\|_{L^2(\Sigma_{t'}^{\epsilon_0})}\\
&\cdot\|\mu^{1/2}\not{d}^{(i_1\cdots i_{l-m})}\rho'^{(1)}_{m,l-m}\|_{L^2(\Sigma_{t'}^{\epsilon_0})}dt'\\
&\leqslant C\delta_0^2\bar{\mu}_m^{-2a}(t)(1+\log(1+t))^{2q}\cdot(\varphi_2(Ca\delta_0)\mathcal{G}'_{1,[l+2];a,q}(t)\\
&+\int_0^t(1+t')^{-2}(1+\log(1+t'))\mathcal{G}'_{1,[l+2];a,q}(t')dt')
\end{aligned}
\tag{14.472}
$$

(similar to (14.362)).

Next, we consider $V'_{1,0}$, given by (14.452). By (14.429) we have

$$|V'_{1,0}| \leqslant C\delta_0 \int_{W^t_{\epsilon_0}} |R_{i_{l-m}} \cdots R_{i_1}(T)^{m+1}\psi_\alpha||(\not{\mathcal{L}}_L + 2\nu)^{(i_1\cdots i_{l-m})} Y_{m,l-m}|dt'dud\mu_{\not{g}}$$

$$\leqslant C\delta_0 \int_0^t \mathcal{W}_{\{l+1\}}(t')\|(\not{\mathcal{L}}_L + 2\nu)^{(i_1\cdots i_{l-m})} Y_{m,l-m}\|_{L^2(\Sigma^{\epsilon_0}_{t'})} dt'dud\mu_{\not{g}} \quad (14.473)$$

Here we need only consider the contribution of $\not{d}^{(i_1\cdots i_{l-m})}\rho'^{(0)}_{m,l-m}$. By (14.464),

$$\|\not{d}^{(i_1\cdots i_{l-m})}\rho'^{(0)}_{m,l-m}\|_{L^2(\Sigma^{\epsilon_0}_t)} \leqslant C\bar{\mu}_m^{-1/2}(1+t)^{-1}\sqrt{\mathcal{E}'_{1,[l+2]}(t)} \quad (14.474)$$

hence by (14.318) the contribution in question is bounded by

$$C\epsilon_0\delta_0 \int_0^t (1+t')^{-1}\bar{\mu}_m^{-2a-1/2}(t')(1+\log(1+t'))^{p+q}\sqrt{\mathcal{G}_{0,[l+2];a,p}(t')\mathcal{G}'_{1,[l+2];a,q}(t')}dt' \quad (14.475)$$

This is similar to (14.364), which has already been bounded by (14.366). This completes the estimates for the spacetime integrals $V'_0, V'_1, V'_2$. Thus, the estimate of the contribution of (14.57) to (14.59) is completed.

# 15

# The Top Order Energy Estimates

In this chapter, we consider the energy estimates of top order $l+2$. Of these the most delicate are those corresponding to the variations (14.52) and (14.54), to which the *borderline* error integrals are associated.

## 15.1 Estimates Associated to $K_1$

For each of the variations, we first consider the integral identity associated to $K_1$ and then the integral identity associated to $K_0$. For each variation $\psi$, of any order, there is an additional *borderline* integral associated to $K_1$, contributed by $Q_{1,3}[\psi]$ (see (5.114), (5.190)—(5.194)). This is the integral $L(t,\epsilon_0)[\psi]$ defined by (5.257):

$$L(t,\epsilon_0)[\psi] = \int_0^t (1+t')^{-1}(1+\log(1+t'))^{-1}\mathcal{E}_1'[\psi](t')dt' \tag{15.1}$$

Consider now the integral identities corresponding to the variations (14.52) and the vectorfield $K_1$. In each of these we have the *borderline* hypersurface integral (14.248) bounded by (14.268) and (14.295):

$$C(\frac{1}{a-1/2}+\frac{1}{q+1/2})\bar{\mu}_m^{-2a}(t)(1+\log(1+t))^{2q(i_1\cdots i_l)}\mathcal{G}_{1,l+2;a,q}'(t) \tag{15.2}$$

We also have the *borderline* spacetime integral (14.361) bounded by (14.363):

$$C(\frac{1}{2a}+\frac{1}{2q})\bar{\mu}_m^{-2a}(t)(1+\log(1+t))^{2q(i_1\cdots i_l)}\mathcal{G}_{1,l+2;a,q}'(t) \tag{15.3}$$

We also have the remaining integrals, bounded, in the case of (14.308) by

$$C_q\delta_0^2(1+\log(1+t))^{2q}\bar{\mu}_m^{-2a}(t)\mathcal{G}_{1,[l+2];a,q}' \tag{15.4}$$

and in the case of (14.313), by

$$C_l\epsilon_0\delta_0\bar{\mu}_m^{-2a}(t)(1+\log(1+t))^{2q}(\mathcal{G}_{0,[l+2];a,p}(t)+\mathcal{G}_{1,[l+2];a,q}'(t)) \tag{15.5}$$

provided that

$$q \geqslant p+1 \tag{15.6}$$

and (14.321) is the same as (14.313). In the case of (14.365) we have a bound by

$$C\delta_0^2\bar{\mu}_m^{-2a}(t)(1+\log(1+t))^{2q}\mathcal{G}'_{1,[l+2];a,q}(t) \tag{15.7}$$

and in the case of (14.368) a bound by

$$C_l\epsilon_0\delta_0\bar{\mu}_m^{-2a}(t)(1+\log(1+t))^{2q}(\mathcal{G}'_{1,[l+2];a,q}(t)+\mathcal{G}_{0,[l+2];a,p}(t)) \tag{15.8}$$

provided that (15.6) holds. Finally (14.370) is the same as (14.368). Combining we see that all the remaining integrals associated to (14.52) and to $K_1$ containing the top order spatial derivatives of the acoustical entities are bounded by

$$\begin{aligned}&C_{q,l}\delta_0\bar{\mu}_m^{-2a}(t)(1+\log(1+t))^{2q}\sqrt{\mathcal{G}'_{1,[l+2];a,q}(t)+\mathcal{G}_{0,[l+2];a,p}(t)}\\&\cdot(\delta_0\mathcal{Y}_0(0)+\mathcal{A}'_{[l]}(0)+\mathcal{B}_{[l+1]}(0)+\sqrt{\mathcal{G}'_{1,[l+2];a,q}(t)+\mathcal{G}_{0,[l+2];a,p}(t)})\end{aligned} \tag{15.9}$$

The bounds in terms of initial data come from the initial hypersurface integrals and the lower order terms which we omitted.

Consider now the error integrals associated to the same variations, or in fact any of the top order variations, and to $K_1$ or to $K_0$, which contain the lower order spatial derivatives of the acoustical entities and are contributed by the remaining terms in the sum (14.14). Consider an arbitrary term in this sum, corresponding to some $k \in \{0,\cdots,l\}$:

$$(Y_{I_{l+1}}+{}^{(Y_{I_{l+1}})}\delta)\cdots(Y_{I_{l+2-k}}+{}^{(Y_{I_{l+2-k}})}\delta)^{(Y_{I_{l+1-k}};I_1\cdots I_{l-k})}\sigma_{l+1-k} \tag{15.10}$$

There is a total of $k$ derivatives with respect to the commutation fields acting on ${}^{(Y)}\sigma_{l+1-k}$. In view of the fact that ${}^{(Y)}\sigma_{l+1-k}$ has the structure described in the paragraph following (14.15), in considering the partial contribution of each term in ${}^{(Y)}\sigma_{1,l+1-k}$, if the factor which is a component of ${}^{(Y)}\tilde{\pi}$ receives more than $(l+1)_*$ derivatives with respect to the commutation fields, then the factor which is a 2nd derivative of $\psi_{l+1-k}$ receives at most $k-(l+1)_*-1$ derivatives of the commutation fields, thus corresponds to a derivative of the $\psi_\alpha$ of order at most

$$k-(l+1)_*+1+l-k=l_*+1$$

therefore this factor is bounded in $L^\infty(\Sigma_t^{\epsilon_0})$ by the bootstrap assumption. Also, in considering the partial contribution of each term in ${}^{(Y)}\sigma_{2,l+1-k}$, if the factor which is a 1st derivative of ${}^{(Y)}\tilde{\pi}$ receives more than $(l+1)_*-1$ derivatives with respect to the commutation fields, then the factor which is a 1st derivative of $\psi_{l+1-k}$ receives at most $k-(l+1)_*$ derivatives with respect to the commutation field, thus corresponds to a derivative of the $\psi_\alpha$ of order at most

$$k-(l+1)_*+1+l-k=l_*+1$$

therefore this factor is again bounded in $L^\infty(\Sigma_t^{\epsilon_0})$ by the bootstrap assumption. Similar considerations apply to ${}^{(Y)}\sigma_{3,l+1-k}$. We conclude that for all the terms in the sum in (14.14) of which one factor is a derivative of the ${}^{(Y)}\tilde{\pi}$ of order more than $(l+1)_*$, the other factor is then a derivative of the $\psi_\alpha$ of order at most $l_*+1$ and is thus bounded in $L^\infty(\Sigma_t^{\epsilon_0})$ by the bootstrap assumption. Of these terms we have already estimated the contribution of those containing the top order spatial derivatives of the acoustical entities. The contributions of the remaining terms are then bounded using Propositions 12.11 and 12.12. We obtain a bound for these contributions to the error integrals associated to $K_1$ and to any of the variations, up to the top order, by (recall (14.59))

$$\begin{aligned}
&C_l\delta_0\bar{\mu}_m^{-a}(t)(1+\log(1+t))^q \\
&\cdot(\delta_0\mathcal{Y}_0(0)+\mathcal{A}'_{[l]}(0)+\mathcal{B}_{[l+1]}(0)+\sqrt{\mathcal{G}'_{1,[l+2];a,q}(t)+\mathcal{G}_{0,[l+2];a,p}(t)}) \\
&\cdot(\int_0^{\epsilon_0}\mathcal{F}'^t_{1,[l+2]}(u')du')^{1/2} \\
&+C_l\delta_0\int_0^{\epsilon_0}\mathcal{F}'^t_{1,[l+2]}(u')du'+C_l\delta_0(\int_0^{\epsilon_0}\mathcal{F}'^t_{1,[l+2]}(u')du')^{1/2}(K_{[l+2]}(t,\epsilon_0))^{1/2} \qquad (15.11)
\end{aligned}$$

The last two terms are contributed by the terms in (14.14) containing the top order derivatives of the acoustical entities of which however one is a derivative with respect to $L$ (hence they are expressible in terms of the top order derivatives of the $\psi_\alpha$).

Here, we have defined

$$\mathcal{F}'^t_{1,[n]}(u)=\sum_{m=1}^{n}\mathcal{F}'^t_{1,m}(u) \qquad (15.12)$$

where $\mathcal{F}'^t_{1,n}(u)$ represents the sum of the fluxes associated to the vectorfield $K_1$ of all the $n$th order variations.

On the other hand, all the other terms in the sum (14.14) contain derivatives of the ${}^{(Y)}\tilde{\pi}$ of order at most $(l+1)_*$, thus spatial derivatives of $\chi'$ of order at most $(l+1)_*$ and spatial derivatives of $\mu$ of order at most $(l+1)_*+1$, which are bounded in $L^\infty(\Sigma_t^{\epsilon_0})$ by virtue of Propositions 12.9 and 12.10 and the bootstrap assumption. So we can use Lemma 7.6 to bound these contributions by

$$\begin{aligned}
&C_l\int_0^{\epsilon_0}\mathcal{F}'^t_{1,[l+2]}(u')du'+C_l(\int_0^{\epsilon_0}\mathcal{F}'^t_{1,[l+2]}(u')du')^{1/2}(K_{[l+1]}(t,\epsilon_0))^{1/2} \\
&+C_l(\int_0^{\epsilon_0}\mathcal{F}'^t_{1,[l+2]}(u')du')^{1/2} \\
&\cdot(\int_0^t(1+t')^{-2}(1+\log(1+t'))^2(\mathcal{E}'_{1,[l+2]}(t')+\epsilon_0^2\mathcal{E}_{0,[l+2]}(t'))dt')^{1/2} \qquad (15.13)
\end{aligned}$$

Here we have defined

$$K_{[n]}(t,u)=\sum_{m=1}^{n}K_m(t,u) \qquad (15.14)$$

where $K_n(t,u)$ represents the sum of the integrals

$$K[\psi](t,u) = -\int_{W_u^t} \frac{\Omega}{2}\omega\nu^{-1}\mu^{-1}(L\mu)_-|\not{d}\psi|^2 d\mu_g \tag{15.15}$$

of all the $n$th order variations.

Finally, for each variation $\psi$ we have the error integrals

$$\int_{W_{\epsilon_0}^t} \sum_{k=1}^{7} Q_{0,k}[\psi]d\mu_g, \quad \int_{W_{\epsilon_0}^t} \sum_{k=1}^{8} Q_{1,k}[\psi]d\mu_g \tag{15.16}$$

from the fundamental energy estimates. According to (5.260), we have

$$\int_{W_{\epsilon_0}^t} \sum_{k=1}^{8} Q_{1,k}[\psi]d\mu_g$$
$$\leqslant -\frac{1}{2}K[\psi](t,\epsilon_0) + CM[\psi](t,\epsilon_0) + \frac{3}{2}L[\psi](t,\epsilon_0) + \int_0^t \tilde{A}(t')\mathcal{E}_1'[\psi](t')dt' \tag{15.17}$$

Here $M[\psi](t,\epsilon_0)$ is given by (5.257):

$$M[\psi](t,\epsilon_0) = \bar{\mathcal{E}}_0[\psi](t)(1+\log(1+t))^4 + \int_0^{\epsilon_0} \mathcal{F}_1'^t[\psi](u')du' \tag{15.18}$$

As in Chapter 5, we will bring the term $-(1/2)K[\psi](t,\epsilon_0)$ on the left-hand side of the integral inequality associated to $\psi$ and $K_1$.

Let us define, for any variation $\psi$, in analogy with (14.109) and (14.111),

$$\mathcal{G}_{0;a,p}[\psi](t) = \sup_{t'\in[0,t]} ((1+\log(1+t'))^{-2p}\bar{\mu}_m^{2a}(t')\mathcal{E}_0[\psi](t')) \tag{15.19}$$

$$\mathcal{G}_{1;a,q}'[\psi](t) = \sup_{t'\in[0,t]} ((1+\log(1+t'))^{-2q}\bar{\mu}_m^{2a}(t')\mathcal{E}_1'[\psi](t')) \tag{15.20}$$

Note that $\mathcal{G}_{0;a,p}[\psi](t)$ and $\mathcal{G}_{1;a,q}'[\psi](t)$ are non-decreasing functions of $t$.

Recalling from Chapter 8 the definition of $\bar{\mu}_m$:

$$\bar{\mu}_m(t) = \min\{\mu_m(t),1\}, \quad \mu_m(t) = \min_{\Sigma_t^{\epsilon_0}} \mu = \min_{(u,\vartheta)\in[0,\epsilon_0]\times S^2} \mu(t,u,\vartheta) \tag{15.21}$$

we define by replacing $[0,\epsilon_0]\times S^2$ by $[0,u]\times S^2$, or $\Sigma_t^{\epsilon_0}$ by $\Sigma_t^u$:

$$\bar{\mu}_{m,u}(t) = \min\{\mu_{m,u}(t),1\}, \quad \mu_{m,u}(t) = \min_{\Sigma_t^u} \mu = \min_{(u',\vartheta)\in[0,u]\times S^2} \mu(t,u',\vartheta) \tag{15.22}$$

Note that $\bar{\mu}_{m,u}(t)$ is a non-increasing function of $u$ at each $t$. We then define

$$\mathcal{H}_{1;a,q}'^t[\psi](u) = \sup_{t'\in[0,t]} ((1+\log(1+t'))^{-2q}\bar{\mu}_{m,u}^{2a}(t')\mathcal{F}_1'^{t'}[\psi](u)) \tag{15.23}$$

$$V_{1;a,q}'[\psi](t,u) = \int_0^u \mathcal{H}_{1;a,q}'[\psi](u')du' \tag{15.24}$$

Note that $\mathcal{H}'^{t}_{1;a,q}[\psi](u)$ is a non-decreasing function of $t$ at each $u$ while $V'_{1;a,q}[\psi](t,u)$ is a non-decreasing function of $u$ at each $t$ as well as a non-decreasing function of $t$ at each $u$. We have, for each $u \in [0, \epsilon_0]$,

$$\begin{aligned}\int_0^u \mathcal{F}'^{t}_1[\psi](u')du' &\leqslant \int_0^u \bar{\mu}^{-2a}_{m,u'}(t)(1+\log(1+t))^{2q}\mathcal{H}'^{t}_{1;a,q}[\psi](u')du' \\ &\leqslant \bar{\mu}^{-2a}_{m,u}(t)(1+\log(1+t))^{2q}V'_{1;a,q}[\psi](t,u) \qquad (15.25)\end{aligned}$$

We also define

$$\mathcal{H}'^{t}_{1,[n];a,q}(u) = \sup_{t'\in[0,t]} ((1+\log(1+t'))^{-2q}\bar{\mu}^{2a}_{m,u}(t')\mathcal{F}'^{t'}_{1,[n]}(u)) \qquad (15.26)$$

$$V'_{1,[n];a,q}(t,u) = \int_0^u \mathcal{H}'^{t}_{1,[n];a,q}(u')du' \qquad (15.27)$$

Then $\mathcal{H}'^{t}_{1,[n];a,q}(u)$ is a non-decreasing function function of $t$ at each $u$, $V'_{1,[n];a,q}(t,u)$ is a non-decreasing function of $u$ at each $t$ as well as a non-decreasing function of $t$ at each $u$, and for each $u \in [0, \epsilon_0]$,

$$\begin{aligned}\int_0^u \mathcal{F}'^{t}_{1,[n]}(u')du' &\leqslant \int_0^u \bar{\mu}^{-2a}_{m,u'}(t)(1+\log(1+t))^{2q}\mathcal{H}'^{t}_{1,[n];a,q}(u')du' \\ &\leqslant \bar{\mu}^{-2a}_{m,u}(t)(1+\log(1+t))^{2q}V'_{1,[n];a,q}(t,u) \qquad (15.28)\end{aligned}$$

Going back to (15.18), we have, in regard to the first term on the right,

$$\begin{aligned}\bar{\mathcal{E}}_0[\psi](t) = \sup_{t'\in[0,t]} \mathcal{E}_0[\psi](t') &\leqslant \sup_{t'\in[0,t]} (\bar{\mu}^{-2a}_m(t')(1+\log(1+t'))^{2p}\mathcal{G}_{0;a,p}[\psi](t')) \\ &\leqslant C(1+\log(1+t))^{2p}\bar{\mu}^{-2a}_m(t)\mathcal{G}_{0;a,p}[\psi](t) \qquad (15.29)\end{aligned}$$

(by Corollary 2 of Lemma 8.11). At this point we set

$$q = p+2 \qquad (15.30)$$

which is consistent with (15.6). Then (15.29) implies that the first term on the right in (15.18) is bounded by

$$C\bar{\mu}^{-2a}_m(t)(1+\log(1+t))^{2q}\mathcal{G}_{0;a,p}[\psi](t) \qquad (15.31)$$

The second term on the right in (15.18) being bounded according to (15.25) with $u = \epsilon_0$, we conclude that

$$\begin{aligned}M[\psi](t,\epsilon_0) &\leqslant \bar{\mu}^{-2a}_m(t)(1+\log(1+t))^{2q}(C\mathcal{G}_{0;a,p}[\psi](t) + V'_{1;a,q}[\psi](t,\epsilon_0)) \\ &\leqslant \bar{\mu}^{-2a}_m(t)(1+\log(1+t))^{2q}(C\mathcal{G}_{0,[l+2];a,p}(t) + V'_{1,[l+2];a,q}(t,\epsilon_0)) \quad (15.32)\end{aligned}$$

for all variations of order not exceeding $l+2$.

In regard to the third term on the right in (15.17), which involves the borderline integral (15.1), we have, by Corollary 2 of Lemma 8.11,

$$L[\psi](t,\epsilon_0) \leqslant C\bar{\mu}_m^{-2a}(t)\mathcal{G}'_{1;a,q}[\psi](t)\int_0^t (1+t')^{-1}(1+\log(1+t'))^{2q-1}dt'$$
$$\leqslant \frac{C}{2q}\bar{\mu}_m^{-2a}(t)(1+\log(1+t))^{2q}\mathcal{G}'_{1;a,q}[\psi](t) \tag{15.33}$$

Finally, in regard to the last term on the right in (15.17) we have

$$\int_0^t \tilde{A}(t')\mathcal{E}'_1[\psi](t')dt'$$
$$\leqslant C\bar{\mu}_m^{-2a}(1+\log(1+t))^{2q}\int_0^t \tilde{A}(t')\mathcal{G}'_{1;a,q}[\psi](t')dt'$$
$$\leqslant C\bar{\mu}_m^{-2a}(t)(1+\log(1+t))^{2q}\int_0^t \tilde{A}(t')\mathcal{G}'_{1,[l+2];a,q}(t')dt' \tag{15.34}$$

for all variations of order not exceeding $l+2$.

We also recall that in (5.73) (with $u=\epsilon_0$) corresponding to the variation $\psi$ and to $K_1$ we have the hypersurface integrals bounded according to (5.267):

$$|\int_{\Sigma_t^{\epsilon_0}}(1/2)\Omega(\underline{L}\omega+\underline{\nu}\omega)\psi^2 - \int_{\Sigma_0^{\epsilon_0}}(1/2)\Omega(\underline{L}\omega+\underline{\nu}\omega)\psi^2|$$
$$\leqslant C\bar{\mathcal{E}}_0[\psi](t)(1+\log(1+t))^4$$
$$\leqslant C\bar{\mu}_m^{-2a}(t)(1+\log(1+t))^{2q}\mathcal{G}_{0;a,p}[\psi](t) \tag{15.35}$$

by (15.29), (15.30).

We turn to the bound (15.13). Let us define

$$I_{[n];a,q}(t,u) = \sup_{t'\in[0,t]}((1+\log(1+t'))^{-2q}\bar{\mu}_{m,u}^{2a}(t')K_{[n]}(t',u)) \tag{15.36}$$

This is a non-decreasing function of $t$ at each $u$. Then, in view of (15.28) with $u=\epsilon_0$, we can bound (15.13) by

$$C_l\bar{\mu}_m^{-2a}(t)(1+\log(1+t))^{2q}$$
$$\cdot(\int_0^t (1+t')^{-2}(1+\log(1+t'))^2(\mathcal{G}_{0,[l+2];a,p}(t')+\mathcal{G}'_{1;[l+2];a,q}(t'))dt'$$
$$+V'_{1,[l+2];a,q}(t,\epsilon_0)+I_{[l+2];a,q}(t,\epsilon_0)^{1/2}V'_{1,[l+2];a,q}(t,\epsilon_0)^{1/2}) \tag{15.37}$$

and the last term in parenthesis is bounded by

$$\frac{\delta}{2}I_{[l+2];a,q}(t,\epsilon_0)+\frac{1}{2\delta}V'_{1,[l+2];a,q}(t,\epsilon_0) \tag{15.38}$$

for any positive constant $\delta$ ($\delta$ will be chosen below).

Finally, (15.11) is bounded by

$$C_l\delta_0\bar{\mu}_m^{-2a}(t)(1+\log(1+t))^{2q}((\mathcal{Y}_0(0)+\mathcal{A}'_{[l]}(0)+\mathcal{B}_{[l+1]}(0))^2+\delta(\mathcal{G}'_{1,[l+2];a,q}(t)$$
$$+\mathcal{G}_{0,[l+2];a,p}(t))+\delta I_{[l+2];a,q}(t,\epsilon_0)+(1+\frac{1}{\delta})V'_{1,[l+2];a,q}(t,\epsilon_0)) \tag{15.39}$$

We now consider the integral identity corresponding to the vectorfield $K_1$ (5.73) with $u=\epsilon_0$ and to the variations (14.52) $R_j R_{i_l}\cdots R_{i_1}\psi_\alpha$, $j=1,2,3$, $\alpha=0,1,2,3$, for a given multi-index $(i_1\cdots i_l)$. Summing over $j$ and $\alpha$, we then obtain from (15.2), (15.3), (15.33), in regard to the borderline integrals, and from (15.9), (15.32), (15.34), (15.35), (15.37), (15.39), in regard to the remaining contributions, the following:

$$
\begin{aligned}
&\Big(\sum_{j,\alpha}\mathcal{E}_1'[R_jR_{i_l}\cdots R_{i_1}\psi_\alpha](t)+\sum_{j,\alpha}\mathcal{F}_1'^t[R_jR_{i_1}\cdots R_{i_1}\psi_\alpha](\epsilon_0)\\
&+\frac{1}{2}\sum_{j,\alpha}K[R_jR_{i_l}\cdots R_{i_1}\psi_\alpha](t,\epsilon_0)\Big)\bar{\mu}_m^{2a}(t)(1+\log(1+t))^{-2q}\\
&\leqslant C(\frac{1}{a}+\frac{1}{q})^{(i_1\cdots i_l)}\mathcal{G}_{1,l+2;a,q}'(t)+C_{q,l}\delta_0\mathcal{G}_{1,[l+2];a,q}'(t)\\
&+C_{q,l}R_{[l+2];a,q}(t,\epsilon_0)+C\int_0^t\bar{A}(t')\mathcal{G}_{1,[l+2];a,q}'(t')dt'
\end{aligned}
\tag{15.40}
$$

Here,

$$
\begin{aligned}
R_{[l+2];a,q}(t,\epsilon_0)=&\mathcal{D}_{[l+2]}+\mathcal{G}_{0,[l+2];a,p}(t)+(1+\frac{1}{\delta})V_{1,[l+2];a,q}'(t,\epsilon_0)\\
&+\delta I_{[l+2];a,q}(t,\epsilon_0)
\end{aligned}
\tag{15.41}
$$

and $\mathcal{D}_{[l+2]}$ is the quantity

$$
\begin{aligned}
\mathcal{D}_{[l+2]}=&\mathcal{E}_{0,[l+2]}(0)+\mathcal{E}_{1,[l+2]}'(0)+(\mathcal{P}_{[l+2]})^2\\
\mathcal{P}_{[l+2]}=&\mathcal{Y}_0(0)+\mathcal{A}_{[l]}'(0)+\mathcal{B}_{\{l+1\}}(0)\\
&+\sum_{i_1\cdots i_l}\|^{(i_1\cdots i_l)}x_l(0)\|_{L^2(\Sigma_0^{\epsilon_0})}\\
&+\sum_{m=0}^{l}\sum_{i_1\cdots i_{l-m}}\|^{(i_1\cdots i_{l-m})}x_{m,l-m}'(0)\|_{L^2(\Sigma_0^{\epsilon_0})}
\end{aligned}
\tag{15.42}
$$

Also,

$$
\bar{A}(t)=\tilde{A}(t)+C_l(1+t)^{-2}(1+\log(1+t))^2 \tag{15.43}
$$

It is crucial that the constant $C$ in front of the first term on the right in (15.40) is independent of $a,q,l$.

Now all the estimates we have made so far depend only on the bootstrap assumption and on assumption **J**, which refer to $W_{\epsilon_0}^s$, on the assumptions of Proposition 12.6, 12.9, 12.10 on the initial data on $\Sigma_0^{\epsilon_0}$, and on the fact that $\epsilon_0\leqslant 1/2$. These assumptions hold if $\epsilon_0$ is replaced by any $u\in(0,\epsilon_0]$; therefore all our estimates hold with $\epsilon_0$ replaced by any $u\in(0,\epsilon_0]$. With this replacement, $\bar{\mu}_m(t)$ is replaced by $\bar{\mu}_{m,u}(t)$ and, for any variation $\psi$, $\mathcal{E}_0[\psi](t)$ and $\mathcal{E}_1'[\psi](t)$ are replaced by $\mathcal{E}_0^u[\psi](t)$ and $\mathcal{E}_1'^u[\psi](t)$ respectively. Thus $\mathcal{G}_{0;a,p}[\psi](t)$, $\mathcal{G}_{1;a,q}'[\psi](t)$ are replaced by

$$\mathcal{G}^u_{0;a,p}[\psi](t) = \sup_{t'\in[0,t]}((1+\log(1+t'))^{-2p}\bar{\mu}^{2a}_{m,u}(t')\mathcal{E}^u_0[\psi](t')) \tag{15.44}$$

$$\mathcal{G}'^u_{1;a,q}[\psi](t) = \sup_{t'\in[0,t]}((1+\log(1+t'))^{-2q}\bar{\mu}^{2a}_{m,u}(t')\mathcal{E}'^u_1[\psi](t')) \tag{15.45}$$

respectively, and ${}^{(i_1\cdots i_l)}\mathcal{G}_{0,l+2;a,p}(t)$, ${}^{(i_1\cdots i_l)}\mathcal{G}_{0,m,l+2;a,p}(t)$, ${}^{(i_1\cdots i_l)}\mathcal{G}'_{1,l+2;a,p}(t)$, ${}^{(i_1\cdots i_l)}\mathcal{G}'_{1,m,l+2;a,p}(t)$, are replaced by

$$\begin{aligned}&{}^{(i_1\cdots i_l)}\mathcal{G}^u_{0,l+2;a,p}(t)\\&=\sup_{t'\in[0,t]}((1+\log(1+t'))^{-2p}\bar{\mu}^{2a}_{m,u}(t')\sum_{j,\alpha}\mathcal{E}^u_0[R_jR_{i_l}\cdots R_{i_1}\psi_\alpha](t'))\end{aligned} \tag{15.46}$$

$$\begin{aligned}&{}^{(i_1\cdots i_{l-m})}\mathcal{G}^u_{0,m,l+2;a,p}(t)\\&=\sup_{t'\in[0,t]}((1+\log(1+t'))^{-2p}\bar{\mu}^{2a}_{m,u}(t')\sum_{\alpha}\mathcal{E}^u_0[R_{i_{l-m}}\cdots R_{i_1}(T)^{m+1}\psi_\alpha](t'))\end{aligned} \tag{15.47}$$

$$\begin{aligned}&{}^{(i_1\cdots i_l)}\mathcal{G}'^u_{1,l+2;a,q}(t)\\&=\sup_{t'\in[0,t]}((1+\log(1+t'))^{-2q}\bar{\mu}^{2a}_{m,u}(t')\sum_{j,\alpha}\mathcal{E}'^u_1[R_jR_{i_l}\cdots R_{i_1}\psi_\alpha](t'))\end{aligned} \tag{15.48}$$

$$\begin{aligned}&{}^{(i_1\cdots i_{l-m})}\mathcal{G}'^u_{1,m,l+2;a,q}(t)\\&=\sup_{t'\in[0,t]}((1+\log(1+t'))^{-2q}\bar{\mu}^{2a}_{m,u}(t')\sum_{\alpha}\mathcal{E}'^u_1[R_{i_{l-m}}\cdots R_{i_1}(T)^{m+1}\psi_\alpha](t'))\end{aligned} \tag{15.49}$$

respectively. Also, $\mathcal{G}_{0,[n];a,p}$, $\mathcal{G}'_{1,[n];a,q}$, are replaced by

$$\mathcal{G}^u_{0,[n];a,p}(t) = \sup_{t'\in[0,t]}((1+\log(1+t'))^{-2p}\bar{\mu}^{2a}_{m,u}(t')\mathcal{E}^u_{0,[n]}(t')) \tag{15.50}$$

$$\mathcal{G}'^u_{1,[n];a,q}(t) = \sup_{t'\in[0,t]}((1+\log(1+t'))^{-2q}\bar{\mu}^{2a}_{m,u}(t')\mathcal{E}'^u_{1,[n]}(t')) \tag{15.51}$$

respectively. Moreover, $\mathcal{D}_{[l+2]}$ is replaced by $\mathcal{D}^u_{[l+2]}$. Thus in place of (15.40) we obtain, for all $u\in[0,\epsilon_0]$, and all $t\in[0,s]$,

$$\begin{aligned}&(\sum_{j,\alpha}\mathcal{E}'^u_1[R_jR_{i_l}\cdots R_{i_1}\psi_\alpha](t)+\sum_{j,\alpha}\mathcal{F}'^t_1[R_jR_{i_l}\cdots R_{i_1}\psi_\alpha](u)\\&\quad+\frac{1}{2}\sum_{j,\alpha}K[R_jR_{i_l}\cdots R_{i_1}\psi_\alpha](t,u))\bar{\mu}^{2a}_{m,u}(t)(1+\log(1+t))^{-2q}\\&\leqslant C(\frac{1}{a}+\frac{1}{q})^{(i_1\cdots i_l)}\mathcal{G}'^u_{1,l+2;a,q}(t)+C_{q,l}\delta_0\mathcal{G}'^u_{1,[l+2];a,q}(t)\\&\quad+C_{q,l}R_{[l+2];a,q}(t,u)+C\int_0^t\bar{A}(t')\mathcal{G}'^u_{1,[l+2];a,q}(t')dt'\end{aligned} \tag{15.52}$$

and we have

$$\begin{aligned}R_{[l+2];a,q}(t,u)=&\mathcal{D}^u_{[l+2]}+\mathcal{G}^u_{0,[l+2];a,p}(t)\\&+(1+\frac{1}{\delta})V'_{1,[l+2];a,q}(t,u)+\delta I_{[l+2];a,q}(t,u)\end{aligned} \tag{15.53}$$

where

$$\begin{aligned}\mathcal{D}^u_{[l+2]} &= \mathcal{E}^u_{0,[l+2]}(0) + \mathcal{E}'^u_{1,[l+2]}(0) + (\mathcal{P}^u_{[l+2]})^2 \\ \mathcal{P}^u_{[l+2]} &= \mathcal{Y}^u_0(0) + \mathcal{A}'^u_{[l]}(0) + \mathcal{B}^u_{\{l+1\}}(0) \\ &+ \sum_{i_1\cdots i_l} \|^{(i_1\cdots i_l)}x_l(0)\|_{L^2(\Sigma^u_0)} \\ &+ \sum_{m=0}^{l} \sum_{i_1\cdots i_{l-m}} \|^{(i_1\cdots i_{l-m})}x'_{m,l-m}(0)\|_{L^2(\Sigma^u_0)}\end{aligned} \tag{15.54}$$

Keeping only the term

$$\sum_{j,\alpha} \mathcal{E}'^u_1[R_j R_{i_l}\cdots R_{i_1}\psi_\alpha](t)$$

on the left of (15.52), we have

$$\begin{aligned}&\bar{\mu}^{2a}_{m,u}(t)(1+\log(1+t))^{-2q} \sum_{j,\alpha} \mathcal{E}'^u_1[R_j R_{i_l}\cdots R_{i_1}\psi_\alpha](t) \\ &\leqslant C(\frac{1}{a}+\frac{1}{q})^{(i_1\cdots i_l)}\mathcal{G}'^u_{1,l+2;a,q}(t) + C_{q,l}\delta_0 \mathcal{G}'^u_{1,[l+2];a,q}(t) \\ &+ C_{q,l}R_{[l+2];a,q}(t,u) + C\int_0^t \bar{A}(t')\mathcal{G}'^u_{1,[l+2];a,q}(t')dt'\end{aligned} \tag{15.55}$$

The same holds with $t$ replaced by $t' \in [0,t]$. Now the right hand side of (15.55) is a non-decreasing function of $t$ at each $u$. The inequality corresponding to $t'$ thus holds a fortiori if we again replace $t'$ by $t$ on the right hand side. Taking the supremum over all $t' \in [0,t]$ on the left hand side we obtain, in view of the definition (15.48),

$$\begin{aligned}{}^{(i_1\cdots i_l)}\mathcal{G}'^u_{1,[l+2];a,q}(t) &\leqslant C(\frac{1}{a}+\frac{1}{q})^{(i_1\cdots i_l)}\mathcal{G}'^u_{1,l+2;a,q}(t) + C_{q,l}\delta_0\mathcal{G}'^u_{1,[l+2];a,q}(t) \\ &+ C_{q,l}R_{[l+2];a,q}(t,u) + C\int_0^t \bar{A}(t')\mathcal{G}'^u_{1,[l+2];a,q}(t')dt'\end{aligned} \tag{15.56}$$

If $a$ and $q$ are chosen suitably large so that

$$C(\frac{1}{a}+\frac{1}{q}) \leqslant \frac{1}{2} \tag{15.57}$$

then (15.56) will imply

$$\begin{aligned}&\frac{1}{2}{}^{(i_1\cdots i_l)}\mathcal{G}'^u_{1,l+2;a,q}(t) \\ &\leqslant C_{q,l}\delta_0\mathcal{G}'^u_{1,[l+2];a,q}(t) + C_{q,l}R_{[l+2];a,q}(t,u) + C\int_0^t \bar{A}(t')\mathcal{G}'^u_{1,[l+2];a,q}(t')dt'\end{aligned} \tag{15.58}$$

The actual choice of $a$ and $q$ shall be specified below.

Consider now the integral identities corresponding to the variations (14.54) and $K_1$. In each of these we have the *borderline* hypersurface integral (14.421) bounded by (14.423):

$$C(\frac{1}{a-1/2}+\frac{1}{q+1/2})\bar{\mu}_m^{-2a}(t)(1+\log(1+t))^{2q(i_1\cdots i_{l-m})}\mathcal{G}'_{1,m,l+2;a,q}(t) \quad (15.59)$$

We also have the *borderline* spacetime integral (14.470) bounded by (14.471):

$$C(\frac{1}{2a}+\frac{1}{2q})\bar{\mu}_m^{-2a}(t)(1+\log(1+t))^{2q(i_1\cdots i_{l-m})}\mathcal{G}'_{1,m,l+2;a,q}(t) \quad (15.60)$$

We also have the remaining integrals, bounded, in the case of (14.429) by (15.4), in the case of (14.435) bounded by (15.5), in the case of (14.474) by (15.7), and in the case of (14.476) by (15.8). Combining we see that all the remaining integrals associated to the variations (14.54) and to $K_1$ containing the top order spatial derivatives of the acoustical entities are bounded by (15.9). On the other hand, by the discussion following (15.9) and (15.13), the error integrals associated to (14.54) and to $K_1$ contributed by all other terms in (14.14) are bounded by (15.11) and (15.13). Finally, we are left with (15.16), which is bounded in a similar way. We thus obtain

$$\begin{aligned}
&(\sum_\alpha \mathcal{E}'_1[R_{i_{l-m}}\cdots R_{i_1}(T)^{m+1}\psi_\alpha](t)+\sum_\alpha \mathcal{F}'^t_1[R_{i_{l-m}}\cdots R_{i_1}(T)^{m+1}\psi_\alpha](\epsilon_0)\\
&+\frac{1}{2}\sum_{j,\alpha}K[R_{i_{l-m}}\cdots R_{i_1}(T)^{m+1}\psi_\alpha](t,\epsilon_0))\bar{\mu}_m^{2a}(t)(1+\log(1+t))^{-2q}\\
&\leqslant C(\frac{1}{a}+\frac{1}{q})^{(i_1\cdots i_{l-m})}\mathcal{G}'_{1,m,l+2;a,q}(t)+C_{q,l}\delta_0\mathcal{G}'_{1,[l+2];a,q}(t) \qquad (15.61)\\
&+C_{q,l}R_{[l+2];a,q}(t,\epsilon_0)+C\int_0^t \bar{A}(t')\mathcal{G}'_{1,[l+2];a,q}(t')dt'
\end{aligned}$$

and also

$$\begin{aligned}
&(\sum_\alpha \mathcal{E}'^u_1[R_{i_{l-m}}\cdots R_{i_1}(T)^{m+1}\psi_\alpha](t)+\sum_\alpha \mathcal{F}'^t_1[R_{i_{l-m}}\cdots R_{i_1}(T)^{m+1}\psi_\alpha](u)\\
&+\frac{1}{2}\sum_{j,\alpha}K[R_{i_{l-m}}\cdots R_{i_1}(T)^{m+1}\psi_\alpha](t,u))\bar{\mu}_{m,u}^{2a}(t)(1+\log(1+t))^{-2q}\\
&\leqslant C(\frac{1}{a}+\frac{1}{q})^{(i_1\cdots i_{l-m})}\mathcal{G}'^u_{1,m,l+2;a,q}(t)+C_{q,l}\delta_0\mathcal{G}'^u_{1,[l+2];a,q}(t) \qquad (15.62)\\
&+C_{q,l}R_{[l+2];a,q}(t,u)+C\int_0^t \bar{A}(t')\mathcal{G}'^u_{1,[l+2];a,q}(t')dt'
\end{aligned}$$

Keeping only the term

$$\sum_\alpha \mathcal{E}'^u_1[R_{i_{l-m}}\cdots R_{i_1}(T)^{m+1}\psi_\alpha](t)$$

on the left of (15.62) we have

$$\bar{\mu}_{m,u}^{2a}(t)(1+\log(1+t))^{-2q}\sum_{\alpha}\mathcal{E}_1'^u[R_{i_{l-m}}\cdots R_{i_1}(T)^{m+1}\psi_\alpha](t)$$
$$\leqslant C(\frac{1}{a}+\frac{1}{q})^{(i_1\cdots i_{l-m})}\mathcal{G}'^u_{1,m,l+2;a,q}(t)+C_{q,l}\delta_0\mathcal{G}'^u_{1,[l+2];a,q}(t)$$
$$+C_{q,l}R_{[l+2];a,q}(t,u)+C\int_0^t\bar{A}(t')\mathcal{G}'^u_{1,[l+2];a,q}(t')dt'$$

By the same reasoning as that leading from (15.55) to (15.58) we then deduce

$$\frac{1}{2}^{(i_1\cdots i_{l-m})}\mathcal{G}'^u_{1,m,l+2;a,q}(t)$$
$$\leqslant C_{q,l}\delta_0\mathcal{G}'^u_{1,[l+2];a,q}(t)+C_{q,l}R_{[l+2];a,q}(t,u)+C\int_0^t\bar{A}(t')\mathcal{G}'^u_{1,[l+2];a,q}(t')dt' \quad (15.63)$$

Consider finally the integral identities corresponding to any of the other variations $\psi$ of order up to $l+2$ and to $K_1$. In each of these we have only one *borderline* integral (15.1) bounded by (15.33). Since all the remaining integrals in these identities are bounded by

$$C_{q,l}\delta_0\mathcal{G}'^u_{1,[l+2];a,q}(t)+C_{q,l}R_{[l+2];a,q}(t,u)+C\int_0^t\bar{A}(t')\mathcal{G}'^u_{1,[l+2];a,q}(t')dt' \quad (15.64)$$

we obtain, for each such variation $\psi$, the inequality

$$(\mathcal{E}_1'^u[\psi](t)+\mathcal{F}_1'^t[\psi](u)+\frac{1}{2}K[\psi](t,u))\bar{\mu}_{m,u}^{2a}(t)(1+\log(1+t))^{-2q}$$
$$\leqslant\frac{C}{2q}\mathcal{G}'^u_{1;a,q}[\psi](t)+C_{q,l}\delta_0\mathcal{G}'^u_{1,[l+2];a,q}(t) \quad (15.65)$$
$$+C_{q,l}R_{[l+2];a,q}(t,u)+C\int_0^t\bar{A}(t')\mathcal{G}'^u_{1,[l+2];a,q}(t')dt'$$

Similarly, keeping only

$$\mathcal{E}_1'^u[\psi](t)$$

on the left and noting that by virtue of the condition (15.57) the coefficient $C/2q$ of the first term on the right is not greater than $1/2$, we have

$$\bar{\mu}_{m,u}^{2a}(t)(1+\log(1+t))^{-2q}\mathcal{E}_1'^u[\psi](t)$$
$$\leqslant\frac{1}{2}\mathcal{G}'^u_{1;a,q}[\psi](t)+C_{q,l}\delta_0\mathcal{G}'^u_{1,[l+2];a,q}(t)$$
$$+C_{q,l}R_{[l+2];a,q}(t,u)+C\int_0^t\bar{A}(t')\mathcal{G}'^u_{1,[l+2];a,q}(t')dt'$$

The same holds with $t$ replaced by any $t'\in[0,t]$. The right hand side being a non-decreasing function of $t$ at each $u$, the inequality corresponding to $t'$ holds fortiori if we again replace $t'$ by $t$ on the right. Taking then the supremum over all $t' \in [0,t]$ on the left we obtain, in view of the definition (15.45),

$$\begin{aligned}&\frac{1}{2}\mathcal{G}'^u_{1;a,q}[\psi](t)\\ &\leqslant C_{q,l}\delta_0\mathcal{G}'^u_{1,[l+2];a,q}(t) + C_{q,l}R_{[l+2];a,q}(t,u) + C\int_0^t \bar{A}(t')\mathcal{G}'^u_{1,[l+2];a,q}(t')dt' \end{aligned}\tag{15.66}$$

Now, according to the definition of $\mathcal{E}'^u_{1,[l+2]}(t)$ we have

$$\begin{aligned}\mathcal{E}'^u_{1,[l+2]}(t) = &\sum_{i_1\cdots i_l}(\sum_{j,\alpha}\mathcal{E}'^u_1[R_jR_{i_l}\cdots R_{i_1}\psi_\alpha](t))\\ &+\sum_{m=0}^{l}\sum_{i_1\cdots i_{l-m}}(\sum_{\alpha}\mathcal{E}'^u_1[R_{i_{l-m}}\cdots R_{i_1}(T)^{m+1}\psi_\alpha](t)) + \sum_{\psi}\mathcal{E}'^u_1[\psi](t)\end{aligned}\tag{15.67}$$

where the last sum is over all the other variations $\psi$ of order up to $l+2$. Noting that for non-negative functions $x_1(t), \cdots, x_N(t)$ we have

$$\sup_{t'\in[0,t]}\sum_{n=1}^{N}x_n(t') \leqslant \sum_{n=1}^{N}\sup_{t'\in[0,t]}x_n(t')\tag{15.68}$$

it follows, in view of the definitions (15.45), (15.48), (15.49), (15.51) that

$$\begin{aligned}\mathcal{G}'^u_{1,[l+2];a,q}(t) \leqslant &\sum_{i_1\cdots i_l}{}^{(i_1\cdots i_l)}\mathcal{G}'^u_{1,l+2;a,q}(t)\\ &+\sum_{m=0}^{l}\sum_{i_1\cdots i_{l-m}}{}^{(i_1\cdots i_{l-m})}\mathcal{G}'^u_{1,m,l+2;a,q}(t) + \sum_{\psi}\mathcal{G}'^u_{1;a,q}[\psi](t)\end{aligned}$$

Thus, summing inequalities (15.58) over $i_1\cdots i_l$, summing inequalities (15.63) over $i_1\cdots i_{l-m}$ and over $m = 0,\cdots,l$, and summing inequalities (15.66) over all the other variations $\psi$ of order up to $l+2$, and then adding the resulting three inequalities we obtain

$$\begin{aligned}&\frac{1}{2}\mathcal{G}'^u_{1,[l+2];a,q}(t)\\ &\leqslant C_{q,l}\delta_0\mathcal{G}'^u_{1,[l+2];a,q}(t) + C_{q,l}R_{[l+2];a,q}(t,u) + C_l\int_0^t \bar{A}(t')\mathcal{G}'^u_{1,[l+2];a,q}(t')dt'\end{aligned}\tag{15.69}$$

for new constants $C_l$, $C_{q,l}$. Requiring then $\delta_0$ to satisfy the smallness condition

$$C_{q,l}\delta_0 \leqslant \frac{1}{4}\tag{15.70}$$

(15.69) implies

$$\mathcal{G}'^u_{1,[l+2];a,q}(t) \leqslant C_{q,l}R_{[l+2];a,q}(t,u) + C_l\int_0^t \bar{A}(t')\mathcal{G}'^u_{1,[l+2];a,q}(t')dt'\tag{15.71}$$

By (15.43), (5.258) and Proposition 13.1, we have

$$\int_0^s \bar{A}(t)dt \leqslant C_l \quad \text{independent of} \quad s \tag{15.72}$$

Since $R_{[l+2];a,q}(t,u)$ is a non-decreasing function of $t$ at each $u$, (15.71) implies

$$\mathcal{G}'^u_{1,[l+2];a,q}(t) \leqslant C_{q,l} R_{[l+2];a,q}(t,u) \tag{15.73}$$

hence also

$$\int_0^t \bar{A}(t')\mathcal{G}'^u_{1,[l+2];a,q}(t')dt' \leqslant C_{q,l} R_{[l+2];a,q}(t,u) \tag{15.74}$$

for a new constant $C_{q,l}$.

Substituting (15.73), (15.74) in the right-hand sides of (15.52), (15.62) and (15.65), omitting in each case the first term in parenthesis on the left, we obtain

$$\begin{aligned}
&(\sum_{j,\alpha} \mathcal{F}'^t_1[R_j R_{i_l} \cdots R_{i_1}\psi_\alpha](u) + \frac{1}{2}\sum_{j,\alpha} K[R_j R_{i_l} \cdots R_{i_1}\psi_\alpha](t,u)) \\
&\cdot \bar{\mu}^{2a}_{m,u}(t)(1+\log(1+t))^{-2q} \\
&\leqslant C_{q,l} R_{[l+2];a,q}(t,u)
\end{aligned} \tag{15.75}$$

$$\begin{aligned}
&(\sum_{\alpha} \mathcal{F}'^t_1[R_{i_{l-m}} \cdots R_{i_1}(T)^{m+1}\psi_\alpha](u) + \frac{1}{2}\sum_{\alpha} K[R_{i_{l-m}} \cdots R_{i_1}(T)^{m+1}\psi_\alpha](t,u)) \\
&\cdot \bar{\mu}^{2a}_{m,u}(t)(1+\log(1+t))^{-2q} \\
&\leqslant C_{q,l} R_{[l+2];a,q}(t,u)
\end{aligned} \tag{15.76}$$

$$\begin{aligned}
&(\mathcal{F}'^t_1[\psi](u) + \frac{1}{2}K[\psi](t,u))\bar{\mu}^{2a}_{m,u}(t)(1+\log(1+t))^{-2q} \\
&\leqslant C_{q,l} R_{[l+2];a,q}(t,u)
\end{aligned} \tag{15.77}$$

Summing then as in (15.67), yields

$$\begin{aligned}
&(\mathcal{F}'^t_{1,[l+2]}(u) + \frac{1}{2}K_{[l+2]}(t,u))\bar{\mu}^{2a}_{m,u}(t)(1+\log(1+t))^{-2q} \\
&\leqslant C_{q,l} R_{[l+2];a,q}(t,u)
\end{aligned} \tag{15.78}$$

for a new constant $C_{q,l}$. Now the definition (15.53) of $R_{[l+2];a,q}(t,u)$ reads

$$R_{[l+2];a,q}(t,u) = \delta I_{[l+2];a,q}(t,u) + N_{[l+2];a,q}(t,u) \tag{15.79}$$

where

$$N_{[l+2];a,q}(t,u) = \mathcal{D}^u_{[l+2]} + \mathcal{G}^u_{0,[l+2];a,p}(t) + (1+\frac{1}{\delta})V'_{1,[l+2];a,q}(t,u) \tag{15.80}$$

Keeping only the term $(1/2)K_{[l+2]}(t,u)$ on the left in (15.78) we have

$$\frac{1}{2}\bar{\mu}_{m,u}^{2a}(t)(1+\log(1+t))^{-2q}K_{[l+2]}(t,u) \leqslant C_{q,l}\delta I_{[l+2];a,q}(t,u) + C_{q,l}N_{[l+2];a,q}(t,u) \tag{15.81}$$

The same holds with $t$ replaced by any $t' \in [0,t]$. The right hand side being a non-decreasing function of $t$ at each $u$, the inequality holds a fortiori if we again replace $t'$ by $t$ on the right. Taking then the supremum over all $t' \in [0,t]$ on the left we obtain, in view of the definition (15.36),

$$\frac{1}{2}I_{[l+2];a,q}(t,u) \leqslant C_{q,l}\delta I_{[l+2];a,q}(t,u) + C_{q,l}N_{[l+2];a,q}(t,u) \tag{15.82}$$

We choose $\delta$ according to

$$C_{q,l}\delta = \frac{1}{4} \tag{15.83}$$

Then (15.82) implies

$$I_{[l+2];a,q}(t,u) \leqslant C_{q,l}N_{[l+2];a,q}(t,u) \tag{15.84}$$

Substituting the estimate (15.84) in (15.79) and the result in (15.78), and keeping only $\mathcal{F}'^{t}_{1,[l+2]}(u)$ on the left of (15.78), we obtain

$$\bar{\mu}_{m,u}^{2a}(t)(1+\log(1+t))^{-2q}\mathcal{F}'^{t}_{1,[l+2]}(u) \leqslant C_{q,l}(Q_{[l+2];a,p}(t,u) + V'_{1,[l+2];a,q}(t,u)) \tag{15.85}$$

for a new $C_{q,l}$, where

$$Q_{[l+2];a,p}(t,u) = \mathcal{D}^{u}_{[l+2]} + \mathcal{G}^{u}_{0,[l+2];a,p}(t) \tag{15.86}$$

Seeing that the right hand side of (15.85) is a non-decreasing function of $t$ at each $u$, the previous reasoning applies and we deduce

$$\mathcal{H}'^{t}_{1,[l+2];a,q}(u) \leqslant C_{q,l}(Q_{[l+2];a,p}(t,u) + V'_{1,[l+2];a,q}(t,u)) \tag{15.87}$$

In view of the definition of $V'_{1,[l+2];a,q}(t,u)$, this reads

$$\mathcal{H}'^{t}_{1,[l+2];a,q}(u) \leqslant C_{q,l}Q_{[l+2];a,p}(t,u) + C_{q,l}\int_0^u \mathcal{H}'^{t}_{1,[l+2];a,q}(u')du' \tag{15.88}$$

Defining the functions

$$\bar{\mathcal{G}}^{u}_{0,[n];a,p}(t) = \sup_{u' \in [0,u]} \mathcal{G}^{u'}_{0,[n];a,p}(t) \tag{15.89}$$

which are non-decreasing functions of $u$ at each $t$ as well as non-decreasing functions of $t$ at each $u$, we may replace $Q_{[l+2];a,p}(t,u)$ on the right in (15.88) by

$$\bar{Q}_{[l+2];a,p}(t,u) = \mathcal{D}^{u}_{[l+2]} + \bar{\mathcal{G}}^{u}_{0,[l+2];a,p}(t) \tag{15.90}$$

which is also a non-decreasing function of $u$ at each $t$ as well as a non-decreasing function of $t$ at each $u$. Recalling that $[0, \epsilon_0]$ is a bounded interval, (15.88) then implies

$$\mathcal{H}'^{t}_{1,[l+2];a,q}(u) \leqslant C_{q,l}\bar{Q}_{[l+2];a,p}(t, u) \tag{15.91}$$

for a new constant $C_{q,l}$. Hence also

$$V'_{1,[l+2];a,q}(t, u) \leqslant C_{q,l}\epsilon_0\bar{Q}_{[l+2];a,p}(t, u) \tag{15.92}$$

Substituting this in (15.80) and the result in (15.84) we obtain

$$I_{[l+2];a,q}(t, u) \leqslant C_{q,l}\bar{Q}_{[l+2];a,p}(t, u) \tag{15.93}$$

for a new constant $C_{q,l}$. Substituting finally this in (15.79) and the result in (15.73) yields

$$\mathcal{G}'^{u}_{1,[l+2];a,q}(t) \leqslant C_{q,l}\bar{Q}_{[l+2];a,p}(t, u) \tag{15.94}$$

for a new constant $C_{q,l}$.

## 15.2 Estimates Associated to $K_0$

Consider now the integral identities corresponding to the variations (14.52) and to the vectorfield $K_0$. In each of these we have the *borderline* integral (14.97), bounded by the sum of (14.102) and (14.107):

$$C(\frac{1}{2a} + \frac{1}{2p})\bar{\mu}_m^{-2a}(t)(1 + \log(1 + t))^{2p(i_1 \cdots i_l)}\mathcal{G}_{0,l+2;a,p}(t) \tag{15.95}$$

We also have the remaining integrals, bounded in the case of (14.115) by (14.132), hence by

$$C_q\delta_0\bar{\mu}_m^{-2a}(t)(1 + \log(1 + t))^{2p}(\mathcal{G}_{0,[l+2];a,p}(t) + \mathcal{G}'_{1,[l+2];a,q}(t)) \tag{15.96}$$

and in the case of (14.157) by (14.158), hence by (15.96) as well. Also, the integral (14.161), is bounded by (14.162), hence by

$$C_p\delta_0\bar{\mu}_m^{-2a}(t)(1 + \log(1 + t))^{2p}\mathcal{P}_{[l+2]}\sqrt{\mathcal{G}_{0,[l+2];a,p}(t)} \tag{15.97}$$

Combining we see that all the remaining integrals associated to (14.52) and to $K_0$ containing the top order spatial derivatives of the acoustical entities are bounded by

$$\begin{aligned}&C_{p,l}\delta_0\bar{\mu}_m^{-2a}(t)(1 + \log(1 + t))^{2p}\sqrt{\mathcal{G}_{0,[l+2];a,p}(t) + \mathcal{G}'_{1,[l+2];a,q}(t)}\\ &\cdot(\mathcal{P}_{[l+2]} + \sqrt{\mathcal{G}_{0,[l+2];a,p}(t) + \mathcal{G}'_{1,[l+2];a,q}(t)})\end{aligned} \tag{15.98}$$

By the discussion following (15.9), the contribution of the terms in (14.14) of which one factor is a derivative of ${}^{(Y)}\tilde{\pi}$ of order more than $(l+1)_*$, but which do not contain top order spatial derivatives of acoustical entities, are bounded using Propositions 12.11 and 12.12 and the bootstrap assumption. We obtain a bound for these contributions to the error integrals associated to $K_0$ and to any of the variations, up to the top order, by

$$\begin{aligned}
&C_l\delta_0\bar{\mu}_m^{-a}(t)(1+\log(1+t))^p(\mathcal{P}_{[l+2]}+\sqrt{\mathcal{G}'_{1,[l+2];a,q}(t)+\mathcal{G}_{0,[l+2];a,p}(t)})\\
&\cdot(\int_0^t(1+t')^{-3/2}\mathcal{E}_{0,[l+2]}(t')dt'+\int_0^{\epsilon_0}\mathcal{F}^t_{0,[l+2]}(u')du')^{1/2}\\
&+C_l\delta_0\int_0^t(1+t')^{-3/2}\mathcal{E}_{0,[l+2]}(t')dt'+C_l\delta_0\int_0^{\epsilon_0}\mathcal{F}^t_{0,[l+2]}(u')du'\\
&+C_l(\bar{\mathcal{F}}'^t_{1,[l+2]}(u')du'+\bar{K}_{[l+2]}(t,\epsilon_0))^{1/2}\\
&\cdot(\int_0^t(1+t')^{-3/2}\mathcal{E}_{0,[l+2]}(t')dt'+\int_0^{\epsilon_0}\mathcal{F}^t_{0,[l+2]}(u')du')^{1/2} \qquad (15.99)
\end{aligned}$$

The last three terms are contributed by the terms in (14.14) containing the top order derivatives of the acoustical entities of which however one is a derivative with respect to $L$ (hence are expressible in terms of the top order derivatives of the $\psi_\alpha$). Here

$$\mathcal{F}^t_{0,[n]}(u)=\sum_{m=1}^n\mathcal{F}^t_{0,m}(u) \qquad (15.100)$$

where $\mathcal{F}^t_{0,n}(u)$ represents the sum of the fluxes associated to the vectorfield $K_0$ of all the $n$th order variations. Also

$$\bar{\mathcal{F}}'^t_{1,[n]}(u)=\sup_{t'\in[0,t]}((1+\log(1+t'))^{-4}\mathcal{F}'^{t'}_{1,[n]}(u)) \qquad (15.101)$$

$$\bar{K}_{[n]}(t,u)=\sup_{t'\in[0,t]}((1+\log(1+t'))^{-4}K_{[n]}(t',u)) \qquad (15.102)$$

On the other hand, all the other terms in the sum (14.14) contain derivatives of the ${}^{(Y)}\tilde{\pi}$ of order at most $(l+1)_*$, thus spatial derivatives of $\chi'$ of order at most $(l+1)_*$ and spatial derivatives of $\mu$ of order at most $(l+1)_*+1$, which are bounded in $L^\infty(\Sigma_t^{\epsilon_0})$ by Propositions 12.9 and 12.10 and the bootstrap assumption. The contributions of these terms are bounded according to Lemma 7.6:

$$\begin{aligned}
&C_l\int_0^t(1+t')^{-3/2}\mathcal{E}_{0,[l+2]}(t')dt'+C_l\int_0^{\epsilon_0}\mathcal{F}^t_{0,[l+2]}(u')du'\\
&+C_l(\int_0^{\epsilon_0}\bar{\mathcal{F}}'^t_{1,[l+2]}(u')du'+\bar{K}_{[l+2]}(t,\epsilon_0))^{1/2}\\
&\cdot(\int_0^t(1+t')^{-3/2}\mathcal{E}_{0,[l+2]}(t')dt'+\int_0^{\epsilon_0}\mathcal{F}^t_{0,[l+2]}(u')du')^{1/2} \qquad (15.103)
\end{aligned}$$

From the definitions (15.26), (15.36) and also the relation (15.30), we have, for all $t' \in [0,t]$,

$$(1+\log(1+t'))^{-4}\mathcal{F}'^{t'}_{1,[n]}(u) \leqslant \bar{\mu}^{-2a}_{m,u}(t')(1+\log(1+t'))^{2p}\mathcal{H}'^{t}_{1,[n];a,q}(u) \quad (15.104)$$

$$(1+\log(1+t'))^{-4}K_{[n]}(t',u) \leqslant \bar{\mu}^{-2a}_{m,u}(t')(1+\log(1+t'))^{2p}I_{[l+2];a,q}(t,u) \quad (15.105)$$

taking the supremum over $t' \in [0,t]$, we obtain, in view of Corollary 2 of Lemma 8.11,

$$\bar{\mathcal{F}}'^{t}_{1,[n]}(u) \leqslant C\bar{\mu}^{-2a}_{m,u}(t)(1+\log(1+t))^{2p}\mathcal{H}'^{t}_{1,[n];a,q}(u) \quad (15.106)$$

$$\bar{K}_{[n]}(t,u) \leqslant C\bar{\mu}^{-2a}_{m,u}(t)(1+\log(1+t))^{2p}I_{[n];a,q}(t,u) \quad (15.107)$$

In view of the definition (15.27) and the fact that $\bar{\mu}_{m,u}(t)$ is a non-increasing function of $u$ at each $t$ we then have

$$\int_0^u \bar{\mathcal{F}}'^{t}_{1,[n]}(u')du' \leqslant C\bar{\mu}^{-2a}_{m,u}(t)(1+\log(1+t))^{2p}V'_{1,[n];a,q}(t,u) \quad (15.108)$$

We also define

$$\mathcal{H}^t_{0,[n];a,p}(u) = \sup_{t'\in[0,t]} ((1+\log(1+t'))^{-2p}\bar{\mu}^{2a}_{m,u}(t')\mathcal{F}^{t'}_{0,[n]}(u)) \quad (15.109)$$

$$V_{0,[n];a,p}(t,u) = \int_0^u \mathcal{H}^t_{0,[n];a,p}(u')du' \quad (15.110)$$

Then $\mathcal{H}^t_{0,[n];a,p}(u)$ is a non-decreasing function of $t$ at each $u$, $V_{0,[n];a,p}(t,u)$ is a non-decreasing function of $u$ at each $t$ as well as a non-decreasing function of $t$ at each $u$, and for each $u \in [0,\epsilon_0]$

$$\begin{aligned}\int_0^u \mathcal{F}^t_{0,[n]}(u')du' &\leqslant \int_0^u \bar{\mu}^{-2a}_{m,u'}(t)(1+\log(1+t))^{2p}\mathcal{H}^t_{0,[n];a,p}(u')du' \\ &\leqslant \bar{\mu}^{-2a}_{m,u}(t)(1+\log(1+t))^{2p}V_{0,[n];a,p}(t,u) \quad (15.111)\end{aligned}$$

So we can bound (15.99) and (15.103) by

$$\begin{aligned}&C_l\delta_0\bar{\mu}_m^{-2a}(t)(1+\log(1+t))^{2p}\cdot((\mathcal{P}_{[l+2]})^2+\delta'(\mathcal{G}'_{1,[l+2];a,q}(t)+\mathcal{G}_{0,[l+2];a,p}(t)) \\ &+(1+\frac{1}{\delta'})(\int_0^t (1+t')^{-3/2}\mathcal{G}_{0,[l+2];a,p}(t')dt' + V_{0,[l+2];a,p}(t,\epsilon_0)) \\ &+\delta'(V'_{1,[l+2];a,q}(t,\epsilon_0)+I_{[l+2];a,q}(t,\epsilon_0))) \quad (15.112)\end{aligned}$$

for any positive constant $\delta'$ ($\delta'$ will be chosen below).

Finally, we have to deal with the first of (15.16). According to (5.282) we have

$$\begin{aligned}&\int_{W^t_{\epsilon_0}} \sum_{k=1}^{7} Q_0[\psi]d\mu_g \\ &\leqslant \int_0^t (1+t')^{-2}(1+\log(1+t'))^4 B_s(t')\mathcal{E}'_1[\psi](t')dt' \quad (15.113)\end{aligned}$$

$$+C\int_0^t(1+t')^{-1}(1+\log(1+t'))^{-2}(\bar{\mathcal{E}}_1'[\psi](t')+\bar{\mathcal{E}}_0[\psi](t'))dt'+CV_0[\psi](t,\epsilon_0)$$
$$+C(V_0[\psi](t,\epsilon_0)+\int_0^t(1+t')^{-1}(1+\log(1+t'))^{-2}\bar{\mathcal{E}}_0[\psi](t')dt')^{1/2}(V_1'[\psi](t,\epsilon_0))^{1/2}$$
$$+C\int_0^t(1+t')^{-2}(1+\log(1+t'))^2V_1'[\psi](t',\epsilon_0)dt'$$
$$+C\bar{K}[\psi](t,\epsilon_0)^{1/2}(\int_0^t(1+t')^{-1}(1+\log(1+t'))^{-2}\bar{\mathcal{E}}_0[\psi](t')dt')^{1/2}$$
$$+C\bar{K}[\psi](t,\epsilon_0)^{1/2}(V_0[\psi](t,\epsilon_0))^{1/2}$$

It follows that

$$\begin{aligned}&\int_{W^t_{\epsilon_0}}\sum_{k=1}^7 Q_0[\psi]d\mu_g\\ &\leqslant\int_0^t\tilde{B}_s(t')\bar{\mathcal{E}}_1'[\psi](t')dt'\\ &\quad+C(1+\frac{1}{\delta'})(\int_0^t(1+t')^{-1}(1+\log(1+t'))^{-2}\bar{\mathcal{E}}_0[\psi](t')dt'\\ &\quad+V_0[\psi](t,\epsilon_0))+\delta'V_1'[\psi](t,\epsilon_0)\\ &\quad+C\int_0^t(1+t')^{-2}(1+\log(1+t'))^2V_1'[\psi](t',\epsilon_0)dt'+\delta'\bar{K}[\psi](t,\epsilon_0)\end{aligned}\tag{15.114}$$

where

$$\tilde{B}_s(t)=(1+t)^{-2}(1+\log(1+t))^4B_s(t)+C(1+t)^{-1}(1+\log(1+t))^{-2}\tag{15.115}$$

Since

$$\begin{aligned}\bar{\mathcal{E}}_1'[\psi](t)&=\sup_{t'\in[0,t]}((1+\log(1+t'))^{-4}\mathcal{E}_1'[\psi](t'))\\ &\leqslant\sup_{t'\in[0,t]}(\bar{\mu}_m^{-2a}(t')(1+\log(1+t'))^{2p}\mathcal{G}_{1;a,q}'[\psi](t'))\\ &\leqslant C\bar{\mu}_m^{-2a}(t)(1+\log(1+t))^{2p}\mathcal{G}_{1;a,q}'[\psi](t)\end{aligned}\tag{15.116}$$

the first term on the right in (15.114) is bounded by

$$C\bar{\mu}_m^{-2a}(t)(1+\log(1+t))^{2p}\int_0^t\tilde{B}_s(t')\mathcal{G}_{1;a,q}'[\psi](t')dt'\tag{15.117}$$

In regard to the second term on the right in (15.114) we have

$$\begin{aligned}&\int_0^t(1+t')^{-1}(1+\log(1+t'))^{-2}\bar{\mathcal{E}}_0[\psi](t')dt'\\ &\leqslant C\bar{\mu}_m^{-2a}(t)(1+\log(1+t))^{2p}\int_0^t(1+t')^{-1}(1+\log(1+t'))^{-2}\mathcal{G}_{0;a,p}[\psi](t')dt'\end{aligned}\tag{15.118}$$

Moreover, defining

$$\mathcal{H}^t_{0;a,p}[\psi](u) = \sup_{t'\in[0,t]} ((1+\log(1+t'))^{-2p}\bar{\mu}^{2a}_{m,u}(t')\mathcal{F}^{t'}_0[\psi](u)) \tag{15.119}$$

$$V_{0;a,p}[\psi](t,u) = \int_0^u \mathcal{H}^t_{0;a,p}[\psi](u')du' \tag{15.120}$$

we have, in analogy with (15.111),

$$V_0[\psi](t,u) = \int_0^u \mathcal{F}^t_0(u')du' \leqslant \bar{\mu}^{-2a}_{m,u}(t)(1+\log(1+t))^{2p}V_{0;a,p}[\psi](t,u) \tag{15.121}$$

In regard to the third and fourth term on the right of (15.114), recalling from Chapter 5 that

$$V'_1[\psi](t,u) = \int_0^u \bar{\mathcal{F}}'^t_1[\psi](u')du' \tag{15.122}$$

we have, using an inequality similar to (15.106),

$$V'_1[\psi](t,u) \leqslant C\bar{\mu}^{-2a}_{m,u}(t)(1+\log(1+t))^{2p}V'_{1;a,q}[\psi](t,u) \tag{15.123}$$

hence also

$$\int_0^t (1+t')^{-2}(1+\log(1+t'))^2 V'_1[\psi](t',\epsilon_0)dt'$$

$$\leqslant C\bar{\mu}^{-2a}_m(t)(1+\log(1+t))^{2p}\int_0^t (1+t')^{-2}(1+\log(1+t'))^2 V'_{1;a,q}[\psi](t',\epsilon_0)dt' \tag{15.124}$$

Finally, in regard to the last term, we have, in analogy with (15.107),

$$\bar{K}[\psi](t,\epsilon_0) \leqslant C\bar{\mu}^{-2a}_m(t)(1+\log(1+t))^{2p}I_{a,q}[\psi](t,\epsilon_0) \tag{15.125}$$

where

$$I_{a,q}[\psi](t,u) = \sup_{t'\in[0,t]} ((1+\log(1+t))^{-2q}\bar{\mu}^{2a}_{m,u}(t')K[\psi](t',u)) \tag{15.126}$$

Substituting the above in (15.114) we obtain

$$\int_{W^t_{\epsilon_0}} \sum_{k=1}^{7} Q_0[\psi]d\mu_g$$

$$\leqslant C\bar{\mu}^{-2a}_m(t)(1+\log(1+t))^{2p}\Big(\int_0^t \tilde{B}_s(t')\mathcal{G}'_{1;a,q}[\psi](t')dt' + (1+\frac{1}{\delta'}) \tag{15.127}$$

$$\cdot\Big(\int_0^t (1+t')^{-1}(1+\log(1+t'))^{-2}\mathcal{G}_{0;a,p}[\psi](t')dt' + V_{0;a,p}[\psi](t,\epsilon_0)\Big)$$

$$+\delta' V'_{1;a,q}[\psi](t,\epsilon_0)$$

$$+\int_0^t (1+t')^{-2}(1+\log(1+t'))^2 V'_{1;a,q}[\psi](t',\epsilon_0)dt' + \delta' I_{a,q}[\psi](t,\epsilon_0)\Big)$$

We now consider the energy identity corresponding to $K_0$ ((5.74) with $u = \epsilon_0$) and to the variations (14.52) $R_j R_{i_l} \cdots R_{i_1}\psi_\alpha$, $j = 1,2,3$, $\alpha = 0,1,2,3$, for a given multi-index $(i_1 \cdots i_l)$.

Summing over $j$ and $\alpha$ we then obtain from (15.95), (15.98), (15.112) and (15.127)

$$
\begin{aligned}
&(\sum_{j,\alpha}\mathcal{E}_0[R_j R_{i_l}\cdots R_{i_1}\psi_\alpha](t)+\sum_{j,\alpha}\mathcal{F}_0^t[R_j R_{i_l}\cdots R_{i_1}\psi_\alpha](\epsilon_0))\bar{\mu}_m^{2a}(t)(1+\log(1+t))^{-2p}\\
&\leqslant C(\frac{1}{a}+\frac{1}{p})^{(i_1\cdots i_l)}\mathcal{G}_{0,l+2;a,p}(t)+C_{p,l}\mathcal{D}_{[l+2]} \qquad (15.128)\\
&+C_{p,l}\delta_0(\mathcal{G}_{0,[l+2];a,p}(t)+\mathcal{G}'_{1,[l+2];a,q}(t))\\
&+C\int_0^t \tilde{B}_s(t')\mathcal{G}'_{1,[l+2];a,q}(t')dt'\\
&+C\int_0^t (1+t')^{-2}(1+\log(1+t'))^2 V'_{1,[l+2];a,q}(t',\epsilon_0)dt'\\
&+C_l(1+\frac{1}{\delta'})(\int_0^t (1+t')^{-1}(1+\log(1+t'))^{-2}\mathcal{G}_{0,[l+2];a,p}(t')dt'+V_{0,[l+2];a,p}(t,\epsilon_0))\\
&+C_l\delta'(V'_{1,[l+2];a,q}(t,\epsilon_0)+I_{[l+2];a,q}(t,\epsilon_0))
\end{aligned}
$$

We substitute (15.92)—(15.94) to obtain

$$
\begin{aligned}
&(\sum_{j,\alpha}\mathcal{E}_0[R_j R_{i_l}\cdots R_{i_1}\psi_\alpha](t)+\sum_{j,\alpha}\mathcal{F}_0^t[R_j R_{i_l}\cdots R_{i_1}\psi_\alpha](\epsilon_0))\bar{\mu}_m^{2a}(t)(1+\log(1+t))^{-2p}\\
&\leqslant C(\frac{1}{a}+\frac{1}{p})^{(i_1\cdots i_l)}\mathcal{G}_{0,l+2;a,p}(t)+C_{p,l}(\delta_0+\delta')\bar{\mathcal{G}}_{0,[l+2];a,p}(t) \qquad (15.129)\\
&+C_{p,l}(1+\frac{1}{\delta'})(\mathcal{D}_{[l+2]}+\int_0^t \tilde{B}_s(t')\bar{\mathcal{G}}_{0,[l+2];a,p}(t')dt'+V_{0,[l+2];a,p}(t,\epsilon_0))
\end{aligned}
$$

Here we have used the fact in Proposition 13.2:

$$\int_0^s \tilde{B}_s(t)dt \leqslant C \quad \text{independent of} \quad s \qquad (15.130)$$

It is crucial that the constant $C$ in front of the first term on the right of (15.129) is independent of $a,p,l$. Moreover, (15.129) holds with $\epsilon_0$ replaced by any $u \in (0,\epsilon_0]$, that is, we have, for all $u \in (0,\epsilon_0]$ and all $t \in [0,s]$,

$$
\begin{aligned}
&(\sum_{j,\alpha}\mathcal{E}_0^u[R_j R_{i_l}\cdots R_{i_1}\psi_\alpha](t)+\sum_{j,\alpha}\mathcal{F}_0^t[R_j R_{i_l}\cdots R_{i_1}\psi_\alpha](u))\bar{\mu}_{m,u}^{2a}(t)(1+\log(1+t))^{-2p}\\
&\leqslant C(\frac{1}{a}+\frac{1}{p})^{(i_1\cdots i_l)}\mathcal{G}^u_{0,l+2;a,p}(t)+C_{p,l}(\delta_0+\delta')\bar{\mathcal{G}}^u_{0,[l+2];a,p}(t) \qquad (15.131)\\
&+C_{p,l}(1+\frac{1}{\delta'})(\mathcal{D}^u_{[l+2]}+\int_0^t \tilde{B}_s(t')\bar{\mathcal{G}}^u_{0,[l+2];a,p}(t')dt'+V_{0,[l+2];a,p}(t,u))
\end{aligned}
$$

Keeping only the term

$$\sum_{j,\alpha} \mathcal{E}_0^u[R_j R_{i_l} \cdots R_{i_1} \psi_\alpha] \tag{15.132}$$

on the left of (15.131), we have

$$\begin{aligned}
&\bar{\mu}_{m,u}^{2a}(t)(1+\log(1+t))^{-2p} \sum_{j,\alpha} \mathcal{E}_0^u[R_j R_{i_l} \cdots R_{i_1} \psi_\alpha](t) \\
&\leqslant C(\frac{1}{a}+\frac{1}{p})^{(i_1\cdots i_l)}\mathcal{G}_{0,l+2;a,p}^u(t) + C_{p,l}(\delta_0+\delta')\bar{\mathcal{G}}_{0,[l+2];a,p}^u(t) \\
&\quad + C_{p,l}(1+\frac{1}{\delta'})(\mathcal{D}_{[l+2]}^u + \int_0^t \tilde{B}_s(t')\bar{\mathcal{G}}_{0,[l+2];a,p}^u(t')dt' + V_{0,[l+2];a,p}(t,u))
\end{aligned} \tag{15.133}$$

The same holds with $t$ replaced by any $t' \in [0,t]$. Now the right hand side of (15.132) is a non-decreasing function of $t$ at each $u$. The inequality corresponding to $t'$ thus holds a fortiori if we again replace $t'$ by $t$ on the right hand side. Taking then the supremum over all $t' \in [0,t]$ on the left we obtain, in view of the definition (15.46),

$$\begin{aligned}
&{}^{(i_1\cdots i_l)}\mathcal{G}_{0,l+2;a,p}^u(t) \\
&\leqslant C(\frac{1}{a}+\frac{1}{p})^{(i_1\cdots i_l)}\mathcal{G}_{0,l+2;a,p}^u(t) + C_{p,l}(\delta_0+\delta')\bar{\mathcal{G}}_{0,[l+2];a,p}^u(t) \\
&\quad + C_{p,l}(1+\frac{1}{\delta'})(\mathcal{D}_{[l+2]}^u + \int_0^t \tilde{B}_s(t')\bar{\mathcal{G}}_{0,[l+2];a,p}^u(t')dt' + V_{0,[l+2];a,p}(t,u))
\end{aligned} \tag{15.134}$$

If we choose $a$ and $p$ suitably large so that

$$C(\frac{1}{a}+\frac{1}{p}) \leqslant \frac{1}{2} \tag{15.135}$$

then (15.133) implies

$$\begin{aligned}
&\frac{1}{2}{}^{(i_1\cdots i_l)}\mathcal{G}_{0,l+2;a,p}^u(t) \\
&\leqslant C_{p,l}(\delta_0+\delta')\bar{\mathcal{G}}_{0,[l+2];a,p}^u(t) \\
&\quad + C_{p,l}(1+\frac{1}{\delta'})(\mathcal{D}_{[l+2]}^u + \int_0^t \tilde{B}_s(t')\bar{\mathcal{G}}_{0,[l+2];a,p}^u(t')dt' + V_{0,[l+2];a,p}(t,u))
\end{aligned} \tag{15.136}$$

Consider now the integral identities corresponding to the variations (14.54) and $K_0$. In each of these we have the *borderline* integral (14.178), bounded by (14.179),

$$C(\frac{1}{2a}+\frac{1}{2p})\bar{\mu}_m^{-2a}(t)(1+\log(1+t))^{2p(i_1\cdots i_{l-m})}\mathcal{G}_{0,m,l+2;a,p}(t) \tag{15.137}$$

We also have the remaining integrals, bounded in the case of (14.185) by (15.96) and in the case of (14.198) by the sum of (14.158) and (14.199), hence also by

(15.96). We also have the integral (14.199) bounded by (14.200), hence by (15.97). Combining we see that all the remaining integrals associated to (14.54) and $K_0$ containing the top order spatial derivatives of the acoustical entities are bounded by (15.98). On the other hand, the error integrals associated to (14.54) contributed by all the other terms in (14.14) are bounded by the sum of (15.99) and (15.103), hence by (15.112). Finally, the contribution from the first of (15.16) is bounded by (15.127).

Consequently, the integral identities in question yield

$$\begin{aligned}
&\Big(\sum_\alpha \mathcal{E}_0[R_{i_{l-m}}\cdots R_{i_1}(T)^{m+1}\psi_\alpha](t)+\sum_\alpha \mathcal{F}_0^t[R_{i_{l-m}}\cdots R_{i_1}(T)^{m+1}\psi_\alpha](\epsilon_0)\Big)\\
&\cdot\bar{\mu}_m^{2a}(t)(1+\log(1+t))^{-2p}\\
&\leqslant C(\frac{1}{a}+\frac{1}{p})\,{}^{(i_1\cdots i_{l-m})}\mathcal{G}_{0,m,[l+2];a,p}(t)+C_{p,l}(\delta_0+\delta')\bar{\mathcal{G}}_{0,[l+2];a,p}(t) \qquad (15.138)\\
&\quad+C_{p,l}(1+\frac{1}{\delta'})\Big(\mathcal{D}_{[l+2]}+\int_0^t \tilde{B}_s(t')\bar{\mathcal{G}}_{0,[l+2];a,p}(t')dt'+V_{0,[l+2];a,p}(t,\epsilon_0)\Big)
\end{aligned}$$

and

$$\begin{aligned}
&\Big(\sum_\alpha \mathcal{E}_0^u[R_{i_{l-m}}\cdots R_{i_1}(T)^{m+1}\psi_\alpha](t)+\sum_\alpha \mathcal{F}_\alpha^t[R_{i_{l-m}}\cdots R_{i_1}(T)^{m+1}\psi_\alpha](u)\Big)\\
&\cdot\bar{\mu}_{m,u}^{2a}(t)(1+\log(1+t))^{-2p}\\
&\leqslant C(\frac{1}{a}+\frac{1}{p})\,{}^{(i_1\cdots i_{l-m})}\mathcal{G}^u_{0,m,[l+2];a,p}(t)+C_{p,l}(\delta_0+\delta')\bar{\mathcal{G}}^u_{0,[l+2];a,p}(t) \qquad (15.139)\\
&\quad+C_{p,l}(1+\frac{1}{\delta'})\Big(\mathcal{D}^u_{[l+2]}+\int_0^t \tilde{B}_s(t')\bar{\mathcal{G}}^u_{0,[l+2];a,p}(t')dt'+V_{0,[l+2];a,p}(t,u)\Big)
\end{aligned}$$

Keeping only the term

$$\sum_\alpha \mathcal{E}_0^u[R_{i_{l-m}}\cdots R_{i_1}(T)^{m+1}\psi_\alpha]$$

on the left in (15.139), we have

$$\begin{aligned}
&\bar{\mu}_{m,u}^{2a}(t)(1+\log(1+t))^{-2p}\sum_\alpha \mathcal{E}_0^u[R_{i_{l-m}}\cdots R_{i_1}(T)^{m+1}\psi_\alpha](t)\\
&\leqslant C(\frac{1}{a}+\frac{1}{p})\,{}^{(i_1\cdots i_{l-m})}\mathcal{G}^u_{0,m,l+2;a,p}(t)+C_{p,l}(\delta_0+\delta')\bar{\mathcal{G}}^u_{0,[l+2];a,p}(t) \qquad (15.140)\\
&\quad+C_{p,l}(1+\frac{1}{\delta'})\Big(\mathcal{D}^u_{[l+2]}+\int_0^t \tilde{B}_s(t')\bar{\mathcal{G}}^u_{0,[l+2];a,p}(t')dt'+V_{0,[l+2];a,p}(t,u)\Big)
\end{aligned}$$

Seeing that the right hand side of (15.140) is a non-decreasing function of $t$ at each $u$, the previous reasoning applies and we deduce

$$\begin{aligned}
&{}^{(i_1\cdots i_{l-m})}\mathcal{G}^u_{0,m,l+2;a,p}(t)\\
&\leqslant C(\frac{1}{a}+\frac{1}{p})\,{}^{(i_1\cdots i_{l-m})}\mathcal{G}^u_{0,m,l+2;a,p}(t)+C_{p,l}(\delta_0+\delta')\bar{\mathcal{G}}^u_{0,[l+2];a,p}(t) \qquad (15.141)\\
&\quad+C_{p,l}(1+\frac{1}{\delta'})\Big(\mathcal{D}^u_{[l+2]}+\int_0^t \tilde{B}_s(t')\bar{\mathcal{G}}^u_{0,[l+2];a,p}(t')dt'+V_{0,[l+2];a,p}(t,u)\Big)
\end{aligned}$$

Choosing $a$ and $p$ sufficiently large so that (15.135) holds, this implies

$$\begin{aligned}&\frac{1}{2}{}^{(i_1\cdots i_{l-m})}\mathcal{G}^u_{0,m,l+2;a,p}(t)\\ &\leqslant C_{p,l}(\delta_0+\delta')\bar{\mathcal{G}}^u_{0,[l+2];a,p}(t)\\ &\quad+C_{p,l}(1+\frac{1}{\delta'})(\mathcal{D}^u_{[l+2]}+\int_0^t\tilde{B}_s(t')\bar{\mathcal{G}}^u_{0,[l+2];a,p}(t')dt'+V_{0,[l+2];a,p}(t,u))\end{aligned} \tag{15.142}$$

Consider finally the integral identities corresponding to any of the other variations $\psi$ of order up to $l+2$, and to $K_0$. These identities contain no borderline integrals and all error integrals involved are bounded by

$$\begin{aligned}&C_{p,l}(\delta_0+\delta')\bar{\mathcal{G}}^u_{0,[l+2];a,p}(t)\\ &\quad+C_{p,l}(1+\frac{1}{\delta'})(\mathcal{D}^u_{[l+2]}+\int_0^t\tilde{B}_s(t')\bar{\mathcal{G}}^u_{0,[l+2];a,p}(t')dt'+V_{0,[l+2];a,p}(t,u))\end{aligned} \tag{15.143}$$

We thus obtain, for each such variation $\psi$ the inequality

$$\begin{aligned}&(\mathcal{E}^u_0[\psi](t)+\mathcal{F}^t_0[\psi](u))\bar{\mu}^{2a}_{m,u}(t)(1+\log(1+t))^{-2p}\\ &\leqslant C_{p,l}(\delta_0+\delta')\bar{\mathcal{G}}^u_{0,[l+2];a,p}(t)\\ &\quad+C_{p,l}(1+\frac{1}{\delta'})(\mathcal{D}^u_{[l+2]}+\int_0^t\tilde{B}_s(t')\bar{\mathcal{G}}^u_{0,[l+2];a,p}(t')dt'+V_{0,[l+2];a,p}(t,u))\end{aligned} \tag{15.144}$$

Keeping only the term

$$\mathcal{E}^u_0[\psi]$$

on the left in (15.144) and applying the above reasoning we deduce

$$\begin{aligned}&\mathcal{G}^u_{0;a,p}[\psi](t)\\ &\leqslant C_{p,l}(\delta_0+\delta')\bar{\mathcal{G}}^u_{0,[l+2];a,p}(t)\\ &\quad+C_{p,l}(1+\frac{1}{\delta'})(\mathcal{D}^u_{[l+2]}+\int_0^t\tilde{B}_s(t')\bar{\mathcal{G}}^u_{0,[l+2];a,p}(t')dt'+V_{0,[l+2];a,p}(t,u))\end{aligned} \tag{15.145}$$

Now according to the definition of $\mathcal{E}^u_{0,[l+2]}(t)$ we have

$$\begin{aligned}\mathcal{E}^u_{0,[l+2]}(t)=&\sum_{i_1\cdots i_l}(\sum_{j,\alpha}\mathcal{E}^u_0[R_jR_{i_l}\cdots R_{i_1}\psi_\alpha](t))\\ &+\sum_{m=0}^{l}\sum_{i_1\cdots i_{l-m}}(\sum_{\alpha}\mathcal{E}^u_0[R_{i_{l-m}}\cdots R_{i_1}(T)^{m+1}\psi_\alpha](t))+\sum_{\alpha}\mathcal{E}^u_0[\psi](t)\end{aligned} \tag{15.146}$$

where the last sum is over all the other variations $\psi$ up to order $l+2$. By (15.68) we then have

$$\begin{aligned}\mathcal{G}^u_{0,[l+2];a,p}(t)\leqslant&\sum_{i_1\cdots i_l}{}^{(i_1\cdots i_l)}\mathcal{G}^u_{0,l+2;a,p}(t)\\ &+\sum_{m=0}^{l}\sum_{i_1\cdots i_{l-m}}{}^{(i_1\cdots i_{l-m})}\mathcal{G}^u_{0,m,l+2;a,p}(t)+\sum_{\psi}\mathcal{G}^u_{0;a,p}[\psi](t)\end{aligned} \tag{15.147}$$

Thus, summing (15.135) over $i_1 \cdots i_l$, summing (15.142) over $i_1 \cdots i_{l-m}$ and over $m = 0, \cdots, l$, and summing (15.145) over all the other variations $\psi$ of order up to $l+2$, we obtain

$$\begin{aligned}
&\frac{1}{2}\mathcal{G}^u_{0,[l+2];a,p}(t) \\
&\leqslant C_{p,l}(\delta_0 + \delta')\bar{\mathcal{G}}^u_{0,[l+2];a,p}(t) \qquad (15.148)\\
&\quad + C_{p,l}(1 + \frac{1}{\delta'})(\mathcal{D}^u_{[l+2]} + \int_0^t \tilde{B}_s(t')\bar{\mathcal{G}}^u_{0,[l+2];a,p}(t')dt' + V_{0,[l+2];a,p}(t,u))
\end{aligned}$$

for a new constant $C_{p,l}$. Requiring $\delta_0$ and $\delta'$ to satisfy the smallness condition and equation respectively:

$$C_{p,l}\delta_0 \leqslant \frac{1}{8}, \quad C_{p,l}\delta' = \frac{1}{8} \qquad (15.149)$$

(15.148) implies

$$\begin{aligned}
\frac{1}{2}\mathcal{G}^u_{0,[l+2];a,p}(t) &\leqslant \frac{1}{4}\bar{\mathcal{G}}^u_{0,[l+2];a,p}(t) + C_{p,l}(\mathcal{D}^u_{[l+2]} \qquad (15.150)\\
&\quad + \int_0^t \tilde{B}_s(t')\bar{\mathcal{G}}^u_{0,[l+2];a,p}(t')dt' + V_{0,[l+2];a,p}(t,u))
\end{aligned}$$

for a new constant $C_{p,l}$. The same holds with $u$ replaced by any $u' \in [0,u]$. Now the right hand side of (15.150) is a non-decreasing function of $u$ at each $t$. The inequality corresponding to $u'$ thus holds a fortiori if we again replace $u'$ by $u$ on the right. Taking then the supremum over all $u' \in [0,u]$ on the left we obtain, in view of the definition (15.89),

$$\frac{1}{4}\bar{\mathcal{G}}^u_{0,[l+2];a,p}(t) \leqslant C_{p,l}(\mathcal{D}^u_{[l+2]} + \int_0^t \tilde{B}_s(t')\bar{\mathcal{G}}^u_{0,[l+2];a,p}(t')dt' + V_{0,[l+2];a,p}(t,u)) \qquad (15.151)$$

In view of (15.130), since $V_{0,[l+2];a,p}(t,u)$ is a non-decreasing function of $t$ at each $u$, (15.151) implies

$$\bar{\mathcal{G}}^u_{0,[l+2];a,p}(t) \leqslant C_{p,l}(\mathcal{D}^u_{[l+2]} + V_{0,[l+2];a,p}(t,u)) \qquad (15.152)$$

hence also

$$\int_0^t \tilde{B}_s(t')\bar{\mathcal{G}}^u_{0,[l+2];a,p}(t')dt' \leqslant C_{p,l}(\mathcal{D}^u_{[l+2]} + V_{0,[l+2];a,p}(t,u)) \qquad (15.153)$$

Substituting (15.152) and (15.153) in the right-hand sides of (15.131), (15.138) and (15.144), omitting in each case the first term in parenthesis on the left we obtain

$$\begin{aligned}
&\bar{\mu}^{2a}_{m,u}(t)(1+\log(1+t))^{-2p}\sum_{j,\alpha}\mathcal{F}^t_0[R_j R_{i_l}\cdots R_{i_1}\psi_\alpha](u) \\
&\leqslant C_{p,l}(\mathcal{D}^u_{[l+2]} + V_{0,[l+2];a,p}(t,u)) \qquad (15.154)
\end{aligned}$$

$$\begin{aligned}&\bar{\mu}_{m,u}^{2a}(t)(1+\log(1+t))^{-2p}\sum_{j,\alpha}\mathcal{F}_0^t[R_j R_{i_{l-m}}\cdots R_{i_1}(T)^{m+1}\psi_\alpha](u)\\&\leqslant C_{p,l}(\mathcal{D}_{[l+2]}^u+V_{0,[l+2];a,p}(t,u))\end{aligned}\tag{15.155}$$

$$\begin{aligned}&\bar{\mu}_{m,u}^{2a}(t)(1+\log(1+t))^{-2p}\sum_{j,\alpha}\mathcal{F}_0^t[\psi](u)\\&\leqslant C_{p,l}(\mathcal{D}_{[l+2]}^u+V_{0,[l+2];a,p}(t,u))\end{aligned}\tag{15.156}$$

Summing as in (15.146) yields

$$\bar{\mu}_{m,u}^{2a}(t)(1+\log(1+t))^{-2p}\mathcal{F}_{0,[l+2]}^t(u)\leqslant C_{p,l}(\mathcal{D}_{[l+2]}^u+V_{0,[l+2];a,p}(t,u))\tag{15.157}$$

Since the inequality holds with $t$ replaced by any $t'\in[0,t]$ and the right hand side is a non-decreasing function of $t$ at each $u$, the inequality corresponding to $t'$ holds a fortiori if we again replace $t'$ by $t$ on the right. Taking then the supremum over all $t'\in[0,t]$ on the left we obtain, in view of the definition (15.109),

$$\mathcal{H}_{0,[l+2];a,p}^t(u)\leqslant C_{p,l}(\mathcal{D}_{[l+2]}^u+V_{0,[l+2];a,p}(t,u))\tag{15.158}$$

Recalling the definition of $V_{0,[l+2];a,p}(t,u)$, this is

$$\mathcal{H}_{0,[l+2];a,p}^t(u)\leqslant C_{p,l}\mathcal{D}_{[l+2]}^u+C_{p,l}\int_0^u\mathcal{H}_{0,[l+2];a,p}^t(u')du'\tag{15.159}$$

Since $[0,\epsilon_0]$ is a bounded interval and $\mathcal{D}_{[l+2]}^u$ is a non-decreasing function of $u$, (15.159) implies

$$\mathcal{H}_{0,[l+2];a,p}^t\leqslant C_{p,l}\mathcal{D}_{[l+2]}^u\tag{15.160}$$

hence also

$$V_{0,[l+2];a,p}(t,u)\leqslant C_{p,l}\epsilon_0\mathcal{D}_{[l+2]}^u\tag{15.161}$$

Substituting this in (15.152) we obtain

$$\bar{\mathcal{G}}_{0,[l+2];a,p}^u(t)\leqslant C_{p,l}\mathcal{D}_{[l+2]}^u\tag{15.162}$$

Substituting finally (15.161) in (15.90) and the result in (15.91), (15.93), (15.94) yields

$$\mathcal{G}_{1,[l+2];a,q}^{\prime u}(t),\quad\mathcal{H}_{1,[l+2];a,q}^{\prime t}(u),\quad I_{[l+2];a,q}(t,u)\leqslant C_{p,l}\mathcal{D}_{[l+2]}^u\tag{15.163}$$

This completes the top order energy estimates.

At this point we shall specify the choice of $a$ and $p$ (recall from (15.30) that $q=p+2$). We also take $a$ to be the form

$$a=[a]+\frac{3}{4}\tag{15.164}$$

where $[a]$ denotes the integral part of $a$. We then choose $[a]$ to be the smallest positive integer so that with $C$ the largest of the two constants in (15.57) and (15.134), we have

$$\frac{C}{a} \leqslant \frac{3}{8} \tag{15.165}$$

We then choose $p$ so that with the same constant $C$, we have

$$\frac{C}{p} = \frac{1}{8} \tag{15.166}$$

Then the conditions (15.57) and (15.134) are both satisfied and thereafter $a, p$ and $q$ are fixed. We then fix $l$ so that

$$l_* \geqslant [a] + 4 \tag{15.167}$$

We shall see the reason for this in the following. So $l$ is thenceforth fixed as well, therefore we shall not denote the dependence of the constants on $a, p, q, l$. Actually, the constants shall now depend only on the the function $H$. We shall still require $\delta_0$ to be suitably small in relation to these constants.

# 16

# The Descent Scheme

We now consider again the estimates for the energies $\mathcal{E}_{0,[l+1]}$ and $\mathcal{E}'_{1,[l+1]}$, associated to $K_0$ and $K_1$ respectively. We shall presently obtain improved estimates for these next to the top order energies, using the estimates for the top order energies just obtained. We shall show that the next to the top order quantities satisfy estimates similar to (15.160), (15.162) and (15.163), but with $a$ replaced by

$$b = a - 1 \tag{16.1}$$

$p$ replaced by 0, and, accordingly, $q$ replaced by 2.

Let us first consider the error integrals associated to the variations of order $l+1$ and to $K_1$, involving the highest spatial derivatives of the acoustical entities. The leading contributions are:

1) The contribution of (14.56) with $l+1$ replaced by $l$ to the corresponding integral (14.59), namely the integral (14.202) with $l$ replaced by $l-1$:

$$-\int_{W_u^t}(\omega/\nu)(R_{i_l}\cdots R_{i_1}\mathrm{tr}\chi')(T\psi_\alpha)((L+\nu)R_{i_l}\cdots R_{i_1}\psi_\alpha)dt'du'd\mu_{\tilde{g}} \tag{16.2}$$

2) The contribution of (14.57) with $l$ replaced by $l-1$ to the corresponding integral (14.59), namely the integral (14.369) with $l$ replaced by $l-1$:

$$\begin{aligned}-\int_{W_u^t}&(\omega/\nu)(T\psi_\alpha)(R_{i_{l-1-m}}\cdots R_{i_1}(T)^m\Delta\mu)\\ &((L+\nu)R_{i_{l-1-m}}\cdots R_{i_1}(T)^{m+1}\psi_\alpha)dt'du'd\mu_{\tilde{g}}\end{aligned} \tag{16.3}$$

for $m = 0,\cdots,l-1$.

By the bounds (14.294) the integral (16.2) is bounded by

$$\begin{aligned}&C\delta_0\int_{W_u^t}(1+t')|R_{i_l}\cdots R_{i_1}\mathrm{tr}\chi'||(L+\nu)R_{i_l}\cdots R_{i_1}\psi_\alpha|dt'du'd\mu_{g}\\ &\leqslant C\delta_0\Big(\int_{W_u^t}|R_{i_l}\cdots R_{i_1}\mathrm{tr}\chi'|^2dt'du'd\mu_{g}\Big)^{1/2}\\ &\quad\cdot\Big(\int_{W_u^t}(1+t')^2|(L+\nu)R_{i_l}\cdots R_{i_1}\psi_\alpha|^2dt'du'd\mu_{g}\Big)^{1/2}\end{aligned} \tag{16.4}$$

Now the last factor is bounded by

$$C(\int_0^u \mathcal{F}'^t_{1,[l+1]}(u')du')^{1/2} \tag{16.5}$$

while the first factor is

$$(\int_0^t \|R_{i_l}\cdots R_{i_1}\mathrm{tr}\chi'\|^2_{L^2(\Sigma^u_{t'})}dt')^{1/2} \tag{16.6}$$

which is bounded in terms of

$$(\int_0^t (\mathcal{A}'^u_{[l]}(t'))^2 dt')^{1/2} \tag{16.7}$$

Here $\mathcal{A}'^u_{[l]}(t)$ is defined by replacing the $L^2$ norms over $\Sigma^{\epsilon_0}_t$ in the definition of $\mathcal{A}'_{[l]}(t)$ by $L^2$ norms over $\Sigma^u_t$. Proposition 12.11 applies with $\epsilon_0$ replaced by $u$, giving a bound for $\mathcal{A}'^u_{[l]}(t)$ by

$$C(1+t)^{-1}\int_0^t (1+t')^{-1}(\mathcal{W}^u_{[l+2]}(t') + \mathcal{W}^{Qu}_{[l+1]}(t'))dt' \tag{16.8}$$

Now we have

$$\begin{aligned}\mathcal{W}^u_{[l+2]}(t) + \mathcal{W}^{Qu}_{[l+1]}(t) &\leqslant C\bar{\mu}^{-1/2}_{m,u}\sqrt{\mathcal{E}'^u_{1,[l+2]}(t)}\\ &\leqslant C\bar{\mu}^{-a-1/2}_{m,u}(t)(1+\log(1+t))^q\sqrt{\mathcal{G}'^u_{1,[l+2];a,q}(t)}\end{aligned} \tag{16.9}$$

Therefore (16.8) is bounded by

$$\begin{aligned}&C(1+t)^{-1}\sqrt{\mathcal{G}'^u_{1,[l+2];a,q}(t)}\int_0^t (1+t')^{-1}\bar{\mu}^{-a-1/2}_{m,u}(t')(1+\log(1+t'))^q dt'\\ &\leqslant C(1+t)^{-1}\sqrt{\mathcal{G}'^u_{1,[l+2];a,q}(t)}J^u_{a,q-1}(t)\\ &\leqslant C(1+t)^{-1}(1+\log(1+t))^{q+1}\bar{\mu}^{-a+1/2}_{m,u}(t)\sqrt{\mathcal{D}^u_{[l+2]}}\end{aligned} \tag{16.10}$$

Here $J^u_{a,q}(t)$ is the integral (14.147) with $\epsilon_0$ replaced by $u$ and we have used (14.153) with $\epsilon_0$ replaced by $u$ as well as the bounds (15.163). It follows that the contribution of (16.8) to (16.7) is bounded by

$$C\sqrt{\mathcal{D}^u_{[l+2]}}(\int_0^t (1+t')^{-2}(1+\log(1+t'))^{2q+2}\bar{\mu}^{-2a+1}_{m,u}(t')dt')^{1/2} \tag{16.11}$$

the last integral being estimated by following an argument similar to that used to estimate (14.115) (with $\epsilon_0$ replaced by $u$). By (15.28) with $a$ replaced by $b$ and $q$ by 2 we have

$$\int_0^u \mathcal{F}'^t_{1,[l+1]}(u')du' \leqslant \bar{\mu}^{-2b}_{m,u}(t)(1+\log(1+t))^4 V'_{1,[l+1];b,2}(t,u) \tag{16.12}$$

we conclude that (16.4) is bounded by

$$C\delta_0\bar{\mu}_{m,u}^{-2b}(t)(1+\log(1+t))^2\sqrt{\mathcal{D}^u_{[l+2]}}\sqrt{V'_{1,[l+1];b,2}(t,u)} \tag{16.13}$$

where $b$ is defined by (16.1).

By (14.294), (16.3) is bounded by

$$\begin{aligned}&C\delta_0\int_{W^t_u}(1+t')|R_{i_{l-1-m}}\cdots R_{i_1}(T)^m\not{\Delta}\mu|\\&\cdot|(L+\nu)R_{i_{l-1-m}}\cdots R_{i_1}(T)^{m+1}\psi_\alpha|dt'du'd\mu_{\not{g}}\\&\leqslant C\delta_0(\int_{W^t_u}|R_{i_{l-1-m}}\cdots R_{i_1}(T)^m\not{\Delta}\mu|^2dt'du'd\mu_{\not{g}})^{1/2}\\&\cdot(\int_{W^t_u}(1+t')^2|(L+\nu)R_{i_{l-1-m}}\cdots R_{i_1}(T)^{m+1}\psi_\alpha|^2dt'du'd\mu_{\not{g}})^{1/2}\end{aligned} \tag{16.14}$$

Now the last factor is bounded by (16.5) while the first factor is

$$(\int_0^t\|R_{i_{l-1-m}}\cdots R_{i_1}(T)^m\not{\Delta}\mu\|^2_{L^2(\Sigma^u_{t'})}dt')^{1/2} \tag{16.15}$$

which is bounded by

$$(\int_0^t(1+t')^{-4}(\mathcal{B}^u_{[m,l+1]}(t'))^2dt')^{1/2} \tag{16.16}$$

Here the quantity $\mathcal{B}^u_{[m,l+1]}(t)$ is defined by replacing the $L^2$ norms over $\Sigma^{\epsilon_0}_t$ in the definition of the quantity $\mathcal{B}_{[m,l+1]}(t)$ by $L^2$ norms over $\Sigma^u_t$. Proposition 12.12 applies with $\epsilon_0$ replaced by $u$, giving a bound for $\mathcal{B}^u_{[m,l+1]}(t)$ in terms of

$$C(1+t)\int_0^t(1+t')^{-1}(\mathcal{W}^u_{\{l+2\}}(t')+\mathcal{W}^{Qu}_{\{l+1\}}(t'))dt' \tag{16.17}$$

Now we have

$$\begin{aligned}&\mathcal{W}^u_{\{l+2\}}(t)+\mathcal{W}^{Qu}_{\{l+1\}}(t)\\&\leqslant C\bar{\mu}_{m,u}^{-1/2}(\sqrt{\mathcal{E}'^u_{1,[l+2]}(t)}+\sqrt{\mathcal{E}^u_{0,[l+2]}(t)})\\&\leqslant C\bar{\mu}_{m,u}^{-a-1/2}(t)(1+\log(1+t))^q(\sqrt{\mathcal{G}'^u_{1,[l+2];a,q}(t)}+\sqrt{\mathcal{G}^u_{0,[l+2];a,p}(t)})\end{aligned} \tag{16.18}$$

Therefore (16.17) is bounded by

$$\begin{aligned}&C(1+t)(\sqrt{\mathcal{G}'^u_{1,[l+2];a,q}(t)}+\sqrt{\mathcal{G}^u_{0,[l+2];a,p}(t)})\\&\cdot\int_0^t(1+t')^{-1}\bar{\mu}_{m,u}^{-a-1/2}(t')(1+\log(1+t'))^qdt'\\&\leqslant C(1+t)(\sqrt{\mathcal{G}'^u_{1,[l+2];a,q}(t)}+\sqrt{\mathcal{G}^u_{0,[l+2];a,p}(t)})J^u_{a,q-1}\\&\leqslant C(1+t)(1+\log(1+t))^{q+1}\bar{\mu}_{m,u}^{-a+1/2}(t)\sqrt{\mathcal{D}^u_{[l+2]}}\end{aligned} \tag{16.19}$$

where we have used (15.162), (15.163). It follows that the contribution of (16.17) to (16.16) is bounded by

$$C\sqrt{\mathcal{D}^u_{[l+2]}}\Big(\int_0^t(1+t')^{-2}(1+\log(1+t'))^{2q+2}\bar{\mu}_{m,u}^{-2a+1}(t')dt'\Big)^{1/2}$$
$$\leqslant\sqrt{\mathcal{D}^u_{[l+2]}}\bar{\mu}_{m,u}^{-a+1}(t) \tag{16.20}$$

In view of (16.12), (16.14) is bounded by

$$C\delta_0\bar{\mu}_{m,u}^{-2b}(t)(1+\log(1+t))^2\sqrt{\mathcal{D}^u_{[l+2]}}\sqrt{V'_{1,[l+1];b,2}(t,u)} \tag{16.21}$$

where $b$ is defined by (16.1).

We proceed to consider the error integrals associated to the variations of order $l+1$ and to $K_0$, involving the highest spatial derivatives of the acoustical entities. The leading contributions are:

1) The contribution of (14.56) with $l+1$ replaced by $l$ to the corresponding integral (14.62), namely the integral on the left in (14.63) with $l+1$ replaced by $l$ (and $\epsilon_0$ by $u$):

$$\int_{W^t_u}|R_{i_l}\cdots R_{i_1}\mathrm{tr}\chi'||T\psi_\alpha||\underline{L}R_{i_l}\cdots R_{i_1}\psi_\alpha|dt'du'd\mu_{\tilde{\not{g}}} \tag{16.22}$$

2) The contribution of (14.57) with $l$ replaced by $l-1$ to the corresponding integral (14.62), namely the integral (14.162) with $l$ replaced by $l-1$:

$$\int_{W^t_u}|R_{i_{l-1-m}}\cdots R_{i_1}(T)^m\not{\Delta}\mu||T\psi_\alpha||\underline{L}R_{i_{l-1-m}}\cdots R_{i_1}(T)^{m+1}\psi_\alpha|dt'du'd\mu_{\tilde{\not{g}}} \tag{16.23}$$

By (14.294), (16.22) is bounded by

$$C\delta_0\int_{W^t_u}(1+t')^{-1}|R_{i_l}\cdots R_{i_1}\mathrm{tr}\chi'||\underline{L}R_{i_l}\cdots R_{i_1}\psi_\alpha|dt'du'd\mu_{\not{g}}$$
$$\leqslant C\delta_0\int_0^t(1+t')^{-1}\|R_{i_l}\cdots R_{i_1}\mathrm{tr}\chi'\|_{L^2(\Sigma^u_{t'})}\|\underline{L}R_{i_l}\cdots R_{i_1}\psi_\alpha\|_{L^2(\Sigma^u_{t'})}dt' \tag{16.24}$$

Substituting (16.10) for the first factor in the integrand on the right and noting the definition (16.1) of $b$ and the fact that

$$\|\underline{L}R_{i_l}\cdots R_{i_1}\psi_\alpha\|_{L^2(\Sigma^u_t)}\leqslant C\sqrt{\mathcal{E}^u_{0,[l+1]}(t)}\leqslant C\bar{\mu}_{m,u}^{-b}(t)\sqrt{\mathcal{G}^u_{0,[l+1];b,0}(t)} \tag{16.25}$$

we obtain that (16.24) is bounded by

$$C\delta_0\sqrt{\mathcal{D}^u_{[l+2]}}\sqrt{\mathcal{G}^u_{0,[l+1];b,0}(t)}\int_0^t(1+t')^{-2}(1+\log(1+t'))^{q+1}\bar{\mu}_{m,u}^{-2b-1/2}(t')dt'$$
$$\leqslant C\delta_0\bar{\mu}_{m,u}^{-2b+1/2}(t)\sqrt{\mathcal{D}^u_{[l+2]}}\sqrt{\mathcal{G}^u_{0,[l+1];b,0}(t)} \tag{16.26}$$

the last integral being estimated in a similar manner as (14.115).

By (14.294), (15.23) is bounded by

$$\begin{aligned} & C\delta_0 \int_{W_u^t} (1+t')^{-1} |R_{i_{l-1-m}} \cdots R_{i_1}(T)^m \Delta\mu| |\underline{L} R_{i_{l-1-m}} \cdots R_{i_1}(T)^{m+1} \psi_\alpha| dt' du' d\mu_{\not g} \\ \leqslant & C\delta_0 \int_0^t (1+t')^{-1} \|R_{i_{l-1-m}} \cdots R_{i_1}(T)^m \Delta\mu\|_{L^2(\Sigma_{t'}^u)} \\ & \cdot \|\underline{L} R_{i_{l-1-m}} \cdots R_{i_1}(T)^{m+1} \psi_\alpha\|_{L^2(\Sigma_{t'}^u)} dt' \end{aligned} \tag{16.27}$$

Substituting (16.19) (multiplied by $C(1+t)^{-2}$) for the first factor in the integrand and noting again the definition (16.1) and the fact that

$$\|\underline{L} R_{i_{l-1-m}} \cdots R_{i_1}(T)^{m+1} \psi_\alpha\|_{L^2(\Sigma_{t'}^u)} \leqslant C\sqrt{\mathcal{E}_{0,[l+1]}^u(t)} \leqslant C\bar{\mu}_{m,u}^{-b}(t)\sqrt{\mathcal{G}_{0,[l+1];b,0}^u(t)} \tag{16.28}$$

we obtain that (16.27) is bounded by

$$\begin{aligned} & C\delta_0 \sqrt{\mathcal{D}_{[l+2]}^u} \sqrt{\mathcal{G}_{0,[l+1];b,0}^u(t)} \int_0^t (1+t')^{-2}(1+\log(1+t'))^{q+1} \bar{\mu}_{m,u}^{-2b-1/2}(t') dt' \\ \leqslant & C\delta_0 \bar{\mu}_{m,u}^{-2b+1/2}(t) \sqrt{\mathcal{D}_{[l+2]}^u} \sqrt{\mathcal{G}_{0,[l+1];b,0}^u(t)} \end{aligned} \tag{16.29}$$

a bound of the same form as (16.26).

Consider now the error integrals associated to the variations of order $l+1$ and to $K_1$ or $K_0$, which contain the lower order spatial derivatives of the acoustical entities and are contributed by the remaining terms in the sum (14.14) with $l+2$ replaced by $l+1$. By the discussion following (15.9), those contributed by the terms in which one factor is a derivative of the ${}^{(Y)}\tilde{\pi}$ of order more than $(l+1)_*$ are bounded, using Propositions 12.11 and 12.12 and the bootstrap assumption, by (15.11), hence by (15.39), with $([l+2];a,q)$ replaced by $([l+1];b,2)$ (and $\epsilon_0$ by $u$), thus by

$$\begin{aligned} & C_l \delta_0 \bar{\mu}_{m,u}^{-2b}(t)(1+\log(1+t))^4 (\mathcal{D}_{[l+2]}^u + \delta(\mathcal{G}_{1,[l+1];b,2}^{\prime u}(t) + \mathcal{G}_{0,[l+1];b,0}^u(t)) \\ & + \delta I_{[l+1];b,2}(t,u) + (1+\frac{1}{\delta}) V_{1,[l+1];b,2}^{\prime}(t,u)) \end{aligned} \tag{16.30}$$

in the case of the error integrals associated to $K_1$, by (15.99), hence by (15.112), with $([l+2];a,p)$ replaced by $([l+1];b,0)$ (and $\epsilon_0$ by $u$), thus by

$$\begin{aligned} & C_l \delta_0 \bar{\mu}_{m,u}^{-2b}(t)(\mathcal{D}_{[l+2]}^u + \delta'(\mathcal{G}_{1,[l+1];b,2}^{\prime u}(t) + \mathcal{G}_{0,[l+1];b,0}^u(t)) \\ & + (1+\frac{1}{\delta'})(\int_0^t (1+t')^{-3/2} \mathcal{G}_{0,[l+1];b,0}^u(t') dt' + V_{0,[l+1];b,0}(t,u)) \\ & + \delta'(V_{1,[l+1];b,2}^{\prime}(t,u) + I_{[l+1];b,2}(t,u))) \end{aligned} \tag{16.31}$$

in the case of the error integrals associated to $K_0$. In (16.30), (16.31), the arbitrary positive constants $\delta$, $\delta'$ shall be chosen appropriately below. In fact, (16.30) and

(16.31) also bound the error integrals associated to $K_1$ and to $K_0$ respectively, and to any of the variations of order up to $l$ in which one factor is a derivative of ${}^{(Y)}\tilde{\pi}$ of order more than $(l+1)_*$.

On the other hand, the error integrals associated to any of the variations of order up to $l+1$ contributed by the terms in the sum (14.14) containing derivatives of the ${}^{(Y)}\tilde{\pi}$ of order at most $(l+1)_*$ are bounded by (15.13) with $[l+2]$ replaced by $[l+1]$ (and $\epsilon_0$ by $u$):

$$\begin{aligned}
&C\int_0^u \mathcal{F}'^t_{1,[l+1]}(u')du' + C(\int_0^u \mathcal{F}'^t_{1,[l+1]}(u')du')^{1/2}(K_{[l+1]}(t,u))^{1/2} \\
&+ C(\int_0^u \mathcal{F}'^t_{1,[l+1]}(u')du')^{1/2} \\
&\cdot(\int_0^t (1+t')^{-2}(1+\log(1+t'))^2(\mathcal{E}'^u_{1,[l+1]}(t') + u^2\mathcal{E}^u_{0,[l+1]}(t'))dt')^{1/2} \\
\leqslant\; &C\bar{\mu}^{-2b}_{m,u}(t)(1+\log(1+t))^4(V'_{1,[l+1];b,2} \qquad\qquad (16.32)\\
&+ (V'_{1,[l+1];b,2}(t,u))^{1/2}(I_{[l+1];b,2}(t,u))^{1/2} + (V'_{1,[l+1];b,2}(t,u))^{1/2} \\
&\cdot(\int_0^t (1+t')^{-2}(1+\log(1+t'))^2(\mathcal{G}'^u_{1,[l+1];b,2}(t') + \mathcal{G}^u_{0,[l+1];b,0}(t'))dt')^{1/2})
\end{aligned}$$

in the case of $K_1$, and by (15.103) with $[l+2]$ replaced by $[l+1]$ (and $\epsilon_0$ by $u$):

$$\begin{aligned}
&C\int_0^t (1+t')^{-3/2}\mathcal{E}^u_{0,[l+1]}(t')dt' + C\int_0^u \mathcal{F}^t_{0,[l+1]}(u')du' \\
&+ C(\int_0^u \bar{\mathcal{F}}'^t_{1,[l+1]}(u')du')^{1/2}(\int_0^t (1+t')^{-3/2}\mathcal{E}^u_{0,[l+1]}(t')dt' + \int_0^u \mathcal{F}^t_{0,[l+1]}(u')du')^{1/2} \\
&+ C(\bar{K}_{[l+1]}(t,u))^{1/2}(\int_0^t (1+t')^{-3/2}\mathcal{E}^u_{0,[l+1]}(t')dt' + \int_0^u \mathcal{F}^t_{0,[l+1]}(u')du')^{1/2} \\
\leqslant\; &C\bar{\mu}^{-2b}_{m,u}(t)(\int_0^t (1+t')^{-3/2}\mathcal{G}^u_{0,[l+1];b,0}(t')dt' + V_{0,[l+1];b,0}(t,u) \qquad\qquad (16.33)\\
&+ (V'_{1,[l+1];b,2}(t,u) + I_{[l+1];b,2}(t,u))^{1/2} \\
&\cdot(\int_0^t (1+t')^{-3/2}\mathcal{G}^u_{0,[l+1];b,0}(t')dt' + V_{0,[l+1];b,0}(t,u))^{1/2})
\end{aligned}$$

in the case of $K_0$.

Finally, for each variation $\psi$ of order up to $l+1$, we have the error integrals (15.16). Consider first the second of these integrals, which is associated to $K_1$. Summing (15.17) over all such variations we obtain

$$\begin{aligned}
\sum_\psi \int_{W^t_u} \sum_{k=1}^{8} Q_{1,k}[\psi]d\mu_g \leqslant -\frac{1}{2}K_{[l+1]}(t,u) + CM_{[l+1]}(t,u) + \frac{3}{2}L_{[l+1]}(t,u) \qquad (16.34)\\
+ \int_0^t \tilde{A}(t')\mathcal{E}'^u_{1,[l+1]}(t')dt'
\end{aligned}$$

Here,

$$M_{[l+1]}(t,u) = \sum_{\psi} M[\psi](t,u), \quad L_{[l+1]}(t,u) = \sum_{\psi} L[\psi](t,u) \tag{16.35}$$

From (15.18) and (15.29) with $(a,p)$ replaced by $(b,0)$, noting the fact that there is a fixed number of variations under consideration, we obtain

$$\begin{aligned} M_{[l+1]}(t,u) &\leqslant C(1+\log(1+t))^4 \sup_{t'\in[0,t]} \mathcal{E}^u_{0,[l+1]}(t') + \int_0^u \mathcal{F}'^t_{1,[l+1]}(u')du' \\ &\leqslant \bar{\mu}^{-2b}_{m,u}(t)(1+\log(1+t))^4(C\mathcal{G}^u_{0;b,0}(t) + V'_{1,[l+1];b,2}(t,u)) \end{aligned} \tag{16.36}$$

Also, from (15.1),

$$L_{[l+1]}(t,u) = \int_0^t (1+t')^{-1}(1+\log(1+t'))^{-1}\mathcal{E}'^u_{1,[l+1]}(t')dt' \tag{16.37}$$

To estimate this borderline integral appropriately, we must use the following variant of Corollary 2 of Lemma 8.11:

**Variant of Corollary 2** Let $b$ be a positive constant and $k$ any constant greater than 1. Then if $\delta_0$ is suitably small, depending on an upper bound for $b$ and a lower bound greater than 1 for $k$, we have, for all $t' \in [0,t]$, $t \in [1,s]$, and $u \in (0,\epsilon_0]$,

$$\bar{\mu}^{-b}_{m,u}(t') \leqslant k\bar{\mu}^{-b}_{m,u}(t)$$

*Proof.* To prove this, we revisit the proof of Corollary 2. In the following, we shall take the constant $a$ in the proof of that corollary to be a suitably large constant, having nothing to do with the choice of $a$ which we made above in relation to the top order energy estimates. In Case 1 of this proof (8.322) holds, hence

$$\frac{\bar{\mu}^{-b}_{m,u}(t')}{\bar{\mu}^{-b}_{m,u}(t)} \leqslant (1-C\delta_0)^{-b} \leqslant k \tag{16.38}$$

provided that $\delta_0$ is suitably small, depending on an upper bound for $b$ and a lower bound greater than 1 for $k$. In Subcase 2a the lower bound (8.273) holds (with $\epsilon_0$ replaced by $u$) for any given $a$, provided that $a\delta_0$ is suitably small, hence

$$\frac{\bar{\mu}^{-b}_{m,u}(t')}{\bar{\mu}^{-b}_{m,u}(t)} \leqslant (1-\frac{2}{a})^{-b} \leqslant k \tag{16.39}$$

if $a$ is suitably large, depending on an upper bound for $b$ and a lower bound greater than 1 for $k$. In Subcase 2b the lower bound (8.303) and the upper bound (8.312) hold (with $\epsilon_0$ replaced by $u$) for any given $a$, provided that $a\delta_0$ is suitably small, hence

$$\frac{\bar{\mu}^{-b}_{m,u}(t')}{\bar{\mu}^{-b}_{m,u}(t)} \leqslant (\frac{1-\frac{2}{a}}{1+\frac{2}{a}})^{-b}(\frac{1-\delta_1\tau}{1-\delta_1\tau'})^b \leqslant (\frac{1-\frac{2}{a}}{1+\frac{2}{a}})^{-b} \leqslant k \tag{16.40}$$

if $a$ is suitably large, depending on an upper bound for $b$ and a lower bound greater than 1 for $k$. We have thus established the variant of Corollary 2. □

We now set

$$k = \sqrt{\frac{4}{3}} \tag{16.41}$$

in the above variant of Corollary 2, and we proceed to estimate $L_{[l+1]}(t,u)$. We have, from (16.38) and the definition (14.111) with $(a,q)$ replaced by $(b,2)$,

$$\begin{aligned} L_{[l+1]}(t,u) &\leqslant \int_0^t (1+t')^{-1}(1+\log(1+t'))^3 \bar{\mu}_{m,u}^{-2b} \mathcal{G}'^u_{1,[l+1];b,2}(t')dt' \\ &\leqslant \frac{4}{3}\bar{\mu}_{m,u}^{-2b}(t)\mathcal{G}'^u_{1,[l+1];b,2}(t)\int_0^t (1+t')^{-1}(1+\log(1+t'))^3 dt' \\ &\leqslant \frac{1}{3}\bar{\mu}_{m,u}^{-2b}(t)(1+\log(1+t))^4 \mathcal{G}'^u_{1,[l+1];b,2}(t) \end{aligned} \tag{16.42}$$

Consider now the integral identity (5.73) corresponding to $K_1$ and to the variations $\psi$ of order up to $l+1$. In each of these identities we also have the hypersurface integrals bounded according to (15.35) with $(a,p)$ replaced by $(b,0)$ (and $\epsilon_0$ by $u$) by

$$C\bar{\mu}_{m,u}^{-2b}(t)(1+\log(1+t))^4 \mathcal{G}^u_{0,[l+1];b,0}(t) \tag{16.43}$$

Summing over all such variations we then obtain, from (16.42), and from (16.13), (16.21), (16.30), (16.32), (16.34), (16.36), the following:

$$\begin{aligned} &(\mathcal{E}'^u_{1,[l+1]}(t) + \mathcal{F}'^t_{1,[l+1]}(u) + \frac{1}{2}K_{[l+1]}(t,u))\bar{\mu}_{m,u}^{2b}(t)(1+\log(1+t))^{-4} \\ &\leqslant \frac{1}{2}\mathcal{G}'^u_{1,[l+1];b,2}(t) + C\delta_0\delta\mathcal{G}'^u_{1,[l+1];b,2}(t) \\ &\quad + CR'_{[l+1];b,2}(t,u) + C\int_0^t \bar{A}(t')\mathcal{G}'^u_{1,[l+1];b,2}(t')dt' \end{aligned} \tag{16.44}$$

Here, $\delta$ is any positive constant ($\delta$ shall be chosen below), and

$$R'_{[l+1];b,2} = \mathcal{D}^u_{[l+2]} + \mathcal{G}^u_{0,[l+1];b,0}(t) + (1+\frac{1}{\delta})V'_{1,[l+1];b,2}(t,u) + \delta I_{[l+1];b,2}(t,u) \tag{16.45}$$

Keeping only the term $\mathcal{E}'^u_{1,[l+1]}(t)$ in parenthesis on the left in (16.44) we have

$$\begin{aligned} &\bar{\mu}_{m,u}^{2b}(t)(1+\log(1+t))^{-4}\mathcal{E}'^u_{1,[l+1]}(t) \\ &\leqslant \frac{1}{2}\mathcal{G}'^u_{1,[l+1];b,2}(t) + C\delta_0\delta\mathcal{G}'^u_{1,[l+1];b,2}(t) \\ &\quad + CR'_{[l+1];b,2}(t,u) + C\int_0^t \bar{A}(t')\mathcal{G}'^u_{1,[l+1];b,2}(t')dt' \end{aligned} \tag{16.46}$$

Since the right-hand side of the above is non-decreasing in $t$, applying the previous reasoning we deduce

$$\frac{1}{3}\mathcal{G}'^{u}_{1,[l+1];b,2}(t) \leqslant CR'_{[l+1];b,2}(t,u) + C\int_0^t \bar{A}(t')\mathcal{G}'^{u}_{1,[l+1];b,2}(t')dt' \tag{16.47}$$

provided that

$$C\delta_0\delta \leqslant \frac{1}{6} \tag{16.48}$$

From this point we proceed as in the argument leading from (15.71) to (15.73), (15.74), to obtain

$$\mathcal{G}'^{u}_{1,[l+1];b,2}(t) \leqslant CR'_{[l+1];b,2}(t,u) \tag{16.49}$$

$$\int_0^t \bar{A}(t')\mathcal{G}'^{u}_{1,[l+1];b,2}(t')dt' \leqslant CR'_{[l+1];b,2}(t,u) \tag{16.50}$$

Substituting these in the right-hand side of (16.44) and omitting the first term in parenthesis on the left, we obtain

$$(\mathcal{F}'^{t}_{1,[l+1]}(u) + \frac{1}{2}K_{[l+1]}(t,u))\bar{\mu}^{2b}_{m,u}(t)(1+\log(1+t))^{-4} \leqslant CR'_{[l+1];b,2}(t,u) \tag{16.51}$$

From this point we proceed as in the argument leading from (15.78) to (15.91)—(15.94), choosing $\delta$ according to

$$C\delta = \frac{1}{4} \tag{16.52}$$

We obtain in this way the estimates

$$\mathcal{G}'^{u}_{1,[l+1];b,2}(t),\quad \mathcal{H}'^{t}_{1,[l+1];b,2}(u),\quad I_{[l+1];b,2}(t,u) \leqslant C\bar{Q}'_{[l+1];b,2}(t,u) \tag{16.53}$$

$$V'_{1,[l+1];b,2}(t,u) \leqslant C\epsilon_0\bar{Q}'_{[l+1];b,2}(t,u) \tag{16.54}$$

where

$$\bar{Q}'_{[l+1];b,2}(t,u) = \mathcal{D}^{u}_{[l+2]} + \bar{\mathcal{G}}^{u}_{0,[l+1];b,0}(t) \tag{16.55}$$

Consider finally the integral identity corresponding to $K_0$ and to the variations $\psi$ of order up to $l+1$. In each of these identities we have the first of (15.16). Summing (15.127) over all such variations we obtain (there is a fixed number of variations under consideration)

$$\begin{aligned}
&\bar{\mu}^{2b}_{m,u}(t)\sum_{\psi}\int_{W^t_u}\sum_{k=1}^{7}Q_0[\psi]d\mu_g \\
&\leqslant C\int_0^t \tilde{B}_s(t')\mathcal{G}'^{u}_{1,[l+1];b,2}(t')dt' \\
&\quad + (1+\frac{1}{\delta'})(\int_0^t (1+t')^{-1}(1+\log(1+t'))^{-2}\mathcal{G}^{u}_{0,[l+1];b,0}(t')dt' + V_{0,[l+1];b,0}(t,u)) \\
&\quad + C\delta' V'_{1,[l+1];b,2}(t,u) + C\int_0^t (1+t')^{-2}(1+\log(1+t'))^{-2}V'_{1,[l+1];b,2}(t',u)dt'
\end{aligned} \tag{16.56}$$

$$+ C\delta' I_{[l+1];b,2}(t,u)$$

Here, as in (16.31), $\delta'$ is an arbitrary positive constant which shall be chosen appropriately below. Summing over all such variations we then obtain, from (16.26), (16.29), (16.31), (16.33), (16.56), the following:

$$\begin{aligned}
&(\mathcal{E}^u_{0,[l+1]}(t) + \mathcal{F}^t_{0,[l+1]}(t))\bar{\mu}^{2b}_{m,u}(t) \\
&\leqslant C\delta_0\delta'(\mathcal{G}^u_{0,[l+1];b,0}(t) + \mathcal{G}'^u_{1,[l+1];b,2}(t)) + C(1+\frac{\delta_0}{\delta'})\mathcal{D}^u_{[l+2]} \\
&\quad + C\int_0^t \tilde{B}_s(t')\mathcal{G}'^u_{1,[l+1];b,2}(t')dt' \\
&\quad + C(1+\frac{1}{\delta'})\int_0^t (1+t')^{-1}(1+\log(1+t'))^{-2}\mathcal{G}^u_{0,[l+1];b,0}(t')dt' \\
&\quad + C\delta' V'_{1,[l+1];b,2}(t,u) + C\int_0^t (1+t')^{-2}(1+\log(1+t'))^2 V'_{1,[l+1];b,2}(t',u)dt' \\
&\quad + C(1+\frac{1}{\delta'})V_{0,[l+1];b,0}(t,u) + C\delta' I_{[l+1];b,2}(t,u)
\end{aligned} \tag{16.57}$$

Substituting on the right (16.53)—(16.55) yields

$$\begin{aligned}
&(\mathcal{E}^u_{0,[l+1]}(t) + \mathcal{F}^t_{0,[l+1]}(u))\bar{\mu}^{2b}_{m,u}(t) \\
&\leqslant C\delta'\bar{\mathcal{G}}^u_{0,[l+1];b,0}(t) + C(1+\frac{\delta_0}{\delta'})\mathcal{D}^u_{[l+2]} \\
&\quad + C(1+\frac{1}{\delta'})\int_0^t \tilde{B}_s(t')\bar{\mathcal{G}}^u_{0,[l+1];b,0}(t')dt' + C(1+\frac{1}{\delta'})V_{0,[l+1];b,0}(t,u)
\end{aligned} \tag{16.58}$$

Keeping only the term $\mathcal{E}^u_{0,[l+1]}(t)$ in parenthesis on the left in (16.58), and noting that the right-hand side is a non-decreasing function of $t$ at each $u$, we may replace $t$ by any $t' \in [0,t]$ on the left while keeping $t$ on the right, thus, taking the supremum over all $t' \in [0,t]$ we obtain, in view of the definition (15.50),

$$\begin{aligned}
&\mathcal{G}^u_{0,[l+1];b,0}(t) \\
&\leqslant C\delta'\bar{\mathcal{G}}^u_{0,[l+1];b,0}(t) + C(1+\frac{\delta_0}{\delta'})\mathcal{D}^u_{[l+2]} \\
&\quad + C(1+\frac{1}{\delta'})\int_0^t \tilde{B}_s(t')\bar{\mathcal{G}}^u_{0,[l+1];b,0}(t')dt' + C(1+\frac{1}{\delta'})V_{0,[l+1];b,0}(t,u)
\end{aligned} \tag{16.59}$$

We now choose $\delta'$ such that

$$C\delta' = \frac{1}{2} \tag{16.60}$$

Then in view of the fact that the right-hand side of (16.59) is non-decreasing in $u$, (16.59) implies

$$\frac{1}{2}\bar{\mathcal{G}}^u_{0,[l+1];b,0}(t) \leqslant C\mathcal{D}^u_{[l+2]} + C\int_0^t \tilde{B}_s(t')\bar{\mathcal{G}}^u_{0,[l+1];b,0}(t')dt' + CV_{0,[l+1];b,0}(t,u) \tag{16.61}$$

for new constants $C$. From this point we proceed as in the argument leading from (15.151) to (15.160)—(15.162) to obtain the estimates

$$\bar{\mathcal{G}}^u_{0,[l+1];b,0}(t), \quad \mathcal{H}^t_{0,[l+1];b,0}(u) \leqslant C\mathcal{D}^u_{[l+2]} \tag{16.62}$$

Substituting finally (16.62) in (16.55) and the result in (16.53) yields

$$\mathcal{G}'^u_{1,[l+1];b,2}(t), \quad \mathcal{H}'^t_{1,[l+1];b,2}(u), \quad I_{[l+1];b,2}(t,u) \leqslant C\mathcal{D}^u_{[l+2]} \tag{16.63}$$

We have thus obtained improved energy estimates of the next to the top order, namely of order $l+1$.

We proceed in this way, taking at the $n$th step

$$a_n = a - n \quad \text{and} \quad b_n = a_n - 1 = a - (n+1) = a_{n+1} \tag{16.64}$$

in the role of $a$ and $b$ respectively, the argument beginning in the paragraph containing (16.1) and concluding with (16.62), (16.63), being step 0. For $n \geqslant 1$ we have 2 and 0 in the role of $q$ and $p$ respectively. The $n$th step is otherwise exactly the same as the 0th step as above, as long as $b_n > 0$, that is as long as $n \leqslant [a] - 1$. In estimating the integrals $J^u_{a_n,1}$ in (16.10), (16.19), as well as the integrals

$$\int_0^t (1+t')^{-2}(1+\log(1+t'))^6 \bar{\mu}_{m,u}^{-2a_n+1}(t')dt' \tag{16.65}$$

for $n \geqslant 1$, we follow the argument leading from (14.147) to (14.153) in the case of $J^u_{a_n,1}$, the argument leading from (14.115) to (14.131) in the case of (16.65). When referring to Lemma 8.11 in these arguments, we now mean Lemma 8.11 with $a$ set equal to 4 in that lemma. Then in Case 1, (14.116) holds which implies

$$\bar{\mu}_{m,u}^{-2a_n-1}(t) \leqslant C \tag{16.66}$$

for all $t \in [0,s]$, as the $a_n$ have a fixed upper bound $a$. Also, the subcases of Case 2 are defined according to

$$t_1 = e^{\frac{1}{8\delta_1}} - 1 \tag{16.67}$$

and in Subcase 2a we have the lower bound

$$\bar{\mu}_{m,u}(t') \geqslant \frac{1}{2} \tag{16.68}$$

in place of (14.121), hence again

$$\bar{\mu}_{m,u}^{-2a_n-1}(t') \leqslant C \tag{16.69}$$

while in Subcase 2b we have the lower bound

$$\bar{\mu}_{m,u} \geqslant \frac{1}{2}(1-\delta_1\tau'), \quad \tau' = \log(1+t') \tag{16.70}$$

in place of (14.124), as well the upper bound (8.312)

$$\bar{\mu}_m(t) \leqslant \frac{7}{4}(1-\delta_1\tau) \tag{16.71}$$

from the proof of Lemma 8.11 with $a$ set to equal to 4, which implies

$$\bar{\mu}_{m,u}^{-2a_n} \geqslant \frac{1}{C}(1-\delta_1\tau)^{-2a_n} \quad \tau = \log(1+t) \tag{16.72}$$

Noting again that the $a_n$ have a fixed upper bound. We thus obtain in Subcase 2b, in place of (14.151),

$$\begin{aligned} J^u_{a_n,1}(t) - J^u_{a_n,1}(t_1) &\leqslant C(1+\tau)^2 \int_{\tau_1}^{\tau} (1-\delta_1\tau')^{-a_n-1/2} d\tau' \\ &\leqslant \frac{C}{\delta_1} \frac{(1+\tau)^2}{a_n - \frac{1}{2}} (1-\delta_1\tau)^{-a_n+1/2} \\ &\leqslant C \frac{(1+\log(1+t))^3}{a_n - \frac{1}{2}} \bar{\mu}_{m,u}^{-a_n+1/2} \end{aligned} \tag{16.73}$$

noting that the role of $a_n$ is different from that of $a$.

Since for any $n \leqslant [a]$,

$$a_n - \frac{1}{2} \geqslant a_{[a]} - \frac{1}{2} = \frac{3}{4} - \frac{1}{2} = \frac{1}{4} \tag{16.74}$$

the denominator in (16.74) is bounded from below by a positive constant even for $n = [a]$, hence combining with the results for Subcase 2a and for Case 1 we conclude that

$$J^u_{a_n,1}(t) \leqslant C(1+\log(1+t))^3 \bar{\mu}_{m,u}^{-a_n+1/2} \tag{16.75}$$

for all $n = 1, \cdots, [a]$

Also, for $n \leqslant [a]-1$, so that $a_n \geqslant 7/4$, we have in Subcase 2b, in regard to the integral (16.65), in place of (14.126),

$$\begin{aligned} &\int_{t_1}^{t} (1+t')^{-2}(1+\log(1+t'))^6 (1-\delta_1\tau')^{-2a_n+1} dt' \\ &\leqslant C(1+t_1)^{-1}(1+\log(1+t_1))^6 \int_{\tau_1}^{\tau} (1-\delta_1\tau')^{-2a_n+1} d\tau' \\ &\leqslant C(1+t_1)^{-1} \frac{(1+\log(1+t_1))^6}{\delta_1} \frac{(1-\delta_1\tau)^{-2a_n+2}}{2(a_n-1)} \\ &\leqslant C(1+t_1)^{-1}(1+\log(1+t_1))^7 \frac{(1-\delta_1\tau)^{-2(a_n-1)}}{2(a_n-1)} \\ &\leqslant C\bar{\mu}_{m,u}^{-2(a_n-1)} \end{aligned} \tag{16.76}$$

hence combining with the results for Subcase 2a and for Case 1 (where (16.65) is simply bounded by a constant) we conclude that (16.65) is bounded by

$$C\bar{\mu}_{m,u}^{-2(a_n-1)} \tag{16.77}$$

for all $n = 1, \cdots, [a] - 1$.

Obviously, in (16.26) and (16.29) we may replace the integrals on the left by the integral (16.65) and we obtain the bounds at the $n$th step by

$$C\delta_0 \bar{\mu}_{m,u}^{-2b_n}(t)\sqrt{\mathcal{D}^u_{[l+2]}}\sqrt{\mathcal{G}^u_{0,[l+1-n];b_n,0}(t)} \tag{16.78}$$

Following then the argument of step 0, we obtain at the $n$th step, for $n = 1, \cdots, [a] - 1$, the estimates

$$\begin{aligned} \bar{\mathcal{G}}^u_{0,[l+1-n];b_n,0}(t), \quad \mathcal{H}^t_{0,[l+1-n];b_n,0}(u) &\leqslant C\mathcal{D}^u_{[l+2]} \\ \mathcal{G}'^u_{1,[l+1-n];b_n,2}(t), \quad \mathcal{H}'^t_{1,[l+1-n];b_n,2}(u), \quad I_{[l+1-n];b_n,2}(t,u) &\leqslant C\mathcal{D}^u_{[l+2]} \end{aligned} \tag{16.79}$$

for all $(t, u) \in [0, s] \times [0, \epsilon_0]$. Here the constants $C$ may depend on $n$, but $n$ is in any case bounded.

We now make the last step $n = [a]$. In this case, we have $a_n = a_{[a]} = \frac{3}{4}$ and (16.75) reads

$$J^u_{3/4,1}(t) \leqslant C(1 + \log(1 + t))^3 \bar{\mu}_{m,u}^{-1/4} \tag{16.80}$$

while in regard to (16.65), we now have, from (16.76), in Subcase 2b, since $-2a_{[a]} + 1 = -1/2$,

$$\begin{aligned} &\int_{t_1}^{t} (1 + t')^{-2}(1 + \log(1 + t'))^6 (1 - \delta_1 \tau')^{-1/2} dt' \\ &\leqslant C(1 + t_1)^{-1}(1 + \log(1 + t_1))^6 \int_{\tau_1}^{\tau} (1 - \delta_1 \tau')^{-1/2} d\tau' \\ &\leqslant C(1 + t_1)^{-1} \frac{(1 + \log(1 + t_1))^6}{\delta_1} \frac{(1 - \delta_1 \tau_1)^{1/2}}{1/2} \\ &\leqslant C(1 + t_1)^{-1}(1 + \log(1 + t_1))^7 \frac{(7/8)^{1/2}}{1/2} \leqslant C \end{aligned} \tag{16.81}$$

hence combining with the results for Subcase 2a and for Case 1 (where (16.65) is in any case bounded by a constant) we conclude that (16.65) is in the case $n = [a]$ simply bounded by a constant. We can thus set

$$b_n = b_{[a]} = 0 \tag{16.82}$$

in the last step, and proceed exactly as in the preceding steps. We thus arrive at the estimates

$$\begin{aligned} \bar{\mathcal{G}}^u_{0,[l+1-[a]];0,0}(t), \quad \mathcal{H}^t_{0,[l+1-[a]];0,0}(u) &\leqslant C\mathcal{D}^u_{[l+2]} \\ \mathcal{G}'^u_{1,[l+1-[a]];0,2}(t), \quad \mathcal{H}'^t_{1,[l+1-[a]];0,2}(u), \quad I_{[l+1-[a]];0,2}(t,u) &\leqslant C\mathcal{D}^u_{[l+2]} \end{aligned} \tag{16.83}$$

for all $(t, u) \in [0, s] \times [0, \epsilon_0]$.

These are the desired estimates. Because from the definitions,

$$\bar{\mathcal{G}}^u_{0,[l+1-[a]];0,0}(t) = \sup_{t'\in[0,t]} (\mathcal{E}^u_{0,[l+1-[a]]}(t')) \tag{16.84}$$

$$\mathcal{H}^t_{0,[l+1-[a]];0,0}(u) = \sup_{t'\in[0,t]} (\mathcal{F}^{t'}_{0,[l+1-[a]]}(u)) \tag{16.85}$$

$$\mathcal{G}'^u_{1,[l+1-[a]];0,2}(t) = \sup_{t'\in[0,t]} ((1+\log(1+t'))^{-4}\mathcal{E}'^u_{1,[l+1-[a]]}(t')) \tag{16.86}$$

$$\mathcal{H}'^t_{1,[l+1-[a]];0,2}(u) = \sup_{t'\in[0,t]} ((1+\log(1+t))^{-4}\mathcal{F}'^{t'}_{1,[l+1-[a]]}(u)) \tag{16.87}$$

$$I_{[l+1-[a]];0,2}(t,u) = \sup_{t'\in[0,t]} ((1+\log(1+t))^{-4}K_{[l+1-[a]]}(t',u)) \tag{16.88}$$

and the weights $\bar{\mu}_{m,u}$ have been eliminated.

# 17

# The Isoperimetric Inequality. Recovery of Assumption J. Recovery of the Bootstrap Assumption. Proof of the Main Theorem

## 17.1 Recovery of J—Preliminary

We now establish assumption **J** on the basis of the bootstrap assumption. We consider the first order variations

$$ {}^{(\alpha)}\tilde{\psi}_1 = \begin{cases} S\phi, & \alpha = 0, \\ \mathring{R}_i\phi, & \alpha = i = 1,2,3 \end{cases} \tag{17.1}$$

To these Theorem 5.1 applies and we obtain, in particular,

$$ \mathcal{E}_0^u[{}^{(\alpha)}\tilde{\psi}_1](t) \leqslant C\mathcal{E}_0^u[{}^{(\alpha)}\tilde{\psi}_1](0) \tag{17.2}$$

for all $(t, u) \in [0, s] \times [0, \epsilon_0]$.

We also consider the higher order variations, corresponding to the first order variations ${}^{(\alpha)}\tilde{\psi}_1$,

$$ {}^{(\alpha;I_1\cdots I_{n-1})}\tilde{\psi}_n = Y_{I_{n-1}}\cdots Y_{I_1}\,{}^{(\alpha)}\tilde{\psi}_1 \tag{17.3}$$

We require the multi-indices $(I_1 \cdots I_{n-1})$ to be of the form specified in the paragraph following (14.4) and concluding with (14.5). We consider all such variations of order up to 4. We denote by $\tilde{\mathcal{E}}_{0,[4]}^u(t)$ the sum of energies corresponding to $K_0$, $\Sigma_t^u$, and to all such variations of order up to 4. Now the integral identities corresponding to the variations in question and to $K_0$ and $K_1$ contain derivatives of ${}^{(Y)}\tilde{\pi}$ of order at most 3, which, in view of (15.167), are bounded in $L^\infty(\Sigma_t^{\epsilon_0})$ by Propositions 12.9 and 12.10 and the bootstrap assumption. The error integrals involved are then bounded exactly as in Chapter 7 and in accordance with the remark following Lemma 7.6 can be all absorbed in the error integrals of the fundamental energy estimate. Consequently, the conclusions of Theorem 5.1 hold for the variations in question and we obtain, in particular,

$$ \tilde{\mathcal{E}}_{0,[4]}^u(t) \leqslant C\tilde{D}_{[4]}^u \tag{17.4}$$

for all $(t,u) \in [0,s] \times [0,\epsilon_0]$. Here we denote

$$\tilde{D}^u_{[4]} = \tilde{\mathcal{E}}^u_{0,[4]}(0) \tag{17.5}$$

Applying Lemma 5.1 we then conclude that

$$\tilde{\mathcal{S}}_{[4]}(t,u) \leqslant C\epsilon_0 \tilde{D}^u_{[4]} \tag{17.6}$$

for all $(t,u) \in [0,s] \times [0,\epsilon_0]$. Here we denote by $\tilde{\mathcal{S}}_{[4]}$ the integral on $S_{t,u}$ (with respect to $d\mu_{\not g}$) of the sum of the squares of all the variations under consideration. In particular, we have

$$\int_{S_{t,u}} (|S\phi|^2 + \sum_{j_1} |R_{j_1} S\phi|^2 + \sum_{j_1,j_2} |R_{j_2} R_{j_1} S\phi|^2) d\mu_{\not g} \leqslant \tilde{\mathcal{S}}_{[4]}(t,u) \tag{17.7}$$

$$\int_{S_{t,u}} (|\mathring{R}_i\phi|^2 + \sum_{j_1} |R_{j_1} \mathring{R}_i\phi|^2 + \sum_{j_1,j_2} |R_{j_2} R_{j_1} \mathring{R}_i\phi|^2) d\mu_{\not g} \leqslant \tilde{\mathcal{S}}_{[4]}(t,u), \quad i = 1,2,3 \tag{17.8}$$

$$\int_{S_{t,u}} (|TS\phi|^2 + \sum_{j_1} |R_{j_1} TS\phi|^2 + \sum_{j_1,j_2} |R_{j_2} R_{j_1} TS\phi|^2) d\mu_{\not g} \leqslant \tilde{\mathcal{S}}_{[4]}(t,u) \tag{17.9}$$

$$\int_{S_{t,u}} (|T\mathring{R}_i\phi|^2 + \sum_{j_1} |R_{j_1} T\mathring{R}_i\phi|^2 + \sum_{j_1,j_2} |R_{j_2} R_{j_1} T\mathring{R}_i\phi|^2) d\mu_{\not g} \leqslant \tilde{\mathcal{S}}_{[4]}(t,u), \quad i = 1,2,3 \tag{17.10}$$

We are now in a position to apply the following lemma.

**Lemma 17.1** Let $f$ be a function on $S_{t,u}$ with square-integrable derivatives up to 2nd order. We denote

$$\mathcal{S}_{[2]}[f] = \int_{S_{t,u}} (|f|^2 + \sum_{j_1} |R_{j_1} f|^2 + \sum_{j_1,j_2} |R_{j_2} R_{j_1} f|^2) d\mu_{\not g}$$

Then $f \in L^\infty(S_{t,u})$ and there is a positive numerical constant $C$ such that

$$\sup_{S_{t,u}} |f| \leqslant C(1+t)^{-1} (\mathcal{S}_{[2]}[f])^{1/2}$$

*Proof*. By (8.332) we have

$$C^{-1}(1+t)^2 \leqslant A(t,u) \leqslant C(1+t)^2 \tag{17.11}$$

where $A(t,u)$ is the area of $S_{t,u}$ (by (8.332)). By Corollary 12.2.a we have

$$\int_{S_{t,u}} (|f|^2 + \sum_i |R_i f|^2) d\mu_{\not g} + A(t,u) \int_{S_{t,u}} (|\not d f|^2 + \sum_i |\not d R_i f|^2) d\mu_{\not g} \leqslant C\mathcal{S}_{[2]}[f] \tag{17.12}$$

To continue with the proof of the lemma, we need the *isoperimetric Sobolev inequality*.

## 17.2 The Isoperimetric Inequality

(See [15], [2]) If $g$ is an arbitrary function which is integrable and with integrable derivative on $S_{t,u}$, then $g$ is square-integrable on $S_{t,u}$ and, denoting by $\bar{g}$ the mean value of $g$ on $S_{t,u}$:

$$\bar{g} = \frac{1}{A(t,u)} \int_{S_{t,u}} g d\mu_{\not{g}} \tag{17.13}$$

we have

$$\int_{S_{t,u}} (g - \bar{g})^2 d\mu_{\not{g}} \leqslant I(t,u) \left( \int_{S_{t,u}} |\not{d} g| d\mu_{\not{g}} \right)^2 \tag{17.14}$$

where $I(t,u)$ is the *isoperimetric constant* of $S_{t,u}$:

$$I = \sup_U \frac{\min\{\text{Area}(U), \text{Area}(U^c)\}}{(\text{Perimeter}(\partial U))^2} \tag{17.15}$$

where the supremum is over all domains $U$ with $C^1$ boundary $\partial U$ in $S_{t,u}$, and $U^c = S_{t,u} \setminus U$ denotes the complement of $U$ in $S_{t,u}$.

Since

$$\|\bar{g}\|_{L^2(S_{t,u})} = |\bar{g}| A^{1/2}, \quad |\bar{g}| \leqslant A^{-1} \|g\|_{L^1(S_{t,u})} \tag{17.16}$$

It follows that

$$\|g\|_{L^2(S_{t,u})} \leqslant \sqrt{I'} \|g\|_{W_1^1(S_{t,u})} \tag{17.17}$$

where

$$\|g\|_{W_1^1(S_{t,u})} = \|\not{d} g\|_{L^1(S_{t,u})} + A^{-1/2} \|g\|_{L^1(S_{t,u})} \tag{17.18}$$

and

$$I' = \max\{I, 1\} \tag{17.19}$$

We have

$$\|f^2\|_{W_1^1(S_{t,u})} = \int_{S_{t,u}} |\not{d}(f^2)| d\mu_{\not{g}} + A^{-1/2} \int_{S_{t,u}} |f|^2 d\mu_{\not{g}} \tag{17.20}$$

and, since $\not{d}(f^2) = 2f \not{d} f$,

$$\int_{S_{t,u}} |\not{d}(f^2)| d\mu_{\not{g}} \leqslant 2 \left( \int_{S_{t,u}} |f|^2 d\mu_{\not{g}} \right)^{1/2} \left( \int_{S_{t,u}} |\not{d} f|^2 d\mu_{\not{g}} \right)^{1/2} \leqslant C A^{-1/2} S_{[2]}[f] \tag{17.21}$$

by (17.12). Since the second term on the right of (17.20) is similarly bounded, we obtain

$$\|f^2\|_{W_1^1(S_{t,u})} \leqslant CA^{-1/2}\mathcal{S}_{[2]}[f] \tag{17.22}$$

We have

$$\|\sum_i |R_i f|^2\|_{W_1^1(S_{t,u})} = \int_{S_{t,u}} |\not{d}(\sum_i |R_i f|^2)|^2 d\mu_{\not{g}} + A^{-1/2}\int_{S_{t,u}} \sum_i |R_i f|^2 d\mu_{\not{g}} \tag{17.23}$$

and, since $\not{d}(\sum_i |R_i f|^2) = 2\sum_i (R_i f)\not{d}(R_i f)$,

$$\int_{S_{t,u}} |\not{d}(\sum_i |R_i f|^2)| d\mu_{\not{g}}$$
$$\leqslant 2\Big(\int_{S_{t,u}} \sum_i |R_i f|^2 d\mu_{\not{g}}\Big)^{1/2}\Big(\int_{S_{t,u}} \sum_i |\not{d}R_i f|^2 d\mu_{\not{g}}\Big)^{1/2} \leqslant CA^{-1/2}\mathcal{S}_{[2]}[f] \tag{17.24}$$

by (17.12). Since the second term on the right in (17.23) is similarly bounded, we obtain

$$\|\sum_i |R_i f|^2\|_{W_1^1(S_{t,u})} \leqslant CA^{-1/2}\mathcal{S}_{[2]}[f] \tag{17.25}$$

We now apply (17.17) first taking $g = f^2$ and then taking $g = \sum_i |R_i f|^2$ to obtain, in view of (17.22), (17.25),

$$\Big(\int_{S_{t,u}} (|f|^4 + (\sum_i |R_i f|^2)^2) d\mu_{\not{g}}\Big)^{1/2} \leqslant C\sqrt{I'}A^{-1/2}\mathcal{S}_{[2]}[f] \tag{17.26}$$

Then by Corollary 12.2.a,

$$\|f\|^2_{W_1^4(S_{t,u})} \leqslant C\sqrt{I'}A^{-3/2}\mathcal{S}_{[2]}[f] \tag{17.27}$$

where for an arbitrary function $f$ on $S_{t,u}$ we denote

$$\|f\|_{W_1^4(S_{t,u})} = \|\not{d}f\|_{L^4(S_{t,u})} + A^{-1/2}\|f\|_{L^4(S_{t,u})} \tag{17.28}$$

We shall now use an argument found in [11]. We rescale the metric $\not{g}$ on $S_{t,u}$, setting

$$\hat{\not{g}} = A^{-1}\not{g} \quad \text{so that} \quad d\mu_{\hat{\not{g}}} = A^{-1}d\mu_{\not{g}} \tag{17.29}$$

to a metric $\hat{\not{g}}$ of unit area. Taking account of the fact that relative to the new metric we have

$$|\not{d}f|^2_{\hat{\not{g}}} = (\hat{\not{g}}^{-1})^{BC}\partial_B f\partial_C f = A(\not{g}^{-1})^{BC}\partial_B f\partial_C f = A|\not{d}f|^2_{\not{g}} \tag{17.30}$$

we have

$$\|f\|_{W_1^4(S_{t,u},\not{g})} = A^{-1/4}\|f\|_{W_1^4(S_{t,u},\hat{\not{g}})} \tag{17.31}$$

where

$$\|f\|_{W_1^p(S_{t,u},\hat{\not g})} = \|\not d f\|_{L^p(S_{t,u},\hat{\not g})} + \|f\|_{L^p(S_{t,u},\hat{\not g})} \tag{17.32}$$

hence (17.27) reads

$$\|f\|^2_{W_1^4(S_{t,u},\hat{\not g})} \leqslant C\sqrt{I'}A^{-1}\mathcal{S}_{[2]}[f] \tag{17.33}$$

Moreover, from (17.17), (17.18) relative to $\hat{\not g}$,

$$\|g\|_{L^2(S_{t,u},\hat{\not g})} \leqslant \sqrt{I'}\|g\|_{W_1^1(S_{t,u},\hat{\not g})} \tag{17.34}$$

We now set

$$\tilde{f} = \frac{1}{\sqrt{I'}}\frac{|f|}{\|f\|_{W_1^4(S_{t,u},\hat{\not g})}} \tag{17.35}$$

Then $\tilde{f} \geqslant 0$ and taking $g = \tilde{f}^k$, $k > 1$, in (17.34) we obtain

$$\|\tilde{f}^k\|_{L^2(S_{t,u},\hat{\not g})} \leqslant \sqrt{I'}\|\tilde{f}^k\|_{W_1^1(S_{t,u},\hat{\not g})} \tag{17.36}$$

Since $\not d(\tilde{f}^k) = k\tilde{f}^{k-1}\not d\tilde{f}$ we have,

$$\|\not d(\tilde{f}^k)\|_{L^1(S_{t,u},\hat{\not g})} \leqslant k\|\tilde{f}^{k-1}\|_{L^{4/3}(S_{t,u},\hat{\not g})}\|\not d\tilde{f}\|_{L^4(S_{t,u},\hat{\not g})} \tag{17.37}$$

and

$$\|\tilde{f}^k\|_{L^1(S_{t,u},\hat{\not g})} = \|\tilde{f}^{k-1}\tilde{f}\|_{L^1(S_{t,u},\hat{\not g})} \leqslant \|\tilde{f}^{k-1}\|_{L^{4/3}(S_{t,u},\hat{\not g})}\|\tilde{f}\|_{L^4(S_{t,u},\hat{\not g})} \tag{17.38}$$

hence

$$\|\tilde{f}^k\|_{W_1^1(S_{t,u},\hat{\not g})} \leqslant k\|\tilde{f}^{k-1}\|_{L^{4/3}(S_{t,u},\hat{\not g})}\|\tilde{f}\|_{W_1^4(S_{t,u},\hat{\not g})} \tag{17.39}$$

Now by (17.35) we have

$$\|\tilde{f}\|_{W_1^4(S_{t,u},\hat{\not g})} = \frac{1}{\sqrt{I'}} \tag{17.40}$$

Substituting (17.39) and using (17.36) yields

$$\|\tilde{f}^k\|_{L^2(S_{t,u},\hat{\not g})} \leqslant k\|\tilde{f}^{k-1}\|_{L^{4/3}(S_{t,u},\hat{\not g})} \tag{17.41}$$

which is equivalent to

$$\|\tilde{f}\|_{L^{2k}(S_{t,u},\hat{\not g})} \leqslant k^{1/k}\|\tilde{f}\|^{1-(1/k)}_{L^{(4/3)(k-1)}(S_{t,u},\hat{\not g})} \tag{17.42}$$

This implies

$$\|\tilde{f}\|_{L^{2k}(S_{t,u},\hat{\not g})} \leqslant k^{1/k}\|\tilde{f}\|^{1-(1/k)}_{L^{(4/3)k}(S_{t,u},\hat{\not g})} \tag{17.43}$$

This is because, by virtue of the fact that $S_{t,u}$ has unit area with respect to $\hat{\not{g}}$, the norm $\|g\|_{L^p(S_{t,u},\hat{\not{g}})}$ is a non-decreasing function of the exponent $p$, for any given function $g$. The ratio of the exponent on the left in (17.43) to the exponent on the right is $3/2$. We now set

$$k = (\frac{3}{2})^n, \quad n = 1, 2, 3, \cdots \tag{17.44}$$

For $n = 1$ the exponent on the right in (17.43) is 2, and taking $g = \tilde{f}$ in (17.34) we obtain, by (17.35),

$$\|\tilde{f}\|_{L^2(S_{t,u},\hat{\not{g}})} \leqslant \sqrt{I'}\|\tilde{f}\|_{W_1^1(S_{t,u},\hat{\not{g}})} = \frac{\|\tilde{f}\|_{W_1^1(S_{t,u},\hat{\not{g}})}}{\|\tilde{f}\|_{W_1^4(S_{t,u},\hat{\not{g}})}} \leqslant 1 \tag{17.45}$$

Then by (17.45) and induction on $n$ we get

$$\|\tilde{f}\|_{L^{2(3/2)^n}(S_{t,u},\hat{\not{g}})} \leqslant (\frac{3}{2})^{\sum_{m=1}^{n} m(3/2)^{-m}} \tag{17.46}$$

Taking the limit $n \to \infty$ we obtain

$$\sup_{S_{t,u}} \tilde{f} \leqslant c \tag{17.47}$$

where

$$c = (\frac{3}{2})^{\sum_{m=1}^{\infty} m(3/2)^{-m}} = (\frac{3}{2})^6$$

In view of (17.35), (17.47) is equivalent to

$$\sup_{S_{t,u}} |f| \leqslant c\sqrt{I'}\|f\|_{W_1^4(S_{t,u},\hat{\not{g}})} \tag{17.48}$$

Substituting finally (17.33) yields

$$\sup_{S_{t,u}} |f| \leqslant cCI'^{3/4}A^{-1/2}(\mathcal{S}_{[2]}[f])^{1/2} \tag{17.49}$$

To complete the proof of the Lemma 17.1, what remains to be done is to obtain an upper bound for $I(S_{t,u})$, the isoperimetric constant of $S_{t,u}$. The integral curves of $T$ on a given $\Sigma_t^{\epsilon_0}$ define a diffeomorphism of $S_{t,0}$ onto each $S_{t,u}$, $u \in [0, \epsilon_0]$. A domain $U_u \subset S_{t,u}$ with $C^1$ boundary $\partial U_u$ is mapped by the inverse onto a domain $U_0 \subset S_{t,0}$ with $C^1$ boundary $\partial U_0$. Consider the image $U_{u'}$ of the domain $U_0$ on each $S_{t,u'}$, $u' \in [0, u]$, under the diffeomorphism. Then from the definition,

$$\not{\mathcal{L}}_T \not{g} = 2\kappa\theta$$

we have

$$\frac{d}{du'}\text{Area}(U_{u'}) = \int_{U_{u'}} \kappa \text{tr}\theta d\mu_{\not{g}} \tag{17.50}$$

and

$$\frac{d}{du'}\text{Perimeter}(\partial U_{u'}) = \int_{\partial U_{u'}} \kappa\theta(V,V)ds \tag{17.51}$$

where $V$ is the unit tangent vectorfield and $ds$ is the element of arc length of $\partial U_{u'}$. Here, $u$ is the parameter of the integral curves of $T$. Since the bootstrap assumption implies

$$\kappa|\theta| \leqslant C(1+t)^{-1}(1+\log(1+t)) \leqslant C \tag{17.52}$$

from (17.50) and (17.51) we obtain

$$|\frac{d}{du'}\text{Area}(U_{u'})| \leqslant C\text{Area}(U_{u'}) \tag{17.53}$$

$$|\frac{d}{du'}\text{Perimeter}(\partial U_{u'})| \leqslant C\text{Perimeter}(\partial U_{u'}) \tag{17.54}$$

Therefore integrating with respect to $u'$ on $[0,u]$ yields

$$\begin{gathered} C^{-1}\text{Area}(U_0) \leqslant \text{Area}(U_u) \leqslant C\text{Area}(U_0) \\ C^{-1}\text{Perimeter}(\partial U_0) \leqslant \text{Perimeter}(\partial U_u) \leqslant C\text{Perimeter}(\partial U_0) \end{gathered} \tag{17.55}$$

for all $u \in [0, \epsilon_0]$.

It follows that

$$C^{-1}I(S_{t,0}) \leqslant I(S_{t,u}) \leqslant CI(S_{t,0}) \tag{17.56}$$

for all $u \in [0, \epsilon_0]$.

Since $S_{t,0}$ is a round sphere in Euclidean space, we have

$$I(S_{t,0}) = \frac{1}{2\pi}$$

So we obtain an upper bound for $I(S_{t,u})$ by a numerical constant. In view of (17.11), the lemma follows from (17.49). □

## 17.3 Recovery of J—Completion

Noting from (17.7)—(17.10) that

$$\begin{aligned} &\mathcal{S}_{[2]}[S\phi] \leqslant \tilde{\mathcal{S}}_{[4]}(t,u), \quad &&\mathcal{S}_{[2]}[\mathring{R}_i\phi] \leqslant \tilde{\mathcal{S}}_{[4]}(t,u), \quad i=1,2,3 \\ &\mathcal{S}_{[2]}[TS\phi] \leqslant \tilde{\mathcal{S}}_{[4]}(t,u), \quad &&\mathcal{S}_{[2]}[T\mathring{R}_i\phi] \leqslant \tilde{\mathcal{S}}_{[4]}(t,u), \quad i=1,2,3 \end{aligned} \tag{17.57}$$

then by Lemma 17.1 and (17.6) we have

$$\sup_{\Sigma_t^{\epsilon_0}} |S\phi| \leqslant C(1+t)^{-1}\sqrt{\tilde{D}^u_{[4]}}, \quad \sup_{\Sigma_t^{\epsilon_0}} |\mathring{R}_i\phi| \leqslant C(1+t)^{-1}\sqrt{\tilde{D}^u_{[4]}}, \quad i=1,2,3$$

$$\sup_{\Sigma_t^{\epsilon_0}} |TS\phi| \leqslant C(1+t)^{-1}\sqrt{\tilde{D}^u_{[4]}}, \quad \sup_{\Sigma_t^{\epsilon_0}} |T\mathring{R}_i\phi| \leqslant C(1+t)^{-1}\sqrt{\tilde{D}^u_{[4]}}, \quad i=1,2,3 \tag{17.58}$$

it follows that if

$$\sqrt{\tilde{D}^u_{[4]}} \leqslant C\delta_0 \tag{17.59}$$

then assumption **J** holds on $W^s_{\epsilon_0}$. We thus have established assumption **J**.

## 17.4 Recovery of the Final Bootstrap Assumption

We are now ready to recover the bootstrap assumption and proceed to the proof of the main theorem of this monograph. Let us denote by $\mathcal{S}_{[n]}(t,u)$ the integral on $S_{t,u}$ (with respect to $d\mu_{\not{g}}$) of the sum of the squares of all the variations (14.4), of order up to $n$. By Lemma 5.1 we have

$$\mathcal{S}_{[l+1-[a]]}(t,u) \leqslant C\epsilon_0 \mathcal{E}^u_{0,[l+1-[a]]}(t) \quad \text{for all} \quad (t,u) \in [0,s]\times[0,\epsilon_0] \tag{17.60}$$

Hence, in view of (16.83) and (16.84),

$$\mathcal{S}_{[l+1-[a]]}(t,u) \leqslant C\epsilon_0 \mathcal{D}^u_{[l+2]} \quad \text{for all} \quad (t,u) \in [0,s]\times[0,\epsilon_0] \tag{17.61}$$

Now, for any variation $\psi$ of order up to $l-1-[a]$ we have

$$\mathcal{S}_{[2]}[\psi] \leqslant \mathcal{S}_{[l+1-[a]]} \tag{17.62}$$

This is because $\psi, R_{j_1}\psi, j_1 = 1,2,3; R_{j_2}R_{j_1}\psi, j_1, j_2 = 1,2,3$ are themselves variations included in $\mathcal{S}_{[l+1-[a]]}$. Then by Lemma 17.1 and (17.61), (17.62), we obtain

$$\sup_{S_{t,u}} |\psi| \leqslant C(1+t)^{-1}\sqrt{\epsilon_0}\sqrt{\mathcal{D}^u_{[l+2]}} \quad \text{for all} \quad (t,u) \in [0,s]\times[0,\epsilon_0] \tag{17.63}$$

That is, we obtain

$$\max_{\alpha} \max_{i_1\cdots i_n} \|R_{i_n}\cdots R_{i_1}(T)^m(Q)^{p(\alpha)}\psi_1\|_{L^\infty(\Sigma_t^{\epsilon_0})} \leqslant C(1+t)^{-1}\sqrt{\epsilon_0}\sqrt{\mathcal{D}_{[l+2]}} \tag{17.64}$$

for all $p+m+n \leqslant l-2-[a]$ and all $t \in [0,s]$.

From (15.167), we have

$$l-2-[a] \geqslant (l+1)_* + 2$$

since $l = l_* + (l+1)_*$. So if

$$C\sqrt{\epsilon_0}\sqrt{\mathcal{D}_{[l+2]}} < \delta_0 \tag{17.65}$$

then we recover the bootstrap assumption $\mathbf{E}_{\{\{(l+1)_*+2\}\}}$.

## 17.5 Completion of the Proof of the Main Theorem

Let now $s_*$ be the least upper bound of the set of values of $s$ in the interval $[0, t_{*\epsilon_0}]$ such that the bootstrap assumption holds on $W^s_{\epsilon_0}$. We recall from Chapter 2 that $t_{*\epsilon_0}$ is defined by

$$t_{*\epsilon_0} = \inf_{u\in[0,\epsilon_0]} t_*(u) \tag{17.66}$$

where $t_*(u)$ is the greatest lower bound of the extent of the generators of $C_u$, in the parameter $t$, in the domain of the maximal solution. We note here that $C_u$ do not contain cut loci. This follows from the fact that the angle between the outward unit normal $-\hat{T}$ to the surface $S_{t,u}$ with respect to the Euclidean metric on $\Sigma_t$ and the outward unit normal $N$ to the Euclidean coordinate spheres does not exceed a fixed constant times $\delta_0$, as follows from the estimate (12.19). In view of the bound on $\chi$, the second fundamental form of the sections $S_{t,u}$ relative to $C_u$, the $C_u$ do not contain focal points (that is, points along a generator of $C_u$ which are conjugate to $S_{0,u}$) either. The absence of cut locus or focal points implies that a bicharacteristic generator of $C_u$ cannot leave the boundary of the domain of dependence, in the domain of maximal solution, of the exterior of the surface $S_{0,u}$ in the initial hyperplane $\Sigma_0$. (For the notion of cut locus and of focal or conjugate points in Riemannian geometry, see [3]. For the corresponding notions in Lorentzian geometry see [16].) Thus, unless $t_{*\epsilon_0} = \infty$, there is on $\Sigma^{\epsilon_0}_{t_{*\epsilon_0}}$ at least one point which belongs to the boundary of the domain of maximal solution and not to the domain itself. We shall presently show that in fact $s_*$ coincides with $t_{*\epsilon_0}$ under the condition (17.65). For, otherwise, that is if $s_* < t_{*\epsilon_0}$, then by continuity the bootstrap assumption holds on $W^{s_*}_{\epsilon_0}$ as well, hence by the above, (17.64) holds with $l - 2 - [a]$ replaced by $(l+1)_* + 2$ and $s$ replaced by $s_*$. By (17.65) and continuity however the bootstrap assumption must also hold for some $s > s_*$, contradicting the definition of $s_*$. We conclude that $s_* = t_{*\epsilon_0}$ and (17.64) holds for all $t \in [0, t_{*\epsilon_0})$. We thus have uniform pointwise estimates for the $\psi_\alpha$ on $W^*_{\epsilon_0}$ up to order $(l+1)_* + 2$. Propositions 12.9 and 12.10 then give uniform pointwise estimates for $\chi'$ up to order $(l+1)_*$ and for $\mu$ up to order $(l+1)_* + 1$. It follows that the $\psi_\alpha$, $\chi'$ and $\mu$ thus also $\kappa$ and the induced acoustical metric $\not{g}$, extend smoothly in *acoustical coordinates* to $\Sigma^{\epsilon_0}_{t_{*\epsilon_0}}$ and $W^{t_{*\epsilon_0}}_{\epsilon_0}$. Also the rectangular components $g_{\mu\nu}$ of the acoustical metric, being functions of $\psi_\alpha$, likewise extend. Moreover, since the bootstrap assumption holds on $W^*_{\epsilon_0}$, all the estimates we have derived, in particular the energy estimates (16.83), (16.79), (15.160)—(15.163), and the $L^2$ acoustical estimates of Propositions 12.11, 12.12, as well as the top order acoustical estimates, hold for all $t \in [0, t_{*\epsilon_0})$. The estimates not containing the weights $\bar{\mu}_{m,u}$, such as the estimates (16.83), then extend to $t = t_{*\epsilon_0}$ as well. Now, we must have

$$\bar{\mu}_m(t_{*\epsilon_0}) = 0 \tag{17.67}$$

For otherwise the Jacobian determinant $\Delta$ of the transformation from acoustical to rectangular coordinates has a positive minimum on $\Sigma^{\epsilon_0}_{t_{*\epsilon_0}}$, therefore the inverse transform is regular and the $\psi_\alpha$ extend smoothly in *rectangular coordinates* to

$\Sigma^{\epsilon_0}_{t_{*\epsilon_0}}$. However, once the $\psi_\alpha$ extend to functions of the rectangular coordinates on $\Sigma^{\epsilon_0}_{t_{*\epsilon_0}}$ which belong to the Sobolev space $H_3$, then the standard local existence theorem applies and we obtain an extension of the solution to a development containing an extension of all the characteristic hypersurfaces $C_u$, $u \in [0, \epsilon_0]$, up to a value $t_1$ of $t$ for some $t_1 > t_*$, in contradiction with the definition of $t_{*\epsilon_0}$. We conclude that there is at least one point on $\Sigma^{\epsilon_0}_{t_{*\epsilon_0}}$ where $\mu$ vanishes. By the same argument, at each point $x_* \in \Sigma^{\epsilon_0}_{t_{*\epsilon_0}}$ which lies on the boundary of the domain of the maximal solution, $\mu(x_*) = 0$. For, otherwise, $\mu$ has a positive minimum in a suitable neighborhood of $x_*$, hence the solution is locally extendible at $x_*$ in contradiction with the fact that $x_*$ lies on the boundary of the domain of the maximal solution.

Now, from Proposition 8.6, taking $t = s$ we have

$$\hat{\mu}_s(s, u, \vartheta) = 1 + \hat{E}_s(u, \vartheta) \log(1 + s) \tag{17.68}$$

and we recall that

$$\hat{\mu}_s(s, u, \vartheta) = \frac{\mu(s, u, \vartheta)}{\mu_{[1],s}(u, \vartheta)}, \quad \hat{E}_s(u, \vartheta) = \frac{E_s(u, \vartheta)}{\mu_{[1],s}(u, \vartheta)} \tag{17.69}$$

and that $\mu_{[1],s}(u, \vartheta)$ is bounded from below by 1/2. We also recall that according to (8.226),

$$|E_s(u, \vartheta) - \frac{1}{4}\ell P_s(u, \vartheta)| \leqslant C\delta_0(1 + s)^{-1}(1 + \log(1 + s)) \tag{17.70}$$

where, from Lemma 8.10,

$$P_s(u, \vartheta) = (1 + s)(\underline{L}\psi_0)(s, u, \vartheta) \tag{17.71}$$

Now by (17.64) we have

$$|P_s(u, \vartheta)| \leqslant C\sqrt{\epsilon_0}\sqrt{\mathcal{D}_{[l+2]}} \tag{17.72}$$

Hence we obtain

$$|E_s(u, \vartheta)| \leqslant C|\ell|\sqrt{\epsilon_0}\sqrt{\mathcal{D}_{[l+2]}} + C\delta_0(1 + s)^{-1}(1 + \log(1 + s)) \tag{17.73}$$

Since

$$\frac{\log(1 + s)(1 + \log(1 + s))}{1 + s} \leqslant C$$

this implies

$$\log(1 + s)|E_s(u, \vartheta)| \leqslant C|\ell|\sqrt{\epsilon_0}\sqrt{\mathcal{D}_{[l+2]}} \log(1 + s) + C\delta_0 \tag{17.74}$$

Substituting in (17.68) then yields

$$\hat{\mu}_s(s,u,\vartheta) \geqslant 1 - C\delta_0 - C|\ell|\sqrt{\epsilon_0}\sqrt{\mathcal{D}_{[l+2]}}\log(1+s) \tag{17.75}$$

It follows that

$$\log(1+t_{*\epsilon_0}) \geqslant \frac{1}{C|\ell|\sqrt{\epsilon_0}\sqrt{\mathcal{D}_{[l+2]}}} \tag{17.76}$$

(for a new constant $C$), because otherwise $\mu$ would be bounded from below by a positive constant on $\Sigma_{t_{*\epsilon_0}}^{\epsilon_0}$ contradicting the conclusions of the previous paragraph.

It remains for us to analyze the requirements on the initial data. We recall from Chapter 2 that $u$ on $\Sigma_0$ is defined by

$$u = 1 - r \tag{17.77}$$

Thus, in rectangular coordinates we have on $\Sigma_0$

$$\partial_i u = -N^i, \quad N^i = \frac{x^i}{r} \tag{17.78}$$

where $N$ is the Euclidean outward unit normal to the Euclidean coordinate spheres. Since $\kappa$ is defined by

$$\kappa^{-2} = (\bar{g}^{-1})^{ij}\partial_i u \partial_j u \tag{17.79}$$

and $\bar{g}_{ij} = \delta_{ij}$, on $\Sigma_0$ we have simply

$$\kappa = 1 \tag{17.80}$$

and

$$\mu = \eta \tag{17.81}$$

According to (2.27) we have

$$\hat{T}^i = \kappa \partial_i u \tag{17.82}$$

then on $\Sigma_0$,

$$\hat{T}^i = -N^i = T^i \tag{17.83}$$

Then by (2.55), $L$ is given on $\Sigma_0$ by

$$L = \frac{\partial}{\partial x^0} + (\eta N^i - \psi_i)\frac{\partial}{\partial x^i} \tag{17.84}$$

On $\Sigma_0$, $Q = L$. Consider next the orthogonal projection to $S_{t,u}$ in $\Sigma_t$, relative to $\bar{g}$. It is given in rectangular coordinates by

$$\Pi_b^a = \delta_b^a - \hat{T}^a \hat{T}^b \tag{17.85}$$

The level surfaces $S_{0,u}$ of $u$ on $\Sigma_0$ being Euclidean spheres centered at the origin and the vectorfields $\mathring{R}_i$ being tangential to these spheres, the commutation fields $R_i$ coincide on $\Sigma_0$ with $\mathring{R}_i$:

$$R_i = \Pi \mathring{R}_i = \mathring{R}_i = \epsilon_{ijk} x^j \frac{\partial}{\partial x^k} \tag{17.86}$$

Moreover, the functions $\lambda_i$ defined by (12.12) vanishes on $\Sigma_0$:

$$\lambda_i = -\kappa N^j \epsilon_{ijk} x^k = 0 \tag{17.87}$$

in view of (17.83).

Since the $S_{0,u}$ are Euclidean spheres of radius $r$ with inward unit normal $\hat{T}$, we have

$$\theta_{ab} = -\frac{1}{r}(\delta_{ab} - \hat{T}^a \hat{T}^b) \tag{17.88}$$

Then from (3.27),

$$\chi_{ab} = -\eta(\theta_{ab} - \not{k}_{ab}) \tag{17.89}$$

where

$$\not{k}_{ab} = -\eta^{-1} \Pi^i_a \Pi^j_b \partial_i \psi_j \tag{17.90}$$

Consider now the initial data for the nonlinear wave equation (1.23). The functions $\psi_\alpha$, $\alpha = 0, 1, 2, 3$ are given in the exterior of the sphere of radius $1 - \epsilon_0$ with center at the origin and satisfy $\partial_i \psi_j = \partial_j \psi_i$. Outside the unit sphere with center at the origin, the initial data coincide with those of a constant state $\psi_\alpha = 0$, $\alpha = 0, 1, 2, 3$. We assume that $\psi_\alpha$, $\alpha = 0, 1, 2, 3$ are, in rectangular coordinates, functions belonging to the Sobolev space $H_{l+2}(\Sigma_0^{\epsilon_0})$ with vanishing traces on the unit sphere. We set

$$D_{[l+2]} = \sum_{\alpha,i} \|\partial_i \psi_\alpha\|^2_{H_{l+1}(\Sigma_0^{\epsilon_0})} \tag{17.91}$$

The nonlinear wave equation in the form

$$(g^{-1})^{\mu\nu} \partial_\mu \psi_\nu = 0, \quad \partial_\mu \psi_\nu = \partial_\nu \psi_\mu \tag{17.92}$$

allows us to express $\partial_0 \psi_0$ and $\partial_0 \psi_i$ in terms of $\partial_i \psi_0$ and $\partial_i \psi_j$, for, $\partial_0 \psi_i = \partial_i \psi_0$ and we have

$$(g^{-1})^{00} = -\eta^{-2} \tag{17.93}$$

We can then express recursively $\partial_0^k \psi_\alpha$, $\alpha = 0, 1, 2, 3$ for $k = 1, \cdots, l+2$ in terms of the data $\partial_{i_k} \cdots \partial_{i_1} \psi_\alpha$, $\alpha = 0, 1, 2, 3$; $i_1, \cdots, i_n = 1, 2, 3$. Now the standard Sobolev inequalities yield

$$\max_{\alpha} \max_{i_1 \cdots i_n} \|\partial_{i_n} \cdots \partial_{i_1} \psi_\alpha\|_{L^\infty(\Sigma_0^{\epsilon_0})}$$
$$\leqslant C\sqrt{\epsilon_0}\sqrt{D_{[l+2]}} \quad \text{for} \quad n = 1, \cdots, l-1 \tag{17.94}$$

It then follows that also

$$\max_{\alpha} \max_{i_1 \cdots i_n} \|\partial_{i_n} \cdots \partial_{i_1} \partial_0^k \psi_j\|_{L^\infty(\Sigma_0^{\epsilon_0})}$$
$$\leqslant C\sqrt{\epsilon_0}\sqrt{D_{[l+2]}} \quad \text{for} \quad n, k \geqslant 0, n+k \leqslant l-1 \tag{17.95}$$

Here, the constant $\sqrt{\epsilon_0}$ comes from using Lemma 5.1.

Now from (17.83), (17.85) and (17.87) the rectangular components of $T$, $L$ and $R_i$ are on $\Sigma_0^{\epsilon_0}$ smooth functions of the rectangular coordinates and the $\psi_\alpha$. It then follows that

$$\mathcal{E}_{0,[l+2]}(0), \quad \mathcal{E}'_{1,[l+2]}(0) \leqslant C D_{[l+2]} \tag{17.96}$$

From (17.94) and (17.95) we have

$$\|\chi'\|_{\infty,\{l-2\},\Sigma_0^{\epsilon_0}} \leqslant C\sqrt{\epsilon_0}\sqrt{D_{[l+2]}} \tag{17.97}$$

The assumptions of Propositions 12.3, 12.6, 12.9, 12.10, on the initial conditions are then recovered provided that

$$\sqrt{\epsilon_0}\sqrt{D_{[l+2]}} \leqslant C^{-1}\delta_0 \tag{17.98}$$

On $\Sigma_0^{\epsilon_0}$, we have $y^i = 0$, hence

$$\mathcal{Y}_0(0) = 0 \tag{17.99}$$

Moreover,

$$\mathcal{B}_{\{l+1\}}(0) \leqslant C\sqrt{D_{[l+2]}} \tag{17.100}$$
$$\mathcal{A}_{[l]}(0) \leqslant C\sqrt{D_{[l+2]}} \tag{17.101}$$

and by (8.44), (9.60) we have

$$\sum_{i_1 \cdots i_l} \|{}^{(i_1 \cdots i_l)}x_l(0)\|_{L^2(\Sigma_0^{\epsilon_0})} \leqslant C\sqrt{D_{[l+2]}} \tag{17.102}$$

$$\sum_{m=0}^{l} \sum_{i_1 \cdots i_{l-m}} \|{}^{(i_1 \cdots i_{l-m})}x'_{m,l-m}(0)\|_{L^2(\Sigma_0^{\epsilon_0})} \leqslant C\sqrt{D_{[l+2]}} \tag{17.103}$$

Combining the above then yields

$$\mathcal{D}_{[l+2]} \leqslant C D_{[l+2]} \tag{17.104}$$

and we can replace $\epsilon_0$ by any $u \in (0, \epsilon_0]$ to obtain

$$\mathcal{D}^u_{[l+2]} \leqslant C D^u_{[l+2]} \tag{17.105}$$

for all $u \in (0, \epsilon_0]$, where

$$D^u_{[l+2]} = \sum_{\alpha,i} \|\partial_i \psi_\alpha\|^2_{H_{l+1}(\Sigma_0^u)} \tag{17.106}$$

In conclusion we have proved the following theorem, which is the main theorem of this monograph.

**Theorem 17.1** Let $(p, s, v^i, i = 1,2,3)$ be initial data for Equations (1.4), (1.10) and (1.12) on $\Sigma_0$ which correspond to the initial data of a constant state

$$p = p_0, \quad s = s_0, \quad v^i = 0, \quad i = 1,2,3$$

outside the unit sphere with center at the origin in $\Sigma_0$. We can adjust the zero-point of the enthalpy $h$ so that $h_0 = 0$, that is, the enthalpy vanishes in the constant state. We can also choose the relation of the unit of time to the unit of length so that $\eta_0 = 1$, that is, the sound speed in the constant state is equal to unity. Let also the initial data be irrotational and isentropic outside the sphere of radius $1 - \epsilon_0$, $0 < \epsilon_0 \leqslant 1/2$ with center at the origin in $\Sigma_0$. Then we have initial data $(\phi, \partial_t \phi)$ for the nonlinear wave equation (1.23) outside the sphere of radius $1 - \epsilon_0$ with center at the origin in $\Sigma_0$, where $\partial_i \phi = -v^i$ and $\partial_t \phi = h + \frac{1}{2}\sum_{i=1}^3 (\partial_i \phi)^2$. The initial data $(\phi, \partial_t \phi)$ vanish outside the unit sphere with center at the origin in $\Sigma_0$. Consider the annular region $\Sigma_0^{\epsilon_0}$ in $\Sigma_0$ bounded by the two concentric spheres. Then there is a positive integer $[a]$ such that the following hold. Let $l$ be a positive integer such that

$$l_* \geqslant [a] + 4$$

and suppose that the functions $\psi_\alpha$, $\alpha = 0,1,2,3$ corresponding to the initial data on $\Sigma_0^{\epsilon_0}$ are functions of the rectangular coordinates belonging to the Sobolev space $H_{l+2}(\Sigma_0^{\epsilon_0})$ with vanishing traces on the unit sphere. Then setting

$$D_{[l+2]} = \sum_{\alpha,i} \|\partial_i \psi_\alpha\|^2_{H_{l+1}(\Sigma_0^{\epsilon_0})}$$

there is a suitably small positive constant $\bar{\delta}_0$ and a positive constant $C$, such that for any $\delta_0 \in (0, \bar{\delta}_0]$, if

$$C\sqrt{\epsilon_0}\sqrt{D_{[l+2]}} < \delta_0$$

the following conclusions hold:

(i) Let $u$ be the function $1 - r$ on $\Sigma_0$ and $S_{0,u}$ the spheres of radius $1 - u$ with center at the origin, the level surfaces of $u$ in $\Sigma_0$. We consider, in the domain of the maximal solution corresponding to the given initial data, the family

$\{C_u, u \in [0, \epsilon_0]\}$ of outgoing characteristic hypersurfaces corresponding to the family $\{S_{0,u}, u \in [0, \epsilon_0]\}$:

$$C_u \bigcap \Sigma_0 = S_{0,u}, \quad \forall u \in [0, \epsilon_0]$$

with each bicharacteristic generator of each $C_u$ extending in the domain of the maximal solution as long as it remains on the boundary of the domain of dependence of the exterior of $S_{0,u}$ in $\Sigma_0$. Then the bicharacteristic generators of each $C_u$ have no future end points except on the boundary of the domain of the maximal solution. Let $t_*(u)$ be the least upper bound of the extent of the generators of $C_u$, in the parameter $t$, in the domain of the maximal solution, and let

$$t_{*\epsilon_0} = \inf_{u \in [0,\epsilon_0]} t_*(u)$$

We define for each $(t, u) \in [0, t_{*\epsilon_0}) \times [0, \epsilon_0]$ the closed surface

$$S_{t,u} = C_u \bigcap \Sigma_t$$

Then either $t_{*\epsilon_0} = \infty$ or there is on $\Sigma^{\epsilon_0}_{t_{*\epsilon_0}}$ at least one point which belongs to the boundary of the domain of the maximal solution and not to the domain itself.

(ii) We have the lower bound

$$\log(1 + t_{*\epsilon_0}) \geqslant \frac{1}{C|\ell|\sqrt{\epsilon_0}\sqrt{D_{[l+2]}}}$$

where $\ell$ is the constant

$$\ell = \frac{dH}{dh}(h_0)$$

In particular, if $\ell = 0$ we have $t_{*\epsilon_0} = \infty$.

(iii) For all $t \in [0, t_{*\epsilon_0})$ we have the $L^\infty$ bounds

$$\max_\alpha \max_{i_1 \cdots i_n} \|R_{i_n} \cdots R_{i_1} (T)^m (Q)^p \psi_\alpha\|_{L^\infty(\Sigma^{\epsilon_0}_t)} \leqslant C(1+t)^{-1}\sqrt{\epsilon_0}\sqrt{D_{[l+2]}}$$
$$\text{for all} \quad p + m + n \leqslant l - 2 - [a]$$

and with $\mu$ and $\chi$ the acoustical entities defined by the family $\{C_u, u \in [0, \epsilon_0]\}$ and for each $C_u : u \in [0, \epsilon_0]$ by the family of sections $\{S_{t,u}, t \in [0, t_{*\epsilon_0})\}$:

$$\|\mu - 1\|_{\infty,\{l-3-[a]\},\Sigma^{\epsilon_0}_t} \leqslant C\delta_0(1 + \log(1 + t))$$
$$\|\chi'\|_{\infty,\{l-4-[a]\},\Sigma^{\epsilon_0}_t} \leqslant C\delta_0(1 + t)^{-2}(1 + \log(1 + t))$$

where

$$\chi' = \chi - \frac{\not{g}}{1 - u + t}$$

and $\not{g}$ is the induced acoustical metric on $S_{t,u}$.

(iv) The $\psi_\alpha$, $\mu$, $\not{g}$ and $\chi$, extend continuously with their first $l-2-[a]$ derivatives in the case of $\psi_\alpha$, $l-3-[a]$ derivatives in the case of $\mu$, $\not{g}$ and $l-4-[a]$ derivatives in the case of $\chi$, in acoustical coordinates to $\Sigma_{t_{*\epsilon_0}}$. The rectangular components $\hat{T}^i$ and $L^i$ of the vectorfields $T$ and $L$ likewise extend continuously with their first $l-3-[a]$ derivatives in acoustical coordinates to $\Sigma_{t_{*\epsilon_0}}$, and so do the rectangular components $g_{\mu\nu}$ of the acoustical spacetime metric. The function $\mu$ so extended vanishes at each point in $\Sigma_{t*\epsilon_0}^{\epsilon_0}$ which lies on the boundary of the domain of the maximal solution, being positive everywhere else. The Jacobian determinant $\Delta$ of the transformation from acoustical to rectangular coordinates also vanishes at the same points, being positive everywhere else. The first derivatives $\hat{T}^i\partial_i\psi_\alpha$ blow up at these points. More precisely what blows up is the component $\hat{T}^i\hat{T}^j\partial_i\psi_j$. Moreover, there is a positive constant $C$ such that in the subdomain $\mathcal{U}$ of $W^*_{\epsilon_0}$ where $\mu < 1/4$, which contains a spacetime neighborhood of each of the points in question, we have

$$L\mu \leqslant -C^{-1}(1+t)^{-1}(1+\log(1+t))^{-1}$$

(v) The following energy estimates hold, for all $u \in [0, \epsilon_0]$,

$$\sup_{t\in[0,t_{*\epsilon_0}]} (\mathcal{E}^u_{0,[l+1-[a]]}(t)) \leqslant CD^u_{[l+2]}$$

$$\mathcal{F}^{t_{*\epsilon_0}}_{0,[l+1-[a]]}(u) \leqslant CD^u_{[l+2]}$$

$$\sup_{t\in[0,t_{*\epsilon_0}]} ((1+\log(1+t))^{-4}\mathcal{E}'^u_{1,[l+1-[a]]}(t)) \leqslant CD^u_{[l+2]}$$

$$\sup_{t\in[0,t_{*\epsilon_0}]} ((1+\log(1+t))^{-4}\mathcal{F}'^t_{1,[l+1-[a]]}(u)) \leqslant CD^u_{[l+2]}$$

$$\sup_{t\in[0,t_{*\epsilon_0}]} ((1+\log(1+t))^{-4}K_{[l+1-[a]]}(t,u)) \leqslant CD^u_{[l+2]}$$

where

$$D^u_{[l+2]} = \sum_{\alpha,i} \|\partial_i\psi_\alpha\|^2_{H_{l+1}(\Sigma^u_t)}$$

and $\Sigma^u_0$ is the annular region on $\Sigma_0$ bounded by $S_{0,u}$ and the unit sphere $S_{0,0}$. Moreover, setting

$$a = [a] + \frac{3}{4}$$

we have, for each $n = 0, \cdots, [a]-1$, and for all $u \in [0, \epsilon_0]$, the estimates

$$\sup_{t\in[0,t_{*\epsilon_0})} (\bar{\mu}^{2(a-n-1)}_{m,u}(t)\mathcal{E}^u_{0,[l+1-n]}(t)) \leqslant CD^u_{[l+2]}$$

$$\sup_{t\in[0,t_{*\epsilon_0})} (\bar{\mu}^{2(a-n-1)}_{m,u}(t)\mathcal{F}^t_{0,[l+1-n]}(u)) \leqslant CD^u_{[l+2]}$$

$$\sup_{t\in[0,t_{*\epsilon_0})} (\bar{\mu}^{2(a-n-1)}_{m,u}(t)(1+\log(1+t))^{-4}\mathcal{E}'^u_{1,[l+1-n]}(t)) \leqslant CD^u_{[l+2]}$$

$$\sup_{t\in[0,t_{*\epsilon_0})}(\bar{\mu}_{m,u}^{2(a-n-1)}(t)(1+\log(1+t))^{-4}\mathcal{F}'^{t}_{1,[l+1-n]}(u))\leqslant CD^u_{[l+2]}$$

$$\sup_{t\in[0,t_{*\epsilon_0})}(\bar{\mu}_{m,u}^{2(a-n-1)}(t)(1+\log(1+t))^{-4}K_{[l+1-n]}(t,u))\leqslant CD^u_{[l+2]}$$

Furthermore, there is a positive real number $p$ such that with $q=p+2$ we have the top order energy estimates

$$\sup_{t\in[0,t_{*\epsilon_0})}(\bar{\mu}_{m,u}^{2a}(t)(1+\log(1+t))^{-2p}\mathcal{E}^u_{0,[l+2]}(t))\leqslant CD^u_{[l+2]}$$

$$\sup_{t\in[0,t_{*\epsilon_0})}(\bar{\mu}_{m,u}^{2a}(t)(1+\log(1+t))^{-2p}\mathcal{F}^t_{0,[l+2]}(u))\leqslant CD^u_{[l+2]}$$

$$\sup_{t\in[0,t_{*\epsilon_0})}(\bar{\mu}_{m,u}^{2a}(t)(1+\log(1+t))^{-2q}\mathcal{E}'^u_{1,[l+2]}(t))\leqslant CD^u_{[l+2]}$$

$$\sup_{t\in[0,t_{*\epsilon_0})}(\bar{\mu}_{m,u}^{2a}(t)(1+\log(1+t))^{-2q}\mathcal{F}'^t_{1,[l+2]}(u))\leqslant CD^u_{[l+2]}$$

$$\sup_{t\in[0,t_{*\epsilon_0})}(\bar{\mu}_{m,u}^{2a}(t)(1+\log(1+t))^{-2q}K_{[l+2]}(t,u))\leqslant CD^u_{[l+2]}$$

(vi) The following $L^2$ acoustical estimates hold, for all $u\in[0,\epsilon_0]$ and all $t\in[0,t_{*\epsilon_0}]$,

$$\mathcal{A}'^u_{[l-[a]-1]}(t)\leqslant C(1+t)^{-1}(1+\log(1+t))^3\sqrt{D^u_{[l+2]}}$$

$$\mathcal{B}^u_{[l-[a]]}(t)\leqslant C(1+t)(1+\log(1+t))^3\sqrt{D^u_{[l+2]}}$$

Moreover, for each $n=1,\cdots,[a]$ and for all $u\in[0,\epsilon_0]$ and all $t\in[0,t_{*\epsilon_0})$,

$$\mathcal{A}'^u_{[l-n]}(t)\leqslant C(1+t)^{-1}(1+\log(1+t))^3\bar{\mu}_{m,u}^{-a+n+1/2}(t)\sqrt{D^u_{[l+2]}}$$

$$\mathcal{B}^u_{\{l-n+1\}}(t)\leqslant C(1+t)(1+\log(1+t))^3\bar{\mu}_{m,u}^{-a+n+1/2}(t)\sqrt{D^u_{[l+2]}}$$

and for $n=0$ and for all $u\in[0,\epsilon_0]$ and all $t\in[0,t_{*\epsilon_0})$,

$$\mathcal{A}'^u_{[l]}(t)\leqslant C(1+t)^{-1}(1+\log(1+t))^{q+1}\bar{\mu}_{m,u}^{-a+1/2}(t)\sqrt{D^u_{[l+2]}}$$

$$\mathcal{B}^u_{\{l+1\}}(t)\leqslant C(1+t)(1+\log(1+t))^{q+1}\bar{\mu}_{m,u}^{-a+1/2}(t)\sqrt{D^u_{[l+2]}}$$

Finally, for all $u\in[0,\epsilon_0]$ and all $t\in[0,t_{*\epsilon_0})$ we have the top order acoustical estimates

$$\max_{i_1\cdots i_{l+1}}\|\mu R_{i_{l+1}}\cdots R_{i_1}\mathrm{tr}\chi'\|_{L^2(\Sigma^u_t)}\leqslant C\bar{\mu}_{m,u}^{-2a}(t)(1+\log(1+t))^{2p}\sqrt{D^u_{[l+2]}}$$

$$\sum_{m=0}^{l}\max_{i_1\cdots i_{l-m}}\|\mu R_{i_{l-m}}\cdots R_{i_1}(T)^m\Delta\mu\|_{L^2(\Sigma^u_t)}\leqslant C\bar{\mu}_{m,u}^{-2a}(t)(1+\log(1+t))^{2p}\sqrt{D^u_{[l+2]}}$$

# 18

# Sufficient Conditions on the Initial Data for the Formation of a Shock in the Evolution

In the present chapter we shall establish sharp sufficient conditions on the initial data for the formation of a shock in the evolution. We just investigate the problem in the isentropic irrotational case.

The set up is as in Theorem 17.1. Following this theorem we must find sharp sufficient conditions on the initial data which will guarantee that the function $\mu$ becomes zero somewhere on $\Sigma_t^{\epsilon_0}$ at some finite $t$. That value of $t$ shall then be $t_{*\epsilon_0}$. We shall use Lemma 8.10, Proposition 8.5, Proposition 8.6. Taking $t = s$ in Proposition 8.6 we obtain

$$\mu(s, u, \vartheta) = \mu_{[1],s}(u, \vartheta)\hat{\mu}_s(s, u, \vartheta) \tag{18.1}$$

where $\mu_{[1],s}(u, \vartheta)$ satisfies

$$\frac{1}{2} \leqslant \mu_{[1],s}(u, \vartheta) \leqslant \frac{3}{2} \tag{18.2}$$

while $\hat{\mu}_s(s, u, \vartheta)$ is given by

$$\hat{\mu}_s(s, u, \vartheta) = 1 + \hat{E}_s(u, \vartheta) \log(1 + s) \tag{18.3}$$

Here

$$\hat{E}_s(u, \vartheta) = \frac{E_s(u, \vartheta)}{\mu_{[1],s}(u, \vartheta)} \tag{18.4}$$

and according to (8.226),

$$|E_s(u, \vartheta) - \frac{1}{4}\ell P_s(u, \vartheta)| \leqslant C\delta_0(1 + s)^{-1}(1 + \log(1 + s)) \tag{18.5}$$

where $P_s(u, \vartheta)$ is the function defined by Lemma 8.10:

$$P_s(u, \vartheta) = (1 + s)(\underline{L}\psi_0)(s, u, \vartheta) \tag{18.6}$$

In (18.5) and in all the following the positive constant $\delta_0$ shall be as in Theorem 17.1. In view of the above, a sharp sufficient condition for $\mu$ to become zero somewhere on $\Sigma_s^{\epsilon_0}$ at some finite $s$, is an upper bound for $\min_{(u,\vartheta)\in[0,\epsilon_0]\times S^2} P_s(u, \vartheta)$ by a negative constant for all sufficiently large $s$ in the case that $\ell > 0$, a lower bound for $\max_{(u,\vartheta)\in[0,\epsilon_0]\times S^2} P_s(u,\vartheta)$ by a positive constant for all sufficiently large $s$ in the case that $\ell < 0$. As was already shown in Theorem 17.1 no shocks can form in the case $\ell = 0$ as in this case we have $t_{*\epsilon_0} = \infty$. Thus in the following we assume $\ell \neq 0$.

The sharp sufficient condition just stated is not a condition on the initial data. Let us revisit the wave equation for $\psi_0$ in the form encountered in the proof of Lemma 8.10 ((8.180) and (8.181)):

$$L(\underline{L}\psi_0) + \nu\underline{L}\psi_0 + \underline{\nu}L\psi_0 = \rho_0' \tag{18.7}$$

where

$$\rho_0' = \mu\not{\Delta}\psi_0 + \mu\not{g}^{-1}(\not{d}\log\Omega, \not{d}\psi_0) - 2\not{g}^{-1}(\zeta, \not{d}\psi_0) \tag{18.8}$$

Consider the function

$$\underline{T} = (1-u+t)\underline{L}\psi_0 - \psi_0 \tag{18.9}$$

We have

$$L\underline{T} = (1-u+t)L(\underline{L}\psi_0) - L\psi_0 + \underline{L}\psi_0 \tag{18.10}$$

Substituting for $L(\underline{L}\psi_0)$ from (18.7) yields

$$L\underline{T} = \omega \tag{18.11}$$

where

$$\omega = -((1-u+t)\nu - 1)\underline{L}\psi_0 - ((1-u+t)\underline{\nu}+1)L\psi_0 + (1-u+t)\rho_0' \tag{18.12}$$

From the conclusions (iii) of Theorem 17.1 we have, recalling

$$\nu = \frac{1}{2}(\mathrm{tr}\chi + L\log\Omega)$$
$$\underline{\nu} + \alpha^{-1}\kappa\nu = \frac{1}{2}\alpha^{-1}\kappa L\log\Omega + \frac{1}{2}\underline{L}\log\Omega + \kappa\mathrm{tr}\not{k}$$

the bounds

$$|(1-u+t)\nu - 1| \leqslant C\delta_0(1+t)^{-1}(1+\log(1+t)) \tag{18.13}$$
$$|(1-u+t)\underline{\nu} + 1| \leqslant C\delta_0(1+\log(1+t)) \tag{18.14}$$

Let now $f$ be an arbitrary function defined on $W^s_{\epsilon_0}$. Let us denote by $\bar{f}(t,u)$ the mean value of $f$ on $S_{t,u}$ with respect to $d\mu_{\tilde{\not{g}}}$:

$$\bar{f}(t,u) = \frac{1}{\tilde{\mathrm{Area}}(t,u)}\int_{S_{t,u}} f d\mu_{\tilde{\not{g}}}, \quad \tilde{\mathrm{Area}}(t,u) = \int_{S_{t,u}} d\mu_{\tilde{\not{g}}} \tag{18.15}$$

From the facts established in Chapter 5,

$$\frac{\partial}{\partial t}\int_{S_{t,u}} f d\mu_{\tilde{g}} = \int_{S_{t,u}} (Lf + 2\nu f) d\mu_{\tilde{g}} \tag{18.16}$$

$$\frac{\partial}{\partial t}\tilde{\text{Area}}(t,u) = \int_{S_{t,u}} 2\nu d\mu_{\tilde{g}} = 2\bar{\nu}\tilde{\text{Area}}(t,u) \tag{18.17}$$

Hence by direct calculation,

$$\frac{\partial \bar{f}}{\partial t}(t,u) = \frac{1}{\tilde{\text{Area}}(t,u)}\int_{S_{t,u}} (Lf + 2\nu f - 2\bar{\nu}\bar{f}) d\mu_{\tilde{g}} \tag{18.18}$$

Noting that for any pair of functions $f, g$ on $S_{t,u}$ we have

$$\int_{S_{t,u}} (fg - \bar{f}\bar{g}) d\mu_{\tilde{g}} = \int_{S_{t,u}} (f - \bar{f})(g - \bar{g}) d\mu_{\tilde{g}} \tag{18.19}$$

we can write (18.18) in the form

$$\frac{\partial \bar{f}}{\partial t}(t,u) = \overline{Lf} + 2\overline{(\nu - \bar{\nu})(f - \bar{f})} \tag{18.20}$$

We apply the formula (18.20) to the function $\underline{\tau}$ obtaining, in view of (18.11),

$$\frac{\partial \bar{\underline{\tau}}}{\partial t}(t,u) = \bar{\omega} + 2\overline{(\nu - \bar{\nu})(\underline{\tau} - \bar{\underline{\tau}})} \tag{18.21}$$

Integrating this with respect to $t$ on $[0, s]$, we obtain

$$\bar{\underline{\tau}}(s,u) = \bar{\underline{\tau}}(0,u) + \int_0^s (\bar{\omega} + 2\overline{(\nu - \bar{\nu})(\underline{\tau} - \bar{\underline{\tau}})})(t,u) dt \tag{18.22}$$

Finally we replace $u$ by $u' \in [0, u]$, multiply by $\tilde{\text{Area}}(0, u')$, and integrate with respect to $u'$ on $[0, u]$ to obtain

$$\int_0^u \bar{\underline{\tau}}(s,u')\tilde{\text{Area}}(0,u') du'$$
$$= \int_0^u \int_{S_{0,u'}} \underline{\tau} d\mu_{\tilde{g}} du' + \int_0^u \tilde{\text{Area}}(0,u')(\int_0^s (\bar{\omega} + 2\overline{(\nu - \bar{\nu})(\underline{\tau} - \bar{\underline{\tau}})})(t,u') dt) du' \tag{18.23}$$

We shall estimate the second integral on the right in (18.23). Interchanging the order of integration this integral is

$$I(s,u) = \int_0^s (\int_0^u \tilde{\text{Area}}(0,u')(\bar{\omega} + 2\overline{(\nu - \bar{\nu})(\underline{\tau} - \bar{\underline{\tau}})})(t,u') du') dt \tag{18.24}$$

Since $\tilde{\text{Area}}(0,u') \leqslant \tilde{\text{Area}}(0,0) = 4\pi$, we have

$$|I(s,u)| \leqslant C \int_0^s J(t,u) dt \tag{18.25}$$

where $J(t,u)$ is the integral

$$J(t,u) = \int_0^u (|\bar{\omega}| + 2\overline{|\nu - \bar{\nu}||\underline{\tau} - \bar{\underline{\tau}}|})(t,u')du' \tag{18.26}$$

We consider first the contribution of the term $|\bar{\omega}|$, through (18.26), to the integral on the right in (18.25). Now $\omega$ is given by (18.12), and we consider first the third term on the right in (18.12). We have

$$(1-u+t)\bar{\rho}_0'(t,u) = \frac{1-u+t}{\tilde{\text{Area}}(t,u)}\int_{S_{t,u}} \rho_0' d\mu_{\tilde{g\!\!/}} \tag{18.27}$$

Now for an arbitrary function $f$ on $S_{t,u}$, we have

$$\begin{aligned}
\tilde{\Delta\!\!\!/} f &= \frac{1}{\sqrt{\det \tilde{g\!\!/}}}\frac{\partial}{\partial \vartheta^A}((\tilde{g\!\!/}^{-1})^{AB}\sqrt{\det \tilde{g\!\!/}}\frac{\partial f}{\partial \vartheta^A}) \\
&= \frac{1}{\Omega\sqrt{\det g\!\!/}}\frac{\partial}{\partial \vartheta^A}((g\!\!/^{-1})^{AB}\sqrt{\det g\!\!/}\frac{\partial f}{\partial \vartheta^A}) = \Omega^{-1}\Delta\!\!\!/ f
\end{aligned} \tag{18.28}$$

Thus, integrating by parts on $S_{t,u}$ we obtain

$$\begin{aligned}
\int_{S_{t,u}} \mu \Delta\!\!\!/ \psi_0 d\mu_{\tilde{g\!\!/}} &= \int_{S_{t,u}} \mu\Omega\tilde{\Delta\!\!\!/}\psi_0 d\mu_{\tilde{g\!\!/}} \\
&= -\int_{S_{t,u}} (\mu\tilde{g\!\!/}^{-1}(d\!\!/\Omega, d\!\!/\psi_0) + \Omega\tilde{g\!\!/}^{-1}(d\!\!/\mu, d\!\!/\psi_0))d\mu_{\tilde{g\!\!/}} \\
&= -\int_{S_{t,u}} (\mu g\!\!/^{-1}(d\!\!/\log\Omega, d\!\!/\psi_0) + g\!\!/^{-1}(d\!\!/\mu, d\!\!/\psi_0))d\mu_{\tilde{g\!\!/}}
\end{aligned} \tag{18.29}$$

Recalling that

$$-(d\!\!/\mu + 2\zeta)\cdot g\!\!/^{-1} = -(\eta+\zeta)\cdot g\!\!/^{-1} = \Lambda$$

we then conclude that

$$\int_{S_{t,u}} \rho_0' d\mu_{\tilde{g\!\!/}} = \int_{S_{t,u}} \Lambda \cdot d\!\!/\psi_0 d\mu_{\tilde{g\!\!/}} \tag{18.30}$$

Now from conclusions (iii) of Theorem 17.1 we have

$$|\Lambda| \leqslant C\delta_0(1+t)^{-1}(1+\log(1+t)) \tag{18.31}$$

Thus (18.30) implies, through (18.27),

$$\begin{aligned}
(1-u+t)|\bar{\rho}_0'(t,u)| &\leqslant \frac{1-u+t}{(\text{Area}(t,u))^{1/2}}\sup_{S_{t,u}}|\Lambda|(\int_{S_{t,u}} |d\!\!/\psi_0|^2 d\mu_{g\!\!/})^{1/2} \\
&\leqslant C\delta_0(1+t)^{-1}(1+\log(1+t))(\int_{S_{t,u}} |d\!\!/\psi_0|^2 d\mu_{\tilde{g\!\!/}})^{1/2}
\end{aligned} \tag{18.32}$$

This contribution to (18.25) is then bounded by

$$C\delta_0\sqrt{u}\int_0^s(1+t)^{-1}(1+\log(1+t))(\int_{\Sigma_t^u}|\not{d}\psi_0|^2)^{1/2}dt$$

$$\leqslant C\delta_0\sqrt{u}(\int_0^s\int_{\Sigma_t^u}|\not{d}\psi_0|^2dt)^{1/2} \tag{18.33}$$

The integral on the right in (18.33) is

$$\int_{W_u^s}|\not{d}\psi_0|^2dtdu'd\mu_{\not{g}} \tag{18.34}$$

This integral has already been estimated in Chapter 7 by (7.146) for any variation. In particular for the first order variation $\psi_0$ we have

$$\int_{W_u^s}|\not{d}\psi_0|^2dtdu'd\mu_{\not{g}}\leqslant C(\overline{K}[\psi_0](s,u)+\int_0^u\mathcal{F}_0^s[\psi_0](u')du')\leqslant C\mathcal{E}_0^u[\psi_0](0) \tag{18.35}$$

The last step is by virtue of Theorem 5.1. Then (18.33) is bounded by

$$C\delta_0\sqrt{u}\sqrt{\mathcal{E}_0^u[\psi_0](0)} \tag{18.36}$$

Next, we consider the first two terms on the right in (18.12). By (18.13) the contribution of the first term is bounded by

$$\frac{C\delta_0(1+t)^{-1}(1+\log(1+t))}{\tilde{\text{Area}}(t,u)}\int_{S_{t,u}}|\underline{L}\psi_0|d\mu_{\tilde{\not{g}}}$$

$$\leqslant\frac{C\delta_0(1+t)^{-1}(1+\log(1+t))}{(\tilde{\text{Area}}(t,u))^{1/2}}(\int_{S_{t,u}}|\underline{L}\psi_0|^2d\mu_{\tilde{\not{g}}})^{1/2}$$

$$\leqslant C\delta_0(1+t)^{-2}(1+\log(1+t))(\int_{S_{t,u}}|\underline{L}\psi_0|^2d\mu_{\tilde{\not{g}}})^{1/2} \tag{18.37}$$

hence the corresponding contribution to $J(t,u)$ is bounded by

$$C\delta_0\sqrt{u}(1+t)^{-2}(1+\log(1+t))\sqrt{\mathcal{E}_0^u[\psi_0](t)} \tag{18.38}$$

and to the integral on the right in (18.25) by

$$C\delta_0\sqrt{u}(\int_0^s(1+t)^{-2}(1+\log(1+t))dt)\sqrt{\mathcal{E}_0^u[\psi_0](0)}\leqslant C\delta_0\sqrt{u}\sqrt{\mathcal{E}_0^u[\psi_0](0)} \tag{18.39}$$

by Theorem 5.1. The factor $\sqrt{u}$ comes from Holder Inequality when we obtain (18.38). By (18.14) the contribution of the second term on the right in (18.12) to $\tilde{\omega}$ is bounded by

$$\frac{C\delta_0(1+\log(1+t))}{\tilde{\text{Area}}(t,u)}\int_{S_{t,u}}|L\psi_0|d\mu_{\tilde{\not{g}}}$$

$$\leqslant\frac{C\delta_0(1+\log(1+t))}{(\tilde{\text{Area}}(t,u))^{1/2}}(\int_{S_{t,u}}|L\psi_0|^2d\mu_{\tilde{\not{g}}})^{1/2}$$

$$\leqslant C\delta_0(1+t)^{-1}(1+\log(1+t))(\int_{S_{t,u}}|L\psi_0|^2 d\mu_{\tilde{\not g}})^{1/2} \tag{18.40}$$

The contribution of this to the right of (18.25) is then bounded by

$$\begin{aligned}
&C\delta_0\sqrt{u}\int_0^s (1+t)^{-1}(1+\log(1+t))(\int_{\Sigma_t^u}|L\psi_0|^2)^{1/2}dt \\
&\leqslant C\delta_0\sqrt{u}(\int_0^s\int_{\Sigma_t^u}|L\psi_0|^2 dt)^{1/2} \\
&\leqslant C\delta_0\sqrt{u}(\int_0^u \mathcal{F}_0^s[\psi_0](u')du')^{1/2} \\
&\leqslant C\delta_0\sqrt{u}\sqrt{\mathcal{E}_0^u[\psi_0](0)}
\end{aligned} \tag{18.41}$$

by Theorem 5.1. Combining (18.36), (18.39), (18.41), we conclude that the contribution of $|\overline{\omega}|$ to the right of (18.25) is bounded by

$$C\delta_0\sqrt{u}\sqrt{\mathcal{E}_0^u[\psi_0](0)} \tag{18.42}$$

Finally, we consider the contribution of the term $2\overline{|\nu-\overline{\nu}||\underline{\tau}-\overline{\underline{\tau}}|}$ to the right of (18.25). We have

$$\overline{|\nu-\overline{\nu}||\underline{\tau}-\overline{\underline{\tau}}|} = \frac{1}{\tilde{\text{Area}}(t,u)}\int_{S_{t,u}}|\nu-\overline{\nu}||\underline{\tau}-\overline{\underline{\tau}}|d\mu_{\tilde{\not g}} \tag{18.43}$$

By (18.13),

$$\begin{aligned}
|\nu-\overline{\nu}| &\leqslant |\nu-\frac{1}{1-u+t}|+|\overline{\nu}-\frac{1}{1-u+t}| \\
&\leqslant C\delta_0(1+t)^{-2}(1+\log(1+t))
\end{aligned} \tag{18.44}$$

Therefore

$$\begin{aligned}
\overline{|\nu-\overline{\nu}||\underline{\tau}-\overline{\underline{\tau}}|} &\leqslant \frac{C\delta_0(1+t)^{-2}(1+\log(1+t))}{\tilde{\text{Area}}(t,u)}\int_{S_{t,u}}|\underline{\tau}-\overline{\underline{\tau}}|d\mu_{\tilde{\not g}} \\
&\leqslant \frac{C\delta_0(1+t)^{-2}(1+\log(1+t))}{(\tilde{\text{Area}}(t,u))^{1/2}}(\int_{S_{t,u}}|\underline{\tau}-\overline{\underline{\tau}}|^2 d\mu_{\tilde{\not g}})^{1/2}
\end{aligned} \tag{18.45}$$

We have

$$\begin{aligned}
\int_{S_{t,u}}|\underline{\tau}-\overline{\underline{\tau}}|^2 d\mu_{\tilde{\not g}} &= \int_{S_{t,u}}\underline{\tau}^2 d\mu_{\tilde{\not g}} - \int_{S_{t,u}}\overline{\underline{\tau}}^2 d\mu_{\tilde{\not g}} \\
&\leqslant \int_{S_{t,u}}\underline{\tau}^2 d\mu_{\tilde{\not g}} \leqslant C(1+t)^2\int_{S_{t,u}}|\underline{L}\psi_0|^2 d\mu_{\not g} + C\int_{S_{t,u}}|\psi_0|^2 d\mu_{\not g}
\end{aligned} \tag{18.46}$$

by (18.9). The contribution of the first integral on the right in (18.46) to the right-hand side of (18.45) is bounded by

$$C\delta_0(1+t)^{-2}(1+\log(1+t))(\int_{S_{t,u}}|\underline{L}\psi_0|^2 d\mu_{\not{g}})^{1/2} \tag{18.47}$$

This coincides with (18.37), hence the corresponding contribution is bounded by (18.39). Finally, the contribution of the second term on the right in (18.46) to (18.45) is bounded by

$$\begin{aligned} &C\delta_0(1+t)^{-3}(1+\log(1+t))(\int_{S_{t,u}}|\psi_0|^2 d\mu_{\not{g}})^{1/2} \\ &\leqslant C\delta_0(1+t)^{-3}(1+\log(1+t))\sqrt{u}\sqrt{\mathcal{E}_0^u[\psi_0](t)} \end{aligned} \tag{18.48}$$

by Lemma 5.1, hence the corresponding contribution to (18.25) is also bounded by (18.39). Combining the above results we conclude that the contribution of $2|\nu-\overline{\nu}||\underline{\tau}-\overline{\underline{\tau}}|$ to the right of (18.25) is also bounded by (18.42).

Then we have

$$|I(s,u)| \leqslant C\delta_0\sqrt{u}\sqrt{\mathcal{E}_0^u[\psi_0](0)} \tag{18.49}$$

It follows from (18.23) that

$$|\int_0^u \overline{\underline{\tau}}(s,u')\tilde{\text{Area}}(0,u')du' - \int_{\Sigma_0^u}\underline{\tau}d\mu_{\tilde{\not{g}}}du'| \leqslant C\delta_0\sqrt{u}\sqrt{\mathcal{E}_0^u[\psi_0](0)} \tag{18.50}$$

From (18.6) and (18.9) we have

$$P_s(u,\vartheta) = \underline{\tau} + \frac{1}{1-u+s}(u\underline{\tau} + (1+s)\psi_0) \tag{18.51}$$

Let us denote

$$Q(u) = \int_{\Sigma_0^u}\underline{\tau}d\mu_{\tilde{\not{g}}}du' \tag{18.52}$$

Consider the case $\ell > 0$. If the initial data satisfy

$$Q(u) \leqslant -2C\delta_0\sqrt{u}\sqrt{\mathcal{E}_0^u[\psi_0](0)} < 0 \tag{18.53}$$

for some $u \in (0,\epsilon_0]$ with the constant $C$ as in (18.50), by (18.50) we have

$$\int_0^u \overline{\underline{\tau}}(s,u')\tilde{\text{Area}}(0,u')du' \leqslant \frac{1}{2}Q(u) < 0 \tag{18.54}$$

It follows that

$$\min_{u' \in [0,\epsilon_0]} \overline{\underline{\tau}}(s,u') \leqslant \frac{(1/2)Q(u)}{\int_0^u \tilde{\text{Area}}(0,u')du'} < 0 \tag{18.55}$$

hence also

$$\min_{(u',\vartheta)\in[0,u]\times S^2} \underline{T}(s,u',\vartheta) \leqslant \frac{(1/2)Q(u)}{\int_0^u \tilde{\mathrm{Area}}(0,u')du'} \tag{18.56}$$

It then follows from (18.51), in view of (iii) of Theorem 17.1, that

$$\min_{(u',\vartheta)\in[0,u]\times S^2} P_s(u',\vartheta) \leqslant \frac{(1/4)Q(u)}{\int_0^u \tilde{\mathrm{Area}}(0,u')du'} < 0 \tag{18.57}$$

for all $s$ large enough so that

$$\frac{1}{2} + s \geqslant -C\delta_0 \frac{\int_0^u \tilde{\mathrm{Area}}(0,u')du'}{(1/8)Q(u)} \tag{18.58}$$

Thus by the discussion in the beginning of the chapter $t_{*\epsilon_0}$ is finite. In fact, (18.57) implies, in view of (18.5), that

$$\min_{(u',\vartheta)\in[0,u]\times S^2} E_s(u',\vartheta) \leqslant \frac{(1/32)\ell Q(u)}{\int_0^u \tilde{\mathrm{Area}}(0,u')du'} < 0 \tag{18.59}$$

for all $s$ large enough so that

$$\frac{1+s}{1+\log(1+s)} \geqslant -C\delta_0 \frac{\int_0^u \tilde{\mathrm{Area}}(0,u')du'}{(1/32)\ell Q(u)} \tag{18.60}$$

where the constant $C$ is the one in (18.5). It follows from (18.3) in view of (18.2) that $s$ is bounded from above according to

$$\log(1+s) \leqslant -\frac{\int_0^u \tilde{\mathrm{Area}}(0,u')du'}{(1/32)\ell Q(u)} \tag{18.61}$$

Obviously, the upper bound in (18.61) exceeds the lower bounds in (18.58) and (18.60), it follows that $t_{*\epsilon_0}$ satisfies the upper bound

$$\log(1+t_{*\epsilon_0}) \leqslant -\frac{\int_0^u \tilde{\mathrm{Area}}(0,u')du'}{(1/32)\ell Q(u)} \tag{18.62}$$

In the case $\ell < 0$ a similar argument shows that if the initial data satisfy .

$$Q(u) \geqslant 2C\delta_0\sqrt{u}\sqrt{\mathcal{E}_0^u[\psi_0](0)} > 0 \tag{18.63}$$

for some $u \in (0,\epsilon_0]$ with the constant $C$ as in (18.50), then $t_{*\epsilon_0}$ is finite, in fact satisfies the upper bound

$$\log(1+t_{*\epsilon_0}) \leqslant -\frac{\int_0^u \tilde{\mathrm{Area}}(0,u')du'}{(1/32)\ell Q(u)} \tag{18.64}$$

In fact, we have

$$\min_{(u',\vartheta)\in[0,u]\times S^2} E_s(u',\vartheta) \leqslant \frac{1}{4}\ell \max_{(u',\vartheta)\in[0,u]\times S^2} P_s(u',\vartheta) + C\delta_0(1+s)^{-1}(1+\log(1+s))$$

Then we get a similar bound as (18.59).

We obtain the first condition to guarantee the formation of shocks:

**Theorem 18.1.a** The quantity $Q(u)$ satisfies

$$Q(u) \geqslant 2C\delta_0\sqrt{u}\sqrt{\mathcal{E}_0^u[\psi_0](0)} > 0$$

for $\ell < 0$; and

$$Q(u) \leqslant -2C\delta_0\sqrt{u}\sqrt{\mathcal{E}_0^u[\psi_0](0)} < 0$$

for $\ell > 0$.

We can modify this condition so that it is expressed in the Euclidean space.

Let us replace $d\mu_{\tilde{g}}$ by $d\mu_{g}$, then since the difference of $\Omega$ from unity, its value in the constant state, is bounded on $\Sigma_0^u$ by $C\delta_0$, the change in the integral will be bounded by $C\delta_0\sqrt{u}\sqrt{\mathcal{E}_0^u[\psi_0](0)}$. Moreover we may replace $\underline{L}$ by the one corresponding to the constant state:

$$\mathring{\underline{L}} = \frac{\partial}{\partial x^0} - N^i\frac{\partial}{\partial x^i} \tag{18.65}$$

Then let us define

$$Q_1(u) = \int_{\Sigma_0^u} \mathcal{I}_0 d^3x$$

where

$$\mathcal{I}_0 = r\mathring{\underline{L}}\psi_0 - \psi_0 = r(\partial_0\psi_0 - N^i\partial_i\psi_0) - \psi_0$$

Since we are now on the initial hypersurface, the difference of

$$\int_{\Sigma_0^u} \mathcal{I}_0 d\mu_{g} du'$$

from $Q(u)$ will be also bounded by $C\delta_0\sqrt{u}\sqrt{\mathcal{E}_0^u[\psi_0](0)}$.

We obtain the second condition to guarantee the formation of shocks:

**Theorem 18.1.b** The quantity $Q_1(u)$ satisfies

$$Q_1(u) \geqslant 2C'\delta_0\sqrt{u}\sqrt{\mathcal{E}_0^u[\psi_0](0)} > 0$$

for $\ell < 0$; and

$$Q_1(u) \leqslant -2C'\delta_0\sqrt{u}\sqrt{\mathcal{E}_0^u[\psi_0](0)} < 0$$

for $\ell > 0$.

The above two conditions depend only on the variation $\psi_0$, but the quantities $Q(u)$ and $Q_1(u)$ involve the first derivatives of $\psi_0$. Now we shall obtain a third condition which depends on other variations $\psi_i, i = 1, 2, 3$ but involves only the variations themselves (actually on the enthalpy $h$ and the normal component of the fluid speed $v_N$).

Consider the wave equation (1.23) on $\Sigma_0$, we can write it in the following form:

$$\partial_0\psi_0 - \partial_i\psi_i = 2\psi_i\partial_0\psi_i - \psi_i\psi_j\partial_i\psi_j + (\eta^2 - 1)\Delta\phi$$

where the repeated index means summing. By the $L^\infty$ bounds for the first order variations established in the previous chapter, we can bound the $L^1$ norm on $\Sigma_0^u$ of $r$ times the right-hand side by $C\delta_0\sqrt{u}\sqrt{\mathcal{E}_{0,[1]}^u(0)}$. Here the $\sqrt{u}$ in the difference comes from the using of Holder inequality.

Therefore defining

$$Q_0(u) = \int_{\Sigma_0^u} (r(\partial_i\psi_i - N^i\partial_i\psi_0) - \psi_0)d\mu_{\not{g}}du' \tag{18.66}$$

we have

$$|Q_0(u) - Q(u)| \leqslant C'\sqrt{\epsilon_0}\delta_0\sqrt{\mathcal{E}_{0,[1]}(0)} \tag{18.67}$$

Hence if in the case $\ell > 0$ we have

$$Q_0(u) \leqslant -2(C' + C)\sqrt{\epsilon_0}\delta_0\sqrt{\mathcal{E}_{0,[1]}^u(0)} < 0 \quad \text{for some} \quad u \in (0, \epsilon_0] \tag{18.68}$$

where $C$ is in (18.53), then by direct calculation, (18.53) holds and we have

$$Q(u) \leqslant \frac{1}{2}Q_0(u) \tag{18.69}$$

hence, in place of (18.54),

$$\int_0^u \underline{T}(s, u')\tilde{\text{Area}}(0, u')du' \leqslant \frac{1}{4}Q_0(u) < 0 \tag{18.70}$$

Similarly, if in the case $\ell > 0$ we have

$$Q_0(u) \geqslant 2(C' + C)\sqrt{\epsilon_0}\delta_0\sqrt{\mathcal{E}_{0,[1]}^u(0)} > 0 \quad \text{for some} \quad u \in (0, \epsilon_0] \tag{18.71}$$

where $C$ is in (18.63), then (18.63) holds and we have

$$Q(u) \geqslant \frac{1}{2}Q_0(u) \tag{18.72}$$

hence

$$\int_0^u \underline{T}(s, u')\tilde{\text{Area}}(0, u')du' \geqslant \frac{1}{4}Q_0(u) > 0 \tag{18.73}$$

Integrating by parts and using divergence theorem,

$$\int_{\Sigma_0^u} r(\partial_i\psi_i - N^i\partial_i\psi_0)d^3x = \int_{S_{0,u}} r(\psi_0 - \psi_N)d\mu_{\not g} + \int_{\Sigma_0^u}(3\psi_0 - \psi_N)dx^3 \quad (18.74)$$

To see this, we just write

$$\int_{\Sigma_0^u} r\partial_i\psi_i d^3x = \int_{\Sigma_0^u}\partial_i(r\psi_i)d^3x - \int_{\Sigma_0^u}\psi_N d^3x$$

and

$$-\int_{\Sigma_0^u} rN^i\partial_i\psi_0 d^3x = \int_{\Sigma_0^u} r\partial_i N^i\psi_0 d^3x + \int_{\Sigma_0^u}\sum_i (N^i)^2\psi_0 d^3x - \int_{\Sigma_0^u}\partial_i(rN^i\psi_0)d^3x$$

We obtain the following condition to guarantee the formation of shocks:

**Theorem 18.1.c** The quantity

$$Q_0(u) = \int_{S_{0,u}} r(h + v_N) + \int_{\Sigma_0^u}(2h + v_N)$$

satisfies

$$Q_0(u) \geqslant 2C''\delta_0\sqrt{u}\sqrt{\mathcal{E}^u_{0,[1]}(0)}$$

for $\ell < 0$; and

$$Q_0(u) \leqslant -2C''\delta_0\sqrt{u}\sqrt{\mathcal{E}^u_{0,[1]}(0)}$$

for $\ell > 0$.

Under one of these three conditions, the lifespan $t_{*\epsilon_0}$ is finite, and it satisfies one of the following inequalities:

$$\log(1 + t_{*\epsilon_0}) \leqslant \frac{Cu}{|\ell||Q(u)|}$$

$$\log(1 + t_{*\epsilon_0}) \leqslant \frac{C'u}{|\ell||Q_1(u)|}$$

$$\log(1 + t^{*\epsilon_0}) \leqslant \frac{C''u}{|\ell||Q_0(u)|}$$

respectively.

# 19

# The Structure of the Boundary of the Domain of the Maximal Solution

## 19.1 Nature of Singular Hypersurface in Acoustical Differential Structure

### 19.1.1 Preliminary

In this chapter we shall take $\epsilon_0$ to be a variable with range $(0, 1/2]$, in agreement with the setup of Theorem 17.1. To emphasize that $\epsilon_0$ is now taken to be a variable, we denote it simply by $\epsilon$. Let us denote as in Chapter 2 by $t_*(u)$ the greatest lower bound of the parameter $t$ of the generators of $C_u$ in the domain of maximal solution, and by $t_{*\epsilon}$ the greatest lower bound of $t_*(u)$ for $u \in [0, \epsilon]$. Obviously $t_{*\epsilon}$ is a non-increasing function of $\epsilon$. Then Theorem 17.1 applies for each $\epsilon \in (0, \epsilon_0)$, where we now denote by $\epsilon_0$ the maximal real number in the interval $(0, 1/2]$ such that the smallness condition

$$C\sqrt{\epsilon_0}\sqrt{\mathcal{D}_{[l+2]}(\epsilon_0)} \leqslant \bar{\delta}_0, \quad \mathcal{D}_{[l+2]}(\epsilon_0) = \sum_{\alpha,i} \|\partial_i\psi_\alpha\|^2_{H_{l+1}(\Sigma_0^{\epsilon_0})}$$

holds. For any such $\epsilon$ we can find a $\delta_0 \in (0, \bar{\delta}_0]$ such that the smallness condition of Theorem 17.1 holds with $\epsilon$ in the role of $\epsilon_0$, that is, we have

$$C\sqrt{\epsilon}\sqrt{\mathcal{D}_{[l+2]}(\epsilon)} < \delta_0$$

($\sqrt{\epsilon}\sqrt{\mathcal{D}_{[l+2]}(\epsilon)}$ is an increasing function in $\epsilon$.)

Now according to Theorem 17.1, for each $\epsilon \in (0, \epsilon_0)$, the solution, which exists in the classical sense in $W_\epsilon^{t_{*\epsilon}} \setminus \Sigma_{t_{*\epsilon}}^\epsilon$ extends smoothly in acoustical coordinates to $\Sigma_{t_{*\epsilon}}^\epsilon$, but there is a non-empty set of points $K_\epsilon \subset \Sigma_{t_{*\epsilon}}^\epsilon$ where $\mu$ vanishes, being positive on the complement of $K_\epsilon$ in $\Sigma_{t_{*\epsilon}}^\epsilon$. Thus the set $K_\epsilon$ is also the set of minima of the function $\mu$ on $\Sigma_{t_{*\epsilon}}^\epsilon$. If $q$ is a point of $K_\epsilon$, then either $q$ is an interior minimum of $\mu$ on $\Sigma_{t_{*\epsilon}}^\epsilon$, in which case we have

$$\mu(q) = 0, \quad \frac{\partial\mu}{\partial u}(q) = 0, \quad \frac{\partial\mu}{\partial\vartheta^A}(q) = 0, \quad A = 1, 2, \quad \frac{\partial^2\mu}{\partial u^2}(q) \geqslant 0 \tag{19.1}$$

at an interior zero, and we shall assume that the non-degeneracy condition

$$\frac{\partial^2 \mu}{\partial u^2}(q) > 0 \quad \text{at an interior zero} \tag{19.2}$$

holds, or $q$ is a boundary minimum, necessarily on $S_{t_{*\epsilon},\epsilon}$ (for on the other boundary of $\Sigma^{\epsilon}_{t_{*\epsilon}}$, namely on $S_{t_{*\epsilon},0}$, the constant state holds), in which case we have

$$\mu(q) = 0, \quad \frac{\partial \mu}{\partial u}(q) \leqslant 0, \quad \frac{\partial \mu}{\partial \vartheta^A}(q) = 0, \quad A = 1, 2 \tag{19.3}$$

and we shall assume the non-degeneracy condition

$$\frac{\partial \mu}{\partial u}(q) < 0 \quad \text{at a boundary zero} \tag{19.4}$$

holds.

Let now $u_m(\epsilon)$ be the minimal value of $u \in (0, \epsilon)$ for which there is an interior zero of $\mu$ on $\Sigma^{u}_{t_{*\epsilon}}$. Then if we replace $\epsilon$ by $\epsilon' \in (0, \epsilon]$, as we decrease $\epsilon'$, it follows directly from the definitions that $t_{*\epsilon'}$ stays constant up to the point where $\epsilon'$ reaches the value $u_m(\epsilon)$, that is , we have

$$t_{*\epsilon'} = t_{*\epsilon} \quad \text{for all} \quad \epsilon' \in [u_m(\epsilon), \epsilon] \tag{19.5}$$

Thereafter $t_{*\epsilon'}$ increases, $\mu$ being everywhere positive on $\Sigma^{\epsilon'}_{t_{*u_m(\epsilon)}}$, and for a while $K_{\epsilon'}$ consists only of boundary zeros. This situation either persists for all $\epsilon' \in (0, u_m(\epsilon))$ or, as we decrease $\epsilon'$ there is a first $\epsilon_1 \in (0, u_m(\epsilon))$ at which one or more interior zeros of $\mu$ appear on $\Sigma^{\epsilon_1}_{t_{*\epsilon_1}}$, from which the preceding repeat with $\epsilon_1$ in the role of $\epsilon$.

In the following, we shall use the conclusion in (iv) of Theorem 17.1, namely, for every $\epsilon \in (0, \epsilon_0)$ the following upper bound for $L\mu$ holds in the region $\mathcal{U}_\epsilon \subset W^{t_{*\epsilon}}_{\epsilon}$ where $\mu < 1/4$:

$$L\mu \leqslant -C^{-1}(1+t)^{-1}(1+\log(1+t))^{-1} \quad \text{in} \quad \mathcal{U}_\epsilon \tag{19.6}$$

What we shall use is just the fact that for finite $t$, $L\mu$ is bounded from above by a negative constant in the region where $\mu < 1/4$.

Let us denote

$$V_{\epsilon_0} = \bigcup_{\epsilon \in (0,\epsilon_0)} W^{t_{*\epsilon}}_{\epsilon} \tag{19.7}$$

This is the domain covered by Theorem 17.1, applied in the above manner (we are disregarding here the exterior domain where the constant state holds). This domain is contained in the closure of $W_{\epsilon_0}$, the domain of maximal solution, and the set

$$J_{\epsilon_0} = \bigcup_{\epsilon \in (0,\epsilon_0)} K_\epsilon \subset V_{\epsilon_0} \tag{19.8}$$

of zeros of $\mu$ in $V_{\epsilon_0}$ is part of the singular boundary of $W_{\epsilon_0}$.

Our purpose in the following is to describe the acoustical spacetime structure in a neighborhood of a point of the singular boundary of $W_{\epsilon_0}$. Consider now a value $\epsilon \in (0, \epsilon_0)$ for which there is an interior zero of $\mu$ on $\Sigma^{\epsilon}_{t_{*\epsilon}}$, that is, there is a $u_0 \in (0, \epsilon)$ and a zero of $\mu$ belonging to $S_{t_{*\epsilon},u_0}$. By (19.2) the minimum of $\frac{\partial^2 \mu}{\partial u^2}$ over all zeros of $\mu$ on $S_{t_{*\epsilon},u_0}$ is positive. It follows that there are $u_1 \in [0, u_0)$, $u_2 \in (u_0, \epsilon]$ such that in the annular region

$$\bigcup_{u \in (u_1, u_2)} S_{t_{*\epsilon},u} \tag{19.9}$$

in $\Sigma^{\epsilon}_{t_{*\epsilon}}$ there are no zeros of $\mu$ except on $S_{t_{*\epsilon},u_0}$ itself. If $u_0 = u_m(\epsilon)$, there are no zeros of $\mu$ on $S_{t_{*\epsilon},u}$ for any $u \in [0, u_0)$ and we can take $u_1 = 0$.

According to the discussion above, a boundary zero of $\mu$ on $\Sigma^{\epsilon}_{t*\epsilon}$ belongs to the surface $S_{t_{*\epsilon},\epsilon}$ and by (19.4) there is a $u_1 < \epsilon$ such that there are no zeros of $\mu$ on $S_{t_{*\epsilon},u}$ for any $u \in (u_1, \epsilon)$. Moreover, there is a $u_0 > \epsilon$ such that for $\epsilon' \in [\epsilon, u_0]$ there is a corresponding boundary zero of $\mu$ on $\Sigma^{\epsilon'}_{t_{*\epsilon'}}$ belonging to the surface $S_{t_{*\epsilon'},\epsilon'}$ and no interior zeros, and there is a $u_2 > u_0$ such that for $\epsilon' \in (u_0, u_2]$ we have $t_{*\epsilon'} = t_{*u_0}$ and on $\Sigma_{t_{*\epsilon'}}$ we have an interior zero of $\mu$ on $S_{t_{*\epsilon'},u_0}$ and no zeros of $\mu$ on $S_{t_{*\epsilon'},u}$, for any $u \in [0, \epsilon'), u \neq u_0$ (otherwise, the solution can not be extended to $t_{*\epsilon}$, which is larger than $t_{*\epsilon'}$). In this case we consider the annular region

$$\bigcup_{u \in (u_1, u_2)} S_{t_{*u_2},u} \tag{19.10}$$

in $\Sigma^{u_2}_{t_{*u_2}}$.

### 19.1.2 Intrinsic View Point

Consider the manifold

$$\mathcal{M}_{\epsilon_0} = [0, \infty) \times [0, \epsilon_0) \times S^2 \tag{19.11}$$

of all possible $(t, u, \vartheta)$ values with $u \in [0, \epsilon_0)$. Then the domain $V_{\epsilon_0}$ in acoustical spacetime corresponds to the domain

$$\mathcal{V}_{\epsilon_0} = \{(t, u, \vartheta) \in \mathcal{M}_{\epsilon_0}, t \leqslant t_{*u}\} \tag{19.12}$$

in $\mathcal{M}_{\epsilon_0}$ and (19.9) corresponds to

$$\{t_{*\epsilon}\} \times (u_1, u_2) \times S^2 \tag{19.13}$$

The mapping

$$(t, u, \vartheta) \in \mathcal{V}_{\epsilon_0} \mapsto x(t, u, \vartheta) \in V_{\epsilon_0}, \quad x = (x^\alpha, \alpha = 0, 1, 2, 3) \tag{19.14}$$

where $x^\alpha$ are rectangular coordinates in Galilean spacetime, is a homeomorphism of $\mathcal{V}_{\epsilon_0}$ onto $V_{\epsilon_0}$, which is locally also a diffeomorphism except at the points of the subset $\mathcal{J}_{\epsilon_0}$, the image of which is $J_{\epsilon_0}$, where $\mu$ vanishes. On the domain (19.12) we have the acoustical metric $g$, given by (2.41),

$$g = -2\mu dt du + \alpha^{-2}\mu^2 du^2 + \not{g}_{AB}(d\vartheta^A + \Xi^A du)(d\vartheta^B + \Xi^B du) \tag{19.15}$$

Recall that $\alpha(t, u, \vartheta)$ is a positive function near 1 and that at each $(t, u)$, $\not{g}_{AB}(t, u, \vartheta)$ are the components of a positive definite metric on $S^2$. The components of $g$ are smooth functions of the acoustical coordinates in the domain $\mathcal{V}_{\epsilon_0}$, including the boundary of this domain. Moreover, the properties of $\alpha$ and $\not{g}_{AB}$ just mentioned hold on the boundary of $\mathcal{V}_{\epsilon_0}$ as well. However on the subset $\mathcal{J}_{\epsilon_0}$ of this boundary $\mu$ vanishes, hence, in view of the fact that

$$\sqrt{-\det g} = \mu\sqrt{\det \not{g}} \tag{19.16}$$

the metric (19.15) degenerates on $\mathcal{J}_{\epsilon_0}$.

We now consider a smooth extension of the acoustical metric $g$ to $t > t_*(u)$. The extension is to satisfy the following two conditions:

(i) The function $\alpha$ remains near 1, and $\not{g}$ remains a positive definite metric on $S^2$.

(ii) The function $\mu$ extends in such a way that $\frac{\partial\mu}{\partial t}$ is bounded from above by a negative constant where $\mu < 1/4$.

Then for each $(u, \vartheta) \in (0, \epsilon_0) \times S^2$ the following alternative holds: Either there is a first $t_*(u, \vartheta) \geqslant t_*(u)$ where $\mu$ vanishes, or $\mu$ is positive for all $t \geqslant t_*(u)$. Consider the subset $\tilde{\mathcal{D}} \subset (0, \epsilon_0) \times S^2$ consisting of those points where the first alternative holds. Then by Property (ii) of the extension $\tilde{\mathcal{D}}$ is an open set. In the case of an interior zero of $\mu$ on $\Sigma^{\epsilon}_{t_{*\epsilon}}$ belonging to $S_{t_{*\epsilon}, u_0}$ we have $t_{*\epsilon} = t_*(u_0)$, there is a $\vartheta_0 \in S^2$ such that $\mu(t_*(u_0), u_0, \vartheta_0) = 0$. Therefore $(u_0, \vartheta_0) \in \tilde{\mathcal{D}}$. In the case of a boundary zero on $\Sigma^{\epsilon}_{t_{*\epsilon}}$, for each $\epsilon' \in [\epsilon, u_0]$ there is a $\vartheta_*(\epsilon') \in S^2$ such that $\mu(t_*(\epsilon'), \epsilon', \vartheta_*(\epsilon')) = 0$. Therefore in this case $\tilde{\mathcal{D}}$ contains the curve

$$\{(\epsilon', \vartheta_*(\epsilon')), \epsilon' \in [\epsilon, u_0]\} \tag{19.17}$$

Consider

$$\tilde{\mathcal{H}} = \{(t_*(u, \vartheta), u, \vartheta), (u, \vartheta) \in \tilde{\mathcal{D}}\} \tag{19.18}$$

This is the zero level set of the function $\mu$ over the domain $\tilde{\mathcal{D}} \subset (0, \epsilon_0) \times S^2$. Now $\mu$ is a smooth function which by (ii) has no critical points on its zero level set. Therefore $\tilde{\mathcal{H}}$ is a smooth graph.

Since $\tilde{\mathcal{H}}$ is a graph over $\tilde{\mathcal{D}} \subset (0, \epsilon_0) \times S^2$, $(u, \vartheta)$ can be used as coordinates on $\tilde{\mathcal{H}}$, and in these coordinates $g_*$, the metric induced on $\tilde{\mathcal{H}}$, is given by, from (19.15),

$$g_* = (\not{g}_*)_{AB}(d\vartheta^A + \Xi^A_* du)(d\vartheta^B + \Xi^B_* du) \tag{19.19}$$

where

$$(\not{g}_*)_{AB}(u, \vartheta) = \not{g}_{AB}(t_*(u, \vartheta), u, \vartheta) \tag{19.20}$$

are the components of a positive definite metric on $S^2$, and

$$\Xi_*^A(u,\vartheta) = \Xi^A(t_*(u,\vartheta),u,\vartheta) \tag{19.21}$$

Now the metric (19.19) is degenerate:

$$\det g_* = 0 \tag{19.22}$$

### 19.1.3 Invariant Curves

We see that although the hypersurface $\tilde{\mathcal{H}}$ is singular, being a hypersurface where the spacetime metric $g$ degenerates, from the point of view of its intrinsic geometry it is just like a regular null hypersurface in a regular spacetime. At each point $q \in \tilde{\mathcal{H}}$ there is a unique line $L_q \subset T_q\tilde{\mathcal{H}}$ which we may consider to be the linear span of a non-zero null vector $V(q)$. Thus we have a null vectorfield $V$ on $\tilde{\mathcal{H}}$ and we can express it in terms of the coordinates $(u,\vartheta)$ by

$$V = V^u\frac{\partial}{\partial u} + \not{V}, \quad \not{V} = V^A\frac{\partial}{\partial\vartheta^A} \tag{19.23}$$

At each $q = (t_*(u,\vartheta),u,\vartheta) \in \tilde{\mathcal{H}}$, the vector $\not{V}(q)$ is tangent to $S_{*u}$ of constant $u$ through $q$, which projects to the domain $\{u\} \times B_u$ where $B_u$ is the domain

$$B_u = \{\vartheta \in S^2, (u,\vartheta) \in \tilde{\mathcal{D}}\} \subset S^2$$

Without loss of generality, we may set $V_u = 1$. By direct calculation we have

$$\begin{aligned} 0 = g(V,V) = g_*(V,V) &= \not{g}_*(\not{V},\not{V}) + 2\not{g}_*(\Xi_*,\not{V}) + \not{g}_*(\Xi_*,\Xi_*) \\ &= \not{g}_*(\not{V}+\Xi_*,\not{V}+\Xi_*) \end{aligned} \tag{19.24}$$

Since $\not{g}_*$ is positive definite, this holds if and only if

$$\not{V} = -\Xi_* \tag{19.25}$$

Therefore $V$ is expressed in $(u,\vartheta)$ coordinates on $\tilde{\mathcal{H}}$ by

$$V = \frac{\partial}{\partial u} - \Xi_* \tag{19.26}$$

Now, let

$$X(q) = X^t(q)\frac{\partial}{\partial t} + X^u(q)\frac{\partial}{\partial u} + X^A\frac{\partial}{\partial\vartheta^A} \tag{19.27}$$

be an arbitrary vector in $T_q\mathcal{M}_{\epsilon_0}, q \in \tilde{\mathcal{H}}$. Then $X(q) \in T_q\tilde{\mathcal{H}}$ if and only if

$$X(q)(t - t_*(u,\vartheta)) = 0 \tag{19.28}$$

Substituting (19.27) this is equivalent to

$$X^t(q) = \frac{\partial t_*}{\partial u} X^u(q) + \frac{\partial t_*}{\partial \vartheta^A} X^A(q) \tag{19.29}$$

In particular the vectorfield $V$ as a vectorfield in spacetime along $\tilde{\mathcal{H}}$ is expressed by

$$V = (\frac{\partial t_*}{\partial u} - \Xi_*^A \frac{\partial t_*}{\partial \vartheta^A})\frac{\partial}{\partial t} + \frac{\partial}{\partial u} - \Xi_*^A \frac{\partial}{\partial \vartheta^A} \tag{19.30}$$

Here, $\frac{\partial}{\partial u}$ and $\frac{\partial}{\partial \vartheta^A}$ are the acoustical coordinate vectorfields in spacetime, not in $\tilde{\mathcal{H}}$. In fact, we have

$$\frac{\partial}{\partial u} = \frac{\partial t_*}{\partial u}\frac{\partial}{\partial t} + \frac{\partial}{\partial u}$$
$$\frac{\partial}{\partial \vartheta^A} = \frac{\partial t_*}{\partial \vartheta^A}\frac{\partial}{\partial t} + \frac{\partial}{\partial \vartheta^A}$$

The vectorfields on the left-hand side are the coordinate vectorfields in $\tilde{\mathcal{H}}$.

Given any vector $X(q) \in T_q\mathcal{M}_{\epsilon_0}, q \in \tilde{\mathcal{H}}$, there is a unique real number $\lambda$ such that $X(q) - \lambda L(q) \in T_q\tilde{\mathcal{H}}$. In fact it is readily seen from (19.29), recalling that $L = \frac{\partial}{\partial t}$, that

$$\lambda = X^t(q) - \frac{\partial t_*}{\partial u} X^u(q) - \frac{\partial t_*}{\partial \vartheta^A} X^A(q)$$

This defines a projection operator $\Pi_*$ from $T_q\mathcal{M}_{\epsilon_0}$ to $T_q\tilde{\mathcal{H}}$ by $\Pi_* X(q) = X(q) - \lambda L(q)$. We call $\Pi_*$ *the* $L$ *projection to* $\tilde{\mathcal{H}}$. Obviously,

$$V = \Pi_* T \tag{19.31}$$

We call the integral curves of $V$ the *invariant curves*. The singular surface $\tilde{\mathcal{H}}$ is ruled by these curves. The invariant curves as 1-dimensional submanifolds of $\tilde{\mathcal{H}}$ are independent of the choice of acoustical function $u$, being the integral manifolds of the 1-dimensional distribution $\{L_q, q \in \tilde{\mathcal{H}}\}$ on $\tilde{\mathcal{H}}$. The invariant curves have zero arc length.

Now we may adapt the coordinates $(u, \vartheta)$ on $\tilde{\mathcal{H}}$ so that the coordinate lines $\vartheta =$ const coincide with the invariant curves. Obviously, this choice is equivalent to the condition

$$\Xi_* = 0 \tag{19.32}$$

We then call the corresponding acoustical coordinates *canonical*. In these coordinates (19.19) takes the form

$$g_* = (\not{g}_*)_{AB} d\vartheta^A d\vartheta^B \tag{19.33}$$

By (19.32), we have, for $t \leqslant t_*(u, \vartheta)$,

$$\Xi^A(t, u, \vartheta) = -\int_t^{t_*(u,\vartheta)} (\frac{\partial \Xi^A}{\partial t})(t', u, \vartheta) dt'$$

Since also

$$\mu(t,u,\vartheta) = -\int_t^{t_*(u,\vartheta)} (\frac{\partial \mu}{\partial t})(t',u,\vartheta)dt'$$

it follows that the functions

$$\hat{\Xi}^A = \mu^{-1}\Xi^A \tag{19.34}$$

which a priori are defined only for $t < t_*(u,\vartheta)$, are actually given by

$$\hat{\Xi}^A(t,u,\vartheta) = \frac{\text{mean value on}[t,t_*(u,\vartheta)]\text{of}\{(\partial\Xi^A/\partial t)(,u,\vartheta)\}}{\text{mean value on}[t,t_*(u,\vartheta)]\text{of}\{(\partial\mu/\partial t)(,u,\vartheta)\}} \tag{19.35}$$

hence extend smoothly to $t = t_*(u,\vartheta)$, that is, to $\tilde{\mathcal{H}}$.

### 19.1.4 Extrinsic View Point

We now consider the character of the singular hypersurface $\tilde{\mathcal{H}}$ from the extrinsic point of view. To do this, we consider the reciprocal acoustical metric, a quadratic form in the cotangent space to the spacetime manifold at each regular point, given by

$$g^{-1} = -(1/2\mu)(L\otimes \underline{L} + \underline{L}\otimes L) + (\not{g}^{-1})^{AB}X_A\otimes X_B \tag{19.36}$$

Since

$$L = \frac{\partial}{\partial t}, \quad \underline{L} = \alpha^{-1}\kappa L + 2T = \alpha^{-2}\mu\frac{\partial}{\partial t} + 2(\frac{\partial}{\partial u} - \mu\hat{\Xi}^A\frac{\partial}{\partial\vartheta^A})$$

(19.36) takes the form in canonical acoustical coordinates:

$$\begin{aligned}\mu g^{-1} = &-(\frac{\partial}{\partial t}\otimes\frac{\partial}{\partial u} + \frac{\partial}{\partial u}\otimes\frac{\partial}{\partial t}) - \mu\alpha^{-2}\frac{\partial}{\partial t}\otimes\frac{\partial}{\partial t}\\ &+\mu(\frac{\partial}{\partial t}\otimes\hat{\Xi}^A\frac{\partial}{\partial\vartheta^A} + \hat{\Xi}^A\frac{\partial}{\partial\vartheta^A}\otimes\frac{\partial}{\partial t}) + \mu(\not{g}^{-1})^{AB}\frac{\partial}{\partial\vartheta^A}\otimes\frac{\partial}{\partial\vartheta^B}\end{aligned} \tag{19.37}$$

We see that although, due to the degeneracy of $\not{g}$ on $\tilde{\mathcal{H}}$, $g^{-1}$ blows up at $\tilde{\mathcal{H}}$, $\mu g^{-1}$ in fact extends smoothly to $\tilde{\mathcal{H}}$. The character of $\tilde{\mathcal{H}}$ from the extrinsic point of view is then determined by the sign of the invariant $\mu(g^{-1})^{\alpha\beta}\partial_\alpha\mu\partial_\beta\mu$ on $\tilde{\mathcal{H}}$. We have

$$\begin{aligned}&\mu(g^{-1})^{\alpha\beta}\partial_\alpha\mu\partial_\beta\mu\\ &= -2\frac{\partial\mu}{\partial t}\frac{\partial\mu}{\partial u} + \mu(-\alpha^{-2}(\frac{\partial\mu}{\partial t})^2 + 2\frac{\partial\mu}{\partial t}\frac{\partial\mu}{\partial\vartheta^A}\hat{\Xi}^A + (\not{g}^{-1})^{AB}\frac{\partial\mu}{\partial\vartheta^A}\frac{\partial\mu}{\partial\vartheta^B})\end{aligned} \tag{19.38}$$

On $\tilde{\mathcal{H}}$, $\mu$ vanishes and this reduces to, simply

$$\mu(g^{-1})^{\alpha\beta}\partial_\alpha\mu\partial_\beta\mu = -2\frac{\partial\mu}{\partial t}\frac{\partial\mu}{\partial u} \tag{19.39}$$

In view of condition (ii), $\mu(g^{-1})^{\alpha\beta}\partial_\alpha\mu\partial_\beta\mu$ is $>0$, $=0$, $<0$, at a point $q\in\tilde{\mathcal{H}}$ according as to whether $(\partial\mu/\partial u)(q)$ is $>0$, $=0$, $<0$. We conclude that $\tilde{\mathcal{H}}$ is spacelike, null, or timelike, at $q$ according as to whether $(\partial\mu/\partial u)(q)$ is $<0$, $=0$, $>0$. Now the boundary of the domain of the maximal solution cannot be timelike at any point. Therefore the timelike part of $\tilde{\mathcal{H}}$ cannot be part of the boundary of the domain of maximal solution. So we shall focus on the spacelike part of $\tilde{\mathcal{H}}$, which we denote by $\mathcal{H}$ and its boundary $\partial\mathcal{H}$, where $\tilde{\mathcal{H}}$ is null.

Consider then the open subset $\mathcal{D}\subset\tilde{\mathcal{D}}$ defined by

$$\mathcal{D}=\{(u,\vartheta)\in\tilde{\mathcal{D}},(\partial\mu/\partial u)(t_*(u,\vartheta),u,\vartheta)<0\} \tag{19.40}$$

(canonical acoustical coordinates). The boundary of $\mathcal{D}$ in $\tilde{\mathcal{D}}$ is given by

$$\partial\mathcal{D}=\{(u,\vartheta)\in\tilde{\mathcal{D}},(\partial\mu/\partial u)(t_*(u,\vartheta),u,\vartheta)=0\} \tag{19.41}$$

Then $\mathcal{H}$ is the graph (19.18) over $\mathcal{D}$:

$$\mathcal{H}=\{(t_*(u,\vartheta),u,\vartheta),(u,\vartheta)\in\mathcal{D}\} \tag{19.42}$$

and $\partial\mathcal{H}$ is the graph (19.18) over $\partial\mathcal{D}$:

$$\partial\mathcal{H}=\{(t_*(u,\vartheta),u,\vartheta),(u,\vartheta)\in\partial\mathcal{D}\} \tag{19.43}$$

Assuming that the non-degeneracy condition

$$(u,\vartheta)\in\partial\mathcal{D}\quad\text{implies}\quad(\partial^2\mu/\partial u^2)(t_*(u,\vartheta),u,\vartheta)\neq 0 \tag{19.44}$$

holds, $\partial\mathcal{D}$ splits into the disjoint union $\partial\mathcal{D}=\partial_-\mathcal{D}\bigcup\partial_+\mathcal{D}$ where

$$\begin{aligned}\partial_-\mathcal{D}&=\{(u,\vartheta)\in\partial\mathcal{D},(\partial^2\mu/\partial u^2)(t_*(u,\vartheta),u,\vartheta)>0\}\\ \partial_+\mathcal{D}&=\{(u,\vartheta)\in\partial\mathcal{D},(\partial^2\mu/\partial u^2)(t_*(u,\vartheta),u,\vartheta)<0\}\end{aligned} \tag{19.45}$$

The boundary $\partial\mathcal{H}$ of $\mathcal{H}$ in $\tilde{\mathcal{H}}$ similarly splits into the disjoint union $\partial\mathcal{H}=\partial_-\mathcal{H}\bigcup\partial_+\mathcal{H}$ where $\partial_-\mathcal{H}$ and $\partial_+\mathcal{H}$ are the graphs (19.18) over $\partial_-\mathcal{D}$ and $\partial_+\mathcal{D}$ respectively:

$$\begin{aligned}\partial_-\mathcal{H}&=\{(t_*(u,\vartheta),u,\vartheta),(u,\vartheta)\in\partial_-\mathcal{D}\}\\ \partial_+\mathcal{H}&=\{(t_*(u,\vartheta),u,\vartheta),(u,\vartheta)\in\partial_+\mathcal{D}\}\end{aligned} \tag{19.46}$$

If there is an interior zero of $\mu$ on $\Sigma^\epsilon_{t_{*\epsilon}}$ corresponding to $(u_0,\vartheta_0)$, then $(u_0\vartheta_0)$ belongs to $\partial_-\mathcal{D}$. If there is a boundary zero of $\mu$ on $\Sigma^\epsilon_{t_{*\epsilon}}$, then by (19.4), $\mathcal{D}$ contains the curve (19.17) except for the point $(u_0,\vartheta_*(u_0))$ which belongs to $\partial_-\mathcal{D}$.

Now on $\tilde{\mathcal{H}}$ we have $\mu=0$, that is, we have

$$\mu(t_*(u,\vartheta),u,\vartheta)=0\quad\text{for all}\quad(u,\vartheta)\in\tilde{\mathcal{D}} \tag{19.47}$$

Differentiating this equation implicitly with respect to $u$ we obtain

$$\frac{\partial t_*}{\partial u}(u,\vartheta)=-\frac{\partial\mu/\partial u}{\partial\mu/\partial t}(t_*(u,\vartheta),u,\vartheta) \tag{19.48}$$

By virtue of property (ii) we have

$$\mathcal{D} = \{(u, \vartheta) \in \tilde{\mathcal{D}}, (\partial t_*/\partial u)(u, \vartheta) < 0\} \tag{19.49}$$

and

$$\partial\mathcal{D} = \{(u, \vartheta) \in \tilde{\mathcal{D}}, (\partial t_*/\partial u)(u, \vartheta) = 0\} \tag{19.50}$$

(canonical acoustical coordinates).

Moreover, differentiating again (19.48) with respect to $u$ and evaluating the result on $\partial\mathcal{D}$, we obtain, in view of (19.41) and (19.50),

$$\frac{\partial^2 t_*}{\partial u^2}(u, \vartheta) = -\frac{\partial^2 \mu/\partial u^2}{\partial \mu/\partial t}(t_*(u, \vartheta), u, \vartheta) \quad \text{on} \quad \partial\mathcal{D} \tag{19.51}$$

Comparing with (19.45) we conclude that

$$\begin{aligned} \partial_-\mathcal{D} &= \{(u, \vartheta) \in \partial\mathcal{D}, (\partial^2 t_*/\partial u^2)(u, \vartheta) > 0\} \\ \partial_+\mathcal{D} &= \{(u, \vartheta) \in \partial\mathcal{D}, (\partial^2 t_*/\partial u^2)(u, \vartheta) < 0\} \end{aligned} \tag{19.52}$$

Thus if we consider a connected component of $\mathcal{H}$ and the corresponding components of $\partial_-\mathcal{H}$, $\partial_+\mathcal{H}$, then the component of $\partial_-\mathcal{H}$, which is not empty (in fact, the point on $\partial_-\mathcal{H}$ corresponds the interior zero. If there is no interior zero, this point corresponds to $u = \epsilon_0$), is the past boundary of $\mathcal{H}$, the component of $\partial_+\mathcal{H}$, which may be empty ($\mathcal{H}$ could be the asymptote of $C_0$), its future boundary, the function $t_*$ reaching a minimum, along each invariant curve, at $\partial_-\mathcal{H}$, a maximum at $\partial_+\mathcal{H}$.

Consider now the function $f = \partial\mu/\partial u$ on $\tilde{\mathcal{H}}$. In canonical acoustical coordinates on $\tilde{\mathcal{H}}$ we have

$$f(u, \vartheta) = \frac{\partial \mu}{\partial u}(t_*(u, \vartheta), u, \vartheta) \tag{19.53}$$

Differentiating with respect to $u$, and evaluating the result on $\partial\mathcal{H}$, we obtain, in view of (19.50),

$$\frac{\partial f}{\partial u}(u, \vartheta) = \frac{\partial^2 \mu}{\partial u^2}(t_*(u, \vartheta), u, \vartheta) \quad \text{on} \quad \partial\mathcal{H} \tag{19.54}$$

Thus $\partial\mathcal{H}$ is the zero level set of a smooth function, namely $f$, on the smooth manifold $\tilde{\mathcal{H}}$, and by (19.54) and the non-degeneracy condition (19.44) this is a non-critical level set. It follows that $\partial\mathcal{H}$ and its two components $\partial_-\mathcal{H}$ and $\partial_+\mathcal{H}$ are smooth. Moreover, since $\partial_-\mathcal{H}$ and $\partial_+\mathcal{H}$ intersect the invariant curves, generators of $\tilde{\mathcal{H}}$ transversally they are both spacelike surfaces in acoustical spacetime.

Finally, we note that $\partial_+\mathcal{H}$ cannot be part of the boundary of the domain of the maximal solution. This is seen as follows. For any given component of $\partial_+\mathcal{H}$, there is a component of $\tilde{\mathcal{H}} \setminus \bar{\mathcal{H}}$, the timelike part of $\tilde{\mathcal{H}}$, whose future boundary is the given component of $\partial_+\mathcal{H}$ (at $\partial_+\mathcal{D}$, $\partial^2 t_*/\partial u^2 < 0$, so as $u$ decreases, $\partial t_*/\partial u$ becomes positive beyond $\partial_+\mathcal{D}$, thus $t_*$ decreases as $u$ decreases). This, or in fact any, component of $\tilde{\mathcal{H}}\backslash\bar{\mathcal{H}}$ has a past boundary, for, $\lim_{u\to 0} t_*(u) = \infty$, therefore $t_*(u,\vartheta)$ must have intervals of increase along each invariant line as we approach $u = 0$. The past boundary of the component in question is then also the past boundary of the next outward (that is, in the direction of decreasing $u$ along the invariant curves) component of $\mathcal{H}$, a component of $\partial_-\mathcal{H}$. But the domain of the maximal solution terminates at the incoming characteristic hypersurface $\underline{C}$ generated by the incoming null normals to this component of $\partial_-\mathcal{H}$. So the maximal solution cannot reach $\partial_+\mathcal{H}$.

Summarizing, we get

**Proposition 19.1** Consider a smooth extension of the acoustical metric $g$ to $t > t_*(u)$, satisfying the following two conditions:

(i) The function $\alpha$ remains near 1, and the metric $\not{g}$ remains a positive definite metric on $S^2$.

(ii) The function $\mu$ is extended in such a way that $\partial\mu/\partial t$ is bounded from above by a negative constant where $\mu < 1/4$.

Then there is an open subset $\tilde{\mathcal{D}} \subset (0,\epsilon_0) \times S^2$ and a smooth graph

$$\tilde{\mathcal{H}} = \{(t_*(u,\vartheta), u, \vartheta), (u,\vartheta) \in \tilde{\mathcal{D}}\}$$

where $\mu$ vanishes, being positive below this graph. The singular hypersurface $\tilde{\mathcal{H}}$ with its induced metric $g_*$ is smooth and has the intrinsic geometry of a regular null hypersurface in a regular spacetime. It is ruled by invariant curves of vanishing arc length. The tangent line to an invariant curve at a point $q$ is the linear span of the vector $\Pi_* T(q)$, the $L$ projection of $T(q)$ to $\tilde{\mathcal{H}}$. Canonical acoustical coordinates are defined by taking the $\vartheta =$ const, coordinate lines on $\tilde{\mathcal{H}}$ to be the invariant curves. On the other hand, from the point of view of how $\tilde{\mathcal{H}}$ is embedded in the acoustical spacetime, the extrinsic point of view, $\tilde{\mathcal{H}}$ is at a point $q$ spacelike, null, or timelike, according as to whether, in canonical acoustical coordinates, $\partial\mu/\partial u$ is $< 0, = 0, > 0$, at $q$, or equivalently, as to whether $\partial t_*/\partial u$ is $< 0, = 0, > 0$, at $q$. The timelike part of $\tilde{\mathcal{H}}$ cannot be part of the boundary of the domain of the maximal solution. Moreover, denoting by $\mathcal{H}$ the spacelike part of $\tilde{\mathcal{H}}$ and by $\partial\mathcal{H}$ its boundary, under the non-degeneracy condition

$$\partial^2\mu/\partial u^2 \neq 0 \quad \text{on} \quad \partial\mathcal{H}$$

the boundary splits into the disjoint union of $\partial_-\mathcal{H}$ and $\partial_+\mathcal{H}$, where $\partial_-\mathcal{H}$ and $\partial_+\mathcal{H}$ correspond to $\partial^2\mu/\partial u^2 > 0$ and $< 0$ respectively, or equivalently $\partial^2 t_*/\partial u^2 > 0$ and $< 0$ respectively. Each of $\partial_-\mathcal{H}$, $\partial_+\mathcal{H}$, is a smooth spacelike surface in the acoustical spacetime. For each connected component of $\mathcal{H}$ the corresponding components of $\partial_-\mathcal{H}$ and $\partial_+\mathcal{H}$ are respectively its past and future boundaries, the sets of past and future end points of its invariant curves. Finally, $\partial_+\mathcal{H}$, which, in contrast to $\partial_-\mathcal{H}$, may be empty, cannot be part of the boundary of the domain of the maximal solution.

## 19.2 The Trichotomy Theorem for Past Null Geodesics Ending at Singular Boundary

We now turn to the investigation of the past null geodesic cone of an arbitrary point $q$ on the singular boundary of the domain of the maximal solution. According to the above, such a point belongs either to $\mathcal{H}$, or to $\partial_-\mathcal{H}$. The domain of the maximal solution or maximal development of the initial data on $\Sigma_0^{\epsilon_0} \bigcup \Sigma_0^E$, where $\Sigma_0^E$ denotes the exterior of the unit sphere in $\Sigma_0$, where the constant state holds, has, like any development of the initial data in question, the property that for each point $q$ in this domain each past directed curve issuing at $q$ which is causal with respect to the acoustical metric terminates in the past at a point of $\Sigma_0^{\epsilon_0} \bigcup \Sigma_0^E$. In particular, each past directed null geodesic issuing at $q$ terminates in the past at a point of $\Sigma_0^{\epsilon_0} \bigcup \Sigma_0^E$. Suppose now that $q$ is a zero of $\mu$ on $\Sigma_{t_{*\epsilon}}^{\epsilon}$ for some $\epsilon \in (0, \epsilon_0)$. Then $q \in W_\epsilon^{t_{*\epsilon}}$ and the union of $W_\epsilon^{t_{*\epsilon}}$ with the domain $E^{t_{*\epsilon}}$ in Galilean spacetime exterior to the cone $C_0$ and bounded by the hyperplanes $\Sigma_0$ and $\Sigma_{t_{*\epsilon}}$, where the constant state holds, is a development of the restriction of the initial data to $\Sigma_0^{\epsilon} \bigcup \Sigma_0^E$, therefore each past directed null geodesic issuing at $q$ remains in $W_\epsilon^{t_{*\epsilon}} \bigcup E^{t_{*\epsilon}}$ and terminates in the past at a point of $\Sigma_0^{\epsilon} \bigcup \Sigma_0^E$.

### 19.2.1 Hamiltonian Flow

Now, the null geodesic flow of a Lorentzian manifold $(\mathcal{M}, g)$ is from the point view of cotangent bundle the Hamiltonian flow on $T^*\mathcal{M}$ generated by the Hamiltonian

$$H = \frac{1}{2}(g^{-1})^{\mu\nu}(q)p_\mu p_\nu \tag{19.55}$$

on the surface $H = 0$. Here $(q^\mu, \mu = 1, \cdots, n)$, $n = \dim\mathcal{M}$, are local coordinates on $\mathcal{M}$ (called in the Hamiltonian context *canonical coordinates*), and expanding $p \in T_q^*\mathcal{M}$ in the basis $(dq^1(q), \cdots, dq^n(q))$:

$$p = p_\mu dq^\mu(q)$$

the coefficients $(p_\mu, \mu = 1, \cdots, n)$ of the expansion are linear coordinates on $T_q^*\mathcal{M}$ (called in the Hamiltonian context *canonical momenta*). We thus have local coordinates $(q^\mu, \mu = 1, \cdots, n; p_\mu, \mu = 1, \cdots, n)$ on $T^*\mathcal{M}$ and the equations defining a Hamiltonian flow are the canonical equations:

$$\frac{dq^\mu}{d\tau} = \frac{\partial H}{\partial p_\mu}, \quad \frac{dp_\mu}{d\tau} = -\frac{\partial H}{\partial q_\mu} \tag{19.56}$$

Each level set of the Hamiltonian $H$ is invariant by the Hamiltonian flow. We may thus consider the restriction of the flow to any given level set of $H$, in particular to the surface $H = 0$. Set now

$$H(q,p) = \Omega(q)\tilde{H}(q,p) \tag{19.57}$$

where $\Omega$ is a positive function on $\mathcal{M}$. Then the Hamiltonian flow of $\tilde{H}$ on its zero level set is equivalent, up to reparametrization, to the Hamiltonian flow of $H$ on its zero level set. More precisely, let $\tau \mapsto (q(\tau), p(\tau))$ be a solution of the canonical equations (19.56) on the surface $H = 0$. Then defining a new parameter $\tilde{\tau}$ by

$$\frac{d\tilde{\tau}}{d\tau} = \Omega(q(\tau)) \tag{19.58}$$

$\tilde{\tau} \mapsto (\tilde{q}(\tilde{\tau}), \tilde{p}(\tilde{\tau})) = (q(\tau), p(\tau))$ is a solution of (19.56) with $H$ replaced by $\tilde{H}$ and $\tau$ by $\tilde{\tau}$ on the surface $\tilde{H} = 0$. This is so because

$$\begin{aligned} \frac{d\tilde{q}^\mu}{d\tilde{\tau}} &= \frac{dq^\mu}{d\tau}\frac{d\tau}{d\tilde{\tau}} = \frac{\partial H}{\partial p_\mu}\Omega^{-1} = \frac{\partial \tilde{H}}{\partial p_\mu} \\ \frac{d\tilde{p}_\mu}{d\tilde{\tau}} &= \frac{dp_\mu}{d\tau}\frac{d\tau}{d\tilde{\tau}} = -\frac{\partial H}{\partial q^\mu}\Omega^{-1} = -\frac{\partial(\tilde{H}\Omega)}{\partial q^\mu}\Omega^{-1} \\ &= -\frac{\partial \tilde{H}}{\partial q^\mu} - \tilde{H}\Omega^{-1}\frac{\partial \Omega}{\partial q^\mu} = -\frac{\partial \tilde{H}}{\partial q^\mu} \quad \text{on} \quad \tilde{H} = 0 \end{aligned}$$

In the case that $H$ is the Hamiltonian (19.55), then

$$\tilde{H} = \frac{1}{2}(\tilde{g}^{-1})^{\mu\nu}(q)p_\mu p_\nu, \quad \tilde{g}_{\mu\nu} = \Omega g_{\mu\nu}$$

and the above corresponds to the fact that null geodesics are invariant, up to reparametrization, under conformal transformations of the metric.

In the case of our acoustical spacetime $(\mathcal{M}_{\epsilon_0}, g)$, the reciprocal metric $g^{-1}$ is as we have seen singular at $\mathcal{H}$, however $\mu g^{-1}$, given in canonical acoustical coordinates by (19.37), extends smoothly to $\tilde{\mathcal{H}}$. So we take the Hamiltonian to be

$$H = -p_u p_t + \mu\left(-\frac{1}{2}\alpha^{-2}p_t^2 + p_t\hat{\Xi}^A \not{p}_A + \frac{1}{2}(\not{g}^{-1})^{AB}\not{p}_A\not{p}_B\right) \tag{19.59}$$

Here $p_t$, $p_u$, and $\not{p}_A, A = 1, 2$ are the momenta conjugate to the coordinates $t$, $u$, $\vartheta^A, A = 1, 2$, respectively. Thus the $\not{p}_A$ are the components of the *angular momentum* $\not{p}$. Given a solution $\tau \mapsto (q(\tau), p(\tau)), q = (t, u, \vartheta), p = (p_t, p_u, \not{p})$ (this is an affinely parametrized null geodesic of the conformal acoustical metric $\mu^{-1}g$), then defining $s$ by

$$\frac{ds}{d\tau} = \mu(q(\tau)) \tag{19.60}$$

and taking the inverse of the mapping $\tau \mapsto s(\tau)$, then $s \mapsto (q(\tau(s)), p(\tau(s)))$ is an affinely parametrized null geodesic of the acoustical metric $g$, $s$ being the affine parameter, and conversely.

The canonical equations (19.56) take for the Hamiltonian (19.59) the following form:

$$\begin{aligned} \frac{dt}{d\tau} &= \frac{\partial H}{\partial p_t} = -p_u + \mu(-\alpha^{-2}p_t + \hat{\Xi}^A\not{p}_A) \\ \frac{du}{d\tau} &= \frac{\partial H}{\partial p_u} = -p_t \\ \frac{d\vartheta^A}{d\tau} &= \frac{\partial H}{\partial \not{p}_A} = \mu((\not{g}^{-1})^{AB}\not{p}_B + p_t\hat{\Xi}^A) \end{aligned} \tag{19.61}$$

$$\begin{aligned}\frac{dp_t}{d\tau} &= -\frac{\partial H}{\partial t} = -\frac{\partial \mu}{\partial t}(-\frac{1}{2}\alpha^{-2}p_t^2 + p_t\hat{\Xi}^A \not{p}_A + \frac{1}{2}(\not{g}^{-1})^{AB}\not{p}_A\not{p}_B) \\ &\quad -\mu(\alpha^{-3}\frac{\partial \alpha}{\partial t}p_t^2 + p_t\frac{\partial \hat{\Xi}^A}{\partial t}\not{p}_A + \frac{1}{2}\frac{\partial (\not{g}^{-1})^{AB}}{\partial t}\not{p}_A\not{p}_B) \\ \frac{dp_u}{d\tau} &= -\frac{\partial H}{\partial u} = -\frac{\partial \mu}{\partial u}(-\frac{1}{2}\alpha^{-2}p_t^2 + p_t\hat{\Xi}^A \not{p}_A + \frac{1}{2}(\not{g}^{-1})^{AB}\not{p}_A\not{p}_B) \\ &\quad -\mu(\alpha^{-3}\frac{\partial \alpha}{\partial u}p_t^2 + p_t\frac{\partial \hat{\Xi}^A}{\partial u}\not{p}_A + \frac{1}{2}\frac{\partial (\not{g}^{-1})^{AB}}{\partial u}\not{p}_A\not{p}_B) \\ \frac{d\not{p}_A}{d\tau} &= -\frac{\partial H}{\partial \vartheta^A} = -\frac{\partial \mu}{\partial \vartheta^A}(-\frac{1}{2}\alpha^{-2}p_t^2 + p_t\hat{\Xi}^B \not{p}_B + \frac{1}{2}(\not{g}^{-1})^{BC}\not{p}_B\not{p}_C) \\ &\quad -\mu(\alpha^{-3}\frac{\partial \alpha}{\partial \vartheta^A}p_t^2 + p_t\frac{\partial \hat{\Xi}^B}{\partial \vartheta^A}\not{p}_B + \frac{1}{2}\frac{\partial (\not{g}^{-1})^{BC}}{\partial \vartheta^A}\not{p}_B\not{p}_C)\end{aligned} \tag{19.62}$$

### 19.2.2 Asymptotic Behavior

Given a point $q_0 \in \mathcal{H}\bigcup\partial_-\mathcal{H}$, we shall study the solutions of the system (19.61)—(19.62), subject to the condition $H = 0$, which end at $q_0 = (t_0, u_0, \vartheta_0)$, $t_0 = t_*(u_0, \vartheta_0)$, at $\tau = 0$. The solution are then studied for $\tau \leqslant 0$. These are the null geodesics of the acoustical metric ending at $q_0$, that is, the past null geodesic cone of $q_0$, with the orientation of each geodesic reversed so that they become future-directed ending at $q_0$ instead of past directed issuing at $q_0$.

Now the condition $H = 0$ defines at a regular point $p \in \mathcal{M}_{\epsilon_0}$, where $\mu(p) > 0$, a double cone in $T_p^*\mathcal{M}_{\epsilon_0}$ ((19.59) is a quadratic homogeneous polynomial in $p_\mu$, the coordinates of $T_p^*\mathcal{M}_{\epsilon_0}$) and we are considering the backward part of this cone. However at the singular point $q_0$, where $\mu(q_0) = 0$, the condition $H = 0$ reduces to

$$p_u p_t = 0 \tag{19.63}$$

Thus the double cone degenerates to the two hyperplanes $p_t = 0$ and $p_u = 0$ and the backward part of the cone degenerates to the following three pieces:

(i) The negative half-hyperplane: $p_t = 0, \quad p_u < 0$.
(ii) The negative half-hyperplane: $p_u = 0, \quad p_t < 0$.
(iii) The plane $p_t = p_u = 0$.

We thus have a *trichotomy* of the past null geodesic cone of $q_0$. The null geodesics ending at $q_0$ such that their momentum covector at $q_0$ belongs to (i) are the *outgoing* null geodesics. These contain the generator of the characteristic hypersurface $C_{u_0}$ through $q_0$, parametrized by $t$, which is the following solution of (19.61)—(19.62) and the condition $H = 0$:

$$t = t_0 + \tau, \quad u = u_0, \quad \vartheta = \vartheta_0; \quad p_t = 0, \quad p_u = -1, \quad \not{p} = 0 \tag{19.64}$$

The null geodesic ending at $q_0$ such that their momentum vector at $q_0$ belongs to (ii) are the *incoming* null geodesics. Finally, the null geodesics ending at $q_0$ such that their momentum vector at $q_0$ belongs to (iii) we simply call the *other* null geodesics.

To obtain an intuitive picture of how the null cone in $T^*_p\mathcal{M}_{\epsilon_0}$ degenerates as $p$, a regular point, approaches $q_0$, a point on the singular boundary, we consider the picture in 4-dimensional Euclidean space with the regular coordinates $x_1$, $x_2$, $y$, $z$, taking

$$x_1^2 + x_2^2 = (\not{g}^{-1})^{AB}\not{p}_A\not{p}_B, \quad y = p_u, \quad z = \alpha^{-1}p_t \tag{19.65}$$

An adequate picture of the degeneration is obtained if we assume for simplicity that $\hat{\Xi} = 0$ at a given $\Sigma_t$. The equation of the double cone $H = 0$ then becomes

$$2yz = \kappa(x_1^2 + x_2^2 - z^2) \, . \tag{19.66}$$

Note that the double cone contains the $y$ axis as well as the line

$$x_1 = x_2 = 0, \quad y = -\frac{1}{2}\kappa z \tag{19.67}$$

Consider the point $P_0$ on the positive $y$ axis at Euclidean distance 1 from the origin. The coordinates of $P_0$ are then $(0, 0, 1, 0)$. Consider also the point $P_1$ on the line (19.67) at Euclidean distance 1 from the origin in the positive $z$ direction. The coordinates of $P_1$ are

$$(0, 0, -(\kappa/2)/\sqrt{(\kappa/2)^2 + 1}, 1/\sqrt{(\kappa/2)^2 + 1})$$

Let $H_1$ be the hyperplane

$$z = -\lambda(y - 1), \quad \lambda = \frac{1}{\sqrt{(\kappa/2)^2 + 1} + (\kappa/2)} \tag{19.68}$$

passing through the points $P_0$, $P_1$, and ruled by planes parallel to the $(x_1, x_2)$ plane. Then the intersection of $H_1$ with the double cone is the spheroid on $H_1$ which projects to the following spheroid on the $(x_1, x_2, y)$ hyperplane:

$$\begin{gathered} \frac{x_1^2 + x_2^2}{a^2} + \frac{(y - y_0)^2}{b^2} = 1; \\ a = \sqrt{\frac{\lambda}{\kappa(2 - \kappa\lambda)}}, b = \frac{1}{2 - \kappa\lambda}, \quad y_0 = \frac{1 - \kappa\lambda}{2 - \kappa\lambda} \end{gathered} \tag{19.69}$$

with semimajor axis $a$, semiminor axis $b$, centered at $y_0$ on the $y$ axis. The spheroid on $H_1$ is centered at the point $(0, 0, y_0, z_0)$, $z_0 = -\lambda(y_0 - 1)$, and has semimajor axis $a' = a$ in a plane through this center parallel to the $(x_1, x_2)$ plane and semiminor axis $b' = \sqrt{1 + \lambda^2}b$ in a line through the center in $H_1$ orthogonal to this plane. As the regular point $p$ approaches the singular point $q_0$, $\kappa \to 0$, hence $\lambda \to 1$, $y_0 \to 1/2$, $z_0 \to 1/2$, $b' \to 1/\sqrt{2}$, but $a' \to \infty$, in fact $\sqrt{2\kappa}a' \to 1$. The hyperplane $H_1$ becomes the hyperplane $z = -(y - 1)$, and the spheroid on $H_1$, the intersection with the double cone, degenerates to the planes $y = 1$, $z = 0$, and $y = 0$, $z = 1$, parallel to the $(x_1, x_2)$ plane, at which the hyperplane $z = -(y - 1)$ intersects the hyperplanes $z = 0$ and $y = 0$. This shows how the double cone degenerates to the two hyperplanes $z = 0$ and $y = 0$.

We return to the study of the null geodesics ending at $q_0$. The value of $p_0 = ((p_t)_0, (p_u)_0, \not{p}_0)$ determines in which of the three classes the null geodesic belongs. Note that $p_0 \neq 0$, so that $(q_0, p_0)$ is a regular non-critical point of the Hamiltonian system (19.61)—(19.62). Now we can impose in each class a normalization condition by making use of the following remark. Let $\tau \mapsto (q(\tau), p(\tau))$ be a solution of the canonical equations (19.56) associated to the Hamiltonian (19.55) and corresponding to the conditions $q(0) = q_0$, $p(0) = p_0$, at $\tau = 0$. Then for any positive constant $a$, $\tau \mapsto (q(a\tau), ap(a\tau))$ is the solution corresponding to the conditions $q(0) = q_0$, $p(0) = ap_0$. Thus the new conditions give a null geodesic which is simply a reparametrization of the null geodesic arising from the original conditions. This is so because if $\tilde{q}(\tau) = q(a\tau)$, $\tilde{p}(\tau) = ap(a\tau)$, we have

$$\begin{aligned}\frac{d\tilde{q}^\mu}{d\tau}(\tau) &= a\frac{dq^\mu}{d\tau}(a\tau) = a\frac{\partial H}{\partial p_\mu}(q(a\tau), p(a\tau)) \\ &= a(g^{-1})^{\mu\nu}(q(a\tau))p_\nu(a\tau) = (g^{-1})^{\mu\nu}(\tilde{q}(\tau))\tilde{p}_\nu(\tau) = (\frac{\partial H}{\partial p_\mu})(\tilde{q}(\tau), \tilde{p}(\tau)) \\ \frac{d\tilde{p}_\mu}{d\tau}(\tau) &= a^2\frac{dp_\mu}{d\tau}(a\tau) = -a^2(\frac{\partial H}{\partial q^\mu})(q(a\tau), p(a\tau)) \\ &= -\frac{1}{2}a^2\frac{\partial(g^{-1})^{\lambda\nu}}{\partial q^\mu}(q(a\tau))p_\lambda(a\tau)p_\nu(a\tau) \\ &= -\frac{1}{2}\frac{\partial(g^{-1})^{\lambda\nu}}{\partial q^\mu}(\tilde{q}(\tau))\tilde{p}_\lambda(\tau)\tilde{p}_\nu(\tau) = -(\frac{\partial H}{\partial q^\mu})(\tilde{q}(\tau), \tilde{p}(\tau))\end{aligned}$$

Given an *outgoing* null geodesic ending at $q_0$ we can, by suitable choice of the scale factor $a$, set $(p_0)_u = -1$. Thus, the outgoing null geodesics ending at $q_0$ correspond to the plane

$$(p_0)_t = 0, \quad (p_0)_u = -1 \quad \text{in} \quad T^*_{q_0}\mathcal{M}_{\epsilon_0} \tag{19.70}$$

Given an *incoming* null geodesic ending at $q_0$ we can, by suitable choice of the scale factor $a$, set $(p_0)_t = -1$. Thus, the incoming null geodesics ending at $q_0$ correspond to the plane

$$(p_0)_u = 0, \quad (p_0)_t = -1 \quad \text{in} \quad T^*_{q_0}\mathcal{M}_{\epsilon_0} \tag{19.71}$$

Finally, given one of the *other* null geodesics ending at $q_0$ we can, by suitable choice of the scale factor $a$, set $|(\not{p})_0| = \sqrt{(\not{g}^{-1})^{AB}(q_0)(\not{p}_0)_A(\not{p}_0)_B} = 1$. Thus, the *other* null geodesics ending at $q_0$ correspond to the circle

$$(p_0)_t = 0, \quad (p_0)_u = 0, \quad |\not{p}_0| = 1 \quad \text{in} \quad T^*_{q_0}\mathcal{M}_{\epsilon_0} \tag{19.72}$$

Consider the class of *outgoing* null geodesics ending at $q_0$. For this class Equations (19.61) evaluated at $\tau = 0$ give

$$\frac{dt}{d\tau}(0) = 1, \quad \frac{du}{d\tau}(0) = 0, \quad \frac{d\vartheta^A}{d\tau}(0) = 0 \tag{19.73}$$

Also, (19.62) give

$$\begin{aligned}\frac{dp_t}{d\tau}(0) &= -\frac{1}{2}\frac{\partial\mu}{\partial t}(q_0)|\not{p}_0|^2\\ \frac{dp_u}{d\tau}(0) &= -\frac{1}{2}\frac{\partial\mu}{\partial u}(q_0)|\not{p}_0|^2\\ \frac{d\not{p}_A}{d\tau}(0) &= -\frac{1}{2}\frac{\partial\mu}{\partial\vartheta^A}(q_0)|\not{p}_0|^2\end{aligned} \tag{19.74}$$

Differentiating (19.61) with respect to $\tau$ and evaluating the result at $\tau = 0$ using (19.74) and the fact that by (19.73),

$$\left(\frac{d\mu(q(\tau))}{d\tau}\right)_{\tau=0} = \frac{\partial\mu}{\partial t}(q_0) \tag{19.75}$$

we obtain

$$\begin{aligned}\frac{d^2t}{d\tau^2}(0) &= \frac{1}{2}\frac{\partial\mu}{\partial u}(q_0)|\not{p}_0|^2 + \frac{\partial\mu}{\partial t}(q_0)\hat{\Xi}^A(q_0)(\not{p}_0)_A\\ \frac{d^2u}{d\tau^2}(0) &= \frac{1}{2}\frac{\partial\mu}{\partial t}(q_0)|\not{p}_0|^2\\ \frac{d^2\vartheta^A}{d\tau^2}(0) &= \frac{\partial\mu}{\partial t}(q_0)(\not{g}^{-1})^{AB}(q_0)(\not{p}_0)_B\end{aligned} \tag{19.76}$$

In view of (19.73), (19.76) the tangent vector to an outgoing null geodesic ending at $q_0$ has the following expansion as $\tau \to 0$:

$$\begin{aligned}\frac{dt}{d\tau}(\tau) &= 1 + \left(\frac{1}{2}\frac{\partial\mu}{\partial u}(q_0)|\not{p}_0|^2 + \frac{\partial\mu}{\partial t}(q_0)\hat{\Xi}^A(q_0)(\not{p}_0)_A\right)\tau + O(\tau^2)\\ \frac{du}{d\tau}(\tau) &= \frac{1}{2}\frac{\partial\mu}{\partial t}(q_0)|\not{p}_0|^2\tau + O(\tau^2)\\ \frac{d\vartheta^A}{d\tau}(\tau) &= \frac{\partial\mu}{\partial t}(q_0)(\not{g}^{-1})^{AB}(q_0)(\not{p}_0)_B\tau + O(\tau^2)\end{aligned} \tag{19.77}$$

We conclude that in canonical acoustical coordinates (see (19.37)) the tangent vectors to all *outgoing* null geodesics ending at $q_0$ tend as we approach $q_0$ to the tangent vector of the generator of $C_{u_0}$ through $q_0$.

Consider the class of *incoming* null geodesics ending at $q_0$. For this class Equations (19.61) evaluated at $\tau = 0$ give

$$\frac{dt}{d\tau}(0) = 0, \quad \frac{du}{d\tau}(0) = 1, \quad \frac{d\vartheta^A}{d\tau}(0) = 0 \tag{19.78}$$

Also Equations (19.62) give

$$\begin{aligned}\frac{dp_t}{d\tau}(0) &= -\frac{1}{2}\frac{\partial\mu}{\partial t}(q_0)(-\alpha^{-2}(q_0) - 2\hat{\Xi}^A(q_0)(\not{p}_0)_A + |\not{p}_0|^2)\\ \frac{dp_u}{d\tau}(0) &= -\frac{1}{2}\frac{\partial\mu}{\partial u}(q_0)(-\alpha^{-2}(q_0) - 2\hat{\Xi}^A(q_0)(\not{p}_0)_A + |\not{p}_0|^2)\\ \frac{d\not{p}_A}{d\tau}(0) &= -\frac{1}{2}\frac{\partial\mu}{\partial\vartheta^A}(q_0)(-\alpha^{-2}(q_0) - 2\hat{\Xi}^B(q_0)(\not{p}_0)_B + |\not{p}_0|^2)\end{aligned} \tag{19.79}$$

Differentiating (19.61) with respect to $\tau$ and evaluating the result at $\tau = 0$ using (19.79) and the fact that by (19.78),

$$\left(\frac{d\mu(q(\tau))}{d\tau}\right)_{\tau=0} = \frac{\partial\mu}{\partial u}(q_0) \tag{19.80}$$

we obtain

$$\begin{aligned} \frac{d^2t}{d\tau^2}(0) &= \frac{1}{2}\frac{\partial\mu}{\partial u}(q_0)(\alpha^{-2}(q_0) + |\not{p}_0|^2) \\ \frac{d^2\vartheta^A}{d\tau^2}(0) &= \frac{\partial\mu}{\partial u}(q_0)((\not{g}^{-1})^{AB}(q_0)(\not{p}_0)_B - \hat{\Xi}^A(q_0)) \end{aligned} \tag{19.81}$$

The right-hand side of the first equation in (19.81) is negative when $q_0 \in \mathcal{H}$. However in the case that $q_0 \in \partial_-\mathcal{H}$, we have

$$\left(\frac{d\mu(q(\tau))}{d\tau}\right)_{\tau=0} = 0, \quad \frac{d^2t}{d\tau^2}(0) = 0, \quad \frac{d^2\vartheta^A}{d\tau^2}(0) = 0 \tag{19.82}$$

We must then proceed to the third derivatives. Using (19.78) and (19.82) we find in this case

$$\left(\frac{d^2\mu(q(\tau))}{d\tau^2}\right)_{\tau=0} = \frac{\partial^2\mu}{\partial u^2}(q_0) \tag{19.83}$$

and differentiating the second equation of (19.62) with respect to $\tau$ and evaluating the result at $\tau = 0$ gives

$$\frac{d^2p_u}{d\tau^2}(0) = \frac{1}{2}\frac{\partial^2\mu}{\partial u^2}(q_0)(\alpha^{-2}(q_0) + 2\hat{\Xi}^A(q_0)(\not{p}_0)_A - |\not{p}_0|^2) \tag{19.84}$$

Differentiating then the second and third equations of (19.61) with respect to $\tau$ a second time and using (19.83) and (19.84) we then obtain that in the case $q_0 \in \partial_-\mathcal{H}$,

$$\begin{aligned} \frac{d^3t}{d\tau^3}(0) &= \frac{1}{2}\frac{\partial^2\mu}{\partial u^2}(q_0)(\alpha^{-2}(q_0) + |\not{p}_0|^2) \\ \frac{d^3\vartheta^A}{d\tau^3}(0) &= \frac{\partial^2\mu}{\partial u^2}(q_0)((\not{g}^{-1})^{AB}(q_0)(\not{p}_0)_B - \hat{\Xi}^A(q_0)) \end{aligned} \tag{19.85}$$

Note that the right-hand side of the first equation of (19.85) is positive. The tangent vector to an incoming null geodesic ending at $q_0$ has the following expansion as $\tau \to 0$:

$$\begin{aligned} \frac{du}{d\tau}(\tau) &= 1 + O(\tau) \\ \frac{dt}{d\tau}(\tau) &= \frac{1}{2}\frac{\partial\mu}{\partial u}(q_0)(\alpha^{-2}(q_0) + |\not{p}_0|^2)\tau + O(\tau^2) \\ \frac{d\vartheta^A}{d\tau}(\tau) &= \frac{\partial\mu}{\partial u}(q_0)((\not{g}^{-1})^{AB}(q_0)(\not{p}_0)_B - \hat{\Xi}^A(q_0))\tau + O(\tau^2) \end{aligned} \tag{19.86}$$

by (19.78), (19.81), in the case that $q_0 \in \mathcal{H}$, and

$$\frac{du}{d\tau}(\tau) = 1 + O(\tau)$$
$$\frac{dt}{d\tau}(\tau) = \frac{1}{4}\frac{\partial^2\mu}{\partial u^2}(q_0)(\alpha^{-2}(q_0) + |\not{p}_0|^2)\tau^2 + O(\tau^3) \tag{19.87}$$
$$\frac{d\vartheta^A}{d\tau}(\tau) = \frac{1}{2}\frac{\partial^2\mu}{\partial u^2}(q_0)((\not{g}^{-1})^{AB}(q_0)(\not{p}_0)_B - \hat{\Xi}^A(q_0))\tau^2 + O(\tau^3)$$

by (19.78), (19.82), (19.85), in the case that $q_0 \in \partial_-\mathcal{H}$.

Consider finally the class of the *other* null geodesics ending at $q_0$. For this class Equations (19.61) and (19.62) evaluated at $\tau = 0$ give

$$\frac{dt}{d\tau}(0) = 0, \quad \frac{du}{d\tau}(0) = 0, \quad \frac{d\vartheta^A}{d\tau}(0) = 0 \tag{19.88}$$

$$\frac{dp_t}{d\tau}(0) = -\frac{1}{2}\frac{\partial\mu}{\partial t}(q_0)$$
$$\frac{dp_u}{d\tau}(0) = -\frac{1}{2}\frac{\partial\mu}{\partial u}(q_0) \tag{19.89}$$
$$\frac{d\not{p}_A}{d\tau}(0) = -\frac{1}{2}\frac{\partial\mu}{\partial\vartheta^A}(q_0)$$

Note that the right-hand side of the first equation of (19.89) is positive. Differentiating (19.61) with respect to $\tau$ and evaluating the result at $\tau = 0$ using (19.89) and the fact that by (19.88),

$$\left(\frac{d\mu(q(\tau))}{d\tau}\right)_{\tau=0} = 0 \tag{19.90}$$

we obtain

$$\frac{d^2t}{d\tau^2}(0) = \frac{1}{2}\frac{\partial\mu}{\partial u}(q_0)$$
$$\frac{d^2u}{d\tau^2}(0) = \frac{1}{2}\frac{\partial\mu}{\partial t}(q_0) \tag{19.91}$$
$$\frac{d^2\vartheta^A}{d\tau^2}(0) = 0$$

Note that the right-hand side of the second equation of (19.91) is negative, while the right-hand side of the first equation of (19.91) is negative when $q_0 \in \mathcal{H}$ and vanishes when $q_0 \in \partial_-\mathcal{H}$. Differentiating the third equation of (19.61) a second time with respect to $\tau$ and evaluating the result at $\tau = 0$ yields

$$\frac{d^3\vartheta^A}{d\tau^3}(0) = \left(\frac{d^2\mu(q(\tau))}{d\tau^2}\right)_{\tau=0}(\not{g}^{-1})^{AB}(q_0)(\not{p}_0)_B \tag{19.92}$$

Now from (19.88) and (19.91) we obtain

$$\left(\frac{d^2\mu(q(\tau))}{d\tau^2}\right)_{\tau=0} = \frac{d^2t}{d\tau^2}(0)\frac{\partial\mu}{\partial t}(q_0) + \frac{d^2u}{d\tau^2}(0)\frac{\partial\mu}{\partial u}(q_0) + \frac{d^2\vartheta^A}{d\tau^2}(0)\frac{\partial\mu}{\partial\vartheta^A}(q_0)$$
$$= \frac{\partial\mu}{\partial t}\frac{\partial\mu}{\partial u}(q_0) \tag{19.93}$$

Substituting in (19.92) then yields

$$\frac{d^3\vartheta^A}{d\tau^3}(0) = \frac{\partial\mu}{\partial t}\frac{\partial\mu}{\partial u}(q_0)(\not{g}^{-1})^{AB}(q_0)(\not{p}_0)_B \tag{19.94}$$

In the case that $q_0 \in \partial_-\mathcal{H}$ the right-hand side of the first equation of (19.91), of (19.93), and of (19.94) all vanish:

$$\left(\frac{d^2\mu(q(\tau))}{d\tau^2}\right)_{\tau=0} = 0, \quad \frac{d^2t}{d\tau^2}(0) = 0, \quad \frac{d^2\vartheta^A}{d\tau^2}(0) = \frac{d^3\vartheta^A}{d\tau^3}(0) = 0 \tag{19.95}$$

Differentiating the first equation of (19.61) a second time with respect to $\tau$ and evaluating the result at $\tau = 0$ then yields

$$\frac{d^3t}{d\tau^3}(0) = -\frac{d^2p_u}{d\tau^2}(0) \tag{19.96}$$

Also, differentiating the second equation of (19.62) with respect to $\tau$ and evaluating the result at $\tau = 0$ yields

$$\frac{d^2p_u}{d\tau^2}(0) = -\frac{1}{2}\left(\frac{d(\partial\mu/\partial u)(q(\tau))}{d\tau}\right)_{\tau=0} \tag{19.97}$$

However, from (19.88),

$$\left(\frac{d(\partial\mu/\partial u)(q(\tau))}{d\tau}\right)_{\tau=0}$$
$$= \frac{\partial^2\mu}{\partial t\partial u}(q_0)\frac{dt}{d\tau}(0) + \frac{\partial^2\mu}{\partial u^2}(q_0)\frac{du}{d\tau}(0) + \frac{\partial^2\mu}{\partial\vartheta^A\partial u}(q_0)\frac{d\vartheta^A}{d\tau}(0) = 0 \tag{19.98}$$

hence

$$\frac{d^3t}{d\tau^3}(0) = \frac{d^2p_u}{d\tau^2}(0) = 0 \tag{19.99}$$

We must proceed to the fourth derivatives. Differentiating the first equation of (19.61) a third time with respect to $\tau$ and evaluating the result at $\tau = 0$ yields

$$\frac{d^4t}{d\tau^4}(0) = -\frac{d^3p_u}{d\tau^3}(0) + \left(\frac{d^3\mu(q(\tau))}{d\tau^3}\right)_{\tau=0}\hat{\Xi}^A(q_0)(\not{p}_0)_A \tag{19.100}$$

Also, differentiating the second equation of (19.62) a second time with respect to $\tau$ and evaluating the result at $\tau = 0$ yields

$$\frac{d^3p_u}{d\tau^3}(0) = -\frac{1}{2}\left(\frac{d^2(\partial\mu/\partial u)(q(\tau))}{d\tau^2}\right)_{\tau=0} \tag{19.101}$$

Now, in view of (19.88) and (19.91)

$$\begin{aligned}\left(\frac{d^2(\partial\mu/\partial u)(q(\tau))}{d\tau^2}\right)_{\tau=0} &= \frac{\partial^2\mu}{\partial t\partial u}(q_0)\frac{d^2t}{d\tau^2}(0) \\ &\quad + \frac{\partial^2\mu}{\partial u^2}(q_0)\frac{d^2u}{d\tau^2}(0) + \left(\frac{\partial^2\mu}{\partial\vartheta^A\partial u}\right)(q_0)\left(\frac{d^2\vartheta^A}{d\tau^2}\right)(0) \\ &= \frac{\partial^2\mu}{\partial u^2}(q_0)\frac{d^2u}{d\tau^2}(0) = \frac{1}{2}\left(\frac{\partial\mu}{\partial t}\frac{\partial^2\mu}{\partial u^2}\right)(q_0)\end{aligned} \tag{19.102}$$

On the other hand, by (19.88), (19.91), (19.95), (19.98), (19.99), and the fact that $(\partial\mu/\partial u)(q_0) = 0$,

$$\begin{aligned}(\frac{d^3\mu(q(\tau))}{d\tau^3})_{\tau=0} &= \frac{\partial\mu}{\partial t}(q_0)\frac{d^3 t}{d\tau^3}(0) + \frac{\partial\mu}{\partial u}(q_0)\frac{d^3 u}{d\tau^3}(0) \\ &+ \frac{\partial\mu}{\partial\vartheta^A}(q_0)\frac{d^3\vartheta^A}{d\tau^3}(0) + 2(\frac{d(\partial\mu/\partial t)(q(\tau))}{d\tau})_{\tau=0}\frac{d^2 t}{d\tau^2}(0) \\ &+ 2(\frac{d(\partial\mu/\partial u)(q(\tau))}{d\tau})_{\tau=0}\frac{d^2 u}{d\tau^2}(0) \\ &+ 2(\frac{d(\partial\mu/\partial\vartheta^A)(q(\tau))}{d\tau})_{\tau=0}(\frac{d^2\vartheta^A}{d\tau^2})(0) = 0\end{aligned} \tag{19.103}$$

Substituting then (19.102) in (19.101) and the result in (19.100), yields, in view of (19.103),

$$\frac{d^4 t}{d\tau^4}(0) = \frac{1}{4}\frac{\partial\mu}{\partial t}\frac{\partial^2\mu}{\partial u^2}(q_0) \tag{19.104}$$

Differentiating the third equation of (19.61) a third time and evaluating the result at $\tau = 0$ we obtain, taking into account (19.103),

$$\frac{d^4\vartheta^A}{d\tau^4}(0) = (\frac{d^3\mu(q(\tau))}{d\tau^3})_{\tau=0}(\not{g}^{-1})^{AB}(q_0)(\not{p}_0)_B = 0 \tag{19.105}$$

Finally, we differentiate the third equation of (19.61) a fourth time and evaluate the result at $\tau = 0$ to obtain

$$\frac{d^5\vartheta^A}{d\tau^5}(0) = (\frac{d^4\mu(q(\tau))}{d\tau^4})_{\tau=0}(\not{g}^{-1})^{AB}(q_0)(\not{p}_0)_B \tag{19.106}$$

and we have

$$\begin{aligned}(\frac{d^4\mu(q(\tau))}{d\tau^4})_{\tau=0} &= \frac{\partial\mu}{\partial t}(q_0)\frac{d^4 t}{d\tau^4}(0) + 3(\frac{d^2(\partial\mu/\partial u)(q(\tau))}{d\tau^2})_{\tau=0}\frac{d^2 u}{d\tau^2}(0) \\ &= ((\frac{\partial\mu}{\partial t})^2\frac{\partial^2\mu}{\partial u^2})(q_0)\end{aligned} \tag{19.107}$$

by (19.102), (19.104), and the second of (19.91). Substituting then in (19.106) yields

$$\frac{d^5\vartheta^A}{d\tau^5}(0) = ((\frac{\partial\mu}{\partial t})^2\frac{\partial^2\mu}{\partial u^2})(q_0)(\not{g}^{-1})^{AB}(q_0)(\not{p}_0)_B \tag{19.108}$$

According to the above results the tangent vector to any of the other null geodesics ending at $q_0$ has the following expansion as $\tau \to 0$:

$$\begin{aligned}\frac{dt}{d\tau}(\tau) &= \frac{1}{2}\frac{\partial\mu}{\partial u}(q_0)\tau + O(\tau^2) \\ \frac{du}{d\tau}(\tau) &= \frac{1}{2}\frac{\partial\mu}{\partial t}(q_0)\tau + O(\tau^2) \\ \frac{d\vartheta^A}{d\tau}(\tau) &= \frac{1}{2}\frac{\partial\mu}{\partial t}\frac{\partial\mu}{\partial u}(q_0)(\not{g}^{-1})^{AB}(\not{p}_0)_B\tau^2 + O(\tau^3)\end{aligned} \tag{19.109}$$

by (19.91), (19.94), in the case $q_0 \in \mathcal{H}$, and

$$\begin{aligned}\frac{dt}{d\tau}(\tau) &= \frac{1}{24}\frac{\partial\mu}{\partial t}\frac{\partial^2\mu}{\partial u^2}(q_0)\tau^3 + O(\tau^4)\\ \frac{du}{d\tau}(\tau) &= \frac{1}{2}\frac{\partial\mu}{\partial t}(q_0)\tau + O(\tau^2)\\ \frac{d\vartheta^A}{d\tau}(\tau) &= \frac{1}{24}((\frac{\partial\mu}{\partial t})^2\frac{\partial^2\mu}{\partial u^2})(q_0)(\not{g}^{-1})^{AB}(q_0)(\not{\psi}_0)_B\tau^4 + O(\tau^5)\end{aligned} \tag{19.110}$$

in the case that $q_0 \in \partial_-\mathcal{H}$.

To obtain the picture in Galilean spacetime, we now investigate the behavior of the tangent vector to a null geodesic ending at $q_0$, in each of the three classes, as $t \to t_0$, in rectangular coordinates. The rectangular components $\tilde{L}'^\mu$ of the tangent vector $\tilde{L}'$ of a null geodesic ending at $q_0$, parametrized by $\tau$, are given by

$$\tilde{L}'^\mu = \frac{dx^\mu}{d\tau} = \frac{\partial x^\mu}{\partial t}\frac{dt}{d\tau} + \frac{\partial x^\mu}{\partial u}\frac{du}{d\tau} + \frac{\partial x^\mu}{\partial \vartheta^A}\frac{d\vartheta^A}{d\tau} \tag{19.111}$$

If the null geodesic is parametrized by $t$ instead of $\tau$ the tangent vector is

$$L' = \frac{\tilde{L}'}{dt/d\tau} \tag{19.112}$$

and its rectangular components are given by

$$L'^\mu = \frac{\partial x^\mu}{\partial t} + \frac{\partial x^\mu}{\partial u}\frac{du}{dt} + \frac{\partial x^\mu}{\partial \vartheta^A}\frac{d\vartheta^A}{dt} \tag{19.113}$$

where

$$\frac{du}{dt} = \frac{du/d\tau}{dt/d\tau}, \quad \frac{d\vartheta^A/d\tau}{dt/d\tau} \tag{19.114}$$

Recalling from Chapter 2 that

$$\frac{\partial x^\mu}{\partial t} = L^\mu, \quad \frac{\partial x^0}{\partial u} = \frac{\partial x^0}{\partial \vartheta^A} = 0, \quad \frac{\partial x^i}{\partial u} = T^i + \Xi^A X^i_A, \quad \frac{\partial x^i}{\partial \vartheta^A} = X^i_A \tag{19.115}$$

we then obtain

$$L'^0 = 1, \quad L'^i = L^i + \mu(\alpha^{-1}\hat{T}^i + \hat{\Xi}^A X^i_A)\frac{du}{dt} + X^i_A\frac{d\vartheta^A}{dt} \tag{19.116}$$

Consider the class of the *outgoing* null geodesics ending at $q_0 \in \mathcal{H} \bigcup \partial_-\mathcal{H}$. Then from (19.77) we have

$$\lim_{t\to t_0}\frac{du}{dt} = 0, \quad \lim_{t\to t_0}\frac{d\vartheta^A}{dt} = 0 \tag{19.117}$$

Since $\lim_{t\to t_0}\mu = 0$, it follows that for all *outgoing* null geodesics ending at $q_0$ we have

$$\lim_{t\to t_0} L'^i = L^i(q_0) \tag{19.118}$$

We conclude that in Galilean spacetime the tangent vectors to all *outgoing* null geodesics ending at $q_0$ tend as we approach $q_0$ to the tangent vector $L(q_0)$ of the generator of $C_{u_0}$ through $q_0$. The vector $L(q_0)$ is therefore an *invariant vector* independent of the choice of acoustical function $u$. From the point of view of Galilean spacetime this vector is equal, up to a multiplicative constant, to the invariant vector $V(q_0)$, defined by (19.23), since the vector $T(q_0)$ vanishes from the point of view of Galilean spacetime (recalling the definition of $\Pi_*$).

Consider the class of the *incoming* null geodesics ending at $q_0 \in \mathcal{H}\bigcup\partial_-\mathcal{H}$. If $q_0 \in \mathcal{H}$, then taking into account the fact that by (19.80),

$$\mu(q(\tau)) = \frac{\partial\mu}{\partial u}(q_0)\tau + O(\tau^2) \tag{19.119}$$

we obtain from (19.86)

$$\lim_{t\to t_0} \mu\frac{du}{dt} = \frac{1}{(1/2)(\alpha^{-2}(q_0) + |\not{p}_0|^2)}, \quad \lim_{t\to t_0} \frac{d\vartheta^A}{dt} = \frac{(\not{g}^{-1})^{AB}(\not{p})_B - \hat{\Xi}^A(q_0)}{(1/2)(\alpha^{-2}(q_0) + |\not{p}_0|^2)} \tag{19.120}$$

If $q_0 \in \partial_-\mathcal{H}$, then taking into account the fact that by (19.82), (19.83),

$$\mu(q(\tau)) = \frac{1}{2}\frac{\partial^2\mu}{\partial u^2}(q_0)\tau^2 + O(\tau^3) \tag{19.121}$$

we obtain from (19.87)

$$\lim_{t\to t_0} \mu\frac{du}{dt} = \frac{1}{(1/2)(\alpha^{-2}(q_0) + |\not{p}_0|^2)}, \quad \lim_{t\to t_0} \frac{d\vartheta^A}{dt} = \frac{(\not{g}^{-1})^{AB}(\not{p})_B - \hat{\Xi}^A(q_0)}{(1/2)(\alpha^{-2}(q_0) + |\not{p}_0|^2)} \tag{19.122}$$

which coincides with (19.120). It follows that in both cases we have

$$\lim_{t\to t_0} L'^i = L^i(q_0) + \frac{\alpha^{-1}(q_0)\hat{T}^i(q_0)}{(1/2)(\alpha^{-2}(q_0) + |\not{p}_0|^2)} + \frac{(\not{g}^{-1})^{AB}(q_0)X^i_A(q_0)(\not{p}_0)_B}{(1/2)(\alpha^{-2}(q_0) + |\not{p}_0|^2)} \tag{19.123}$$

Note that this limit depends on $\not{p}_0$. For $\not{p}_0 = 0$ the limit vector is $(L + 2\alpha\hat{T})(q_0) = \alpha^2(q_0)\underline{\hat{L}}(q_0)$, where $\underline{\hat{L}}(q_0)$ is the conjugate to $L(q_0)$, the null vector in the orthogonal complement of $T_{q_0}S_{t_0,u_0}$. As $\not{p}_0$ tends to $\infty$ the limit vector tends to $L(q_0)$.

Consider finally the class of the *other* null geodesics ending at $q_0 \in \mathcal{H}\bigcup\partial_-\mathcal{H}$. If $q_0 \in \mathcal{H}$, then taking into account the fact that by (19.90), (19.93),

$$\mu(q(\tau)) = \frac{1}{2}(\frac{\partial\mu}{\partial t}\frac{\partial\mu}{\partial u})(q_0)\tau^2 + O(\tau^3) \tag{19.124}$$

we obtain from (19.109)

$$\lim_{t \to t_0} \mu \frac{du}{dt} = 0, \quad \lim_{t \to t_0} \frac{d\vartheta^A}{dt} = 0 \tag{19.125}$$

If $q_0 \in \partial_-\mathcal{H}$, then taking into account the fact that by (19.90), (19.95), (19.103), (19.107),

$$\mu(q(\tau)) = \frac{1}{24}((\frac{\partial\mu}{\partial t})^2 \frac{\partial^2\mu}{\partial u^2})(q_0)\tau^4 + O(\tau^5) \tag{19.126}$$

we obtain from (19.110)

$$\lim_{t \to t_0} \mu \frac{du}{dt} = 0, \quad \lim_{t \to t_0} \frac{d\vartheta^A}{dt} = 0 \tag{19.127}$$

as well. It follows that in both cases we have simply

$$\lim_{t \to t_0} L'^i = L^i(q_0) \tag{19.128}$$

Thus, from the point of view of Galilean spacetime, the tangent vectors to all the other null geodesics ending at $q_0$ tend to the vector $L(q_0)$ as we approach $q_0$.

We now consider, for a fixed point $q_0 \in \mathcal{H} \bigcup \partial_-\mathcal{H}$ and each of the three classes of null geodesics ending at $q_0$, the mapping

$$(\tau; \not{p}_0) \mapsto (t, u, \vartheta) = F(\tau; \not{p}_0), \quad F = (F^t, F^u, \not{F}) \tag{19.129}$$

where for the *outgoing* and *incoming* null geodesics: $\not{p}_0 \in T_{\vartheta_0}S^2$ which may be identified with $\mathbb{R}^2$, while for the *other* null geodesics $\not{p}_0$ takes values in the unit circle in $T_{\vartheta_0}S^2$, which may be identified with $S^1 \subset \mathbb{R}^2$. In each of the three cases the mapping (19.129) is smooth (by standard ordinary differential equation theory), the range of $F^t$ being restricted to $t \geqslant 0$.

Let us consider first the mapping

$$\tau \mapsto F^t(\tau; \not{p}_0) \tag{19.130}$$

for fixed $\not{p}_0$. We shall presently show that this mapping is a diffeomorphism of $[\tau_0, 0]$ onto $[0, t_0]$ in the case of the *outgoing* null geodesic, a diffeomorphism of $[\tau_0, 0)$ onto $[0, t_0)$ in the case of *incoming* null geodesics as well as in the case of the *other* null geodesics, the difference of the latter from the former being that $(dt/d\tau)(0) = 1$ for the *outgoing* (see (19.77)), while $(dt/d\tau)(0) = 0$ for the *incoming* and the *other* (see (19.86), (19.87), (19.109), (19.110)). Here $\tau_0 = \tau_0(\not{p}_0)$ ($p_t$ and $p_u$ are constants). Now the expressions (19.77) and (19.86), (19.87), (19.109), (19.110), show that $(dt/d\tau)(\tau) > 0$ for $\tau \in [\tau_1, 0)$, $\tau_1$ suitably small. Thus either $(dt/d\tau)(\tau) > 0$ for all $\tau \in [\tau_0, 0)$, or there is a first $\tau_*$ where $(dt/d\tau)(\tau_*) = 0$. Now in general, for a Hamiltonian of the form (19.55), the condition $H = 0$ is equivalent to the following condition on the tangent vector:

$$\frac{1}{2} g_{\mu\nu} \frac{dq^\mu}{d\tau} \frac{dq^\nu}{d\tau} = 0 \tag{19.131}$$

In the present case, for our Hamiltonian (19.59), the condition $H = 0$ is equivalent to the condition

$$-\frac{dt}{d\tau}\frac{du}{d\tau} + \frac{1}{2}\alpha^{-2}\mu(\frac{du}{d\tau})^2 + \frac{1}{2}\not{g}_{AB}(\frac{d\vartheta^A}{d\tau} + \mu\hat{\Xi}^A\frac{du}{d\tau})(\frac{d\vartheta^B}{d\tau} + \mu\hat{\Xi}^B\frac{du}{d\tau}) = 0 \quad (19.132)$$

Obviously, at $\tau_*$ we have

$$\frac{dt}{d\tau}(\tau_*) = \frac{du}{d\tau}(\tau_*) = \frac{d\vartheta^A}{d\tau}(\tau_*) = 0$$

By (19.61), we have

$$p_t(\tau_*) = p_u(\tau_*) = \not{p}_A(\tau_*) = 0$$

Then with $(t_*, u_*, \vartheta_*) = (t(\tau_*), u(\tau_*), \vartheta(\tau_*))$,

$$((t, u, \vartheta); (p_t, p_u, \not{p})) = ((t_*, u_*, \vartheta_*); (0, 0, 0))$$

is a solution of the Hamiltonian system (19.61)—(19.62) coinciding with our solution at $\tau = \tau_*$. By the uniqueness theorem for the Hamiltonian system (19.61)—(19.62) our null geodesic must coincide with the above constant solution for all $\tau$, in particular we must have $t(0) = t_*$ in contradiction with the fact that $t(0) = t_0 > t_*$. This rules out the second alternative above and we conclude that $(dt/d\tau)(\tau) > 0$ for all $\tau \in [\tau_0, 0)$. It follows that the mapping (19.130) is a diffeomorphism of $[\tau_0, 0]$ onto $[0, t_0]$ in the case of the *outgoing* null geodesics, a diffeomorphism of $[\tau_0, 0)$ onto $[0, t_0)$ in the cases of the *incoming* and the *other* null geodesics. We can thus invert the mapping (19.130) obtaining a continuous mapping of $[0, t_0]$ onto $[\tau_0, 0]$ by

$$t \mapsto (F^t)^{-1}(t; \not{p}_0) \quad (19.133)$$

which is smooth on $[0, t_0]$ in the case of the *outgoing* null geodesics, on $[0, t_0)$ in the case of the *incoming* and the *other* null geodesics.

Consider the mapping

$$(t; \not{p}_0) \mapsto (t, u, \vartheta) = G(t; \not{p}_0) = F((F^t)^{-1}(t; \not{p}_0); \not{p}_0) \quad (19.134)$$

and for fixed $t \in [0, t_0]$ the mapping

$$\not{p}_0 \mapsto (t, u, \vartheta) = G_t(\not{p}_0) \quad \text{where} \quad G_t(\not{p}_0) = G(t; \not{p}_0) \quad (19.135)$$

where, as above, $\not{p}_0 \in \mathbb{R}^2$ for the *outgoing* and *incoming* null geodesics, while $\not{p}_0 \in S^1$ for the *other* null geodesics. This is the domain of the mapping (19.135), while its range is the hypersurface $\{t\} \times (-\infty, \epsilon_0) \times S^2$ in the extension of the manifold $\mathcal{M}_{\epsilon_0}$ to

$$\mathcal{M}^e_{\epsilon_0} = \mathcal{M}_{\epsilon_0} \bigcup \mathcal{M}^E = [0, \infty) \times (-\infty, \epsilon_0) \times S^2$$

where

$$\mathcal{M}^E = [0,\infty) \times (-\infty, 0) \times S^2$$

corresponds to the domain of the surrounding constant state. The hypersurface $\{t\} \times (-\infty, \epsilon_0) \times S^2$ corresponds to the domain in the hyperplane $\Sigma_t$ in Galilean spacetime where $u > \epsilon_0$.

The mapping (19.135) is smooth for each $t \in [0, t_0)$, and for $t = t_0$, $G_{t_0}$ is the constant mapping

$$G_{t_0}(\not{p}_0) = G(t_0; \not{p}_0) = q_0 \tag{19.136}$$

Also, writing

$$G = (G^t, G^u, \not{G}), \quad G_t = (G_t^t, G_t^u, \not{G}_t) \tag{19.137}$$

we have

$$G_t^t = G^t = t \tag{19.138}$$

We have

$$\frac{\partial \not{G}^A}{\partial t}(t; \not{p}_0) = \frac{d\vartheta^A}{dt}(t) = \frac{(d\vartheta^A/d\tau)(\tau)}{(dt/d\tau)(\tau)} \tag{19.139}$$

and

$$\begin{aligned}\frac{\partial^2 \not{G}^A}{\partial t^2}(t; \not{p}_0) &= (\frac{d^2\vartheta^A}{dt^2})(t) \\ &= (\frac{(d^2\vartheta^A/d\tau^2)(dt/d\tau) - (d^2t/d\tau^2)(d\vartheta^A/d\tau)}{(dt/d\tau)^3})(\tau)\end{aligned} \tag{19.140}$$

In the case of the *outgoing* null geodesics we obtain, from (19.73),

$$\frac{\partial \not{G}^A}{\partial t}(t_0; \not{p}_0) = 0 \tag{19.141}$$

and, from (19.76),

$$\frac{\partial^2 \not{G}^A}{\partial t^2}(t_0; \not{p}_0) = \frac{\partial \mu}{\partial t}(q_0)(\not{g}^{-1})^{AB}(q_0)(\not{p}_0)_B \tag{19.142}$$

Let us consider the matrix $\partial \not{G}_t^A / \partial (\not{p}_0)_B$. By (19.141), (19.142),

$$\begin{aligned}&\frac{\partial}{\partial t}(\frac{\partial \not{G}^A}{\partial (\not{p}_0)_B})(t_0; \not{p}_0) = 0 \\ &\frac{\partial^2}{\partial t^2}(\frac{\partial \not{G}^A}{\partial (\not{p}_0)_B})(t_0; \not{p}_0) = (\frac{\partial \mu}{\partial t})(q_0)(\not{g}^{-1})^{AB}(q_0)\end{aligned} \tag{19.143}$$

It follows that

$$\frac{\partial \not{G}_t^A}{\partial(\not{p}_0)_B}(\not{p}_0) = \frac{1}{2}\frac{\partial \mu}{\partial t}(q_0)(\not{g}^{-1})^{AB}(q_0)((t-t_0)^2) + O((t-t_0)^3) \tag{19.144}$$

and

$$(\det \partial \not{G}_t/\partial \not{p}_0) = \frac{1}{4}(\frac{\partial \mu}{\partial t})^2(q_0)\det \not{g}^{-1}(t-t_0)^4 + O((t-t_0)^5) \tag{19.145}$$

Therefore in the case of the *outgoing* null geodesics the mapping (19.135) is of maximal rank at each $\not{p}_0 \in \mathbb{R}^2$ for $t$ suitably close to $t_0$. By the implicit function theorem we conclude that in the case of the *outgoing* null geodesics the image of $\mathbb{R}^2$ by the mapping (19.135) is, for $t$ suitably close to $t_0$, an immersed disk in the hypersurface $\{t\} \times (-\infty, \epsilon_0) \times S^2 \subset \mathcal{M}^e_{\epsilon_0}$, which corresponds to the hyperplane $\Sigma_t$ in Galilean spacetime. To show that this image is in fact an embedded disk we show that the mapping (19.135) is one-to-one for $t$ suitably close to $t_0$. For $|\not{p}_0| \leqslant B$ and any given $B > 0$, this follows from the fact that by (19.141), (19.142),

$$\not{G}_t^A(\not{p}_0) = \frac{1}{2}\frac{\partial \mu}{\partial t}(q_0)(\not{g}^{-1})^{AB}(q_0)(\not{p}_0)_B(t-t_0)^2 + O((t-t_0)^3) \tag{19.146}$$

provided that $t$ is suitably close to $t_0$ depending on $B$ (this guarantees the first term on the right of (19.146) is the main part of $\not{G}_t^A(\not{p}_0)$). To handle the case that $|\not{p}_0| > B$ (now, $B$ is fixed), we choose a different renormalization of the momentum covector at $q_0$ for the class of *outgoing* geodesics ending at $q_0$ than that given by (19.70). We now choose the scale factor $a$ so that

$$|\not{p}_0| = 1 \tag{19.147}$$

Then $p_u = -1/|\not{p}_0|$. This corresponds to the diffeomorphism

$$f : \not{p}_0 \mapsto (-\frac{1}{|\not{p}_0|}, \frac{\not{p}_0}{|\not{p}_0|}) \tag{19.148}$$

which maps the exterior of the disk of radius $B$ in $\mathbb{R}^2$ onto $(0, -1/B) \times S^1$. Using results analogous to (19.73), (19.76), substituted in (19.139), (19.140), we can show that the mapping $G_t \circ f^{-1}$ is one-to-one on $(0, -1/B) \times S^1$. Actually, what we have done here is just putting the "radial variable" outside the Taylor expansion, so we can proceed as in the case $|\not{p}_0| < B$. Still, we have to show that the image of $(0, -1/B) \times S^1$ by $G_t \circ f^{-1}$ does not intersect the image of the disk of radius $B$ by $G_t$. In fact, although we have made a coordinate transformation, but the actual Taylor expansion is the same, so the conclusion mentioned above is obvious.

In the case of the *other* null geodesics, $(F^t)^{-1}$ is not differentiable at $t = t_0$, so we first investigate the mapping $F$. We choose a basis $\omega^1$, $\omega^2$ for $T_{\vartheta_0}S^2$ which is orthonormal relative to $\not{g}^{-1}(q_0)$ and express $\not{p}_0$, which belongs to the unit circle in $T_{\vartheta_0}S^2$ in the form

$$\not{p}_0 = \omega^1 \cos\varphi + \omega^2 \sin\varphi \tag{19.149}$$

In the case that $q_0 \in \mathcal{H}$, we have, from (19.88), (19.91), (19.94),

$$F^t(0;\varphi) = t_0, \quad \frac{\partial F^t}{\partial \tau}(0;\varphi) = 0, \quad \frac{\partial^2 F^t}{\partial \tau^2}(0;\varphi) = \frac{1}{2}\frac{\partial \mu}{\partial u}(q_0) \tag{19.150}$$

and

$$\begin{aligned} \not{F}^A(0;\varphi) &= \vartheta_0^A, \quad \frac{\partial \not{F}^A}{\partial \tau}(0;\varphi) = (\frac{\partial^2 \not{F}^A}{\partial \tau^2})(0;\varphi) = 0 \\ \frac{\partial^3 \not{F}^A}{\partial \tau^3}(0;\varphi) &= \frac{\partial \mu}{\partial t}\frac{\partial \mu}{\partial u}(q_0)(\not{g}^{-1})^{AB}(q_0)(\not{p}_0)_B(\varphi) \\ (\not{p}_0)_A &= (\omega^1)_A \cos\varphi + (\omega^2)_A \sin\varphi \end{aligned} \tag{19.151}$$

It follows that in the case that $q_0 \in \mathcal{H}$ we have

$$\begin{aligned} F^t(\tau;\varphi) &= t_0 + \frac{1}{4}\frac{\partial \mu}{\partial u}(q_0)\tau^2 + O(\tau^3) \\ \not{F}^A(\tau;\varphi) &= \vartheta_0^A + \frac{1}{6}\frac{\partial \mu}{\partial u}\frac{\partial \mu}{\partial t}(q_0)(\not{g}^{-1})^{AB}(q_0)(\not{p}_0)_B(\varphi)\tau^3 + O(\tau^4) \end{aligned} \tag{19.152}$$

hence, since $\not{G}_t^A(\varphi) = \not{F}^A((F^t)^{-1}(t;\varphi);\varphi)$,

$$\not{G}_t^A(\varphi) = \vartheta_0^A - \frac{4}{3}\frac{(\partial\mu/\partial t)(q_0)}{\sqrt{-(\partial\mu/\partial u)(q_0)}}(\not{g}^{-1})^{AB}(q_0)(\not{p}_0)_B(\varphi)(t_0 - t)^{3/2} + O((t_0 - t)^2) \tag{19.153}$$

(We just use the first of (19.152) to express $\tau$ in terms of $t - t_0$.)

Moreover, (19.150) implies

$$\frac{\partial F^t}{\partial \varphi}(0;\varphi) = \frac{\partial}{\partial \tau}(\frac{\partial F^t}{\partial \varphi})(0;\varphi) = \frac{\partial^2}{\partial \tau^2}(\frac{\partial F^t}{\partial \varphi})(0;\varphi) = 0 \tag{19.154}$$

hence

$$\frac{\partial F^t}{\partial \varphi}(\tau;\varphi) = O(\tau^3) \tag{19.155}$$

Also, (19.151) implies

$$\begin{aligned} \frac{\partial \not{F}^A}{\partial \varphi}(0;\varphi) &= \frac{\partial}{\partial \tau}(\frac{\partial \not{F}^A}{\partial \varphi})(0;\varphi) = (\frac{\partial^2}{\partial \tau^2}(\frac{\partial \not{F}^A}{\partial \varphi}))(0;\varphi) = 0 \\ \frac{\partial^3}{\partial \tau^3}(\frac{\partial \not{F}^A}{\partial \varphi})(0;\varphi) &= \frac{\partial \mu}{\partial t}\frac{\partial \mu}{\partial u}(q_0)(\not{g}^{-1})^{AB}(q_0)(-(\omega^1)_B \sin\varphi + (\omega^2)_B \cos\varphi) \end{aligned} \tag{19.156}$$

hence

$$\frac{\partial \not{F}^A}{\partial \varphi}(\tau;\varphi) = \frac{1}{6}\frac{\partial \mu}{\partial t}\frac{\partial \mu}{\partial u}(q_0)(\not{g}^{-1})^{AB}(q_0)(-(\omega^1)_B \sin\varphi + (\omega^2)_B \cos\varphi)\tau^3 + O(\tau^4) \tag{19.157}$$

Differentiating the equation $F^t((F^t)^{-1}(t;\varphi);\varphi) = t$ implicitly with respect to $\varphi$ yields

$$\frac{\partial (F^t)^{-1}}{\partial \varphi} = -\frac{\partial F^t/\partial \varphi}{\partial F^t/\partial \tau} \tag{19.158}$$

Now (19.150) implies

$$\frac{\partial F^t}{\partial \tau}(\tau;\varphi) = \frac{1}{2}\frac{\partial \mu}{\partial u}(q_0)\tau + O(\tau^2) \tag{19.159}$$

Substituting (19.155) and (19.159) in (19.158) yields

$$\frac{\partial (F^t)^{-1}}{\partial \varphi} = O(\tau^2) \tag{19.160}$$

Differentiating $\not{G}^A(t;\varphi) = \not{F}^A((F^t)^{-1}(t;\varphi);\varphi)$ with respect to $\varphi$ we obtain

$$\frac{\partial \not{G}_t^A}{\partial \varphi} = \frac{\partial \not{F}^A}{\partial \tau}\frac{\partial (F^t)^{-1}}{\partial \varphi} + \frac{\partial \not{F}^A}{\partial \varphi} \tag{19.161}$$

Now (19.151) implies

$$\frac{\partial \not{F}^A}{\partial \tau} = \frac{1}{2}\frac{\partial \mu}{\partial t}\frac{\partial \mu}{\partial u}(q_0)(\not{g}^{-1})^{AB}(q_0)(\not{p}_0)_B(\varphi)\tau^2 + O(\tau^3) \tag{19.162}$$

In view of (19.160) and (19.162) the first term on the right in (19.161) is $O(\tau^4)$, while the second term is given by (19.157). We conclude that

$$\frac{\partial \not{G}_t^A}{\partial \varphi} = a(\not{g}^{-1})^{AB}(q_0)(-(\omega^1)_B \sin\varphi + (\omega^2)_B \cos\varphi)\tau^3 + O(\tau^4) \tag{19.163}$$

where

$$a = \frac{1}{6}\frac{\partial \mu}{\partial t}\frac{\partial \mu}{\partial u}(q_0) \tag{19.164}$$

which gives

$$\begin{aligned}
&\not{g}_{AB}(q_0)\frac{\partial \not{G}_t^A}{\partial \varphi}\frac{\partial \not{G}_t^B}{\partial \varphi} \\
&= a^2(\not{g}^{-1})^{AB}(-(\omega^1)_A \sin\varphi + (\omega^2)_A \cos\varphi)(-(\omega^1)_B \sin\varphi + (\omega^2)_B \cos\varphi)\tau^6 + O(\tau^7) \\
&= a^2\tau^6 + O(\tau^7)
\end{aligned} \tag{19.165}$$

The above hold in the case that $q_0 \in \mathcal{H}$. In the case $q_0 \in \partial_-\mathcal{H}$ similar calculations using (19.88), (19.95), (19.99), (19.104), (19.105), (19.108), give

$$\not{G}_t^A(\varphi) = \vartheta_0^A - \frac{8}{5}\frac{(-(\partial\mu/\partial t)(q_0))^{3/4}}{((1/6)(\partial^2\mu/\partial u^2)(q_0))^{1/4}}(t_0 - t)^{5/4} + O((t_0 - t)^{3/2}) \tag{19.166}$$

$$\frac{\partial \not{G}_t^A}{\partial \varphi} = b(\not{g}^{-1})^{AB}(q_0)(-(\omega^1)_B \sin\varphi + (\omega^2)_B \cos\varphi)\tau^5 + O(\tau^6) \tag{19.167}$$

where

$$b = \frac{1}{120}((\frac{\partial\mu}{\partial t})^2\frac{\partial^2\mu}{\partial u^2})(q_0) \tag{19.168}$$

and

$$\not{g}_{AB}(q_0)\frac{\partial \not{\mathcal{G}}_t^A}{\partial\varphi}\frac{\partial \not{\mathcal{G}}_t^B}{\partial\varphi} = b^2\tau^{10} + O(\tau^{11}) \tag{19.169}$$

The formulas (19.165) and (19.169) imply in the case of the *other* null geodesics ending at $q_0$ that the mapping (19.135) is of maximal rank at each $\not{p}_0 \in S^1$ for $t$ suitably close to $t_0$. By the implicit function theorem we conclude that in the case of the other null geodesics the image of $S^1$ by the mapping (19.135) is, for $t$ suitably close to $t_0$, an immersed circle in the hypersurface $\{t\}\times(-\infty,\epsilon_0)\times S^2 \subset \mathcal{M}^e_{\epsilon_0}$, which corresponds to the hyperplane $\Sigma_t$ in Galilean spacetime. Moreover, we deduce from (19.153) and (19.166) that the mapping (19.135) is one-to-one for $t$ suitably close to $t_0$. It then follows that the image of $S^1$ by the mapping (19.135) is actually an embedded circle. This circle is the boundary of the embedded disk corresponding to the outgoing null geodesics.

Similar arguments show that in the case of the *incoming* null geodesics the image of $\mathbb{R}^2$ by the mapping (19.135) is an embedded disk for $t$ suitably close to $t_0$. The boundary of this disk is the circle corresponding to the *other* null geodesics. We have thus proved the following theorem.

**Theorem 19.1** Let $q_0 = (t_0, u_0, \vartheta_0)$ be a point belonging to $\mathcal{H}\bigcup\partial_-\mathcal{H}$, the singular boundary of the domain of the maximal solution. Then the backward past null cone of $q_0$ in $T^*_{q_0}\mathcal{M}_{\epsilon_0}$ degenerates to the three pieces:
(i) The negative half-hyperplane: $p_t = 0$, $p_u < 0$.
(ii) The negative half hyperplane: $p_u = 0$, $p_t < 0$.
(iii) The plane $p_t = p_u = 0$.

There is a corresponding trichotomy of the past null geodesic cone of $q_0$ into three classes of null geodesics ending at $q_0$: the *outgoing* null geodesics, the *incoming* null geodesics, and the *other* null geodesics, corresponding to (i), (ii), (iii), respectively. Let all these null geodesics be parametrized by $t$. Then, in Galilean spacetime, the following hold:

The tangent vectors to all *outgoing* null geodesics ending at $q_0$ tend as we approach $q_0$ to the tangent vector $L(q_0)$ of the generator of $C_{u_0}$ through $q_0$. The vector $L(q_0)$ is thus an invariant null vector associated to the singular point $q_0$.

The tangent vector to an *incoming* null geodesic ending at $q_0$ with angular momentum $\not{p}_0$ at $q_0$, tends as $t \to t_0$ to the following limit:

$$L^i(q_0) + \frac{\alpha^{-1}(q_0)\hat{T}^i(q_0)}{(1/2)(\alpha^{-2}(q_0) + |\not{p}_0|^2)} + \frac{(\not{g}^{-1})^{AB}(q_0)X^i_A(q_0)(\not{p}_0)_B}{(1/2)(\alpha^{-2}(q_0) + |\not{p}_0|^2)}$$

which, for $\not{p}_0 = 0$ is equal to $\alpha^2(q_0)\underline{\hat{L}}(q_0)$, while as $\not{p}_0 \to \infty$ tends to $L(q_0)$.

The tangent vector to all the *other* null geodesics ending at $q_0$ likewise tend as $t \to t_0$ to $L(q_0)$.

Consider for each of the pieces (i), (ii), (iii), above, and for each $t \in [0, t_0]$, the mapping

$$\not{p}_0 \mapsto G_t(\not{p}_0)$$

where $G_t(\not{p}_0)$ is the point at parameter value $t$ along the null geodesic ending at parameter value $t_0$ at the point $q_0$ with momentum covector $\not{p}_0$. Then for $t \neq t_0$ and suitably close to $t_0$ the following hold:

The image of (i), which corresponds to the points of intersection of the outgoing null geodesics ending at $q_0$ with the hyperplane $\Sigma_t$, is an embedded disk.

The image of (ii), which corresponds to the points of intersection of the incoming null geodesics ending at $q_0$ with the hyperplane $\Sigma_t$, is also an embedded disk.

The image of (iii), which corresponds to the points of intersection of the other null geodesics ending at $q_0$ with the hyperplane $\Sigma_t$, is an embedded circle, the common boundary of the two disks.

## 19.3 Transformation of Coordinates

Consider now again the mapping (19.129) for the class of *outgoing* null geodesics ending at $q_0 \in \mathcal{H} \bigcup \partial_- \mathcal{H}$. We now wish to make the dependence on the point $q_0 \in \mathcal{H} \bigcup \partial_- \mathcal{H}$ explicit, so we denote $F(\tau; \not{p}_0)$ by $F(\tau; q_0; \not{p}_0)$. What we wish to investigate presently is the change from one acoustical function $u$ to another $u'$ (arising from different initial data for the eikonal equation), such that the null geodesic generators of the level sets $C'_{u'}$ of $u'$, the characteristic hypersurfaces corresponding to $u'$, which have future end points on the singular boundary of the domain of the maximal solution, belong to the class of *outgoing* null geodesics ending at these points. We can view this change as generated by a coordinate transformation:

$$u_0 = u_0(u'_0, \vartheta'_0), \quad \vartheta_0 = \vartheta_0(\vartheta'_0) \tag{19.170}$$

on $\mathcal{H} \bigcup \partial_- \mathcal{H}$, preserving the invariant curves. We also assume

$$\frac{\partial u_0}{\partial u'_0} > 0, \quad \det(\frac{\partial \vartheta_0}{\partial \vartheta'_0}) > 0 \tag{19.171}$$

If the coordinates of a point $q_0 \in \mathcal{H} \bigcup \partial_- \mathcal{H}$ in the unprimed system are $(t_0, u_0, \vartheta_0)$, $t_0 = t_*(u_0, \vartheta_0)$, then the coordinates of the same point in the primed system are $(t'_0, u'_0, \vartheta'_0)$, $t'_0 = t'_*(u'_0, \vartheta'_0)$, where

$$t'_*(u'_0, \vartheta'_0) = t_*(u_0, \vartheta_0) \tag{19.172}$$

hence $t'_0 = t_0$.

We shall determine $\not{p}_0$ as a function of $(u'_0, \vartheta'_0)$, such that the hypersurfaces, given in parametric form by

$$\begin{aligned} t &= M^t(\tau, u'_0, \vartheta'_0) = F^t(\tau; (u'_0, \vartheta'_0); \not{p}_0(u'_0, \vartheta'_0)) \\ u &= M^u(\tau, u'_0, \vartheta'_0) = F^u(\tau; (u'_0, \vartheta'_0); \not{p}_0(u'_0, \vartheta'_0)) \\ \vartheta &= \not{M}(\tau, u'_0, \vartheta'_0) = \not{F}(\tau; (u'_0, \vartheta'_0); \not{p}_0(u'_0, \vartheta'_0)) \end{aligned} \tag{19.173}$$

for constant values of $u'_0$, are null hypersurfaces with respect to the acoustical metric, namely the level sets $C'_{u'_0}$ of the new acoustical function $u'$.

In (19.173), $(t'_*(u'_0, \vartheta'_0), u'_0, \vartheta'_0)$ are the coordinates of a point $q_0 \in \mathcal{H} \bigcup \partial_- \mathcal{H}$ in the primed system, hence we have

$$\begin{aligned} F^t(0; (u'_0, \vartheta'_0); \not{p}_0) &= t_*(u_0, \vartheta_0) \\ F^u(0; (u'_0, \vartheta'_0); \not{p}_0) &= u_0 \\ \not{F}(0; (u'_0, \vartheta'_0); \not{p}_0) &= \vartheta_0 \end{aligned} \tag{19.174}$$

The hypersurfaces in question are generated by the outgoing null geodesics corresponding to constant values of $(u'_0, \vartheta'_0)$, and the hypersurfaces are null hypersurfaces if and only if these null geodesics are orthogonal, with respect to the acoustical metric, to the sections $S'_{\tau, u'_0}$ corresponding to conatant values of $\tau$. Let then $X'_A$ be the tangent vectorfield to the $\vartheta'^A_0$ coordinate lines on these sections and, as before, $\tilde{L}'$ be the tangent vectorfield to the null geodesics, as parametrized by $\tau$. Then the orthogonality condition reads

$$g(\tilde{L}', X'_A) = 0, \quad A = 1, 2 \tag{19.175}$$

and we have

$$\tilde{L}' = \frac{dF^t}{d\tau}\frac{\partial}{\partial t} + \frac{dF^u}{d\tau}\frac{\partial}{\partial u} + \frac{d\not{F}^A}{d\tau}\frac{\partial}{\partial \vartheta^A} \tag{19.176}$$

and

$$X'_A = X'^t_A \frac{\partial}{\partial t} + X'^u_A \frac{\partial}{\partial u} + X'^B_A \frac{\partial}{\partial \vartheta^B} \tag{19.177}$$

where

$$\begin{aligned} X'^t_A &= \frac{\partial F^t}{\partial \vartheta'^A_0} + \frac{\partial F^t}{\partial (\not{p}_0)_B}\frac{\partial (\not{p}_0)_B}{\partial \vartheta'^A_0} \\ X'^u_A &= \frac{\partial F^u}{\partial \vartheta'^A_0} + \frac{\partial F^u}{\partial (\not{p}_0)_B}\frac{\partial (\not{p}_0)_B}{\partial \vartheta'^A_0} \\ X'^C_A &= \frac{\partial \not{F}^C}{\partial \vartheta'^A_0} + \frac{\partial \not{F}^C}{\partial (\not{p}_0)_B}\frac{\partial (\not{p}_0)_B}{\partial \vartheta'^A_0} \end{aligned} \tag{19.178}$$

In fact, (19.173) defines a change of coordinates in the spacetime domain covered by the hypersurfaces in question, from $(t, u, \vartheta)$ coordinates to $(\tau, u'_0, \vartheta'_0)$ coordinates, and in the new coordinates we have, simply,

$$\tilde{L}' = \frac{\partial}{\partial \tau}, \quad X'_A = \frac{\partial}{\partial \vartheta_0'^A} \tag{19.179}$$

It follows that

$$[\tilde{L}', X'_A] = 0 \tag{19.180}$$

Now by (19.60) the integral curves of the vectorfield $\mu^{-1}\tilde{L}'$ are affinely parametrized null geodesics, hence

$$D_{\tilde{L}'}\tilde{L}' = \mu^{-1}(\tilde{L}'\mu)\tilde{L}', \quad g(\tilde{L}', \tilde{L}') = 0 \tag{19.181}$$

Let us then define

$$\iota_A = g(\tilde{L}', X'_A) \tag{19.182}$$

What we shall do is to prove $\iota_A = 0$.

We have, along the integral curves of $\tilde{L}'$,

$$\frac{d\iota_A}{d\tau} = \tilde{L}'(g(\tilde{L}', X'_A)) = g(D_{\tilde{L}'}\tilde{L}', X'_A) + g(\tilde{L}', D_{\tilde{L}'}X'_A)$$

and from (19.180), $D_{\tilde{L}'}X'_A = D_{X'_A}\tilde{L}'$, hence

$$g(\tilde{L}', D_{\tilde{L}'}X'_A) = \frac{1}{2}X'_A(g(\tilde{L}', \tilde{L}')) = 0$$

while by (19.181),

$$g(D_{\tilde{L}'}\tilde{L}', X'_A) = \mu^{-1}\frac{d\mu}{d\tau}\iota_A$$

It follows that

$$\frac{d}{d\tau}(\mu^{-1}\iota_A) = 0 \tag{19.183}$$

In view of (19.15) and (19.176),

$$\begin{aligned} \iota_A = & -\mu\frac{dF^t}{d\tau}X_A'^u - \mu\frac{dF^u}{d\tau}X_A'^t + \alpha^{-2}\mu^2\frac{dF^u}{d\tau}X_A'^u \\ & + \not{g}_{BC}(\frac{d\not{F}^B}{d\tau} + \mu\hat{\Xi}^B\frac{dF^u}{d\tau})(X_A'^C + \mu\hat{\Xi}^C X_A'^u) \end{aligned} \tag{19.184}$$

At $\tau = 0$ this reduces to

$$(\iota_A)_{\tau=0} = (\not{g}_{BC}\frac{d\not{F}^B}{d\tau}X_A'^C)_{\tau=0} = 0 \tag{19.185}$$

by (19.73). Moreover, we have

$$\lim_{\tau\to 0}(\mu^{-1}\iota_A) = -(\frac{dF^t}{d\tau}X'^u_A + \frac{dF^u}{d\tau}X'^t_A)_{\tau=0} + (\not{g}_{BC}X'^C_A)_{\tau=0}\lim_{\tau\to 0}(\mu^{-1}\frac{d\not{F}^B}{d\tau}) \tag{19.186}$$

Now by (19.73),

$$(\frac{dF^t}{d\tau})_{\tau=0} = 1, \quad (\frac{dF^u}{d\tau})_{\tau=0} = 0 \tag{19.187}$$

and by (19.75), (19.77),

$$\lim_{\tau\to 0}(\mu^{-1}\frac{d\not{F}^A}{d\tau}) = (\not{g}^{-1})^{AB}(q_0)(\not{p}_0)_B \tag{19.188}$$

while by (19.174) and (19.178),

$$(X'^u_A)_{\tau=0} = \frac{\partial u_0}{\partial\vartheta'^A_0}, \quad (X'^B_A)_{\tau=0} = \frac{\partial\vartheta^B_0}{\partial\vartheta'^A_0} \tag{19.189}$$

Note that we view $F^u$ and $\not{F}$ as the functions of $(u'_0, \vartheta'_0)$.

Substituting in (19.186) we obtain

$$\lim_{\tau\to 0}(\mu^{-1}\iota_A) = -\frac{\partial u_0}{\partial\vartheta'^A_0} + (\not{p}_0)_B\frac{\partial\vartheta^B_0}{\partial\vartheta'^A_0} \tag{19.190}$$

Therefore the condition $\lim_{\tau\to 0}(\mu^{-1}\iota_A) = 0$ is equivalent to the following definition of $\not{p}_0$ as a function of $(u'_0, \vartheta'_0)$:

$$(\not{p}_0)_A = \frac{\partial u_0}{\partial\vartheta'^B_0}\frac{\partial\vartheta'^B_0}{\partial\vartheta^A_0} \tag{19.191}$$

Once $\not{p}_0$ is defined according to (19.191), Equation (19.183) implies that $\iota_A = 0$ for all $\tau$, that is, the orthogonality condition (19.175) is everywhere satisfied and the hypersurfaces of constant $u'_0$, given by (19.173) are outgoing characteristic hypersurfaces.

In the following we denote $(u'_0, \vartheta'_0)$ simply by $(u', \vartheta')$. At fixed $q_0$, $\not{p}_0$ we invert the mapping (19.130), to obtain the mapping (19.133). Let us denote

$$f(t', u', \vartheta') = (F^t)^{-1}(t'; (u', \vartheta'); \not{p}_0(u', \vartheta')) \tag{19.192}$$

We then substitute in (19.173) to obtain the following transformation from the original acoustical coordinates $(t, u, \vartheta)$ to the new acoustical coordinates $(t', u', \vartheta')$:

$$\begin{aligned} t &= t' \\ u &= N^u(t', u', \vartheta') = F^u(f(t', u', \vartheta'); (u', \vartheta'); \not{p}_0(u', \vartheta')) \\ \vartheta &= \not{N}(t', u', \vartheta') = \not{F}(f(t', u', \vartheta'); (u', \vartheta'); \not{p}_0(u', \vartheta')) \end{aligned} \tag{19.193}$$

The acoustical metric has in the new acoustical coordinates a form similar to (19.15):

$$g = -2\mu' dt' du' + \alpha'^{-2}\mu'^2 du'^2 + \not{g}'_{AB}(d\vartheta'^A + \Xi'^A du')(d\vartheta'^B + \Xi'^B du') \quad (19.194)$$

where we have the new coefficients $\alpha'$, $\mu'$, $\Xi'^A$, and $\not{g}'_{AB}$. We have, in arbitrary local coordinates,

$$\alpha'^{-2} = -(g^{-1})^{\mu\nu}\partial_\mu t'\partial_\nu t' = -(g^{-1})^{\mu\nu}\partial_\mu t\partial_\nu t = \alpha^{-2} \quad (19.195)$$

so the functions $\alpha$ and $\alpha'$ coincide. The function $\mu'$ is given by (see Chapter 2):

$$L' = \mu'\hat{L}', \quad \hat{L}'^\mu = -(g^{-1})^{\mu\nu}\partial_\nu u' \quad (19.196)$$

and we have (in arbitrary local coordinates)

$$\frac{1}{\mu'} = -(g^{-1})^{\mu\nu}\partial_\mu t'\partial_\nu u' = -(g^{-1})^{\mu\nu}\partial_\mu t\partial_\nu u' \quad (19.197)$$

Here $L'$, related to $\tilde{L}'$ by (19.112), has the same integral curves as $\tilde{L}'$, but parametrized by $t$ instead of $\tau$, while the integral curves of $\hat{L}'$ are also the same but affinely parametrized. As we have shown above, the integral curves of the vectorfield $\mu^{-1}\tilde{L}'$ are likewise the same and affinely parametrized. Now two different affine parameters $s$ and $s'$ along the same geodesic, giving the same orientation, satisfy $s' = as + b$ where $a$ and $b$ are constants along the geodesic, $a > 0$. It follows that there is a function $a(u', \vartheta') > 0$ such that

$$\mu^{-1}\tilde{L}' = a\hat{L}' \quad (19.198)$$

On the other hand, from (19.37) and the second of (19.196) we obtain, working in the original acoustical coordinates,

$$\lim_{\tau\to 0}(\mu\hat{L}') = (\frac{\partial u'}{\partial u})_{\tau=0}\frac{\partial}{\partial t} + (\frac{\partial u'}{\partial t})_{\tau=0}\frac{\partial}{\partial u} \quad (19.199)$$

while by the first of (19.73),

$$\lim_{\tau\to 0}\tilde{L}' = \frac{\partial}{\partial t} \quad (19.200)$$

Now, from (19.193) we obtain

$$\begin{aligned}
&\frac{\partial t}{\partial t'} = 1, \quad \frac{\partial t}{\partial u'} = 0, \quad \frac{\partial t}{\partial \vartheta'^A} = 0 \\
&\frac{\partial u}{\partial t'} = \frac{\partial F^u}{\partial \tau}\frac{\partial f}{\partial t'}, \quad \frac{\partial u}{\partial u'} = \frac{\partial F^u}{\partial \tau}\frac{\partial f}{\partial u'} + \frac{\partial F^u}{\partial u'}, \quad \frac{\partial u}{\partial \vartheta'^A} = \frac{\partial F^u}{\partial \tau}\frac{\partial f}{\partial \vartheta'^A} + \frac{\partial F^u}{\partial \vartheta'^A} \quad (19.201) \\
&\frac{\partial \vartheta^A}{\partial t'} = \frac{\partial \not{F}^A}{\partial \tau}\frac{\partial f}{\partial t'}, \quad \frac{\partial \vartheta^A}{\partial u'} = \frac{\partial \not{F}^A}{\partial \tau}\frac{\partial f}{\partial u'} + \frac{\partial \not{F}^A}{\partial u'}, \quad \frac{\partial \vartheta^A}{\partial \vartheta'^B} = \frac{\partial \not{F}^A}{\partial \tau}\frac{\partial f}{\partial \vartheta'^B} + \frac{\partial \not{F}^A}{\partial \vartheta'^B}
\end{aligned}$$

In the above, we are considering $F^t$, $F^u$ and $\not{F}$ as functions of $\tau$, $u'$, $\vartheta'$, according to

$$\begin{aligned} F^t(\tau, u', \vartheta') &= F^t(\tau; (u', \vartheta'); \not\phi_0(u', \vartheta')) \\ F^u(\tau, u', \vartheta') &= F^u(\tau; (u', \vartheta'); \not\phi_0(u', \vartheta')) \\ \not{F}(\tau, u', \vartheta') &= \not{F}(\tau; (u', \vartheta'); \not\phi_0(u', \vartheta')) \end{aligned} \tag{19.202}$$

At $\tau = 0$, which corresponds to $t' = t'_*(u', \vartheta')$, so that

$$f(t'_*(u', \vartheta'), u', \vartheta') = 0 \tag{19.203}$$

we have, from (19.73),

$$\frac{\partial F^t}{\partial \tau} = 1, \qquad \frac{\partial F^u}{\partial \tau} = 0, \qquad \frac{\partial \not{F}^A}{\partial \tau} = 0 \tag{19.204}$$

hence the second and the third of (19.201) reduce at $\tau = 0$ to

$$\begin{aligned} &\frac{\partial u}{\partial t'} = 0, \quad \frac{\partial u}{\partial u'} = \frac{\partial F^u}{\partial u'}, \quad \frac{\partial u}{\partial \vartheta'^A} = \frac{\partial F^u}{\partial \vartheta'^A} \\ &\frac{\partial \vartheta^A}{\partial t'} = 0, \quad \frac{\partial \vartheta^A}{\partial u'} = \frac{\partial \not{F}^A}{\partial u'}, \quad \frac{\partial \vartheta^A}{\partial \vartheta'^B} = \frac{\partial \not{F}^A}{\partial \vartheta'^B} \end{aligned} \tag{19.205}$$

Now at $\tau = 0$ we have, according to (19.170) and (19.174),

$$\begin{aligned} F^t(0, u', \vartheta') &= t'_*(u', \vartheta') = t_*(u_0(u', \vartheta'), \vartheta_0(\vartheta')) \\ F^u(0, u', \vartheta') &= u_0(u', \vartheta') \\ \not{F}(0, u', \vartheta') &= \vartheta_0(\vartheta') \end{aligned} \tag{19.206}$$

Hence, at $\tau = 0$,

$$\begin{aligned} &\frac{\partial F^u}{\partial u'} = \frac{\partial u_0}{\partial u'}, \quad \frac{\partial F^u}{\partial \vartheta'^A} = \frac{\partial u_0}{\partial \vartheta'^A} \\ &\frac{\partial \not{F}^A}{\partial u'} = 0, \quad \frac{\partial \not{F}^A}{\partial \vartheta'^B} = \frac{\partial \vartheta_0^A}{\partial \vartheta'^B} \end{aligned} \tag{19.207}$$

Substituting in (19.205) we conclude that the Jacobian matrix of the transformation (19.193) is at $\tau = 0$ given by

$$\frac{\partial(t, u, \vartheta)}{\partial(t', u', \vartheta')} = \begin{bmatrix} 1 & 0 & 0 & 0 \\ 0 & \partial u_0/\partial u' & \partial u_0/\partial \vartheta'^1 & \partial u_0/\partial \vartheta'^2 \\ 0 & 0 & \partial \vartheta_0^1/\partial \vartheta'^1 & \partial \vartheta_0^1/\partial \vartheta'^2 \\ 0 & 0 & \partial \vartheta_0^2/\partial \vartheta'^1 & \partial \vartheta_0^2/\partial \vartheta'^2 \end{bmatrix} \tag{19.208}$$

Computing the reciprocal matrix we then find

$$\left(\frac{\partial u'}{\partial u}\right)_{\tau=0} = \frac{1}{\partial u_0/\partial u'}, \qquad \left(\frac{\partial u'}{\partial t}\right)_{\tau=0} = 0 \tag{19.209}$$

Substituting in (19.199) and comparing with (19.198) and (19.200) we conclude that

$$a(u',\vartheta') = \frac{\partial u_0}{\partial u'} > 0 \tag{19.210}$$

by (19.171). And from (19.37), (19.112) and (19.198),

$$\frac{\mu'}{\mu} = \frac{a}{dF^t/d\tau} \tag{19.211}$$

Then we know that $\mu'/\mu$ is bounded from above and below by positive constants. Also, since $\alpha' = \alpha$ we have

$$\frac{\kappa'}{\kappa} = \frac{\mu'}{\mu} \tag{19.212}$$

Next we consider the new coefficients $\Xi'^A$ and $\not{g}'_{AB}$. We have, in view of (19.194),

$$\begin{aligned}\Xi'_A = \not{g}'_{AB}\Xi'^B &= g'_{Au'} = g_{\mu\nu}\frac{\partial x^\mu}{\partial \vartheta'^A}\frac{\partial x^\nu}{\partial u'} \\ &= \mu^2(\alpha^{-2} + |\hat{\Xi}|^2)\frac{\partial u}{\partial \vartheta'^A}\frac{\partial u}{\partial u'} \\ &+ \mu\hat{\Xi}_B(\frac{\partial u}{\partial \vartheta'^A}\frac{\partial \vartheta^B}{\partial u'} + \frac{\partial \vartheta^B}{\partial \vartheta'^A}\frac{\partial u}{\partial u'}) \\ &+ \not{g}_{BC}\frac{\partial \vartheta^B}{\partial \vartheta'^A}\frac{\partial \vartheta^C}{\partial u'}\end{aligned} \tag{19.213}$$

Here,

$$|\hat{\Xi}|^2 = \not{g}_{AB}\hat{\Xi}^A\hat{\Xi}^B, \quad \hat{\Xi}_A = \not{g}_{AB}\hat{\Xi}^B$$

From (19.208) we have

$$(\frac{\partial \vartheta^A}{\partial u'})_{\tau=0} = 0$$

In view of (19.213) this implies that $\Xi'$ vanishes at $\tau = 0$, that is, on the singular boundary $\mathcal{H}\bigcup\partial_-\mathcal{H}$, a reflection of the fact that the new acoustical coordinates $(t', u', \vartheta')$ are also canonical. Since, by (19.211) $\partial\mu'/\partial t' < 0$ at $\tau = 0$, it follows as in (19.35) that

$$\hat{\Xi}' = \mu'^{-1}\Xi' \tag{19.214}$$

is regular at $\tau = 0$. ($a$ does not depend on $t'$.)

We have

$$\begin{aligned}\not{g}'_{AB} &= g_{\mu\nu}\frac{\partial x^\mu}{\partial \vartheta'^A}\frac{\partial x^\nu}{\partial \vartheta'^B} \\ &= \mu^2(\alpha^{-2} + |\hat{\Xi}|^2)\frac{\partial u}{\partial \vartheta'^A}\frac{\partial u}{\partial \vartheta'^B} \\ &+ \mu\hat{\Xi}_C(\frac{\partial \vartheta^C}{\partial \vartheta'^B}\frac{\partial u}{\partial \vartheta'^A} + \frac{\partial \vartheta^C}{\partial \vartheta'^A}\frac{\partial u}{\partial \vartheta'^B}) \\ &+ \not{g}_{CD}\frac{\partial \vartheta^C}{\partial \vartheta'^A}\frac{\partial \vartheta^D}{\partial \vartheta'^B}\end{aligned} \tag{19.215}$$

Let

$$Y' = Y'^A \frac{\partial}{\partial \vartheta'^A}, \quad \not{g}'(Y', Y') = \not{g}'_{AB} Y'^A Y'^B \tag{19.216}$$

Let also

$$\tilde{\not{g}}_{AB} = \not{g}_{CD} \frac{\partial \vartheta^C}{\partial \vartheta'^A} \frac{\partial \vartheta^D}{\partial \vartheta'^B}, \quad \tilde{\not{g}}(Y', Y') = \tilde{\not{g}}_{AB} Y'^A Y'^B \tag{19.217}$$

We then deduce from (19.215) the following lower bound for $\not{g}'$ in terms of $\not{g}$: for any vector $Y'$ as in (19.216) it holds that

$$\not{g}'(Y', Y') \geqslant \frac{\tilde{\not{g}}(Y', Y')}{1 + \alpha^2 \not{g}(\hat{\Xi}, \hat{\Xi})} \tag{19.218}$$

In fact, by (19.215),

$$\not{g}'(Y', Y') = \mu^2(\alpha^{-2} + |\hat{\Xi}|^2)(Y'u)^2 + 2\mu(\hat{\Xi}' \cdot Y')(Y'u) + \tilde{\not{g}}(Y', Y')$$

where

$$\hat{\Xi}'_B = \hat{\Xi}_C \frac{\partial \vartheta^C}{\partial \vartheta'_B}$$

then by Cauchy-Schwarz inequality,

$$|\hat{\Xi}' \cdot Y'| \leqslant \sqrt{\tilde{\not{g}}^{-1}(\hat{\Xi}', \hat{\Xi}')} \sqrt{\tilde{\not{g}}(Y', Y')}$$

Since

$$\tilde{\not{g}}^{-1}(\hat{\Xi}', \hat{\Xi}') = \not{g}^{-1}(\hat{\Xi}, \hat{\Xi}) - |\hat{\Xi}|^2$$

we obtain

$$|\hat{\Xi}' \cdot Y'| \leqslant |\hat{\Xi}| \sqrt{\tilde{\not{g}}(Y', Y')}$$

Setting

$$a = \mu^2(\alpha^{-2} + |\hat{\Xi}|^2), \quad b = \mu|\hat{\Xi}|, \quad x = |Y'u|, \quad y = \sqrt{\tilde{\not{g}}(Y', Y')}$$

then

$$\not{g}'(Y', Y') \geqslant ax^2 - 2bxy + y^2$$

Setting

$$\rho = \frac{x}{y}$$

we have

$$\frac{\not g'(Y',Y')}{\tilde{\not g}(Y',Y')} \geqslant a\rho^2 - 2b\rho + 1 \geqslant 1 - \frac{b^2}{a} = \frac{\mu^2\alpha^{-2}}{\mu^2(\alpha^{-2} + |\hat{\Xi}|^2)}$$

Reversing the roles of the coordinates $(t, u, \vartheta)$ and $(t', u', \vartheta')$, we obtain in a similar manner a lower bound for $\not g$ in terms of $\not g'$. Let

$$Y = Y^A \frac{\partial}{\partial \vartheta^A}, \quad \not g(Y, Y) = \not g_{AB} Y^A Y^B \tag{19.219}$$

Let also

$$\tilde{\not g}'_{AB} = \not g'_{CD} \frac{\partial \vartheta'^C}{\partial \vartheta^A} \frac{\partial \vartheta'^D}{\partial \vartheta^B}, \quad \tilde{\not g}'(Y, Y) = \tilde{\not g}'_{AB} Y^A Y^B \tag{19.220}$$

Then for any $Y$ as in (19.219) it holds that

$$\not g(Y, Y) \geqslant \frac{\tilde{\not g}'(Y, Y)}{1 + \alpha^2 \not g'(\hat{\Xi}', \hat{\Xi}')} \tag{19.221}$$

We summarize the above results in the following proposition.

**Proposition 19.2** Let $u'$ be any other acoustical function such that the null geodesic generators of the level surfaces $C'_{u'}$ of $u'$ which have future end points on the singular boundary of the domain of the maximal solution belong to the class of *outgoing* null geodesics ending at this points. The change from the original acoustical function $u$ to the new acoustical function $u'$ can be viewed as generated by a coordinate transformation

$$u_0 = u_0(u'_0, \vartheta'_0), \quad \vartheta_0 = \vartheta_0(\vartheta'_0) \tag{19.222}$$

on $\mathcal{H} \bigcup \partial_- \mathcal{H}$ preserving the invariant curves, so that both systems of acoustical coordinates are canonical. We require that the transformation $\vartheta_0 \mapsto \vartheta'_0$ be orientation preserving and that on each invariant curve the transformation $u_0 \mapsto u'_0$ be also orientation preserving. The characteristic hypersurfaces $C'_{u'_0}$ are given in parametric form in terms of the original acoustical coordinates $(t, u, \vartheta)$ by

$$\begin{aligned} t &= F^t(\tau, u'_0, \vartheta'_0) \\ u &= F^u(\tau, u'_0, \vartheta'_0) \\ \vartheta &= \not F(\tau, u'_0, \vartheta'_0) \end{aligned}$$

where

$$\begin{aligned} F^t(\tau, u'_0, \vartheta'_0) &= F^t(\tau; q_0; \not p_0(q_0)) \\ F^u(\tau, u'_0, \vartheta'_0) &= F^u(\tau; q_0; \not p_0(q_0)) \\ \not F(\tau, u'_0, \vartheta'_0) &= \not F(\tau; q_0; \not p_0(q_0)) \end{aligned}$$

Here $q_0$ is the point on $\mathcal{H} \bigcup \partial_- \mathcal{H}$ with coordinates $(t_0, u_0, \vartheta_0)$ in the original system and $(t'_0, u'_0, \vartheta'_0)$ in the new system, where

$$t_0 = t_*(u_0, \vartheta_0) = t'_*(u'_0, \vartheta'_0) = t'_0$$

Also,

$$\not{p}_0(q_0) = \frac{\partial u_0}{\partial \vartheta'^A_0} d\vartheta'^A_0$$

is the angular momentum at $q_0$. In the above

$$\tau \mapsto (F^t(\tau; q_0; \not{p}_0), F^u(\tau; q_0; \not{p}_0), \not{F}(\tau; q_0; \not{p}_0))$$

is the *outgoing* null geodesic, parametrized by $\tau$, ending at $\tau = 0$ at $q_0 \in \mathcal{H} \bigcup \partial_- \mathcal{H}$ with angular momentum $\not{p}_0$. Denoting by $\tau = f(t', u', \vartheta')$ the solution of

$$F^t(\tau, u', \vartheta') = t'$$

for fixed $(u', \vartheta')$, the transformation between the original system of acoustical coordinates $(t, u, \vartheta)$ and the new system of acoustical coordinates $(t', u', \vartheta')$ is given by

$$\begin{aligned} t &= t' \\ u &= N^u(t', u', \vartheta') \\ \vartheta &= \not{N}(t', u', \vartheta') \end{aligned}$$

where

$$\begin{aligned} N^u(t', u', \vartheta') &= F^u(f(t', u', \vartheta'), u', \vartheta') \\ \not{N}(t', u', \vartheta') &= \not{F}(f(t', u', \vartheta'), u', \vartheta') \end{aligned}$$

The acoustical metric has the same form in the new acoustical coordinates:

$$g = -2\mu dt du + \alpha^{-2}\mu^2 du^2 + \not{g}_{AB}(d\vartheta^A + \mu\hat{\Xi}^A du)(d\vartheta^B + \mu\hat{\Xi}^B du)$$
$$g = -2\mu' dt' du' + \alpha'^{-2}\mu'^2 du'^2 + \not{g}'_{AB}(d\vartheta'^A + \mu\hat{\Xi}'^A du')(d\vartheta'^B + \mu\hat{\Xi}'^B du')$$

where $\alpha = \alpha'$. The ratio $\mu'/\mu$ is bounded above and below by positive constants, while the metric $\not{g}'$ and $\not{g}$ are uniformly equivalent in a neighborhood of the singular boundary.

We now revisit the mapping (19.135) for a fixed point $q_0 \in \mathcal{H} \bigcup \partial_- \mathcal{H}$ and for the class of *outgoing* null geodesics ending at $q_0$. Setting $t = 0$ we have the mapping of $\mathbb{R}^2$ into $\Sigma_0$ by

$$\begin{aligned} \not{p}_0 &\mapsto (0, u, \vartheta) = G_0(\not{p}_0) \\ G_0(\not{p}_0) &= G(0, \not{p}_0) = (0, F^u(f(\not{p}_0); \not{p}_0), \not{F}(f(\not{p}_0); \not{p}_0)) \\ \text{where} \quad \tau &= f(\not{p}_0) \quad \text{is the solution of} \quad F^t(\tau; \not{p}_0) = 0 \end{aligned} \tag{19.223}$$

Let $u'$ be a function defined on $\Sigma_0$, which has a unique maximum $p_0$ on $\Sigma_0$, and is smooth and without critical points on $\Sigma_0 \setminus \{p_0\}$. We may consider $u'$ as a function of the acoustical coordinates $(0, u, \vartheta)$ on $\Sigma_0$, defined on $(-\infty, 1) \times S^2 \setminus \{(u_0, \vartheta_0)\}$ where $(u_0, \vartheta_0)$ are the coordinates of the maximum point $p_0$.

Consider then the following function on $\mathbb{R}^2$:

$$g(\not{p}_0) = u'(G_0(\not{p}_0)) \tag{19.224}$$

Consider now a maximum of the function $g(\not{p}_0)$ which is achieved at some point $\not{p}_{0,M} \in \mathbb{R}^2$. Such a maximum point corresponds to a point $p_M \in \Sigma_0$, where

$$p_M = (0, u_M, \vartheta_M), \quad u_M = G_0^u(\not{p}_{0,M}), \quad \vartheta_M = \not{G}_0(\not{p}_{0,M}) \tag{19.225}$$

and we have

$$\frac{\partial g}{\partial(\not{p}_0)_A} = 0 \quad \text{at} \quad \not{p}_{0,M} \tag{19.226}$$

that is

$$\frac{\partial u'}{\partial u}\frac{\partial G_0^u}{\partial(\not{p}_0)_A} + \frac{\partial u'}{\partial \vartheta^B}\frac{\partial \not{G}_0^B}{\partial(\not{p}_0)_A} = 0 \quad \text{at} \quad \not{p}_{0,M} \tag{19.227}$$

Consider the level surface $S'_{0,u'}$ of $u'$ on $\Sigma_0$ through the point $p_M$. We shall show that $L'$ is $g$-orthogonal to $S_{0,u'}$ at $p_M$.

This fact is equivalent to the following. Denoting by $C^-(q_0)$ the past null geodesic cone of the point $q_0$ on the singular boundary, and by $C^-_{out}(q_0)$ the part of $C^-(q_0)$ corresponding to the *outgoing* null geodesics ending at $q_0$, then the statement is that the surface $C^-_{out}(q_0) \bigcap \Sigma_0$ and $S'_{0,u'}$ have the same tangent plane at $p_M$:

$$T_{p_M}(C^-_{out}(q_0) \bigcap \Sigma_0) = T_{p_M} S'_{0,u'} \tag{19.228}$$

To show this, we shall produce a basis for the left-hand side and show that this is also a basis for the right-hand side.

Consider the vectorfields

$$Y^A = \frac{\partial F^t}{\partial(\not{p}_0)_A}\frac{\partial}{\partial t} + \frac{\partial F^u}{\partial(\not{p}_0)_A}\frac{\partial}{\partial u} + \frac{\partial \not{F}^B}{\partial(\not{p}_0)_A}\frac{\partial}{\partial \vartheta^B}, \quad A = 1, 2 \tag{19.229}$$

These are tangential to the sections of $C^-_{out}(q_0)$ which correspond to constant values of $\tau$ (recalling the mapping $F$, we view $\tau$ and $\not{p}_0$ are independent coordinates), hence are $g$-orthogonal to $\tilde{L}'$. For each point $p \in C^-_{out}(q_0)$, $(Y^1(p), Y^2(p))$ is a basis for the tangent plane at $p$ to the $\tau =$ const section of $C^-_{out}(q_0)$ through $p$, provided that the vectors $Y^1(p)$ , $Y^2(p)$ are linearly independent. We project the $Y^A$ to the $\Sigma_t$ as follows. We find suitable functions $\lambda^A$, $A = 1, 2$, such that the vectorfields

$$Y'^A = Y^A - \lambda^A L' \tag{19.230}$$

are tangential to the $\Sigma_t$. The vectorfields $Y'^A$, being $g$-orthogonal to $L'$ hence tangential to $C^-_{out}(q_0)$, shall then be tangential to the sections $C^-_{out}(q_0)\bigcap\Sigma_t$. As $L'$ is given by

$$L' = \frac{\partial}{\partial t} + \frac{\partial F^u/\partial\tau}{\partial F^t/\partial\tau}\frac{\partial}{\partial u} + \frac{\partial \not{F}^A/\partial\tau}{\partial F^t/\partial\tau}\frac{\partial}{\partial\vartheta^A} \tag{19.231}$$

we find

$$\lambda^A = \frac{\partial F^t}{\partial(\not{p}_0)_A} \tag{19.232}$$

and

$$\begin{aligned}\frac{\partial F^t}{\partial\tau}Y'^A = &\left(\frac{\partial F^u}{\partial(\not{p}_0)_A}\frac{\partial F^t}{\partial\tau} - \frac{\partial F^t}{\partial(\not{p}_0)_A}\frac{\partial F^u}{\partial\tau}\right)\frac{\partial}{\partial u}\\ &+\left(\frac{\partial\not{F}^B}{\partial(\not{p}_0)_A}\frac{\partial F^t}{\partial\tau} - \frac{\partial F^t}{\partial(\not{p}_0)_A}\frac{\partial\not{F}^B}{\partial\tau}\right)\frac{\partial}{\partial\vartheta^B}\end{aligned} \tag{19.233}$$

Then $(Y'^1(p_M), Y'^2(p_M))$ is a basis for $T_{p_M}(C^-_{out}(q_0)\bigcap\Sigma_0)$, provided that $p_M$ is not conjugate to $q_0$. Now, we have, from (19.223),

$$\begin{aligned}\frac{\partial G^u_0}{\partial(\not{p}_0)_A} &= \frac{\partial F^u}{\partial\tau}\frac{\partial f}{\partial(\not{p}_0)_A} + \frac{\partial F^u}{\partial(\not{p}_0)_A}\\ \frac{\partial\not{G}^B_0}{\partial(\not{p}_0)_A} &= \frac{\partial\not{F}^B}{\partial\tau}\frac{\partial f}{\partial(\not{p}_0)_A} + \frac{\partial\not{F}^B}{\partial(\not{p}_0)_A}\end{aligned} \tag{19.234}$$

Also, differentiating the equation $F^t(f(\not{p}_0);\not{p}_0) = 0$ implicitly with respect to $\not{p}_0$ yields

$$\frac{\partial f}{\partial(\not{p}_0)_A} = -\frac{\partial F^t/\partial(\not{p}_0)_A}{\partial F^t/\partial\tau} \tag{19.235}$$

Substituting in (19.234) we then obtain

$$\begin{aligned}\frac{\partial G^u_0}{\partial(\not{p}_0)_A} &= \left(-\frac{\partial F^u}{\partial\tau}\frac{\partial F^t}{\partial(\not{p}_0)_A} + \frac{\partial F^t}{\partial\tau}\frac{\partial F^u}{\partial(\not{p}_0)_A}\right)\frac{1}{\partial F^t/\partial\tau}\\ \frac{\partial\not{G}^B_0}{\partial(\not{p}_0)_A} &= \left(-\frac{\partial\not{F}^B}{\partial\tau}\frac{\partial F^t}{\partial(\not{p}_0)_A} + \frac{\partial F^t}{\partial\tau}\frac{\partial\not{F}^B}{\partial(\not{p}_0)_A}\right)\frac{1}{\partial F^t/\partial\tau}\end{aligned} \tag{19.236}$$

Comparing with (19.233) we see that

$$Y'^A = \frac{\partial G^u_0}{\partial(\not{p}_0)_A}\frac{\partial}{\partial u} + \frac{\partial\not{G}^B_0}{\partial(\not{p}_0)_A}\frac{\partial}{\partial\vartheta^B} \tag{19.237}$$

hence (19.227) reads, simply,

$$Y'^A(p_M)u' = 0, \quad A = 1,2 \tag{19.238}$$

We conclude that $(Y'^1(p_M), Y'^2(p_M))$ is a basis for $T_{p_M} S'_{0,u'}$, provided that the vectors $Y'^1(p_M)$, $Y'^2(p_M)$ are linearly independent. We have thus demonstrated (19.228) under the condition that $p_M$ is not conjugate to $q_0$. However, (19.226) holds even in the case that $p_M$ is conjugate to $q_0$, but in this case we must approach $p_M$, which corresponds to $\tau = f(\not{p}_{0,M})$ along the outgoing geodesic ending at $q_0$ with angular momentum $\not{p}_{0,M}$, by a sequence of non-conjugate points corresponding to $\tau = \tau_n \to f(\not{p}_{0,M})$ along the same geodesic.

The equality (19.228) is equivalent to $L'$ being $g$-orthogonal to $S'_{0,u'}$ at $p_M$. Thus, extending $u'$ to an acoustical function so that its level sets are outgoing characteristic hypersurface $C'_{u'}$ with $C'_{u'} \bigcap \Sigma_0 = S'_{0,u'}$, the generator of the outgoing characteristic hypersurfaces $C'_{u'}$ through $p_M$ coincides with the *outgoing* null geodesic ending at $q_0$ with angular momentum $\not{p}_{0,M}$.

Let us now consider again our main problem from a more general point of view. Our problem is the initial value problem for the equations of motion of a perfect fluid, with initial data on the hyperplane $\Sigma_0$ which coincide with those of a constant state outside a compact subset $P_0 \subset \Sigma_0$. We may take $S_{0,0}$ to be any sphere in $\Sigma_0$ containing $P_0$. The construction in Chapter 2 then gives an acoustical function $u$ based on $S_{0,0}$. If the initial data is isentropic and irrotational, Theorem 17.1 can be applied after suitable translation and rescaling, giving a development of the initial data in a domain $V_{\epsilon_0}$ as in (19.7) corresponding to the annular region on $\Sigma_0$ bounded by $S_{0,0}$ and the concentric sphere of radius $1 - \epsilon_0$ that of $S_{0,0}$. Then by the local uniqueness theorem the union of all the developments each of which corresponds to a sphere $S_{0,0}$ containing $P_0$ is also a development of the initial data, contained in the maximal development, the union of the domains $V_{\epsilon_0}$ being contained in the closure of the domain of the maximal solution. Consider now a point $q_0$ belonging to the singular boundary of the domain of the maximal solution such that $q_0 \in V'_{\epsilon_0}$ where $V'_{\epsilon_0}$ is the domain associated to the acoustical function $u'$, corresponding to the sphere $S'_{0,0}$. Thus $q_0 \in \Sigma^{\epsilon}_{t'_{*\epsilon}}$ for some $\epsilon \in (0, \epsilon_0)$, where $t'_{*\epsilon} = \inf_{u' \in [0,\epsilon]} t'_*(u')$ and $t'_*(u')$ is the least upper bound of the extent of the generators of $C'_{u'}$ in the domain of maximal solution. Then for any given acoustical function $u$, corresponding to a given sphere $S_{0,0}$, the above argument (with the roles of $u$ and $u'$ reversed) shows that there is an outgoing characteristic hypersurface $C_u$ associated to $u$ and a generator of $C_u$ reaching $q_0$, thus of extent $t'_{*\epsilon}$, contained in $V'_{\epsilon_0}$, hence in the union of the domains $V_{\epsilon_0}$. In view of the properties of the transformation from one acoustical function to another, expounded by Proposition 19.2, Theorem 17.1 covers the union of the domains $V_{\epsilon_0}$.

However we can extend the domain covered by Theorem 17.1 still further in the following manner. We can replace the hyperplanes $\Sigma_t$, by the hyperplanes $\Sigma_{\alpha;t}$ where for any fixed $\alpha = (\alpha_1, \alpha_2, \alpha_3) \in \mathbb{R}^3$ such that

$$\|\alpha\| = \sqrt{\alpha_1^2 + \alpha_2^2 + \alpha_3^2} < 1$$

and any $t \in \mathbb{R}$, $\Sigma_{\alpha;t}$ is the hyperplane in Galilean spacetime given by

$$x^0 = t + \sum_{i=1}^{3} \alpha_i x^i$$

The hyperplane $\Sigma_{\alpha;0}$ is then spacelike relative to the actual acoustical metric $g$ provided that the departure of the data induced on $\Sigma_{\alpha;0}$ from those of the constant state is suitably small. We then consider as above a sphere $S_{0,0}$ in $\Sigma_0$ such that the initial data on $\Sigma_0$ coincide with those of the constant state outside $S_{0,0}$, and proceed to construct the corresponding acoustical function $u$. The surfaces $S_{\alpha;t,0} = C_0 \bigcap \Sigma_{\alpha;t}$ are then round spheres. The first variations are defined exactly as before, for, these are adapted to the rest frame of the constant state. Of the commutation fields $T$, $Q$, $R_i, i = 1,2,3$, which define the higher order variations, $Q$ is defined exactly as before, $T$ is now defined to be tangential to the hyperplanes $\Sigma_{\alpha;t}$, $g$-orthogonal to the surfaces $S_{\alpha;t,u} = C_u \bigcap \Sigma_{\alpha;t}$ and satisfying $Tu = 1$, while the $R_i, i = 1,2,3$ are defined to be the $g$-orthogonal projections to the surfaces $S_{\alpha;t,u}$, of the rotations associated to $g_0$ which leave $\Sigma_{\alpha;t}$ and $S_{\alpha;t,0}$ invariant. We note that, similarly, $\Sigma_{\alpha;t}$ remain spacelike relative to the acoustical metric provided that the initial departure from the constant state is suitably small. We obtain in this way an extension of Theorem 17.1 and a domain $V_{\alpha;\epsilon_0}$ defined in a similar manner as the domain $V_{\epsilon_0}$, but with the hyperplanes $\Sigma_{\alpha;t}$ in the role of the hyperplanes $\Sigma_t$. The domain covered by Theorem 17.1 is then maximally extended to be the union of all the domains $V_{\alpha;\epsilon_0}$, and the remarks of the previous paragraph apply to this maximal extension.

## 19.4 How $\mathcal{H}$ Looks Like in Rectangular Coordinates in Galilean Spacetime

We now consider a given component of $\mathcal{H}$ and its past boundary, the corresponding component of $\partial_-\mathcal{H}$. By Proposition 19.1 these are smooth 3-dimensional and 2-dimensional submanifolds of $\mathcal{M}_{\epsilon_0}$, respectively. We shall now investigate the image of the component of $\mathcal{H}$ as well as $\partial_-\mathcal{H}$ in Galilean spacetime. For clarity we distinguish these images from the sets themselves and we denote by $H$ and $\partial_- H$ the image in Galilean spacetime of $\mathcal{H}$ and $\partial_-\mathcal{H}$ respectively.

Now by Proposition 19.1 $\partial_-\mathcal{H}$ is a spacelike surface in $(\mathcal{M}_{\epsilon_0}, g)$ intersecting the invariant curves transversally. Thus there is a smooth function $u_*$ defined on a connected domain $\mathcal{B} \subset S^2$, such that the component of $\partial_-\mathcal{H}$ is given in canonical acoustical coordinates by

$$\{(t_*(u_*(\vartheta), \vartheta), u_*(\vartheta), \vartheta), \vartheta \in \mathcal{B}\} \tag{19.239}$$

and the corresponding component of $\mathcal{H}$ is given by

$$\{(t_*(u, \vartheta), u, \vartheta), u_* > u > \underline{u}(\vartheta), \vartheta \in \mathcal{B}\} \tag{19.240}$$

where a given invariant curve, parametrized by $u$, either extends to any $u > 0$, in which case $\underline{u}(\vartheta) = 0$, or ends at $u = \underline{u}(\vartheta)$, the point of intersection with the incoming characteristic hypersurface $\underline{C}$ generated by the incoming null normals to the next outward component of $\partial_-\mathcal{H}$.

Consider then the component of $\partial_- H$, which is the image of (19.239) in Galilean spacetime. This is the image in Galilean spacetime of the domain $\mathcal{B} \subset S^2$ under the mapping

$$\vartheta \longmapsto x = q(\vartheta), \quad \text{where} \quad q^\mu(\vartheta) = x^\mu(t_*(u_*(\vartheta), \vartheta), \quad u_*(\vartheta), \vartheta), \mu = 0, 1, 2, 3 \tag{19.241}$$

This mapping is one-to-one (it is simply the transformation between rectangular coordinates and acoustical coordinates). We shall presently show that the mapping is of maximal rank. Consider the 2-dimensional matrix $m$ with entries:

$$m_{AB} = g_{\mu\nu} X^\mu_A \frac{\partial q^\nu}{\partial \vartheta^B} \tag{19.242}$$

Now, we have (see (19.115))

$$\begin{aligned} &\frac{\partial x^0}{\partial t} = 1, \quad \frac{\partial x^i}{\partial t} = L^i \\ &\frac{\partial x^0}{\partial u} = \frac{\partial x^0}{\partial \vartheta^A} = 0, \quad \frac{\partial x^i}{\partial u} = \mu(\alpha^{-1}\hat{T}^i + \hat{\Xi}^A X^i_A), \quad \frac{\partial x^i}{\partial \vartheta^A} = X^i_A \end{aligned} \tag{19.243}$$

and from (19.241),

$$\begin{aligned} \frac{\partial q^0}{\partial \vartheta^B} &= \frac{\partial t_*}{\partial u}\frac{\partial u_*}{\partial \vartheta^B} + \frac{\partial t_*}{\partial \vartheta^B} \\ \frac{\partial q^j}{\partial \vartheta^B} &= \frac{\partial x^j}{\partial t}\left(\frac{\partial t_*}{\partial u}\frac{\partial u_*}{\partial \vartheta^B} + \frac{\partial t_*}{\partial \vartheta^B}\right) + \frac{\partial x^j}{\partial u}\frac{\partial u_*}{\partial \vartheta^B} + \frac{\partial x^j}{\partial \vartheta^B} \end{aligned} \tag{19.244}$$

Now on $\mathcal{H} \bigcup \partial_- \mathcal{H}$ we have $\mu = 0$, therefore, from (19.243),

$$\frac{\partial x^i}{\partial u} = 0 \quad \text{on} \quad \mathcal{H} \bigcup \partial_- \mathcal{H} \tag{19.245}$$

Moreover by Proposition 19.1,

$$\frac{\partial t_*}{\partial u} = 0 \quad \text{on} \quad \partial_- \mathcal{H} \tag{19.246}$$

Thus substituting in (19.244) from (19.245) and taking account of (19.245) and (19.246), we obtain

$$\frac{\partial q^0}{\partial \vartheta^B} = \frac{\partial t_*}{\partial \vartheta^B}, \quad \frac{\partial q^j}{\partial \vartheta^B} = L^j \frac{\partial t_*}{\partial \vartheta^B} + X^j_B$$

or, simply,

$$\frac{\partial q^\nu}{\partial \vartheta^B} = L^\nu \frac{\partial t_*}{\partial \vartheta^B} + X^\nu_B \tag{19.247}$$

Substituting this in (19.242) we obtain simply

$$m_{AB} = g(X_A, X_B) = \not{g}_{AB} \tag{19.248}$$

Hence $\det m = \det \not{g} > 0$ and the mapping (19.241) is of maximal rank. It follows that this mapping is a diffeomorphism onto its image and its image, the component of $\partial_- H$, is a smooth embedded 2-dimensional submanifold in Galilean spacetime.

The component of $H$ can similarly be shown to be a smooth embedded 3-dimensional submanifold in Galilean spacetime except along its past boundary (just using the definition of linearly independent, the three linearly independent vectors are $L\frac{\partial t_*}{\partial u}$ and the $L\frac{\partial t_*}{\partial \vartheta^A} + X_A$), the corresponding component of $\partial_- H$. To analyze the behavior near $\partial_- H$, we shall derive an analytic expression for an arbitrary invariant curve of $H$ in a neighborhood of $\partial_- H$. Now the parametric equations of an invariant curve of $H$, in rectangular coordinates in Galilean spacetime, the parameter being $u$, are (see (19.240))

$$x^\mu = x^\mu(t_*(u,\vartheta), u, \vartheta), \quad u_*(\vartheta) > u > \underline{u}(\vartheta), \quad \vartheta = \text{const} \tag{19.249}$$

or, since $x^0 = t$,

$$\begin{aligned} x^0 &= t_*(u,\vartheta) \\ x^i &= x^i(t_*(u,\vartheta), u, \vartheta), \quad u_*(\vartheta) > u > \underline{u}(\vartheta), \quad \vartheta = \text{const} \end{aligned} \tag{19.250}$$

We have

$$\begin{aligned} \frac{dx^0}{du}(u) &= \frac{\partial t_*}{\partial u}(u,\vartheta) \\ \frac{dx^i}{du}(u) &= \frac{\partial x^i}{\partial t}(t_*(u,\vartheta), u, \vartheta)\frac{\partial t_*}{\partial u}(u,\vartheta) + \frac{\partial x^i}{\partial u}(t_*(u,\vartheta), u, \vartheta) \end{aligned}$$

Substituting from (19.243) and taking into account of (19.245) we obtain

$$\begin{aligned} \frac{dx^0}{du}(u) &= \frac{\partial t_*}{\partial u}(u,\vartheta) \\ \frac{dx^i}{du}(u) &= L^i(t_*(u,\vartheta), u, \vartheta)\frac{\partial t_*}{\partial u}(u,\vartheta) \end{aligned} \tag{19.251}$$

or simply

$$\frac{dx^\mu}{du}(u) = L^\mu(t_*(u,\vartheta), u, \vartheta)\frac{\partial t_*}{\partial u}(u,\vartheta) \tag{19.252}$$

We see again that in Galilean spacetime the vectorfield $L$ is tangential to the invariant curves. We also see that at $u = u_*(\vartheta)$, that is on $\partial_- H$, the right-hand sides vanish by (19.246), so the past end point of an invariant curve is a singular point.

To analyze the behavior at the past end point we consider the second and third derivatives at this point. Differentiating (19.251) with respect to $u$ we obtain

$$\begin{aligned}\frac{d^2x^0}{du^2}(u) &= \frac{\partial^2 t_*}{\partial u^2}(u,\vartheta)\\ \frac{d^2x^i}{du^2}(u) &= \frac{\partial L^i}{\partial t}(t_*(u,\vartheta),u,\vartheta)(\frac{\partial t_*}{\partial u})^2(u,\vartheta)\\ &+\frac{\partial L^i}{\partial u}(t_*(u,\vartheta),u,\vartheta)\frac{\partial t_*}{\partial u}(u,\vartheta)\\ &+L^i(t_*(u,\vartheta),u,\vartheta)(\frac{\partial^2 t_*}{\partial u^2})(u,\vartheta)\end{aligned} \tag{19.253}$$

Also, differentiating (19.48) with respect to $u$ we obtain

$$\begin{aligned}\frac{\partial^2 t_*}{\partial u^2}(u,\vartheta) = &-(\frac{\partial\mu}{\partial t})^{-2}(t_*(u,\vartheta),u,\vartheta)\\ &\cdot(\frac{\partial\mu}{\partial t}(\frac{\partial^2\mu}{\partial u\partial t}\frac{\partial t_*}{\partial u}+\frac{\partial^2\mu}{\partial u^2})-\frac{\partial\mu}{\partial u}(\frac{\partial^2\mu}{\partial t^2}\frac{\partial t_*}{\partial u}+\frac{\partial^2\mu}{\partial u\partial t}))\end{aligned}$$

Substituting for $\partial t_*/\partial u$ from (19.48) then yields

$$\begin{aligned}\frac{\partial^2 t_*}{\partial u^2}(u,\vartheta) = &-(\frac{\partial\mu}{\partial t})^{-3}((\frac{\partial\mu}{\partial t})^2\frac{\partial^2\mu}{\partial u^2}-2\frac{\partial\mu}{\partial t}\frac{\partial\mu}{\partial u}\frac{\partial^2\mu}{\partial u\partial t}\\ &+(\frac{\partial\mu}{\partial u})^2\frac{\partial^2\mu}{\partial t^2})(t_*(u,\vartheta),u,\vartheta)\end{aligned} \tag{19.254}$$

In particular, at $u = u_*(\vartheta)$ we obtain, in accordance with (19.51),

$$\frac{\partial^2 t_*}{\partial u^2}(u_*(\vartheta),\vartheta) = a(\vartheta) \tag{19.255}$$

where

$$a(\vartheta) = -\frac{\partial^2\mu/\partial u^2}{\partial\mu/\partial t}(q(\vartheta)) > 0 \tag{19.256}$$

by the first equation of (19.45), and we denote by

$$q(\vartheta) = (t_*(u_*(\vartheta),\vartheta),u_*(\vartheta),\vartheta) \tag{19.257}$$

the end point on $\partial_- H$. By (19.246) and (19.255), at $u = u_*(\vartheta)$ (19.253) becomes

$$\begin{aligned}\frac{d^2x^0}{du^2}(u_*(\vartheta)) &= a(\vartheta)\\ \frac{d^2x^i}{du^2}(u_*(\vartheta)) &= L^i(q(\vartheta))a(\vartheta)\end{aligned} \tag{19.258}$$

Moreover, differentiating (19.253) with respect to $u$ and evaluating the result at $u = u_*(\vartheta)$ we obtain, in view of (19.246),

$$\begin{aligned}\frac{d^3x^0}{du^3}(u_*(\vartheta)) &= \frac{\partial^3 t_*}{\partial u^3}(u_*(\vartheta),\vartheta)\\ \frac{d^3x^i}{du^3}(u_*(\vartheta)) &= 2\frac{\partial L^i}{\partial u}(q(\vartheta))\frac{\partial^2 t_*}{\partial u^2}(u_*(\vartheta),\vartheta)+L^i(q(\vartheta))(\frac{\partial^3 t_*}{\partial u^3})(u_*(\vartheta),\vartheta)\end{aligned} \tag{19.259}$$

We shall presently determine the coefficients

$$\frac{\partial^3 t_*}{\partial u^3}(u_*(\vartheta),\vartheta) \quad \text{and} \quad \frac{\partial L^i}{\partial u}(q(\vartheta)) = TL^i(q(\vartheta))$$

In view of the fact that for any smooth function $f(t,u,\vartheta)$ we have

$$\frac{d}{du}f(t_*(u,\vartheta),u,\vartheta) = \frac{\partial f}{\partial t}\frac{\partial t_*}{\partial u} + \frac{\partial f}{\partial u}$$

hence by (19.246),

$$(\frac{d}{du}f(t_*(u,\vartheta),u,\vartheta))_{u=u_*(\vartheta)} = \frac{\partial f}{\partial u}(q(\vartheta))$$

differentiating (19.254) with respect to $u$ and evaluating the result at $u = u_*(\vartheta)$ yields

$$\frac{\partial^3 t_*}{\partial u^3}(u_*(\vartheta),\vartheta) = b(\vartheta) \tag{19.260}$$

where

$$b(\vartheta) = (\frac{\partial\mu}{\partial t})^{-2}(3\frac{\partial^2\mu}{\partial u^2}\frac{\partial^2\mu}{\partial t\partial u} - \frac{\partial\mu}{\partial t}\frac{\partial^3\mu}{\partial u^3})(q(\vartheta)) \tag{19.261}$$

Next, $TL^i$ is given by (3.184):

$$TL^i = a_T\hat{T}^i + b_T^i \tag{19.262}$$

where $a_T$ is given by (3.187) and $b_T^i$ are the rectangular components of $S_{t,u}$-tangential vectorfield given by (3.189). On $H\bigcup\partial_-H$, in particular at $q(\vartheta)$, these formulas reduce to

$$\begin{aligned} a_T &= \alpha^{-1}L\mu < 0 \\ b_T^i &= (\not{g}^{-1}\cdot \not{d}\mu)^i \end{aligned} \tag{19.263}$$

Substituting the above as well as (19.255) in (19.259) we obtain

$$\begin{aligned} &\frac{d^3x^0}{du^3}(u_*(\vartheta)) = b(\vartheta) \\ &\frac{d^3x^i}{du^3}(u_*(\vartheta)) = L^i(q(\vartheta))b(\vartheta) + 2a(\vartheta)(a_T\hat{T}^i + b_T^i)(q(\vartheta)) \end{aligned} \tag{19.264}$$

In view of (19.258) and (19.264) we obtain the following analytic expression for an invariant curve of $H$ in a neighborhood of its origin $q(\vartheta) \in \partial_-H$:

$$\begin{aligned} x^0(u) =& x^0(q(\vartheta)) + \frac{1}{2}a(\vartheta)(u-u_*(\vartheta))^2 + \frac{1}{6}b(\vartheta)(u-u_*(\vartheta))^3 + O((u-u_*(\vartheta))^4) \\ x^i(u) =& x^i(q(\vartheta)) + \frac{1}{2}a(\vartheta)L^i(q(\vartheta))(u-u_*(\vartheta))^2 \\ &+ \frac{1}{6}(b(\vartheta)L^i(q(\vartheta)) + 2a(\vartheta)(a_T\hat{T}^i + b_T^i)(q(\vartheta)))(u-u_*(\vartheta))^3 + O((u-u_*(\vartheta))^4) \end{aligned}$$

We summarize the above results in the following proposition.

**Proposition 19.3** The singular boundary of the domain of the maximal solution in Galilean spacetime is $H \bigcup \partial_- H$. Consider a given component of $H$ and its past boundary, the corresponding component of $\partial_- H$. The component of $\partial_- H$ is a smooth 2-dimensional embedded submanifold in Galilean spacetime, which is spacelike with respect to the acoustical metric. The component of $H$ is a smooth embedded 3-dimensional submanifold in Galilean spacetime ruled by invariant curves of vanishing arc length with respect to the acoustical metric, having past end points on the corresponding component of $\partial_- H$. In a neighborhood of its origin at the point $q(\vartheta) \in \partial_- H$ an invariant curve is given, in parametric form, by

$$\begin{aligned} x^0(u) &= x^0(q(\vartheta)) + \frac{1}{2}a(\vartheta)(u - u_*(\vartheta))^2 + \frac{1}{6}b(\vartheta)(u - u_*(\vartheta))^3 + O((u - u_*(\vartheta))^4) \\ x^i(u) &= x^i(q(\vartheta)) + \frac{1}{2}a(\vartheta)L^i(q(\vartheta))(u - u_*(\vartheta))^2 \\ &\quad + \frac{1}{6}(b(\vartheta)L^i(q(\vartheta)) + 2a(\vartheta)(a_T\hat{T}^i + b_T^i)(q(\vartheta)))(u - u_*(\vartheta))^3 + O((u - u_*(\vartheta))^4) \end{aligned}$$

where

$$\begin{aligned} a(\vartheta) &= -\frac{\partial^2\mu/\partial u^2}{\partial\mu/\partial t}(q(\vartheta)) > 0 \\ b(\vartheta) &= \left(\frac{\partial\mu}{\partial t}\right)^{-2}\left(3\frac{\partial^2\mu}{\partial u^2}\frac{\partial^2\mu}{\partial t\partial u} - \frac{\partial\mu}{\partial t}\frac{\partial^3\mu}{\partial u^3}\right)(q(\vartheta)) \\ a_T(q(\vartheta)) &= (\alpha^{-1}L\mu)(q(\vartheta)) < 0 \\ b_T^i(q(\vartheta)) &= (\not{g}^{-1} \cdot \not{d}\mu)^i(q(\vartheta)) \end{aligned}$$

# References


[1] D. Bernoulli, *Hydrodynamica*, Argentorati, 1738.

[2] I. Chavel, *Eigenvalues in Riemannian Geometry*, Academic Press, 1984.

[3] J. Cheeger and D. Ebin, *Comparison Theorems in Riemannian Geometry*, North Holland Publ. Comp., 1975.

[4] D. Christodoulou, *The Action Principle and Partial Differential Equations*, Ann. Math. Stud., vol. 146, Princeton University Press, 2000.

[5] D. Christodoulou, *The Formation of Shocks in 3-Dimensional Fluids*, EMS Monographs in Mathematics, EMS Publishing House, 2007.

[6] D. Christodoulou and S. Klainerman, *The Global Nonlinear Stability of the Minkowski Space*, Princeton Mathematical Series, vol. 41, Princeton University Press, 1993.

[7] R. Clausius, *Über verschiedene für die Anwendung bequeme Formen der Hauptgleichungen der mechanischen Wärmatheorie*, Annalen der Physik und Chemie **125** (1865), 353–400.

[8] J.-B. le R. D'Alembert, *Recherches sur la courbe que forme une corde tendue mise en vibration*, Mém. Acad. Sci. Berlin **3** (1749), 214–253.

[9] L. Euler, *Principes généraux du mouvement des fluides*, Hist. de l'Acad. de Berlin (1755).

[10] L. Euler, *De principiis motus fluidorum*, Novi Comm. Acad. Petrop. xiv, vi (1759, 1761).

[11] D. Glibarg and N.S. Trudinger, *Elliptic Partial Differential Equations of Second Order*, Springer-Verlag, 1977.

[12] S. Klainerman, *Uniform decay and Lorentz invariance of the classical wave equation*, Comm. Pure Appl. Math. **38** (1985), 321–332.

[13] P.S. Laplace, *Sur la vitesse du son dans l'air et dans l'eau*, Ann. de Chim. et de Phys.iii **238** (1816).

[14] E. Noether, *Invariante Variationsprobleme*, Nachr. D. König. Gesellsch. D. Wiss. Zu Göttingen, Math-phys. (1918), 235–257.

[15] R. Osserman, *The isoperimetric inequality*, Bull.A.M.S. **84** (1978), 1182–1238.

[16] R. Penrose, *Techniques of Differential Topology in Relativity*, Regional Conf. Series in Appl. Math., vol. 7, SIAM, 1972.

[17] B. Riemann, *Über die Fortpflanzung ebener Luftvellen von endlicher Schwingungsweite*, Mathematisch-physikalische Klasse **8** (1858—1859), 43.

[18] T. Sideris, *Formation of sigularities in three-dimensional fluids*, Comm. Math. Phys. **101** (1985), 475–485.

# *About the Authors*

**Demetrios Christodoulou** is a professor of mathematics and physics at Eidgenössische Technische Hochschule (ETH) Zürich. His areas of research include partial differential equations, geometric analysis, general relativity, and fluid mechanics. His honors include the MacArthur Fellows Award, the Bocher Memorial Prize, and the Shaw Prize in Mathematics. He is a member of the American Academy of Arts and Sciences, the European Academy of Sciences, and the National Academy of Sciences (U.S.A.)

**Shuang Miao** is a post-doctoral assistant professor of mathematics at the University of Michigan. He has worked as a Research Fellow at ETH Zürich, and was a PhD student at the Chinese Academy of Sciences. His areas of research include partial differential equations, fluid dynamics, and mathematical physics.

